Quaternion and Clifford Fourier Transforms

Quaternion and Clifford Fourier Transforms

Eckhard Hitzer
International Christian University, College of Liberal Arts

CRC Press
Taylor & Francis Group
Boca Raton London New York

CRC Press is an imprint of the
Taylor & Francis Group, an **informa** business
A CHAPMAN & HALL BOOK

First edition published 2022
by CRC Press
6000 Broken Sound Parkway NW, Suite 300, Boca Raton, FL 33487-2742

and by CRC Press
2 Park Square, Milton Park, Abingdon, Oxon, OX14 4RN

CRC Press is an imprint of Taylor & Francis Group, LLC

ISBN: 978-0-367-77466-0 (hbk)
ISBN: 978-1-032-02658-9 (pbk)
ISBN: 978-1-003-18447-8 (ebk)

DOI: 10.1201/9781003184478

Typeset in LatinModern
by KnowledgeWorks Global Ltd.

Access the Support Material (color figures) at www.routledge.com/9780367774660

Contents

Foreword

I have known the author of this book since 2010, and worked with him in person and remotely throughout the decade since. Eckhard Hitzer is an expert on Clifford algebras and in this book he has brought together a wide body of mathematical knowledge on the subject of hypercomplex Fourier transforms including those based on the algebra of quaternions (a special case of Clifford algebra), and on Clifford algebras in general. This represents a comprehensive collection of ideas and results by many authors over many years, as evidenced by the extensive list of references. They are brought together here in a consistent style and notation which will make this book a useful reference.

This book should provide a sound mathematical basis for further research on Clifford Fourier transforms, both theoretical, and applied to problems in science and engineering.

Stephen J. Sangwine
Colchester, United Kingdom
October 2020

Preface

Dedicated to:
Eriko, Jesaja and Joshua

Lift up your eyes to the heavens,
look at the earth beneath;
the heavens will vanish like smoke,
the earth will wear out like a garment
and its inhabitants die like flies.
But my salvation will last forever,
my righteousness will never fail.
Isaiah 51:6 (NIV)

Jesus said, *Heaven and earth will pass away, but My words will by no means pass away.*
Luke 21:33 (NKJV)

Der Mensch lebt und bestehet nur eine kleine Zeit,
und alle Welt vergehet mit ihrer Herrlichkeit.
Es ist nur Einer ewig und an allen Enden und wir in seinen Händen.
Matthias Claudius

This book describes the development of quaternion and Clifford Fourier transforms in Clifford geometric algebra over the last 30 years.

It begins with a historical introduction, Chapter 1, to quaternion Fourier transforms, Clifford Fourier transforms and their refinement in the form of wavelets.

Then, Chapter 2 begins with the axiomatic formulation of Clifford algebras, the important role of quadratic forms, the fundamental Clifford (geometric) product of vectors and multivectors and convenient products derived from it, determinants in Clifford algebra, the method of Gram-Schmidt orthonormalization, a brief guide to the Clifford algebras of the plane, Euclidean space, spacetime and conformal geometric algebra. Chapter 2 concludes with the study of imaginary numbers in real Clifford algebras generalized to multivector square roots of -1, and the introduction to quaternions, of interest in its own and frequently appearing as subalgebra of higher dimensional Clifford algebras.

Chapter 3 provides an elementary introduction to the basics of differential calculus with vectors.

Chapter 4 is devoted to quaternion Fourier transforms (QFT). In many respects, due to its low dimensionality of four, and its nature as division algebra, quaternions are easier to handle as higher dimensional Clifford algebras, but they already display the essential feature of non-commutativity. While QFTs are by themselves of great interest, their knowledge helps to better grasp the treatment of Clifford Fourier transforms in Chapter 5. Chapter 4 first outlines the fundamentals of QFTs, including their important orthogonal two-dimensional planes split. It then describes properties of the QFT such as uncertainty, convolution, Wiener Khinchine type theorems, special types of QFTs (windowed, Fourier-Mellin, quaternion-domain), and the physics-related generalization to volume-time and spacetime Fourier transforms.

Chapter 5 first provides an overview of Clifford Fourier transforms (CFT), studies the important case of CFTs in three-dimensional Euclidean space, continues with generalization to one-sided CFTs in quadratic space Clifford algebras $Cl(p,q)$, and generalization to two-sided CFTs. This is followed by the treatment of properties of CFTs such as uncertainty and convolution. Finally in this chapter, special CFTs (windowed, Fourier-Mellin, conformal geometric algebra) are expounded.

The last chapter, Chapter 6, endeavors to provide a holistic view on quaternionic and Clifford Fourier transforms, some important applied forms, and further extensions such as to Fourier Stieltjes and linear canonical transforms that appear in the recent literature.

Appendix A provides a summary of Clifford square roots of -1, proofs the multivector Cauchy-Schwartz inequality and the uncertainty equality in case of Gaussian multivector functions.

Even though Chapters 2 and 3 lay the foundations of Clifford algebra and quaternions, and geometric calculus, we try to keep later chapters on more special topics reasonably self contained, so that readers with basic familiarity with quaternions and Clifford algebra will hopefully be able to begin reading directly in the chapter and section of their particular interest, without frequently needing to skip back and forth. Finally, Fourier transforms are a prime example of integral transforms. Readers who look for more background on integral transforms are referred to textbooks like [10, 87, 278].

Acknowlegements

I gratefully acknowledge the fruitful scientific collaboration with the following colleagues:

- R. Abłamowicz
- S. Adji
- R. Ashino
- F. Brackx
- R. Bujack
- D. Eelbode
- A. Hayashi
- J. Helmstetter
- M. Hlawitschka
- I. Hotz
- B. Mawardi
- R. Nagaoka
- S. J. Sangwine
- G. Scheuermann
- R. Vaillancourt

Many results of our joint work are also described in this book.

I thank my colleagues at the Department of Natural Science of International Christian University, especially in the Physics Major, for strongly supporting my current sabbatical research year. During this year with Airborne Hydro Mapping (AHM) GmbH in Innsbruck, Austria, I thank for excellent collaboration my colleagues at AHM; and my precious neighbours Evi and Hermann Jakschitz, Cinzia Zotti, Leopoldo d'Agostino, Johann Salchner and his family, the family of Benedikt Mair and Brigitte Mader, and Helga managing the Ladele, who provide a sort of home away from home.

CHAPTER 1

Introduction

...for geometry, you know, is the gate of science, and the gate is so low and small that one can only enter it as a little child. (*William K. Clifford* [152])

But the gate to life is narrow and the way that leads to it is hard, and there are few people who find it. ...I assure you that unless you change and become like children, you will never enter the Kingdom of heaven. (*Jesus Christ* [224])

At the beginning, we survey the historical development of quaternion and Clifford Fourier transforms and wavelets. This chapter is mainly based on [51].

The first quotation above stems from Clifford himself, who initially was a theologian and then became an atheist. But somehow his view of science was strongly colored by what Jesus taught as the Gospel about the Kingdom of God.[1] To agree or disagree on what Clifford believed is a matter of faith and not of science. But I quite like his point, that geometry is like a gateway to a new understanding of science.

The development of hypercomplex Fourier transforms and wavelets has taken place in several different threads, reflected in the overview of the subject presented in this chapter. We give in Section 1.2 an overview of the development of quaternion Fourier transforms, then in Section 1.3 the development of Clifford Fourier transforms. Additionally, wavelets are a more recent refinement of Fourier transforms, and the distinction between their quaternion and Clifford algebra approach has been much

[1]It is interesting to note that a parallel to this exists even in the Japanese tea ceremony: "...enter the teahouse. The sliding door is only thirty six inches high. Thus all who enter must bow their heads and crouch. This door points to the reality that all are equal in tea, irrespective of status or social position."[223]. The present form of the tea ceremony was established by Sen Rikyu in the 16th century. His wife is said to have been a secret Christian (Kirishitan), some think even Sen Rikyu was.

DOI: 10.1201/9781003184478-1

less pronounced than in the case of Fourier transforms, Section 1.4 reviews the history of both quaternion and Clifford wavelets, although within this book we will not be able to treat wavelets in any depth.

We recognize that the history presented here is bound to be incomplete, and that work by some authors will have been overlooked, for which we can only offer our humble apologies.

1.1 BRIEF HISTORICAL NOTES

W.K. Clifford (1845–1879), a young English Goldsmid professor of applied mathematics at the University College of London, published in 1878 in the American Journal of Mathematics Pure and Applied a nine page long paper on *Applications of Grassmann's Extensive Algebra* [77]. In this paper, the young genius Clifford, standing on the shoulders of two giants of algebra: W.R. Hamilton (1805–1865), the inventor of *quaternions* [155, 156], and H.G. Grassmann (1809–1877), the inventor of *extensive algebra* [146], added the measurement of length and angle to Grassmann's abstract and coordinate free algebraic methods for computing with a space and all its subspaces. Clifford thus unified and generalized in his geometric algebras (= Clifford algebras) the works of Hamilton and Grassmann by finalizing the fundamental concept of *directed numbers* [304].

Any Clifford algebra $Cl(V)$ is generated from an inner-product[2] vector space $(V, a \cdot b : a, b \in V \mapsto \mathbb{R})$ by Clifford's geometric product setting[3] the geometric product[4] of any vector with itself equal to their inner product: $aa = a \cdot a$. We indeed have the *universal property* [244, 283] that any isometry[5] from the vector space V into an inner-product algebra[6] $\mathcal{A}$ over the field[7] $\mathbb{K}$ can be uniquely extended to an isometry[8] from the Clifford algebra $Cl(V)$ into $\mathcal{A}$. The Clifford algebra $Cl(V)$ is the *unique* associative and multilinear algebra with this property. Thus if we wish to generalize methods from algebra, analysis, calculus, differential geometry (etc.) of real numbers, complex numbers and quaternion algebra to vector spaces and multivector spaces (which include additional elements representing 2D up to nD subspaces, i.e. plane elements up to hypervolume elements), the study of Clifford algebras becomes *unavoidable*. Indeed, repeatedly and independently a long list of Clifford algebras, their subalgebras and in Clifford algebras embedded algebras (like *octonions* [247]) of many spaces have been studied and applied historically, often under different names.

[2]The inner product defines the measurement of length and angle.

[3]This setting amounts to an algebra generating relationship.

[4]No product sign will be introduced, simple juxtaposition implies the geometric product just like $2x = 2 \times x$.

[5]A $\mathbb{K}$-isometry between two inner-product spaces is a $\mathbb{K}$-linear mapping preserving the inner products.

[6]A $\mathbb{K}$-algebra is a $\mathbb{K}$-vector space equipped with an associative and multilinear product. An inner-product $\mathbb{K}$-algebra is a $\mathbb{K}$-algebra equipped with an inner product structure when taken as $\mathbb{K}$-vector space.

[7]Important fields are real $\mathbb{R}$ and complex numbers $\mathbb{C}$, etc.

[8]That is a $\mathbb{K}$-linear homomorphism preserving the inner products, i.e. a $\mathbb{K}$-linear mapping preserving both the products of the algebras when taken as rings, and the inner products of the algebras when taken as inner-product vector spaces.

Some of these algebras are *complex numbers (and the complex number plane), hyperbolic numbers (split complex numbers, real tessarines), dual numbers, quaternions, biquaternions (complex quaternions), dual quaternions, Plücker coordinates, bicomplex numbers (commutative quaternions, tessarines, Segre quaternions), Pauli algebra (space algebra), Dirac algebra (space-time algebra, Minkowski algebra), algebra of physical space, para-vector algebra, spinor algebra, Lie algebras, Cartan algebra, versor algebra, rotor algebra, motor algebra, Clifford bracket algebra, conformal algebra, algebra of differential forms, etc.*

Section 1.1.1 can also be skipped by readers less interested in mathematical definitions.

1.1.1 Definitions related to Clifford's geometric algebras

Definition of an algebra [316]: Let $\mathcal{A}$ be a vector space over the reals $\mathbb{R}$ with an additional binary operation from $\mathcal{A}\times\mathcal{A}$ to $\mathcal{A}$, denoted here by $\circ$ ($x\circ y$ is the product of any $x,y\in\mathcal{A}$). Then $\mathcal{A}$ is an algebra over $\mathbb{R}$ if the following identities hold $\forall\, x,y,z\in\mathcal{A}$, and "scalars" $\alpha,\beta\in\mathbb{R}$: (1,2) Left and right distributivity: $(x+y)\circ z = x\circ z + y\circ z$, $x\circ(y+z) = x\circ y + x\circ z$. (3) Compatibility with scalars: $(\alpha x)\circ(\beta y) = (\alpha\beta)(x\circ y)$. This means that $x\circ y$ is bilinear. The binary operation is often referred to as multiplication in $\mathcal{A}$, which is not necessarily associative.

Definition of inner product space [2]: An inner product space is a vector space V over $\mathbb{R}$ together with an inner product map $\langle .,.\rangle : V\times V\to\mathbb{R}$, that satisfies $\forall\, x,y,z\in V$ and $\forall\,\alpha\in\mathbb{R}$: (1) Symmetry: $\langle x,y\rangle = \langle y,x\rangle$. (2) Linearity in the first argument: $\langle \alpha x,y\rangle = \alpha\langle x,y\rangle,\ \langle x+y,z\rangle = \langle x,z\rangle + \langle y,z\rangle$.

Note: We do not assume positive definiteness.

Definition of inner product algebra: An inner product algebra is an algebra equipped with an inner product $\mathcal{A}\times\mathcal{A}\to\mathbb{R}$.

Definition of Clifford's geometric algebra (GA) [131,247]: Let $\{e_1, e_2, \ldots, e_p, e_{p+1}, \ldots, e_{p+q}, e_{p+q+1}, \ldots, e_n\}$, with $n = p+q+r$, $e_k^2 = \varepsilon_k$, $\varepsilon_k = +1$ for $k = 1,\ldots,p$, $\varepsilon_k = -1$ for $k = p+1,\ldots,p+q$, $\varepsilon_k = 0$ for $k = p+q+1,\ldots,n$, be an *orthonormal base* of the inner product vector space $\mathbb{R}^{p,q,r}$ with a geometric product according to the multiplication rules

$$e_k e_l + e_l e_k = 2\varepsilon_k\delta_{k,l}, \qquad k,l = 1,\ldots n, \tag{1.1.1}$$

where $\delta_{k,l}$ is the Kronecker symbol with $\delta_{k,l} = 1$ for $k = l$, and $\delta_{k,l} = 0$ for $k \neq l$. This non-commutative product and the additional axiom of *associativity* generate the 2^n-dimensional Clifford geometric algebra $Cl(p,q,r) = Cl(\mathbb{R}^{p,q,r}) = Cl_{p,q,r} = \mathcal{G}_{p,q,r} = \mathbb{R}_{p,q,r}$ over $\mathbb{R}$. The set $\{e_A : A\subseteq\{1,\ldots,n\}\}$ with $e_A = e_{h_1}e_{h_2}\cdots e_{h_k}$, $1\leq h_1 < \ldots < h_k \leq n$, $e_\emptyset = 1$, forms a graded (blade) basis of $Cl(p,q,r)$. The grades k range from 0 for scalars, 1 for vectors, 2 for bivectors, s for s-vectors, up to n for pseudoscalars. The vector space $\mathbb{R}^{p,q,r}$ is included in $Cl(p,q,r)$ as the subset of 1-vectors. The general elements of $Cl(p,q,r)$ are real linear combinations of basis blades e_A, called Clifford numbers, multivectors or hypercomplex numbers.

Note: The definition of Clifford's GA is fundamentally *coordinate system independent*, i.e. *coordinate free*. Equation (1.1.1) is fully equivalent to the coordinate independent definition: $aa = a \cdot a, \forall a \in \mathbb{R}^{p,q,r}$ [161]. This even applies to Clifford analysis (Geometric Calculus). Clifford algebra is thus ideal for computing with geometrical *invariants* [244]. A review of five different ways to define GA, including one definition based on vector space basis element multiplication rules, and one definition focusing on $Cl(p,q,r)$ as a *universal* associative algebra, is given in chapter 14 of the textbook [247].

In general $\langle A \rangle_k$ denotes the grade k part of $A \in Cl(p,q,r)$. The parts of grade 0, $(s-k)$, $(k-s)$, and $(s+k)$, respectively, of the geometric product of a k-vector $A_k \in Cl(p,q,r)$ with an s-vector $B_s \in Cl(p,q,r)$

$$A_k * B_s := \langle A_k B_s \rangle_0, \quad A_k \rfloor B_s := \langle A_k B_s \rangle_{s-k}, \tag{1.1.2}$$
$$A_k \lfloor B_s := \langle A_k B_s \rangle_{k-s}, \quad A_k \wedge B_s := \langle A_k B_s \rangle_{k+s},$$

are called *scalar product*, *left contraction* (zero for $s < k$), *right contraction* (zero for $k < s$), and (associative) *outer product*, respectively. These definitions extend by linearity to the corresponding products of general multivectors. The various derived products of (1.1.2) are related, e.g. by

$$(A \wedge B) \rfloor C = A \rfloor (B \rfloor C), \quad \forall A, B, C \in Cl(p,q,r). \tag{1.1.3}$$

Note that for vectors a, b in $\mathbb{R}^{p,q,r} \subset Cl(p,q,r)$ we have

$$ab = a \rfloor b + a \wedge b, \qquad a \rfloor b = a \lfloor b = a \cdot b = a * b, \tag{1.1.4}$$

where $a \cdot b$ is the usual inner product of $\mathbb{R}^{p,q,r}$.

For $r = 0$ we often denote $\mathbb{R}^{p,q} = \mathbb{R}^{p,q,0}$, and $Cl(p,q) = Cl(p,q,0)$. For Euclidean vector spaces $(n = p)$ we use $\mathbb{R}^n = \mathbb{R}^{n,0} = \mathbb{R}^{n,0,0}$, and $Cl(n) = Cl(n,0) = Cl(n,0,0)$. The *even* grade subalgebra of $Cl(p,q,r)$ is denoted by $Cl^+(p,q,r)$, the k-vectors of its basis have only even grades k. Every k-vector B that can be written as the outer product $B = b_1 \wedge b_2 \wedge \ldots \wedge b_k$ of k vectors $b_1, b_2, \ldots, b_k \in \mathbb{R}^{p,q,r}$ is called a *simple* k-vector or *blade*.

Definition of outermorphism [161,248]: An *outermorphism* is the unique extension to $Cl(V)$ of a vector space map for all $a \in V$, $f : a \mapsto f(a) \in V'$, and is given by the mapping $B = b_1 \wedge b_2 \wedge \ldots \wedge b_k \to f(b_1) \wedge f(b_2) \wedge \ldots \wedge f(b_k)$ in $Cl(V')$, for every blade B in $Cl(V)$.

1.2 QUATERNION FOURIER TRANSFORMS (QFT)

1.2.1 Major developments in the history of the quaternion Fourier transform

Quaternions were first applied to Fourier transforms by Ernst [130, Section 6.4.2] and Delsuc [102, Eqn. 20] in the late 1980s, seemingly without knowledge of the earlier work of Sommen [306, 307] on Clifford Fourier and Laplace transforms further explained in Section 1.3.2. Ernst and Delsuc's quaternion transforms were two-dimensional (that is they had two independent variables) and proposed for application to nuclear magnetic resonance (NMR) imaging. Written in terms of two independent time variables[9] t_1 and t_2, the forward transforms were of the following form:

$$F(\omega_1, \omega_2) = \int_{-\infty}^{\infty} \int_{-\infty}^{\infty} f(t_1, t_2) e^{\boldsymbol{i}\omega_1 t_1} e^{\boldsymbol{j}\omega_2 t_2} \, dt_1 \, dt_2 \,. \tag{1.2.1}$$

Notice the use of different quaternion basis units $\boldsymbol{i}$ and $\boldsymbol{j}$ in each of the two exponentials, a feature that was essential to maintain the separation between the two dimensions (the prime motivation for using a quaternion Fourier transformation was to avoid the mixing of information that occurred when using a complex Fourier transform – something that now seems obvious, but must have been less so in the 1980s). The signal waveforms/samples measured in NMR are complex, so the quaternion aspect of this transform was essential only for maintaining the separation between the two dimensions. As we will see below, there was some unused potential here.

The fact that exponentials in the above formulation do not commute (with each other, or with the 'signal' function f), means that other formulations are possible, and indeed Ell in 1992 [123, 124] formulated a transform with the two exponentials positioned either side of the signal function:

$$F(\omega_1, \omega_2) = \int_{-\infty}^{\infty} \int_{-\infty}^{\infty} e^{\boldsymbol{i}\omega_1 t_1} f(t_1, t_2) \, e^{\boldsymbol{j}\omega_2 t_2} \, dt_1 \, dt_2 \,. \tag{1.2.2}$$

Ell's transform was a theoretical development, but it was soon applied to the practical problem of computing a holistic Fourier transform of a color image [291] in which the signal samples (discrete image pixels) had three-dimensional values (represented as quaternions with zero scalar parts). This was a major change from the previously intended application in nuclear magnetic resonance, because now the two-dimensional nature of the transform mirrored the two-dimensional nature of the image, and the four-dimensional nature of the algebra used followed naturally from the three-dimensional nature of the image pixels.

Other researchers in signal and image processing have followed Ell's formulation (with trivial changes of basis units in the exponentials) [56, 57, 72], but as with the NMR transforms, the quaternion nature of the transforms was applied essentially to separation of the two independent dimensions of an image (Bülow's work [56, 57] was based on greyscale images, that is with one-dimensional pixel values). Two new ideas emerged in 1998 in a paper by Sangwine and Ell [292]. These were, firstly, the choice of a general root $\boldsymbol{\mu}$ of -1 (a unit quaternion with zero scalar part) rather than a basis unit ($\boldsymbol{i}$, $\boldsymbol{j}$ or $\boldsymbol{k}$) of the quaternion algebra, and secondly, the choice of a single exponential rather than two (giving a choice of ordering relative to the quaternionic signal function):

$$F(\omega_1, \omega_2) = \int_{-\infty}^{\infty} \int_{-\infty}^{\infty} e^{\boldsymbol{\mu}(\omega_1 t_1 + \omega_2 t_2)} f(t_1, t_2) \, dt_1 \, dt_2 \,. \tag{1.2.3}$$

[9]The two independent time variables arise naturally from the formulation of two-dimensional NMR spectroscopy.

This made possible a quaternion Fourier transform of a one-dimensional signal:

$$F(\omega) = \int_{-\infty}^{\infty} e^{\mu \omega t} f(t)\, dt\,. \tag{1.2.4}$$

Such a transform makes sense only if the signal function has quaternion values, suggesting applications where the signal has three or four independent components. (An example is vibrations in a solid, such as rock, detected by a sensor with three mutually orthogonal transducers, such as a vector geophone.)

Very little has appeared in print about the interpretation of the Fourier coefficients resulting from a quaternion Fourier transform. One interpretation are components of different symmetry, as explained by Ell in [128]. Sangwine and Ell in 2007 published a paper about quaternion Fourier transforms applied to color images, with a detailed explanation of the Fourier coefficients in terms of elliptical paths in color space (the n-dimensional space of the values of the image pixels in a color image) [127].

1.2.2 Towards splitting quaternions and the QFT

Following the earlier works of Ernst, Ell, Sangwine (see Section 1.2.1) and Bülow [56, 57], Hitzer thoroughly studied the quaternion Fourier transform (QFT) applied to quaternion-valued functions in [184]. As part of this work a quaternion split

$$q_{\pm} = \frac{1}{2}(q \pm \boldsymbol{i} q \boldsymbol{j}), \qquad q \in \mathbb{H}, \tag{1.2.5}$$

was devised and applied, which lead to a better understanding of $GL(\mathbb{R}^2)$ transformation properties of the QFT spectrum of 2D images, including color images, and opened the way to a generalization of the QFT concept to a full spacetime Fourier transformation (SFT) for spacetime algebra $Cl(3,1)$-valued signals.

This was followed up by the establishment of a fully *directional* (opposed to component wise) uncertainty principle for the QFT and the SFT [187]. Independently Mawardi et al. [255] established a componentwise uncertainty principle for the QFT.

The QFT with a Gabor window was treated by Bülow [56], a study which has been continued by Mawardi et al. in [256].

Hitzer reports in [194] initial results (obtained in co-operation with Sangwine) about a further generalization of the QFT to a general form of orthogonal 2D planes split (OPS-) QFT, where the split (1.2.5) with respect to two orthogonal pure quaternion units $\boldsymbol{i}, \boldsymbol{j}$ is generalized to a steerable split with respect to any two pure unit quaternions. This approach is fully elaborated upon in [195, 205].

1.3 DEVELOPMENT OF CLIFFORD FOURIER TRANSFORMATIONS IN CLIFFORD'S GEOMETRIC ALGEBRA

Regarding the literature, an introduction to the vector and multivector calculus used in the field of Clifford Fourier transforms (CFT) can be found in [174, 175]. A tutorial introduction to CFTs and Clifford wavelet transforms can be found in [183]. The Clifford algebra application survey [202] contains an up to date section on applications of Clifford algebra integral transforms, including CFTs, QFTs and wavelet transforms.

1.3.1 On the role of Clifford algebra square roots of -1 for Clifford Fourier transformations

In 1990, Jancewicz [222] defined a trivector Fourier transformation

$$\mathcal{F}_3\{g\}(\boldsymbol{\omega}) = \int_{\mathbb{R}^3} g(\mathbf{x}) e^{-i_3 \mathbf{x}\cdot\boldsymbol{\omega}} d^3\mathbf{x}, \quad i_3 = \boldsymbol{e}_1\boldsymbol{e}_2\boldsymbol{e}_3, \quad g : \mathbb{R}^3 \to Cl(3,0), \qquad (1.3.1)$$

for the electromagnetic field replacing the imaginary unit $i \in \mathbb{C}$ by the trivector i_3 of the geometric algebra $Cl(3,0)$ of three-dimensional Euclidean space $\mathbb{R}^3 = \mathbb{R}^{3,0}$ with ortho-normal vector basis $\{\boldsymbol{e}_1, \boldsymbol{e}_2, \boldsymbol{e}_3\}$.

In [132] Felsberg makes use of signal embeddings in low dimensional Clifford algebras $\mathbb{R}_{2,0}$ and $\mathbb{R}_{3,0}$ to define his Clifford-Fourier transform (CFT) for one-dimensional signals as

$$\mathcal{F}_1^{fe}[f](\underline{u}) = \int_{\mathbb{R}} \exp\left(-2\pi i_2 \underline{u}\, \underline{x}\right) f(\underline{x})\, d\underline{x}, \quad i_2 = \boldsymbol{e}_1\boldsymbol{e}_2, \quad f : \mathbb{R} \to \mathbb{R}, \qquad (1.3.2)$$

where he uses the pseudoscalar $i_2 \in Cl(2,0)$. For two-dimensional signals he defines the CFT as

$$\mathcal{F}_2^{fe}[f](\underline{u}) = \int_{\mathbb{R}^2} \exp\left(-2\pi i_3 < \underline{u}, \underline{x} >\right) f(\underline{x})\, d\underline{x}, \quad f : \mathbb{R}^2 \to \mathbb{R}^2, \qquad (1.3.3)$$

where he uses the pseudoscalar $i_3 \in Cl(3,0)$. It is used a.o. to introduce a concept of two-dimensional analytic signal. Together with Bülow, and Sommer, Felsberg applied these CFTs to image structure processing (key-notion: structure multivector) [56, 132].

Ebling and Scheuermann [117,118] consequently applied to vector signal processing in two- and three dimensions, respectively, the following two-dimensional CFT

$$\mathcal{F}_2\{f\}(\boldsymbol{\omega}) = \int_{\mathbb{R}^2} f(\mathbf{x}) e^{-i_2 \mathbf{x}\cdot\boldsymbol{\omega}} d^2\mathbf{x}, \quad f : \mathbb{R}^2 \to \mathbb{R}^2, \qquad (1.3.4)$$

with Clifford Fourier kernel

$$\exp\left(-\boldsymbol{e}_1\boldsymbol{e}_2(\omega_1 x_1 + \omega_2 x_2)\right), \qquad (1.3.5)$$

and the three-dimensional CFT (1.3.1) of Jancewicz with Clifford Fourier kernel

$$\exp\left(-\boldsymbol{e}_1\boldsymbol{e}_2\boldsymbol{e}_3(\omega_1 x_1 + \omega_2 x_2 + \omega_3 x_3)\right). \qquad (1.3.6)$$

An important integral operation defined and applied in this context by Ebling and Scheuermann was the Clifford convolution. These Clifford-Fourier transforms and the corresponding convolution theorems allow Ebling and Scheuermann for a.o. the analysis of vector-valued patterns in the frequency domain.

Note that the latter Fourier kernel (1.3.6) has also been used by Mawardi and Hitzer in [181,251,251] to define their Clifford-Fourier transform of three-dimensional multivector signals: that means, they researched the properties of $\mathcal{F}_3\{g\}(\boldsymbol{\omega})$ in detail when applied to full multivector signals $g : \mathbb{R}^3 \to Cl(3,0)$. This included an investigation of the uncertainty inequality for this type of CFT. They subsequently generalized $\mathcal{F}_3\{g\}(\boldsymbol{\omega})$ to dimensions $n = 3 (\text{mod } 4)$, *i.e.* $n = 3, 7, 11, \ldots$,

$$\mathcal{F}_n\{g\}(\boldsymbol{\omega}) = \int_{\mathbb{R}^n} g(\mathbf{x}) e^{-i_n \mathbf{x}\cdot\boldsymbol{\omega}} d^n\mathbf{x}, \quad g : \mathbb{R}^n \to Cl(n,0), \tag{1.3.7}$$

which is straight forward, since for these dimensions the pseudoscalar $i_n = \boldsymbol{e}_1 \ldots \boldsymbol{e}_n$ squares to -1 and is central [182], i.e. it commutes with every other multivector belonging to $Cl(n, 0)$. A little less trivial is the generalization of $\mathcal{F}_2\{f\}(\boldsymbol{\omega})$ to

$$\mathcal{F}_n\{f\}(\boldsymbol{\omega}) = \int_{\mathbb{R}^n} f(\mathbf{x}) e^{-i_n \mathbf{x}\cdot\boldsymbol{\omega}} d^n\mathbf{x}, \quad f : \mathbb{R}^n \to Cl(n,0), \tag{1.3.8}$$

with $n = 2 (\text{mod } 4)$, *i.e.* $n = 2, 6, 10 \ldots$, because in these dimensions the pseudoscalar $i_n = \boldsymbol{e}_1 \ldots \boldsymbol{e}_n$ squares to -1, but it ceases to be central. So the relative order of the factors in $\mathcal{F}_n\{f\}(\boldsymbol{\omega})$ becomes important, see [185] for a systematic investigation and comparison.

In the context of generalizing quaternion Fourier transforms (QFT) via algebra isomorphisms to higher dimensional Clifford algebras, Hitzer [184] constructed a spacetime Fourier transform (SFT) in the full algebra of spacetime $Cl(3, 1)$, which includes the CFT (1.3.1) as a partial transform of space. Implemented analogous (isomorphic) to the orthogonal 2D planes split of quaternions, the SFT permits a natural spacetime split, which algebraically splits the SFT into right- and left propagating multivector wave packets. This analysis allows to compute the effect of Lorentz transformations on the spectra of these wavepackets, as well as a 4D directional spacetime uncertainty formula [187] for spacetime signals.

Mawardi et al. extended the CFT $\mathcal{F}_2\{f\}(\boldsymbol{\omega})$ to a windowed CFT in [254]. Fu et al. establish in [136] a strong version of Heisenberg's uncertainty principle for Gabor-windowed CFTs.

In [61] Bujack, Scheuermann and Hitzer expand the notion of Clifford Fourier transform to include multiple left and right exponential kernel factors, in which commuting (or anticommuting) blades, that square to -1, replace the complex unit $i \in \mathbb{C}$, thus managing to include most practically used CFTs in a single comprehensive framework. Based on this, they have also constructed a general CFT convolution theorem [59].

Spurred by the systematic investigation of (complex quaternion) biquaternion square roots of -1 in $Cl(3, 0)$ by Sangwine [294], Hitzer and Ablamowicz [193] systematically investigated the explicit equations and solutions for square roots of -1 in all real Clifford algebras $Cl(p, q), p + q \leq 4$. This investigation has been continued by Hitzer, Helmstetter and Ablamowicz [203] for all square roots of -1 in all real Clifford algebras $Cl(p, q)$ without restricting the value of $n = p + q$. One important motivation for this is the relevance of the Clifford algebra square roots of -1 for the general construction of CFTs, where the imaginary unit $i \in \mathbb{C}$ is replaced by a $\sqrt{-1} \in Cl(p, q)$, without restriction to pseudoscalars or blades.

Based on the knowledge of square roots of -1 in real Clifford algebras $Cl(p, q)$, [201] develops a general CFT in $Cl(p, q)$, wherein the complex unit $i \in \mathbb{C}$ is replaced by any square root of -1 chosen from any component and (or) conjugation class of the submanifold of square roots of -1 in $Cl(p,q)$, and details its properties, including a convolution theorem. A similar general approach is taken in [196] for the construction of two-sided CFTs in real Clifford algebras $Cl(p,q)$, freely choosing two square roots from any one or two components and (or) conjugation classes of the submanifold of square roots of -1 in $Cl(p,q)$. These transformations are therefore generically steerable.

This algebraically motivated approach may eventually be favourably combined with group theoretic, operator theoretic and spinorial approaches, to be reviewed in the following.

1.3.2 Clifford Fourier transform in the light of Clifford analysis

Two robust tools used in image processing and computer vision for the analysis of scalar fields are convolution and Fourier transformation. Several attempts have been made to extend these methods to two- and three-dimensional vector fields and even multivector fields. Let us give an overview of those generalized Fourier transforms.

In [57] Bülow and Sommer define a so-called quaternionic Fourier transform of two-dimensional signals $f(x_1, x_2)$ taking their values in the algebra $\mathbb{H}$ of real quaternions. Note that the quaternion algebra $\mathbb{H}$ is nothing else but (isomorphic to) the Clifford algebra $\mathbb{R}_{0,2} = Cl(0,2)$ where, traditionally, the basis vectors are denoted by $\boldsymbol{i}$ and $\boldsymbol{j}$, with $\boldsymbol{i}^2 = \boldsymbol{j}^2 = -1$, and the bivector by $\boldsymbol{k} = \boldsymbol{ij}$. In terms of these basis vectors, this quaternionic Fourier transform takes the form

$$\mathcal{F}^q[f](u_1, u_2) = \int_{\mathbb{R}^2} \exp(-2\pi \boldsymbol{i} u_1 x_1)\ f(x_1, x_2)\ \exp(-2\pi \boldsymbol{j} u_2 x_2)\ d\underline{x} \tag{1.3.9}$$

Due to the non-commutativity of the multiplication in $\mathbb{H}$, the convolution theorem for this quaternionic Fourier transform is rather complicated, see also [59].

This is also the case for its higher dimensional analogue, the so-called Clifford Fourier transform given by

$$\mathcal{F}^{cl}[f](\underline{u}) = \int_{\mathbb{R}^m} f(\underline{x})\ \exp(-2\pi \boldsymbol{e}_1 u_1 x_1)\ \ldots\ \exp(-2\pi \boldsymbol{e}_m u_m x_m)\ d\underline{x} \tag{1.3.10}$$

Note that for $m = 1$ and interpreting the Clifford basis vector $\boldsymbol{e}_1$ as the imaginary unit i, the Clifford Fourier transform (1.3.10) reduces to the standard Fourier transform on the real line, while for $m = 2$ the quaternionic Fourier transform (1.3.9) is recovered when restricting to real signals.

Finally, Bülow and Sommer also introduce a so-called commutative hypercomplex Fourier transform given by

$$\mathcal{F}^h[f](\underline{u}) = \int_{\mathbb{R}^m} f(\underline{x}) \exp\left(-2\pi \textstyle\sum_{j=1}^m \tilde{\boldsymbol{e}}_j u_j x_j\right)\ d\underline{x} \tag{1.3.11}$$

where the basis vectors $(\tilde{\boldsymbol{e}}_1, \ldots, \tilde{\boldsymbol{e}}_m)$ obey the commutative multiplication rules $\tilde{\boldsymbol{e}}_j \tilde{\boldsymbol{e}}_k = \tilde{\boldsymbol{e}}_k \tilde{\boldsymbol{e}}_j$, $j, k = 1, \ldots, m$, while still retaining $\tilde{\boldsymbol{e}}_j^2 = -1$, $j = 1, \ldots, m$. This commutative hypercomplex Fourier transform offers the advantage of a simple convolution theorem.

The hypercomplex Fourier transforms $\mathcal{F}^q$, $\mathcal{F}^{cl}$ and $\mathcal{F}^h$ enable Bülow and Sommer to establish a theory of multi-dimensional signal analysis and in particular to introduce the notions of multi-dimensional analytic signal, Gabor filter, instantaneous and local amplitude and phase, *etc.*

In this context the Clifford Fourier transformations by Felsberg [132] for one- and two-dimensional signals, by Ebling and Scheuermann for two- and three-dimensional vector signal processing [117,118], and by Mawardi and Hitzer for general multivector signals in $Cl(3,0)$ [181, 251, 251], and their respective kernels, as already reviewed in Section 1.3.1, should also be considered.

The above mentioned Clifford Fourier kernel of Bülow and Sommer

$$\exp\left(-2\pi \boldsymbol{e}_1 u_1 x_1\right) \cdots \exp\left(-2\pi \boldsymbol{e}_m u_m x_m\right) \tag{1.3.12}$$

was in fact already introduced in [33] and [305] as a theoretical concept in the framework of Clifford analysis. This generalized Fourier transform was further elaborated by Sommen in [306, 307] in connection with similar generalizations of the Cauchy, Hilbert and Laplace transforms. In this context also the work of Li, McIntosh and Qian should be mentioned; in [243] they generalize the standard multi-dimensional Fourier transform of a function in $\mathbb{R}^m$, by extending the Fourier kernel $\exp\left(i\left\langle \underline{\boldsymbol{\xi}}, \underline{x}\right\rangle\right)$ to a function which is holomorphic in $\mathbb{C}^m$ and monogenic in $\mathbb{R}^{m+1}$.

In [40,42,46] Brackx, De Schepper and Sommen follow another philosophy in their construction of a Clifford Fourier transform. One of the most fundamental features of Clifford analysis is the factorization of the Laplace operator. Indeed, whereas in general the square root of the Laplace operator is only a pseudo-differential operator, by embedding Euclidean space into a Clifford algebra, one can realize $\sqrt{-\Delta_m}$ as the Dirac operator $\partial_{\underline{x}}$. In this way Clifford analysis spontaneously refines harmonic analysis. In the same order of ideas, Brackx et al. decided to not replace nor to improve the classical Fourier transform by a Clifford analysis alternative, since a refinement of it automatically appears within the language of Clifford analysis. The key step to making this refinement apparent is to interpret the standard Fourier transform as an operator exponential:

$$\mathcal{F} = \exp\left(-i\frac{\pi}{2}\mathcal{H}\right) = \sum_{k=0}^{\infty} \frac{1}{k!}\left(-i\frac{\pi}{2}\right)^k \mathcal{H}^k, \tag{1.3.13}$$

where $\mathcal{H}$ is the scalar operator

$$\mathcal{H} = \frac{1}{2}\left(-\Delta_m + r^2 - m\right) \tag{1.3.14}$$

This expression links the Fourier transform with the Lie algebra $\mathfrak{sl}_2$ generated by Δ_m and $r^2 = |x|^2$ and with the theory of the quantum harmonic oscillator determined by the Hamiltonian $-\frac{1}{2}\left(\Delta_m - r^2\right)$. Splitting the operator $\mathcal{H}$ into a sum of Clifford algebra–valued second order operators containing the angular Dirac operator Γ, one is led, in a natural way, to a *pair* of transforms $\mathcal{F}_{\mathcal{H}^\pm}$, the harmonic average of which is precisely the standard Fourier transform:

$$\mathcal{F}_{\mathcal{H}^\pm} = \exp\left(\frac{i\pi m}{4}\right)\exp\left(\mp\frac{i\pi\Gamma}{2}\right)\exp\left(\frac{i\pi}{4}\left(\Delta_m - r^2\right)\right) \tag{1.3.15}$$

For the special case of dimension 2, Brackx et al. obtain a closed form for the kernel of the integral representation of this Clifford Fourier transform leading to its internal representation

$$\mathcal{F}_{\mathcal{H}^\pm}[f](\underline{\boldsymbol{\xi}}) = \mathcal{F}_{\mathcal{H}^\pm}[f](\boldsymbol{\xi}_1, \boldsymbol{\xi}_2) = \frac{1}{2\pi} \int_{\mathbb{R}^2} \exp\left(\pm \boldsymbol{e}_{12}(\boldsymbol{\xi}_1 x_2 - \boldsymbol{\xi}_2 x_1)\right) f(\underline{x})\, d\underline{x} \tag{1.3.16}$$

which enables the generalization of the calculation rules for the standard Fourier transform both in the L_1 and in the L_2 context. Moreover, the Clifford-Fourier transform of Ebling and Scheuermann

$$\mathcal{F}^e[f](\underline{\boldsymbol{\xi}}) = \int_{\mathbb{R}^2} \exp\left(-\boldsymbol{e}_{12}(x_1 \boldsymbol{\xi}_1 + x_2 \boldsymbol{\xi}_2)\right) f(\underline{x})\, d\underline{x} \tag{1.3.17}$$

can be expressed in terms of the Clifford-Fourier transform:

$$\mathcal{F}^e[f](\underline{\boldsymbol{\xi}}) = 2\pi\ \mathcal{F}_{\mathcal{H}^\pm}[f](\mp\boldsymbol{\xi}_2, \pm\boldsymbol{\xi}_1) = 2\pi\ \mathcal{F}_{\mathcal{H}^\pm}[f](\pm\boldsymbol{e}_{12}\underline{\boldsymbol{\xi}}), \tag{1.3.18}$$

taking into account that, under the isomorphism between the Clifford algebras $\mathbb{R}_{2,0}$ and $\mathbb{R}_{0,2}$, both pseudoscalars are isomorphic images of each other.

The question whether $\mathcal{F}_{\mathcal{H}^\pm}$ can be written as an integral transform is answered positively in the case of even dimension by De Bie and Xu in [90]. The integral kernel of this transform is not easy to obtain and looks quite complicated. In the case of odd dimension the problem is still open.

Later, in [92], De Bie and De Schepper have studied the fractional Clifford Fourier transform as a generalization of both the standard fractional Fourier transform and the Clifford Fourier transform. It is given as an operator exponential by

$$\mathcal{F}_{\alpha,\beta} = \exp\left(\frac{i\alpha m}{2}\right) \exp\left(\imath\beta\Gamma\right) \exp\left(\frac{i\alpha}{2}\left(\Delta_m - r^2\right)\right). \tag{1.3.19}$$

For the corresponding integral kernel a series expansion is obtained, and, in the case of dimension 2, an explicit expression in terms of Bessel functions.

The above, more or less chronological, overview of generalized Fourier transforms in the framework of quaternionic and Clifford analysis gives the impression of a medley of ad hoc constructions. However, there is a structure behind some of these generalizations, which becomes apparent when, as already slightly touched upon above, the Fourier transform is linked to group representation theory, in particular the Lie algebras $\mathfrak{sl}_2$ and $\mathfrak{osp}(1|2)$. This unifying character is beautifully demonstrated by De Bie in the overview paper [94], where, next to an extensive bibliography, also new results on some of the transformations mentioned below can be found. It is shown that using realizations of the Lie algebra $\mathfrak{sl}_2$, one is lead to scalar generalizations of the Fourier transform, such as:

(i) The fractional Fourier transform, which is, as the standard Fourier transform, invariant under the orthogonal group; this transform has been reinvented several times as well in mathematics as in physics, and is attributed to Namias [272], Condon [81], Bargmann [16], Collins [79], Moshinsky and Quesne [266]; for a detailed overview of the theory and recent applications of the fractional Fourier transform we refer the reader to [276];

(ii) The Dunkl transform, see e.g. [113], where the symmetry is reduced to that of a finite reflection group;

(iii) The radially deformed Fourier transform, see e.g. [235], which encompasses both the fractional Fourier and the Dunkl transform;

(iv) The super Fourier transform, see e.g. [83, 88], which is defined in the context of superspaces and is invariant under the product of the orthogonal with the symplectic group.

Realizations of the Lie algebra $\mathfrak{osp}(1|2)$, on the contrary, need the framework of Clifford analysis, and lead to:

(v) The Clifford–Fourier transform and the fractional Clifford–Fourier transform, both already mentioned above; meanwhile an entire class of Clifford–Fourier transforms has been thoroughly studied in [91];

(vi) The radially deformed hypercomplex Fourier transform, which appears as a special case in the theory of radial deformations of the Lie algebra $\mathfrak{osp}(1|2)$, see [93, 96], and is a topic of current research, see [98].

1.4 DEVELOPMENT OF QUATERNION AND CLIFFORD WAVELETS

1.4.1 Towards Clifford wavelets in Clifford analysis

The interest of the Belgian Ghent Clifford Research Group for generalizations of the Fourier transform in the framework of Clifford analysis, grew out from the study of the multidimensional Continuous Wavelet Transform in this particular setting. Clifford wavelet theory, however restricted to the continuous wavelet transform, was initiated by Brackx and Sommen in [34] and further developed by N. De Schepper in her PhD thesis [105]. The Clifford wavelets originate from a mother wavelet not only by translation and dilation, but also by rotation, making the Clifford wavelets appropriate for detecting directional phenomena. Rotations are implemented as specific actions on the variable by a spin element, since, indeed, the special orthogonal group $\mathrm{SO}(m)$ is doubly covered by the spin group $\mathrm{Spin}(m)$ of the real Clifford algebra $\mathbb{R}_{0,m}$. The mother wavelets themselves are derived from intentionally devised orthogonal polynomials in Euclidean space. It should be noted that these orthogonal polynomials are not tensor products of one–dimensional ones, but genuine multidimensional ones satisfying the usual properties such as a Rodrigues formula, recurrence relations and differential equations. In this way multidimensional Clifford wavelets were constructed grafted on the Hermite polynomials [35], Laguerre polynomials [38], Gegenbauer polynomials [39], Jacobi polynomials [43] and Bessel functions [36].

Taking the dimension m to be even, say $m = 2n$, introducing a complex structure, i.e. an $\mathrm{SO}(2n)$–element squaring up to -1, and considering functions with values in the complex Clifford algebra $\mathbb{C}_{2n}$, so-called Hermitian Clifford analysis originates as a refinement of standard or Euclidean Clifford analysis. It should be noticed that the traditional holomorphic functions of several complex variables appear as a special case of Hermitian Clifford analysis, when the function values are restricted to a specific homogeneous part of spinor space. In this Hermitian setting the standard Dirac operator, which is invariant under the orthogonal group $O(m)$, is split into two Hermitian Dirac operators, which are now invariant under the unitary group $U(n)$. Also in this Hermitian Clifford analysis framework, multidimensional wavelets have been introduced by Brackx, H. De Schepper and Sommen [41,44], as kernels for a Hermitian Continuous Wavelet Transform, and (generalized) Hermitian Clifford Hermite polynomials have been devised to generate the corresponding Hermitian wavelets [45,47].

1.4.2 Further developments in quaternion and Clifford wavelet theory

Clifford algebra multiresolution analysis (MRA) has been pioneered by M. Mitrea [264]. Important are also the electromagnetic signal application oriented developments of Clifford algebra wavelets by G. Kaiser [228–230, 232].

Quaternion MRA wavelets with applications to image analysis have been developed in [314] by Traversoni. Clifford algebra multiresolution analysis has been applied by Bayro-Corrochano [22–24] to: Clifford wavelet neural networks (information processing), also considering quaternionic MRA, a quaternionic wavelet phase concept, as well as applications to (e.g. robotic) motion estimation and image processing.

Beyond this Zhao and Peng [335] established a theory of quaternion-valued admissible wavelets. Zhao [336] studied Clifford algebra-valued admissible (continuous) wavelets using the complex Fourier transform for the spectral representation. Mawardi and Hitzer [252,253] extended this to continuous Clifford and Clifford-Gabor wavelets in $Cl(3,0)$ using the CFT of (1.3.1) for the spectral representation. They also studied a corresponding Clifford wavelet transform uncertainty principle. Hitzer [188,189] generalized this approach to continuous admissible Clifford algebra wavelets in real Clifford algebras $Cl(n,0)$ of dimensions $n = 2,3 (\text{mod } 4)$, i.e. $n = 2,3,6,7,10,11,\ldots$ Restricted to $Cl(n,0)$ of dimensions $n = 2 (\text{mod } 4)$ this approach has also been taken up in [257].

Kähler et al. [227] treated monogenic (Clifford) wavelets over the unit ball. Bernstein studied Clifford continuous wavelet transforms in $L_{0,2}$ and $L_{0,3}$ [27], as well as monogenic kernels and wavelets on the 3D sphere [28]. Bernstein et al. [29] further studied Clifford diffusion wavelets on conformally flat cylinders and tori. Soulard and Carré extend in [309] the theory and application of monogenic wavelets to color image denoising. De Martino and Diki [103] establish a special one dimensional quaternion short-time Fourier transform (QSTFT). Its construction is based on a so called slice hyperholomorphic Segal-Bargmann transform.

CHAPTER 2

Clifford algebra

We study fundamental aspects of Clifford algebra, survey its modern formulation, look at how imaginary numbers are dealt with in real multivector algebra, and introduce useful aspects of quaternion based geometry.

Sections 2.1 to 2.5 aim to highlight several fundamental aspects of Clifford's geometric algebra (GA). They are mainly based on unpublished notes for a linear algebra course with some geometric algebra content at the University of the Air in Japan broadcast between 2004 and 2008 [177], and cover questions which often arise when first dealing with Clifford's geometric algebra, and on how familiar linear algebra concepts are dealt with. Because Clifford's geometric algebra serves later in this book as the algebraic grammar for hypercomplex integral transforms, it is important to become familiar with this essential extension of Grassmann algebra. Section 2.6 provides an overview of important geometric algebra examples of the Euclidean plane, Euclidean three-dimensional space, spacetime and of conformal geometric algebra. Then Section 2.7 studies how real multivector elements of geometric algebra replace the many roles (e.g. in Fourier transform kernels) of imaginary numbers, providing a real geometric interpretation at the same time. Finally, Section 2.8 studies in detail the role quaternions play for rotations in three and four dimensions.

2.1 AXIOMS OF CLIFFORD'S GEOMETRIC ALGEBRA

2.1.1 Axioms for geometric algebra $\mathcal{R}_{p,q} = Cl(p,q)$

2.1.1.1 Algebra over field of real numbers

The set $\mathbb{R}$ of real numbers forms a *field*. For the addition of $a, b, c \in \mathbb{R}$ we have the properties of

$a + b = b + a$	commutativity
$(a + b) + c = a + (b + c)$	associativity
$a + 0 = a$	zero 0
$a + (-a) = 0$	opposite $-a$ of a.

DOI: 10.1201/9781003184478-2

For the *multiplication* of $a, b, c \in \mathbb{R}$ we have the properties of

$(a+b)c = bc + ac$	
$a(b+c) = ab + ac$	distributivity
$(ab)c = a(bc)$	associativity
$1a = a$	unity $1 \neq 0$
$aa^{-1} = 1$	inverse a^{-1} for $a \neq 0$.
$ab = ba$	commutativity.

Definition 2.1.1. *An algebra over $\mathbb{R}$ is a linear space A over $\mathbb{R}$ together with a bilinear map (implying distributivity) $A \times A \to A, (a,b) \to ab$.*

Hence, for $\mathcal{G}_n$ to be a Clifford geometric algebra over the real n-dimensional Euclidean vector space $\mathbb{R}^n$, the *geometric product* of elements A, B, C $\in \mathcal{G}_n$ must satisfy the following axioms:

Axiom 2.1.2. *Addition is commutative:*

$$A + B = B + A. \tag{2.1.1}$$

Axiom 2.1.3. *Addition and the geometric product are associative:*

$$(A+B)+C = A+(B+C), \qquad A(BC) = (AB)C, \tag{2.1.2}$$

and distributive:

$$A(B+C) = AB + AC, \qquad (A+B)C = AC + BC. \tag{2.1.3}$$

Axiom 2.1.4. *There exist unique additive and multiplicative identities 0 and 1 such that:*

$$A + 0 = A, \qquad 1A = A. \tag{2.1.4}$$

Axiom 2.1.5. *Every A in $\mathcal{G}_n$ has an additive inverse:*

$$A + (-A) = 0. \tag{2.1.5}$$

Axiom 2.1.6. *For any nonzero vector $\boldsymbol{a}$ in $\mathcal{G}_n$ the square of $\boldsymbol{a}$ is equal to a unique positive scalar $|\boldsymbol{a}|^2$, that is*

$$\boldsymbol{aa} = \boldsymbol{a^2} = |\boldsymbol{a}|^2 > 0. \tag{2.1.6}$$

Depending on the signature of the underlying vector space $\mathbb{R}^{p,q}$, for defining $Cl(p,q)$, zero and negative squares of vectors will also occur.

Axiom 2.1.7. *Every simple k-vector, $A_k = \boldsymbol{a}_1\boldsymbol{a}_2...\boldsymbol{a}_k$, can be factorized into pairwise orthogonal vector factors, which satisfy:*

$$\boldsymbol{a}_i\boldsymbol{a}_j = -\boldsymbol{a}_j\boldsymbol{a}_i, \qquad i,j = 1,2,...,k \quad \text{and} \quad i \neq j. \tag{2.1.7}$$

2.1.1.2 *Definition (1): using quadratic form*

Let $V = (\mathbb{R}^n, \mathcal{Q})$ be a real vector space with a non-degenerate quadratic form $\mathcal{Q}$: $V \to \mathbb{R}$ of signature (p, q) with $p + q = n$. It is conventional to use the abbreviation $\mathbb{R}^{p,q} = Cl(p, q)$. For vectors $\mathbf{a}, \mathbf{b} \in \mathbb{R}^{p,q}$ we associate with $\mathcal{Q}$ the symmetric bilinear form $\langle \mathbf{ab} \rangle = \frac{1}{2}[\mathcal{Q}(\mathbf{a} + \mathbf{b}) - \mathcal{Q}(\mathbf{a}) - \mathcal{Q}(\mathbf{b})]$.

Definition 2.1.8. *An associative algebra over the field $\mathbb{R}$ with unity 1 is the geometric algebra $\mathbb{R}_{p,q}$ of the non-degenerate quadratic form $\mathcal{Q}$ (signature (p, q) and $p+q = n$) on $\mathbb{R}^n$, which contains copies of the field $\mathbb{R}$ and of the vector space $\mathbb{R}^{p,q}$ as distinct subspaces so that*

(1) $\mathbf{a}^2 = \mathcal{Q}(\mathbf{a})$ for any $\mathbf{a} \in \mathbb{R}^{p,q}$;

(2) $\mathbb{R}^{p,q}$ generates $\mathbb{R}_{p,q}$ as an algebra over the field $\mathbb{R}$;

(3) $\mathbb{R}_{p,q} = Cl(p, q)$ is not generated by any proper subspace of $\mathbb{R}^{p,q}$.

Note the deliberate choice of unity 1! If $q = 0$ the second index is often omitted

$$\mathbb{R}_n = \mathbb{R}_{n,0} \quad \text{or} \quad Cl_n = Cl(n, 0). \tag{2.1.8}$$

Using an orthonormal basis $\{\mathbf{e}_1, \mathbf{e}_2, \ldots, \mathbf{e}_n\}$ for $\mathbb{R}^{p,q}$, the condition (1) can be expressed as

$$\mathbf{e}_k^2 = 1, \quad 1 \le k \le p, \quad \mathbf{e}_k^2 = -1, \quad p < k \le n, \quad \mathbf{e}_k\mathbf{e}_l = -\mathbf{e}_l\mathbf{e}_k, \quad k < l. \tag{2.1.9}$$

(3) is only needed for signatures $p - q = 1 \mod 4$ where $(\mathbf{e}_1\mathbf{e}_2 \ldots \mathbf{e}_n)^2 = 1$.

The basic multiplication rules for the basis vectors of an orthonormal basis can also be used to define a geometric algebra. This was the approach taken by Clifford himself in 1878 [77] and 1882 [78].

2.1.1.3 *Definition (2): by basic multiplication rules*

Historically the first multiplication rule for vectors of the linear space $\mathbb{R}^n$ of importance for us is the bilinear outer (exterior) product of vectors. For a fixed orthonormal basis $\{\mathbf{e}_1, \mathbf{e}_2, \ldots, \mathbf{e}_n\}$ of $\mathbb{R}^n$ Grassmann introduced "bivectors"

$$\mathbf{e}_k \wedge \mathbf{e}_l = -\mathbf{e}_l \wedge \mathbf{e}_k, k \neq l$$

$$\mathbf{e}_k \wedge \mathbf{e}_l = 0, k = l.$$

The set of all bivectors $\{\mathbf{e}_k \wedge \mathbf{e}_l | k < l\}$ forms a basis of a new linear space $\bigwedge^2 \mathbb{R}^n$ of dimension $\begin{pmatrix} n \\ 2 \end{pmatrix}$. The name outer (or exterior) product stems from the fact, that it assigns to any pair of vectors of the original space $\mathbb{R}^n$ an element of a different vector space $\bigwedge^2 \mathbb{R}^n$.

Grassmann then defines multivectors. To any r-tuple $\{\mathbf{v}_1, \mathbf{v}_2, \ldots, \mathbf{v}_r\}$ of vectors $\mathbf{v}_k \in \mathbb{R}^n$, he assigns the multivector of grade r (or rank r, German: Stufe r)

$$\mathbf{v}_1 \wedge \mathbf{v}_2 \wedge \ldots \wedge \mathbf{v}_r.$$

$\mathbf{v}_1 \wedge \mathbf{v}_2 \wedge \ldots \wedge \mathbf{v}_r$ is set zero if the $\mathbf{v}_k$ are linearly dependent. The grade r multivectors change sign, under the permutation of any two adjacent vectors

$$\mathbf{v}_1 \wedge \ldots \wedge \mathbf{v}_k \wedge \mathbf{v}_{k+1} \wedge \ldots \wedge \mathbf{v}_r = -\mathbf{v}_1 \wedge \ldots \wedge \mathbf{v}_{k+1} \wedge \mathbf{v}_k \wedge \ldots \wedge \mathbf{v}_r$$

and form their own vector space $\bigwedge^r \mathbb{R}^n$. It follows that the set of all r-vectors of the form

$$\mathbf{e}_{k_1} \wedge \mathbf{e}_{k_2} \wedge \ldots \wedge \mathbf{e}_{k_r}, \quad \{k_1, k_2, \ldots, k_r\} \ r\text{-subset of } \{1, 2, \ldots, n\}$$

forms a basis of the new vector space $\bigwedge^r \mathbb{R}^n$ of dimension $\begin{pmatrix} n \\ r \end{pmatrix}$.

The full exterior algebra (or Grassmann algebra) is the vector space $\bigwedge \mathbb{R}^n$ of dimension 2^n with the r-vectors $\mathbf{e}_{k_1} \wedge \mathbf{e}_{k_2} \wedge \ldots \wedge \mathbf{e}_{k_r}, 1 \leq r \leq n$ as a basis.

For defining the associative bilinear *geometric product* of vectors, Clifford used Grassmann's multiplication rule of orthogonal vectors,

$$\mathbf{e}_k \mathbf{e}_l = -\mathbf{e}_l \mathbf{e}_k, k \neq l,$$

but set the geometric product of a vector with itself to be a real number

$$\mathbf{e}_k \mathbf{e}_k = \pm 1.$$

The associative algebra of dimension 2^n so defined is the geometric algebra $\mathbb{R}_{p,q} = Cl(p,q)$. p says how many basis vectors of an orthogonal basis of the linear space $\mathbb{R}^n$ have positive square and the rest $q = n - p$ have negative square.

The geometric product operating on multivectors $u, v, w \in \bigwedge \mathbb{R}^{p,q}$ can be expressed with the help of the *left contraction* $u \rfloor v \in \bigwedge \mathbb{R}^{p,q}$

(a) $\mathbf{x} \rfloor \mathbf{y} = \frac{1}{2}(\mathbf{x}\mathbf{y} + \mathbf{y}\mathbf{x}) \in \mathbb{R}$

(b) $\mathbf{x} \rfloor (u \wedge v) = (\mathbf{x} \rfloor u) \wedge v + \hat{u} \wedge (\mathbf{x} \rfloor v)$

(c) $(u \wedge v) \rfloor w = u \rfloor (v \rfloor w)$

for $\mathbf{x}, \mathbf{y} \in \mathbb{R}^{p,q}$ and the grade involution

$$\hat{u} = \sum_r (-1)^r \langle u \rangle_r,$$

with $\langle u \rangle_r$ the grade r-vector parts of u. (a) shows that the left contraction generalizes the usual inner product of 1-vectors. (b) indicates that the vector $\mathbf{x}$ acts like a "derivation". With the help of the left contraction, the geometric product of a vector $\mathbf{x} \in \mathbb{R}^{p,q}$ and a general multivector $u \in \bigwedge \mathbb{R}^{p,q}$ can be written as

$$\mathbf{x}u = \mathbf{x} \rfloor u + x \wedge u.$$

Demanding associativity and linearity, the geometric product extends to all of $\bigwedge \mathbb{R}^{p,q}$. This gives the bilinear map which equips Grassmann's exterior algebra vector space $\bigwedge \mathbb{R}^{p,q}$ to become the geometric *algebra* $\mathbb{R}_{p,q} = Cl(p,q)$.

2.1.1.4 Definition (3): compact form

Definition 2.1.9 (Clifford's geometric algebra [131, 247]). *Let* $\{e_1, e_2, \ldots, e_p, e_{p+1}, \ldots, e_n\}$, *with* $n = p + q$, $e_k^2 = \varepsilon_k$, $\varepsilon_k = +1$ *for* $k = 1, \ldots, p$, $\varepsilon_k = -1$ *for* $k = p + 1, \ldots, n$, *be an orthonormal base of the inner product vector space* $\mathbb{R}^{p,q}$ *with a geometric product according to the multiplication rules*

$$e_k e_l + e_l e_k = 2\varepsilon_k \delta_{k,l}, \qquad k, l = 1, \ldots n, \tag{2.1.10}$$

where $\delta_{k,l}$ *is the Kronecker symbol with* $\delta_{k,l} = 1$ *for* $k = l$, *and* $\delta_{k,l} = 0$ *for* $k \neq l$. *This non-commutative product and the additional axiom of associativity generate the* 2^n-*dimensional Clifford geometric algebra* $Cl(p, q) = Cl(\mathbb{R}^{p,q}) = Cl_{p,q} = \mathcal{G}_{p,q} = \mathbb{R}_{p,q}$ *over* $\mathbb{R}$. *The set* $\{e_A : A \subseteq \{1, \ldots, n\}\}$ *with* $e_A = e_{h_1} e_{h_2} \ldots e_{h_k}$, $1 \leq h_1 < \ldots < h_k \leq n$, $e_\emptyset = 1$, *forms a graded (blade) basis of* $Cl(p, q)$. *The grades* k *range from* 0 *for scalars,* 1 *for vectors,* 2 *for bivectors,* s *for* s*-vectors, up to* n *for pseudoscalars. The vector space* $\mathbb{R}^{p,q}$ *is included in* $Cl(p, q)$ *as the subset of 1-vectors. The general elements of* $Cl(p, q)$ *are real linear combinations of basis blades* e_A, *called Clifford numbers, multivectors or hypercomplex numbers.*

2.1.1.5 Grade r subspaces

All elements $u \in \mathbb{R}_{p,q} = Cl(p, q)$ are defined as sums of terms of length r, summing over all $r = 0 \ldots n$. Terms of constant length r

$$\langle u \rangle_r = \sum_{1 \leq k_1 < \ldots < k_r \leq n} u_{k_1 \ldots k_r} \mathbf{e}_{k_1} \ldots \mathbf{e}_{k_r}$$

are said to be of degree (or grade) r or $r-$vectors. This notion is independent of the choice of the orthonormal basis $\{\mathbf{e}_1, \mathbf{e}_2, \ldots, \mathbf{e}_n\}$. Linear combinations of degree $r = 2r'$ are called *even*, those with a degree $r = 2r' + 1$ are called *odd*. According to the definition, $\mathbb{R}^{p,q}$ forms the subspace of all $1-$vectors and $\mathbb{R}$ the subspace of all scalars (that is $0-$vectors or elements of grade 0). The geometric algebra $\mathbb{R}_{p,q} = Cl(p, q)$ is the direct sum of its (even and odd) subspaces $\bigwedge^r \mathbb{R}^{p,q}$ of $r-$vectors:

$$\mathbb{R}_{p,q} = \mathbb{R} \oplus \mathbb{R}^{p,q} \oplus \bigwedge^2 \mathbb{R}^{p,q} \oplus \ldots \oplus \bigwedge^n \mathbb{R}^{p,q}.$$

As mentioned earlier, the dimension of each subspace $\bigwedge^r \mathbb{R}^{p,q}$ of $r-$vectors is $\binom{n}{r}$.

Therefore the dimension of $\bigwedge^n \mathbb{R}^{p,q}$ is $\binom{n}{n} = 1$. The n-vectors are all scalar multiples of the pseudoscalar (oriented n-volume)

$$I = I_n = \mathbf{e}_1 \wedge \mathbf{e}_2 \wedge \ldots \wedge \mathbf{e}_n.$$

2.1.1.6 Reverse and principal reverse

The *reverse* of $M \in Cl(p, q)$ is defined as

$$\widetilde{M} = \sum_{k=0}^{n} (-1)^{\frac{k(k-1)}{2}} \langle M \rangle_k. \tag{2.1.11}$$

Taking the *reverse* is equivalent to reversing the order of products of basis vectors in the basis blades e_A.

The *principal reverse* [7, 54, 55, 239] of $M \in Cl(p,q)$ is defined as

$$\widetilde{M}^P = \sum_{k=0}^{n} (-1)^{\frac{k(k-1)}{2}} \langle \overline{M} \rangle_k. \tag{2.1.12}$$

The operation $\overline{M}$ means here to additionally change in the basis decomposition of M the sign of every vector of negative square $\overline{e_A} = \varepsilon_{h_1} e_{h_1} \varepsilon_{h_2} e_{h_2} \ldots \varepsilon_{h_k} e_{h_k}$, $1 \le h_1 < \ldots < h_k \le n$. Note that the notation $\overline{M}$ is usually reserved for the Clifford conjugation (composition of reversion and grade involution). Reversion, $\overline{M}$ in (2.1.12), and principal reversion are all involutions.

Note that both often replace complex conjugation and quaternion conjugation. The upper index P in the notation for principal reversion $\widetilde{M}^P$ in (2.1.12) is usually omitted, as both definitions agree on Clifford algebras $Cl(n,0)$, and whenever needed it is pointed out which of the two is employed in the context of a particular Clifford Fourier transform.

2.1.2 Geometric algebra $\mathcal{R}_2 = Cl(2,0)$

2.1.2.1 *Complex numbers*

$\mathbb{R}_2 = \mathbb{R}_{2,0} = Cl(2,0)$ is a 4−dimensional real algebra with a basis $\{1, \mathbf{e}_1, \mathbf{e}_2, \mathbf{e}_{12} = \mathbf{e}_1\mathbf{e}_2\}$. The multiplication table is Table 2.3. $\mathbb{R}_2$ has grade 0, grade 1 and grade 2 subspaces spanned by

1	$\mathbb{R}$	scalars (even)
$\mathbf{e}_1, \mathbf{e}_2$	$\mathbb{R}^2$	vectors (odd)
$\mathbf{e}_{12}$	$\bigwedge^2 \mathbb{R}^2$	bivectors (even)

The geometric algebra $\mathbb{R}_2 = Cl(2,0)$ can also be written as the direct sum $\mathbb{R}_2 = \mathbb{R}_2^+ \oplus \mathbb{R}_2^- = Cl^+(2,0) + Cl^-(2,0)$ of its even and odd parts:
$Cl^+(2,0) = \mathbb{R}_2^+ = \mathbb{R} \oplus \bigwedge^2 \mathbb{R}^2$ (even)
$Cl^-(2,0) = \mathbb{R}_2^- = \mathbb{R}^2$ (odd).
The even part is not only a subspace but also a subalgebra isomorphic to the field $\mathbb{C}$ of complex numbers.

2.1.2.2 *Reflections and rotations*

To learn more about the geometric meaning of Clifford algebra, we look at how it elegantly expresses reflections and rotations. In a Euclidean space, we can fix one point as origin O. All other points are then defined by their position vectors $\mathbf{x}$. Each straight line through O can be expressed in terms of a (unit length)direction vector $\mathbf{a}$ with $\mathbf{aa} = 1$. Using the fact that in the geometric product of vectors, parallel components commute and orthogonal components anticommute, we can reverse the sign of the orthogonal (to $\mathbf{a}$) component of a general vector $\mathbf{x}$ by

$$\mathbf{x}' = \mathbf{axa} = \mathbf{aa}(\mathbf{x}_\parallel - \mathbf{x}_\perp) = \mathbf{x}_\parallel - \mathbf{x}_\perp.$$

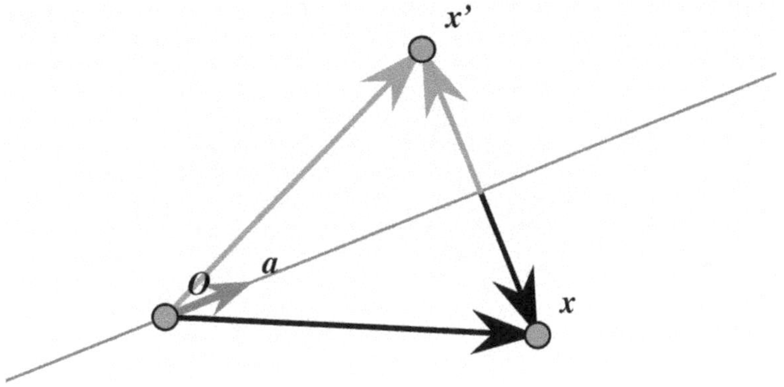

Figure 2.1 Reflection at a line through O in direction $\mathbf{a}$. Source: [180, Fig. 1].

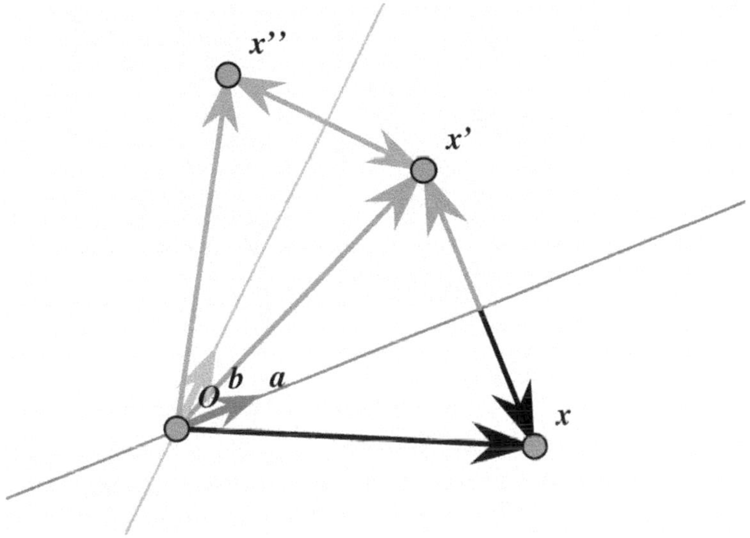

Figure 2.2 Rotation by two reflections. Source: [180, Fig. 2].

This describes a *reflection* at the straight line through O in the direction $\mathbf{a}$ (Figure 2.1).

Two successive reflections at two lines of angle $\vartheta/2$ produce a *rotation* by the angle ϑ. If the two lines intersect in O and have unit length direction vectors $\mathbf{a}$ and $\mathbf{b}$ with angle $\vartheta/2$, the rotation by ϑ is described by (compare Figure 2.2.)

$$\mathbf{x}'' = \mathbf{b}\mathbf{x}'\mathbf{b}2re = \mathbf{baxab}.$$

The product $R = \mathbf{ba}$ is the *rotation operator*, the *rotor*. The reverse product $\widetilde{R} = \mathbf{ab}$ is often just referred to as the *reverse*. A second successive rotation with rotor $R' = \mathbf{cb}$ by twice the angle $\vartheta'/2$ between the vectors $\mathbf{c}$ and $\mathbf{b}$ combines as expected to give the rotation by $\vartheta + \vartheta'$

$$\mathbf{x}''' = \mathbf{cbbaxabbc} = \mathbf{caxac}$$

with rotor $R'' = R'R$. The multiplication of two rotors gives therefore a new rotor.

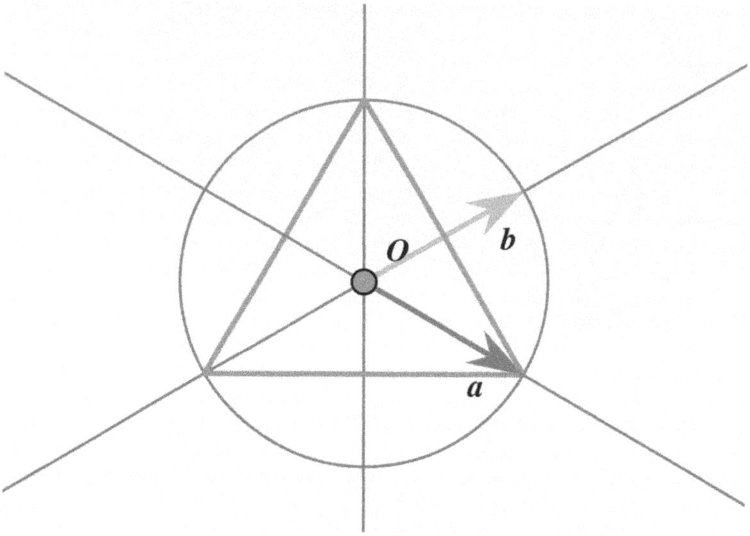

Figure 2.3 Vectors $\mathbf{a}, \mathbf{b}$ generate $2\mathcal{H}_3$. Source: [180, Fig. 3].

This yields the dirotation group of rotations. (The prefix *di* indicates that the rotors $\pm R$ describe *oriented* equivalent rotations with opposite senses. Physicists call this representation also the *spin-$\frac{1}{2}$ representation* of the rotation group.)

2.1.2.3 Two-dimensional point groups

A regular polygon with k sides in two dimensions has $2k$ oriented lines of reflections, that leave the polygon invariant. These are the k lines through the k corners and the k lines through the middles of each side. (Each line is counted twice giving two orientations to each reflection, by the two orientations $\pm\mathbf{a}$ of the direction vectors.) Combining an even number of these reflections, we get the symmetry rotations of the polygon. Odd numbers of reflections simply combine to new reflections. All combinations of reflections together generate the symmetry group of the polygon, the *dihedral group* $2\mathcal{H}_k$. Compare Table 2.1 for $k = 3$. The subgroup of symmetry rotations is simply the even subgroup $2\mathcal{C}_k$ of the dihedral group, also called *dicyclic group*. If we don't distinguish between the two "orientations" of each reflection, we have only k distinct symmetry reflections of the k-polygon, generating the group $\mathcal{H}_k$. Its even subgroup $\mathcal{C}_k$ comprises k distinct symmetry rotations.

Each dihedral group can be generated from only two reflections at straight lines passing through the center O of a k-polygon. The first straight line passes through a corner (direction vector $\mathbf{a}$) and the second straight line through the midpoint of a side (direction vector $\mathbf{b}$). Compare Figure 2.3 for $k = 3$. Repeated reflections at both lines generate all symmetry transformations of the dihedral group. The kth power of the elementary rotor $R = \mathbf{ba}$ leads to a full rotation by 360 degree

$$(\mathbf{ba})^k = -1, \quad R^k\mathbf{x}\widetilde{R}^k = (-1)^2\mathbf{x} = \mathbf{x}.$$

Table 2.1 The 12 distinct elements of the dihedral group $2\mathcal{H}_3$. Source: [180, Tab. 1].

3 distinct "positive sense" rotations with rotors	3 distinct "negative sense" rotations with rotors	6 distinct reflections at "oriented" lines with directions
$1 = \mathbf{a}^2 = \mathbf{b}^2$	$-1 = (\mathbf{ba})^3 = (\mathbf{ab})^3$	$\pm\mathbf{a}$
$\mathbf{ba}$	$-\mathbf{ba} = \mathbf{ba}(\mathbf{ab})^3 = (\mathbf{ab})^2$	$\pm\mathbf{b}$
$(\mathbf{ba})^2$	$-(\mathbf{ba})^2 = \mathbf{ab}$	$\pm\mathbf{bab} = \pm\mathbf{aba}$

Table 2.2 Multiplication table of $\mathcal{G}_3 = Cl(3,0)$ basis elements. Source: [251, Tab. 1].

	1	e_1	e_2	e_3	e_{12}	e_{31}	e_{23}	e_{123}
1	1	e_1	e_2	e_3	e_{12}	e_{13}	e_{23}	e_{123}
e_1	e_1	1	e_{12}	$-e_{31}$	e_2	$-e_3$	e_{123}	e_{23}
e_2	e_2	$-e_{12}$	1	e_{23}	$-e_1$	e_{123}	e_3	e_{31}
e_3	e_3	e_{31}	$-e_{23}$	1	e_{123}	e_1	$-e_2$	e_{12}
e_{12}	e_{12}	$-e_2$	e_1	e_{123}	-1	e_{23}	$-e_{31}$	$-e_3$
e_{31}	e_{31}	e_3	e_{123}	$-e_1$	$-e_{23}$	-1	e_{12}	$-e_2$
e_{23}	e_{23}	e_{123}	$-e_3$	e_2	e_{31}	$-e_{12}$	-1	$-e_1$
e_{123}	e_{123}	e_{23}	e_{31}	e_{12}	$-e_3$	$-e_2$	$-e_1$	-1

This relation for the kth power of R together with the unity conditions $\mathbf{a}^2 = \mathbf{b}^2 = 1$ determines the dicyclic rotor group $2\mathcal{C}_k$ completely.

2.1.2.4 *Clifford's geometric algebra $\mathcal{G}_3 = Cl(3,0)$ of $\mathbb{R}^3$*

Here we only provide the multiplication table (Table 2.2) for $Cl(3,0)$, which shows that $Cl(2,0)$ can simply be extended to $Cl(3,0)$ by introducing a third orthonormal basis vector $\mathbf{e}_3$. A detailed discussion of $Cl(3,0)$ will follow in Section 2.6.3.

2.2 QUADRATIC FORMS IN CLIFFORD'S GEOMETRIC ALGEBRA

2.2.1 Definition of geometric algebra with quadratic form

Quadratic forms allow a basis-free definition[1] of Clifford geometric algebra [53, 247].

Definition 2.2.1. *An associative algebra over a field F with unity 1 is the Clifford geometric algebra $\mathcal{G}(\mathcal{Q})$ of a non-degenerate quadratic form $\mathcal{Q}$ on a linear space V over the field F, if $\mathcal{G}(\mathcal{Q})$ contains the linear space V itself and the field $F = F \cdot 1$ as distinct subspaces so that*

[1]The following definition is not the only one using quadratic forms. C. Chevalley gave another definition in terms of a tensor algebra divided by an ideal. For the generation of this ideal a quadratic form becomes again essential ([71], chapter 11).

- $\mathbf{x}^2 = \mathcal{Q}(\mathbf{x}) \ \forall \mathbf{x} \in V$
- *V generates $\mathcal{G}(\mathcal{Q})$ as a algebra over the field F*
- *$\mathcal{G}(\mathcal{Q})$ is not generated by any proper subspace of V.*

The third condition guarantees the universal property for odd dimensions n with signatures 1 mod 4 of the quadratic form $\mathcal{Q}$, and the dimension of $\mathcal{G}(\mathcal{Q})$ to be 2^n. In general, the quadratic form $\mathcal{Q}$ is associated to the following symmetric bilinear form:

$$\begin{aligned} \mathbf{x} * \mathbf{y} &\equiv \frac{1}{2}[\mathcal{Q}(\mathbf{x}+\mathbf{y}) - \mathcal{Q}(\mathbf{x}) - \mathcal{Q}(\mathbf{y})] = \frac{1}{2}[(\mathbf{x}+\mathbf{y})^2 - \mathbf{x}^2 - \mathbf{y}^2] \\ &= \frac{1}{2}[\mathbf{xy} + \mathbf{xy}] \in \mathbb{R} \ \forall \mathbf{x}, \mathbf{y} \in V. \end{aligned}$$

As an example, let us look at the real ($F = \mathbb{R}$) linear quadratic space $V = \mathbb{R}^{p,q}$, which generates the geometric algebra $\mathbb{R}_{p,q} = \mathcal{G}(\mathbb{R}^{p,q}) = Cl(p,q)$ of signature $p - q$. With the help of $\mathcal{Q}$ (or the scalar $*$ product[110]), we can introduce an orthonormal basis $\{\mathbf{e}_1, \mathbf{e}_2, \ldots, \mathbf{e}_n\}$ or $\mathbb{R}^{p,q}$.

The quadratic form applied to the orthonormal basis vectors gives explicitly

$$\mathbf{e}_i^2 = 1, \ \forall 1 \le i \le p, \quad \mathbf{e}_i^2 = -1, \ \forall p < i \le n.$$

For distinct orthonormal basis vectors we must have

$$\mathbf{e}_i * \mathbf{e}_j = \frac{1}{2}[\mathbf{e}_i\mathbf{e}_j + \mathbf{e}_j\mathbf{e}_i] = 0 \ \forall 1 \le i < j \le n.$$

This yields

$$\mathbf{e}_i\mathbf{e}_j = -\mathbf{e}_j\mathbf{e}_i \ \forall 1 \le i < j \le n,$$

i.e. the geometric product of orthogonal basis vectors is antisymmetric – exactly like the exterior product of Grassmann. New for Clifford's geometric product $\mathbf{xy}$ of vectors $\mathbf{x}$ and $\mathbf{y}$ is the symmetric scalar part $\mathbf{x} * \mathbf{y} \equiv \frac{1}{2}[\mathbf{xy} + \mathbf{xy}]$.

2.2.2 Examples of quadratic forms and associated geometric algebras

A number of examples appear in Section 2.1. These were:

- The geometric algebra $\mathbb{R}_2 = \mathbb{R}_{2,0} = Cl(2,0)$ of the Euclidean plane R^2 with $n = p = 2, q = 0$ and signature $p - q = 2 - 0 = 2$. The even subalgebra $\mathbb{R}_2^+$ of $\mathbb{R}_2$ was found to be isomorphic to the complex numbers.
- The geometric algebra $\mathbb{R}_3 = \mathbb{R}_{3,0} = Cl(3,0)$ of the three-dimensional Euclidean space $\mathbb{R}^3$ with $n = p = 3, q = 0$ and signature 3. Its even subalgebra $\mathbb{R}_3^+$ is found to be isomorphic to Hamilton's famous quaternion algebra, which allows the most elegant spinorial description of rotations.
- The geometric algebra $\mathbb{R}_{3,1} = Cl(3,1)$ of the four-dimensional ($n = 4$) Minkowski space $\mathbb{R}^{3,1}$, whose quadratic form is characterized by $p = 3, q = 1$ and signature $p - q = 2$. This particular geometric algebra is named Space Time

Algebra (STA), because it is of great use for uniformly describing physics. Its even part $\mathbb{R}_{3,1}^{+}$ is isomorphic to the geometric algebra of Euclidean space. This isomorphism depends on the particular choice of the (time) vector with negative square and singles out a laboratory frame for measurements.

- The geometric algebra $\mathbb{R}_{4,1} = Cl(4,1)$ of the five-dimensional ($n = 5$) quadratic linear space $\mathbb{R}^{4,1}$, whose quadratic form is characterized by $p = 4, q = 1$ with signature $p - q = 3$. $\mathbb{R}_{4,1}$ is very versatile as an algebraic model for three-dimensional Euclidean space, called homogeneous or conformal model of Euclidean space. Simple multivectors in $\mathbb{R}_{4,1}$ are in one-to-one correspondence with points, lines, planes, circles and spheres in Euclidean space. The operations of translation, rotation, join and intersection of these elements all become simple exception-free, monomial multivector product expressions.

2.2.3 Geometric algebras with quadratic form of signature $p = q$

2.2.3.1 *A new interpretation of the geometric algebra of the Minkowski plane*

Given a certain basis of n linearly independent vectors we have the freedom replace it by another set of n linearly independent vectors, performing a basis transformation. The expression for the quadratic form in the new basis depends on this basis transformation.

Let us look at the non-trivial example of the Minkowski plane $\mathbb{R}^{1,1}$ and its geometric algebra $\mathbb{R}_{1,1}$, with $n = 2$, $p = q = 1$, and signature $p - q = 1 - 1 = 0$. In the *orthonormal* basis $\{\mathbf{e}_0, \mathbf{e}_1\}$ the quadratic form relationship $\mathbf{x}^2 = \mathcal{Q}(\mathbf{x})$, $\forall \mathbf{x} \in \mathbb{R}^{1,1}$ can be expressed as

$$\mathbf{e}_0^2 = -1, \quad \mathbf{e}_1^2 = 1, \quad \mathbf{e}_0\mathbf{e}_1 = -\mathbf{e}_1\mathbf{e}_0 \equiv N, \tag{2.2.1}$$

and the orthonormality relation

$$\mathbf{e}_0 * \mathbf{e}_1 = 0, \tag{2.2.2}$$

where N is just another name for the bivector pseudoscalar I_2 of the geometric algebra $\mathbb{R}_{1,1}$ of the Minkowski plane.

A particular choice of a new basis, is the *null* basis $\{\mathbf{n}, \bar{\mathbf{n}}\}$, defined by the basis transformation[2]

$$\bar{\mathbf{n}} = \frac{1}{2}(\mathbf{e}_0 + \mathbf{e}_1), \quad \mathbf{n} = \frac{1}{2}(\mathbf{e}_0 - \mathbf{e}_1). \tag{2.2.3}$$

Because the new basis is no longer orthonormal, the same quadratic form relationship $\mathbf{x}^2 = \mathcal{Q}(\mathbf{x})$, $\forall \mathbf{x} \in \mathbb{R}^{1,1}$ looks now a little unfamiliar

$$\mathbf{n}^2 = \bar{\mathbf{n}}^2 = 0, \quad \mathbf{n} \wedge \bar{\mathbf{n}} = \frac{1}{2}N, \quad \text{and} \quad \mathbf{n} * \bar{\mathbf{n}} = \frac{1}{2}[\mathbf{n}\bar{\mathbf{n}} + \bar{\mathbf{n}}\mathbf{n}] = -\frac{1}{2}. \tag{2.2.4}$$

The last relationship $\mathbf{n} * \bar{\mathbf{n}} = \frac{1}{2}[\mathbf{n}\bar{\mathbf{n}} + \bar{\mathbf{n}}\mathbf{n}] = -\frac{1}{2}$ can be interpreted as a "duality"

---

[2]The factor $\frac{1}{2}$ in the definition of $\bar{\mathbf{n}}$ is the only major difference to the treatment in Section 2.1.

between the two one-dimensional spaces V^1 and V^{1*} spanned by the vectors $\mathbf{n}$ and $\bar{\mathbf{n}}$, respectively [106]. The duality condition being, that V^{1*} is the space of

$$\forall \mathbf{x} = x\mathbf{n} \in V^1, \;\; x \in \mathbb{R}: \;\; \exists_1 \bar{\mathbf{x}} = \frac{1}{x}\bar{\mathbf{n}} \in V^{1*} : \mathbf{x} * \bar{\mathbf{x}} = -\frac{1}{2}. \tag{2.2.5}$$

The inverse basis transformation is given by

$$\mathbf{e}_0 = \bar{\mathbf{n}} + \mathbf{n}, \quad \mathbf{e}_1 = \bar{\mathbf{n}} - \mathbf{n}. \tag{2.2.6}$$

2.2.3.2 Generalizing to the geometric mother algebra with $p = q = n$

We can now generalize the interpretation of the last subsection. We extend the one-dimensional space V^1 to an n-dimensional space V^n with an orthogonal null-vector basis $\{\mathbf{n}_1, \mathbf{n}_2, \ldots, \mathbf{n}_n\}$ and the *dual* space V^{1*} to the n-dimensional space V^{n*} with the orthogonal null-vector basis $\{\bar{\mathbf{n}}_1, \bar{\mathbf{n}}_2, \ldots, \bar{\mathbf{n}}_n\}$. The analogous duality conditions on the two basis are now that the basis vectors $\mathbf{n}_i$ are related to dual map vectors $\bar{\mathbf{n}}_j$ by

$$\mathbf{n}_i * \bar{\mathbf{n}}_j = -\frac{1}{2}\delta_{i,j} \;\; \forall i, j = 1, 2, \ldots, n. \tag{2.2.7}$$

$\delta_{i,j}$ is the usual Kronecker delta symbol. Assuming that all vectors $\mathbf{n}_i$ and $\bar{\mathbf{n}}_j$ are linearly independent, the direct sum of the two vector spaces V^n and V^{n*} is a $2n$-dimensional vector space

$$\mathbb{R}^{n,n} = V^n \oplus V^{n*}. \tag{2.2.8}$$

The geometric algebra of $\mathbb{R}^{n,n}$ is denoted by

$$\mathbb{R}_{n,n} = \mathcal{G}(\mathbb{R}^{n,n}) = Cl(n,n), \tag{2.2.9}$$

and called (geometric) *mother algebra*. The name stems from the fact that it is very useful for working with (multi)linear functions on n-dimensional vector spaces [106, 107].

Continuing the analogy, we change the combined null-vector basis of the vector space $\mathbb{R}^{n,n}$ into an orthonormal basis, that clearly shows the quadratic form to which the geometric mother algebra is associated

$$\bar{\mathbf{e}}_i = \bar{\mathbf{n}}_i + \mathbf{n}_i, \;\; \mathbf{e}_i = \bar{\mathbf{n}}_i - \mathbf{n}_i, \;\; 1 \leq i \leq n. \tag{2.2.10}$$

Based on the null-vector properties of the vectors $\mathbf{n}_i$ and $\bar{\mathbf{n}}_j$ and on the scalar product condition $\mathbf{n}_i * \bar{\mathbf{n}}_j = -\frac{1}{2}\delta_{i,j} \; \forall i, j = 1, 2, \ldots, n$, we can calculate the orthonormality relationships

$$\bar{\mathbf{e}}_i * \bar{\mathbf{e}}_j = -\delta_{i,j}, \;\; \mathbf{e}_i * \bar{\mathbf{e}}_j = 0, \;\; \mathbf{e}_i * \mathbf{e}_j = \delta_{i,j} \;\; \forall i, j = 1, 2, \ldots, n. \tag{2.2.11}$$

The orthonormal set of vectors $\{\mathbf{e}_1, \mathbf{e}_2, \ldots, \mathbf{e}_n\}$ is seen to span a real Euclidean vector space $\mathbb{R}^n$, and the orthonormal set of vectors $\{\bar{\mathbf{e}}_1, \bar{\mathbf{e}}_2, \ldots, \bar{\mathbf{e}}_n\}$ is seen to span an anti-Euclidean vector space $\bar{\mathbb{R}}^n$. Hence the $2n$-dimensional vector space $\mathbb{R}^{n,n}$ can also be written as the direct sum of

$$\mathbb{R}^{n,n} = \mathbb{R}^n \oplus \bar{\mathbb{R}}^n. \tag{2.2.12}$$

Finally the simple homogeneous grade $(p+q)$ multivectors ($(p+q)$-blades)

$$\mathbf{e}_1\mathbf{e}_2\ldots\mathbf{e}_p\,\bar{\mathbf{e}}_1\bar{\mathbf{e}}_2\ldots\bar{\mathbf{e}}_q \tag{2.2.13}$$

can be used to project out any desired subspace $\mathbb{R}^{p,q}$, because each $(p+q)$-blade is in one-to-one correspondence a subspace of $\mathbb{R}^{n,n}$([161], p. 19).

We therefore now understand in general, how any null-vector (sub)space with even dimension $2n'$ can be reinterpreted by a change of basis as a vector space $\mathbb{R}^{n',n'}$. In the same way that vector spaces and quadratic forms are invariant under a change of basis, the geometric algebra generated by a vector space and its quadratic form according to the general definition on page 23 is also invariant. This gives a strategy how to deal with degenerate geometric algebras, i.e. geometric algebras with quadratic forms that result in subspaces whose vectors square to zero.

A general notation used for vector spaces with arbitrary signatures is $\mathbb{R}^{p,q,r}$, where p,q,r mean the dimensions of the subspaces whose vectors have positive, negative or zero squares, respectively [164]. The third index r is often omitted, if $r=0$. According to the main argument of this subsection, any degenerate n-dimensional vector space is a subspace of the comprehensive vector space $\mathbb{R}^{n,n}$ and hence any degenerate geometric algebra can also be embedded in a larger non-degenerate geometric algebra $\mathbb{R}_{n,n}$, called the mother algebra of n-space.

As an example, the algebra $\mathbb{R}_{0,0,2}$ is a subalgebra of the mother algebra $\mathbb{R}_{2,2,0}$ and the degenerate algebra $\mathbb{R}_{3,0,1}$ is a subalgebra of conformal model $\mathbb{R}_{4,1,0}$. These two examples are closely related, because

$$R^{0,0,2} \subset R^{2,2,0}, \quad \mathbb{R}^{4,1,0} = \mathbb{R}^{3,0,0} \oplus \mathbb{R}^{1,1,0}. \tag{2.2.14}$$

Both relationships correspond to simple changes of basis.

2.3 CLIFFORD'S GEOMETRIC PRODUCT AND DERIVED PRODUCTS

2.3.1 The geometric product of multivectors

The aim of this passage is to show how the geometric product of multivectors is defined in general, extending the basic geometric product of vectors given by Clifford. An alternative definition of Clifford geometric algebra, that guarantees existence as quotient algebra of the tensor algebra was given by Chevalley in 1954 [71].

In line with [161], the approach I take here is to introduce how vectors and homogeneous simple multivectors are multiplied. By linearity this extends first to the products of vectors and homogeneous multivectors and second to the products of vectors and arbitrary multivectors. Third, using the factorization of homogeneous simple multivectors into a geometric product of anticommuting vectors, we can repeatedly apply the product formula for vectors and arbitrary multivectors to calculate products of homogeneous simple multivectors and arbitrary multivectors. Fourth, by linearity this then extends to fully general geometric products of arbitrary multivectors. For brevity I always refer to a linear space of the reals $\mathbb{R}^n$ and its geometric algebra $\mathbb{R}_n = Cl(n,0)$, but the formulas and definitions apply also to the geometric algebras $\mathbb{R}_{p,q} = Cl(p,q)$ of linear spaces $\mathbb{R}^{p,q}$ of arbitrary signature $\{p,q\}$.

Following Grassmann's definition, which was also adopted by Clifford, the anti-symmetric outer product of two vectors $\mathbf{a}, \mathbf{b}_1 \in \mathbb{R}^n$ is given by

$$\mathbf{a} \wedge \mathbf{b}_1 = -\mathbf{b}_1 \wedge \mathbf{a}, \tag{2.3.1}$$

mapping the pair $\mathbf{a}, \mathbf{b}_1$ to a bivector.

Following Clifford's definition, the symmetric contractions (or scalar product) of $\mathbf{a}, \mathbf{b}_1$ are given by the following grade 0 scalar

$$\mathbf{a} \rfloor \mathbf{b}_1 = \mathbf{a} \lfloor \mathbf{b}_1 = \mathbf{a} * \mathbf{b}_1 \in \mathbb{R}, \tag{2.3.2}$$

which also corresponds to the conventional inner product of vectors.

The full geometric product of the two vectors $\mathbf{a}, \mathbf{b}_1$ is the sum

$$\mathbf{a}\mathbf{b}_1 = \mathbf{a} \rfloor \mathbf{b}_1 + \mathbf{a} \wedge \mathbf{b}_1. \tag{2.3.3}$$

We now define a homogeneous simple grade r multivector $B_r \in \mathbb{R}_n$ to be the product of r anticommuting vectors

$$B_r = \mathbf{b}_1\mathbf{b}_2 \dots \mathbf{b}_r, \quad \mathbf{b}_j\mathbf{b}_k = -\mathbf{b}_k\mathbf{b}_j, \quad 1 \leq j < k \leq r \leq n. \tag{2.3.4}$$

The outer product of Grassmann maps $\mathbf{a}$ and B_r to a grade $r+1$ multivector which represents the grade $r+1$ part (where grade selection is indicated by angular brackets) of the geometric product $\mathbf{a}B_r$

$$\mathbf{a} \wedge B_r = \langle \mathbf{a}B_r \rangle_{r+1}. \tag{2.3.5}$$

The (left) contraction of a vector $\mathbf{a}$ and B_r is then given by the grade $r-1$ part of the geometric product $\mathbf{a}B_r$

$$\mathbf{a} \rfloor B_r = \langle \mathbf{a}B_r \rangle_{r-1}. \tag{2.3.6}$$

It can be calculated explicitly as a linear combination of homogeneous simple grade $r-1$ vectors

$$\mathbf{a} \rfloor B_r = \mathbf{a} \rfloor (\mathbf{b}_1\mathbf{b}_2 \dots \mathbf{b}_r) = \sum_{k=1}^{r} (-1)^{k+1} \mathbf{a} \rfloor \mathbf{b}_k (\mathbf{b}_1\mathbf{b}_2 \dots \cancel{\mathbf{b}_k}\, \mathbf{b}_{k+1} \dots \mathbf{b}_r) \tag{2.3.7}$$

where $\cancel{\mathbf{b}_k}$ means that $\mathbf{b}_k$ is to be omitted from the geometric product in round brackets. Because of the anticommutativity (i.e. orthogonality) of the vectors $\{\mathbf{b}_1, \mathbf{b}_2, \dots, \mathbf{b}_r\}$ we can rewrite $\mathbf{a} \rfloor B_r$ also as

$$\begin{aligned} \mathbf{a} \rfloor B_r &= \mathbf{a} \rfloor (\mathbf{b}_1 \wedge \mathbf{b}_2 \wedge \dots \wedge \mathbf{b}_r) \\ &= \sum_{k=1}^{r} (-1)^{k+1} \mathbf{a} \rfloor \mathbf{b}_k (\mathbf{b}_1 \wedge \mathbf{b}_2 \wedge \dots \wedge \cancel{\mathbf{b}_k} \wedge \dots \wedge \mathbf{b}_r), \end{aligned}$$

which will be of use later on.

The full geometric product $\mathbf{a}B_r$ is the sum of the (left) contraction and the outer product

$$\mathbf{a}B_r = \langle \mathbf{a}B_r \rangle_{r-1} + \langle \mathbf{a}B_r \rangle_{r+1} = \mathbf{a} \rfloor B_r + \mathbf{a} \wedge B_r. \tag{2.3.8}$$

Notice that we get two new grade parts: one from the left contraction resulting in a part of grade $r-1$ and one from the outer product resulting in a part of grade $r+1$.

Every multivector in a given geometric algebra can be represented as a linear combination of its homogeneous grade parts. Every grade part can be represented by a linear combination of simple multivectors of the same grade. This second summation will only become necessary for $n > 3$, because for $n \leq 3$ every homogeneous grade multivector can be factorized into a product of grade 1 vectors. For brevity I will omit the second summation over simple multivectors of the same grade.

$$B = \sum_{r=1}^{n} \langle B \rangle_r. \tag{2.3.9}$$

Applying linearity the product of a grade 1 vector $\mathbf{a}$ and a general multivector B is then given as

$$\mathbf{a}B = \mathbf{a} \sum_{r=1}^{n} \langle B \rangle_r = \sum_{r=1}^{n} \mathbf{a} \langle B \rangle_r = \sum_{r=1}^{n} (\mathbf{a} \rfloor \langle B \rangle_r + \mathbf{a} \wedge \langle B \rangle_r). \tag{2.3.10}$$

Next let us consider the geometric product of a simple grade s multivector A_s with a general multivector B. Because A_s is considered to be simple and of grade s, it can be factorized into a product of s anticommuting vectors

$$A_s = \mathbf{a}_1 \ldots \mathbf{a}_{s-1} \mathbf{a}_s, \quad \mathbf{a}_j \mathbf{a}_k = -\mathbf{a}_k \mathbf{a}_j, \quad 1 \leq j < k \leq s \leq n. \tag{2.3.11}$$

Making use of the associativity of the geometric product, we can therefore rewrite the product $A_s B$ as

$$A_s B = \mathbf{a}_1 \ldots \mathbf{a}_{s-1} \mathbf{a}_s B = \mathbf{a}_1 (\ldots (\mathbf{a}_{s-1} (\mathbf{a}_s B) \ldots). \tag{2.3.12}$$

This can now be explicitly calculated by repeatedly applying the previous formula for the product of a vector $\mathbf{a}$ and a general multivector B. Because each geometric multiplication of a vector $\mathbf{a}_k$, $1 \leq k \leq s$ with a homogeneous multivector part $\langle B \rangle_r$ yields two new parts of one grade lower and one grade higher, the result of $A_s \langle B \rangle_r$ will be a linear combination of parts with grades ranging from grade $r - s$ in steps of two up to the highest grade part $r + s$ (The left contraction will automatically be zero for the contraction of a vector with a scalar, therefore hypothetical negative grade parts will not occur, i.e. they will simply be zero.)

$$A_s \langle B \rangle_r = \langle A_s \langle B \rangle_r \rangle_{r-s} + \langle A_s \langle B \rangle_r \rangle_{r-s+2} + \ldots + \langle A_s \langle B \rangle_r \rangle_{r+s}. \tag{2.3.13}$$

To treat the most general case of the geometric product of two arbitrary multivectors A and B we represent A also as a linear combination of its homogeneous grade parts. (Each homogeneous grade part can in turn be represented as a linear combination of simple multivectors of the same grade. But again we omit this second summation for brevity.)

$$A = \sum_{s=1}^{n} \langle A \rangle_s. \tag{2.3.14}$$

The general product of two arbitrary multivectors is then by linearity

$$AB = \left(\sum_{s=1}^{n} \langle A \rangle_s \right) B = \sum_{s=1}^{n} (\langle A \rangle_s B). \tag{2.3.15}$$

To the expressions $\langle A \rangle_s B$ we can in turn apply factorization of the simple homogeneous grade parts of $\langle A \rangle_s$, etc. and break down the whole general product into simple elementary geometric products of grade 1 vectors (or in linear combinations of expressions just given in terms of left contractions and outer products of vectors).

To illustrate the method of explicit calculation, let us conclude this section with an example. For ease of calculation, we write the multivectors in terms of orthonormal basis vectors $\{\mathbf{e}_1, \mathbf{e}_2, \mathbf{e}_3\}$ of the linear space $\mathbb{R}^n$ and their geometric products. This corresponds already to decomposing the multivector factors A and B into linear combinations of homogeneous simple grade parts. The scalar coefficients for the magnitude of each homogeneous simple grade component are represented by Greek letters:

$$\begin{aligned} A &= \alpha + \alpha_1 \mathbf{e}_1 + \alpha_{12} \mathbf{e}_1 \mathbf{e}_2 + \alpha_{123} \mathbf{e}_1 \mathbf{e}_2 \mathbf{e}_3, \\ B &= \beta + \beta_2 \mathbf{e}_2 + \beta_{23} \mathbf{e}_2 \mathbf{e}_3 + \beta_{123} \mathbf{e}_1 \mathbf{e}_2 \mathbf{e}_3. \end{aligned} \tag{2.3.16}$$

The geometric product AB is

$$\begin{aligned} AB &= (\alpha + \alpha_1 \mathbf{e}_1 + \alpha_{12} \mathbf{e}_1 \mathbf{e}_2 + \alpha_{123} \mathbf{e}_1 \mathbf{e}_2 \mathbf{e}_3) \\ &\quad (\beta + \beta_2 \mathbf{e}_2 + \beta_{23} \mathbf{e}_2 \mathbf{e}_3 + \beta_{123} \mathbf{e}_1 \mathbf{e}_2 \mathbf{e}_3) \\ &= \alpha\beta + \alpha\beta_2 \mathbf{e}_2 + \alpha\beta_{23} \mathbf{e}_2 \mathbf{e}_3 + \alpha\beta_{123} \mathbf{e}_1 \mathbf{e}_2 \mathbf{e}_3 + \beta\alpha_1 \mathbf{e}_1 + \beta\alpha_{12} \mathbf{e}_1 \mathbf{e}_2 \\ &\quad + \beta\alpha_{123} \mathbf{e}_1 \mathbf{e}_2 \mathbf{e}_3 + \alpha_1\beta_2 \mathbf{e}_1 \mathbf{e}_2 + \alpha_1\beta_{23} \mathbf{e}_1 \mathbf{e}_2 \mathbf{e}_3 + \alpha_1\beta_{123} \mathbf{e}_1 \mathbf{e}_1 \mathbf{e}_2 \mathbf{e}_3 \\ &\quad + \alpha_{12}\beta_2 \mathbf{e}_1 \mathbf{e}_2 \mathbf{e}_2 + \alpha_{12}\beta_{23} \mathbf{e}_1 \mathbf{e}_2 \mathbf{e}_2 \mathbf{e}_3 + \alpha_{12}\beta_{123} \mathbf{e}_1 \mathbf{e}_2 \mathbf{e}_1 \mathbf{e}_2 \mathbf{e}_3 \\ &\quad + \alpha_{123}\beta_2 \mathbf{e}_1 \mathbf{e}_2 \mathbf{e}_3 \mathbf{e}_2 + \alpha_{123}\beta_{23} \mathbf{e}_1 \mathbf{e}_2 \mathbf{e}_3 \mathbf{e}_2 \mathbf{e}_3 \\ &\quad + \alpha_{123}\beta_{123} \mathbf{e}_1 \mathbf{e}_2 \mathbf{e}_3 \mathbf{e}_1 \mathbf{e}_2 \mathbf{e}_3. \end{aligned} \tag{2.3.17}$$

Because $\{\mathbf{e}_1, \mathbf{e}_2, \mathbf{e}_3\}$ is orthonormal, the square of each vector is unity $\mathbf{e}_1\mathbf{e}_1 = \mathbf{e}_1^2 = \mathbf{e}_2^2 = \mathbf{e}_3^2 = 1$ and vectors of different index anticommute, e.g. $\mathbf{e}_1\mathbf{e}_2 = -\mathbf{e}_2\mathbf{e}_1$, etc. We therefore have

$$
\begin{aligned}
AB &= \alpha\beta - \alpha_{123}\beta_{123}\mathbf{e}_2\mathbf{e}_1\mathbf{e}_1\mathbf{e}_3\mathbf{e}_3\mathbf{e}_2 \\
&\quad + \alpha\beta_2\mathbf{e}_2 + (\alpha_1\beta + \alpha_{12}\beta_2)\mathbf{e}_1 - \alpha_{12}\beta_{123}\mathbf{e}_1^2\mathbf{e}_2^2\mathbf{e}_3 - \alpha_{123}\beta_{23}\mathbf{e}_1\mathbf{e}_2^2\mathbf{e}_3^2 \\
&\quad + (\alpha_{12}\beta + \alpha_1\beta_2)\mathbf{e}_1\mathbf{e}_2 + (\alpha\beta_{23} + \alpha_1\beta_{123})\mathbf{e}_2\mathbf{e}_3 + \alpha_{12}\beta_{23}\mathbf{e}_1\mathbf{e}_3 \\
&\quad - \alpha_{123}\beta_2\mathbf{e}_1\mathbf{e}_2\mathbf{e}_2\mathbf{e}_3 + (\alpha\beta_{123} + \beta\alpha_{123} + \alpha_1\beta_{23})\mathbf{e}_1\mathbf{e}_2\mathbf{e}_3 \\
&= \alpha\beta - \alpha_{123}\beta_{123} \\
&\quad + (\alpha_1\beta + \alpha_{12}\beta_2 - \alpha_{123}\beta_{23})\mathbf{e}_1 + \alpha\beta_2\mathbf{e}_2 - \alpha_{12}\beta_{123}\mathbf{e}_3 \\
&\quad + (\alpha_{12}\beta + \alpha_1\beta_2)\mathbf{e}_1\mathbf{e}_2 + (\alpha\beta_{23} + \alpha_1\beta_{123})\mathbf{e}_2\mathbf{e}_3 \\
&\quad + (\alpha_{12}\beta_{23} - \alpha_{123}\beta_2)\mathbf{e}_1\mathbf{e}_3 \\
&\quad + (\alpha\beta_{123} + \beta\alpha_{123} + \alpha_1\beta_{23})\mathbf{e}_1\mathbf{e}_2\mathbf{e}_3,
\end{aligned}
\tag{2.3.18}
$$

where we have listed the result line by line in terms of grade 0 scalars, grade 1 vectors, grade 2 bivectors and grade 3 trivectors.

2.3.2 The scalar product

A useful product of multivectors $A, B \in \mathbb{R}_n$ that can be derived from their general geometric product is the so-called *scalar product* defined as the scalar part of the geometric product and indicated by an asterisk

$$A * B \equiv \langle AB \rangle_0 \in \mathbb{R}. \tag{2.3.19}$$

Note that the index 0 is sometimes dropped, so that angular brackets without an index come to mean the scalar part of the enclosed expression.

The scalar product of homogeneous multivectors A_s of grade s and B_r of grade r will only be different from zero, if $s = r$. Given that $s \leq r$ we can understand this restriction due to the fact that as described in the previous section, the geometric product A_sB_r has as its lowest grade part a term of grade $r - s$. For the case that $s > r$ we can reverse the order of all vector factors in the geometric product AB and get

$$\langle \widetilde{A_sB_r} \rangle = \langle \widetilde{B}_r\widetilde{A}_s \rangle = (-1)^{s(s-1)/2}(-1)^{r(r-1)/2}\langle B_rA_s \rangle. \tag{2.3.20}$$

The powers of (-1) are due to the anticommutativity of the vector factors of the simple grade components of the homogeneous multivectors A_s and B_r. Now we can apply the same argument as for the case $s \leq r$ and see in general that the scalar product is only nonvanishing if $r = s$. But if $r = s$ the powers of (-1) cancel each other and we find that

$$A_s * B_r = B_r * A_s. \tag{2.3.21}$$

That is the scalar product of two homogeneous mulitvectors is symmetric. By linearity, this extends to the scalar product of arbitrary multivectors

$$
\begin{aligned}
A * B &= \left(\sum_{s=1}^{n}\langle A\rangle_s\right) * \left(\sum_{r=1}^{n}\langle B\rangle_r\right) = \sum_{s=1}^{n}\langle A\rangle_s * \langle B\rangle_s = \sum_{s=1}^{n}\langle B\rangle_s * \langle A\rangle_s \\
&= B * A.
\end{aligned}
\tag{2.3.22}
$$

The scalar product inherits linearity from the geometric product.

As an example let us consider the scalar product of a simple grade $s-$vector A_s with its reverse $\widetilde{A}_s$

$$\widetilde{A}_s * A_s = \langle \widetilde{A}_s A_s \rangle = \langle \mathbf{a}_s \dots \mathbf{a}_2 \mathbf{a}_1 \mathbf{a}_1 \dots \mathbf{a}_{s-1} \mathbf{a}_s \rangle = \mathbf{a}_1^2 \mathbf{a}_2^2 \dots \mathbf{a}_s^2, \tag{2.3.23}$$

where we used the associativity of the geometric product, which allows us convenient pairwise multiplication of vector factors. Remember that the square of each vector is a scalar. In case that the linear space $\mathbb{R}^n$ has positive signature ($q = 0$) the above result will be positive and by linearity we can define a positive magnitude for a general multivector A as

$$|A|^2 = \sum_{s=1}^{n} \langle \widetilde{A} \rangle_s * \langle A \rangle_s \geq 0. \tag{2.3.24}$$

2.3.3 The outer product

The outer product in the Grassmann algebra [146] historically precedes the definition of geometric algebras by Clifford [77, 78]. But Clifford did not intend to do away with it, he rather wanted to unify Grassmann algebra and Hamilton's algebra of quaternions in a single algebraic framework. It is therefore not surprising, that the geometric product of two arbitrary multivectors comprises the outer product as the sum over the maximum grade parts of the result

$$A \wedge B = (\sum_{s=1}^{n} \langle A \rangle_s) \wedge (\sum_{r=1}^{n} \langle B \rangle_r) = \sum_{s=1}^{n} \sum_{r=1}^{n} \langle \langle A \rangle_s \langle B \rangle_r \rangle_{r+s}. \tag{2.3.25}$$

The outer product inherits both linearity and associativity from the geometric product.

As defined by Grassmann, the outer product of two vectors $\mathbf{a}, \mathbf{b}$ is antisymmetric

$$\mathbf{a} \wedge \mathbf{b} = -\mathbf{b} \wedge \mathbf{a}. \tag{2.3.26}$$

Calculating the reverse of two homogeneous simple grade s and grade r multivectors A_s and B_r we see that

$$\begin{aligned} \widetilde{A_s \wedge B_r} &= \widetilde{\langle A_s B_r \rangle}_{r+s} = \langle \widetilde{B}_r \widetilde{A}_s \rangle_{r+s} \\ &= (-1)^{s(s-1)/2} (-1)^{r(r-1)/2} \langle B_r A_s \rangle_{r+s}. \end{aligned} \tag{2.3.27}$$

On the other hand, we also have

$$\widetilde{\langle A_s B_r \rangle}_{r+s} = (-1)^{(s+r)(s+r-1)/2} \langle A_s B_r \rangle_{r+s}. \tag{2.3.28}$$

Equating both right sides again and taking care of the powers of (-1) we end up with the symmetry formula for the outer product of homogeneous multivectors

$$A_s \wedge B_r = (-1)^{rs} B_r \wedge A_s, \tag{2.3.29}$$

which includes the antisymmetry of the outer product of vectors as a special case for $r = s = 1$.

2.3.3.1 The cross product of three dimensions

In conventional three-dimensional vector analysis, frequent use is made of an anti-symmetric product of vectors, which results in a third vector perpendicular to the two vector factors, with the length equal to the area of the parallelogram spanned by these two vectors and with the orientation given by the so-called right hand rule. The names used for this product are: vector product, cross product or outer product and it is often indicated by an x-shaped product sign

$$\mathbf{a} \times \mathbf{b} = (a_2b_3 - a_3b_2)\mathbf{e}_1 + (a_3b_1 - a_1b_3)\mathbf{e}_2 + (a_1b_2 - a_2b_1)\mathbf{e}_3. \tag{2.3.30}$$

Historically, this product has actually been derived from Grassmann's outer product of vectors by mapping the resulting grade 2 bivector area element to the *dual* vector perpendicular to it. This is done by multiplication with the pseudoscalar volume element I_3.

$$\mathbf{a} \times \mathbf{b} \equiv -I_3\mathbf{a} \wedge \mathbf{b}, \quad \mathbf{a} \wedge \mathbf{b} = I_3\mathbf{a} \times \mathbf{b}. \tag{2.3.31}$$

For an orthonormal basis $\{\mathbf{e}_1, \mathbf{e}_2, \mathbf{e}_3\}$ of the Euclidean space $\mathbb{R}^3$ we have $I_3 = \mathbf{e}_1\mathbf{e}_2\mathbf{e}_3$. Let us conclude this subsection with an example:

$$\mathbf{a} = \mathbf{e}_1 + 3\mathbf{e}_2, \quad \mathbf{b} = \mathbf{e}_2 + 4\mathbf{e}_3, \tag{2.3.32}$$

$$\begin{aligned} \mathbf{a} \times \mathbf{b} &= -I_3(\mathbf{a} \wedge \mathbf{b}) \\ &= -I_3(\mathbf{e}_1 + 3\mathbf{e}_2) \wedge (\mathbf{e}_2 + 4\mathbf{e}_3) \\ &= -(\mathbf{e}_1\mathbf{e}_2\mathbf{e}_3)(\mathbf{e}_1\mathbf{e}_2 + 4\mathbf{e}_1\mathbf{e}_3 + 12\mathbf{e}_2\mathbf{e}_3) \\ &= -\mathbf{e}_1\mathbf{e}_2\mathbf{e}_3\mathbf{e}_1\mathbf{e}_2 - 4\mathbf{e}_1\mathbf{e}_2\mathbf{e}_3\mathbf{e}_1\mathbf{e}_3 - 12\mathbf{e}_1\mathbf{e}_2\mathbf{e}_3\mathbf{e}_2\mathbf{e}_3 \\ &= \mathbf{e}_3 - 4\mathbf{e}_2 + 12\mathbf{e}_1, \end{aligned} \tag{2.3.33}$$

where we have used orthonormality of the vectors $\{\mathbf{e}_1, \mathbf{e}_2, \mathbf{e}_3\}$ and that $\mathbf{e}_j\mathbf{e}_k = -\mathbf{e}_k\mathbf{e}_j$ for $k \neq j$. To recover the outer product bivector we simply multiply again with I_3 to get

$$\mathbf{a} \wedge \mathbf{b} = I_3(\mathbf{a} \times \mathbf{b}) = \mathbf{e}_1\mathbf{e}_2\mathbf{e}_3(\mathbf{e}_3 - 4\mathbf{e}_2 + 12\mathbf{e}_1) = \mathbf{e}_1\mathbf{e}_2 + 4\mathbf{e}_1\mathbf{e}_3 + 12\mathbf{e}_2\mathbf{e}_3. \tag{2.3.34}$$

It is important to note that the cross product vector construction is limited to three dimensions. It does not work in two dimensions, because no third perpendicular dimension is available and it does not work in four and higher dimensions, because the perpendicular space is then two or higher dimensional, i.e. a unique perpendicular vector can no longer be defined. In contrast to this, Grassmann's original bivector construction is not bound to the dimensionality of the space.

2.3.3.2 Linear dependence and independence

The nonzero outer product of a set of r linearly independent vectors uniquely determines an r-dimensional subspace of $\mathbb{R}^n$ spanned by these vectors [161]. Let us therefore proof the following *proposition*:

$$\mathbf{a}_1 \wedge \mathbf{a}_2 \wedge \ldots \wedge \mathbf{a}_r = 0 \iff \mathbf{a}_1, \mathbf{a}_2, \ldots, \mathbf{a}_r \in \mathbb{R}^n \text{ linearly dependent.} \tag{2.3.35}$$

($\Leftarrow$) Let $\mathbf{a}_1, \mathbf{a}_2, \ldots, \mathbf{a}_r \in \mathbb{R}^n$ be a set of linearly dependent vectors. Without loss of generality we assume that

$$\mathbf{a}_1 = \sum_{k=2}^{r} \alpha_k \mathbf{a}_k, \tag{2.3.36}$$

with at least one of the coefficients $\alpha_k \neq 0$. Inserting the sum for $\mathbf{a}_1$ we get

$$\mathbf{a}_1 \wedge \mathbf{a}_2 \wedge \ldots \wedge \mathbf{a}_r = \left(\sum_{k=2}^{r} \alpha_k \mathbf{a}_k\right) \wedge \mathbf{a}_2 \wedge \ldots \wedge \mathbf{a}_r = 0, \tag{2.3.37}$$

because each term of the sum contains an expression of the form $\mathbf{a}_k \wedge \mathbf{a}_k = 0$.

($\Rightarrow$) Let the outer product of r vectors $\mathbf{a}_1, \mathbf{a}_2, \ldots, \mathbf{a}_r \in \mathbb{R}^n$ be zero: $\mathbf{a}_1 \wedge \mathbf{a}_2 \wedge \ldots \wedge \mathbf{a}_r = 0$. But let us assume, that $A_{r-1} = \mathbf{a}_2 \wedge \ldots \wedge \mathbf{a}_r \neq 0$, and that the vectors $\{\mathbf{a}_2, \ldots, \mathbf{a}_r\}$ are orthogonal. Then

$$A_{r-1} = \mathbf{a}_2\mathbf{a}_3 \ldots \mathbf{a}_r \tag{2.3.38}$$

The following definitions will be convenient for showing that $\mathbf{a}_1$ is a linear combination of $\mathbf{a}_2, \ldots, \mathbf{a}_r$. The multiplicative inverse of each vector with respect to the geometric product is given by

$$\mathbf{a}_k^{-1} = \frac{\mathbf{a}_k}{\mathbf{a}_k^2}, \quad k = 2, \ldots, r \tag{2.3.39}$$

and the inverse of A_{r-1} is then given by the reversely ordered product of the inverse vectors

$$A_{r-1}^{-1} = \mathbf{a}_r^{-1}\mathbf{a}_{r-1}^{-1} \ldots \mathbf{a}_2^{-1} \tag{2.3.40}$$

We then have

$$\begin{aligned} A_{r-1}A_{r-1}^{-1} &= \mathbf{a}_2 \ldots \mathbf{a}_{r-1}\, \mathbf{a}_r\mathbf{a}_r^{-1}\mathbf{a}_{r-1}^{-1} \ldots \mathbf{a}_2^{-1} = \mathbf{a}_2 \ldots \mathbf{a}_{r-1}\mathbf{a}_{r-1}^{-1} \ldots \mathbf{a}_2^{-1} \\ &= \ldots = 1. \end{aligned} \tag{2.3.41}$$

According to our assumption, the full geometric product of $\mathbf{a}_1$ and A_{r-1} is

$$\mathbf{a}_1 A_{r-1} = \mathbf{a}_1 \rfloor A_{r-1} + \mathbf{a}_1 \wedge A_{r-1} = \mathbf{a}_1 \rfloor A_{r-1} \tag{2.3.42}$$

and therefore

$$\begin{aligned} \mathbf{a}_1 &= \mathbf{a}_1 A_{r-1}\, A_{r-1}^{-1} = (\mathbf{a}_1 \rfloor A_{r-1})A_{r-1}^{-1} \\ &= \sum_{k=2}^{r} (-1)^{r-k} (\mathbf{a}_1 \rfloor \mathbf{a}_k) \mathbf{a}_r^{-1}\mathbf{a}_{r-1}^{-1} \ldots \not{\mathbf{a}_k} \ldots \mathbf{a}_2^{-1}\mathbf{a}_2 \ldots \mathbf{a}_k \ldots \mathbf{a}_r \\ &= \sum_{k=2}^{r} (\mathbf{a}_1 \rfloor \mathbf{a}_k) \mathbf{a}_r^{-1}\mathbf{a}_{r-1}^{-1} \ldots \not{\mathbf{a}_k} \ldots \mathbf{a}_2^{-1}\mathbf{a}_2 \ldots \not{\mathbf{a}_k} \ldots \mathbf{a}_r\mathbf{a}_k \\ &= \sum_{k=2}^{r} (\mathbf{a}_1 \rfloor \mathbf{a}_k) \mathbf{a}_k. \end{aligned} \tag{2.3.43}$$

End of the proof.

A *corollary* of the above proposition is, that

$$\mathbf{a}_1 \wedge \mathbf{a}_2 \wedge \ldots \wedge \mathbf{a}_r \neq 0 \iff \mathbf{a}_1, \mathbf{a}_2, \ldots, \mathbf{a}_r \in \mathbb{R}^n \text{ linearly independent.} \tag{2.3.44}$$

Another *corollary* is that for a set of linearly independent vectors $\mathbf{a}_1, \mathbf{a}_2, \ldots, \mathbf{a}_r \in \mathbb{R}^n$ we have for $\mathbf{a} \in \mathbb{R}^n$:

$$\mathbf{a} \in \operatorname{Span}\{\mathbf{a}_1, \mathbf{a}_2, \ldots, \mathbf{a}_r\} \iff \mathbf{a} \wedge \mathbf{a}_1 \wedge \mathbf{a}_2 \wedge \ldots \wedge \mathbf{a}_r = 0. \tag{2.3.45}$$

Up to an arbitrary nonzero scalar factor, each homogeneous simple multivector is therefore in one to one correspondence with a subspace of $\mathbb{R}^n$. This is the reason, why the operations of join, intersection (meet) and projection on subspaces can be easily be expressed by geometric products of corresponding homogeneous simple multivectors. The pseudoscalar I_n of a geometric algebra $\mathbb{R}_n$ corresponds to the space $\mathbb{R}^n$ itself.

2.3.4 Right and left contraction

We have already encountered the left and right contractions of two grade 1 vectors and the left contraction of a vector and a homogeneous grade r multivector. But the left and right contractions can be generalized to apply to arbitrary multivectors.

The most general approach is to define left and right contractions solely in terms of the outer product and the scalar product. In this way, the *left contraction* can be defined as [110]

$$C * (A \rfloor B) \equiv (C \wedge A) * B, \qquad \forall A, B, C \in \mathbb{R}_n, \tag{2.3.46}$$

and the *right contraction* as

$$(B \lfloor A) * C \equiv B * (A \wedge C), \qquad \forall A, B, C \in \mathbb{R}_n. \tag{2.3.47}$$

Note that both left and right contractions are linear, because the scalar product and the outer product used in these definitions are linear. Decomposing the multivectors A, B, C grade by grade we get

$$A = \sum_{k=1}^{n} A_k, \quad B = \sum_{l} B_l, \quad C = \sum_{m} C_m. \tag{2.3.48}$$

Inserting this into the expression for the left contraction we get by linearity of both the outer and the scalar product

$$\begin{aligned}(C \wedge A) * B &= \sum_{k,l,m} (C_m \wedge A_k) * B_l = \sum_{m,k} \langle C_m A_k B_{l=m+k} \rangle_0 \\ &= \sum_{m,k} \langle C_m \langle A_k B_{m+k} \rangle_m \rangle_0 = \sum_{m} C_m * \left(\sum_{k} \langle A_k B_{m+k} \rangle_m \right) \\ &= C * (A \rfloor B). \end{aligned} \tag{2.3.49}$$

This gives an explicit expression for the left contraction in terms of the grade parts of A and B

$$A \rfloor B = \sum_{k,l} \langle \langle A \rangle_k \langle B \rangle_l \rangle_{l-k}. \tag{2.3.50}$$

In analogy to this we get for the right contraction

$$B\lfloor A = \sum_{k,l}\langle\langle B\rangle_l\langle A\rangle_k\rangle_{l-k}. \tag{2.3.51}$$

Note that in both cases $m = l - k \geq 0$, i.e. combinations of l and k with $l < k$ do not contribute.

It is now straightforward to see that the contractions of two homogeneous multivectors of the same grade give their scalar product

$$\langle A\rangle_k\rfloor\langle B\rangle_k = \langle A\rangle_k\lfloor\langle B\rangle_k = \langle\langle A\rangle_k\langle B\rangle_k\rangle_{k-k=0} = \langle A\rangle_k * \langle B\rangle_k, \tag{2.3.52}$$

and that, e.g. the left contraction of a vector $\mathbf{a}$ with a homogeneous grade r multivector B_r will give a homogeneous grade $r-1$ multivector

$$\mathbf{a}\rfloor B_r = \langle\mathbf{a}B_r\rangle_{r-1}. \tag{2.3.53}$$

In order to derive convenient explicit formulas for the calculation of the contractions of homogeneous simple multivectors, we will show the following formula:

$$A_r\rfloor(B_s\rfloor C_t) = (A_r \wedge B_s)\rfloor C_t, \tag{2.3.54}$$

where A_r, B_s and C_t are supposed to be homogeneous simple multivectors of grade r, s and t, respectively. For the left side to be nonzero, we must have $t \geq s$ and $t - s > r$ which is equivalent to $t > r + s$. To perform the proof, we scalar multiply with an arbitrary multivector $D \in \mathbb{R}_n$ from the left to get by repeated application of the defining relationship of the left contraction

$$\begin{aligned} D * [A_r\rfloor(B_s\rfloor C_t)] &= (D \wedge A_r) * (B_s\rfloor C_t) = (D \wedge A_r \wedge B_s) * C_t \\ &= [D \wedge (A_r \wedge B_s)] * C_t = D * [(A_r \wedge B_s)\rfloor C_t], \\ &\quad \forall D \in \mathbb{R}_n. \end{aligned} \tag{2.3.55}$$

And hence $A_r\rfloor(B_s\rfloor C_t) = (A_r \wedge B_s)\rfloor C_t$. Note that for conducting the proof no conditions on the values of the grades r, s and t had to be made. The left contraction "takes care" of this.

Let us now look in detail at the left contraction of two homogeneous simple multivectors A_r and B_s of grades r and s, respectively, given by the following vector factorizations

$$A_r = \mathbf{a}_1 \wedge \mathbf{a}_2 \wedge \ldots \wedge \mathbf{a}_r\,, \quad B_s = \mathbf{b}_1 \wedge \mathbf{b}_2 \wedge \ldots \wedge \mathbf{b}_s. \tag{2.3.56}$$

With repeated application of the formula, which we have just proved, we can rewrite the left contraction of A_r and B_s as

$$\begin{aligned} A_r\rfloor B_s &= (\mathbf{a}_1 \wedge \mathbf{a}_2 \wedge \ldots \wedge \mathbf{a}_r)\rfloor(\mathbf{b}_1 \wedge \mathbf{b}_2 \wedge \ldots \wedge \mathbf{b}_s) \\ &= ((\mathbf{a}_1 \wedge \mathbf{a}_2 \wedge \ldots \wedge \mathbf{a}_{r-1}) \wedge \mathbf{a}_r)\rfloor(\mathbf{b}_1 \wedge \mathbf{b}_2 \wedge \ldots \wedge \mathbf{b}_s) \\ &= (\mathbf{a}_1 \wedge \mathbf{a}_2 \wedge \ldots \wedge \mathbf{a}_{r-1})\rfloor(\mathbf{a}_r\rfloor(\mathbf{b}_1 \wedge \mathbf{b}_2 \wedge \ldots \wedge \mathbf{b}_s)) \\ &= \ldots = \\ &= \mathbf{a}_1\rfloor(\mathbf{a}_2\rfloor(\ldots\rfloor(\mathbf{a}_r\rfloor(\mathbf{b}_1 \wedge \mathbf{b}_2 \wedge \ldots \wedge \mathbf{b}_s))\ldots)). \end{aligned} \tag{2.3.57}$$

By repeated application of a previously shown formula ($\not{\mathbf{b}}_k$ means to omit $\mathbf{b}_k$)

$$\mathbf{a}\rfloor B_s = \sum_{k=1}^{r}(-1)^{k+1}\mathbf{a}\rfloor\mathbf{b}_k(\mathbf{b}_1 \wedge \mathbf{b}_2 \wedge \ldots \wedge \not{\mathbf{b}}_k \wedge \ldots \wedge \mathbf{b}_s), \tag{2.3.58}$$

and by reordering we finally reach a very explicit formula for $A_r\rfloor B_s$

$$A_r\rfloor B_s = \sum_{j_1<\ldots<j_r} \epsilon(j_1 \ldots j_s) A_r\rfloor(\mathbf{b}_{j_1} \wedge \ldots \wedge \mathbf{b}_{j_r})(\mathbf{b}_{j_{r+1}} \wedge \ldots \wedge \mathbf{b}_{j_s}). \tag{2.3.59}$$

Note that the left contraction is only nonzero for $r \leq s$. For $r > s$ the right side has to be replaced by zero. Each $j_k \leq s$ is a positive integer with $j_1 < \ldots < j_r$ and with $j_{r+1} < \ldots < j_s$. $\epsilon(j_1 \ldots j_s) = 1$ for even permutations of $(1, 2, \ldots, s)$ and -1 for odd permutations, respectively. Compare also formula (1.40) on page 11 of [161].

Reordering and the explicit formula for $\mathbf{a}\rfloor B_s$ further yield for $r = s$ that

$$\begin{aligned}\tilde{A}_r * B_r &= \tilde{A}_r\rfloor B_r \\ &= \sum_{k=1}^{r}(-1)^{j+k}(\mathbf{a}_j\rfloor\mathbf{b}_k)(\mathbf{a}_r \wedge \ldots \mathbf{a}_j \times \ldots \wedge \mathbf{a}_1)\rfloor(\mathbf{b}_1 \wedge \ldots \mathbf{b}_k \times \ldots \wedge \mathbf{b}_r).\end{aligned} \tag{2.3.60}$$

Continuing this expansion for the homogeneous simple grade $r-1$ multivectors $(\mathbf{a}_r \wedge \ldots \not{\mathbf{a}}_j \ldots \wedge \mathbf{a}_1)$ and $(\mathbf{b}_1 \wedge \ldots \not{\mathbf{b}}_k \ldots \wedge \mathbf{b}_r)$ we get the expansion formula for the determinant of the matrix with coefficients $f_{jk} = \mathbf{a}_j\rfloor\mathbf{b}_k = \mathbf{a}_j * \mathbf{b}_k$. This is also precisely the reason why the determinant definition in geometric algebra given previously as

$$\det(f) = \underline{f}(I_n) * \tilde{I}_n = \underline{f}(\tilde{I}_n) * I_n \tag{2.3.61}$$

agrees with the traditional matrix calculus definition. We just need to set $r = n$, $B_r = I_n$ and $A_r = \underline{f}(I_n) \Leftrightarrow \tilde{A}_r = \underline{f}(\tilde{I}_n)$.

2.4 DETERMINANTS IN GEOMETRIC ALGEBRA

2.4.1 Determinant definition

Let f be a linear map[3], of a real linear vector space $\mathbb{R}^n$ into itself, an endomorphism

$$f : \mathbf{a} \in \mathbb{R}^n \rightarrow \mathbf{a}' \in \mathbb{R}^n. \tag{2.4.1}$$

This map is extended by outermorphism (symbol $\underline{f}$) to act linearly on multivectors

$$\underline{f}(\mathbf{a}_1 \wedge \mathbf{a}_2 \ldots \wedge \mathbf{a}_k) = f(\mathbf{a}_1) \wedge f(\mathbf{a}_2) \ldots \wedge f(\mathbf{a}_k), \qquad k \leq n. \tag{2.4.2}$$

By definition $\underline{f}$ is grade preserving and linear, mapping multivectors to multivectors. Examples are the reflections, rotations and translations described earlier. The outermorphism of a product of two linear maps fg is the product of the outermorphisms $\underline{f}\,\underline{g}$

$$\begin{aligned}f[g(\mathbf{a}_1)] \wedge f[g(\mathbf{a}_2)] \ldots \wedge f[g(\mathbf{a}_k)] &= \underline{f}[g(\mathbf{a}_1) \wedge g(\mathbf{a}_2) \ldots \wedge g(\mathbf{a}_k)] \\ &= \underline{f}[\underline{g}(\mathbf{a}_1 \wedge \mathbf{a}_2 \ldots \wedge \mathbf{a}_k)],\end{aligned} \tag{2.4.3}$$

[3]The treatment in this section largely follows [107].

with $k \leq n$. The square brackets can safely be omitted.

The n-grade pseudoscalars of a geometric algebra are unique up to a scalar factor. This can be used to *define* the determinant[4] of a linear map as

$$\det(f) = \underline{f}(I)I^{-1} = \underline{f}(I) * I^{-1}, \text{ and therefore } \underline{f}(I) = \det(f)I. \tag{2.4.4}$$

For an orthonormal basis $\{\mathbf{e}_1, \mathbf{e}_2, \ldots, \mathbf{e}_n\}$ the unit pseudoscalar is $I = \mathbf{e}_1\mathbf{e}_2 \ldots \mathbf{e}_n$ with inverse $I^{-1} = (-1)^q \mathbf{e}_n \mathbf{e}_{n-1} \ldots \mathbf{e}_1 = (-1)^q (-1)^{n(n-1)/2} I$, where q gives the number of basis vectors, that square to -1 (the linear space is then $\mathbb{R}^{p,q}$). According to Grassmann n-grade vectors represent oriented volume elements of dimension n. The determinant therefore shows how these volumes change under linear maps. Composing two linear maps gives the product of these volume factors

$$\underline{f}\,\underline{g}(I) = \underline{f}[\det(g)I] = \det(g)\underline{f}(I) = \det(g)\det(f)I. \tag{2.4.5}$$

Therefore

$$\det(fg) = \det(g)\det(f). \tag{2.4.6}$$

2.4.2 Adjoint and inverse linear maps

For every linear map $f : \mathbb{R}^n \to \mathbb{R}^n$ exists[5] a unique adjoint linear map $\overline{f} : \mathbb{R}^n \to \mathbb{R}^n$, such that

$$\mathbf{b} * \overline{f}(\mathbf{a}) = \underline{f}(\mathbf{b}) * \mathbf{a}, \qquad \forall \mathbf{a}, \mathbf{b} \in \mathbb{R}^n. \tag{2.4.7}$$

The adjoint linear map extends again via outermorphism

$$\overline{f}(\mathbf{a}_1 \wedge \mathbf{a}_2 \ldots \wedge \mathbf{a}_k) = \overline{f}(\mathbf{a}_1) \wedge \overline{f}(\mathbf{a}_2) \ldots \wedge \overline{f}(\mathbf{a}_k), \qquad k \leq n. \tag{2.4.8}$$

In general we have for multivectors A, B that

$$B * \overline{f}(A) = \underline{f}(B) * A, \tag{2.4.9}$$

which can be applied to the defining[6] relationship [110] for the (right) contraction

$$(C \lfloor A) * B = C * (A \wedge B), \qquad \forall \text{ multivectors } A, B, C. \tag{2.4.10}$$

For simple grade c-vectors C and a-vectors A, the right contraction $(C \lfloor A)$ is a grade $c - a$ sub-space multivector of C perpendicular to A. We now get $\forall A, B, C$

$$\begin{aligned} \overline{f}(C \lfloor A) * B &= (C \lfloor A) * \underline{f}(B) = C * (A \wedge \underline{f}(B)) \\ &= C * (\underline{f}(\underline{f}^{-1}(A)) \wedge \underline{f}(B)) = C * \underline{f}(\underline{f}^{-1}(A) \wedge B) \\ &= \overline{f}(C) * (\underline{f}^{-1}(A) \wedge B) = (\overline{f}(C) \lfloor \underline{f}^{-1}(A)) * B, \end{aligned} \tag{2.4.11}$$

[4]The symbol $(*)$ means the (symmetric) scalar product of two multivectors, i.e. the scalar (0 grade) part of their geometric product.

[5]An explicit definition for the adjoint linear map can be given as $\overline{f}(a) = \mathbf{e}^k(f(\mathbf{e}_k) * \mathbf{a})$, with $\mathbf{e}^k * \mathbf{e}_l = \delta_l^k$ (the *Kronecker delta* symbol), where $1 \leq k, l \leq n$. Here the vectors $\{\mathbf{e}_1, \mathbf{e}_2, \ldots, \mathbf{e}_n\}$ form (a not necessarily orthonormal nor orthogonal) basis of $\mathbb{R}^n$.

[6]The symbols $(*)$ and $(\wedge)$ denote the (symmetric) scalar and the antisymmetric outer product parts of the geometric product of multivectors.

and therefore

$$\overline{f}(C\lfloor A) = \overline{f}(C)\lfloor \underline{f}^{-1}(A). \tag{2.4.12}$$

Similarly we obtain

$$\underline{f}(C\lfloor A) = \underline{f}(C)\lfloor \overline{f}^{-1}(A). \tag{2.4.13}$$

Reversion gives two more identities

$$\overline{f}(A\rfloor C) = \underline{f}^{-1}(A)\rfloor \overline{f}(C), \qquad \underline{f}(A\rfloor C) = \overline{f}^{-1}(A)\rfloor \underline{f}(C). \tag{2.4.14}$$

By substituting in $\overline{f}(C\lfloor A)$ the pseudoscalar I for C and left multiplying with the inverse I^{-1} we get a general formula for calculating the inverse of $\underline{f}$

$$\begin{aligned} I^{-1}\overline{f}(IA) &= I^{-1}(\overline{f}(I)\lfloor \underline{f}^{-1}(A)) = I^{-1}\overline{f}(I)\underline{f}^{-1}(A) = \det(f)\underline{f}^{-1}(A), \\ \iff \quad \underline{f}^{-1}(A) &= \frac{1}{\det(f)} I^{-1}\overline{f}(IA) \end{aligned} \tag{2.4.15}$$

where we used the fact that right contraction with a pseudoscalar is nothing but the geometric product and that $\underline{f}$ is grade preserving.

In the derivation of $\underline{f}^{-1}$ we tacitly used the following property of the determinant obtained by applying $B * \overline{f}(A) = \underline{f}(B) * A$

$$\det(f) = \underline{f}(I) * I^{-1} = I * \overline{f}(I^{-1}) = \overline{f}(I) * I^{-1} = \det(f), \tag{2.4.16}$$

because of the symmetry of the scalar product and because $I^{-1} = (-1)^q (-1)^{n(n-1)/2} I$.

An analogous explicit expression can be derived for $\overline{f}^{-1}$

$$\begin{aligned} \underline{f}^{-1}(A) &= \det(f)^{-1}\overline{f}(AI)I^{-1} = \det(f)^{-1}I^{-1}\overline{f}(IA), \\ \overline{f}^{-1}(A) &= \det(f)^{-1}\underline{f}(AI)I^{-1} = \det(f)^{-1}I^{-1}\underline{f}(IA). \end{aligned} \tag{2.4.17}$$

These formulas are very compact and computationally efficient. They show that for invertible maps ($\det(f) \neq 0$) the inverse mappings can be easily constructed as double-dualities. Duality here means multiplication with the pseudoscalar I or I^{-1}.

2.5 GRAM-SCHMIDT ORTHOGONALIZATION IN GEOMETRIC ALGEBRA

Let Cl_n be the Clifford geometric algebra of the real linear space $\mathbb{R}^n$. Let $\{\mathbf{a}_l, \mathbf{a}_2, \ldots, \mathbf{a}_r\}$, $r \leq n$, be a set of r linearly independent vectors [161]. Then the simple r-multivector $A_r = \mathbf{a}_l \wedge \mathbf{a}_2 \wedge \ldots \wedge \mathbf{a}_r$ (blade) will necessarily be different from zero: $A_r \neq 0$, and vice versa, because the r-volume defined by A_r will be different from zero.

The linearly independent set of vectors $\{\mathbf{a}_l, \mathbf{a}_2, \ldots, \mathbf{a}_r\}$, $r \leq n$, can be systematically orthogonalized. We construct the graded sequence of multivectors

$$A_0 = 1, \;\; A_1 = \mathbf{a}_1, \;\; A_2 = \mathbf{a}_1 \wedge \mathbf{a}_2, \;\; \ldots, \;\; A_r = \mathbf{a}_1 \wedge \mathbf{a}_2 \wedge \ldots \wedge \mathbf{a}_r. \tag{2.5.1}$$

We can use $A_0, A_1, \ldots, A_r$ in order to define a new set of vectors

$$\mathbf{c}_k = \widetilde{A}_{k-1} A_k = \widetilde{A}_{k-1} \rfloor A_k, \qquad k = 1, \ldots, r, \tag{2.5.2}$$

where the tilde over A_{k-1} means to reverse the order of vector factors. For example, $\widetilde{A}_2 = \mathbf{a}_2 \wedge \mathbf{a}_1 = -A_2$, $\widetilde{A}_3 = \mathbf{a}_3 \wedge \mathbf{a}_2 \wedge \mathbf{a}_1 = -A_3$, etc. The geometric product $\widetilde{A}_{k-1}A_k$ can be replaced by the left contraction, because by construction, the $(k-1)$-subspace defined by A_{k-1} is fully contained in the k-subspace defined by A_k. Let us remember the meaning of the left contraction: $\widetilde{A}_{k-1} \rfloor A_k$ results in an $k-(k-1) = 1$ dimensional subspace of the k-subspace defined by A_k, which is orthogonal to the $(k-1)$-subspace defined by A_{k-1}. Therefore the set of r vectors $\mathbf{c}_k$, $k = 1, \ldots, r$, must be an orthogonal set, and span the r-subspace defined by A_r. The last property, can be easily verified by calculating the geometric product of all $\mathbf{c}_k$, $k = 1, \ldots, r$:

$$\begin{aligned} \mathbf{c}_1\mathbf{c}_2 \ldots \mathbf{c}_r &= 1A_1\widetilde{A}_1A_2\widetilde{A}_2 \ldots A_{r-1}\widetilde{A}_{r-1}A_r \\ &= A_1 * \widetilde{A}_1A_2 * \widetilde{A}_2 \ldots A_{r-1} * \widetilde{A}_{r-1}A_r \\ &= |A_1|^2|A_2|^2 \ldots |A_{r-1}|^2 A_r \end{aligned} \tag{2.5.3}$$

where the symbol $(*)$ signifies the scalar product, i.e. the scalar part of the geometric product of two multivectors, and $|A|$ is the positive scalar magnitude of the multivector A defined by $|A|^2 = \widetilde{A} * A$. Obviously the product $\mathbf{c}_1\mathbf{c}_2 \ldots \mathbf{c}_r$ constitutes a factorization of A_r into a product of orthogonal vectors.

This result fully corresponds to the conventional Gram-Schmidt orthogonalization process in linear algebra.

2.6 A BRIEF GUIDE TO IMPORTANT CLIFFORD GEOMETRIC ALGEBRAS

This section intends to guide the reader in compact overview form through the history, the current formulation and higher dimensional extensions of Clifford's geometric algebras *relevant for applications*. It is mainly based on [200].

Geometric algebra was initiated by W.K. Clifford over 130 years ago [77, 78]. It unifies all branches of physics, and has found rich applications in robotics, signal processing, ray tracing, virtual reality, computer vision, vector field processing, tracking, geographic information systems and neural computing. We now survey the basics of geometric algebra on a slightly more advanced level than in the first part of this chapter, with concrete examples of the plane, of 3D space, of spacetime, and the popular conformal model. Geometric algebras are ideal to represent geometric transformations in the general framework of Clifford groups (also called versor or Lipschitz groups). Geometric (algebra based) calculus allows, e.g. to optimize learning algorithms of Clifford neurons, etc.

2.6.1 Overview of Clifford's geometric algebra of the Euclidean plane

In order to demonstrate how to compute with Clifford numbers, we begin with a low dimensional example.

2.6.2 Example of $Cl(2,0)$

A Euclidean plane $\mathbb{R}^2 = \mathbb{R}^{2,0} = \mathbb{R}^{2,0,0}$ is spanned by $e_1, e_2 \in \mathbb{R}^2$ with

$$e_1 \cdot e_1 = e_2 \cdot e_2 = 1, \quad e_1 \cdot e_2 = 0. \tag{2.6.1}$$

$\{e_1, e_2\}$ is an *orthonormal* vector basis of $\mathbb{R}^2$.

Under Clifford's *associative* geometric product we set

$$\begin{aligned} e_1^2 &= e_1e_1 := e_1 \cdot e_1 = 1, \\ e_2^2 &= e_2e_2 := e_2 \cdot e_2 = 1, \end{aligned} \tag{2.6.2}$$

$$\text{and} \quad (e_1+e_2)(e_1+e_2) = e_1^2 + e_2^2 + e_1e_2 + e_2e_1$$
$$= 2 + e_1e_2 + e_2e_1 := (e_1+e_2)\cdot(e_1+e_2) = 2. \tag{2.6.3}$$

Therefore

$$e_1e_2 + e_2e_1 = 0 \;\Leftrightarrow\; e_1e_2 = -e_2e_1, \tag{2.6.4}$$

i.e. the geometric product of orthogonal vectors forms a new entity, called unit *bivector* $e_{12} = e_1e_2$ by Grassmann, and is *anti-symmetric*. General bivectors in $Cl(2,0)$ are βe_{12}, $\forall\, \beta \in \mathbb{R} \setminus \{0\}$. For orthogonal vectors the geometric product equals Grassmann's antisymmetric outer product (exterior product, symbol $\wedge$)

$$e_{12} = e_1e_2 = e_1 \wedge e_2 = -e_2 \wedge e_1 = -e_2e_1 = -e_{21}. \tag{2.6.5}$$

Using associativity, we can compute the products

$$e_1e_{12} = e_1e_1e_2 = e_1^2e_2 = e_2, \quad e_2e_{12} = -e_2e_{21} = -e_1, \tag{2.6.6}$$

which represent a mathematically *positive* (anti-clockwise) $90°$ *rotation*. The opposite order gives

$$e_{12}e_1 = -e_{21}e_1 = -e_2, \quad e_{12}e_2 = e_1, \tag{2.6.7}$$

which represents a mathematically *negative* (clockwise) $90°$ *rotation*. The bivector e_{12} acts like a *rotation operator*, and we observe the general anti-commutation property

$$ae_{12} = -e_{12}a, \quad \forall a = a_1e_1 + a_2e_2 \in \mathbb{R}^2, \;\; a_1, a_2 \in \mathbb{R}. \tag{2.6.8}$$

The square of the unit bivector is -1,

$$e_{12}^2 = e_1e_2e_{12} = e_1(-e_1) = -1, \tag{2.6.9}$$

just like the imaginary unit j of complex numbers $\mathbb{C}$.

Table 2.3 is the complete multiplication table of the Clifford algebra $Cl(\mathbb{R}^2) = Cl(2,0,0) = Cl(2,0)$ with algebra basis elements $\{1, e_1, e_2, e_{12}\}$ (which includes the vector basis of $\mathbb{R}^2$). The even subalgebra spanned by $\{1, e_{12}\}$ (closed under geometric multiplication), consisting of even grade scalars (0-vectors) and bivectors (2-vectors), is isomorphic to $\mathbb{C}$.

Table 2.3 Multiplication table of plane Clifford algebra $Cl(2,0)$.

	1	e_1	e_2	e_{12}
1	1	e_1	e_2	e_{12}
e_1	e_1	1	e_{12}	e_2
e_2	e_2	$-e_{12}$	1	$-e_1$
e_{12}	e_{12}	$-e_2$	e_1	-1

2.6.2.1 *Aspects of algebraic unification and vector inverse*

The general geometric product of two vectors $a, b \in \mathbb{R}^2$

$$\begin{aligned} ab &= (a_1e_1 + a_2e_2)(b_1e_1 + b_2e_2) \\ &= a_1b_1 + a_2b_2 + (a_1b_2 - a_2b_1)e_{12} \\ &= \frac{1}{2}(ab + ba) + \frac{1}{2}(ab - ba) = a \cdot b + a \wedge b, \end{aligned} \tag{2.6.10}$$

has therefore a scalar *symmetric* inner product part

$$\begin{aligned} \frac{1}{2}(ab + ba) &= a \cdot b = a_1b_1 + a_2b_2 \\ &= |a||b| \cos\theta_{a,b}, \end{aligned} \tag{2.6.11}$$

and a bi-vector *skew-symmetric* outer product part

$$\frac{1}{2}(ab - ba) = a \wedge b = (a_1b_2 - a_2b_1)e_{12} = |a||b|e_{12} \sin\theta_{a,b}. \tag{2.6.12}$$

We observe that parallel vectors ($\theta_{a,b} = 0$) commute, $ab = a \cdot b = ba$, and orthogonal vectors ($\theta_{a,b} = 90°$) anti-commute, $ab = a \wedge b = -ba$. The outer product part $a \wedge b$ represents the *oriented area* of the parallelogram spanned by the vectors a, b in the plane of $\mathbb{R}^2$, with oriented magnitude

$$\det(a,b) = |a||b| \sin\theta_{a,b} = (a \wedge b)e_{12}^{-1}, \tag{2.6.13}$$

where $e_{12}^{-1} = -e_{12}$, because $e_{12}^2 = -1$.

With the *Euler* formula we can rewrite the geometric product as

$$ab = |a||b|(\cos\theta_{a,b} + e_{12} \sin\theta_{a,b}) = |a||b|e^{\theta_{a,b}e_{12}}, \tag{2.6.14}$$

again because $e_{12}^2 = -1$.

The geometric product of vectors is *invertible* for all vectors with non-zero square $a^2 \neq 0$

$$\begin{aligned} a^{-1} &:= a/a^2, \quad aa^{-1} = aa/a^2 = 1, \\ a^{-1}a &= \frac{a}{a^2}a = a^2/a^2 = 1. \end{aligned} \tag{2.6.15}$$

The inverse vector a/a^2 is a rescaled version (reflected at the unit circle) of the vector a. This invertibility leads to enormous simplifications and ease of practical computations.

2.6.2.2 On geometric operations and transformations

For example, the *projection* of one vector $x \in \mathbb{R}^2$ onto another $a \in \mathbb{R}^2$ is

$$x_{\|} = |x\,| \cos\theta_{a,x} \frac{a}{|a|} = (x \cdot \frac{a}{|a|})\frac{a}{|a|} = (x \cdot a)\frac{a}{|a|^2} = (x \cdot a)a^{-1}. \tag{2.6.16}$$

The *rejection* (perpendicular part) is

$$x_{\perp} = x - x_{\|} = xaa^{-1} - (x \cdot a)a^{-1} = (xa - x \cdot a)a^{-1} = (x \wedge a)a^{-1}. \tag{2.6.17}$$

We can now use $x_{\|}, x_{\perp}$ to compute the reflection[7] of $x = x_{\|} + x_{\perp}$ at the line (hyperplane[8]) with normal vector a, which means to reverse $x_{\|} \to -x_{\|}$

$$\begin{aligned} x' = -x_{\|} + x_{\perp} &= -a^{-1}a\,x_{\|} + a^{-1}a\,x_{\perp} \\ = -a^{-1}x_{\|}a - a^{-1}\,x_{\perp}a &= -a^{-1}(x_{\|} + x_{\perp})a = -a^{-1}xa. \end{aligned} \tag{2.6.18}$$

The combination of two reflections at two lines (hyperplanes) with normals a, b

$$x'' = -b^{-1}x'b = b^{-1}a^{-1}xab = (ab)^{-1}xab = R^{-1}xR, \tag{2.6.19}$$

gives a rotation. The rotation angle is $\alpha = 2\theta_{a,b}$ and the *rotor*

$$R = e^{\theta_{a,b}e_{12}} = e^{\frac{1}{2}\alpha e_{12}}, \tag{2.6.20}$$

where the lengths $|a||b|$ of ab cancel against $|a|^{-1}|b|^{-1}$ in $(ab)^{-1}$. The rotor R gives the *spinor* form of rotations, fully replacing rotation matrices, and introducing the same elegance to *real* rotations in $\mathbb{R}^2$, like in the complex plane.

In 2D, the product of three reflections, i.e. of a rotation and a reflection, leads to another reflection. In 2D, the product of an *odd* number of reflections always results in a *reflection*. That the product of an *even* number of reflections leads to a *rotation* is true in general dimensions. These transformations are in Clifford algebra simply described by the products of the vectors normal to the lines (hyperplanes) of reflection and called versors.

Definition of a versor [244]: A *versor* refers to a Clifford monomial (product expression) composed of invertible vectors. It is called a *rotor*, or *spinor*, if the number of vectors is even. It is called a *unit versor* if its magnitude is 1.

Every versor $A = a_1 \ldots a_r, \quad a_1, \ldots, a_r \in \mathbb{R}^2, r \in \mathbb{N}$ has an inverse

$$A^{-1} = a_r^{-1} \ldots a_1^{-1} = a_r \ldots a_1/(a_1^2 \ldots a_r^2), \tag{2.6.21}$$

such that

$$AA^{-1} = A^{-1}A = 1. \tag{2.6.22}$$

[7] Note that reflections at hyperplanes are nothing but the *Householder transformations* [220] of matrix analysis.

[8] A hyperplane of a nD space is a $(n-1)$D subspace, thus a hyperplane of $\mathbb{R}^2$, $n = 2$, is a 1D $(2 - 1 = 1)$ subspace, i.e. a line. Every hyperplane is characterized by a vector normal to the hyperplane.

This makes the set of all versors in $Cl(2,0)$ a group, the so called *Lipschitz group* with symbol $\Gamma(2,0)$, also called *Clifford group* or *versor group*. Versor transformations apply via *outermorphisms* to all elements of a Clifford algebra. It is the group of all reflections and rotations of $\mathbb{R}^2$. The reverse product order of a versor represents an involution (applying it twice leads to identity) called *reversion*[9]

$$\widetilde{A} = (a_1 \dots a_r)^\sim = a_r \dots a_1. \tag{2.6.23}$$

In the case of $Cl(2,0)$ we have

$$\Gamma(2,0) = Cl^-(2,0) \cup Cl^+(2,0), \tag{2.6.24}$$

where the odd grade vector part $Cl^-(2,0) = \mathbb{R}^2$ generates reflections, and the even grade part of scalars and bivectors $Cl^+(2,0) = \{A \mid A = \alpha + \beta e_{12}, \alpha, \beta \in \mathbb{R}\}$ generates rotations. The normalized subgroup of versors is called *pin group*

$$\mathrm{Pin}(2,0) = \{A \in \Gamma(2,0) \mid A\widetilde{A} = \pm 1\}. \tag{2.6.25}$$

In the case of $Cl(2,0)$ we have

$$\mathrm{Pin}(2,0) = \{a \in \mathbb{R}^2 \mid a^2 = 1\} \cup \{A \mid A = \cos\varphi + e_{12}\sin\varphi,\ \varphi \in \mathbb{R}\}. \tag{2.6.26}$$

The pin group has an even subgroup, called *spin group*

$$\mathrm{Spin}(2,0) = \mathrm{Pin}(2,0) \cap Cl^+(2,0). \tag{2.6.27}$$

As mentioned above $Cl^+(2,0)$ has the basis $\{1, e_{12}\}$ and is thus isomorphic to $\mathbb{C}$. In the case of $Cl(2,0)$ we have explicitly

$$\mathrm{Spin}(2,0) = \{A \mid A = \cos\varphi + e_{12}\sin\varphi,\ \varphi \in \mathbb{R}\}.$$

The spin group has in general a *spin plus subgroup*[10]

$$\mathrm{Spin}_+(2,0) = \{A \in \mathrm{Spin}(2,0) \mid A\widetilde{A} = +1\}. \tag{2.6.28}$$

The groups $\mathrm{Pin}(2,0)$, $\mathrm{Spin}(2,0)$ and $\mathrm{Spin}_+(2,0)$ are two-fold coverings[11] of the orthogonal group $\mathrm{O}(2,0)$ [315], the special orthogonal group $\mathrm{SO}(2,0)$, and the component of the special orthogonal group connected to the identity $\mathrm{SO}_+(2,0)$.

Let us point out, that this natural combination of reflections leading to spinors rightly indicates the way how Clifford's GA is able to give a fully real algebraic description to quantum mechanics, with a clear cut geometric interpretation.

---

[9] Reversion is an anti-automorphism. Often a dagger $A^\dagger$ is used instead of the tilde, as well as the term transpose.

[10] Note, that in general for Clifford algebras $Cl(n,0)$ of Euclidean spaces $\mathbb{R}^{n,0}$ we have the identity $\mathrm{Spin}(n) = \mathrm{Spin}_+(n)$, where $\mathrm{Spin}(n) = \mathrm{Spin}(n,0)$. The reason is that $A\widetilde{A} < 0$ is only possible for non-Euclidean spaces $\mathbb{R}^{p,q}$, with $q > 0$.

[11] Two-fold covering means, that there are always two elements $\pm A$ in $\mathrm{Pin}(2,0)$, $\mathrm{Spin}(2,0)$ and $\mathrm{Spin}_+(2,0)$, representing one element in $\mathrm{O}(2,0)$, $\mathrm{SO}(2,0)$ and $\mathrm{SO}_+(2,0)$, respectively.

2.6.2.3 Concepts of vectors, k-vectors and multivectors

A general element in $Cl(2,0)$, also called *multivector* can be represented as

$$M = m_0 + m_1 e_1 + m_2 e_2 + m_{12} e_{12}, \quad m_0, m_1, m_2, m_{12} \in \mathbb{R}. \tag{2.6.29}$$

Like real and imaginary parts of a complex number, we have a scalar part $\langle M \rangle_0$ of grade 0, a vector part $\langle M \rangle_1$ of grade 1 and a bivector part $\langle M \rangle_2$ of grade 2

$$M = \langle M \rangle_0 + \langle M \rangle_1 + \langle M \rangle_2, \tag{2.6.30}$$

with

$$\langle M \rangle_0 = m_0, \quad \langle M \rangle_1 = m_1 e_1 + m_2 e_2, \quad \langle M \rangle_2 = m_{12} e_{12}\,. \tag{2.6.31}$$

The set of all grade k elements, $0 \leq k \leq 2$, is denoted $Cl^k(2,0)$.

Grade extraction $\langle \ldots \rangle_k$, $0 \leq k \leq 2$, allows to do many useful computations, like e.g. the *angle* between two vectors ($\text{atan} = \tan^{-1} = \arctan$)

$$\theta_{a,b} = \text{atan}\frac{\langle ab \rangle_2 e_{12}^{-1}}{\langle ab \rangle_0} \overset{(2.6.14)}{=} \text{atan}\frac{\sin\theta_{a,b}}{\cos\theta_{a,b}}. \tag{2.6.32}$$

The symmetric scalar part of the geometric product of two multivectors $M, N \in Cl(2,0)$ is also called their *scalar product* and because of its fundamental importance denoted with a special product sign $*$

$$M * N = \langle MN \rangle_0 = m_0 n_0 + m_1 n_1 + m_2 n_2 - m_{12} n_{12} = N * M, \tag{2.6.33}$$

which is easy to compute using the multiplication table Table 2.3, since only the *diagonal* entries of Table 2.3 are scalar. Using the reversion ($\widetilde{e_{12}} = \widetilde{e_1 e_2} = e_2 e_1 = -e_1 e_2 = -e_{12}$)

$$\widetilde{M} = m_0 + m_1 e_1 + m_2 e_2 - m_{12} e_{12}, \tag{2.6.34}$$

we get the *norm* $|M|$ of a multivector as

$$|M|^2 = M * \widetilde{M} = \langle M\widetilde{M} \rangle_0 = m_0^2 + m_1^2 + m_2^2 + m_{12}^2, \tag{2.6.35}$$

which for vectors $M \in Cl^1(2,0)$ is identical to the length of a vector, and for even grade subalgebra elements $M \in Cl^+(2,0)$ to the modulus of complex numbers, and for pure bivectors like $a \wedge b = \langle ab \rangle_2$ gives the area content of the parallelogram spanned by the vectors $a, b \in \mathbb{R}^2$

$$|a \wedge b| = |a|\,|b|\,|\sin\theta_{a,b}| = |\det(a,b)|. \tag{2.6.36}$$

The reversion operation maps $\widetilde{e_{12}} = -e_{12}$, it is therefore the equivalent of complex conjugation in the isomorphism $Cl^+(2,0) \cong \mathbb{C}$, fully consistent with its use in the norm (2.6.35).

2.6.2.4 *Higher dimensional types of inner and outer products for multivectors*

The grade extraction also allows us to generalize the inner product of vectors, which maps two grade one vectors $a, b \in \mathbb{R}^2 = Cl^1(2,0)$ to a grade zero scalar

$$a \cdot b = \langle ab \rangle_{(1-1=0)} \in \mathbb{R} = Cl^0(2,0), \tag{2.6.37}$$

and therefore *lowers* the grade by 1. In contrast, the outer product with a vector *raises* the grade by 1, i.e.

$$a \wedge b = \langle ab \rangle_{(1+1=2)} \in Cl^2(2,0). \tag{2.6.38}$$

In general, the *left contraction* (symbol $\rfloor$) of a k-vector $A_k = \langle A \rangle_k$ with a l-vector $B_l = \langle B \rangle_l$ is defined as

$$A_k \rfloor B_l = \langle A_k B_l \rangle_{(l-k)}, \tag{2.6.39}$$

which is zero if $0 > l - k$, i.e. if $k > l$. Figuratively speaking (like with projections), we can only contract objects of the same or lower dimension from the left onto an object on the right. The *right contraction* (symbol $\lfloor$) is defined as

$$A_k \lfloor B_l = \langle A_k B_l \rangle_{(k-l)}, \tag{2.6.40}$$

which is zero if $k - l < 0$, i.e. if $k < l$. Figuratively speaking, we can only contract objects of the same or lower dimension from the right onto an object on the left. Both contractions are bilinear and can thus be extended to contractions of multivectors

$$A \rfloor B = \sum_{k=0}^{2} \sum_{l=0}^{2} \langle \langle A \rangle_k \langle B \rangle_l \rangle_{(l-k)}, \tag{2.6.41}$$

$$A \lfloor B = \sum_{k=0}^{2} \sum_{l=0}^{2} \langle \langle A \rangle_k \langle B \rangle_l \rangle_{(k-l)}. \tag{2.6.42}$$

The reversion changes the order of factors and therefore relates left and right contractions by

$$(A \rfloor B)^{\sim} = \widetilde{B} \lfloor \widetilde{A}, \qquad (A \lfloor B)^{\sim} = \widetilde{B} \rfloor \widetilde{A}. \tag{2.6.43}$$

In general, the associative outer product of a k-vector $A_k = \langle A \rangle_k$ with a l-vector $B_l = \langle B \rangle_l$ is defined as the maximum grade part of the geometric product

$$A_k \wedge B_l = \langle A_k B_l \rangle_{(l+k)}, \tag{2.6.44}$$

where $A_k \wedge B_l = 0$ for $l + k > 2$, because in $Cl(2,0)$ the highest grade possible is 2. For example, the following outer products are obtained ($\alpha, \beta \in \mathbb{R}, a, b, c \in \mathbb{R}^2$)

$$\alpha \wedge \beta = \alpha\beta, \quad \alpha \wedge a = a \wedge \alpha = \alpha a, \quad a \wedge b \wedge c = 0, \tag{2.6.45}$$

where the last identity is specific to $Cl(2,0)$, it does generally not hold in GAs of higher dimensional vector spaces. The last identity is due to the fact, that in a plane every third vector c can be expressed by linear combination of two linearly independent vectors a, b. If a, b would not be linearly independent, then already $a \wedge b = 0$, i.e. $a \parallel b$. The outer product in 2D is of great advantage, because the cross product of vectors does only exist in 3D, not in 2D. We next treat the GA of $\mathbb{R}^3$.

2.6.3 Overview of Clifford's geometric algebra of the 3D Euclidean space

The Clifford algebra $Cl(\mathbb{R}^3) = Cl(3,0)$ of three-dimensional (3D) Euclidean space $\mathbb{R}^3$ is arguably the by far most thoroughly studied and applied geometric algebra (GA). In physics, it is also known as *Pauli algebra*, since Pauli's spin matrices provide a 2×2 matrix representation. This shows how GA unifies *classical* with *quantum* mechanics.

Given an orthonormal vector basis $\{e_1, e_2, e_3\}$ of $\mathbb{R}^3$, the eight-dimensional ($2^3 = 8$) Clifford algebra $Cl(\mathbb{R}^3) = Cl(3,0)$ has a basis of one scalar, three vectors, three bivectors and one trivector

$$\{1, e_1, e_2, e_3, e_{23}, e_{31}, e_{12}, e_{123}\}, \tag{2.6.46}$$

where as before $e_{23} = e_2e_3, e_{123} = e_1e_2e_3$, etc. All basis bivectors square to -1, and the product of two basis bivectors gives the third

$$e_{23}e_{31} = e_{21} = -e_{12}, \quad \text{etc.} \tag{2.6.47}$$

Therefore the even subalgebra $Cl^+(3,0)$ with basis[12] $\{1, -e_{23}, -e_{31}, -e_{12}\}$ is indeed found to be isomorphic to quaternions $\{1, \mathbf{i}, \mathbf{j}, \mathbf{k}\}$. This isomorphism is not incidental. As we have learned already for $Cl(2,0)$, also in $Cl(3,0)$, the even subalgebra is the algebra of rotors (rotation operators) or spinors, and describes rotations in the same efficient way as do quaternions [322]. We therefore gain a *real geometric* interpretation of quaternions, as the oriented bi-vector side faces of a unit cube, with edge vectors $\{e_1, e_2, e_3\}$.

In $Cl(3,0)$ a reflection at a plane (=hyperplane) is specified by the plane's normal vector $a \in \mathbb{R}^3$

$$x' = -a^{-1}xa, \tag{2.6.48}$$

the proof is identical to the one in (2.6.18) for $Cl(2,0)$. The combination of two such reflections leads to a rotation by $\alpha = 2\theta_{a,b}$

$$x'' = R^{-1}xR, \quad R = ab = |a||b|e^{\theta_{a,b}\mathbf{i}_{a,b}} = |a||b|e^{\frac{1}{2}\alpha\mathbf{i}_{a,b}}, \tag{2.6.49}$$

where $\mathbf{i}_{a,b} = a \wedge b/(|a \wedge b|)$ specifies the oriented unit bivector of the plane spanned by $a, b \in \mathbb{R}^3$.

The unit trivector $i_3 = e_{123}$ also squares to -1

$$\begin{aligned} i_3^2 &= e_1e_2e_3e_1e_2e_3 = -e_1e_2e_1e_3e_2e_3 \\ &= e_1e_2e_1e_2e_3e_3 = (e_1e_2)^2(e_3)^2 = -1, \end{aligned} \tag{2.6.50}$$

where we only used that the permutation of two orthogonal vectors in the geometric product produces a minus sign. Hence $i_3^{-1} = -i_3$. We further find, that i_3 commutes with every vector, e.g.

$$\begin{aligned} &e_1i_3 = e_1e_1e_2e_3 = e_{23}, \\ &i_3e_1 = e_1e_2e_3e_1 = -e_1e_2e_1e_3 = e_1e_1e_2e_3 = e_{23}, \end{aligned} \tag{2.6.51}$$

---

[12]The minus signs are only chosen, to make the product of two bivectors identical to the third, and not minus the third.

and the like for $e_2 i_3 = i_3 e_2$, $e_3 i_3 = i_3 e_3$. If i_3 commutes with every vector, it also commutes with every bivector $a \wedge b = \frac{1}{2}(ab - ba)$, hence i_3 commutes with every element of $Cl(3,0)$, a property which is called *central* in mathematics. The central subalgebra spanned by $\{1, i_3\}$ is isomorphic to complex numbers $\mathbb{C}$. i_3 changes bivectors into orthogonal vectors

$$e_{23} i_3 = e_2 e_3 e_1 e_2 e_3 = e_1 e_{23}^2 = -e_1\,, \quad \text{etc.} \tag{2.6.52}$$

This is why the GA $Cl(3,0)$ is isomorphic to *complex quaternions*, one form of *biquaternions*

$$\left\{1, e_1 = e_{23} i_3^{-1}, e_2 = e_{31} i_3^{-1}, e_3 = e_{12} i_3^{-1}, e_{23}, e_{31}, e_{12}, i_3\right\}. \tag{2.6.53}$$

Yet writing the basis in the simple product form (2.6.46), fully preserves the *geometric interpretation* in terms of scalars, vectors, bivectors and trivectors, and allows to *reduce* all products to elementary geometric products of basis vectors, which is used in *computational optimization* schemes for GA software like Galoop [166].

2.6.3.1 Explicit multiplication table and important subalgebras of $Cl(3,0)$

For the full multiplication table of $Cl(3,0)$ we still need the geometric products of vectors and bivectors. By changing labels in Table 2.3 ($1 \leftrightarrow 3$ or $2 \leftrightarrow 3$), we get that

$$\begin{aligned} e_2 e_{23} &= -e_{23} e_2 = e_3, \quad e_3 e_{23} = -e_{23} e_3 = -e_2, \\ e_1 e_{31} &= -e_{31} e_1 = -e_3, \quad e_3 e_{31} = -e_{31} e_3 = e_1, \end{aligned} \tag{2.6.54}$$

which shows that in general a vector and a bivector, which includes the vector, anti-commute. The products of a vector with its orthogonal bivector always give the trivector i_3

$$e_1 e_{23} = e_{23} e_1 = i_3, \quad e_2 e_{31} = e_{31} e_2 = i_3, \quad e_3 e_{12} = e_{12} e_3 = i_3, \tag{2.6.55}$$

which also shows that in general vectors and orthogonal bivectors necessarily commute. Commutation relationships therefore clearly depend on both *orthogonality* properties and on the *grades* of the factors, which can frequently be exploited for computations even without the explicit use of coordinates.

Table 2.2 gives the *multiplication table* of $Cl(3,0)$. The elements on the left most column are to be multiplied from the left in the geometric product with the elements in the top row. Every subtable of Table 2.2, that is closed under the geometric product represents a *subalgebra* of $Cl(3,0)$.

We naturally find Table 2.3 as a subtable, because with $\mathbb{R}^2 \subset \mathbb{R}^3$ we necessarily have $Cl(2,0) \subset Cl(3,0)$. In general, any pair of orthonormal vectors will generate a 4D *subalgebra* of $Cl(3,0)$ *specific to the plane* spanned by the two vectors.

We also recognize the subtable of the *even subalgebra* with basis $\{1, e_{23}, e_{31}, e_{12}\}$ isomorphic to *quaternions*.

Then there is the subtable of $\{1, e_{123}\}$ of the *central subalgebra* isomorphic to $\mathbb{C}$. Any element of $Cl(3,0)$ squaring to -1 generates a 2D subalgebra isomorphic to $\mathbb{C}$.

This fundamental observation leads to the generalization of the conventional complex Fourier transformation (FT) to Clifford FTs, sometimes called geometric algebra FTs [184]. Moreover, we also obtain generalizations of real and complex (dual) wavelets to Clifford wavelets, which include quaternion wavelets.

We further have closed subtables with elements $\{1, e_1, e_{23}, e_{123} = i_3\}$, which are fully commutative and correspond to *tessarines*, also called *bicomplex numbers*, *Segre quaternions*, *commutative quaternions* or *Cartan subalgebras*. In general, we can specify any unit vector $u \in \mathbb{R}^3$, $u^2 = 1$, and get a commutative tessarine subalgebra with basis $\{1, u, ui_3, i_3\}$, which includes besides 1 and the vector u itself, the bivector ui_3 orthogonal to u, and the oriented unit volume trivector i_3. Again GA provides a clear and useful geometric interpretation of tessarines and their products. Knowledge of the geometric interpretation is fundamental in order to *see* these algebras in nature and in problems at hand to be solved, and to *identify* settings, where a restriction to subalgebra computations may save considerable computation costs.

2.6.3.2 *Concepts of grade structure of $Cl(3,0)$ and duality*

A general multivector in $Cl(3,0)$, can be represented as

$$M = m_0 + m_1 e_1 + m_2 e_2 + m_3 e_3 + m_{23} e_{23} + m_{31} e_{31} + m_{12} e_{12} \\ + m_{123} e_{123}, \qquad m_0, \dots, m_{123} \in \mathbb{R}. \tag{2.6.56}$$

We have a scalar part $\langle M \rangle_0$ of grade 0, a vector part $\langle M \rangle_1$ of grade 1, a bivector part $\langle M \rangle_2$ of grade 2, and a trivector part $\langle M \rangle_3$ of grade 3

$$M = \langle M \rangle_0 + \langle M \rangle_1 + \langle M \rangle_2 + \langle M \rangle_3, \tag{2.6.57}$$
$$\langle M \rangle_0 = m_0, \quad \langle M \rangle_1 = m_1 e_1 + m_2 e_2 + m_3 e_3,$$
$$\langle M \rangle_2 = m_{23} e_{23} + m_{31} e_{31} + m_{12} e_{12}, \quad \langle M \rangle_3 = m_{123} e_{123}.$$

The set of all grade k elements, $0 \le k \le 3$, is denoted $Cl^k(3,0)$.

The multiplication table of $Cl(3,0)$, Table 2.2, reveals that multiplication with i_3 (or $i_3^{-1} = -i_3$) consistently changes an element of grade k, $0 \le k \le 3$, into an element of grade $3-k$, i.e. scalars to trivectors (also called pseudoscalars) and vectors to bivectors, and vice versa. This means that the geometric product of a multivector $M \in Cl(3,0)$ and the pseudoscalar i_3 (or $i_3^{-1} = -i_3$) always results[13] in the left or right contraction

$$M i_3 = M \rfloor i_3, \qquad i_3 M = i_3 \lfloor M. \tag{2.6.58}$$

Mapping grades k, $0 \le k \le 3$, to grades $3-k$ is known as *duality* or *Hodge duality*. Because of its usefulness and importance it gets the symbol $*$ as an upper index

$$M^* := M i_3^{-1} = -m_0 i_3 - m_1 e_{23} - m_2 e_{31} - m_3 e_{12} \\ + m_{23} e_1 + m_{31} e_2 + m_{12} e_3 + m_{123}. \tag{2.6.59}$$

---

[13]In the context of blade subspaces, whenever a blade B contains another blade A as factor, then the geometric product is reduced to left or right contraction: $AB = A \rfloor B$, $BA = B \lfloor A$.

Duality (2.6.59) in $Cl(3,0)$ changes the outer product of vectors into the cross product

$$(e_1 \wedge e_2)^* = e_1 e_2(-e_{123}) = e_3 = e_1 \times e_2,$$
$$(e_2 \wedge e_3)^* = e_1, \qquad (e_3 \wedge e_1)^* = e_2. \tag{2.6.60}$$

Therefore we can use

$$(a \wedge b)^* = a \times b, \qquad a \wedge b = i_3(a \times b), \tag{2.6.61}$$

to *translate* well known results of standard 3D vector algebra into GA. Yet we emphasize, that the cross product $a \times b$ only exists in 3D, where a bivector has a unique (orthogonal) dual vector. The outer product has the advantage, that it can be *universally* used in all dimensions, because geometrically speaking the parallelogram spanned by two linearly independent vectors is always well defined, independent of the dimension of the embedding space.

Duality[14] also relates outer product and contraction of any $A, B \in Cl(3,0)$

$$A \wedge B^* = (A \rfloor B)^*, \qquad A^* \wedge B = (A \lfloor B)^*, \tag{2.6.62}$$
$$A \rfloor B^* = A^* \lfloor B = (A \wedge B)^*. \tag{2.6.63}$$

For vectors $a, b \in \mathbb{R}^3$ this can be rewritten as

$$a \wedge b^* = a^* \wedge b = (a \cdot b)^*, \qquad a \rfloor b^* = a^* \lfloor b = (a \wedge b)^*. \tag{2.6.64}$$

Inserting the duality definition this gives

$$a \wedge (i_3 b) = (i_3 a) \wedge b = i_3(a \cdot b), \quad a \rfloor (i_3 b) = (i_3 a) \lfloor b = i_3(a \wedge b), \tag{2.6.65}$$

which are handy relations in GA computations.

Let us do two simple examples. First, for $a = b = e_1$. Then we have $i_3 a = i_3 b = e_{23}$, $a \cdot b = 1$, and get

$$a \wedge (i_3 b) = e_1 \wedge e_{23} = i_3,$$
$$(i_3 a) \wedge b = e_{23} \wedge e_1 = i_3, \quad i_3(a \cdot b) = i_3. \tag{2.6.66}$$

Second, for $a = e_1, b = e_2$. Then we have $i_3 a = e_{23}$, $i_3 b = e_{31}$, $a \wedge b = e_{12}$, and get

$$a \rfloor (i_3 b) = e_1 \rfloor e_{31} = \langle e_1 e_{31} \rangle_1 = -e_3,$$
$$(i_3 a) \lfloor b = e_{23} \lfloor e_2 = -e_3, \quad i_3(a \wedge b) = i_3 e_{12} = -e_3. \tag{2.6.67}$$

2.6.3.3 *Concept of blade subspaces of* $Cl(3,0)$

A vector $a \in \mathbb{R}^3$ can be used to define a hyperplane $H(a)$ by

$$H(a) = \{x \in \mathbb{R}^3 \mid x \cdot a = x \rfloor a = 0\}, \tag{2.6.68}$$

[14] Equations (2.6.62) apply in *all* Clifford algebras, but (2.6.63) needs some modification if the pseudoscalar is not central.

which is an example of an *inner product null space* (IPNS) definition. It can also be used to define a line $L(a)$ by

$$L(a) = \{x \in \mathbb{R}^3 \mid x \wedge a = 0\}, \tag{2.6.69}$$

which shows an *outer product null space* (OPNS) definition. We can also use two linearly independent vectors to span a plane $P(a \wedge b)$. By that we mean, that $x = \alpha a + \beta b, \alpha, \beta \in \mathbb{R}$, if and only if, $x \wedge a \wedge b = 0$. Let us briefly check that. Assume $x = \alpha a + \beta b, \alpha, \beta \in \mathbb{R}$, then $x \wedge a \wedge b = (\alpha a + \beta b) \wedge a \wedge b = \alpha(a \wedge a) \wedge b - \beta a \wedge (b \wedge b) = 0$. On the other hand, if x has a component not in the plane spanned by a, b, then $x \wedge a \wedge b$ results in a non-zero trivector, the oriented parallelepiped volume spanned by x, a, b. Hence every plane spanned by two linearly independent vectors $a, b \in \mathbb{R}$ is in OPNS

$$\begin{aligned} P = P(a \wedge b) &= \{x \in \mathbb{R}^3 \mid x = \alpha a + \beta b, \forall \alpha, \beta \in \mathbb{R}\} \\ &= \{x \in \mathbb{R}^3 \mid x \wedge a \wedge b = 0\}. \end{aligned} \tag{2.6.70}$$

Three linearly independent vectors $a, b, c \in \mathbb{R}^3$ always span $\mathbb{R}^3$, and their outer product with any fourth vector $x \in \mathbb{R}^3$ is zero by default (because in $Cl(3,0)$ no 4-vectors exist). Therefore formally $\mathbb{R}^3 = \{x \in \mathbb{R}^3 \mid x \wedge a \wedge b \wedge c = 0\}$.

These geometric facts motivate the notion of *blade*. A blade A is an element of $Cl(3,0)$, that can be written as the outer product $A = a_1 \wedge \ldots \wedge a_k$ of k linearly independent vectors $a_1, \ldots, a_k \in \mathbb{R}^3$, $0 \leq k \leq 3$. Another name for k-blade A is *simple* k-vector. As we have just seen, every k-blade fully defines and characterizes a k-dimensional vector subspace $V(A) \subset \mathbb{R}^3$. Such subspaces are therefore often simply called *blade subspaces*. The GA of a blade subspace $Cl(V(A))$ is a natural subalgebra of $Cl(3,0)$. Examples are $Cl(V(e_1)) = Cl(1,0) \subset Cl(3,0)$, $Cl(V(e_{12})) = Cl(2,0) \subset Cl(3,0)$ and $Cl(V(e_{123}) = \mathbb{R}^3) = Cl(3,0) \subseteq Cl(3,0)$, etc.

The formulas for projection and rejection of vectors can now be generalized to the projection and rejection of subspaces, represented by their blades in their OPNS representations, where the grade of a blade shows the dimension of the blade subspace. The *projection* of a blade A onto a blade B is given by

$$P_B(A) = (A \rfloor B)B^{-1}, \tag{2.6.71}$$

where the left contraction automatically takes care of the fact that it would not be meaningful to project a higher dimensional subspace $V(A)$ onto a lower dimensional subspace $V(B)$.

This projection formula is a first example of how GA allows to *solve* geometric problems *by* means of elementary *algebraic products*, rather than the often cumbersome conventional solution of systems of linear equations.

The *rejection* (orthogonal part) of a blade A from a second blade B is given by

$$P_B^{\perp}(A) = (A \wedge B)B^{-1}. \tag{2.6.72}$$

If we interpret a general multivector $A \in Cl(3,0)$ as a *weighted sum of blades*, where the weights can also be interpreted as blade volumes, then by linearity the projection $P_B(A)$ [and rejection $P_B^{\perp}(A)$] formula applied to a multivector A yields the weighted sum of blades, each projected onto [rejected from] the blade B.

Let us consider the example of $A = e_1 + e_3 + e_{12} + e_{23}$ and $B = e_{12}$, $B^{-1} = -e_{12}$. We get

$$\begin{aligned} P_B(A) &= [(e_1 + e_3 + e_{12} + e_{23})\rfloor e_{12}](-e_{12}) \\ &= [e_2 + 0 - 1 + 0](-e_{12}) = e_1 + e_{12}, \end{aligned} \tag{2.6.73}$$

which is exactly what we expect, because the vector e_1 is in the plane $B = e_{12}$, e_3 is orthogonal to e_{12} and is projected out, the bivector component e_{12} is also in the plane $B = e_{12}$, but the bivector component e_{23} is perpendicular to e_{12} and is correctly projected out. Let us also compute the rejection

$$\begin{aligned} P_B^{\perp}(A) &= [(e_1 + e_3 + e_{12} + e_{23}) \wedge e_{12}](-e_{12}) \\ &= [0 + e_3 e_{12} + 0 + 0](-e_{12}) = e_3, \end{aligned} \tag{2.6.74}$$

which is again correct, because only the vector e_3 is orthogonal to the plane $B = e_{12}$, whereas e.g. e_{23} is not fully orthogonal[15], because it has the vector e_2 in common with $B = e_{12}$.

The general duality of the outer product and the left contraction in (2.6.62) and (2.6.63) has the remarkable consequence of *duality of IPNS and OPNS* subspace representations, because for all blades $A \in Cl(3,0)$

$$x \rfloor A = 0 \;\Leftrightarrow\; x \wedge A^* = 0, \quad \forall x \in \mathbb{R}^3, \tag{2.6.75}$$

i.e. because in general $(x \rfloor A)^* = x \wedge A^*$ for all $x, A \in Cl(3,0)$. And because of $(A^*)^* = A(-i_3)^2 = -A$, we also have

$$x \rfloor A^* = 0 \;\Leftrightarrow\; x \wedge A = 0, \quad \forall x \in \mathbb{R}^3. \tag{2.6.76}$$

So we can either represent a subspace in OPNS or IPNS by a blade $A \in Cl(3,0)$ as

$$\begin{aligned} V_{\mathrm{OPNS}}(A) &= \{x \in \mathbb{R}^3 \mid x \wedge A = 0\} \\ &= \{x \in \mathbb{R}^3 \mid x \rfloor A^* = 0\} = V_{\mathrm{IPNS}}(A^*). \end{aligned} \tag{2.6.77}$$

We also have the relationship

$$\begin{aligned} V_{\mathrm{IPNS}}(A) &= \{x \in \mathbb{R}^3 \mid x \rfloor A = 0\} \\ &= \{x \in \mathbb{R}^3 \mid x \wedge A^* = 0\} = V_{\mathrm{OPNS}}(A^*). \end{aligned} \tag{2.6.78}$$

In application contexts both representations are frequently used and it should always be taken care to *clearly specify* if a blade is meant to represent a subspace in the

[15] To include e_{23} one can simply compute $A - P_B(A) = e_3 + e_{23}$.

OPNS or the IPNS representation. In GA software, like CLUCalc or CLUViz [281], this is done by an initial multivector interpretation command.

The outer product itself *joins* (or unifies) disjoint (orthogonal) blade subspaces, e.g., the two lines $L(a)$ and $L(b)$ are unified by the outer product to the plane $P(a \wedge b)$ in (2.6.70). The left contraction has the dual property to *cut out* one subspace contained in a larger one and leave only the orthogonal complement. For example, if $A = e_1$, $B = e_{12}$, then $A \rfloor B = e_1 \rfloor e_{12} = e_2$, the orthogonal complement of $V(A)$ in $V(B)$. In general in the OPNS, the *orthogonal complement* of $V(A)$ in $V(B)$ is therefore given by $V(A \rfloor B)$.

Let us assume two blades $A, B \in Cl(3,0)$, which represent two blade subspaces $V(A), V(B) \subset \mathbb{R}^3$ with intersection $V(M) = V(A) \cap V(B)$, given by the common blade factor $M \in Cl(3,0)$ (called *meet*)

$$A = A'M = A' \wedge M, \qquad B = MB' = M \wedge B', \tag{2.6.79}$$

where A', B' are the orthogonal complement blades of M in A, B, respectively,

$$A' = AM^{\ 1} = A \lfloor M^{\ 1}, \qquad B' = M^{-1}B = M^{-1} \rfloor B, \tag{2.6.80}$$

where we freely use the fact that scalar factors, like $M^{-2} = \pm |M|^{-2} \in \mathbb{R}$ of $M^{-1} = M\,M^{-2}$, are not relevant for determining a subspace (in both OPNS and IPNS), i.e. $V(M) = V(\lambda M)$, $\forall\, \lambda \in \mathbb{R} \setminus \{0\}$.

The *join* (union) J of any two blade subspaces $V(A), V(B)$ can therefore be represented as

$$J = A' \wedge M \wedge B' = A \wedge B' = A' \wedge B. \tag{2.6.81}$$

Inserting the above contraction formulas for A', B' we get

$$J = A \wedge (M^{-1} \rfloor B) = (A \lfloor M^{-1}) \wedge B. \tag{2.6.82}$$

In turn, we can use the join J (or equivalently its inverse blade J^{-1}) to compute M. The argument is from set theory. If we cut out B from J (or J^{-1}), then only A' (or A'^{-1}) will remain as the orthogonal complement of B in J (or J^{-1}). Cutting out A' (or A'^{-1}) from A itself leaves only the *meet*[16] (intersection) M

$$M = A \vee B = A'^{-1} \rfloor A = (B \rfloor J^{-1}) \rfloor A = B \lfloor (J^{-1} \lfloor A), \tag{2.6.83}$$

The right most form $M = B \lfloor (J^{-1} \lfloor A)$ is obtained by cutting A out of from J (or J^{-1}), and using the resulting $B'^{-1} = J^{-1} \lfloor A$, to get M as orthogonal complement of B' in B.

For example, if we assume $A = e_{12}$, $B = e_{23}$, then the join is obviously $J = e_{123} = i_3$. This allows to compute the meet as

$$M = A \vee B = (B \rfloor J^{-1}) \rfloor A = (e_{23} \rfloor (-e_{123})) \rfloor e_{12} = e_1 \rfloor e_{12} = e_2, \tag{2.6.84}$$

which represents the line of intersection $L(e_2)$ of the two planes $P(e_{12})$ and $P(e_{23})$.

[16]The symbol $\vee$ stems from Grassmann-Cayley algebra.

2.6.3.4 Useful algebraic formulas in Clifford's geometric algebras

Let $A, B \in Cl(p,q,r)$. Some authors use angular brackets without index to indicate the scalar part of a multivector, e.g.

$$\langle AB \rangle = \langle AB \rangle_0 = \langle BA \rangle_0. \tag{2.6.85}$$

For reversion (an involution) we have

$$\begin{aligned} &(AB)^{\sim} = \widetilde{B}\widetilde{A}, \quad \tilde{a} = a, \quad \langle \widetilde{A} \rangle_0 = \langle A \rangle_0 = \langle A \rangle_0^{\sim}, \\ &\langle A \rangle_k^{\sim} = (-1)^{k(k-1)/2} \langle A \rangle_k. \end{aligned} \tag{2.6.86}$$

Apart from reversion, there is the automorphism *main involution* (or *grade involution*) which maps all vectors $a \to \hat{a} = -a$, and therefore

$$\widehat{a_1 \dots a_s} = (-1)^s a_1 \dots a_s, \tag{2.6.87}$$

i.e. even blades are invariant under the grade involution, odd grade blades change sign. The composition of reversion and grade involution is called *Clifford conjugation*

$$\begin{aligned} &\overline{A} = \widetilde{(\hat{A})} = \widehat{(\widetilde{A})}, \qquad \overline{a_1 \dots a_s} = (-1)^s a_s \dots a_1, \\ &\overline{\langle A \rangle_k} = (-1)^{k(k+1)/2} \langle A \rangle_k. \end{aligned} \tag{2.6.88}$$

Applied to a multivector $M \in Cl(3,0)$ we would get

$$\begin{aligned} \widetilde{M} &= \langle M \rangle_0 + \langle M \rangle_1 - \langle M \rangle_2 - \langle M \rangle_3, \\ \widehat{M} &= \langle M \rangle_0 - \langle M \rangle_1 + \langle M \rangle_2 - \langle M \rangle_3, \\ \overline{M} &= \langle M \rangle_0 - \langle M \rangle_1 - \langle M \rangle_2 + \langle M \rangle_3. \end{aligned} \tag{2.6.89}$$

The following algebraic identities [162] are frequently applied

$$\begin{aligned} a \rfloor (b \wedge c) &= (a \rfloor b)c - (a \rfloor c)b = (a \cdot b)c - (a \cdot c)b \\ &\overset{Cl(3,0)}{=} -a \times (b \times c), \end{aligned} \tag{2.6.90}$$

$$\begin{aligned} a \rfloor (b \wedge c \wedge d) &= (b \wedge c \wedge d) \lfloor a, \\ &= (b \wedge c)(a \cdot d) - (b \wedge d)(c \cdot a) + (c \wedge d)(b \cdot a), \end{aligned} \tag{2.6.91}$$

$$a \wedge b \wedge c \wedge d \overset{Cl(3,0)}{=} 0. \tag{2.6.92}$$

2.6.4 Extending Clifford's geometric algebra to spacetime (STA)

The GA $Cl(1,3)$ of flat Minkowski *spacetime*[17] $\mathbb{R}^{1,3}$ [108], has the $2^4 = 16$D basis (e_0 represents the time dimension)

$$\begin{aligned} &\{1, e_0, e_1, e_2, e_3, \sigma_1 = e_{10}, \sigma_2 = e_{20}, \sigma_3 = e_{30}, \\ &e_{23}, e_{31}, e_{12}, e_{123}, e_{230}, e_{310}, e_{120}, e_{0123} = I\} \end{aligned} \tag{2.6.93}$$

[17]Note that some authors prefer opposite signature $Cl(3,1)$.

of one scalar, 4 vectors (basis of $\mathbb{R}^{1,3}$), 6 bivectors, 4 trivectors and one pseudoscalar. The 4 vectors fulfill $e_{\mu}\rfloor e_{\nu} = \eta_{\mu,\nu} = \text{diag}(+1,-1,-1,-1)_{\mu,\nu}$, $0 \leq \mu,\nu \leq 3$.

The commutator of two bivectors always gives a third bivector, e.g.,

$$\frac{1}{2}(e_{10}e_{20} - e_{20}e_{10}) = -e_{12}, \ldots \tag{2.6.94}$$

Exponentiating all bivectors gives a group of rotors, the *Lorentz group*, its *Lie algebra* is the commutator algebra of the six bivectors. The even grade subalgebra $Cl^{+}(1,3)$ with 8D basis ($l = 1,2,3$): $\{1, \{\sigma_l\}, \{I\sigma_l\}, I\}$ is isomorphic to $Cl(3,0)$.

Assigning *electromagnetic* field vectors $\vec{E} = E_1\sigma_1 + E_2\sigma_2 + E_3\sigma_3$, $\vec{B} = B_1\sigma_1 + B_2\sigma_2 + B_3\sigma_3$, the Faraday bivector is $F = \vec{E} + I\vec{B}$. All four *Maxwell equations* then *unify* into one

$$\nabla F = J, \tag{2.6.95}$$

with vector derivative $\nabla = \sum_{\mu,\nu=0}^{3} \eta^{\mu,\nu} e_{\nu}\partial_{\mu} = \sum_{\mu=0}^{3} e^{\mu}\partial_{\mu}$.

The *Dirac* equation of *quantum mechanics* for the electron becomes in STA simply

$$\nabla\psi I\sigma_3 = m\psi e_0, \tag{2.6.96}$$

where the spinor ψ is an even grade multivector, and m the mass of the electron. The electron spinor acts as a spacetime rotor resulting in observables, e.g. the current vector

$$J = \psi e_0 \widetilde{\psi}. \tag{2.6.97}$$

2.6.5 Geometric modeling with conformal extension

2.6.5.1 *On points, planes and motors in* $Cl(4,1)$

In order to *linearize* translations as in (2.6.100) we need the conformal model [9, 244, 246] of Euclidean space (in $Cl(4,1)$), which adds to $\boldsymbol{x} \in \mathbb{R}^3 \subset \mathbb{R}^{4,1}$ two null-vector dimensions for the origin $\boldsymbol{e}_0$ and infinity $\boldsymbol{e}_\infty$, $\boldsymbol{e}_0\rfloor\boldsymbol{e}_\infty = -1$, to obtain by a non-linear embedding in $\mathbb{R}^{4,1}$ the homogeneous[18] conformal point

$$X = \boldsymbol{x} + \frac{1}{2}\boldsymbol{x}^2\boldsymbol{e}_\infty + \boldsymbol{e}_0, \qquad \boldsymbol{e}_0^2 = \boldsymbol{e}_\infty^2 = X^2 = 0, \qquad X\rfloor\boldsymbol{e}_\infty = -1. \tag{2.6.98}$$

The condition $X^2 = 0$ restricts the point manifold to the so-called null-cone of $\mathbb{R}^{4,1}$, similar to the light cone of special relativity. The second condition $X\rfloor\boldsymbol{e}_\infty = \boldsymbol{e}_0\rfloor\boldsymbol{e}_\infty = -1$ further restricts all points to a hyperplane of $\mathbb{R}^{4,1}$, similar to points in projective geometry. The remaining point manifold is again 3D. See [108, 111, 190, 244, 248, 281, 282, 304] for details and illustrations. We can always move from the Euclidean representation $\boldsymbol{x} \in \mathbb{R}^3$ to the conformal point $X \in \mathbb{R}^{4,1}$ by adding the $\boldsymbol{e}_0$ and $\boldsymbol{e}_\infty$ components, or by dropping them (projection (2.6.71) with i_3, or rejection (2.6.72) with $E = \boldsymbol{e}_\infty \wedge \boldsymbol{e}_0$).

The contraction of two conformal points gives their *Euclidean distance* and therefore a plane m equidistant from two points $A = \boldsymbol{a} + \frac{1}{2}\boldsymbol{a}^2\boldsymbol{e}_\infty + \boldsymbol{e}_0$, $B = \boldsymbol{b} + \frac{1}{2}\boldsymbol{b}^2\boldsymbol{e}_\infty + \boldsymbol{e}_0$

[18] Unique up to a nonzero scalar factor.

as

$$X \rfloor A = -\frac{1}{2}(\boldsymbol{x} - \boldsymbol{a})^2 \tag{2.6.99}$$
$$\Rightarrow \; X \rfloor (A - B) = 0, \;\; m = A - B \propto \boldsymbol{n} + d\, \boldsymbol{e}_\infty,$$

where $\boldsymbol{n} \in \mathbb{R}^3$ is a unit normal to the plane and d its signed scalar distance from the origin, "$\propto$" means proportional. Reflecting at two parallel planes m, m' with distance $\boldsymbol{t}/2 \in \mathbb{R}^3$ we get the *translation* opera*tor* (translator by $\boldsymbol{t}$)

$$X' = m' m\, X\, m m' = T_{\boldsymbol{t}}^{-1} X T_{\boldsymbol{t}}, \;\; T_{\boldsymbol{t}} := m m' = 1 + \frac{1}{2}\boldsymbol{t}\boldsymbol{e}_\infty. \tag{2.6.100}$$

Reflection at two non-parallel planes m, m' yields the rotation around the m, m'-intersection by twice the angle subtended by m, m'.

Group theoretically the conformal group $C(3)$ [304] is isomorphic to $O(4,1)$ [315] and the Euclidean group $E(3)$ is the subgroup of $O(4,1)$ leaving infinity $\boldsymbol{e}_\infty$ invariant. Now general translations *and* rotations are both *linearly* represented by geometric products of invertible vectors (called Clifford monomials, Lipschitz elements, versors or simply *mo*tion opera*tors* = *motors*). The commutator algebra of the bivectors of $Cl(4,1)$ constitutes the *Lie algebra* of all *conformal transformations* (exponentials of bivectors) of $\mathbb{R}^3$. Derivatives with respect to these bivector motion parameters allow *motor optimization*, used in pose estimation, structure and motion estimation, motion capture and airborne laser strip adjustment (lidar) [219].

2.6.5.2 *Modeling geometric objects in $Cl(4,1)$ with blades*

Computer vision, computer graphics and robotics are interested in the intuitive *geometric* OPNS *meaning* of blades (outer products of conformal points $P_1, \ldots, P_4$) in conformal GA (Pp = point pair)

$$Pp = P_1 \wedge P_2, \qquad Circle = P_1 \wedge P_2 \wedge P_3,$$
$$Sphere = P_1 \wedge P_2 \wedge P_3 \wedge P_4, \tag{2.6.101}$$

homogeneously representing the *point pair* $\{P_1, P_2\}$, the *circle* through $\{P_1, P_2, P_3\}$, and the *sphere* with surface points $\{P_1, P_2, P_3, P_4\}$. Abstractly these are 0D, 1D, 2D and 3D spheres S with center $\boldsymbol{c} \in \mathbb{R}^3$, radius $r \geq 0$, and *Euclidean* carrier blade directions $\mathbf{D}$: 1 for points P, distance $\boldsymbol{d} \in \mathbb{R}^3$ of P_2 from P_1, circle plane bivector $\mathbf{i}_c$, and sphere volume trivector $i_s \propto e_{123} = i_3$, see [190].

$$S = \mathbf{D} \wedge \boldsymbol{c} + [\frac{1}{2}(c^2 + r^2)\mathbf{D} - \boldsymbol{c}(\boldsymbol{c} \rfloor \mathbf{D})]\boldsymbol{e}_\infty + \mathbf{D}\boldsymbol{e}_0 + (\mathbf{D} \lfloor \boldsymbol{c})E, \tag{2.6.102}$$

with the origin-infinity bivector $E = \boldsymbol{e}_\infty \wedge \boldsymbol{e}_0$.

A point $X \in \mathbb{R}^{4,1}$ is on the sphere S, if and only if $S \wedge X = 0$. By duality this is equivalent to $X \rfloor S^* = 0$, which is equivalent in Euclidean terms to $(\boldsymbol{x} - \boldsymbol{c})^2 = r^2$, where $\boldsymbol{x}$ is the Euclidean part of X, $\boldsymbol{c} \in \mathbb{R}^3$ the center of S, and r the radius of S.

These objects are stretched to infinity (flattened) by wedging with $\boldsymbol{e}_\infty$

$$F = S \wedge \boldsymbol{e}_\infty = \mathbf{D} \wedge \boldsymbol{c}\boldsymbol{e}_\infty - \mathbf{D}E = \mathbf{D}\boldsymbol{c}_\perp \boldsymbol{e}_\infty - \mathbf{D}E, \tag{2.6.103}$$

where $\boldsymbol{c}_\perp \in \mathbb{R}^3$ indicates the support vector (shortest distance from the origin): $\boldsymbol{p}$ for finite–infinite point pair $P \wedge \boldsymbol{e}_\infty$, $\boldsymbol{c}_\perp$ for the line $P_1 \wedge P_2 \wedge \boldsymbol{e}_\infty$ and the plane $P_1 \wedge P_2 \wedge P_3 \wedge \boldsymbol{e}_\infty$, and 0 for the 3D space $\mathbb{R}^3$ itself ($F = -i_s E \propto I_5$). All geometric entities can be *extracted* easily (as derived and illustrated in [190])

$$\mathbf{D} = -F \lfloor E, \qquad r^2 = \frac{S\widehat{S}}{\mathbf{D}^2}, \qquad \boldsymbol{c} = \mathbf{D}^{-1}[S \wedge (1+E)]\lfloor E. \tag{2.6.104}$$

The conformal model is therefore also a complete super model for projective geometry.

In the dual IPNS representation spheres (points for $r = 0$) and planes become vectors in $\mathbb{R}^{4,1}$

$$Sphere^* = C - \frac{1}{2} r^2 \boldsymbol{e}_\infty, \qquad Plane^* = \boldsymbol{n} + d\boldsymbol{e}_\infty \tag{2.6.105}$$

where C is the conformal center point, $\boldsymbol{n} := \boldsymbol{c}_\perp / |\boldsymbol{c}_\perp| \propto \mathbf{i}_c i_3$, "$\propto$" means proportional.

A point $X \in \mathbb{R}^{4,1}$ is on the plane $Plane^*$, if and only if $X \rfloor Plane^* = 0$, which is equivalent in Euclidean terms to $\boldsymbol{x} \cdot \boldsymbol{n} = d$, where $\boldsymbol{x}$ is the Euclidean part of X, $\boldsymbol{n} \in \mathbb{R}^3$ the oriented unit normal vector of $Plane^*$, and d its signed scalar distance from the origin.

Points are spheres with $r = 0$, circles and lines are intersection bivectors

$$Circle^* = S_1^* \wedge S_2^*, \qquad Line^* = Plane_1^* \wedge Plane_2^*, \tag{2.6.106}$$

of two sphere, and two plane vectors, respectively. Inversion at a sphere becomes $X \to SXS = S^* X S^*$, inversion at two concentric spheres gives scaling. For example, the conformal center point C of a sphere S is $C = S\boldsymbol{e}_\infty S$.

2.6.6 On Clifford analysis

Within geometric calculus (Clifford analysis), compare Chapter 3, we can define *quaternionic* and *Clifford FTs* and *wavelet transforms*, with applications in image and signal processing, and *multivector wave packet analysis* [184]. We also obtain a single *fundamental theorem of multivector calculus*, which *unifies* a host of classical theorems of integration for path independence, Green's, Stokes' and Gauss' divergence theorems. *Monogenic* functions f, $\nabla f = 0$, generalize complex analytic functions, and allow us to generalize Cauchy's famous integral theorem for analytic functions in the complex plane to n dimensions [303].

2.7 HOW IMAGINARY NUMBERS BECOME REAL IN CLIFFORD ALGEBRAS

2.7.1 What is an imaginary number?

The current subsection tries to provide a light introduction to the fundamental question of the nature of imaginary numbers. It is based on notes that first appeared in [178].

The previous Japanese emperor (Emperor Showa) is said to have asked this question. Today many students and scientists still ask it, but the traditional canon of mathematics at school and university needs to be widened for the answer. We find it in the works of Hamilton, Grassmann and Clifford. Hamilton introduced quaternions $\boldsymbol{i}, \boldsymbol{j}, \boldsymbol{k}$, with

$$\begin{aligned} &\boldsymbol{i}^2 = \boldsymbol{j}^2 = \boldsymbol{k}^2 = \boldsymbol{ijk} = -1, \\ &\boldsymbol{ij} = -\boldsymbol{ji} = \boldsymbol{k}, \qquad \boldsymbol{jk} = -\boldsymbol{kj} = \boldsymbol{i}, \qquad \boldsymbol{ki} = -\boldsymbol{ik} = \boldsymbol{j}, \end{aligned} \tag{2.7.1}$$

for three-dimensional rotations. Grassmann invented the outer product of oriented line segments (vectors) $\boldsymbol{a}, \boldsymbol{b}$ to give the directed oriented area of the enclosed parallelogram:

$$\boldsymbol{a} \wedge \boldsymbol{b} = -\boldsymbol{b} \wedge \boldsymbol{a}. \tag{2.7.2}$$

Clifford unified their work with the geometric product

$$\boldsymbol{ab} = \boldsymbol{a} \cdot \boldsymbol{b} + \boldsymbol{a} \wedge \boldsymbol{b}, \tag{2.7.3}$$

leading to geometric algebras.

In two dimensions we have orthogonal, unit vectors e_1, e_2 as vector space basis with

$$e_1^2 = e_2^2 = 1, \qquad e_1 \cdot e_2 = 0. \tag{2.7.4}$$

The associative geometric multiplication of the oriented directed unit square

$$i = e_1 e_2 \tag{2.7.5}$$

gives:

$$\begin{aligned} ii &= e_1 e_2 e_1 e_2 = e_1 (e_2 e_1) e_2 = e_1 (e_2 \cdot e_1 + e_2 \wedge e_1) e_2 \\ &= e_1 (0 - e_1 \wedge e_2) e_2 = -e_1 e_1 e_2 e_2 = -1. \end{aligned} \tag{2.7.6}$$

Remark 2.7.1. *We only used*

$$e_2 \cdot e_1 = e_1 \cdot e_2 = 0 \quad \textit{and} \quad e_2 e_1 = e_2 \wedge e_1 = -e_1 \wedge e_2 = -e_1 e_2. \tag{2.7.7}$$

So the square of the oriented unit area i is -1. Enough to satisfy the emperor's curiosity!

But today's politicians ask for an application. As an answer we calculate:

$$i e_1 = e_1 e_2 e_1 = -e_1 e_1 e_2 = -e_2, \tag{2.7.8}$$

and

$$i e_2 = e_1 e_2 e_2 = e_1, \tag{2.7.9}$$

which is a clockwise 90°-rotation. We can also calculate (NB: the order!)

$$e_1 i = e_1 e_1 e_2 = e_1, \tag{2.7.10}$$

and

$$e_2 i = e_2 e_1 e_2 = -e_1 e_2 e_2 = -e_1, \tag{2.7.11}$$

which is an anticlockwise (mathematically positive) 90°-rotation. For a general rotation in two dimensions, we simply add trigonometric coefficients:

$$\boldsymbol{a}(\cos\alpha + i\sin\alpha) \tag{2.7.12}$$

rotates the real vector $\boldsymbol{a}$ by α radians. Now even a politician can rotate vectors without using (or even knowing) matrices.

Of what use may the geometric product be for some new advanced technology venture business? As an application to laser beam optics let us imagine a laser beam with direction vector $\boldsymbol{a}$ hitting a mirror surface element approximated with unit normal vector $\boldsymbol{n}$, ($\boldsymbol{n}^2 = 1$). We can write $\boldsymbol{a}$ in components parallel and perpendicular to $\boldsymbol{n}$:

$$\boldsymbol{a} = \boldsymbol{a}_{\|} + \boldsymbol{a}_{\perp}. \tag{2.7.13}$$

Now

$$\boldsymbol{a}_{\|} \wedge \boldsymbol{n} = 0, \tag{2.7.14}$$

because parallel vectors span no parallelogram, and

$$\boldsymbol{a}_{\perp} \cdot \boldsymbol{n} = 0, \tag{2.7.15}$$

because of perpendicularity. So we must have

$$\boldsymbol{a}_{\|}\boldsymbol{n} = \boldsymbol{a}_{\|} \cdot \boldsymbol{n} + 0 = \boldsymbol{n} \cdot \boldsymbol{a}_{\|} + 0 = \boldsymbol{n}\boldsymbol{a}_{\|}, \tag{2.7.16}$$

and

$$\boldsymbol{a}_{\perp}\boldsymbol{n} = 0 + \boldsymbol{a}_{\perp} \wedge \boldsymbol{n} = 0 - \boldsymbol{n} \wedge \boldsymbol{a}_{\perp} = -\boldsymbol{n}\boldsymbol{a}_{\perp}. \tag{2.7.17}$$

Reflection only changes the sign of $\boldsymbol{a}_{\|}$. Therefore

$$\begin{aligned} \boldsymbol{a}' &= -\boldsymbol{a}_{\|} + \boldsymbol{a}_{\perp} = -\boldsymbol{n}\boldsymbol{n}(\boldsymbol{a}_{\|} - \boldsymbol{a}_{\perp}) = -\boldsymbol{n}(\boldsymbol{n}\boldsymbol{a}_{\|} - \boldsymbol{n}\boldsymbol{a}_{\perp}) \\ &= -\boldsymbol{n}(\boldsymbol{a}_{\|}\boldsymbol{n} + \boldsymbol{a}_{\perp}\boldsymbol{n}) = -\boldsymbol{n}(\boldsymbol{a}_{\|} + \boldsymbol{a}_{\perp})\boldsymbol{n} = -\boldsymbol{n}\boldsymbol{a}\boldsymbol{n}, \end{aligned} \tag{2.7.18}$$

is the reflected vector. In a cavity we may want to trace many reflections at a sequence of surface elements with normal vectors $\boldsymbol{n}_1, \boldsymbol{n}_2, \ldots \boldsymbol{n}_s$ which simply results in

$$\boldsymbol{a}' = (-1)^s \boldsymbol{n}_s \ldots \boldsymbol{n}_2\boldsymbol{n}_1\boldsymbol{a}\boldsymbol{n}_1\boldsymbol{n}_2 \ldots \boldsymbol{n}_s. \tag{2.7.19}$$

Nanoscience is a modern buzz word. On this scale mechanics meets quantum mechanics. Geometric algebra provides complete tools for both. From elementary geometry we know that two reflections at planes with normal vectors $\boldsymbol{n}, \boldsymbol{m}$ enclosing the angle $\theta/2$ result in a rotation by angle θ:

$$\boldsymbol{a}' = \boldsymbol{m}\boldsymbol{n}\,\boldsymbol{a}\,\boldsymbol{n}\boldsymbol{m}. \tag{2.7.20}$$

The general rotation operator (rotor) is

$$R = \boldsymbol{n}\boldsymbol{m} = \boldsymbol{n} \cdot \boldsymbol{m} + \boldsymbol{n} \wedge \boldsymbol{m} = \cos(\theta/2) + i\sin(\theta/2) = e^{i\theta/2} \tag{2.7.21}$$

with unit area element i in the $\boldsymbol{n}, \boldsymbol{m}$ rotation plane. Two rotations are given by the geometric product of two rotors RR'. A second $\theta' = 2\pi$ (equivalent to 360°) rotation produces

$$RR' = Re^{2\pi i/2} = Re^{\pi i} = R(\cos\pi + i\sin\pi) = R(-1 + i0) = -R. \tag{2.7.22}$$

The rotor R itself behaves therefore like the first known quantum particle, i.e. the electron described by a Pauli spinor

$$y = \rho^{1/2}R. \tag{2.7.23}$$

We summarize, that Clifford's geometric algebra answers fundamental questions, which the traditional canon of mathematics taught at schools and universities cannot. It further provides great methodological simplifications and geometric insight in applications to physics, molecular geometry, image processing, computer graphics, robotics, quantum computing, etc., compare [202].

2.7.2 What about imaginary eigenvalues and complex eigenvectors?

This subsection is mainly based on [171]. For further reading on the same topic see also [172, 173].

We first review how anti-symmetric matrices in two dimensions yield imaginary eigenvalues and complex eigenvectors. We will see how this carries on to rotations by means of the Cayley transformation. Then a real geometric interpretation is given to the eigenvalues and eigenvectors by means of real geometric algebra. The eigenvectors are seen to be *two component eigenspinors* which can be further reduced to underlying vector duplets. The eigenvalues are interpreted as rotation operators, which rotate the underlying vector duplets. We then extend and generalize the treatment to three dimensions.

The idea for this treatment arose from a linear algebra problem on anti-symmetric matrices for undergraduate engineering students. I wrote it, looking for a real geometric understanding of the imaginary eigenvalues and complex eigenvectors. Being already familiar with geometric algebra [152, 159, 161, 162] it was natural to try to apply it in this context.

2.7.2.1 *Two real dimensions*

2.7.2.1.1 Complex treatment Any anti-symmetric matrix in two real dimensions is proportional to

$$U = \begin{pmatrix} 0 & -1 \\ 1 & 0 \end{pmatrix}. \tag{2.7.24}$$

The characteristic polynomial equation of the matrix U is

$$|U - \lambda E| = \begin{vmatrix} -\lambda & -1 \\ 1 & -\lambda \end{vmatrix} = \lambda^2 + 1 = 0, \quad \text{i.e.} \quad \lambda^2 = -1. \tag{2.7.25}$$

The classical way to solve this equation is to postulate an imaginary entity j to be the root of -1: $j = \sqrt{-1}$. This leads to many interesting consequences, yet any real geometric meaning of this imaginary quantity is left obscure. The two eigenvalues are therefore the imaginary unit j and $-j$.

$$\lambda_1 = j, \qquad \lambda_2 = -j. \tag{2.7.26}$$

The corresponding complex eigenvectors $\mathbf{x}_1$ and $\mathbf{x}_2$ are

$$\begin{aligned} U\mathbf{x}_1 = \lambda_1\mathbf{x}_1 = j\mathbf{x}_1 \quad &\rightarrow \quad \mathbf{x}_1 = \begin{pmatrix} 1 \\ -j \end{pmatrix}, \\ U\mathbf{x}_2 = \lambda_2\mathbf{x}_2 = -j\mathbf{x}_2 \quad &\rightarrow \quad \mathbf{x}_2 = \begin{pmatrix} 1 \\ j \end{pmatrix}. \end{aligned} \tag{2.7.27}$$

The Cayley transformation [271] $C(-kU)$, with $k = (1 - \cos\vartheta)/\sin\vartheta$

$$\begin{aligned} C(-kU) &= (E + (-kU))^{-1}(E - (-kU)) = E - \frac{2}{1+k^2}(-kU - (-kU)^2) \\ &= \begin{pmatrix} \cos\vartheta & -\sin\vartheta \\ \sin\vartheta & \cos\vartheta \end{pmatrix} \end{aligned} \tag{2.7.28}$$

allows to describe two-dimensional rotations.

The third expression of equation (2.7.28) shows that U and $C(-kU)$ must have the same eigenvectors $\mathbf{x}_1$ and $\mathbf{x}_2$. The corresponding eigenvalues of $C(-kU)$ can now easily be calculated from (2.7.28) as

$$\lambda_{c1} = 1 + \frac{2k\lambda_1(1 + k\lambda_1)}{1+k^2}, \qquad \lambda_{c2} = 1 + \frac{2k\lambda_2(1 + k\lambda_2)}{1+k^2}. \tag{2.7.29}$$

Inserting $\lambda_1 = j$ and $\lambda_2 = -j$ we obtain the complex eigenvalues of the two-dimensional rotation $C(-kU)$ as

$$\lambda_{c1} = \cos\vartheta + j\sin\vartheta, \qquad \lambda_{c2} = \cos\vartheta - j\sin\vartheta. \tag{2.7.30}$$

We now face the question what the imaginary and complex eigenvalues and the complex eigenvectors of U and the rotation $C(-kU)$ mean in terms of purely real geometry. In order to do this let us turn to the real geometric algebra $\mathbb{R}_2$ of a real two-dimensional vector space $\mathbb{R}^2$ [152, 161, 162].

2.7.2.1.2 Real explanation Instead of postulating the imaginary unit j we now solve the characteristic polynomial equation (2.7.25) using both orientations of the oriented unit area element $\mathbf{i}$ of $\mathbb{R}_2$:

$$\lambda_1 = \mathbf{i}, \qquad \lambda_2 = -\mathbf{i}. \tag{2.7.31}$$

The corresponding "eigenvectors" $\mathbf{x}_1$ and $\mathbf{x}_2$ will then be:

$$\mathbf{x}_1 = \begin{pmatrix} 1 \\ -\mathbf{i} \end{pmatrix}, \qquad \mathbf{x}_2 = \begin{pmatrix} 1 \\ \mathbf{i} \end{pmatrix}. \tag{2.7.32}$$

As before, the "eigenvectors" of the Cayley transformation $C(-kU)$ will be the same. And the eigenvalues of $C(-kU)$ now become:

$$\lambda_{c1} = \cos\vartheta + \mathbf{i}\sin\vartheta, \qquad \lambda_{c2} = \cos\vartheta - \mathbf{i}\sin\vartheta. \tag{2.7.33}$$

We can now take the first step in our real explanation and identify the two "eigenvectors" $\mathbf{x}_1$ and $\mathbf{x}_2$ as *two-component spinors* with the entries: $x_{11} = 1$, $x_{12} = -\mathbf{i}$ and $x_{21} = 1$, $x_{22} = \mathbf{i}$.

Now we want to better understand what the real-oriented-unit-area-element eigenvalues λ_1, λ_2 as well as λ_{c1} and λ_{c2} do when multiplied with the two-component eigen-spinors $\mathbf{x}_1$ and $\mathbf{x}_2$. Every spinor can be understood to be the geometric product of two vectors. We therefore choose an arbitrary, but fixed reference vector $\mathbf{z}$ from the vector space R^2. For simplicity let us take $\mathbf{z}$ to be $\mathbf{z} = \boldsymbol{\sigma}_1$, assuming $\{\boldsymbol{\sigma}_1, \boldsymbol{\sigma}_2\}$ to be the orthonormal basis of R^2. We can then factorize the spinor components of the eigen-spinors $\mathbf{x}_1$ and $\mathbf{x}_2$ to:

$$x_{11} = x_{21} = 1 = \boldsymbol{\sigma}_1\boldsymbol{\sigma}_1, \quad x_{12} = -\mathbf{i} = \boldsymbol{\sigma}_2\boldsymbol{\sigma}_1, \quad x_{22} = \mathbf{i} = -\boldsymbol{\sigma}_2\boldsymbol{\sigma}_1. \tag{2.7.34}$$

Note that we always factored out $\mathbf{z}$ to the right. In two real dimensions it now seems natural to adopt the following interpretation: The eigen-spinor $\mathbf{x}_1$ corresponds (modulus the geometric multiplication from the right with $\mathbf{z} = \boldsymbol{\sigma}_1$) to the real vector pair $\{\boldsymbol{\sigma}_1, \boldsymbol{\sigma}_2\}$, whereas $\mathbf{x}_2$ corresponds to the real vector pair $\{\boldsymbol{\sigma}_1, -\boldsymbol{\sigma}_2\}$. Multiplication with λ_1 from the left as in

$$U\mathbf{x}_1 = \lambda_1\mathbf{x}_1 = \mathbf{i}\begin{pmatrix} x_{11} \\ x_{12} \end{pmatrix} = \begin{pmatrix} \mathbf{i}x_{11} \\ \mathbf{i}x_{12} \end{pmatrix}, \tag{2.7.35}$$

results in

$$x_{11} \;\rightarrow\; \mathbf{i}x_{11} = (-\boldsymbol{\sigma}_2)\boldsymbol{\sigma}_1, \qquad x_{12} \;\rightarrow\; \mathbf{i}x_{12} = \boldsymbol{\sigma}_1\boldsymbol{\sigma}_1. \tag{2.7.36}$$

That is the multiplication with $\lambda_1 = \mathbf{i}$ from the left transforms the vector pair $\{\boldsymbol{\sigma}_1, \boldsymbol{\sigma}_2\}$ to the new pair $\{$-$\boldsymbol{\sigma}_2, \boldsymbol{\sigma}_1\}$, which is a simple rotation by -90 degrees. Here the non-commutative nature of the geometric product is important.

The analogous calculation for $\lambda_2\mathbf{x}_2 = -\mathbf{i}\mathbf{x}_2$ shows that the pair $\{\boldsymbol{\sigma}_1, -\boldsymbol{\sigma}_2\}$, which corresponds to $\mathbf{x}_2$ is transformed to $\{\boldsymbol{\sigma}_2, -\boldsymbol{\sigma}_1\}$, i.e. it is rotated by $+90$ degree.

I will now treat $C(-kU)\mathbf{x}_1 = \lambda_{c1}\mathbf{x}_1$ and $C(-kU)\mathbf{x}_2 = \lambda_{c2}\mathbf{x}_2$ in the same way.

$$\begin{aligned} x_{11} \;\rightarrow\; \lambda_{c1}x_{11} &= (\cos\vartheta + \mathbf{i}\sin\vartheta)\boldsymbol{\sigma}_1^2 = \boldsymbol{\sigma}_1(\cos\vartheta - \mathbf{i}\sin\vartheta)\boldsymbol{\sigma}_1 \\ &= (\boldsymbol{\sigma}_1 R(-\vartheta))\boldsymbol{\sigma}_1, \end{aligned} \tag{2.7.37}$$

where $R(-\vartheta)$ is the rotation operator by $-\vartheta$. For the second component x_{12} we have

$$x_{12} \;\rightarrow\; \lambda_{c1}x_{12} = (\cos\vartheta + \mathbf{i}\sin\vartheta)(\boldsymbol{\sigma}_2\boldsymbol{\sigma}_1) = (\boldsymbol{\sigma}_2 R(-\vartheta))\boldsymbol{\sigma}_1. \tag{2.7.38}$$

The action of λ_{c1} on $\mathbf{x}_1$ means therefore a rotation of the corresponding vector pair $\{\boldsymbol{\sigma}_1, \boldsymbol{\sigma}_2\}$ by $-\vartheta$.

The analogous calculations for $\lambda_{c2}\mathbf{x}_2$ show that λ_{c1} rotates the vector pair

$\{\boldsymbol{\sigma}_1, -\boldsymbol{\sigma}_2\}$, which corresponds to $\mathbf{x}_2$, into $\{\boldsymbol{\sigma}_1 R(\vartheta), -\boldsymbol{\sigma}_2 R(\vartheta)\}$. This corresponds to a rotation of the vector pair by $+\vartheta$.

Summarizing the two-dimensional situation, we see that the complex eigenvectors $\mathbf{x}_1$ and $\mathbf{x}_2$ may rightfully be interpreted as two-component eigen-spinors with underlying vector pairs. The multiplication of these eigen-spinors with the unit-oriented-area-element eigenvalues λ_1 and λ_2 means a real rotation of the underlying vector pairs by -90 and $+90$ degrees, respectively. Whereas the multiplication with λ_{c1} and λ_{c2} means a real rotation of the underlying vector pairs by $-\vartheta$ and $+\vartheta$, respectively.

Now all imaginary eigenvalues and complex eigenvectors of anti-symmetric matrices in two real dimensions have a real geometric interpretation. Let us examine next how this carries on to three dimensions.

2.7.2.2 Three real dimensions

2.7.2.2.1 Complex treatment of three dimensions Any anti-symmetric matrix in three real dimensions is proportional to a matrix of the form

$$U = \begin{pmatrix} 0 & -c & b \\ c & 0 & -a \\ -b & a & 0 \end{pmatrix}, \tag{2.7.39}$$

with $a^2 + b^2 + c^2 = 1$. The characteristic polynomial equation of the matrix U is

$$|U - \lambda E| = \begin{vmatrix} -\lambda & -c & b \\ c & -\lambda & -a \\ -b & a & -\lambda \end{vmatrix} = \lambda(\lambda^2 + a^2 + b^2 + c^2) = 0. \tag{2.7.40}$$

If we use the condition that $a^2 + b^2 + c^2 = 1$, this simplifies and breaks up into the two equations

$$\lambda_{1,2}^2 = -1, \quad \lambda_3 = 0. \tag{2.7.41}$$

That means we have one eigenvalue λ_3 equal to zero and for the other two eigenvalues λ_1, λ_2 we have the same condition (2.7.25) as in the two-dimensional case for the matrix of equation (2.7.24). It is therefore clear that in the conventional treatment one would again assign[19] $\lambda_1 = j$ and $\lambda_2 = -j$. The corresponding complex eigenvectors are:

$$\mathbf{x}_1 = \begin{pmatrix} 1 - a^2 \\ -ab - jc \\ -ac + jb \end{pmatrix} \doteq \begin{pmatrix} -ab + jc \\ 1 - b^2 \\ -bc - ja \end{pmatrix} \doteq \begin{pmatrix} -ac - jb \\ -bc + ja \\ 1 - c^2 \end{pmatrix},$$
$$\mathbf{x}_2 = \mathrm{cc}(\mathbf{x}_1), \tag{2.7.42}$$

where cc(.) stands for the usual complex conjugation, i.e. $\mathrm{cc}(j) = -j$. The symbol

[19] Here an incompleteness of the conventional treatment becomes obvious. A priori there is no reason to assume that the solutions to the characteristic polynomial equations in two and three dimensions (2.7.25) and (2.7.41) must geometrically be the same.

$\doteq$ expresses that all three given forms are equivalent up to the multiplication with a scalar (complex) constant.

The eigenvector that corresponds to λ_3 simply is:

$$\mathbf{x}_3 = \begin{pmatrix} a \\ b \\ c \end{pmatrix}. \tag{2.7.43}$$

The fact that $\lambda_3 = 0$ simply means that the matrix U projects out any component of a vector parallel to $\mathbf{x}_3$. U maps the three-dimensional vector space therefore to a plane perpendicular to $\mathbf{x}_3$ containing the origin.

The Cayley transformation [271] $C(-kU)$ with $k = \frac{1-\cos\vartheta}{\sin\vartheta}$ now describes rotations in three dimensions:

$$\begin{aligned} C(-kU) &= (E + (-kU))^{-1}(E - (-kU)) \\ &= E - \frac{2}{1 + k^2(a^2+b^2+c^2)}(-kU - (-kU)^2) = \end{aligned} \tag{2.7.44}$$

$$\begin{pmatrix} 1+(1-\cos\vartheta)(1-a^2) & -c\sin\vartheta + ab(1-\cos\vartheta) & b\sin\vartheta + ac(1-\cos\vartheta) \\ c\sin\vartheta + ab(1-\cos\vartheta) & 1+(1-\cos\vartheta)(1-b^2) & -a\sin\vartheta + bc(1-\cos\vartheta) \\ -b\sin\vartheta + ac(1-\cos\vartheta) & a\sin\vartheta + bc(1-\cos\vartheta) & 1+(1-\cos\vartheta)(1-b^2) \end{pmatrix}.$$

The vector $\mathbf{x}_3$ is the rotation axis.

The expression for $C(-kU)$ after the second equal sign in (2.7.44) clearly shows that the eigenvectors of U and $C(-kU)$ agree in three dimensions as well. The general formula for calculating the eigenvalues λ_c of $C(-kU)$ from the eigenvalues λ of U reads:

$$\lambda_c = 1 + \frac{2k\lambda(1+k\lambda)}{1+k^2(a^2+b^2+c^2)} \overset{a^2+b^2+c^2=1}{=} 1 + \frac{2k\lambda(1+k\lambda)}{1+k^2}. \tag{2.7.45}$$

Inserting λ_1, λ_2 and λ_3 in this formula yields:

$$\lambda_{c1} = \cos\vartheta + j\sin\vartheta, \quad \lambda_{c2} = \cos\vartheta - j\sin\vartheta, \quad \lambda_{c3} = 1. \tag{2.7.46}$$

We see that in three dimensions the complex eigenvectors (2.7.42) contain more structure and the explicit form of the Cayley transformation (2.7.44) gets rather unwieldy.

2.7.2.2.2 Real explanation for three dimensions If we follow the treatment of the two-dimensional case given in Section 2.7.2.1.2, then we need to replace the imaginary unit j in the eigenvalues λ_1, λ_2 and in the eigenvectors $\mathbf{x}_1$, $\mathbf{x}_2$ by an element of the real three-dimensional geometric algebra $\mathbb{R}_3$ [152,159,161]. In principle, there are two different choices: The volume element i or any two-dimensional unit area element like e.g. $\mathbf{i}_1 = \boldsymbol{\sigma}_2\boldsymbol{\sigma}_3$, $\mathbf{i}_2 = \boldsymbol{\sigma}_3\boldsymbol{\sigma}_1$ or $\mathbf{i}_3 = \boldsymbol{\sigma}_1\boldsymbol{\sigma}_2$. ($\{\boldsymbol{\sigma}_1, \boldsymbol{\sigma}_2, \boldsymbol{\sigma}_3\}$ denotes an orthonormal basis in R^3.)

While both interpretations are possible, let me argue for the second possibility: We have seen in Section 2.7.2.2.1 that the multiplication of U with a vector always projects out the component of this vector parallel to $\mathbf{x}_3$ so that the $\mathbf{y}$ on the right hand side of equations like $U\mathbf{x} = \mathbf{y}$ is necessarily a vector in the two-dimensional plane perpendicular to $\mathbf{x}_3$ containing the origin. Thus it seems only natural to interpret the square root of -1 in the solution of equation (2.7.41) to be the oriented unit area element $\mathbf{i} = a\mathbf{i}_1 + b\mathbf{i}_2 + c\mathbf{i}_3$ characteristic for the plane perpendicular to $\mathbf{x}_3$ containing the origin as opposed to the volume element element i or any other two-dimensional unit area element. I will show in the following, that this leads indeed to a consistent interpretation.

Using this area element $\mathbf{i}$ we have $\lambda_1 = \mathbf{i}$, $\lambda_2 = \widetilde{\lambda}_1 = -\mathbf{i}$ and

$$\mathbf{x}_1 = \begin{pmatrix} 1-a^2 \\ -ab - \mathbf{i}c \\ -ac + \mathbf{i}b \end{pmatrix} \doteq \begin{pmatrix} -ab + \mathbf{i}c \\ 1-b^2 \\ -bc - \mathbf{i}a \end{pmatrix} \doteq \begin{pmatrix} -ac - \mathbf{i}b \\ -bc + \mathbf{i}a \\ 1-c^2 \end{pmatrix},$$
$$\mathbf{x}_2 = \widetilde{\mathbf{x}}_1, \tag{2.7.47}$$

where the tilde operation marks the reverse of geometric algebra. As in the two-dimensional case I interpret the three components of each "eigenvector" as spinorial components, i.e. elementary geometric products of two vectors. (In the following we will therefore use the expression *three-component eigenspinor* instead of "eigenvector".) I again arbitrarily fix one vector from the $\mathbf{i}$ plane (the plane perpendicular to $\mathbf{x}_3$) as a reference vector $\mathbf{z}$ with respect to which I will factorize the three component eigenspinors $\mathbf{x}_1$ and $\mathbf{x}_2$. With regard to the first representation of the eigenspinors $\mathbf{x}_1$ and $\mathbf{x}_2$ we choose to set

$$\mathbf{z} = \boldsymbol{\sigma}_{1\|} = \boldsymbol{\sigma}_1 \cdot \mathbf{i}\mathbf{i}^{-1} = (1-a^2)\boldsymbol{\sigma}_1 - ab\boldsymbol{\sigma}_2 - ac\boldsymbol{\sigma}_3. \tag{2.7.48}$$

Using $a^2+b^2+c^2 = 1$, the square $\mathbf{z}^2 = 1 - a^2$ is seen to be the first component spinor of the first representation of $\mathbf{x}_1$ and $\mathbf{x}_2$ as given in (2.7.47).

Next we will use the inverse of $\mathbf{z}$

$$\mathbf{z}^{-1} = \boldsymbol{\sigma}_1 - \frac{ab}{1-a^2}\boldsymbol{\sigma}_2 - \frac{ac}{1-a^2}\boldsymbol{\sigma}_3 \tag{2.7.49}$$

in order to factorize the two other component spinors $\mathbf{n}_2\mathbf{z} = -ab - \mathbf{i}c$ and $\mathbf{n}_3\mathbf{z} = -ac + \mathbf{i}b$ of $\mathbf{x}_1$ in (2.7.47) as well. A somewhat cumbersome calculation[20] renders

$$\mathbf{n}_2 = \mathbf{n}_2\mathbf{z}\mathbf{z}^{-1} = \boldsymbol{\sigma}_{2\|}, \quad \mathbf{n}_3 = \mathbf{n}_3\mathbf{z}\mathbf{z}^{-1} = \boldsymbol{\sigma}_{3\|}. \tag{2.7.50}$$

Summarizing these calculations we have (setting $\mathbf{n}_1 = \mathbf{z} = \boldsymbol{\sigma}_{1\|}$):

$$\mathbf{x}_1 = \begin{pmatrix} 1-a^2 \\ -ab - \mathbf{i}c \\ -ac + \mathbf{i}b \end{pmatrix} = \begin{pmatrix} \mathbf{n}_1\mathbf{z} \\ \mathbf{n}_2\mathbf{z} \\ \mathbf{n}_3\mathbf{z} \end{pmatrix} = \begin{pmatrix} \boldsymbol{\sigma}_{1\|}\boldsymbol{\sigma}_{1\|} \\ \boldsymbol{\sigma}_{2\|}\boldsymbol{\sigma}_{1\|} \\ \boldsymbol{\sigma}_{3\|}\boldsymbol{\sigma}_{1\|} \end{pmatrix}. \tag{2.7.51}$$

[20] A good way to speed up and verify such calculations is geometric algebra software, such as [14] and others.

The other two equivalent representations of $\mathbf{x}_1$ given in (2.7.47) can be written as:

$$\begin{pmatrix} -ab + \mathbf{i}c \\ 1 - b^2 \\ -bc - \mathbf{i}a \end{pmatrix} = \begin{pmatrix} \boldsymbol{\sigma}_{1\|}\boldsymbol{\sigma}_{2\|} \\ \boldsymbol{\sigma}_{2\|}\boldsymbol{\sigma}_{2\|} \\ \boldsymbol{\sigma}_{3\|}\boldsymbol{\sigma}_{2\|} \end{pmatrix}, \quad \begin{pmatrix} -ac - \mathbf{i}b \\ -bc + \mathbf{i}a \\ 1 - c^2 \end{pmatrix} = \begin{pmatrix} \boldsymbol{\sigma}_{1\|}\boldsymbol{\sigma}_{3\|} \\ \boldsymbol{\sigma}_{2\|}\boldsymbol{\sigma}_{3\|} \\ \boldsymbol{\sigma}_{3\|}\boldsymbol{\sigma}_{3\|} \end{pmatrix}. \tag{2.7.52}$$

We see that this simply corresponds to a different choice of the reference vector $\mathbf{z}$, as $\mathbf{z}' = \boldsymbol{\sigma}_{2\|}$ and as $\mathbf{z}'' = \boldsymbol{\sigma}_{3\|}$, respectively. In general all possible ways to write $\mathbf{x}_1$ correspond to different choices of $\mathbf{z}$ from the $\mathbf{i}$ plane. The geometric product $R_{\mathbf{z}^{-1}\mathbf{z}'} = \mathbf{z}^{-1}\mathbf{z}'$ for any two such reference vectors $\mathbf{z}$ and $\mathbf{z}'$ gives the rotation operation to rotate one choice of three-component eigenspinor representation $\mathbf{x}_1(\mathbf{z})$ into the other $\mathbf{x}_1(\mathbf{z}') = \mathbf{x}_1(\mathbf{z})R_{\mathbf{z}^{-1}\mathbf{z}'}$.

As for two dimensions on page 62 we could also try to interpret $\mathbf{x}_2$ by factoring out a reference vector $\mathbf{z}$ to the right. But since according to equation (2.7.47) $\mathbf{x}_2$ is simply the reverse of $\mathbf{x}_1$, it seems not really needed for a real interpretation. Doing it nevertheless, yields less handy expressions.

So all we need to give a real geometric interpretation for the three-component eigenspinors $\mathbf{x}_1$ (and $\mathbf{x}_2$) is the triplet $(\boldsymbol{\sigma}_{1\|}, \boldsymbol{\sigma}_{2\|}, \boldsymbol{\sigma}_{3\|})$ of projections of the three basis vectors $\boldsymbol{\sigma}_1$, $\boldsymbol{\sigma}_2$ and $\boldsymbol{\sigma}_3$ onto the $\mathbf{i}$ plane. Multiplying this triplet with any vector $\mathbf{z}$, element of the $\mathbf{i}$ plane, from the right (from the left) yields all representations of $\mathbf{x}_1$ (and $\mathbf{x}_2$).

After successfully clarifying the real interpretation of the "complex eigenvectors" in terms of a real vector space R^3 vector triplet, we turn briefly to the interpretation of the eigenvalues. The real oriented plane unit area element eigenvalues $\lambda_1 = \mathbf{i}$ and $\lambda_2 = \tilde{\lambda}_1 = -\mathbf{i}$ yield via equation (2.7.45) the eigenvalues of the Cayley transformation $C(-kU)$ as:

$$\lambda_{c1} = \cos\vartheta + \mathbf{i}\sin\vartheta, \quad \lambda_{c2} = \tilde{\lambda}_{c1} = \cos\vartheta - \mathbf{i}\sin\vartheta, \quad \lambda_{c3} = 1. \tag{2.7.53}$$

The action of λ_{c1} on $\mathbf{x}_1$ and λ_{c2} on $\mathbf{x}_2$, respectively, give

$$C(-kU)\mathbf{x}_1 = \lambda_{c1}\mathbf{x}_1 = \lambda_{c1}\begin{pmatrix} \mathbf{n}_1\mathbf{z} \\ \mathbf{n}_2\mathbf{z} \\ \mathbf{n}_3\mathbf{z} \end{pmatrix} = \begin{pmatrix} \lambda_{c1}\mathbf{n}_1\mathbf{z} \\ \lambda_{c1}\mathbf{n}_2\mathbf{z} \\ \lambda_{c1}\mathbf{n}_3\mathbf{z} \end{pmatrix}, \tag{2.7.54}$$

and

$$C(-kU)\mathbf{x}_2 = \lambda_{c2}\mathbf{x}_2 = \lambda_{c2}\begin{pmatrix} \mathbf{z}\mathbf{n}_1 \\ \mathbf{z}\mathbf{n}_2 \\ \mathbf{z}\mathbf{n}_3 \end{pmatrix} = \begin{pmatrix} \mathbf{z}\lambda_{c1}\mathbf{n}_1 \\ \mathbf{z}\lambda_{c1}\mathbf{n}_2 \\ \mathbf{z}\lambda_{c1}\mathbf{n}_3 \end{pmatrix}. \tag{2.7.55}$$

In equation (2.7.55), we have used the facts that $\mathbf{x}_2 = \tilde{\mathbf{x}}_1$ and that $\lambda_{c2}\mathbf{z} = \mathbf{z}\tilde{\lambda}_{c2} = \mathbf{z}\lambda_{c1}$, since $\mathbf{z}$ is element of the $\mathbf{i}$ plane.

We can therefore consistently interpret the both $\lambda_{c1}\mathbf{x}_1$ and $\lambda_{c2}\mathbf{x}_2$ as one and the same rotation of the vector triplet $(\mathbf{n}_1, \mathbf{n}_2, \mathbf{n}_3) = (\boldsymbol{\sigma}_{1\|}, \boldsymbol{\sigma}_{2\|}, \boldsymbol{\sigma}_{3\|})$ by the angle $-\vartheta$ in a right handed sense around the axis $\mathbf{x}_3$ in the $\mathbf{i}$ plane. But we are equally free to alternatively view it as a $+\vartheta$ rotation (in the $\mathbf{i}$ plane) of the reference vector $\mathbf{z}$

instead. No further discussion for the eigenvalues λ_1, λ_2 of U is needed, since these are special cases of λ_{c1}, λ_{c2} with $\vartheta = \pi/2$.

The third eigenvalue of the Cayley transformation $C(-kU)$ is $\lambda_{c3} = 1$, which means that any component parallel to $\mathbf{x}_3$ will be invariant under multiplication with $C(-kU)$.

So far geometric algebra has served us as an investigative tool in order to gain a consistent real geometric vector space interpretation of imaginary eigenvalues and complex eigenvectors of antisymmetric matrices in two and three dimensions. But it is equally possible to pretend not to know about the antisymmetric matrices and their eigenvalues and eigenvectors in the first place, and synthetically construct relationships in geometric algebra which give all the counterparts found in our investigative (analytical) approach so far. As shown in [172] this necessitates in three dimensions the use of the two-sided spinorial description [159, 162] of rotations.

2.7.3 From imaginary numbers to multivector square roots of -1 in Clifford geometric algebras $Cl(p,q)$ with $p+q \leq 4$

It is known that Clifford (geometric) algebra[21] offers a geometric interpretation for square roots of -1 in the form of blades that square to -1, e.g. unit bivectors and the unit pseudoscalar in $Cl(3,0)$. Furthermore, this extends to a geometric interpretation of quaternions as the side face bivectors of a unit cube $\mathbb{H} \cong Cl^+(3,0)$. Research has been done [294] on the biquaternion roots of -1, abandoning the restriction to blades. Biquaternions are isomorphic to the Clifford (geometric) algebra $Cl(3,0)$ of $\mathbb{R}^3$. All these roots of -1 find immediate applications in the construction of new types of geometric Clifford Fourier transformations.

In the current section we show how to explicitly extend this research to general algebras $\mathcal{G}_{p,q} = Cl(p,q)$. We fully present the geometric roots of -1 for the Clifford (geometric) algebras with $p+q \leq 4$. The description is mainly based on [193], which contains fully detailed algebraic proofs.

Historically, Grassmann invented the antisymmetric outer product of vectors, that regards the *oriented parallelogram area* spanned by two vectors as a new type of number, commonly called *bivector*. The bivector represents its own plane, because outer products with vectors in the plane vanish. In three dimensions the outer product of three linearly independent vectors defines a so-called *trivector* with the magnitude of the volume of the *parallelepiped* spanned by the vectors. Its orientation (sign) depends on the handedness of the three vectors.

In the Clifford algebra [162] of $\mathbb{R}^3$ the three bivector side faces of a unit cube $\{\mathbf{e}_1\mathbf{e}_2, \mathbf{e}_2\mathbf{e}_3, \mathbf{e}_3\mathbf{e}_1\}$ oriented along the three coordinate directions $\{\mathbf{e}_1, \mathbf{e}_2, \mathbf{e}_3\}$ correspond to the three quaternion units $\boldsymbol{i}$, $\boldsymbol{j}$ and $\boldsymbol{k}$. Like quaternions, these three bivectors square to minus one and generate the rotations in their respective planes.

Beyond that Clifford algebra allows to extend complex numbers to higher

[21] In his original publication [77] Clifford first used the term *geometric algebras*. Subsequently in mathematics the new term *Clifford algebras* [247] has become the proper mathematical term. For emphasizing the *geometric* nature of the algebra, some researchers continue [58, 161, 162] to use the original term geometric algebra(s).

dimensions [33, 161] and systematically generalize our knowledge of complex numbers, holomorphic functions and quaternions. It has found rich applications in symbolic computation, physics, robotics, computer graphics, etc. [56, 58, 132, 244]. Since bivectors and trivectors in the Clifford algebras of Euclidean vector spaces square to minus one, we can use them to create new geometric kernels for Fourier transformations. This leads to a large variety of new Fourier transformations, which all deserve to be studied in their own right [58, 117, 182, 184, 185, 243, 251, 254, 256, 259, 286, 290]. We will treat both Euclidean (positive definite metric) and non-Euclidean (indefinite metric) vector spaces. We know from Einstein's special theory of relativity that non-Euclidean vector spaces are of fundamental importance in nature [158]. Therefore this section is about finding square roots of -1 in a non-degenerate Clifford algebra $\mathcal{G}_{p,q} = Cl(p,q)$.

2.7.3.1 Notation for multivector square root computations

The associative geometric product of two vectors $\boldsymbol{a}, \boldsymbol{b} \in \mathbb{R}^{p,q}, p+q=n$ is defined as the sum of their symmetric inner product (scalar) and their antisymmetric outer product (bivector)

$$\boldsymbol{a}\boldsymbol{b} = \boldsymbol{a}\cdot\boldsymbol{b} + \boldsymbol{a}\wedge\boldsymbol{b}. \tag{2.7.56}$$

We recall [247] a real Clifford algebra $\mathcal{G}_{p,q} = Cl(p,q)$ as the linear space of all elements generated by the associative (and distributive) bilinear geometric product of vectors of an inner product vector space $\mathbb{R}^{p,q}$, $p+q=n$ over the field of reals $\mathbb{R}$. A Clifford algebra includes the field of reals $\mathbb{R}$ and the vector space $\mathbb{R}^{p,q}$ as grade zero and grade one elements, respectively.

Clifford algebras in one, two and three dimensions have the following basis blades of grade 0 (scalars), grade 1 (vectors), grade 2 (bivectors) and grade 3 (trivectors)

$$\{1, \boldsymbol{e}_1, \boldsymbol{e}_2, \boldsymbol{e}_3, \mathbf{e}_{23}, \mathbf{e}_{31}, \mathbf{e}_{12}, \mathbf{e}_{123}\}, \tag{2.7.57}$$

where we use abbreviations $\mathbf{e}_{12} = \boldsymbol{e}_1\boldsymbol{e}_2, \mathbf{e}_{23} = \boldsymbol{e}_2\boldsymbol{e}_3, \mathbf{e}_{31} = \boldsymbol{e}_3\boldsymbol{e}_1, \mathbf{e}_{123} = \boldsymbol{e}_1\boldsymbol{e}_2\boldsymbol{e}_3$. Every multivector can be expanded in terms of these basis blades with real coefficients. We give examples for $M \in \mathcal{G}_{p,q}$, $n = p+q = 1,2,3$:

$$M = \alpha + \beta\boldsymbol{e}_1, \tag{2.7.58}$$

$$M' = \alpha + b_1\boldsymbol{e}_1 + b_2\boldsymbol{e}_2 + \beta\mathbf{e}_{12}, \tag{2.7.59}$$

$$M'' = \alpha + b_1\boldsymbol{e}_1 + b_2\boldsymbol{e}_2 + b_3\boldsymbol{e}_3 + c_1\mathbf{e}_{23} + c_2\mathbf{e}_{31} + c_3\mathbf{e}_{12} + \beta\mathbf{e}_{123}. \tag{2.7.60}$$

The general notation for the quadratic form of basis vectors in $\mathbb{R}^{p,q}$ is:

$$\vec{e}_k^{\,2} = \varepsilon_k = \begin{cases} +1 & \text{for } 1 \le k \le p, \\ -1 & \text{for } p+1 \le k \le p+q = n. \end{cases} \tag{2.7.61}$$

We therefore always have $\vec{e}_k^{\,4} = \varepsilon_k^2 = 1$, and we abbreviate $\mathcal{G}_p = \mathcal{G}_{p,0}$. We follow the convention that inner and outer products have priority over the geometric product, which saves writing a number of brackets. Therefore, $\vec{a}\cdot\vec{b}\vec{c}$ equals $(\vec{a}\cdot\vec{b})\vec{c}$ and not $\vec{a}\cdot(\vec{b}\vec{c})$, etc.

We will frequently use the following basic formulas of Clifford algebra in the rest of this work. The symmetric part of the geometric product of any two vectors $\vec{a}$, $\vec{b}$ is the inner product (contraction, scalar product)

$$\frac{1}{2}(\vec{a}\vec{b}+\vec{b}\vec{a}) = \vec{a}\cdot\vec{b} = \langle\vec{a}\vec{b}\rangle_0 = \langle\vec{a}\vec{b}\rangle. \tag{2.7.62}$$

Likewise the inner product (contraction, scalar product) of any two bivectors $\underline{c}$, $\underline{c}'$ is symmetric

$$\frac{1}{2}(\underline{c}\,\underline{c}'+\underline{c}'\underline{c}) = \underline{c}\cdot\underline{c}' = \langle\underline{c}\,\underline{c}'\rangle_0 = \langle\underline{c}\,\underline{c}'\rangle. \tag{2.7.63}$$

The antisymmetric part of the geometric product of any two vectors $\vec{a}$, $\vec{b}$ is the outer product (bivector)

$$\frac{1}{2}(\vec{a}\vec{b}-\vec{b}\vec{a}) = \vec{a}\wedge\vec{b} = \langle\vec{a}\vec{b}\rangle_2. \tag{2.7.64}$$

The inner product (left contraction) of a vector $\vec{a}$ with a bivector $\underline{c}$ is antisymmetric

$$\vec{a}\cdot\underline{c} = \frac{1}{2}(\vec{a}\underline{c}-\underline{c}\vec{a}) = \langle\vec{a}\underline{c}\rangle_1. \tag{2.7.65}$$

Let $I_n = \Pi_{k=1}^n \vec{e}_k$ be the unit oriented pseudoscalar of $\mathcal{G}_{p,q}, n = p+q$. Let $A_r, B_s \in \mathcal{G}_{p,q}$ be two blades of grade r and s, respectively. Then we have the following general rules [244]. The inner product (left contraction [110]) is related to the outer product by[22]

$$(A_r\cdot B_s)I_n = A_r\wedge(B_sI_n), \quad \text{if } r\leq s; \tag{2.7.66}$$
$$(A_r\wedge B_s)I_n = A_r\cdot(B_sI_n), \quad \text{if } r+s\leq n, \quad r,s>0. \tag{2.7.67}$$

Two blades $A_r, B_s \in \mathcal{G}_{p,q}$ are called orthogonal if and only if their inner product is zero

$$A_r\perp B_s \quad\Longleftrightarrow\quad A_r\cdot B_s = 0. \tag{2.7.68}$$

With (2.7.66) follows that for $r\leq s$,

$$A_r\perp B_s \quad\Longleftrightarrow\quad A_r\wedge(B_sI_n)=0 \quad\Longleftrightarrow\quad A_r\wedge\widetilde{B_s}=0, \tag{2.7.69}$$

where $\widetilde{B_s} = B_sI_n^{-1}$ is the *dual* of B_s with $I_n^{-1} = \pm I_n$. Likewise (2.7.67) shows that for $r+s\leq n,\ \ r,s>0$

$$A_r\perp\widetilde{B_s} \quad\Longleftrightarrow\quad A_r\wedge B_s = 0. \tag{2.7.70}$$

Example 2.7.2. Let $\vec{b},\underline{c}\in\mathcal{G}_{p,q}, p+q=3$ be a vector $\vec{b}$ and a bivector $\underline{c}$ with vanishing outer product. Then by (2.7.70) the dual vector $\vec{c}=\widetilde{\underline{c}}$ is always perpendicular to $\vec{b}$ independent of the signature of the underlying vector space $\mathbb{R}^{p,q}$, $p+q=3$,

$$\vec{b}\wedge\underline{c}=0 \quad\Longleftrightarrow\quad \vec{b}\cdot\vec{c}=0 \quad\Longleftrightarrow\quad \vec{b}\perp\vec{c}. \tag{2.7.71}$$

[22] In order to avoid a discussion of deviating definitions of the inner product for $r = 0$ or $s = 0$, we exclude scalars in (2.7.67), but depending on the definition of the inner product (or contraction), a single general formula for all grades exists. For example, for the left contraction [110] $A, B \in \mathcal{G}_{p,q}, A\rfloor B = \sum_{r,s}\langle\langle A\rangle_r\langle B\rangle_s\rangle_{s-r}$ we have the two formulas $(A\wedge B)I_n = A\rfloor(BI_n)$ and $(A\rfloor B)I_n = A\wedge(BI_n)$.

2.7.3.2 Geometric multivector square roots of -1

Definition 2.7.3 (Geometric root of -1). *A geometric multivector square root (geometric root) of -1 is a multivector $A \in \mathcal{G}_{p,q}$ with*

$$A^2 = AA = -1. \tag{2.7.72}$$

An immediate application of this definition is the generalization of the famous Euler formula to geometric roots A of -1

$$e^{\varphi A} = \cos\varphi + A\sin\varphi. \tag{2.7.73}$$

For example, Lounesto considers $\cos\varphi + \mathbf{e}_{12}\sin\varphi$ in $\mathcal{G}_2$ in [247, Page 29].

Theorem 2.7.4. *Every multivector square root A of -1 is subject to $n+1 = p+q+1$ grade-wise constraints:*

$$A^2 = \langle AA\rangle = -1, \tag{2.7.74}$$

and

$$\langle AA\rangle_k = 0, \qquad 1 \leq k \leq n, \tag{2.7.75}$$

where $\langle AA\rangle_k$ denotes the k-th vector part of AA, and $\langle AA\rangle = \langle AA\rangle_0$.

We point out that $\langle AA\rangle$ is identical to the scalar product $A * A$ of [161]. In the following we call the scalar equation (2.7.74) the *root equation* of $\mathcal{G}_{p,q}$ and (2.7.75) the *constraints*. Depending on the value of k, each k-vector constraint represents $\binom{n}{k}$ scalar equations. We will sometimes conveniently split up a k-vector constraint equation and still call the resulting partial equations *constraints*.

Table 2.5 on page 72 lists all geometric roots of -1 of $\mathcal{G}_{p,q}$, $n = p+q = 4$. We point out, that similar to Table 2.4, also in Table 2.5 all root equations of the third column result from the general $n = 4$ root equation $A^2 = \langle A^2\rangle = -1$, simply by inserting the case conditions and constraints of columns one and two.

2.7.3.3 Summary on multivector square roots of -1 in Clifford algebras $Cl(p,q)$, $n = p+q \leq 4$

Table 2.4 lists all geometric roots of -1 for Clifford algebras $\mathcal{G}_{p,q}$, $n = p+q \leq 3$, and Table 2.5 does the same for Clifford algebras $\mathcal{G}_{p,q}$, $n = p+q = 4$. The content of both tables has been checked with the MAPLE package CLIFFORD [5]. The solutions for $\mathcal{G}_3$ included in Table 2.4 correspond to the biquaternion roots of -1 found in [294].

Overall the calculations in [193] and the results demonstrate how in Clifford algebras extensive calculations can be done without referring to coordinates [1, 161]. In the case of $\mathcal{G}_{p,q}$, $n = p+q = 4$, one arbitrarily selects one non-isotropic vector $\vec{e}_4$ for suitably splitting the algebra in order to use well developed techniques for algebras $\mathcal{G}_{p,q}$, $n = p+q = 3$. In the end it is always possible to express the results in coordinates as in Table 2.4. However, this considerably blows up the expressions and blurs the mostly (p,q)-signature independent form of the root equations of the

Table 2.4 Geometric roots of -1 for Clifford algebras $\mathcal{G}_{p,q}$, $n = p + q \leq 3$. The multivectors are denoted for $n = 1$ by $\alpha + \beta \mathbf{e}_1$, for $n = 2$ by $\alpha + b_1 \mathbf{e}_1 + b_2 \mathbf{e}_2 + \beta \mathbf{e}_{12}$ and for $n = 3$ by $\alpha + b_1 \mathbf{e}_1 + b_2 \mathbf{e}_2 + b_3 \mathbf{e}_3 + c_1 \mathbf{e}_{23} + c_2 \mathbf{e}_{31} + c_3 \mathbf{e}_{12} + \beta \mathbf{e}_{123}$. Source: [193, Tab. 1].

n	Cases	Solutions A and root equations
1		No solution for $\mathcal{G}_1$ $A = \pm \mathbf{c}_1$ for $\mathcal{G}_{0,1}$
2	$\alpha = 0$	$\beta^2 = b_1^2 \varepsilon_2 + b_2^2 \varepsilon_1 + \varepsilon_1 \varepsilon_2$ $\beta^2 = \begin{cases} b_1^2 + b_2^2 + 1 & \text{for } \mathcal{G}_2 \\ -b_1^2 + b_2^2 - 1 & \text{for } \mathcal{G}_{1,1} \\ -b_1^2 - b_2^2 + 1 & \text{for } \mathcal{G}_{0,2} \end{cases}$
	$\alpha \neq 0$	No solution
3	Constraint: $\alpha = \beta = 0$	$0 = \boldsymbol{b} \wedge \underline{c} = b_1 c_1 + b_2 c_2 + b_3 c_3$ $-1 = \vec{b}^2 + \underline{c}^2$ $-1 = b_1^2 \varepsilon_1 + b_2^2 \varepsilon_2 + b_3^2 \varepsilon_3 - c_1^2 \varepsilon_2 \varepsilon_3 - c_2^2 \varepsilon_3 \varepsilon_1 - c_3^2 \varepsilon_1 \varepsilon_2$ $-1 = \begin{cases} b_1^2 + b_2^2 + b_3^2 - (c_1^2 + c_2^2 + c_3^2) & \text{for } \mathcal{G}_3 \\ b_1^2 - b_2^2 - b_3^2 - (c_1^2 - c_2^2 - c_3^2) & \text{for } \mathcal{G}_{1,2} \\ b_1^2 + b_2^2 - b_3^2 + (c_1^2 + c_2^2 - c_3^2) & \text{for } \mathcal{G}_{2,1} \\ -(b_1^2 + b_2^2 + b_3^2) - (c_1^2 + c_2^2 + c_3^2) & \text{for } \mathcal{G}_{0,3} \end{cases}$
	$\alpha = 0$, $\beta \neq 0$	$A = \pm \mathbf{e}_{123}$ for $\mathcal{G}_3$, $\mathcal{G}_{1,2}$ No solution for $\mathcal{G}_{2,1}$, $\mathcal{G}_{0,3}$
	$\alpha \neq 0$	No solution

Table 2.5 Geometric roots of -1 for Clifford algebras $\mathcal{G}_{p,q}$, $n = p + q = 4$. The multivectors are denoted by $\alpha + \vec{b} + \underline{c} + \beta\,\mathbf{e}_{123} + (\alpha' + \vec{b}' + \underline{c}' + \beta'\,\mathbf{e}_{123})\,\mathbf{e}_4$, for details see equ. (68) in [193]. Source: [193, Tab. 2].

Case	Subcase / Constraints	Solutions and root equations
$\alpha \neq 0$		No solution
$\alpha = 0$, $\alpha' \neq 0$	$\underline{c} = \frac{1}{\alpha'}\vec{b}' \wedge \vec{b}$, $\underline{c}' = \frac{1}{\alpha'}(\beta'\vec{b}' - \varepsilon_4\beta\vec{b})\,\mathbf{e}_{123}$	$\vec{b}^2 + \frac{1}{\alpha'^2}(\vec{b}' \wedge \vec{b})^2 + \beta^2\,\mathbf{e}_{123}^2 + \varepsilon_4\alpha'^2 - \varepsilon_4\vec{b}'^2$ $+\varepsilon_4\frac{1}{\alpha'^2}(\beta'\vec{b}' - \varepsilon_4\beta\vec{b})^2\,\mathbf{e}_{123}^2 - \varepsilon_4\beta'^2\,\mathbf{e}_{123}^2$ $= -1$
$\alpha = 0$, $\alpha' = 0$, $\vec{b}' = 0$	$\beta = \beta' = 0$ $\underline{c} \cdot \underline{c}' = 0$, $\vec{b} \cdot \underline{c}' = 0$, $\vec{b} \wedge \underline{c} = 0$	$\vec{b}^2 + \underline{c}^2 + \varepsilon_4\underline{c}'^2 = -1$
	$\beta = 0$, $\beta' \neq 0$ $\underline{c} = -\frac{1}{\beta'}\vec{b} \cdot \underline{c}'\,\mathbf{e}_{123}^{-1}$	$\vec{b}^2 + \frac{1}{\beta'^2}(\vec{b} \cdot \underline{c}')^2\,\mathbf{e}_{123}^2 + \varepsilon_4\underline{c}'^2 - \varepsilon_4\beta'^2\,\mathbf{e}_{123}^2$ $= -1$
	$\beta \neq 0$ $\vec{b} = 0$, $\underline{c} = 0$	$\beta^2\,\mathbf{e}_{123}^2 + \varepsilon_4\underline{c}'^2 - \varepsilon_4\beta'^2\,\mathbf{e}_{123}^2 = -1$
$\alpha = 0$, $\alpha' = 0$, $\vec{b}' \neq 0$	$\vec{b} = 0$, $\beta = 0$ $\beta' = 0$, $\underline{c} \cdot \underline{c}' = 0$, $\vec{b}' \cdot \underline{c}' = 0$, $\vec{b}' \wedge \underline{c} = 0$	$\underline{c}^2 - \varepsilon_4\vec{b}'^2 + \varepsilon_4\underline{c}'^2 = -1$
	$\vec{b} = 0$, $\beta \neq 0$ $\underline{c} = -\frac{\varepsilon_4}{\beta}\vec{b}' \cdot \underline{c}'\,\mathbf{e}_{123}^{-1}$, $\beta' = 0$	$\frac{1}{\beta^2}(\vec{b}' \cdot \underline{c}')^2\,\mathbf{e}_{123}^2 + \beta^2\,\mathbf{e}_{123}^2 - \varepsilon_4\vec{b}'^2 + \varepsilon_4\underline{c}'^2$ $= -1$
	$\vec{b} \neq 0$, $\beta = 0$ $\underline{c} \cdot \underline{c}' = 0$, $\beta' = 0$, $\vec{b} \wedge \underline{c} = 0$, $\vec{b} \cdot \underline{c}' = 0$, $\vec{b}' = \gamma\vec{b}$, $\gamma \in \mathbb{R} \setminus \{0\}$	$(1 - \varepsilon_4\gamma^2)\,\vec{b}^2 + \underline{c}^2 + \varepsilon_4\,\underline{c}'^2 = -1$
	$\vec{b} \neq 0$, $\beta \neq 0$ $\beta' \neq 0$, $\vec{b}' = \varepsilon_4\frac{\beta}{\beta'}\vec{b}$, $\underline{c} = -\frac{1}{\beta'}\vec{b} \cdot \underline{c}'\,\mathbf{e}_{123}^{-1}$	$(1 - \varepsilon_4\frac{\beta^2}{\beta'^2})\vec{b}^2 + \frac{1}{\beta'^2}(\vec{b} \cdot \underline{c}')^2\,\mathbf{e}_{123}^2 + \beta^2\,\mathbf{e}_{123}^2$ $+\varepsilon_4\underline{c}'^2 - \varepsilon_4\beta'^2\,\mathbf{e}_{123}^2 = -1$

families of geometric roots of -1. In the case of $\mathcal{G}_{p,q}$, $n = p + q = 4$, in Table 2.5, we have not expressed the results in coordinates, because then the table would extend over several pages.

Open questions that are partly answered in the next Section 2.7.4 are:

- How can the graded structure of $\mathcal{G}_{p,q}$ be used best in the calculation of higher order geometric multivector square roots of -1? This also includes a question how to best use, for this type of computation, invariance of the equation $AA = -1$ under Clifford algebra (anti) automorphisms such as grade involution, reversion or conjugation, and under symmetries of the root equation. For example, under the grade involution,

$$AA = -1 \iff \hat{A}\hat{A} = -1. \tag{2.7.76}$$

  Another example would be a rotor R symmetry

$$\begin{aligned} &AA = \langle AA \rangle = -1 \iff \\ &R^{-1}ARR^{-1}AR = \langle R^{-1}ARR^{-1}AR \rangle = \langle R^{-1}AAR \rangle = \langle AA \rangle = -1. \end{aligned} \tag{2.7.77}$$

- The interesting relationship with families of idempotents of Clifford geometric algebras [3].

- What is the relationship with combinatorics?

- Expansion of this work to Clifford algebras $\mathcal{G}_{p,q}$ in arbitrary dimension $n = p+q$. For this purpose, it will be appropriate to use the modulo eight periodicity of Clifford algebras and the isomorphisms with matrix rings. Central elements squaring to -1 would be of particular importance as then they can be used in place of the imaginary i.

- The further use of Clifford algebra computation software like CLIFFORD for MAPLE and other packages [4, 5, 169].

Of special interest in physics are the Clifford algebras of Minkowski space-time, sometimes called [158] *space-time algebras* $\mathcal{G}_{3,1}$ and $\mathcal{G}_{1,3}$. Table 2.5 contains the complete set of all geometric roots of -1 for these algebras, so in particular all possible geometric multivector elements that may take on the role of the imaginary unit i in quantum mechanics, which is e.g. fundamental for the description of spin and for wave propagation.

Finally, the door is now wide open to construct all possible new types of Clifford Fourier transformations (CFT) [290] for multivector fields with domains and image domains ranging over the full Clifford algebras involved or subalgebras and subspaces thereof. In particular, all known Fourier transformations will find their place in this new general framework. The close relationship of wavelet transformations [188, 189, 253] and windowed transformations [254] to Fourier transformations shows that also in these fields new mathematics is to be expected.

Examples of CFTs working with non-central replacements of the imaginary unit i are the quaternion FT (QFT) [56, 58, 184, 256], and the CFT [117, 185] where i is replaced by pseudoscalars in $\mathcal{G}_n, n = 2 \,(\text{mod}\, 4)$. This shows that in principle every geometric root of -1, be it central or not, gives rise to its own geometric FT. Regarding the non-central geometric roots of -1, the example of the QFT shows that the non-commutativity may indeed be of advantage for obtaining more information about the symmetry and the physical nature of signals thus processed.

2.7.4 General theory of square roots of -1 in real Clifford algebras

It is well known that Clifford (geometric) algebra offers a geometric interpretation for square roots of -1 in the form of blades that square to minus 1. This extends to a geometric interpretation of quaternions as the side face bivectors of a unit cube. Systematic research has been done [294] on the biquaternion roots of -1, abandoning the restriction to blades. Biquaternions are isomorphic to the Clifford (geometric) algebra $C\ell(3,0)$ of $\mathbb{R}^3$. Further research on general algebras $C\ell(p,q)$ has explicitly derived the geometric roots of -1 for $p+q \leq 4$, as we just discussed in the Subsection 2.7.3, [193]. The current Subsection overcomes this dimension limit and by applying the Clifford algebra to matrix algebra isomorphisms in order to algebraically characterize the continuous manifolds of square roots of -1 found in the different types of Clifford algebras, depending on the type of associated ring ($\mathbb{R}$, $\mathbb{H}$, $\mathbb{R}^2$, $\mathbb{H}^2$ or $\mathbb{C}$). At the end of the book in Appendix A.1 explicit computer generated tables of representative square roots of -1 are given for all Clifford algebras with $n = 5, 7$, and $s = 3 \,(\text{mod}\, 4)$ with the associated ring $\mathbb{C}$. This includes, e.g., $C\ell(0,5)$ important in Clifford analysis, and $C\ell(4,1)$ which in applications is at the foundation of conformal geometric algebra. All these roots of -1 are immediately useful in the construction of new types of geometric Clifford Fourier transformations. This section is mainly based on [203].

2.7.4.1 *Introductory remarks*

Beyond bivectors, quaternions and rotations, Clifford algebra allows to extend complex numbers to higher dimensions [33, 161] and systematically generalize our knowledge of complex numbers, holomorphic functions and quaternions into the realm of Clifford analysis. It has found rich applications in symbolic computation, physics, robotics, computer graphics, etc. [56, 58, 112, 132, 244]. Since bivectors and trivectors in the Clifford algebras of Euclidean vector spaces square to minus one, we can use them to create new geometric kernels for Fourier transformations. This leads to a large variety of new Fourier transformations, which all deserve to be studied in their own right [58, 117, 182, 184, 185, 194, 243, 251, 254, 256, 259, 286, 290].

In this section, we will treat square roots of -1 in Clifford algebras $C\ell(p,q)$ of both Euclidean (positive definite metric) and non-Euclidean (indefinite metric) non-degenerate vector spaces, $\mathbb{R}^n = \mathbb{R}^{n,0}$ and $\mathbb{R}^{p,q}$, respectively. We know from Einstein's special theory of relativity that non-Euclidean vector spaces are of fundamental importance in nature [158]. They are further, e.g., used in computer vision and robotics [112] and for general algebraic solutions to contact problems [244]. Therefore this section is about characterizing square roots of -1 in all Clifford algebras $C\ell(p,q)$, extending previous limited research on $C\ell(3,0)$ in [294] and $C\ell(p,q), n = p+q \leq 4$

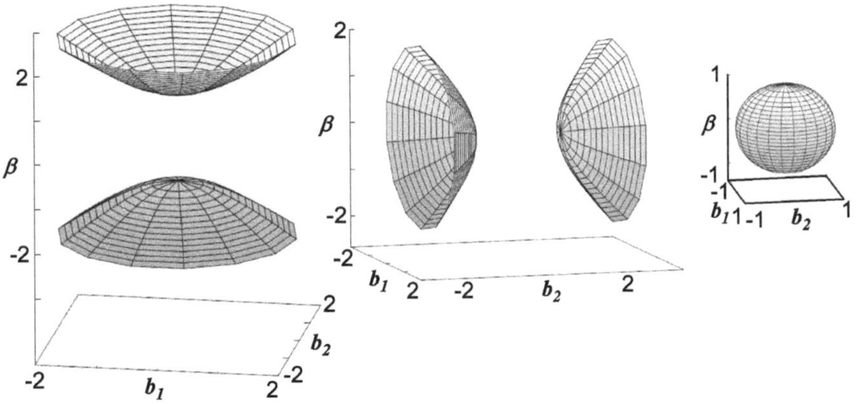

Figure 2.4 Manifolds of square roots f of -1 in $C\ell(2,0)$ (left), $C\ell(1,1)$ (center), and $C\ell(0,2) \cong \mathbb{H}$ (right). The square roots are $f = \alpha + b_1 e_1 + b_2 e_2 + \beta e_{12}$, with $\alpha, b_1, b_2, \beta \in \mathbb{R}$, $\alpha = 0$ and $\beta^2 = b_1^2 e_2^2 + b_2^2 e_1^2 + e_1^2 e_2^2$. Source: [203, Fig. 1].

in [193]. The manifolds of square roots of -1 in $C\ell(p,q)$, $n = p+q = 2$, compare Table 1 of [193], are visualized in Figure 2.4.

First, we introduce necessary background knowledge of Clifford algebras and matrix ring isomorphisms and explain in more detail how we will characterize and classify the square roots of -1 in Clifford algebras in Section 2.7.4.2. Next, we treat section by section (in Sections 2.7.4.3 to 2.7.4.7) the square roots of -1 in Clifford algebras which are isomorphic to matrix algebras with associated rings $\mathbb{R}$, $\mathbb{H}$, $\mathbb{R}^2$, $\mathbb{H}^2$ and $\mathbb{C}$, respectively. The term *associated* means that the isomorphic matrices will only have matrix elements from the associated ring. The square roots of -1 in Section 2.7.4.7 with associated ring $\mathbb{C}$ are of particular interest, because of the existence of classes of *exceptional* square roots of -1, which all include a nontrivial term in the central element of the respective algebra different from the identity. Section 2.7.4.7 therefore includes a detailed discussion of all classes of square roots of -1 in the algebras $C\ell(4,1)$, the isomorphic $C\ell(0,5)$, and in $C\ell(7,0)$. Finally, we add Appendix A.1 with tables of square roots of -1 for all Clifford algebras with $n = 5, 7$, and $s = 3 \pmod 4$. The square roots of -1 in Section 2.7.4.7 and in Appendix A.1 were all computed with the Maple package CLIFFORD [5], as explained in [204].

2.7.4.2 Background and problem formulation

Let $C\ell(p,q)$ be the algebra (associative with unit 1) generated over $\mathbb{R}$ by $p+q$ elements e_k (with $k = 1, 2, \ldots, p+q$) with the relations $e_k^2 = 1$ if $k \le p$, $e_k^2 = -1$ if $k > p$ and $e_h e_k + e_k e_h = 0$ whenever $h \ne k$, see [247]. We set the vector space dimension $n = p+q$ and the signature $s = p-q$. This algebra has dimension 2^n, and its even subalgebra $C\ell_0(p,q)$ has dimension 2^{n-1} (if $n > 0$). We are concerned with square roots of -1 contained in $C\ell(p,q)$ or $C\ell_0(p,q)$. If the dimension of $C\ell(p,q)$ or $C\ell_0(p,q)$ is ≤ 2, it is isomorphic to $\mathbb{R} \cong C\ell(0,0)$, $\mathbb{R}^2 \cong C\ell(1,0)$ or $\mathbb{C} \cong C\ell(0,1)$, and it is clear that there is no square root of -1 in $\mathbb{R}$ and $\mathbb{R}^2 = \mathbb{R} \times \mathbb{R}$, and that there are two squares roots i and $-i$ in $\mathbb{C}$. Therefore we only consider algebras of dimension ≥ 4. Square roots of -1 have been computed explicitly in [294] for $C\ell(3,0)$, and in [193] for algebras of dimensions $2^n \leq 16$.

An algebra $C\ell(p,q)$ or $C\ell_0(p,q)$ of dimension ≥ 4 is isomorphic to one of the five matrix algebras: $\mathcal{M}(2d,\mathbb{R})$, $\mathcal{M}(d,\mathbb{H})$, $\mathcal{M}(2d,\mathbb{R}^2)$, $\mathcal{M}(d,\mathbb{H}^2)$ or $\mathcal{M}(2d,\mathbb{C})$. The integer d depends on n. According to the parity of n, it is either $2^{(n-2)/2}$ or $2^{(n-3)/2}$ for $C\ell(p,q)$, and, either $2^{(n-4)/2}$ or $2^{(n-3)/2}$ for $C\ell_0(p,q)$. The associated ring (either $\mathbb{R}$, $\mathbb{H}$, $\mathbb{R}^2$, $\mathbb{H}^2$ or $\mathbb{C}$) depends on s in this way[23]:

Table 2.6 Associated rings in matrix algebra isomorphisms. Source: [203, p. 4].

s mod 8	0	1	2	3	4	5	6	7
Associated ring for $C\ell(p,q)$	$\mathbb{R}$	$\mathbb{R}^2$	$\mathbb{R}$	$\mathbb{C}$	$\mathbb{H}$	$\mathbb{H}^2$	$\mathbb{H}$	$\mathbb{C}$
Associated ring for $C\ell_0(p,q)$	$\mathbb{R}^2$	$\mathbb{R}$	$\mathbb{C}$	$\mathbb{H}$	$\mathbb{H}^2$	$\mathbb{H}$	$\mathbb{C}$	$\mathbb{R}$

Therefore we shall answer this question: What can we say about the square roots of -1 in an algebra $\mathcal{A}$ that is isomorphic to $\mathcal{M}(2d,\mathbb{R})$, $\mathcal{M}(d,\mathbb{H})$, $\mathcal{M}(2d,\mathbb{R}^2)$, $\mathcal{M}(d,\mathbb{H}^2)$, or, $\mathcal{M}(2d,\mathbb{C})$? They constitute an algebraic submanifold in $\mathcal{A}$; how many connected components[24] (for the usual topology) does it contain? Which are their dimensions? This submanifold is invariant by the action of the *group* $\mathrm{Inn}(\mathcal{A})$ of *inner automorphisms*[25] of $\mathcal{A}$, i.e. for every $r \in \mathcal{A}, r^2 = -1 \Rightarrow f(r)^2 = -1 \ \forall f \in \mathrm{Inn}(\mathcal{A})$. The orbits of $\mathrm{Inn}(\mathcal{A})$ are called conjugacy classes[26]; how many conjugacy classes are there in this submanifold? If the associated ring is $\mathbb{R}^2$ or $\mathbb{H}^2$ or $\mathbb{C}$, the group $\mathrm{Aut}(\mathcal{A})$ of all automorphisms of $\mathcal{A}$ is larger than $\mathrm{Inn}(\mathcal{A})$, and the action of $\mathrm{Aut}(\mathcal{A})$ in this submanifold shall also be described.

We recall some properties of $\mathcal{A}$ that do not depend on the associated ring. The group $\mathrm{Inn}(\mathcal{A})$ contains as many connected components as the *group* $\mathcal{G}(\mathcal{A})$ of *invertible elements* in $\mathcal{A}$. We recall that this assertion is true for $\mathcal{M}(2d,\mathbb{R})$ but not for $\mathcal{M}(2d+1,\mathbb{R})$ which is not one of the relevant matrix algebras. If f is an element of $\mathcal{A}$, let $\mathrm{Cent}(f)$ be the centralizer of f, that is, the subalgebra of all $g \in \mathcal{A}$ such that $fg = gf$. The conjugacy class of f contains as many connected components[27] as $\mathcal{G}(\mathcal{A})$ if (and

[23] Compare chapter 16 on *matrix representations and periodicity of 8*, as well as Table 1 on p. 217 of [247].

[24] Two points are in the same connected component of a manifold, if they can be joined by a continuous path inside the manifold under consideration. (This applies to all topological spaces satisfying the property that each neighbourhood of any point contains a neighbourhood in which every pair of points can always be joined by a continuous path.)

[25] An inner automorphism f of $\mathcal{A}$ is defined as $f : \mathcal{A} \to \mathcal{A}, f(x) = a^{-1}xa, \forall x \in \mathcal{A}$, with given fixed $a \in \mathcal{A}$. The composition of two inner automorphisms $g(f(x)) = b^{-1}a^{-1}xab = (ab)^{-1}x(ab)$ is again an inner automorphism. With this operation the inner automorphisms form the group $\mathrm{Inn}(\mathcal{A})$, compare [323].

[26] The conjugacy class (similarity class) of a given $r \in \mathcal{A}, r^2 = -1$ is $\{f(r) : f \in \mathrm{Inn}(\mathcal{A})\}$, compare [321]. Conjugation is transitive, because the composition of inner automorphisms is again an inner automorphism.

[27] According to the general theory of groups acting on sets, the conjugacy class (as a topological space) of a square root f of -1 is isomorphic to the *quotient* of $\mathcal{G}(\mathcal{A})$ and $\mathrm{Cent}(f)$ (the subgroup

only if) $\mathrm{Cent}(f) \bigcap \mathcal{G}(\mathcal{A})$ is contained in the neutral[28] connected component of $\mathcal{G}(\mathcal{A})$, and the dimension of its conjugacy class is

$$\dim(\mathcal{A}) - \dim(\mathrm{Cent}(f)). \tag{2.7.78}$$

Note that for invertible $g \in \mathrm{Cent}(f)$ we have $g^{-1}fg = f$.

Besides, let $\mathrm{Z}(\mathcal{A})$ be the center of $\mathcal{A}$, and let $[\mathcal{A}, \mathcal{A}]$ be the subspace spanned by all $[f, g] = fg - gf$. In all cases $\mathcal{A}$ is the direct sum of $\mathrm{Z}(\mathcal{A})$ and $[\mathcal{A}, \mathcal{A}]$. For example,[29] $\mathrm{Z}(\mathcal{M}(2d, \mathbb{R})) = \{a\mathbf{1} \mid a \in \mathbb{R}\}$ and $\mathrm{Z}(\mathcal{M}(2d, \mathbb{C})) = \{c\mathbf{1} \mid c \in \mathbb{C}\}$. If the associated ring is $\mathbb{R}$ or $\mathbb{H}$ (that is for even n), then $\mathrm{Z}(\mathcal{A})$ is canonically isomorphic to $\mathbb{R}$, and from the projection $\mathcal{A} \to \mathrm{Z}(\mathcal{A})$ we derive a linear form Scal : $\mathcal{A} \to \mathbb{R}$. When the associated ring[30] is $\mathbb{R}^2$ or $\mathbb{H}^2$ or $\mathbb{C}$, then $\mathrm{Z}(\mathcal{A})$ is spanned by $\mathbf{1}$ (the unit matrix[31]) and some element ω such that $\omega^2 = \pm\mathbf{1}$. Thus, we get two linear forms ScalixScal!scalar coefficient and Spec such that $\mathrm{Scal}(f)\mathbf{1} + \mathrm{Spec}(f)\omega$ is the projection of f in $\mathrm{Z}(\mathcal{A})$ for every $f \in \mathcal{A}$. Instead of ω we may use $-\omega$ and replace Spec with $-$Spec. The following assertion holds for every $f \in \mathcal{A}$: The trace of each multiplication[32] $g \mapsto fg$ or $g \mapsto gf$ is equal to the product

$$\mathrm{tr}(f) = \dim(\mathcal{A})\,\mathrm{Scal}(f). \tag{2.7.79}$$

The word "trace" (when nothing more is specified) means a matrix trace in $\mathbb{R}$, which is the sum of its diagonal elements. For example, the matrix $M \in \mathcal{M}(2d, \mathbb{R})$ with elements $m_{kl} \in \mathbb{R}, 1 \leq k, l \leq 2d$ has the trace $\mathrm{tr}(M) = \sum_{k=1}^{2d} m_{kk}$ [220].

We shall prove that in all cases $\mathrm{Scal}(f) = 0$ for every square root of -1 in $\mathcal{A}$. Then, we may distinguish *ordinary* square roots of -1, and *exceptional* ones. In all cases the ordinary square roots of -1 constitute a unique[33] conjugacy class of dimension $\dim(\mathcal{A})/2$ which has as many connected components as $\mathcal{G}(\mathcal{A})$, and they satisfy the equality $\mathrm{Spec}(f) = 0$ if the associated ring is $\mathbb{R}^2$ or $\mathbb{H}^2$ or $\mathbb{C}$. The exceptional square roots of -1 only exist[34] if $\mathcal{A} \cong \mathcal{M}(2d, \mathbb{C})$. In $\mathcal{M}(2d, \mathbb{C})$ there are $2d$ conjugacy classes of exceptional square roots of -1, each one characterized by an equality $\mathrm{Spec}(f) = k/d$ with $\pm k \in \{1, 2, \ldots, d\}$ [see Section 2.7.4.7], and their dimensions are $< \dim(\mathcal{A})/2$ [see equation (2.7.99)]. For instance, ω (mentioned

of stability of f). Quotient means here the set of left handed classes modulo the subgroup. If the subgroup is contained in the neutral connected component of $\mathcal{G}(\mathcal{A})$, then the number of connected components is the same in the quotient as in $\mathcal{G}(\mathcal{A})$. See also [70].

[28] *Neutral* means to be connected to the identity element of $\mathcal{A}$.

[29] A matrix algebra based proof is, e.g., given in [312].

[30] This is the case for n (and s) odd. Then the pseudoscalar $\omega \in C\ell(p, q)$ is also in $\mathrm{Z}(C\ell(p, q))$.

[31] The number 1 denotes the unit of the Clifford algebra $\mathcal{A}$, whereas the bold face $\mathbf{1}$ denotes the unit of the isomorphic matrix algebra $\mathcal{M}$.

[32] These multiplications are bilinear over the center of $\mathcal{A}$.

[33] Let $\mathcal{A}$ be an algebra $\mathcal{M}(m, \mathbb{K})$ where $\mathbb{K}$ is a division ring. Thus two elements f and g of $\mathcal{A}$ induce $\mathbb{K}$-linear endomorphisms f' and g' on $\mathbb{K}^m$; if $\mathbb{K}$ is not commutative, $\mathbb{K}$ operates on $\mathbb{K}^m$ on the right side. The matrices f and g are conjugate (or similar) if and only if there are two $\mathbb{K}$-bases B_1 and B_2 of $\mathbb{K}^m$ such that f' operates on B_1 in the same way as g' operates on B_2. This theorem allows us to recognize that in all cases but the last one (with exceptional square roots of $-\mathbf{1}$), two square roots of $-\mathbf{1}$ are always conjugate.

[34] The pseudoscalars of Clifford algebras whose isomorphic matrix algebra has ring $\mathbb{R}^2$ or $\mathbb{H}^2$ square to $\omega^2 = +1$.

above) and $-\omega$ are central square roots of -1 in $\mathcal{M}(2d,\mathbb{C})$ which constitute two conjugacy classes of dimension 0. Obviously, $\mathrm{Spec}(\omega) = 1$.

For symbolic computer algebra systems (CAS), like MAPLE, there exist Clifford algebra packages, e.g., CLIFFORD [5], which can compute idempotents [3] and square roots of -1. This will be of especial interest for the exceptional square roots of -1 in $\mathcal{M}(2d,\mathbb{C})$.

Regarding a square root r of -1, a Clifford algebra is the direct sum of the subspaces $\mathrm{Cent}(r)$ (all elements that commute with r) and the skew-centralizer $\mathrm{SCent}(r)$ (all elements that anticommute with r). Every Clifford algebra multivector has a unique split by this Lemma.

Lemma 2.7.5. *Every multivector $A \in C\ell(p,q)$ has, with respect to a square root $r \in C\ell(p,q)$ of -1, i.e., $r^{-1} = -r$, the unique decomposition*

$$A_{\pm} = \frac{1}{2}(A \pm r^{-1}Ar), \quad A = A_+ + A_-, \quad A_+ r = rA_+, \quad A_- r = -rA_-. \tag{2.7.80}$$

Proof. For $A \in C\ell(p,q)$ and a square root $r \in C\ell(p,q)$ of -1, we compute

$$\begin{aligned} A_{\pm} r = \frac{1}{2}(A \pm r^{-1}Ar)r &= \frac{1}{2}(Ar \pm r^{-1}A(-1)) \overset{r^{-1}=-r}{=} \frac{1}{2}(rr^{-1}Ar \pm rA) \\ &= \pm r\frac{1}{2}(A \pm r^{-1}Ar). \end{aligned}$$

□

For example, in Clifford algebras $C\ell(n,0)$ [185] of dimensions $n = 2 \bmod 4$, $\mathrm{Cent}(r)$ is the even subalgebra $C\ell_0(n,0)$ for the unit pseudoscalar r, and the subspace $C\ell_1(n,0)$ spanned by all k-vectors of odd degree k, is $\mathrm{SCent}(r)$. The most interesting case is $\mathcal{M}(2d,\mathbb{C})$, where a whole range of conjugacy classes becomes available. These results will therefore be particularly relevant for constructing *Clifford Fourier transformations* using the square roots of -1.

2.7.4.3 *Square roots of* -1 *in* $\mathcal{M}(2d,\mathbb{R})$

Here $\mathcal{A} = \mathcal{M}(2d,\mathbb{R})$, whence $\dim(\mathcal{A}) = (2d)^2 = 4d^2$. The group $\mathcal{G}(\mathcal{A})$ has *two* connected components determined by the inequalities $\det(g) > 0$ and $\det(g) < 0$.

For the case $d = 1$ we have, e.g., the algebra $C\ell(2,0)$ isomorphic to $\mathcal{M}(2,\mathbb{R})$. The basis $\{1, e_1, e_2, e_{12}\}$ of $C\ell(2,0)$ is mapped to

$$\left\{ \begin{pmatrix} 1 & 0 \\ 0 & 1 \end{pmatrix}, \begin{pmatrix} 0 & 1 \\ 1 & 0 \end{pmatrix}, \begin{pmatrix} 1 & 0 \\ 0 & -1 \end{pmatrix}, \begin{pmatrix} 0 & -1 \\ 1 & 0 \end{pmatrix} \right\}.$$

The general element $\alpha + b_1 e_1 + b_2 e_2 + \beta e_{12} \in C\ell(2,0)$ is thus mapped to

$$\begin{pmatrix} \alpha + b_2 & -\beta + b_1 \\ \beta + b_1 & \alpha - b_2 \end{pmatrix} \tag{2.7.81}$$

in $\mathcal{M}(2,\mathbb{R})$. Every element f of $\mathcal{A} = \mathcal{M}(2d,\mathbb{R})$ is treated as an $\mathbb{R}$-linear endomorphism of $V = \mathbb{R}^{2d}$. Thus, its scalar component and its trace (2.7.79) are related as follows: $\mathrm{tr}(f) = 2d\mathrm{Scal}(f)$. If f is a square root of $-\mathbf{1}$, it turns V into a vector space over $\mathbb{C}$ (if the complex number i operates like f on V). If $(e_1, e_2, \ldots, e_d)$ is a $\mathbb{C}$-basis of V, then $(e_1, f(e_1), e_2, f(e_2), \ldots, e_d, f(e_d))$ is a $\mathbb{R}$-basis of V, and the $2d \times 2d$ matrix of f in this basis is

$$\mathrm{diag}\underbrace{\left(\begin{pmatrix} 0 & -1 \\ 1 & 0 \end{pmatrix}, \ldots, \begin{pmatrix} 0 & -1 \\ 1 & 0 \end{pmatrix}\right)}_{d}. \tag{2.7.82}$$

Consequently all square roots of $-\mathbf{1}$ in $\mathcal{A}$ are conjugate. The centralizer of a square root f of $-\mathbf{1}$ is the algebra of all $\mathbb{C}$-linear endomorphisms g of V (since i operates like f on V). Therefore, the $\mathbb{C}$-dimension of $\mathrm{Cent}(f)$ is d^2 and its $\mathbb{R}$-dimension is $2d^2$. Finally, the dimension (2.7.78) of the conjugacy class of f is $\dim(\mathcal{A}) - \dim(\mathrm{Cent}(f)) = 4d^2 - 2d^2 = 2d^2 = \dim(\mathcal{A})/2$. The two connected components of $\mathcal{G}(\mathcal{A})$ are determined by the sign of the determinant. Because of the next lemma, the $\mathbb{R}$-determinant of every element of $\mathrm{Cent}(f)$ is ≥ 0. Therefore, the intersection $\mathrm{Cent}(f) \bigcap \mathcal{G}(\mathcal{A})$ is contained in the neutral connected component of $\mathcal{G}(\mathcal{A})$ and, consequently, the conjugacy class of f has two connected components like $\mathcal{G}(\mathcal{A})$. Because of the next lemma, the $\mathbb{R}$-trace of f vanishes (indeed its $\mathbb{C}$-trace is di, because f is the multiplication by the scalar i: $f(v) = iv$ for all v) whence $\mathrm{Scal}(f) = 0$. This equality is corroborated by the matrix written above.

We conclude that the square roots of $-\mathbf{1}$ constitute one conjugacy class with two connected components of dimension $\dim(\mathcal{A})/2$ contained in the hyperplane defined by the equation

$$\mathrm{Scal}(f) = 0. \tag{2.7.83}$$

Before stating the lemma that here is so helpful, we show what happens in the easiest case $d = 1$. The square roots of $-\mathbf{1}$ in $\mathcal{M}(2,\mathbb{R})$ are the real matrices

$$\begin{pmatrix} a & c \\ b & -a \end{pmatrix} \text{ with } \begin{pmatrix} a & c \\ b & -a \end{pmatrix}\begin{pmatrix} a & c \\ b & -a \end{pmatrix} = (a^2 + bc)\,\mathbf{1} = -\mathbf{1}; \tag{2.7.84}$$

hence $a^2 + bc = -1$, a relation between a, b, c which is equivalent to $(b-c)^2 = (b+c)^2 + 4a^2 + 4 \Rightarrow (b-c)^2 \geq 4 \Rightarrow b - c \geq 2$ (one component) or $c - b \geq 2$ (second component). Thus, we recognize the two connected components of square roots of $-\mathbf{1}$: The inequality $b \geq c + 2$ holds in one connected component, and the inequality $c \geq b + 2$ in the other one, compare Figure 2.5.

In terms of $C\ell(2,0)$ coefficients (2.7.81) with $b - c = \beta + b_1 - (-\beta + b_1) = 2\beta$, we get the two component conditions simply as

$$\beta \geq 1 \quad \text{(one component)}, \qquad \beta \leq -1 \quad \text{(second component)}. \tag{2.7.85}$$

Rotations ($\det(g) = 1$) leave the pseudoscalar βe_{12} invariant (and thus preserve the two connected components of square roots of $-\mathbf{1}$), but reflections ($\det(g') = -1$) change its sign $\beta e_{12} \to -\beta e_{12}$ (thus interchanging the two components).

Because of the previous argument involving a complex structure on the real space

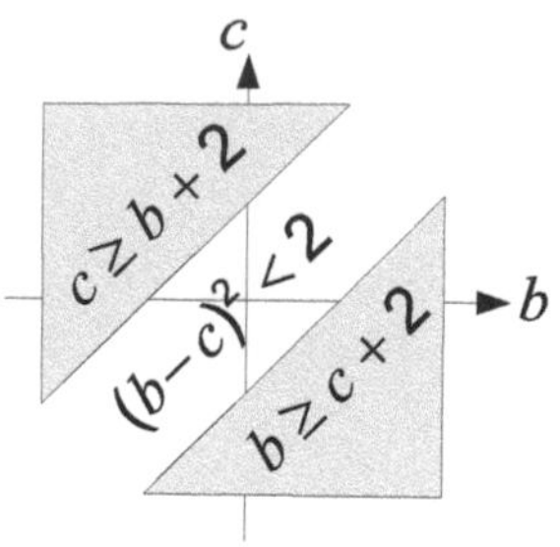

Figure 2.5 Two components of square roots of $-\mathbf{1}$ in $\mathcal{M}(2,\mathbb{R})$. Source: [203, Fig. 2].

V, we conversely consider the complex space $\mathbb{C}^d$ with its structure of vector space over $\mathbb{R}$. If $(e_1, e_2, \ldots, e_d)$ is a $\mathbb{C}$-basis of $\mathbb{C}^d$, then $(e_1, ie_1, e_2, ie_2, \ldots, e_d, ie_d)$ is a $\mathbb{R}$-basis. Let g be a $\mathbb{C}$-linear endomorphism of $\mathbb{C}^d$ (i.e. a complex $d \times d$ matrix), let $\mathrm{tr}_{\mathbb{C}}(g)$ and $\det_{\mathbb{C}}(g)$ be the trace and determinant of g in $\mathbb{C}$, and $\mathrm{tr}_{\mathbb{R}}(g)$ and $\det_{\mathbb{R}}(g)$ its trace and determinant for the real structure of $\mathbb{C}^d$.

Exercise 2.7.6. *For $d = 1$ an endomorphism of $\mathbb{C}^1$ is given by a complex number $g = a + ib$, $a, b \in \mathbb{R}$. Its matrix representation is according to* (2.7.82)

$$\begin{pmatrix} a & -b \\ b & a \end{pmatrix} \text{ with } \begin{pmatrix} a & -b \\ b & a \end{pmatrix}^2 = (a^2 - b^2)\begin{pmatrix} 1 & 0 \\ 0 & 1 \end{pmatrix} + 2ab\begin{pmatrix} 0 & -1 \\ 1 & 0 \end{pmatrix}. \tag{2.7.86}$$

Then we have $\mathrm{tr}_{\mathbb{C}}(g) = a + ib$, $\mathrm{tr}_{\mathbb{R}}\begin{pmatrix} a & -b \\ b & a \end{pmatrix} = 2a = 2\Re(\mathrm{tr}_{\mathbb{C}}(g))$ *and* $\det_{\mathbb{C}}(g) = a + ib$, $\det_{\mathbb{R}}\begin{pmatrix} a & -b \\ b & a \end{pmatrix} = a^2 + b^2 = |\det_{\mathbb{C}}(g)|^2 \geq 0$.

Lemma 2.7.7. *For every $\mathbb{C}$-linear endomorphism g we can write* $\mathrm{tr}_{\mathbb{R}}(g) = 2\Re(\mathrm{tr}_{\mathbb{C}}(g))$ *and* $\det_{\mathbb{R}}(g) = |\det_{\mathbb{C}}(g)|^2 \geq 0$.

Proof. There is a $\mathbb{C}$-basis in which the $\mathbb{C}$-matrix of g is triangular [then $\det_{\mathbb{C}}(g)$ is the product of the entries of g on the main diagonal]. We get the $\mathbb{R}$-matrix of g in the derived $\mathbb{R}$-basis by replacing every entry $a + bi$ of the $\mathbb{C}$-matrix with the elementary matrix $\begin{pmatrix} a & -b \\ b & a \end{pmatrix}$. The conclusion soon follows. The fact that the determinant of a block triangular matrix is the product of the determinants of the blocks on the main diagonal is used. □

2.7.4.4 Square roots of -1 in $\mathcal{M}(2d, \mathbb{R}^2)$

Here $\mathcal{A} = \mathcal{M}(2d, \mathbb{R}^2) = \mathcal{M}(2d, \mathbb{R}) \times \mathcal{M}(2d, \mathbb{R})$, whence $\dim(\mathcal{A}) = 8d^2$. The group $\mathcal{G}(\mathcal{A})$ has four[35] connected components. Every element $(f, f') \in \mathcal{A}$ (with $f, f' \in$

[35] In general, the number of connected components of $\mathcal{G}(\mathcal{A})$ is two if $\mathcal{A} = \mathcal{M}(m, \mathbb{R})$, and one if $\mathcal{A} = \mathcal{M}(m, \mathbb{C})$ or $\mathcal{A} = \mathcal{M}(m, \mathbb{H})$, because in all cases every matrix can be joined by a continuous

$\mathcal{M}(2d, \mathbb{R})$) has a determinant in $\mathbb{R}^2$ which is obviously $(\det(f), \det(f'))$, and the four connected components of $\mathcal{G}(\mathcal{A})$ are determined by the signs of the two components of $\det_{\mathbb{R}^2}(f, f')$.

The lowest dimensional example ($d = 1$) is $C\ell(2,1)$ isomorphic to $\mathcal{M}(2, \mathbb{R}^2)$. Here the pseudoscalar $\omega = e_{123}$ has square $\omega^2 = +1$. The center of the algebra is $\{1, \omega\}$ and includes the idempotents $\epsilon_\pm = (1 \pm \omega)/2$, $\epsilon_\pm^2 = \epsilon_\pm$, $\epsilon_+\epsilon_- = \epsilon_-\epsilon_+ = 0$. The basis of the algebra can thus be written as $\{\epsilon_+, e_1\epsilon_+, e_2\epsilon_+, e_{12}\epsilon_+, \epsilon_-, e_1\epsilon_-, e_2\epsilon_-, e_{12}\epsilon_-\}$, where the first (and the last) four elements form a basis of the subalgebra $C\ell(2,0)$ isomorphic to $\mathcal{M}(2, \mathbb{R})$. In terms of matrices we have the identity matrix $(\mathbf{1}, \mathbf{1})$ representing the scalar part, the idempotent matrices $(\mathbf{1}, 0)$, $(0, \mathbf{1})$, and the ω matrix $(\mathbf{1}, -\mathbf{1})$, with $\mathbf{1}$ the unit matrix of $\mathcal{M}(2, \mathbb{R})$.

The square roots of $(-\mathbf{1}, -\mathbf{1})$ in $\mathcal{A}$ are pairs of two square roots of $-\mathbf{1}$ in $\mathcal{M}(2d, \mathbb{R})$. Consequently they constitute a unique conjugacy class with four connected components of dimension $4d^2 = \dim(\mathcal{A})/2$. This number can be obtained in two ways. First, since every element $(f, f') \in \mathcal{A}$ (with $f, f' \in \mathcal{M}(2d, \mathbb{R})$) has twice the dimension of the components $f \in \mathcal{M}(2d, \mathbb{R})$ of Section 2.7.4.3, we get the component dimension $2 \cdot 2d^2 = 4d^2$. Second, the centralizer $\mathrm{Cent}(f, f')$ has twice the dimension of $\mathrm{Cent}(f)$ of $\mathcal{M}(2d, \mathbb{R})$, therefore $\dim(\mathcal{A}) - \mathrm{Cent}(f, f') = 8d^2 - 4d^2 = 4d^2$. In the above example for $d = 1$ the four components are characterized according to (2.7.85) by the values of the coefficients of $\beta e_{12}\epsilon_+$ and $\beta' e_{12}\epsilon_-$ as

$$\begin{aligned} c_1 &: \quad \beta \geq 1, & \beta' &\geq 1, \\ c_2 &: \quad \beta \geq 1, & \beta' &\leq -1, \\ c_3 &: \quad \beta \leq -1, & \beta' &\geq 1, \\ c_4 &: \quad \beta \leq -1, & \beta' &\leq -1. \end{aligned} \tag{2.7.87}$$

For every $(f, f') \in \mathcal{A}$ we can with (2.7.79) write $\mathrm{tr}(f) + \mathrm{tr}(f') = 2d\mathrm{Scal}(f, f')$ and

$$\mathrm{tr}(f) - \mathrm{tr}(f') = 2d\mathrm{Spec}(f, f') \quad \text{if} \quad \omega = (\mathbf{1}, -\mathbf{1}); \tag{2.7.88}$$

whence $\mathrm{Scal}(f, f') = \mathrm{Spec}(f, f') = 0$ if (f, f') is a square root of $(-\mathbf{1}, -\mathbf{1})$, compare (2.7.83).

The group $\mathrm{Aut}(\mathcal{A})$ is larger than $\mathrm{Inn}(\mathcal{A})$, because it contains the swap automorphism $(f, f') \mapsto (f', f)$ which maps the central element ω to $-\omega$, and interchanges the two idempotents ϵ_+ and ϵ_-. The group $\mathrm{Aut}(\mathcal{A})$ has eight connected components which permute the four connected components of the submanifold of square roots of $(-\mathbf{1}, -\mathbf{1})$. The permutations induced by $\mathrm{Inn}(\mathcal{A})$ are the permutations of the *Klein group*. For example, for $d = 1$ of (2.7.87) we get the following $\mathrm{Inn}(\mathcal{M}(2, \mathbb{R}^2))$

path to a diagonal matrix with entries 1 or -1. When an algebra $\mathcal{A}$ is a direct product of two algebras $\mathcal{B}$ and $\mathcal{C}$, then $\mathcal{G}(\mathcal{A})$ is the direct product of $\mathcal{G}(\mathcal{B})$ and $\mathcal{G}(\mathcal{C})$, and the number of connected components of $\mathcal{G}(\mathcal{A})$ is the product of the numbers of connected components of $\mathcal{G}(\mathcal{B})$ and $\mathcal{G}(\mathcal{C})$.

permutations

$$\begin{aligned} &\det(g) > 0, \quad \det(g') > 0: \quad \text{identity}, \\ &\det(g) > 0, \quad \det(g') < 0: \quad (c_1, c_2), (c_3, c_4), \\ &\det(g) < 0, \quad \det(g') > 0: \quad (c_1, c_3), (c_2, c_4), \\ &\det(g) < 0, \quad \det(g') < 0: \quad (c_1, c_4), (c_2, c_3). \end{aligned} \tag{2.7.89}$$

Beside the identity permutation, Inn($\mathcal{A}$) gives the three permutations that permute two elements and also the other two ones.

The automorphisms outside Inn($\mathcal{A}$) are

$$(f, f') \mapsto (gf'g^{-1}, g'fg'^{-1}) \quad \text{for some} \quad (g, g') \in \mathcal{G}(\mathcal{A}). \tag{2.7.90}$$

If $\det(g)$ and $\det(g')$ have opposite signs, it is easy to realize that this automorphism induces a circular permutation on the four connected components of square roots of $(-\mathbf{1}, -\mathbf{1})$: If $\det(g)$ and $\det(g')$ have the same sign, this automorphism leaves globally invariant two connected components, and permutes the other two ones. For example, for $d = 1$ the automorphisms (2.7.90) outside Inn($\mathcal{A}$) permute the components (2.7.87) of square roots of $(-\mathbf{1}, -\mathbf{1})$ in $\mathcal{M}(2, \mathbb{R}^2)$ as follows:

$$\begin{aligned} &\det(g) > 0, \quad \det(g') > 0: \quad (c_1), (c_2, c_3), (c_4), \\ &\det(g) > 0, \quad \det(g') < 0: \quad c_1 \to c_2 \to c_4 \to c_3 \to c_1, \\ &\det(g) < 0, \quad \det(g') > 0: \quad c_1 \to c_3 \to c_4 \to c_2 \to c_1, \\ &\det(g) < 0, \quad \det(g') < 0: \quad (c_1, c_4), (c_2), (c_3). \end{aligned} \tag{2.7.91}$$

Consequently, the quotient of the group Aut($\mathcal{A}$) by its neutral connected component is isomorphic to the group of isometries of a square in a Euclidean plane.

2.7.4.5 *Square roots of* -1 *in* $\mathcal{M}(d, \mathbb{H})$

Let us first consider the easiest case $d = 1$, when $\mathcal{A} = \mathbb{H}$, e.g., of $C\ell(0, 2)$. The square roots of -1 in $\mathbb{H}$ are the quaternions $ai + bj + cij$ with $a^2 + b^2 + c^2 = 1$. They constitute a compact and connected manifold of dimension 2. Every square root f of -1 is conjugate with i, i.e. there exists $v \in \mathbb{H} : v^{-1}fv = i \Leftrightarrow fv = vi$. If we set $v = -fi + 1 = a + bij - cj + 1$ we have

$$fv = -f^2 i + f = f + i = (f(-i) + 1)i = vi.$$

v is invertible, except when $f = -i$. But i is conjugate with $-i$ because $ij = j(-i)$, hence, by transitivity f is also conjugate with $-i$.

Here $\mathcal{A} = \mathcal{M}(d, \mathbb{H})$, whence $\dim(A) = 4d^2$. The ring $\mathbb{H}$ is the algebra over $\mathbb{R}$ generated by two elements i and j such that $i^2 = j^2 = -1$ and $ji = -ij$. We identify $\mathbb{C}$ with the subalgebra generated by[36] i alone.

The group $\mathcal{G}(\mathcal{A})$ has only *one* connected component. We shall soon prove that

[36]This choice is usual and convenient.

every square root of $-\mathbf{1}$ in $\mathcal{A}$ is conjugate with $i\mathbf{1}$. Therefore, the submanifold of square roots of $-\mathbf{1}$ is a conjugacy class, and it is connected. The centralizer of $i\mathbf{1}$ in $\mathcal{A}$ is the subalgebra of all matrices with entries in $\mathbb{C}$. The $\mathbb{C}$-dimension of $\mathrm{Cent}(i\mathbf{1})$ is d^2, its $\mathbb{R}$-dimension is $2d^2$, and, consequently, the dimension (2.7.78) of the submanifold of square roots of $-\mathbf{1}$ is $4d^2 - 2d^2 = 2d^2 = \dim(\mathcal{A})/2$.

Here $V = \mathbb{H}^d$ is treated as a (unitary) *module* over $\mathbb{H}$ on the *right* side: The product of a line vector ${}^t v = (x_1, x_2, \ldots, x_d) \in V$ by $y \in \mathbb{H}$ is ${}^t v\, y = (x_1 y, x_2 y, \ldots, x_d y)$. Thus, every $f \in \mathcal{A}$ determines an $\mathbb{H}$-linear endomorphism of V: The matrix f multiplies the column vector $v = {}^t(x_1, x_2, \ldots, x_d)$ on the left side $v \mapsto fv$. Since $\mathbb{C}$ is a subring of $\mathbb{H}$, V is also a vector space of dimension $2d$ over $\mathbb{C}$. The scalar i always operates on the right side (like every scalar in $\mathbb{H}$). If $(e_1, e_2, \ldots, e_d)$ is an $\mathbb{H}$-basis of V, then $(e_1, e_1 j, e_2, e_2 j, \ldots, e_d, e_d j)$ is a $\mathbb{C}$-basis of V. Let f be a square root of $-\mathbf{1}$, then the eigenvalues of f in $\mathbb{C}$ are $+i$ or $-i$. If we treat V as a $2d$ vector space over $\mathbb{C}$, it is the direct ($\mathbb{C}$-linear) sum of the eigenspaces

$$V^+ = \{v \in V \mid f(v) = vi\} \quad \text{and} \quad V^- = \{v \in V \mid f(v) = -vi\}, \tag{2.7.92}$$

representing f as a $2d \times 2d$ $\mathbb{C}$-matrix w.r.t. the $\mathbb{C}$-basis of V, with $\mathbb{C}$-scalar eigenvalues (multiplied from the right): $\lambda_\pm = \pm i$.

Since $ij = -ji$, the multiplication $v \mapsto vj$ permutes V^+ and V^-, as $f(v) = \pm vi$ is mapped to $f(v)j = \pm vij = \mp(vj)i$. Therefore, if $(e_1, e_2, \ldots, e_r)$ is a $\mathbb{C}$-basis of V^+, then $(e_1 j, e_2 j, \ldots, e_r j)$ is a $\mathbb{C}$-basis of V^-, consequently $(e_1, e_1 j, e_2, e_2 j, \ldots, e_r, e_r j)$ is a $\mathbb{C}$-basis of V, and $(e_1, e_2, \ldots, e_{r=d})$ is an $\mathbb{H}$-basis of V. Since f by $f(e_k) = e_k i$ for $k = 1, 2, \ldots, d$ operates on the $\mathbb{H}$-basis $(e_1, e_2, \ldots, e_d)$ in the same way as $i\mathbf{1}$ on the natural $\mathbb{H}$-basis of V, we conclude that f and $i\mathbf{1}$ are conjugate.

Besides, $\mathrm{Scal}(i\mathbf{1}) = 0$ because $2i\mathbf{1} = [j\mathbf{1}, ij\mathbf{1}] \in [\mathcal{A}, \mathcal{A}]$, thus $i\mathbf{1} \notin Z(\mathcal{A})$. Whence,[37]

$$\mathrm{Scal}(f) = 0 \quad \text{for every square root of} \quad -\mathbf{1}. \tag{2.7.93}$$

These results are easily verified in the above example of $d = 1$ when $\mathcal{A} = \mathbb{H}$.

2.7.4.6 *Square roots of* -1 *in* $\mathcal{M}(d, \mathbb{H}^2)$

Here, $\mathcal{A} = \mathcal{M}(d, \mathbb{H}^2) = \mathcal{M}(d, \mathbb{H}) \times \mathcal{M}(d, \mathbb{H})$, whence $\dim(A) = 8d^2$. The group $\mathcal{G}(\mathcal{A})$ has only one connected component (see Footnote 35).

The square roots of $(-\mathbf{1}, -\mathbf{1})$ in $\mathcal{A}$ are pairs of two square roots of $-\mathbf{1}$ in $\mathcal{M}(d, \mathbb{H})$. Consequently, they constitute a unique conjugacy class which is connected and its dimension is $2 \times 2d^2 = 4d^2 = \dim(\mathcal{A})/2$.

For every $(f, f') \in \mathcal{A}$ we can write $\mathrm{Scal}(f) + \mathrm{Scal}(f') = 2\,\mathrm{Scal}(f, f')$ and, similarly to (2.7.88),

$$\mathrm{Scal}(f) - \mathrm{Scal}(f') = 2\,\mathrm{Spec}(f, f') \quad \text{if} \quad \omega = (\mathbf{1}, -\mathbf{1}); \tag{2.7.94}$$

whence $\mathrm{Scal}(f, f') = \mathrm{Spec}(f, f') = 0$ if (f, f') is a square root of $(-\mathbf{1}, -\mathbf{1})$, compare with (2.7.93).

---

[37] Compare the definition of $\mathrm{Scal}(f)$ in Section 2.7.4.2, remembering that in the current section the associated ring is $\mathbb{H}$.

The group $\mathrm{Aut}(\mathcal{A})$ has two[38] connected components; the neutral component is $\mathrm{Inn}(\mathcal{A})$, and the other component contains the swap automorphism $(f, f') \mapsto (f', f)$.

The simplest example is $d = 1$, $\mathcal{A} = \mathbb{H}^2$, where we have the identity pair $(1, 1)$ representing the scalar part, the idempotents $(1, 0)$, $(0, 1)$, and ω as the pair $(1, -1)$. $\mathcal{A} = \mathbb{H}^2$ is isomorphic to $C\ell(0, 3)$. The pseudoscalar $\omega = e_{123}$ has the square $\omega^2 = +1$. The center of the algebra is $\{1, \omega\}$, and includes the idempotents $\epsilon_\pm = \frac{1}{2}(1 \pm \omega)$, $\epsilon_\pm^2 = \epsilon_\pm$, $\epsilon_+\epsilon_- = \epsilon_-\epsilon_+ = 0$. The basis of the algebra can thus be written as $\{\epsilon_+, e_1\epsilon_+, e_2\epsilon_+, e_{12}\epsilon_+, \epsilon_-, e_1\epsilon_-, e_2\epsilon_-, e_{12}\epsilon_-\}$ where the first (and the last) four elements form a basis of the subalgebra $C\ell(0, 2)$ isomorphic to $\mathbb{H}$.

2.7.4.7 *Square roots of* -1 *in* $\mathcal{M}(2d, \mathbb{C})$

The lowest dimensional example for $d = 1$ is the Pauli matrix algebra $\mathcal{A} = \mathcal{M}(2, \mathbb{C})$ isomorphic to the geometric algebra $C\ell(3, 0)$ of the 3D Euclidean space and $C\ell(1, 2)$. The $C\ell(3, 0)$ vectors e_1, e_2, e_3 correspond one-to-one to the Pauli matrices

$$\sigma_1 = \begin{pmatrix} 0 & 1 \\ 1 & 0 \end{pmatrix}, \qquad \sigma_2 = \begin{pmatrix} 0 & -i \\ i & 0 \end{pmatrix}, \qquad \sigma_3 = \begin{pmatrix} 1 & 0 \\ 0 & -1 \end{pmatrix}, \tag{2.7.95}$$

with $\sigma_1\sigma_2 = i\sigma_3 = \begin{pmatrix} i & 0 \\ 0 & -i \end{pmatrix}$. The element $\omega = \sigma_1\sigma_2\sigma_3 = i\mathbf{1}$ represents the central pseudoscalar e_{123} of $C\ell(3, 0)$ with square $\omega^2 = -\mathbf{1}$. The Pauli algebra has the following idempotents

$$\epsilon_1 = \sigma_1^2 = \mathbf{1}, \qquad \epsilon_0 = \frac{1}{2}(\mathbf{1} + \sigma_3), \qquad \epsilon_{-1} = \mathbf{0}. \tag{2.7.96}$$

The idempotents correspond via

$$f = i(2\epsilon - \mathbf{1}), \tag{2.7.97}$$

to the square roots of $-\mathbf{1}$:

$$f_1 = i\mathbf{1} = \begin{pmatrix} i & 0 \\ 0 & i \end{pmatrix}, \quad f_0 = i\sigma_3 = \begin{pmatrix} i & 0 \\ 0 & -i \end{pmatrix}, \quad f_{-1} = -i\mathbf{1} = \begin{pmatrix} -i & 0 \\ 0 & -i \end{pmatrix}, \tag{2.7.98}$$

where by *complex* conjugation $f_{-1} = \overline{f_1}$. Let the idempotent $\epsilon_0' = \frac{1}{2}(\mathbf{1} - \sigma_3)$ correspond to the matrix $f_0' = -i\sigma_3$. We observe that f_0 is conjugate to $f_0' = \sigma_1^{-1} f_0 \sigma_1 = \sigma_1\sigma_2 = f_0$ using $\sigma_1^{-1} = \sigma_1$ but f_1 is not conjugate to f_{-1}. Therefore, only f_1, f_0, f_{-1} lead to three distinct conjugacy classes of square roots of $-\mathbf{1}$ in $\mathcal{M}(2, \mathbb{C})$. Compare [204] for the corresponding computations with CLIFFORD for Maple.

In general, if $\mathcal{A} = \mathcal{M}(2d, \mathbb{C})$, then $\dim(\mathcal{A}) = 8d^2$. The group $\mathcal{G}(\mathcal{A})$ has one connected component. The square roots of $-\mathbf{1}$ in $\mathcal{A}$ are in bijection with the idempotents ϵ [3] according to (2.7.97). According[39] to (2.7.97) and its inverse $\epsilon = \frac{1}{2}(\mathbf{1} - if)$

[38] Compare Footnote 35.

[39] On the other hand it is clear that complex conjugation always leads to $f_- = \overline{f_+}$, where the overbar means complex conjugation in $\mathcal{M}(2d, \mathbb{C})$ and Clifford conjugation in the isomorphic Clifford algebra $C\ell(p, q)$. So either the trivial idempotent $\epsilon_- = 0$ is included in the bijection (2.7.97) of idempotents and square roots of $-\mathbf{1}$ or alternatively the square root of $-\mathbf{1}$ with $\mathrm{Spec}(f_-) = -1$ is obtained from $f_- = \overline{f_+}$.

the square root of $-\mathbf{1}$ with $\mathrm{Spec}(f_-) = k/d = -1$, i.e. $k = -d$ (see below), always corresponds to the trivial idempotent $\epsilon_- = 0$, and the square root of $-\mathbf{1}$ with $\mathrm{Spec}(f_+) = k/d = +1$, $k = +d$, corresponds to the identity idempotent $\epsilon_+ = \mathbf{1}$.

If f is a square root of $-\mathbf{1}$, then $V = \mathbb{C}^{2d}$ is the direct sum of the eigenspaces[40] associated with the eigenvalues i and $-i$. There is an integer k such that the dimensions of the eigenspaces are respectively $d+k$ and $d-k$. Moreover, $-d \leq k \leq d$. Two square roots of $-\mathbf{1}$ are conjugate if and only if they give the same integer k. Then, all elements of $\mathrm{Cent}(f)$ consist of diagonal block matrices with 2 square blocks of $(d+k) \times (d+k)$ matrices and $(d-k) \times (d-k)$ matrices. Therefore, the $\mathbb{C}$-dimension of $\mathrm{Cent}(f)$ is $(d+k)^2 + (d-k)^2$. Hence the $\mathbb{R}$-dimension (2.7.78) of the conjugacy class of f:

$$8d^2 - 2(d+k)^2 - 2(d-k)^2 = 4(d^2 - k^2). \tag{2.7.99}$$

Also, from the equality $\mathrm{tr}(f) = (d+k)i - (d-k)i = 2ki$ we deduce that $\mathrm{Scal}(f) = 0$ and that $\mathrm{Spec}(f) = (2ki)/(2di) = k/d$ if $\omega = i\mathbf{1}$ (whence $\mathrm{tr}(\omega) = 2di$).

As announced on page 77, we consider that a square root of $-\mathbf{1}$ is *ordinary* if the associated integer k vanishes, and that it is *exceptional* if $k \neq 0$. Thus the following assertion is true in all cases: the ordinary square roots of $-\mathbf{1}$ in $\mathcal{A}$ constitute one conjugacy class of dimension $\dim(\mathcal{A})/2$ which has as many connected components as $\mathcal{G}(\mathcal{A})$, and the equality $\mathrm{Spec}(f) = 0$ holds for every ordinary square root of $-\mathbf{1}$ when the linear form Spec exists. All conjugacy classes of exceptional square roots of $-\mathbf{1}$ have a dimension $< \dim(\mathcal{A})/2$.

All square roots of $-\mathbf{1}$ in $\mathcal{M}(2d, \mathbb{C})$ constitute $(2d+1)$ conjugacy classes[41] which are also the connected components of the submanifold of square roots of $-\mathbf{1}$ because of the equality $\mathrm{Spec}(f) = k/d$, which is conjugacy class specific.

When $\mathcal{A} = \mathcal{M}(2d, \mathbb{C})$, the group $\mathrm{Aut}(\mathcal{A})$ is larger than $\mathrm{Inn}(\mathcal{A})$ since it contains the complex conjugation (that maps every entry of a matrix to the conjugate complex number). It is clear that the class of ordinary square roots of $-\mathbf{1}$ is invariant by complex conjugation. But the class associated with an integer k other than 0 is mapped by complex conjugation to the class associated with $-k$. In particular, the complex conjugation maps the class $\{\omega\}$ (associated with $k = d$) to the class $\{-\omega\}$ associated with $k = -d$.

All these observations can easily verified for the above example of $d = 1$ of the Pauli matrix algebra $\mathcal{A} = \mathcal{M}(2, \mathbb{C})$. For $d = 2$ we have the isomorphism of $\mathcal{A} = \mathcal{M}(4, \mathbb{C})$ with $C\ell(0,5)$, $C\ell(2,3)$ and $C\ell(4,1)$. While $C\ell(0,5)$ is important in Clifford analysis, $C\ell(4,1)$ is both the geometric algebra of the Lorentz space $\mathbb{R}^{4,1}$ and the conformal geometric algebra of 3D Euclidean geometry. Its set of square roots of $-\mathbf{1}$ is therefore of particular practical interest.

Exercise 2.7.8. *Let $C\ell(4,1) \cong \mathcal{A}$ where $\mathcal{A} = \mathcal{M}(4, \mathbb{C})$ for $d = 2$. The $C\ell(4,1)$*

[40]The following theorem is sufficient for a matrix f in $\mathcal{M}(m, \mathbb{K})$, if $\mathbb{K}$ is a (commutative) field. The matrix f is diagonalizable if and only if $P(f) = 0$ for some polynomial P that has only simple roots, all of them in the field $\mathbb{K}$. (This implies that P is a multiple of the minimal polynomial, but we do not need to know whether P is or is not the minimal polynomial.)

[41]Two conjugate (similar) matrices have the same eigenvalues and the same trace. This suffices to recognize that $2d+1$ conjugacy classes are obtained.

1-vectors can be represented[42] *by the following matrices:*

$$e_1 = \begin{pmatrix} 1 & 0 & 0 & 0 \\ 0 & -1 & 0 & 0 \\ 0 & 0 & -1 & 0 \\ 0 & 0 & 0 & 1 \end{pmatrix}, \; e_2 = \begin{pmatrix} 0 & 1 & 0 & 0 \\ 1 & 0 & 0 & 0 \\ 0 & 0 & 0 & 1 \\ 0 & 0 & 1 & 0 \end{pmatrix}, \; e_3 = \begin{pmatrix} 0 & -i & 0 & 0 \\ i & 0 & 0 & 0 \\ 0 & 0 & 0 & -i \\ 0 & 0 & i & 0 \end{pmatrix},$$

$$e_4 = \begin{pmatrix} 0 & 0 & 1 & 0 \\ 0 & 0 & 0 & -1 \\ 1 & 0 & 0 & 0 \\ 0 & -1 & 0 & 0 \end{pmatrix}, \; e_5 = \begin{pmatrix} 0 & 0 & -1 & 0 \\ 0 & 0 & 0 & 1 \\ 1 & 0 & 0 & 0 \\ 0 & -1 & 0 & 0 \end{pmatrix}. \qquad (2.7.100)$$

We find five conjugacy classes of roots f_k of $-\mathbf{1}$ in $C\ell(4,1)$ for $k \in \{0, \pm 1, \pm 2\}$: four exceptional and one ordinary. Since f_k is a root of $p(t) = t^2 + 1$ which factors over $\mathbb{C}$ into $(t-i)(t+i)$, the minimal polynomial $m_k(t)$ of f_k is one of the following: $t - i$, $t + i$ or $(t-i)(t+i)$. Respectively, there are three classes of characteristic polynomial $\Delta_k(t)$ of the matrix $\mathcal{F}_k$ in $\mathcal{M}(4,\mathbb{C})$ which corresponds to f_k, namely, $(t-i)^4$, $(t+i)^4$ and $(t-i)^{n_1}(t+i)^{n_2}$, where $n_1 + n_2 = 2d = 4$ and $n_1 = d + k = 2 + k$, $n_2 = d - k = 2 - k$. As predicted by the above discussion, the ordinary root corresponds to $k = 0$ whereas the exceptional roots correspond to $k \neq 0$.

1. *For $k = 2$, we have $\Delta_2(t) = (t-i)^4$, $m_2(t) = t - i$, and so $\mathcal{F}_2 = \mathrm{diag}(i,i,i,i)$ which in the above representation (2.7.100) corresponds to the non-trivial central element $f_2 = \omega = e_{12345}$. Clearly,* $\mathrm{Spec}(f_2) = 1 = \frac{k}{d}$; $\mathrm{Scal}(f_2) = 0$; *the $\mathbb{C}$-dimension of the centralizer* $\mathrm{Cent}(f_2)$ *is 16; and the $\mathbb{R}$-dimension of the conjugacy class of f_2 is zero as it contains only f_2 since $f_2 \in \mathrm{Z}(\mathcal{A})$. Thus, the $\mathbb{R}$-dimension of the class is again zero in agreement with (2.7.99).*

2. *For $k = -2$, we have $\Delta_{-2}(t) = (t+i)^4$, $m_{-2}(t) = t + i$, and $\mathcal{F}_{-2} = \mathrm{diag}(-i,-i,-i,-i)$ which corresponds to the central element $f_{-2} = -\omega = -e_{12345}$. Again,* $\mathrm{Spec}(f_{-2}) = -1 = \frac{k}{d}$; $\mathrm{Scal}(f_{-2}) = 0$; *the $\mathbb{C}$-dimension of the centralizer* $\mathrm{Cent}(f_{-2})$ *is 16 and the conjugacy class of f_{-2} contains only f_{-2} since $f_{-2} \in \mathrm{Z}(\mathcal{A})$. Thus, the $\mathbb{R}$-dimension of the class is again zero in agreement with (2.7.99).*

3. *For $k \neq \pm 2$, we consider three subcases when $k = 1$, $k = 0$ and $k = -1$. When $k = 1$, then $\Delta_1(t) = (t-i)^3(t+i)$ and $m_1(t) = (t-i)(t+i)$. Then the root $\mathcal{F}_1 = \mathrm{diag}(i,i,i,-i)$ corresponds to*

$$f_1 = \frac{1}{2}(e_{23} + e_{123} - e_{2345} + e_{12345}). \qquad (2.7.101)$$

Note that $\mathrm{Spec}(f_1) = \frac{1}{2} = \frac{k}{d}$ *so f_1 is an exceptional root of $-\mathbf{1}$.*

[42] For the computations of this example in the Maple package CLIFFORD we have used the identification $i = e_{23}$. Yet the results obtained for the square roots of $-\mathbf{1}$ are independent of this setting (we can alternatively use, e.g., $i = e_{12345}$, or the imaginary unit $i \in \mathbb{C}$), as can easily be checked for f_1 of (2.7.101), f_0 of (2.7.102) and f_{-1} of (2.7.103) by only assuming the standard Clifford product rules for e_1 to e_5.

When $k = 0$, *then* $\Delta_0(t) = (t-i)^2(t+i)^2$ *and* $m_0(t) = (t-i)(t+i)$. *Thus the root of* $-\mathbf{1}$ *in this case is* $\mathcal{F}_0 = \mathrm{diag}(i, i, -i, -i)$ *which corresponds to just*

$$f_0 = e_{123}. \tag{2.7.102}$$

Note that $\mathrm{Spec}(f_0) = 0$ *thus* $f_0 = e_{123}$ *is an ordinary root of* $-\mathbf{1}$.

When $k = -1$, *then* $\Delta_{-1}(t) = (t-i)(t+i)^3$ *and* $m_{-1}(t) = (t-i)(t+i)$. *Then, the root of* $-\mathbf{1}$ *in this case is* $\mathcal{F}_{-1} = \mathrm{diag}(i, -i, -i, -i)$ *which corresponds to*

$$f_{-1} = \frac{1}{2}(e_{23} + e_{123} + e_{2345} - e_{12345}). \tag{2.7.103}$$

Since $\mathrm{Scal}(f_{-1}) = -\frac{1}{2} = \frac{k}{d}$, *we gather that* f_{-1} *is an exceptional root.*

As expected, we can also see that the roots ω *and* $-\omega$ *are related via the grade involution whereas* $f_1 = -\tilde{f}_{-1}$ *where* $\tilde{}$ *denotes the reversion in* $C\ell(4,1)$.

Exercise 2.7.9. *Let* $C\ell(0,5) \cong \mathcal{A}$ *where* $\mathcal{A} = \mathcal{M}(4,\mathbb{C})$ *for* $d = 2$. *The* $C\ell(0,5)$ *1-vectors can be represented*[43] *by the following matrices:*

$$e_1 = \begin{pmatrix} 0 & -1 & 0 & 0 \\ 1 & 0 & 0 & 0 \\ 0 & 0 & 0 & -1 \\ 0 & 0 & 1 & 0 \end{pmatrix}, \quad e_2 = \begin{pmatrix} 0 & -i & 0 & 0 \\ -i & 0 & 0 & 0 \\ 0 & 0 & 0 & -i \\ 0 & 0 & -i & 0 \end{pmatrix}, \quad e_3 = \begin{pmatrix} -i & 0 & 0 & 0 \\ 0 & i & 0 & 0 \\ 0 & 0 & i & 0 \\ 0 & 0 & 0 & -i \end{pmatrix},$$
$$e_4 = \begin{pmatrix} 0 & 0 & -1 & 0 \\ 0 & 0 & 0 & 1 \\ 1 & 0 & 0 & 0 \\ 0 & -1 & 0 & 0 \end{pmatrix}, \quad e_5 = \begin{pmatrix} 0 & 0 & -i & 0 \\ 0 & 0 & 0 & i \\ -i & 0 & 0 & 0 \\ 0 & i & 0 & 0 \end{pmatrix}. \tag{2.7.104}$$

Like for $C\ell(4,1)$, *we have five conjugacy classes of the roots* f_k *of* $-\mathbf{1}$ *in* $C\ell(0,5)$ *for* $k \in \{0, \pm 1, \pm 2\}$: *four exceptional and one ordinary. Using the same notation as in Example 2.7.8, we find the following representatives of the conjugacy classes.*

1. *For* $k = 2$, *we have* $\Delta_2(t) = (t-i)^4$, $m_2(t) = t - i$ *and* $\mathcal{F}_2 = \mathrm{diag}(i, i, i, i)$ *which in the above representation (2.7.104) corresponds to the non-trivial central element* $f_2 = \omega = e_{12345}$. *Then,* $\mathrm{Spec}(f_2) = 1 = \frac{k}{d}$; $\mathrm{Scal}(f_2) = 0$; *the* $\mathbb{C}$*-dimension of the centralizer* $\mathrm{Cent}(f_2)$ *is* 16; *and the* $\mathbb{R}$*-dimension of the conjugacy class of* f_2 *is zero as it contains only* f_2 *since* $f_2 \in \mathrm{Z}(\mathcal{A})$. *Thus, the* $\mathbb{R}$*-dimension of the class is again zero in agreement with (2.7.99).*

2. *For* $k = -2$, *we have* $\Delta_{-2}(t) = (t+i)^4$, $m_{-2}(t) = t + i$ *and* $\mathcal{F}_{-2} = \mathrm{diag}(-i, -i, -i, -i)$ *which corresponds to the central element* $f_{-2} = -\omega =$

[43] For the computations of this example in the Maple package CLIFFORD we have used the identification $i = e_3$. Yet the results obtained for the square roots of $-\mathbf{1}$ are independent of this setting (we can alternatively use, e.g., $i = e_{12345}$ or the imaginary unit $i \in \mathbb{C}$), as can easily be checked for f_1 of (2.7.105), f_0 of (2.7.106) and f_{-1} of (2.7.107) by only assuming the standard Clifford product rules for e_1 to e_5.

$-e_{12345}$. *Again,* $\mathrm{Spec}(f_{-2}) = -1 = \frac{k}{d}$; $\mathrm{Scal}(f_{-2}) = 0$; *the* $\mathbb{C}$*-dimension of the centralizer* $\mathrm{Cent}(f_{-2})$ *is* 16 *and the conjugacy class of* f_{-2} *contains only* f_{-2} *since* $f_{-2} \in \mathrm{Z}(\mathcal{A})$. *Thus, the* $\mathbb{R}$*-dimension of the class is again zero in agreement with (2.7.99).*

3. *For* $k \neq \pm 2$*, we consider three subcases when* $k = 1$*,* $k = 0$ *and* $k = -1$*. When* $k = 1$*, then* $\Delta_1(t) = (t - i)^3(t + i)$ *and* $m_1(t) = (t - i)(t + i)$*. Then the root* $\mathcal{F}_1 = \mathrm{diag}(i, i, i, -i)$ *corresponds to*

$$f_1 = \frac{1}{2}(e_3 + e_{12} + e_{45} + e_{12345}). \tag{2.7.105}$$

Since $\mathrm{Spec}(f_1) = \frac{1}{2} = \frac{k}{d}$, f_1 *is an exceptional root of* $-\mathbf{1}$.

When $k = 0$*, then* $\Delta_0(t) = (t-i)^2(t+i)^2$ *and* $m_0(t) = (t-i)(t+i)$*. Thus the root of* $-\mathbf{1}$ *is this case is* $\mathcal{F}_0 = \mathrm{diag}(i, i, -i, -i)$ *which corresponds to just*

$$f_0 = e_{45}. \tag{2.7.106}$$

Note that $\mathrm{Spec}(f_0) = 0$ *thus* $f_0 = e_{45}$ *is an ordinary root of* $-\mathbf{1}$.

When $k = -1$*, then* $\Delta_{-1}(t) = (t-i)(t+i)^3$ *and* $m_{-1}(t) = (t-i)(t+i)$*. Then, the root of* $-\mathbf{1}$ *in this case is* $\mathcal{F}_{-1} = \mathrm{diag}(i, -i, -i, -i)$ *which corresponds to*

$$f_{-1} = \frac{1}{2}(-e_3 + e_{12} + e_{45} - e_{12345}). \tag{2.7.107}$$

Since $\mathrm{Scal}(f_{-1}) = -\frac{1}{2} = \frac{k}{d}$*, we gather that* f_{-1} *is an exceptional root.*

Again we can see that the roots f_2 *and* f_{-2} *are related via the grade involution whereas* $f_1 = -\tilde{f}_{-1}$ *where* $\tilde{}$ *denotes the reversion in* $C\ell(0, 5)$.

Exercise 2.7.10. *Let* $C\ell(7, 0) \cong \mathcal{A}$ *where* $\mathcal{A} = \mathcal{M}(8, \mathbb{C})$ *for* $d = 4$*. We have nine conjugacy classes of roots* f_k *of* $-\mathbf{1}$ *for* $k \in \{0, \pm 1, \pm 2 \pm 3 \pm 4\}$*. Since* f_k *is a root of a polynomial* $p(t) = t^2 + 1$ *which factors over* $\mathbb{C}$ *into* $(t - i)(t + i)$*, its minimal polynomial* $m(t)$ *will be one of the following:* $t - i$*,* $t + i$ *or* $(t - i)(t + i) = t^2 + 1$.

Respectively, each conjugacy class is characterized by a characteristic polynomial $\Delta_k(t)$ *of the matrix* $M_k \in \mathcal{M}(8, \mathbb{C})$ *which represents* f_k*. Namely, we have*

$$\Delta_k(t) = (t - i)^{n_1}(t + i)^{n_2},$$

where $n_1 + n_2 = 2d = 8$ *and* $n_1 = d + k = 4 + k$ *and* $n_2 = d - k = 4 - k$*. The ordinary root of* $-\mathbf{1}$ *corresponds to* $k = 0$ *whereas the exceptional roots correspond to* $k \neq 0$.

1. *When* $k = 4$*, we have* $\Delta_4(t) = (t - i)^8$*,* $m_4(t) = t - i$ *and* $\mathcal{F}_4 = \mathrm{diag}(\overbrace{i, \ldots, i}^{8})$ *which in the representation used by CLIFFORD [5] corresponds to the non-trivial central element* $f_4 = \omega = e_{1234567}$*. Clearly,* $\mathrm{Spec}(f_4) = 1 = \frac{k}{d}$*;* $\mathrm{Scal}(f_4) = 0$*; the* $\mathbb{C}$*-dimension of the centralizer* $\mathrm{Cent}(f_4)$ *is* 64*; and the* $\mathbb{R}$*-dimension of the conjugacy class of* f_4 *is zero since* $f_4 \in \mathrm{Z}(\mathcal{A})$*. Thus, the* $\mathbb{R}$*-dimension of the class is again zero in agreement with (2.7.99).*

2. *When* $k = -4$, *we have* $\Delta_{-4}(t) = (t+i)^8$, $m_{-4}(t) = t+i$ *and* $\mathcal{F}_{-4} = \mathrm{diag}(\overbrace{-i,\ldots,-i}^{8})$ *which corresponds to* $f_{-4} = -\omega = -e_{1234567}$. *Again,* $\mathrm{Spec}(f_{-4}) = -1 = \frac{k}{d}$; $\mathrm{Scal}(f_{-4}) = 0$; *the* $\mathbb{C}$*-dimension of the centralizer* $\mathrm{Cent}(f)$ *is* 64 *and the conjugacy class of* f_{-4} *contains only* f_{-4} *since* $f_{-4} \in \mathrm{Z}(\mathcal{A})$. *Thus, the* $\mathbb{R}$*-dimension of the class is again zero in agreement with (2.7.99).*

3. *When* $k \neq \pm 4$, *we consider seven subcases when* $k = \pm 3$, $k = \pm 2$, $k = \pm 1$ *and* $k = 0$.

   When $k = 3$, *then* $\Delta_3(t) = (t-i)^7(t+i)$ *and* $m_3(t) = (t-i)(t+i)$. *Then the root* $\mathcal{F}_3 = \mathrm{diag}(\overbrace{i,\ldots,i}^{7},-i)$ *corresponds to*

$$f_3 = \frac{1}{4}(e_{23} - e_{45} + e_{67} - e_{123} + e_{145} - e_{167} + e_{234567} + 3e_{1234567}). \quad (2.7.108)$$

   Since $\mathrm{Spec}(f_3) = \frac{3}{4} = \frac{k}{d}$, f_3 *is an exceptional root of* $-\mathbf{1}$.
   When $k = 2$, *then* $\Delta_2(t) = (t-i)^6(t+i)^2$ *and* $m_2(t) = (t-i)(t+i)$. *Then the root* $\mathcal{F}_2 = \mathrm{diag}(\overbrace{i,\ldots,i}^{6},-i,-i)$ *corresponds to*

$$f_2 = \frac{1}{2}(e_{67} - e_{45} - e_{123} + e_{1234567}). \quad (2.7.109)$$

   Since $\mathrm{Spec}(f_2) = \frac{1}{2} = \frac{k}{d}$, f_2 *is also an exceptional root.*
   When $k = 1$, *then* $\Delta_1(t) = (t-i)^5(t+i)^3$ *and* $m_1(t) = (t-i)(t+i)$. *Then the root* $\mathcal{F}_1 = \mathrm{diag}(\overbrace{i,\ldots,i}^{5},-i,-i,-i)$ *corresponds to*

$$f_1 = \frac{1}{4}(e_{23} - e_{45} + 3e_{67} - e_{123} + e_{145} + e_{167} - e_{234567} + e_{1234567}). \quad (2.7.110)$$

   Since $\mathrm{Spec}(f_1) = \frac{1}{4} = \frac{k}{d}$, f_1 *is another exceptional root.*
   When $k = 0$, *then* $\Delta_0(t) = (t-i)^4(t+i)^4$ *and* $m_0(t) = (t-i)(t+i)$. *Then the root* $\mathcal{F}_0 = \mathrm{diag}(i,i,i,i,-i,-i,-i,-i)$ *corresponds to*

$$f_0 = \frac{1}{2}(e_{23} - e_{45} + e_{67} - e_{234567}). \quad (2.7.111)$$

   Since $\mathrm{Spec}(f_0) = 0 = \frac{k}{d}$, *we see that* f_0 *is an ordinary root of* $-\mathbf{1}$.
   When $k = -1$, *then* $\Delta_{-1}(t) = (t-i)^3(t+i)^5$ *and* $m_{-1}(t) = (t-i)(t+i)$. *Then the root* $\mathcal{F}_{-1} = \mathrm{diag}(i,i,i,\overbrace{-i,\ldots,-i}^{5})$ *corresponds to*

$$f_{-1} = \frac{1}{4}(e_{23} - e_{45} + 3e_{67} + e_{123} - e_{145} - e_{167} - e_{234567} - e_{1234567}). \quad (2.7.112)$$

Thus, $\text{Spec}(f_{-1}) = -\frac{1}{4} = \frac{k}{d}$ *and so* f_{-1} *is another exceptional root.*
When $k = -2$, *then* $\Delta_{-2}(t) = (t-i)^2(t+i)^6$ *and* $m_{-2}(t) = (t-i)(t+i)$. *Then the root* $\mathcal{F}_{-2} = \text{diag}(i, i, \overbrace{-i, \ldots, -i}^{6})$ *corresponds to*

$$f_{-2} = \frac{1}{2}(e_{67} - e_{45} + e_{123} - e_{1234567}). \tag{2.7.113}$$

Since $\text{Spec}(f_{-2}) = -\frac{1}{2} = \frac{k}{d}$, *we see that* f_{-2} *is also an exceptional root.*
When $k = -3$, *then* $\Delta_{-3}(t) = (t-i)(t+i)^7$ *and* $m_{-3}(t) = (t-i)(t+i)$. *Then the root* $\mathcal{F}_{-3} = \text{diag}(i, \overbrace{-i, \ldots, -i}^{7})$ *corresponds to*

$$f_{-3} = \frac{1}{4}(e_{23} - e_{45} + e_{67} + e_{123} - e_{145} + e_{167} + e_{234567} - 3e_{1234567}). \tag{2.7.114}$$

Again, $\text{Spec}(f_{-3}) = -\frac{3}{4} = \frac{k}{d}$ *and so* f_{-3} *is another exceptional root of* $\mathbf{-1}$.
As expected, we can also see that the roots ω *and* $-\omega$ *are related via the reversion whereas* $f_3 = -\bar{f}_{-3}$, $f_2 = -\bar{f}_{-2}$, $f_1 = -\bar{f}_{-1}$ *where* $\bar{\ }$ *denotes the conjugation in* $C\ell(7,0)$.

2.7.4.8 Square roots of -1 in conformal geometric algebra $Cl(4,1)$

We pay special attention to the square roots of -1 in conformal geometric algebra $Cl(4,1)$, because of the enormous practical importance of this algebra in applications to robotics, computer graphics, robot and computer vision, virtual reality, visualization and the like [202]. See Table 2.7 for representative exceptional ($k \neq 0$) square roots of $\mathbf{-1}$ in conformal geometric algebra $Cl(4,1)$ of three-dimensional Euclidean space [203].

k	f_k	$\Delta_k(t)$
2	$\omega = e_{12345}$	$(t-i)^4$
1	$\frac{1}{2}(e_{23} + e_{123} - e_{2345} + e_{12345})$	$(t-i)^3(t+i)$
0	e_{123}	$(t-i)^2(t+i)^2$
-1	$\frac{1}{2}(e_{23} + e_{123} + e_{2345} - e_{12345})$	$(t-i)(t+i)^3$
-2	$-\omega = -e_{12345}$	$(t+i)^4$

Table 2.7 Square roots of $\mathbf{-1}$ in conformal geometric algebra $Cl(4,1) \cong \mathcal{M}(4,\mathbb{C})$, $d = 2$, with characteristic polynomials $\Delta_k(t)$. See [203] for details. Source: modified version of [203, Tab. 2].

2.7.4.8.1 Ordinary square roots of -1 in $Cl(4,1)$ with $k = 0$

In the algebra basis of $Cl(4,1)$ there are nine blades which represent ordinary square roots of -1:

$$e_5, \qquad e_{234}, e_{134}, e_{124}, e_{123}, \qquad e_{2345}, e_{1345}, e_{1245}, e_{1235}. \tag{2.7.115}$$

But remembering the work in [193], we know that even if we only look at the subalgebras $Cl(4,0)$ or $Cl(3,1)$, which do not contain the pseudoscalar e_{12345}, and contain therefore only ordinary square roots of -1 for $Cl(4,1)$, we have long parametrized expressions for ordinary square roots of -1. But because of the high dimensionality it may not be easy to compute a complete expression for the whole 16D submanifold of ordinary square roots of -1 in $Cl(4,1)$ by hand.

2.7.4.8.2 Exceptional square roots of -1 in $Cl(4,1)$ with $k = 1$ In this case we can generalize Table 2.7 to patches of the 12-dimensional submanifold of exceptional square roots of -1 in $Cl(4,1)$. In the future a complete parametrized expression obtained, e.g., with Clifford for Maple would be very desirable.

We begin with the general expression

$$f_1 = (\frac{1+u}{2}E + \frac{1-u}{2})\,\omega, \qquad \omega = e_{12345}, \tag{2.7.116}$$

where we assume that $E, u \in Cl(4,1)$, $E^2 = u^2 = +1$. This makes the expressions $\frac{1\pm u}{2}$ become idempotents $(\frac{1\pm u}{2})^2 = \frac{1\pm u}{2}$. In the following we put forward certain values for E and u which will yield linearly independent patches of the 12-dimensional submanifold of $\sqrt{-1}$.

- $E = ve_5$, $v \in \mathbb{R}^4$, $v^2 = 1$, $u \in \mathbb{R}^3_{\perp v}$, $u^2 = 1$ gives a 3D $\times$ 2D = 6D submanifold. As a concrete example in this submanifold we can e.g. set $v = e_4$, $u = e_1$ and get

$$f_1 = \frac{1}{2}[(1+e_1)e_{45} + 1 - e_1]\omega = \frac{1}{2}[e_{45} + e_{145} + 1 - e_1]\omega. \tag{2.7.117}$$

- $E = e_{1234}$, $u \in \mathbb{R}^4$, $u^2 = 1$ gives a 3D submanifold of $\sqrt{-1}$. A concrete example is e.g. $u = e_1$, then

$$f_1 = \frac{1}{2}[(1+e_1)e_{1234} + 1 - e_1]\omega = \frac{1}{2}[e_{1234} + e_{234} + 1 - e_1]\omega. \tag{2.7.118}$$

- $E = v$, $v \in \mathbb{R}^4$, $v^2 = 1$, $u = e_{1234}$ gives another 3D submanifold. A concrete example is e.g. $v = e_1$ and gives

$$f_1 = \frac{1}{2}[(1+e_{1234})e_1 + 1 - e_{1234}]\omega = \frac{1}{2}[e_1 - e_{234} + 1 - e_{1234}]\omega. \tag{2.7.119}$$

2.7.4.8.3 Exceptional square roots of -1 in $Cl(4,1)$ with $k = -1$ This is completely analogous to $k = +1$ by starting with

$$f_{-1} = (\frac{1+u}{2}E - \frac{1-u}{2})\omega, \qquad \omega = e_{12345}. \tag{2.7.120}$$

2.7.4.8.4 Exceptional square roots of -1 in $Cl(4,1)$ with $k = \pm 2$ The exceptional square roots of -1 are zero-dimensional in this case and therefore uniquely given by

$$f_{\pm 2} = \pm e_{12345}. \tag{2.7.121}$$

2.7.4.9 *Summary on general multivector square roots of* -1

We proved that in all cases $\mathrm{Scal}(f) = 0$ for every square root of $-\mathbf{1}$ in $\mathcal{A}$ isomorphic to $C\ell(p,q)$. We distinguished *ordinary* square roots of $-\mathbf{1}$, and *exceptional* ones.

In all cases the ordinary square roots f of $-\mathbf{1}$ constitute a unique conjugacy class of dimension $\dim(\mathcal{A})/2$ which has as many connected components as the group $\mathcal{G}(\mathcal{A})$ of invertible elements in $\mathcal{A}$. Furthermore, we have $\mathrm{Spec}(f) = 0$ (zero pseudoscalar part) if the associated ring is $\mathbb{R}^2$, $\mathbb{H}^2$ or $\mathbb{C}$. The exceptional square roots of $-\mathbf{1}$ *only* exist if $\mathcal{A} \cong \mathcal{M}(2d, \mathbb{C})$ (see Section 2.7.4.7).

For $\mathcal{A} = \mathcal{M}(2d, \mathbb{R})$ of Section 2.7.4.3, the centralizer and the conjugacy class of a square root f of $-\mathbf{1}$ both have $\mathbb{R}$-dimension $2d^2$ with two connected components, pictured in Figure 2.5 for $d = 1$.

For $\mathcal{A} = \mathcal{M}(2d, \mathbb{R}^2) = \mathcal{M}(2d, \mathbb{R}) \times \mathcal{M}(2d, \mathbb{R})$ of Section 2.7.4.4, the square roots of $(-\mathbf{1}, -\mathbf{1})$ are pairs of two square roots of $-\mathbf{1}$ in $\mathcal{M}(2d, \mathbb{R})$. They constitute a unique conjugacy class with four connected components, each of dimension $4d^2$. Regarding the four connected components, the group $\mathrm{Inn}(\mathcal{A})$ induces the permutations of the Klein group whereas the quotient group $\mathrm{Aut}(\mathcal{A})/\mathrm{Inn}(\mathcal{A})$ is isomorphic to the group of isometries of a Euclidean square in 2D.

For $\mathcal{A} = \mathcal{M}(d, \mathbb{H})$ of Section 2.7.4.5, the submanifold of the square roots f of $-\mathbf{1}$ is a single connected conjugacy class of $\mathbb{R}$-dimension $2d^2$ equal to the $\mathbb{R}$-dimension of the centralizer of every f. The easiest example is $\mathbb{H}$ itself for $d = 1$.

For $\mathcal{A} = \mathcal{M}(d, \mathbb{H}^2) = \mathcal{M}(2d, \mathbb{H}) \times \mathcal{M}(2d, \mathbb{H})$ of Section 2.7.4.6, the square roots of $(-\mathbf{1}, -\mathbf{1})$ are pairs of two square roots (f, f') of $-\mathbf{1}$ in $\mathcal{M}(2d, \mathbb{H})$ and constitute a unique connected conjugacy class of $\mathbb{R}$-dimension $4d^2$. The group $\mathrm{Aut}(\mathcal{A})$ has two connected components: the neutral component $\mathrm{Inn}(\mathcal{A})$ connected to the identity and the second component containing the swap automorphism $(f, f') \mapsto (f', f)$. The simplest case for $d = 1$ is $\mathbb{H}^2$ isomorphic to $C\ell(0,3)$.

For $\mathcal{A} = \mathcal{M}(2d, \mathbb{C})$ of Section 2.7.4.7, the square roots of $-\mathbf{1}$ are in bijection to the idempotents. First, the ordinary square roots of $-\mathbf{1}$ (with $k = 0$) constitute a conjugacy class of $\mathbb{R}$-dimension $4d^2$ of a single connected component which is invariant under $\mathrm{Aut}(\mathcal{A})$. Second, there are $2d$ conjugacy classes of exceptional square roots of $-\mathbf{1}$, each composed of a single connected component, characterized by equality $\mathrm{Spec}(f) = k/d$ (the pseudoscalar coefficient) with $\pm k \in \{1, 2, \dots, d\}$, and their $\mathbb{R}$-dimensions are $4(d^2 - k^2)$. The group $\mathrm{Aut}(\mathcal{A})$ includes conjugation of the pseudoscalar $\omega \mapsto -\omega$ which maps the conjugacy class associated with k to the class associated with $-k$. The simplest case for $d = 1$ is the Pauli matrix algebra isomorphic to the geometric algebra $C\ell(3,0)$ of 3D Euclidean space $\mathbb{R}^3$, and to complex biquaternions [294].

Section 2.7.4.7 includes explicit examples for $d = 2$: $C\ell(4,1)$ and $C\ell(0,5)$, and for $d = 4$: $C\ell(7,0)$. Appendix A.1 summarizes the square roots of $-\mathbf{1}$ in all $C\ell(p,q) \cong \mathcal{M}(2d, \mathbb{C})$ for $d = 1, 2, 4$. [204] contains details on how square roots of $-\mathbf{1}$ can be computed using the package CLIFFORD for Maple.

Among the many possible *applications* of these findings, the possibility of *new integral transformations* in Clifford analysis is very promising. This field thus obtains essential algebraic information, which can e.g., be used to create *steerable* transformations, which may be steerable within a connected component of a submanifold of square roots of $-\mathbf{1}$.

2.8 QUATERNIONS AND THE GEOMETRY OF ROTATIONS IN THREE AND FOUR DIMENSIONS

2.8.1 Introduction to quaternions

The electromagnetic field equations were originally formulated by J. C. Maxwell [258] in the language of Hamilton's quaternions [155]. Later, among many other applications, quaternions began to play an important role in aerospace engineering [238], color signal processing [127], and in material science for texture analysis (understood as the distribution of crystallographic orientations of a polycrystalline sample [328]) [12, 261].

During the last decade the general orthogonal planes split (OPS) with respect to any two pure unit quaternions $f, g \in \mathbb{H}$, $f^2 = g^2 = -1$, including the case $f = g$, has proved extremely useful for the construction and geometric interpretation of general classes of double-kernel quaternion Fourier transformations (QFT) [205]. Applications include color image processing, where the OPS with $f = g =$ the greyline, naturally splits a pure quaternionic three-dimensional color signal into luminance and chrominance components. Yet it is found independently in the quaternion geometry of rotations [261], that the pure quaternion units f, g and the analysis planes, which they define, play a key role in the geometry of rotations, and the geometrical interpretation of integrals related to the spherical Radon transform of probability density functions of unit quaternions, as relevant for texture analysis in crystallography. In our current section, we further investigate these connections.

Gauss, Rodrigues and Hamilton's [273, 274] four-dimensional (4D) quaternion algebra $\mathbb{H}$ is defined over $\mathbb{R}$ with three imaginary units $\boldsymbol{i}$, $\boldsymbol{j}$, $\boldsymbol{k}$ and multiplication laws:

$$\begin{aligned} &\boldsymbol{ij} = -\boldsymbol{ji} = \boldsymbol{k}, \quad \boldsymbol{jk} = -\boldsymbol{kj} = \boldsymbol{i}, \quad \boldsymbol{ki} = -\boldsymbol{ik} = \boldsymbol{j}, \\ &\boldsymbol{i}^2 = \boldsymbol{j}^2 = \boldsymbol{k}^2 = \boldsymbol{ijk} = -1. \end{aligned} \tag{2.8.1}$$

The explicit form of a quaternion $q \in \mathbb{H}$ is

$$q = q_r + q_i\boldsymbol{i} + q_j\boldsymbol{j} + q_k\boldsymbol{k} \in \mathbb{H}, \qquad q_r, q_i, q_j, q_k \in \mathbb{R}. \tag{2.8.2}$$

Quaternions are isomorphic to the Clifford geometric algebra $Cl_{0,2}$ of $\mathbb{R}^{0,2}$, and to the even subalgebra $Cl^+_{3,0}$ of the Clifford geometric algebra $Cl_{3,0}$ of $\mathbb{R}^3$, i.e. $\mathbb{H}$ is isomorphic to the algebra of rotation operators in $Cl(3,0)$:

$$\mathbb{H} \cong Cl_{0,2} \cong Cl^+_{3,0}. \tag{2.8.3}$$

$Cl^+_{3,0}$ has, with an orthonormal basis $\{\boldsymbol{e}_1, \boldsymbol{e}_2, \boldsymbol{e}_3\}$ of $\mathbb{R}^3$, the four dimensional basis

$$\{1, \boldsymbol{e}_{32} = \boldsymbol{e}_3\boldsymbol{e}_2, \boldsymbol{e}_{13} = \boldsymbol{e}_1\boldsymbol{e}_3, \boldsymbol{e}_{21} = \boldsymbol{e}_2\boldsymbol{e}_1\}. \tag{2.8.4}$$

The quaternion conjugate (equivalent to Clifford conjugation in $Cl(3,0)^+$ and $Cl(0,2)$) is defined as

$$\overline{q} = q_r - q_i\boldsymbol{i} - q_j\boldsymbol{j} - q_k\boldsymbol{k}, \qquad \overline{pq} = \overline{q}\,\overline{p}, \tag{2.8.5}$$

which leaves the scalar part q_r unchanged. This leads to the norm of $q \in \mathbb{H}$

$$|q| = \sqrt{q\overline{q}} = \sqrt{q_r^2 + q_i^2 + q_j^2 + q_k^2}, \qquad |pq| = |p|\,|q|\,. \tag{2.8.6}$$

The part

$$\boldsymbol{q} = V(q) = q - q_r = \frac{1}{2}(q - \overline{q}) = q_i\boldsymbol{i} + q_j\boldsymbol{j} + q_k\boldsymbol{k}, \tag{2.8.7}$$

is called a *pure* quaternion or vector part, it squares to the *negative* number $-(q_i^2 + q_j^2 + q_k^2)$. Every unit quaternion $\in S^3$ (i.e. $|q| = 1$) can be written as:

$$\begin{aligned} q &= q_r + q_i\boldsymbol{i} + q_j\boldsymbol{j} + q_k\boldsymbol{k} = q_r + \sqrt{q_i^2 + q_j^2 + q_k^2}\,\widehat{\boldsymbol{q}} = \cos\alpha + \widehat{\boldsymbol{q}}\sin\alpha \\ &= \exp(\alpha\,\widehat{\boldsymbol{q}}), \end{aligned} \tag{2.8.8}$$

where

$$\cos\alpha = q_r, \qquad \sin\alpha = \sqrt{q_i^2 + q_j^2 + q_k^2},$$
$$\widehat{\boldsymbol{q}} = \frac{\boldsymbol{q}}{|\boldsymbol{q}|} = \frac{q_i\boldsymbol{i} + q_j\boldsymbol{j} + q_k\boldsymbol{k}}{\sqrt{q_i^2 + q_j^2 + q_k^2}}, \tag{2.8.9}$$

and

$$\widehat{\boldsymbol{q}}^2 = -1, \qquad \widehat{\boldsymbol{q}} \in S^2. \tag{2.8.10}$$

The left and right *inverse* of a non-zero quaternion is

$$q^{-1} = \overline{q}/\,|q|^2 = \overline{q}/(q\overline{q}). \tag{2.8.11}$$

The real scalar part (grade zero selection [161] in Clifford geometric algebra) is

$$\mathrm{S}(q) = \langle q\rangle_0 = q_r = \frac{1}{2}(q + \overline{q}), \tag{2.8.12}$$

with *symmetries* $\forall p, q \in \mathbb{H}$:

$$\mathrm{S}(pq) = \mathrm{S}(qp) = p_rq_r - p_iq_i - p_jq_j - p_kq_k, \qquad \mathrm{S}(q) = \mathrm{S}(\overline{q})\,, \tag{2.8.13}$$

leads to a cyclic multiplication symmetry

$$\mathrm{S}(pqs) = \mathrm{S}(spq) = \mathrm{S}(qsp)\,, \quad \forall q, r, s \in \mathbb{H}. \tag{2.8.14}$$

We further have *linearity*

$$\mathrm{S}(\alpha p + \beta q) = \alpha\,\mathrm{S}(p) + \beta\,\mathrm{S}(q) = \alpha p_r + \beta q_r, \quad \forall p, q \in \mathbb{H},\ \alpha, \beta \in \mathbb{R}. \tag{2.8.15}$$

The scalar part and the quaternion conjugate allow the definition of the $\mathbb{R}^4$ *inner product* of two quaternions p, q as

$$p \cdot q = \mathrm{S}(p\overline{q}) = p_r q_r + p_i q_i + p_j q_j + p_k q_k \in \mathbb{R}. \tag{2.8.16}$$

Accordingly, we interpret in this section the four quaternion coefficients as coordinates in $\mathbb{R}^4$. In this interpretation selecting any two-dimensional plane subspace[44] and its orthogonal complement two-dimensional subspace allows to split four-dimensional quaternions $\mathbb{H}$ into pairs of orthogonal two-dimensional planes (compare Theorem 3.5 of [205]). Dealing with rotations we include general rotations in $\mathbb{R}^4$.

Definition 2.8.1 (Orthogonality of quaternions). *Two quaternions* $p, q \in \mathbb{H}$ *are* orthogonal $p \perp q$, *if and only if* $\mathrm{S}(p\overline{q}) = 0$.

2.8.2 Motivation for quaternion split

2.8.2.1 Splitting quaternions and knowing what it means

We deal with a split of quaternions, motivated by the consistent appearance of two terms in the *quaternion Fourier transform* [184]

$$\mathcal{F}\{f\}(u,v) = \int_{\mathbb{R}^2} e^{-\boldsymbol{i}xu} f(x,y) e^{-\boldsymbol{j}yv} dxdy. \tag{2.8.17}$$

This observation[45] (note that in the following always $\boldsymbol{i}$ is on the left, and $\boldsymbol{j}$ is on the right) and that every quaternion can be rewritten as

$$q = q_r + q_i\boldsymbol{i} + q_j\boldsymbol{j} + q_k\boldsymbol{k} = q_r + q_i\boldsymbol{i} + q_j\boldsymbol{j} + q_k\boldsymbol{i}\boldsymbol{j}, \tag{2.8.18}$$

motivated the quaternion *split*[46] with respect to the pair of orthonormal pure unit quaternions $\boldsymbol{i}, \boldsymbol{j}$

$$q = q_+ + q_-, \quad q_\pm = \frac{1}{2}(q \pm \boldsymbol{i}q\boldsymbol{j}). \tag{2.8.19}$$

Using (2.8.1), the detailed results of this split can be expanded in terms of real components $q_r, q_i, q_j, q_k \in \mathbb{R}$, as

$$q_\pm = \{q_r \pm q_k + \boldsymbol{i}(q_i \mp q_j)\}\frac{1 \pm \boldsymbol{k}}{2} = \frac{1 \pm \boldsymbol{k}}{2}\{q_r \pm q_k + \boldsymbol{j}(q_j \mp q_i)\}. \tag{2.8.20}$$

The analysis of these two components leads to the following Pythagorean *modulus identity* [187].

---

[44]The notion of two-dimensional plane employed here is thus *different* from a two-dimensional plane in $\mathbb{R}^3$. The latter can be characterized by a unit bivector area element of the plane, which corresponds via the isomorphism $Cl(3,0)^+ \cong \mathbb{H}$ to a pure unit quaternion. This difference in interpretation means also that despite of the isomorphism $Cl(3,0)^+ \cong \mathbb{H}$, the notion and expression of rotations in $\mathbb{R}^4$ cannot be automatically carried over to rotation operators of $Cl(3,0)$. Only in the case when rotations are restricted to the three-dimensional subset of pure quaternions, then Hamilton's original $\mathbb{R}^3$ interpretation of these rotations is obvious.

[45]Replacing e.g. $\boldsymbol{i} \to \boldsymbol{j}$, $\boldsymbol{j} \to \boldsymbol{k}$ throughout would merely change the notation, but not the fundamental observation.

[46]Also called OPS as explained below.

Lemma 2.8.2 (Modulus identity). *For $q \in \mathbb{H}$, $|q|^2 = |q_-|^2 + |q_+|^2$.*

Lemma 2.8.3 (Orthogonality of OPS split parts [187]). *Given any two quaternions $p, q \in \mathbb{H}$ and applying the OPS split of (2.8.19) the resulting two parts are* orthogonal, *i.e. $p_+ \perp q_-$ and $p_- \perp q_+$,*

$$\mathrm{S}(p_+\overline{q_-}) = 0, \qquad \mathrm{S}(p_-\overline{q_+}) = 0. \tag{2.8.21}$$

Next, we discuss the map $\boldsymbol{i}(\,)\boldsymbol{j}$, which will lead to an adapted orthogonal basis of $\mathbb{H}$. We observe, that $\boldsymbol{i}q\boldsymbol{j} = q_+ - q_-$, i.e. under the map $\boldsymbol{i}(\,)\boldsymbol{j}$ the q_+ part is *invariant*, but the q_- part *changes sign*. Both parts are *two-dimensional* (2.8.20), and by Lemma 2.8.3 they span *two completely orthogonal planes*, therefore also the name *OPS*. The q_+ plane has the orthogonal quaternion basis $\{\boldsymbol{i}-\boldsymbol{j} = \boldsymbol{i}(1+\boldsymbol{ij}), 1+\boldsymbol{ij} = 1+\boldsymbol{k}\}$, and the q_- plane has orthogonal basis $\{\boldsymbol{i}+\boldsymbol{j} = \boldsymbol{i}(1-\boldsymbol{ij}), 1-\boldsymbol{ij} = 1-\boldsymbol{k}\}$. All four basis quaternions (if normed: $\{q_1, q_2, q_3, q_4\}$)

$$\{\boldsymbol{i}-\boldsymbol{j}, 1+\boldsymbol{ij},\ \boldsymbol{i}+\boldsymbol{j}, 1-\boldsymbol{ij}\}, \tag{2.8.22}$$

form an orthogonal basis of $\mathbb{H}$ interpreted as $\mathbb{R}^4$. Moreover, we obtain the following geometric picture on the left side of Figure 2.6. The map $\boldsymbol{i}()\boldsymbol{j}$ *rotates* the q_- plane by 180° around the two-dimensional q_+ axis plane. This interpretation of the map $\boldsymbol{i}()\boldsymbol{j}$ is in perfect agreement with Coxeter's notion of *half-turn* [84]. In agreement with its geometric interpretation, the map $\boldsymbol{i}(\,)\boldsymbol{j}$ is an *involution*, because applying it twice leads to identity

$$\boldsymbol{i}(\boldsymbol{i}q\boldsymbol{j})\boldsymbol{j} = \boldsymbol{i}^2 q \boldsymbol{j}^2 = (-1)^2 q = q. \tag{2.8.23}$$

We have the important exponential factor identity

$$e^{\alpha \boldsymbol{i}} q_\pm e^{\beta \boldsymbol{j}} = q_\pm e^{(\beta \mp \alpha)\boldsymbol{j}} = e^{(\alpha \mp \beta)\boldsymbol{i}} q_\pm. \tag{2.8.24}$$

This equation should be compared with the kernel construction of the quaternion Fourier transform (QFT). The equation is also often used in our present context for values $\alpha = \pi/2$ or $\beta = \pi/2$.

Finally, we note the interpretation [205] of the QFT integrand $e^{-\boldsymbol{i}x_1\omega_1} h(\boldsymbol{x})\, e^{-\boldsymbol{j}x_2\omega_2}$ as a *local rotation* by phase angle $-(x_1\omega_1 + x_2\omega_2)$ of $h_-(\boldsymbol{x})$ in the two-dimensional q_- plane, spanned by $\{\boldsymbol{i}+\boldsymbol{j}, 1-\boldsymbol{ij}\}$, and a *local rotation* by phase angle $-(x_1\omega_1 - x_2\omega_2)$ of $h_+(\boldsymbol{x})$ in the two-dimensional q_+ plane, spanned by $\{\boldsymbol{i}-\boldsymbol{j}, 1+\boldsymbol{ij}\}$. This concludes the geometric picture of the OPS of $\mathbb{H}$ (interpreted as $\mathbb{R}^4$) with respect to two orthonormal pure quaternion units.

2.8.2.2 Even one pure unit quaternion can do a nice split

Let us now analyze the involution $\boldsymbol{i}(\,)\boldsymbol{i}$. The map $\boldsymbol{i}(\,)\boldsymbol{i}$ gives

$$\boldsymbol{i}q\boldsymbol{i} = \boldsymbol{i}(q_r + q_i\boldsymbol{i} + q_j\boldsymbol{j} + q_k\boldsymbol{k})\boldsymbol{i} = -q_r - q_i\boldsymbol{i} + q_j\boldsymbol{j} + q_k\boldsymbol{k}. \tag{2.8.25}$$

The following OPS with respect to the *single quaternion unit* $\boldsymbol{i}$ gives

$$q_\pm = \frac{1}{2}(q \pm \boldsymbol{i}q\boldsymbol{i}), \quad q_+ = q_j\boldsymbol{j} + q_k\boldsymbol{k} = (q_j + q_k\boldsymbol{i})\boldsymbol{j}, \quad q_- = q_r + q_i\boldsymbol{i}, \tag{2.8.26}$$

where the q_+ plane is two-dimensional and manifestly *orthogonal* to the two-dimensional q_- plane. The basis of the two planes are (if normed: $\{q_1, q_2\}$, $\{q_3, q_4\}$)

$$q_+\text{-basis: } \{\boldsymbol{j}, \boldsymbol{k}\}, \qquad q_-\text{-basis: } \{1, \boldsymbol{i}\}. \tag{2.8.27}$$

The geometric interpretation of $\boldsymbol{i}(\,)\boldsymbol{i}$ as Coxeter *half-turn* is perfectly analogous to the case $\boldsymbol{i}(\,)\boldsymbol{j}$. This form (2.8.26) of the OPS is identical to the quaternionic *simplex/perplex* split applied in quaternionic signal processing, which leads in color image processing to the *luminosity/chrominance* split [127].

2.8.3 General orthogonal two-dimensional planes split (OPS)

Assume in the following an arbitrary pair of pure unit quaternions f, g, $f^2 = g^2 = -1$. The *orthogonal 2D planes split (OPS)* is then defined with respect to *any two pure unit quaternions* f, g as

$$q_\pm = \frac{1}{2}(q \pm fqg) \qquad \Longrightarrow \qquad fqg = q_+ - q_-, \tag{2.8.28}$$

i.e. under the map $f()g$ the q_+ part is invariant, but the q_- part changes sign.

Both parts are two-dimensional, and span two completely orthogonal planes. For $f \neq \pm g$ the q_+ plane is spanned by two orthogonal quaternions $\{f - g, 1 + fg = -f(f-g)\}$, the q_- plane is, e.g., spanned by $\{f+g, 1-fg = -f(f+g)\}$. For $g = f$ a fully *orthonormal* four-dimensional basis of $\mathbb{H}$ is (R acts as rotation operator (rotor))

$$\{1, f, \boldsymbol{j}', \boldsymbol{k}'\} = R^{-1}\{1, \boldsymbol{i}, \boldsymbol{j}, \boldsymbol{k}\}R, \qquad R = \boldsymbol{i}(\boldsymbol{i} + f), \tag{2.8.29}$$

and the two orthogonal two-dimensional planes basis:

$$q_+\text{-basis: } \{\boldsymbol{j}', \boldsymbol{k}'\}, \qquad q_-\text{-basis: } \{1, f\}. \tag{2.8.30}$$

Note the notation for normed vectors in [261] $\{q_1, q_2, q_3, q_4\}$ for the resulting total *orthonormal basis of* $\mathbb{H}$.

Lemma 2.8.4 (Orthogonality of two OPS planes). *Given any two quaternions q, p and applying the OPS with respect to any two pure unit quaternions f, g we get zero for the scalar part of the mixed products*

$$Sc(p_+\widetilde{q}_-) = 0, \qquad Sc(p_-\widetilde{q}_+) = 0. \tag{2.8.31}$$

Note, that the two parts $x_\pm$ can be *represented* as

$$x_\pm = x_{+f}\,\frac{1 \pm fg}{2} + x_{-f}\,\frac{1 \mp fg}{2} = \frac{1 \pm fg}{2}\,x_{+g} + \frac{1 \mp fg}{2}\,x_{-g}, \tag{2.8.32}$$

with commuting and anticommuting parts $x_{\pm f}f = \pm f x_{\pm f}$, etc.

Next we mention the possibility to perform a split along any given set of two (two-dimensional) analysis planes. It has been found, that any two-dimensional plane in $\mathbb{R}^4$ determines in an elementary way an OPS split and vice versa, compare Theorem 3.5 of [205].

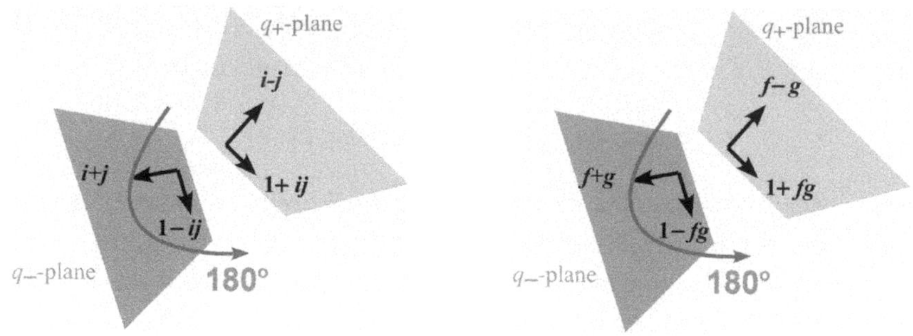

Figure 2.6 Geometric pictures of the involutions $\boldsymbol{i}(\,)\boldsymbol{j}$ and $f()g$ as half turns. Source: [212, Fig. 1].

Let us turn to the geometric interpretation of the map $f()g$. It *rotates* the q_- plane by 180° around the q_+ axis plane. This is in perfect agreement with Coxeter's notion of *half-turn* [84], see the right side of Figure 2.6. The following *identities* hold

$$e^{\alpha f} q_\pm e^{\beta g} = q_\pm e^{(\beta \mp \alpha) g} = e^{(\alpha \mp \beta) f} q_\pm. \tag{2.8.33}$$

This leads to a straightforward geometric interpretation of the integrands of the quaternion Fourier transform (OPS-QFT) with two pure quaternions f, g, and of the orthogonal 2D planes phase rotation Fourier transform [205].

We can further incorporate quaternion conjugation, which consequently provides a geometric interpretation of the QFT involving quaternion conjugation of the signal function. For $d = e^{\alpha g}, t = e^{\beta f}$ the map $d\,\widetilde{(\,)}\,t$ represents a *rotary-reflection* in four dimensions with pointwise *invariant line* $d + t$, a rotary-reflection *axis* $d - t$: $d\,\widetilde{(d - t)}\,t = -(d - t)$, and rotation *angle* $\Gamma = \pi - \arccos \mathrm{S}\big(\widetilde{dt}\big)$ in the plane $\perp \{d + t, d - t\}$. (The derivation of Γ will be shown later.) We obtain the following Lemma.

Lemma 2.8.5. *For OPS* $q_\pm = \frac{1}{2}(q \pm fqg)$, *and left and right exponential factors we have the identity*

$$e^{\alpha g}\, \overline{q_\pm} e^{\beta f} = \overline{q_\pm} e^{(\beta \mp \alpha) f} = e^{(\alpha \mp \beta) g}\, \overline{q_\pm}. \tag{2.8.34}$$

2.8.4 Coxeter on quaternions and reflections

The four-dimensional *angle* Θ between two unit quaternions $p, q \in \mathbb{H}$, $|p| = |q| = 1$, is defined by

$$\cos\Theta = Sc(p\widetilde{q}). \tag{2.8.35}$$

The right and left *Clifford translations* are defined by Coxeter [84] as

$$q \to q' = qa, \quad q \to q'' = aq, \quad a = e^{\widehat{a}\Theta}, \quad \widehat{a}^2 = -1. \tag{2.8.36}$$

Both Clifford translations represent *turns by constant angles* $\Theta_{q,q'} = \Theta_{q,q''} = \Theta$. We analyze the following special cases, assuming the split $q_\pm$ w.r.t. $f = g = \widehat{a}$:

- For $\widehat{a} = \boldsymbol{i}$, $aq_- = q_-a = (q_-a)_-$, is a mathematically *positive* (anti clockwise) *rotation* in the q_- plane $\{1, \boldsymbol{i}\}$.
- Similarly, $aq_+ = (aq_+)_+$, is a mathematically *positive rotation* in the q_+ plane $\{\boldsymbol{j}, \boldsymbol{k}\}$.
- Finally, $q_+a = \tilde{a}q_+ = (q_+a)_+$ is a mathematically *negative rotation* (clockwise) by Θ in the q_+ plane $\{\boldsymbol{j}, \boldsymbol{k}\}$.

Next, we compose Clifford translations, assuming the *split* $q_\pm$ w.r.t. $f = g = \widehat{a}$. For the unit quaternion $a = e^{\widehat{a}\Theta}$, $\widehat{a}^2 = -1$ we find that

$$q \to aqa = a^2 q_- + q_+ \tag{2.8.37}$$

is a *rotation* only in the q_- plane by the *angle* 2Θ, and

$$q \to aq\tilde{a} = q_- + a^2 q_+ \tag{2.8.38}$$

is a *rotation* only in the q_+ plane by the *angle* 2Θ.

Let us now revisit Coxeter's **Lemma 2.2** in [84]: For any two quaternions $a, b \in \mathbb{H}$, $|a| = |b|$, $a_r = b_r$, we can find a $y \in \mathbb{H}$ such that

$$ay = yb. \tag{2.8.39}$$

We now further ask for the set of *all* $y \in \mathbb{H}$ such that $ay = yb$? Based on the OPS, the answer is straightforward. For $a = |a|e^{\Theta\hat{a}}$, $b = |a|e^{\Phi\hat{b}}$, $\Phi = \pm\Theta$, $\hat{a}^2 = \hat{b}^2 = -1$ we use the split $q_\pm = \frac{1}{2}(q \pm \hat{a}q\hat{b})$ to obtain:

- For $\Theta = \Phi$: The set of all y spans the q_- plane. Moreover,

$$\begin{aligned} &aq_+b = q_+, \quad q_+b = \tilde{a}q_+, \quad aq_+ = q_+\tilde{b}, \\ &aq_-b = a^2q_- = q_-b^2, aq_- = q_-b. \end{aligned} \tag{2.8.40}$$

- For $\Theta = -\Phi$: The set of all y spans the q_+ plane. Moreover,

$$\begin{aligned} &aq_-b = q_-, \quad q_-b = \tilde{a}q_-, \quad aq_- = q_-\tilde{b}, \\ &aq_+b = a^2q_+ = q_+b^2, aq_+ = q_+b. \end{aligned} \tag{2.8.41}$$

Let us turn to a reflection in a hyperplane. **Theorem 5.1** in [84] says: The reflection in the hyperplane $\perp a \in \mathbb{H}$: $Sc(a\tilde{q}) = 0$, $|a|^2 = 1$, $a = |a|e^{\Theta\hat{a}}$, $\hat{a}^2 = -1$, is represented by

$$q \to -a\tilde{q}a. \tag{2.8.42}$$

We analyze the situation using the OPS. We define the split $q_\pm = \frac{1}{2}(q \pm \widehat{a}q\widehat{a})$ to obtain

$$q_+ \to -a\widetilde{q_+}a = q_+, \qquad a \to -a\tilde{a}a = -a. \tag{2.8.43}$$

and for $a' = ae^{-\frac{\pi}{2}\hat{a}}$

$$a' \to -a\widetilde{a'}a = a'. \tag{2.8.44}$$

We further consider a general rotation. **Theorem 5.2** in [84] states: The general rotation through 2Φ (about a plane) is $q \to aqb$, $a = |a|e^{\Phi\hat{a}}$, $b = |a|e^{\Theta\hat{b}}$, $\Phi = \pm\Theta$, $\hat{a}^2 = \hat{b}^2 = -1$.

We again apply the OPS. We define the split $q_{\pm} = \frac{1}{2}(q \pm \hat{a}q\hat{b})$ to obtain:

- For $\Theta = \Phi$: Rotation of q_- plane by 2Φ around q_+-plane.
- For $\Theta = -\Phi$: Rotation of q_+ plane by 2Φ around q_--plane.

Let us illustrate this with an *example*: $\hat{a} = \hat{b} = \boldsymbol{i}$, $\Phi = -\Theta$,

$$aq_-b = q_-. \tag{2.8.45}$$

For $q_+ = \boldsymbol{j}$:

$$aq_+b = a\boldsymbol{j}b = \boldsymbol{j}b^2 = \boldsymbol{j}e^{-2\theta\boldsymbol{i}} = \boldsymbol{j}\cos 2\Phi - \boldsymbol{k}\sin 2\Phi, \tag{2.8.46}$$

a rotation in the q_+-plane around the q_- plane. Note, that the detailed analysis of general $q \to aqb$, $q_{\pm} = \frac{1}{2}(q \pm \hat{a}q\hat{b})$, $|a| = |b| = 1$, can be found in [205].

As for the *rotary inversion*, we follow the discussion in [205], sec. 5.1, but add a simple formula for determining the rotation angle. The rotary inversion is given by, $d, t \in \mathbb{H}$, $|d| = |t| = 1$, $q \to d\tilde{q}t$. For $d \neq \pm t$, $[d,t] = dt - td$, we obtain two vectors in the rotation plane $v_{1,2} = [d,t](1 \pm \tilde{d}t)$, with $d\widetilde{v_{1,2}}t = -v_{1,2}\tilde{d}t$. The angle Γ of rotation can therefore be simply found from

$$\cos\Gamma = Sc(\frac{1}{|v_1|^2}\tilde{v_1}d\tilde{v_1}t) = Sc(-\tilde{d}t) = \cos(\pi - \gamma), \tag{2.8.47}$$

with γ the angle between d and t : $\cos\gamma = \tilde{d}t$.

2.8.5 Quaternion geometry of rotations analyzed by 2D OPS

According to [261] the circle $C(q_1, q_2)$ of all unit quaternions,[47] which rotate $g \to f$ $f \neq \pm g$ is given by

$$\begin{aligned} q(t) &= \frac{1 - fg}{|1 - fg|}e^{\frac{t}{2}g} = q_1 e^{\frac{t}{2}g}, \quad t \in [0, 2\pi), \\ q(t)g\widetilde{q(t)} &= f, \quad q_2 = \frac{f+g}{|f+g|}. \end{aligned} \tag{2.8.48}$$

The two-dimensional OPS $q_{\pm}^{f,g} = \frac{1}{2}(q \pm fqg)$ tells us, that all $q(t), t \in [\,0, 2\pi\,)$ are elements of the q_- plane. And indeed, $fq_-g = -q_-$ for all $q_- \in \mathbb{H}$ leads to

$$f = q_- g q_-^{-1}, \tag{2.8.49}$$

[47] Note that apart from the shortest (and longest) rotation(s) along a geodesic connecting two end points of two unit vectors attached to the center of a unit sphere, any circle on the unit sphere, which includes the end points gives another trajectory of (two) rotation(s) between the two endpoints with rotation axis through the center of the unit sphere and the center of the circle on the unit sphere.

for all q_- in the q_--plane. Note, that this is valid for all $f, g \in \mathbb{H}$, $f^2 = g^2 = -1$, even for $f = \pm g$! We therefore get a *one line proof*, which at the same time generalizes from the unit circle to the whole plane.

Meister and Schaeben [261] state that for $q \in C(q_1, q_2)$: $fq, qg, fqg \in C(q_1, q_2)$. This can easily be generalized to the whole q_--plane, because

$$(fq_-)_- = fq_-, \quad (q_-g)_- = q_-g, \quad (fq_-g)_- = fq_-g. \tag{2.8.50}$$

We can use the exponential form, and show that the circle $C(q_1, q_2)$ parametrization of (34), (35) in [261] is a *specialization of the general relation*

$$e^{\frac{t}{2}f}q_- = q_-e^{\frac{t}{2}g} \tag{2.8.51}$$

which means that the two parametrizations are element wise identical.

Now we look at the quaternion circles for the rotations $g \to \pm f$. **Prop. 5** of [261] states: Two circles $C(q_1, q_2) = G(g, f)$ and $C(q_3, q_4) = G(g, -f) = G(-g, f)$, representing all rotations $g \to f$ and $g \to -f$, respectively, are orthonormal to each other. Here four orthogonal unit quaternions are defined as:

$$q_1 = \frac{1 - fg}{|1 - fg|}, \quad q_2 = \frac{f + g}{|f + g|}, \quad q_3 = \frac{1 + fg}{|1 + fg|}, \quad q_4 = \frac{f - g}{|f - g|}. \tag{2.8.52}$$

We provide a simple proof: We already know that all q_1, q_2 span the q_- plane of the split $q_{\pm}^{f,g} = \frac{1}{2}(q \pm fqg)$, and all q_3, q_4 span the q_+-plane. And that

$$fq_{\pm}g = \pm q_{\pm} \quad \iff \quad f = q_{\pm}(\mp g)q_{\pm}^{-1}, \qquad \text{QED.} \tag{2.8.53}$$

Note, the proof is again much faster than in [261]. We see that $G(g, f) = \{q_-^{f,g}/|q_-^{f,g}|, \forall q \in \mathbb{H}\}$, and $G(g, -f) = \{q_+^{f,g}/|q_+^{f,g}|, \forall q \in \mathbb{H}\}$.

For later use, we translate the notation of [261] (38),(39):

$$n_3 = -n_1 = \frac{[f, g]}{|[f, g]|}, \quad n_4 = q_4, \quad n_2 = n_4 n_1, \quad n_4 = n_1 n_2, \quad n_1 = n_2 n_4, \tag{2.8.54}$$

which shows that $\{n_1, n_2, n_4\}$ is a right handed set of three orthonormal pure quaternions, obtained by rotating $\{\boldsymbol{i}, \boldsymbol{j}, \boldsymbol{k}\}$.

The two circles $G(g, f), G(g', f)$ do not intersect for $g \neq g'$, see Cor. 1(i) of [261]. We provide a simple proof: Assume $\exists_1 q \in H : fqg = -q, fqg' = -q$ for $g \neq g'$. Then

$$fqg = fqg' \quad \Leftrightarrow \quad g = g' \quad \Rightarrow \quad G(g, f) \bigcap G(g', f) = \emptyset. \tag{2.8.55}$$

QED.

Cor. 1 (iii) of [261] further states that for every 3D rotation R and given g_0, $g_0^2 = -1$ we can always find f, $f^2 = -1$, such that R is represented by a (unit) quaternion q in $G(g_0, f)$. We can equivalently ask for f, such that q representing the rotation R is $\in q_-^{f,g_0}$-plane. We find

$$fqg_0 = -q \quad \Leftrightarrow \quad f = qg_0q^{-1}. \tag{2.8.56}$$

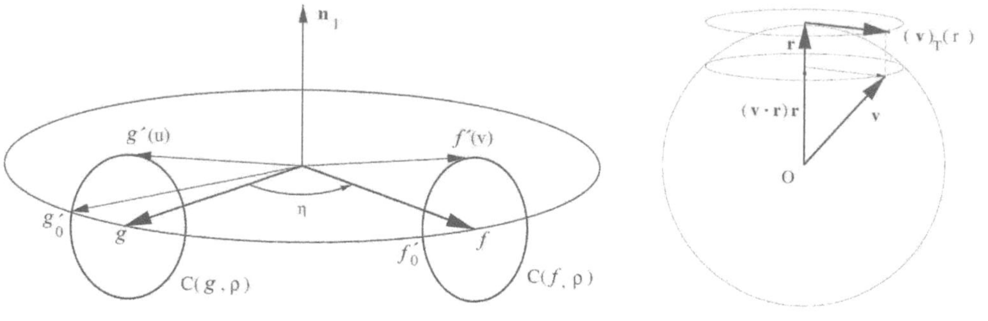

Figure 2.7 Small circles and tangential plane projection. Adapted from Figs. 2 and 3 of [261]. Source: [212, Fig. 2].

The left side of Figure 2.7 shows two small circles $C(g,\rho), C(f,\rho) \subset S^2$ [261]. We now analyze the mapping between pairs of small circles. A small circle with center g and radius ρ is defined as $C(g,\rho) = \{g' \in \S^2 : g \cdot g' = \cos\rho\}$, and all $q \in q_-^{f,g}$-plane map $C(g,\rho)$ to the small circle $C(f,\rho)$ of the same radius (a slight generalization of [261], Prop. 6), with the correspondence

$$
\begin{aligned}
q(t)g'(u)q(t)^{-1} &= f'(u+2t), \\
q(t) = q_1 e^{tg}, \quad g'(u) = e^{\frac{u}{2}g} g'_0 e^{-\frac{u}{2}g}, &\quad f'(u) = e^{\frac{u}{2}f} f'_0 e^{-\frac{u}{2}f}
\end{aligned} \tag{2.8.57}
$$

starting with the corresponding circle points $q_1 g'_0 = f'_0 q_1$.

We provide the following direct proof: We repeatedly apply (2.8.33) to obtain

$$
\begin{aligned}
q_1 g'_0 = f'_0 q_1 &\Leftrightarrow e^{\frac{u}{2}f} q_1 g'_0 e^{-\frac{u}{2}g} = e^{\frac{u}{2}f} f'_0 q_1 e^{-\frac{u}{2}g} \\
&\Leftrightarrow q_1 e^{\frac{u}{2}g} g'_0 e^{-\frac{u}{2}g} = e^{\frac{u}{2}f} f'_0 e^{-\frac{u}{2}f} q_1 \\
&\Leftrightarrow e^{tf} q_1 e^{\frac{u}{2}g} g'_0 e^{-\frac{u}{2}g} = e^{tf} e^{\frac{u}{2}f} f'_0 e^{-\frac{u}{2}f} e^{-tf} e^{tf} q_1 \\
&\Leftrightarrow q_1 e^{tg}\, e^{\frac{u}{2}g} g'_0 e^{-\frac{u}{2}g} = e^{tf} e^{\frac{u}{2}f} f'_0 e^{-\frac{u}{2}f} e^{-tf}\, q_1 e^{tg}.
\end{aligned} \tag{2.8.58}
$$

QED.

Note, that this proof is much shorter than in [261], and we do not need to use addition theorems.

We consider the *projection onto the tangential plane* of a S^2 vector, see the right side of Figure 2.7. Assume $v, r \in S^2$. Note, that $(v)_T(r) = v - (v \cdot r)r$ of [261] can be simplified to $(v)_T(r) = V(vr)r^{-1}$, valid for all pure (non-unit) quaternions r.

Finally we consider the *torus theorem* for all maps $g \to$ small circle $C(f, 2\Theta)$. We slightly reformulate the theorem **Prop. 13** of [261]. We will use the two-dimensional OPS with respect to $f, g \in S^2$, and the corresponding orthonormal basis $\{q_1, q_2, q_3, q_4\}$ of (2.8.52). The theorem says, that the circle $C(q_1, q_2) \in q_-$-plane: $q_-(s) = q_1 \exp(sg/2)$, $s \in [0, 2\pi)$, represents all rotations $g \to f$, while the orthogonal circle $C(q_3, q_4) \in q_+$-plane: $q_+(t) = q_3 \exp(-tg/2)$, $t \in [0, 2\pi)$, represents all rotations $g \to -f$. Then the *spherical torus* $T(q_-(s), q_+(t); \Theta)$ is defined as the set of quaternions

$$q(s,t;\Theta) = q_-(2s)\cos\Theta + q_+(2t)\sin\Theta, \quad s,t \in [0,2\pi), \quad \Theta \in [0,\pi/2], \tag{2.8.59}$$

and represents all rotations $g \to C(f, 2\Theta) \subset S^2$.

In particular, the set $q(s,-s;\Theta)$ maps g for all $s \in [0,2\pi)$ onto f'_0 in the f,g plane with $g \cdot f'_0 = \cos(\eta - 2\Theta)$, $g \cdot f = \cos\eta$,

$$q(s,-s;\Theta)gq(s,-s;\Theta)^{-1} = f'_0 \qquad \forall s \in [0,2\pi). \tag{2.8.60}$$

Moreover, for arbitrary $s_0 \in [0,2\pi)$, the set $q(s_0, t-s_0;\Theta)$ (or equivalently $q(s_0 + t, s_0;\Theta))$ maps $g \to f' \in C(f,2\Theta)$, which results from positive rotation (counter-clockwise) of f'_0 about f by the angle $t \in [0,2\pi)$,

$$q(s_0,t-s_0;\Theta)gq(s_0,t-s_0;\Theta)^{-1} = e^{\frac{t}{2}f} f'_0 e^{-\frac{t}{2}f} \qquad \forall s_0 \in [0,2\pi). \tag{2.8.61}$$

We state the following *direct proof* of the torus theorem.

$$\begin{aligned}
q(s,t;\Theta) &= q_-(2s)\cos\Theta + q_+(2t)\sin\Theta = q_1 e^{sg}\cos\Theta + q_3 e^{-tg}\sin\Theta \\
&= (q_1\cos\Theta + q_3 e^{-(t+s)g}\sin\Theta)e^{sg} \\
&= (\cos\Theta + (q_3/q_1)e^{-(t+s)f}\sin\Theta)q_1 e^{sg} \\
&= (\cos\Theta + (-n_1)e^{-(t+s)f}\sin\Theta)q_1 e^{sg} \\
&= (\cos\Theta + e^{(t+s)f}(-n_1)\sin\Theta)e^{sf} q_1 \\
&= e^{-n'_1\Theta}e^{sf}q_1, \qquad n'_1 = e^{(t+s)f}(-n_1), \quad (n'_1)^2 = -1.
\end{aligned} \tag{2.8.62}$$

We observe, that n'_1 is n_1 rotated around f by angle $s+t$. Application to g gives

$$q(s,t;\Theta)g\; q(s,t;\Theta)^{-1} = e^{-n'_1\Theta} f e^{n'_1\Theta}, \tag{2.8.63}$$

so geometrically g is rotated into $f = e^{sf}q_1 g\, q_1^{-1}e^{-sf}$, which in turn is rotated around n'_1 on the circle $C(f,2\Theta)$. For $t=-s$ obviously

$$q(s,-s;\Theta)gq(s,-s;\Theta)^{-1} = e^{-n_1\Theta} f e^{n_1\Theta} = f'_0, \tag{2.8.64}$$

is a rotation in the f,g plane of g into f'_0, with $g \cdot f'_0 = \cos\eta - 2\Theta$. We further note, that for $s = s_0$, $t \to t - s_0$: $n'_1 = e^{tf}(-n_1)$, such that

$$q(s_0,t-s_0;\Theta)gq(s_0,t-s_0;\Theta)^{-1} = e^{-n'_1\Theta} f e^{n'_1\Theta} = e^{tf} f'_0 e^{-tf}, \tag{2.8.65}$$

describes the small circle $C(f,2\Theta)$. QED.

Our proof is very *compact*, obtained by *direct* computation of *monomial* results, which in turn permit direct *geometric interpretation.*

2.8.6 Summary on quaternions and rotation geometry

We have exposed the geometric understanding of the *general OPS split of quaternions* [205] into two *orthogonal planes* ($\mathbb{R}^4$ interpretation). Moreover, we have consolidated the OPS with the geometric understanding by Altmann [12, 322], Coxeter [84], and Meister and Schaeben [261].

CHAPTER 3

Geometric calculus

> Now faith is being sure of what we hope for and certain of what we do not see. This is what the ancients were commended for. By faith we understand that the universe was formed at God's command, so that what is seen was not made out of what was visible. [157]

The German 19th century mathematician H. Grassmann had the clear vision that his

> ... extension theory (now developed to geometric calculus) ... forms the keystone of the entire structure of mathematics. [146]

We introduce geometric vector differential calculus. This chapter is mainly based on [174]. We treat the fundamentals of the *vector differential calculus* part of *universal geometric calculus.* Geometric calculus is seen to simplify and unify the structure and notation of mathematics for all of science and engineering, and for technological applications. In order to make the treatment reasonably self-contained, we also note the most important *geometric algebra* relationships, which are necessary for vector differential calculus. Readers, who need more than this summary are advised to turn back to Chapter 2, or for a comprehensive less technical introduction to Clifford's geometric algebras we refer to [200]. We further recommend the two excellent college level textbooks [248, 249]. After that *differentiation by vectors* is introduced and a host of major vector differential and vector derivative relationships is proven explicitly in a very elementary step by step approach. The chapter is thus intended to serve as reference material, giving details, which are usually skipped in more advanced discussions of the subject matter.

DOI: 10.1201/9781003184478-3

3.1 INTRODUCTORY NOTES ON VECTOR DIFFERENTIAL CALCULUS

The algebraic *grammar* of the universal form of (geometric) calculus, as envisioned by H. Grassmman, is *geometric algebra* (or Clifford algebra).

A brief overview of vector differential calculus is given in Section 3.2. Next, the basic geometric algebra necessary for this is summarized for notation and reference in Section 3.3. Finally, Section 3.4 develops vector differential calculus with the help of few simple definitions. This approach is generically coordinate free, and fully shows both the concrete and abstract geometric and algebraic beauty of Grassmann's "keystone" of mathematics. Applications of geometric algebra are surveyed in [202].

We demonstrate the proofs for all common formulas of vector differential calculus in an elementary step by step fashion. Where other texts (e.g. [162],[161]) tend both to skip "elementary steps," and to presume, that the reader would be smart enough to fill in the gaps himself. I put the emphasis therefore on thorough proofs and not on comments, interpretations or application. Following the same approach as taken here, the step from vector to multivector differential calculus is taken in [175].

3.2 BRIEF OVERVIEW OF VECTOR DIFFERENTIAL CALCULUS

We start with a brief overview without the proofs (that are given in the following sections). Note that we use in the current section the principal reverse of Section 2.1.1.6.

Multivector valued functions $f : \mathbb{R}^{p,q} \to Cl_{p,q}$, $p+q=n$, have 2^n blade components ($f_A : \mathbb{R}^{p,q} \to \mathbb{R}$)

$$f(\boldsymbol{x}) = \sum_A f_A(\boldsymbol{x})\boldsymbol{e}_A. \tag{3.2.1}$$

We define the *inner product* of two $\mathbb{R}^{p,q} \to Cl_{p,q}$ functions f, g by

$$(f,g) = \int_{\mathbb{R}^{p,q}} f(\boldsymbol{x})\widetilde{g(\boldsymbol{x})}\, d^n\boldsymbol{x} = \sum_{A,B} \boldsymbol{e}_A\widetilde{\boldsymbol{e}_B} \int_{\mathbb{R}^{p,q}} f_A(\boldsymbol{x})g_B(\boldsymbol{x})\, d^n\boldsymbol{x}. \tag{3.2.2}$$

In (3.2.2), the inner product $(\ ,\ )_{L^2(\mathbb{R}^{p,q};Cl_{p,q})}$ satisfies the following conditions[33]

$$\begin{aligned}
(f, g+h)_{L^2(\mathbb{R}^{p,q};Cl_{p,q})} &= (f,g)_{L^2(\mathbb{R}^{p,q};Cl_{p,q})} + (f,h)_{L^2(\mathbb{R}^{p,q};Cl_{p,q})},\\
(f, \lambda g)_{L^2(\mathbb{R}^{p,q};Cl_{p,q})} &= (f,g)_{L^2(\mathbb{R}^{p,q};Cl_{p,q})}\tilde{\lambda},\\
(f\lambda, g)_{L^2(\mathbb{R}^{p,q};Cl_{p,q})} &= (f, g\tilde{\lambda})_{L^2(\mathbb{R}^{p,q};Cl_{p,q})},\\
(f, g)_{L^2(\mathbb{R}^{p,q};Cl_{p,q})} &= \widetilde{(g,f)}_{L^2(\mathbb{R}^{p,q};Cl_{p,q})}.
\end{aligned} \tag{3.2.3}$$

where $f, g \in L^2(\mathbb{R}^{p,q}; Cl_{p,q})$, and the constant multivector $\lambda \in Cl_{p,q}$. The *symmetric scalar part* is

$$\langle f,g\rangle = \int_{\mathbb{R}^{p,q}} f(\boldsymbol{x}) * \widetilde{g(\boldsymbol{x})}\, d^n\boldsymbol{x} = \sum_A \int_{\mathbb{R}^{p,q}} f_A(\boldsymbol{x})g_A(\boldsymbol{x})\, d^n\boldsymbol{x}, \tag{3.2.4}$$

and the *norm* for functions in $L^2(\mathbb{R}^n; Cl_{n,0}) = \{f : \mathbb{R}^n \to Cl_{n,0} \mid \|f\| < \infty\}$ as

$$\|f\|^2 = \langle (f, f) \rangle = \int_{\mathbb{R}^n} |f(\boldsymbol{x})|^2 d^n\boldsymbol{x}.$$
$$= \int_{\mathbb{R}^n} f(\boldsymbol{x}) * \tilde{f}(\boldsymbol{x}) d^3\boldsymbol{x} \overset{(5.2.148)}{=} \int_{\mathbb{R}^n} \sum_A f_A^2(\boldsymbol{x}) d^3\boldsymbol{x}. \tag{3.2.5}$$

Example 3.2.1. *For* $g = \boldsymbol{a}f$, $f, g \in L^2(\mathbb{R}^3; Cl_{3,0})$, $\boldsymbol{a} \in \mathbb{R}^3$ *we get because of* $\langle \boldsymbol{a}f\widetilde{\boldsymbol{a}f}\rangle_0 = \langle \boldsymbol{a}f\tilde{f}\boldsymbol{a}\rangle_0 = \langle \boldsymbol{a}^2 f\tilde{f}\rangle_0 = \boldsymbol{a}^2 f * \tilde{f}$

$$\|\boldsymbol{a}f\|^2_{L^2(\mathbb{R}^3; Cl_{3,0})} = \int_{\mathbb{R}^3} \boldsymbol{a}^2 f(\boldsymbol{x}) * \tilde{f}(\boldsymbol{x}) d^3\boldsymbol{x} = \int_{\mathbb{R}^3} \boldsymbol{a}^2 \sum_A f_A^2(\boldsymbol{x}) d^3\boldsymbol{x}. \tag{3.2.6}$$

Definition 3.2.2 (Clifford module). *Let* $Cl_{p,q}$ *be the real Clifford algebra of the quadratic space* $\mathbb{R}^{p,q}$. *A Clifford algebra module* $L^2(\mathbb{R}^{p,q}; Cl_{p,q})$ *is defined by*

$$L^2(\mathbb{R}^{p,q}; Cl_{p,q}) = \{f : \mathbb{R}^{p,q} \longrightarrow Cl_{p,q} \mid \|f\|_{L^2(\mathbb{R}^{p,q}; Cl_{p,q})} < \infty\}. \tag{3.2.7}$$

An alternative definition of Clifford module uses the principal reverse $\widetilde{N}$ of a multivector $N \in Cl(p,q)$ of Section 2.1.1.6. For $M, N \in Cl(p,q)$ we get $M * \widetilde{N} = \sum_A M_A N_A$. Two multivectors $M, N \in Cl(p,q)$ are *orthogonal* if and only if $M * \widetilde{N} = 0$. The modulus $|M|$ of a multivector $M \in Cl(p,q)$ is defined as

$$|M|^2 = M * \widetilde{M} = \sum_A M_A^2. \tag{3.2.8}$$

The possibility to differentiate with respect to any multivector (representing a geometric object or a transformation operator) is essential for solving *optimization* problems in GA.

We can then define the *vector differential* [161,174] of f for any constant $\boldsymbol{a} \in \mathbb{R}^{p,q}$ as

$$\boldsymbol{a} \cdot \nabla f(\boldsymbol{x}) = \lim_{\epsilon \to 0} \frac{f(\boldsymbol{x} + \epsilon \boldsymbol{a}) - f(\boldsymbol{x})}{\epsilon}, \tag{3.2.9}$$

where $\boldsymbol{a} \cdot \nabla$ is scalar. The *vector* derivative ∇ can be expanded in terms of the basis vectors $\boldsymbol{e}_k$ as

$$\nabla = \nabla_{\boldsymbol{x}} = \sum_{k=1}^{n} \boldsymbol{e}_k \partial_k, \qquad \partial_k = \boldsymbol{e}_k \cdot \nabla = \frac{\partial}{\partial x_k}. \tag{3.2.10}$$

Both $\boldsymbol{a} \cdot \nabla$ and ∇ are *coordinate independent.* Replacing the vectors $\boldsymbol{a}, \boldsymbol{x}$ by multivectors gives the *multivector derivative* [161,175].

Examples. The five multivector functions

$$f_1 = \boldsymbol{x}, \qquad f_2 = \boldsymbol{x}^2, \qquad f_3 = |\boldsymbol{x}|, \qquad f_4 = \boldsymbol{x} \cdot \langle A \rangle_k, \quad 0 \le k \le n,$$
$$f_5 = \log r, \quad \boldsymbol{r} = \boldsymbol{x} - \boldsymbol{x}_0, \quad r = |\boldsymbol{r}|, \tag{3.2.11}$$

have the following vector differentials [174]

$$\boldsymbol{a}\cdot\nabla f_1 = \boldsymbol{a},\ \boldsymbol{a}\cdot\nabla f_2 = 2\boldsymbol{a}\cdot\boldsymbol{x},\ \boldsymbol{a}\cdot\nabla f_3 = \frac{\boldsymbol{a}\cdot\boldsymbol{x}}{|\boldsymbol{x}|},$$
$$\boldsymbol{a}\cdot\nabla f_4 = \boldsymbol{a}\cdot\langle A\rangle_k, \quad \boldsymbol{a}\cdot\nabla f_5 = \frac{\boldsymbol{a}\cdot\boldsymbol{r}}{r^2}. \tag{3.2.12}$$

This leads to the vector derivatives [174]

$$\nabla f_1 = \nabla\cdot\boldsymbol{x} = 3,\ (n=3),\ \nabla f_2 = 2\boldsymbol{x},\ \nabla f_3 = \boldsymbol{x}/|\boldsymbol{x}|,$$
$$\nabla f_4 = k\langle A\rangle_k, \qquad \nabla f_5 = \boldsymbol{r}^{-1} = \boldsymbol{r}/r^2. \tag{3.2.13}$$

We can compute the *derivative from the differential* for $\nabla_{\boldsymbol{a}}$: regard $\boldsymbol{x}$ = constant, $\boldsymbol{a}$ = variable, and compute

$$\nabla f = \nabla_{\boldsymbol{a}}(\boldsymbol{a}\cdot\nabla f) \tag{3.2.14}$$

The vector derivative obeys *sum rules* and *product rules*ixvector derivative!product rule (overdots indicate functions to be differentiated). But non-commutativity leads to modifications, because $\dot\nabla f\dot g \neq f\dot\nabla\dot g$. The *product rule* is

$$\nabla(fg) = (\dot\nabla\dot f)g + \dot\nabla f\dot g = (\dot\nabla\dot f)g + \sum_{k=1}^{n} \boldsymbol{e}_k f(\partial_k g). \tag{3.2.15}$$

The *chain rules* for the vector differential and the vector derivative of $f(\boldsymbol{x}) = g(\lambda(\boldsymbol{x}))$, $\lambda(\boldsymbol{x}) \in \mathbb{R}$, are

$$\boldsymbol{a}\cdot\nabla f = \{\boldsymbol{a}\cdot\nabla\lambda(\boldsymbol{x})\}\frac{\partial g}{\partial\lambda}, \qquad \nabla f = (\nabla\lambda)\frac{\partial g}{\partial\lambda}. \tag{3.2.16}$$

Example: For $\boldsymbol{a} = \boldsymbol{e}_k$ $(1 \le k \le n)$ we have

$$\boldsymbol{e}_k\cdot\nabla f = \partial_k f = (\partial_k\lambda)\frac{\partial g}{\partial\lambda}. \tag{3.2.17}$$

GA thus provides a new formalism for *differentiation on vector manifolds* and for *mappings* between surfaces, including *conformal mapping* [304].

3.3 GEOMETRIC ALGEBRA FOR DIFFERENTIAL CALCULUS

This section is a basic summary (without proofs) of important relationships in geometric algebra. For further details we refer to Chapter 2. This summary mainly serves as a reference section (including notation) for the *vector differential calculus* to be developed. Most of the relationships below can also be found in the synopsis of geometric algebra and in chapters 1 and 2 of [162], as well as in chapter 1 of [161], together with relevant proofs. Beyond that [162] and [161] follow a much more didactic approach for complete newcomers to geometric algebra.

$\mathcal{G}(I)$ is the full *geometric algebra* over all vectors in the n-dimensional unit *pseudoscalar* $I = \vec{e}_1 \wedge \vec{e}_2 \wedge \ldots \wedge \vec{e}_n$. $\mathcal{A}_n \equiv \mathcal{G}^1(I)$ is the n-dimensional vector sub-space of grade-1 elements in $\mathcal{G}(I)$ spanned by $\vec{e}_1, \vec{e}_2, \ldots, \vec{e}_n$. For vectors $\vec{a}, \vec{b}, \vec{c} \in \mathcal{A}_n \equiv \mathcal{G}^1(I)$ and scalars $\alpha, \beta, \lambda, \tau$; $\mathcal{G}(I)$ has the fundamental properties of

- Associativity

$$\vec{a}(\vec{b}\vec{c}) = \vec{a}(\vec{b}\vec{c}), \qquad \vec{a}+\vec{b}+\vec{c} = (\vec{a}+\vec{b})+\vec{c}, \tag{3.3.1}$$

- Commutativity

$$\alpha\vec{a} = \vec{a}\alpha, \qquad \vec{a}+\vec{b} = \vec{b}+\vec{a}, \tag{3.3.2}$$

- Distributivity

$$\vec{a}(\vec{b}+\vec{c}) = \vec{a}\vec{b}+\vec{a}\vec{c}, \qquad (\vec{b}+\vec{c})\vec{a} = \vec{b}\vec{a}+\vec{c}\vec{a}, \tag{3.3.3}$$

- Linearity

$$\alpha(\vec{a}+\vec{b}) = \alpha\vec{a}+\alpha\vec{b} = (\vec{a}+\vec{b})\alpha, \tag{3.3.4}$$

- Scalar square (vector length $|\vec{a}|$)

$$\vec{a}^2 = \vec{a}\vec{a} = \vec{a}\cdot\vec{a} = |\vec{a}|^2. \tag{3.3.5}$$

The *geometric product* $\vec{a}\vec{b}$ is related to the (scalar) *inner product* $\vec{a}\cdot\vec{b}$ and to the (bivector or 2-vector) *outer product* $\vec{a}\wedge\vec{b}$ by

$$\vec{a}\vec{b} = \vec{a}\cdot\vec{b}+\vec{a}\wedge\vec{b}, \tag{3.3.6}$$

with

$$\vec{a}\cdot\vec{b} = \frac{1}{2}(\vec{a}\vec{b}+\vec{b}\vec{a}) = \vec{a}\vec{b}-\vec{a}\wedge\vec{b} = \vec{b}\cdot\vec{a} \overset{(3.3.13)}{=} \langle\vec{a}\vec{b}\rangle_0, \tag{3.3.7}$$

$$\vec{a}\wedge\vec{b} = \frac{1}{2}(\vec{a}\vec{b}-\vec{b}\vec{a}) = -\vec{b}\wedge\vec{a} = \vec{a}\vec{b}-\vec{a}\cdot\vec{b} \overset{(3.3.13)}{=} \langle\vec{a}\vec{b}\rangle_2. \tag{3.3.8}$$

The inner and the outer product are both linear and distributive

$$\vec{a}\cdot(\alpha\vec{b}+\beta\vec{c}) = \alpha\vec{a}\cdot\vec{b}+\beta\vec{a}\cdot\vec{c}, \tag{3.3.9}$$

$$\vec{a}\wedge(\alpha\vec{b}+\beta\vec{c}) = \alpha\vec{a}\wedge\vec{b}+\beta\vec{a}\wedge\vec{c}. \tag{3.3.10}$$

A unit vector $\hat{a}$ in the direction of $\vec{a}$ is

$$\hat{a} = \frac{\vec{a}}{|\vec{a}|} \quad \text{with} \quad \hat{a}^2 = \hat{a}\hat{a} = 1, \qquad \vec{a} = \hat{a}|\vec{a}|. \tag{3.3.11}$$

The inverse of a vector is

$$\vec{a}^{-1} = \frac{1}{\vec{a}} = \frac{\vec{a}}{\vec{a}^2} = \frac{\hat{a}}{|\vec{a}|}. \tag{3.3.12}$$

A multivector A can be uniquely decomposed into its homogeneous grade k parts ($\langle\ \rangle_k$ grade k selector):

$$A = \underbrace{\langle A\rangle_0}_{\text{scalar}} + \underbrace{\langle A\rangle_1}_{\text{vector}} + \underbrace{\langle A\rangle_2}_{\text{bivector}} + \ldots + \underbrace{\langle A\rangle_k}_{k\text{-vector}} + \ldots + \underbrace{\langle A\rangle_n}_{\text{pseudoscalar}} \tag{3.3.13}$$

If A is homogeneous of grade k one often simply writes

$$A = \langle A \rangle_k = A_k. \tag{3.3.14}$$

Grade selection is invariant under scalar multiplication

$$\lambda \langle A \rangle_k = \langle \lambda A \rangle_k. \tag{3.3.15}$$

The consistent definition of inner and outer products of vectors $\vec{a}$ and r-vectors A_r is

$$\vec{a} \cdot A_r \equiv \langle \vec{a} A_r \rangle_{r-1} = \frac{1}{2}(\vec{a} A_r - (-1)^r A_r \vec{a}) \tag{3.3.16}$$

$$\vec{a} \wedge A_r \equiv \langle \vec{a} A_r \rangle_{r+1} = \frac{1}{2}(\vec{a} A_r + (-1)^r A_r \vec{a}). \tag{3.3.17}$$

By linearity the full geometric product of a vector and a multivector A is then

$$\vec{a} A = \vec{a} \cdot A + \vec{a} \wedge A. \tag{3.3.18}$$

This extends to the distributive multiplication with arbitrary multivectors A and B

$$\vec{a}(A + B) = \vec{a} A + \vec{a} B. \tag{3.3.19}$$

The inner and outer products of homogeneous multivectors A_r and B_s are defined ([161], p. 5, (1.21), (1.22)) as

$$A_r \cdot B_s \equiv \langle A_r B_s \rangle_{|r-s|} \text{ for } r, s > 0, \tag{3.3.20}$$

$$A_r \cdot B_s \equiv 0 \text{ for } r = 0 \text{ or } s = 0, \tag{3.3.21}$$

$$A_r \wedge B_s \equiv \langle A_r B_s \rangle_{r+s}, \tag{3.3.22}$$

$$A_r \wedge \lambda = \lambda \wedge A_r = \lambda A_r \text{ for scalar } \lambda. \tag{3.3.23}$$

The inner (and outer) product is again linear and distributive

$$(\lambda A_r) \cdot B_s = A_r \cdot (\lambda B_s) = \lambda (A_r \cdot B_s) = \lambda A_r \cdot B_s, \tag{3.3.24}$$

$$A_r \cdot (B_s + C_t) = A_r \cdot B_s + A_r \cdot C_t, \tag{3.3.25}$$

$$\lambda (B_s + C_t) = \lambda B_s + \lambda C_t. \tag{3.3.26}$$

The *reverse* of a multivector is

$$\widetilde{A} = \sum_{k=1}^{n} (-1)^{k(k-1)/2} \langle A \rangle_k. \tag{3.3.27}$$

[161] uses a *dagger* instead of the *tilde*.

Special examples are

$$\widetilde{\lambda} = \lambda, \quad \widetilde{\vec{a}} = \vec{a}, \quad \widetilde{\vec{a} \wedge \vec{b}} = \vec{b} \wedge \vec{a} = -\vec{a} \wedge \vec{b}, \tag{3.3.28}$$

The scalar *magnitude* $|A|$ of a multivector A is

$$|A|^2 \equiv \underbrace{\widetilde{A} * A}_{\text{scalar product}} = \langle A\rangle_0^2 + \sum_{r=1}^{n} \langle \widetilde{A}\rangle_r \cdot \langle A\rangle_r, \tag{3.3.29}$$

where the separate term $\langle A\rangle_0^2$ is in particular due to the definition of the inner product in [161], p. 5, (1.21). The magnitude allows to define the inverse[1] for simple k-blade vectors

$$A^{-1} \equiv \frac{\widetilde{A}}{|A|^2}, \quad \text{with} \quad A^{-1}A = AA^{-1} = 1. \tag{3.3.30}$$

Alternative ways to express $\vec{a} \in \mathcal{A}_n \equiv \mathcal{G}^1(I)$ are

$$I \wedge \vec{a} = 0 \quad \text{or} \quad I\vec{a} = I \cdot \vec{a}. \tag{3.3.31}$$

The projection of $\vec{a}$ into $\mathcal{A}_n \equiv \mathcal{G}^1(I)$ is

$$P_I(\vec{a}) = P(\vec{a}) \equiv \sum_{k=1}^{n} \vec{a}^k \vec{a}_k \cdot \vec{a} = \sum_{k=1}^{n} \vec{a}_k \vec{a}^k \cdot \vec{a}, \tag{3.3.32}$$

where $\vec{a}^k$ is the *reciprocal frame* defined by

$$\vec{a}^k \vec{a}_j = \delta_j^k = \text{Kronecker delta} = \begin{cases} 1 & \text{if} \quad j = k, \\ 0 & \text{if} \quad j \neq k, \end{cases} . \tag{3.3.33}$$

A general convention is that inner products $\vec{a} \cdot \vec{b}$ and outer products $\vec{a} \wedge \vec{b}$ have priority over geometric products $\vec{a}\vec{b}$, e.g.

$$\vec{a} \cdot \vec{b}\vec{c} \wedge \vec{d}\vec{e} = (\vec{a} \cdot \vec{b})(\vec{c} \wedge \vec{d})\vec{e}. \tag{3.3.34}$$

The projection of a multivector B on a subspace described by a simple m-vector (m-blade) $A_m = \vec{a}_1 \wedge \vec{a}_2 \wedge \ldots \wedge \vec{a}_m$ is

$$P_A(B) \equiv \underbrace{(B \cdot A) \cdot A^{-1}}_{\text{general}} \overset{s \leq m}{\equiv} \underbrace{A^{-1} \cdot (A \cdot B_{(s)})}_{\text{degree dependent}}, \tag{3.3.35}$$

$$P_A(\langle B\rangle_0) \equiv \langle B\rangle_0, \quad P_A(\langle B\rangle_n) \equiv \langle B\rangle_n \cdot AA^{-1}, \tag{3.3.36}$$

the exceptions for scalars $\langle B\rangle_0$ and pseudoscalars $\langle B\rangle_n$ being again due to the definition of the inner product in [161], p. 6, (1.21). A projection of one factor of an inner product has the effect to project the other factor as well

$$\vec{a} \cdot P(\vec{b}) = P(\vec{a}) \cdot P(\vec{b}) = P(\vec{a}) \cdot \vec{b}. \tag{3.3.37}$$

For a multivector $B \in \mathcal{G}(A_m)$, with $A = A_m$ we have

$$(\vec{a} \wedge B) \cdot A = (\vec{a} \wedge B)A = \vec{a} \cdot (BA), \quad \text{if} \quad \vec{a} \wedge A = 0. \tag{3.3.38}$$

[1] Every multivector can be inverted, as long as its determinant is nonzero, see [216, 300].

Reordering rules for products of homogeneous multivector are

$$A_r \cdot B_s = (-1)^{r(s-r)} B_s \cdot A_r \quad \text{for} \quad r \leq s, \tag{3.3.39}$$

$$A_r \wedge B_s = (-1)^{rs} B_s \wedge A_r. \tag{3.3.40}$$

Elementary combinations that occur often are

$$\vec{a} \cdot (\vec{b} \wedge \vec{c}) = (\vec{a} \cdot \vec{b})\vec{c} - (\vec{a} \cdot \vec{c})\vec{b} = \vec{a} \cdot \vec{b}\vec{c} - \vec{a} \cdot \vec{c}\vec{b}, \tag{3.3.41}$$

$$(\vec{a} \wedge \vec{b}) \cdot (\vec{c} \wedge \vec{d}) = \vec{a} \cdot (\vec{b} \cdot (\vec{c} \wedge \vec{d}) = (\vec{a} \cdot \vec{d})(\vec{b} \cdot \vec{c}) - (\vec{a} \cdot \vec{c})(\vec{b} \cdot \vec{d}), \tag{3.3.42}$$

$$(\vec{a} \wedge \vec{b})^2 = (\vec{a} \wedge \vec{b}) \cdot (\vec{a} \wedge \vec{b}) = (\vec{a} \cdot \vec{b})^2 - \vec{a}^2\vec{b}^2 = -(\vec{b} \wedge \vec{a}) \cdot (\vec{a} \wedge \vec{b})$$
$$= -|\vec{a} \wedge \vec{b}|^2, \tag{3.3.43}$$

and the cyclic *Jacobi identity*

$$\vec{a} \cdot (\vec{b} \wedge \vec{c}) + \vec{b} \cdot (\vec{c} \wedge \vec{a}) + \vec{c} \cdot (\vec{a} \wedge \vec{b}) = 0. \tag{3.3.44}$$

The *commutator product* of multivectors A, B is

$$A \times B \equiv \frac{1}{2}(AB - BA). \tag{3.3.45}$$

One useful identity using it is

$$(\vec{a} \wedge \vec{b}) \times A = \vec{a}\vec{b} \cdot A - A \cdot \vec{a}\vec{b} = \vec{a}\vec{b} \wedge A - A \wedge \vec{a}\vec{b}. \tag{3.3.46}$$

The commutator product is to be distinguished from the *cross product*, which is strictly limited to the three-dimensional Euclidean case with unit pseudoscalar I_3 :

$$\vec{a} \times \vec{b} = (\vec{b} \wedge \vec{a})I_3 = -(\vec{a} \wedge \vec{b})I_3 = (\vec{a} \wedge \vec{b})I_3^{-1}. \tag{3.3.47}$$

For more on basic geometric algebra I refer to [162], [161], to section 3 of [173], and to Chapter 2 in the present book.

3.4 VECTOR DIFFERENTIAL CALCULUS

This section shows how to differentiate functions on linear subspaces of the universal geometric algebra $\mathcal{G}$ by vectors. It has wide applications particularly to mechanics and physics in general [162]. Separate concepts of gradient, divergence and curl merge into a single concept of vector derivative, united by the geometric product.

The relationship of differential and derivative is clarified. The *Taylor expansion* (Prop. 116) is applied to important examples, yielding e.g. the *Legendre polynomials* (Prop. 124). The *adjoint* (Def. 3.4.57) and the *integrability* (Prop. 3.4.42, etc.) of multivector functions are defined and discussed. Throughout this section a number of basic differentials and derivations are performed explicitly illustrating ease and power of the calculus.

Since my emphasis here is on explicit step by step proofs, I refer the reader, who is interested in the philosophy, comments and interpretation to the literature in the final Reference section.

As for the notation: Prop. 3.4.7 refers to proposition 3.4.7 of this section. Def. 3.4.13 refers to definition 3.4.13 of this section. (3.3.6) refers to equation number (3.3.6) in the previous section on basic geometric algebra, etc.

Standard definitions of continuity and scalar differentiability apply to multivector-valued functions, because the scalar product determines a unique "distance" $|A - B|$ between two elements $A, B \in \mathcal{G}(I)$.

Definition 3.4.1. *Directional derivative.* $F = F(\vec{x})$ *multivector-valued function of a vector variable* $\vec{x}$ *defined on an* n*-dimensional vector space* $\mathcal{A}_n = \mathcal{G}^1(I)$, I *unit pseudoscalar,* $\vec{a} \in \mathcal{A}_n$:

$$\vec{a} \cdot \vec{\partial} F = \frac{dF(\vec{x} + \vec{a}\tau)}{d\tau} = \lim_{\tau \to 0} \frac{F(\vec{x} + \vec{a}\tau) - F(\vec{x})}{\tau}. \tag{3.4.1}$$

Nomenclature: *derivative* of F in the *direction* $\vec{a}$, $\vec{a}$*-derivative* of F. ([162] uses $\nabla \equiv \vec{\partial}$, [161] uses $\partial \equiv \vec{\partial}$.)

Proposition 3.4.2. *Distributivity w.r.t. vector argument. Assume that the directional derivative of* F *is continuous for all direction vectors of an arbitrary basis of* $\mathcal{A}_n$*."*

$$(\vec{a} + \vec{b}) \cdot \vec{\partial} F = \vec{a} \cdot \vec{\partial} F + \vec{b} \cdot \vec{\partial} F, \quad \forall \vec{a}, \vec{b} \in \mathcal{A}_n. \tag{3.4.2}$$

Proof.

$$\begin{aligned}
(\vec{a} + \vec{b}) \cdot \vec{\partial} F &= \lim_{\tau \to 0} \frac{F(\vec{x} + \tau\vec{a} + \tau\vec{b}) - F(\vec{x})}{\tau} \\
&= \lim_{\tau \to 0} \left\{ \frac{F(\vec{x} + \tau\vec{a} + \tau\vec{b}) - F(\vec{x} + \tau\vec{b}) + F(\vec{x} + \tau\vec{b}) - F(\vec{x})}{\tau} \right\} \\
&= \lim_{\tau \to 0} \vec{a} \cdot \vec{\partial} F(\vec{x} + \tau\vec{b}) + \vec{b} \cdot \vec{\partial} F = \vec{a} \cdot \vec{\partial} F + \vec{b} \cdot \vec{\partial} F.
\end{aligned} \tag{3.4.3}$$

□

Proposition 3.4.3. *For scalar* λ

$$(\lambda\vec{a}) \cdot \vec{\partial} F = \lambda(\vec{a} \cdot \vec{\partial} F). \tag{3.4.4}$$

Proof.

$$(\lambda\vec{a}) \cdot \vec{\partial} F \overset{\text{Def.3.4.1}}{=} \lim_{\tau \to 0} \frac{F(\vec{x} + \lambda\vec{a}\tau) - F(\vec{x})}{\tau}. \tag{3.4.5}$$

Case 1: $\lambda \neq 0$:

$$\begin{aligned}
(\lambda\vec{a}) \cdot \vec{\partial} F &\overset{\text{Def.3.4.1}}{=} \lim_{\tau \to 0} \lambda \frac{F(\vec{x} + \vec{a}(\lambda\tau)) - F(\vec{x})}{\lambda\tau} \\
&\overset{\lambda \neq 0,\ \text{rp.}}{=} \lambda \lim_{\tau' \to 0} \frac{F(\vec{x} + \vec{a}\tau') - F(\vec{x})}{\tau'} \overset{\text{Def.3.4.1}}{=} \lambda(\vec{a} \cdot \vec{\partial} F),
\end{aligned} \tag{3.4.6}$$

with rp. = reparametrization: $\tau \to \tau'$.

Case 2: $\lambda = 0$:

$$(\lambda\vec{a}) \cdot \vec{\partial} F \overset{\text{Def.3.4.1}}{=} \lim_{\tau \to 0} \frac{F(\vec{x}) - F(\vec{x})}{\tau} = 0 = 0\,(\vec{a} \cdot \vec{\partial} F). \tag{3.4.7}$$

□

Proposition 3.4.4. *Distributivity w.r.t. multivector-valued function.*

$$\vec{a}\cdot\vec{\partial}(F+G)=\vec{a}\cdot\vec{\partial}F+\vec{a}\cdot\vec{\partial}G, \tag{3.4.8}$$

with $G=G(\vec{x})$, $F=F(\vec{x})$, *multivector-valued functions of a vector variable* $\vec{x}$. *In the notation of Def. 3.4.13:*

$$\underline{F+G}=\underline{F}+\underline{G}. \tag{3.4.9}$$

Proof.

$$\begin{aligned}\vec{a}\cdot\vec{\partial}(F+G) &\overset{\text{Def.3.4.1}}{=} \lim_{\tau\to 0}\frac{F(\vec{x}+\vec{a}(\tau))+G(\vec{x}+\vec{a}(\tau))-F(\vec{x})-G(\vec{x})}{\tau}\\ &= \lim_{\tau\to 0}\frac{F(\vec{x}+\vec{a}(\tau))-F(\vec{x})}{\tau}+\lim_{\tau\to 0}\frac{G(\vec{x}+\vec{a}(\tau))-G(\vec{x})}{\tau}\\ &\overset{\text{Def.3.4.1}}{=} \vec{a}\cdot\vec{\partial}F+\vec{a}\cdot\vec{\partial}G. \end{aligned}\tag{3.4.10}$$

□

Proposition 3.4.5. *Product rule.*

$$\vec{a}\cdot\vec{\partial}(FG)=(\vec{a}\cdot\vec{\partial}F)G+F(\vec{a}\cdot\vec{\partial}G). \tag{3.4.11}$$

In the notation of Def. 13:

$$\underline{FG}=\underline{F}\,G+F\underline{G}. \tag{3.4.12}$$

Proof.

$$\begin{aligned}&\vec{a}\cdot\vec{\partial}(FG) \overset{\text{Def.3.4.1}}{=} \lim_{\tau\to 0}\frac{F(\vec{x}+\vec{a}\tau)G(\vec{x}+\vec{a}\tau)-F(\vec{x})G(\vec{x})}{\tau}\\ &= \lim_{\tau\to 0}\frac{F(\vec{x}+\vec{a}\tau)G(\vec{x}+\vec{a}\tau)-F(\vec{x})G(\vec{x}+\vec{a}\tau)+F(\vec{x})G(\vec{x}+\vec{a}\tau)-F(\vec{x})G(\vec{x})}{\tau}\\ &= \lim_{\tau\to 0}\left\{\frac{F(\vec{x}+\vec{a}\tau)-F(\vec{x})}{\tau}G(\vec{x}+\vec{a}\tau)+F(\vec{x})\frac{G(\vec{x}+\vec{a}\tau)-G(\vec{x})}{\tau}\right\}\\ &\overset{\text{Def.3.4.1}}{=} (\vec{a}\cdot\vec{\partial}F)\lim_{\tau\to 0}G(\vec{x}+\vec{a}\tau)+F(\vec{a}\cdot\vec{\partial}G)\\ &= (\vec{a}\cdot\vec{\partial}F)G+F(\vec{a}\cdot\vec{\partial}G). \end{aligned}\tag{3.4.13}$$

□

Proposition 3.4.6. *Grade invariance.*

$$\vec{a}\cdot\vec{\partial}\langle F\rangle_k=\langle\vec{a}\cdot\vec{\partial}F\rangle_k. \tag{3.4.14}$$

$\vec{a}\cdot\vec{\partial}$ *is therefore said to be a scalar differential operator.*

Proof.

$$\begin{aligned}\vec{a}\cdot\vec{\partial}\langle F\rangle_k &\overset{\text{Def.3.4.1}}{=} \lim_{\tau\to 0}\frac{\langle F(\vec{x}+\vec{a}\tau)\rangle_k-\langle F(\vec{x})\rangle_k}{\tau}=\lim_{\tau\to 0}\left\langle\frac{F(\vec{x}+\vec{a}\tau)-F(\vec{x})}{\tau}\right\rangle_k\\ &\overset{\text{Def.3.4.1}}{=} \langle\vec{a}\cdot\vec{\partial}F\rangle_k. \end{aligned}\tag{3.4.15}$$

□

Proposition 3.4.7. *Scalar chain rule.*

$$\vec{a}\cdot\vec{\partial}F = (\vec{a}\cdot\vec{\partial}\lambda)\frac{dF}{d\lambda} \tag{3.4.16}$$

with $F = F(\lambda(\vec{x}))$, $\lambda = \lambda(\vec{x})$ *a scalar valued function.*

Proof. Using the Taylor expansions:

$$F(\lambda + \tau\Delta\lambda) = F(\lambda) + \tau\Delta\lambda\frac{dF}{d\lambda} + \frac{\tau^2(\Delta\lambda)^2}{2}\frac{d^2F}{d\lambda^2} + \ldots, \tag{3.4.17}$$

$$\lambda(\vec{x} + \tau\vec{a}) \overset{(3.4.28)}{=} \lambda(\vec{x}) + \tau\vec{a}\cdot\vec{\partial}\lambda(\vec{x}) + \frac{\tau^2}{2}(\vec{a}\cdot\vec{\partial})^2\lambda(\vec{x}) + \ldots, \tag{3.4.18}$$

we have

$$\begin{aligned}
\vec{a}\cdot\vec{\partial}F(\lambda(\vec{x})) &\overset{\text{Def.3.4.1}}{=} \lim_{\tau\to 0}\frac{F(\lambda(\vec{x} + \vec{a}\tau)) - F(\lambda(\vec{x}))}{\tau} \\
&\overset{(3.4.18)}{=} \lim_{\tau\to 0}\frac{F(\lambda(\vec{x}) + \tau\vec{a}\cdot\vec{\partial}\lambda(\vec{x})) - F(\lambda(\vec{x}))}{\tau} \\
&\overset{(3.4.17)}{=} \lim_{\tau\to 0}\frac{F(\lambda(\vec{x})) + \tau\left\{\vec{a}\cdot\vec{\partial}\lambda(\vec{x})\right\}\frac{dF}{d\lambda} - F(\lambda(\vec{x}))}{\tau} \\
&= (\vec{a}\cdot\vec{\partial}\lambda)\frac{dF}{d\lambda}.
\end{aligned} \tag{3.4.19}$$

□

Proposition 3.4.8. *Identity.*

$$\vec{a}\cdot\vec{\partial}\vec{x} = \vec{a}. \tag{3.4.20}$$

Proof.

$$F(\vec{x}) = \vec{x}, \qquad \vec{a}\cdot\vec{\partial}\vec{x} \overset{\text{Def.3.4.1}}{=} \lim_{\tau\to 0}\frac{\vec{x} + \tau\vec{a} - \vec{x}}{\tau} = \lim_{\tau\to 0}\vec{a} = \vec{a}. \tag{3.4.21}$$

□

Proposition 3.4.9. *Constant function. For* A *independent of* $\vec{x}$ *:*

$$\vec{a}\cdot\vec{\partial}A = 0. \tag{3.4.22}$$

Proof.

$$F(\vec{x}) = A, \qquad \vec{a}\cdot\vec{\partial}A \overset{\text{Def.3.4.1}}{=} \lim_{\tau\to 0}\frac{A - A}{\tau} = 0. \tag{3.4.23}$$

□

Proposition 3.4.10. *Vector length.*

$$\vec{a}\cdot\vec{\partial}|\vec{x}| = \frac{\vec{a}\cdot\vec{x}}{|\vec{x}|} = \vec{a}\cdot\hat{x}, \tag{3.4.24}$$

with $\hat{x} = \frac{\vec{x}}{|\vec{x}|}$ *unit vector in the direction of* $\vec{x}$*, see* (3.3.11).

Proof.

$$
\begin{aligned}
\vec{a}\cdot\vec{\partial}\vec{x}^2 &\overset{\text{Prop. 3.4.5}}{=} (\vec{a}\cdot\vec{\partial}\vec{x})\vec{x} + \vec{x}(\vec{a}\cdot\vec{\partial}\vec{x}) \overset{\text{Prop. 3.4.8}}{=} \vec{a}\vec{x} + \vec{x}\vec{a} = 2\vec{a}\cdot\vec{x},\\
\vec{a}\cdot\vec{\partial}|\vec{x}|^2 &\overset{\text{Prop. 3.4.5}}{=} (\vec{a}\cdot\vec{\partial}|\vec{x}|)|\vec{x}| + |\vec{x}|(\vec{a}\cdot\vec{\partial}|\vec{x}|) = 2|\vec{x}|(\vec{a}\cdot\vec{\partial}\,|\vec{x}|),\\
\vec{x}^2 = |\vec{x}|^2 \quad &\Rightarrow \quad \vec{a}\cdot\vec{\partial}\vec{x}^2 = \vec{a}\cdot\vec{\partial}\,|\vec{x}|^2,\\
\Rightarrow \quad \vec{a}\cdot\vec{x} = |\vec{x}|(\vec{a}\cdot\vec{\partial}\,|\vec{x}|) \quad &\Rightarrow \quad \vec{a}\cdot\vec{\partial}\,|\vec{x}| = \frac{\vec{a}\cdot\vec{x}}{|\vec{x}|} = \vec{a}\cdot\hat{x}.
\end{aligned}
\tag{3.4.25}
$$

□

Proposition 3.4.11. *Direction function.*

$$
\vec{a}\cdot\vec{\partial}\hat{x} = \frac{\vec{a}-\vec{a}\cdot\hat{x}\hat{x}}{|\vec{x}|} = \frac{\hat{x}\hat{x}\wedge\vec{a}}{|\vec{x}|}.
\tag{3.4.26}
$$

Proof.

$$
\begin{aligned}
\vec{a}\cdot\vec{\partial}\hat{x} &= \vec{a}\cdot\vec{\partial}\frac{\vec{x}}{|\vec{x}|} \overset{\text{Prop. 3.4.5}}{=} \frac{\vec{a}\cdot\vec{\partial}\vec{x}}{|\vec{x}|} + \vec{x}\vec{a}\cdot\vec{\partial}\frac{1}{|\vec{x}|}\\
&\overset{\text{Props. 3.4.7, 3.4.8}}{=} \frac{\vec{a}}{|\vec{x}|} + \vec{x}(\vec{a}\cdot\vec{\partial}|\vec{x}|)\frac{d\frac{1}{|\vec{x}|}}{d|\vec{x}|} \overset{\text{Prop. 3.4.10}}{=} \frac{\vec{a}}{|\vec{x}|} + \vec{x}(\vec{a}\cdot\hat{x})\frac{-1}{|\vec{x}|^2}\\
&= \frac{\vec{a}}{|\vec{x}|} - \hat{x}(\vec{a}\cdot\hat{x})\frac{1}{|\vec{x}|} = \frac{\vec{a}-\vec{a}\cdot\hat{x}\hat{x}}{|\vec{x}|} = \frac{\hat{x}\hat{x}\vec{a}-\hat{x}\hat{x}\cdot\vec{a}}{|\vec{x}|}\\
&= \frac{\hat{x}(\hat{x}\cdot\vec{a}+\hat{x}\wedge\vec{a})-\hat{x}\hat{x}\cdot\vec{a}}{|\vec{x}|} = \frac{\hat{x}\hat{x}\wedge\vec{a}}{|\vec{x}|}.
\end{aligned}
\tag{3.4.27}
$$

□

Proposition 3.4.12. *Taylor expansion.*

$$
F(\vec{x}+\vec{a}) = \exp(\vec{a}\cdot\vec{\partial})F(\vec{x}) = \sum_{k=0}^{\infty}\frac{(\vec{a}\cdot\vec{\partial})^k}{k!}F(\vec{x}).
\tag{3.4.28}
$$

Proof. Note that this proof is done without referring to Props. 3.4.7 to 3.4.11!

$$
\begin{aligned}
&G(\tau) \equiv F(\vec{x}+\vec{a}\tau)\\
\Rightarrow \quad & \frac{dG(0)}{d\tau} = \frac{dF(\vec{x}+\vec{a}\tau)}{d\tau}|_{\tau=0} \overset{\text{Def.3.4.1}}{=} \vec{a}\cdot\vec{\partial}F\\
\Rightarrow \quad & \frac{d^2G(0)}{d\tau^2} = \frac{d}{d\tau}\frac{dF(\vec{x}+\vec{a}\tau)}{d\tau}|_{\tau=0} \overset{\text{Def.3.4.1}}{=} \vec{a}\cdot\vec{\partial}\frac{dF(\vec{x}+\vec{a}\tau)}{d\tau}|_{\tau=0}\\
& \overset{\text{Def.3.4.1}}{=} \vec{a}\cdot\vec{\partial}(\vec{a}\cdot\vec{\partial}F(\vec{x})) = (\vec{a}\cdot\vec{\partial})^2F(\vec{x}).
\end{aligned}
\tag{3.4.29}
$$

General: $\frac{d^kG(0)}{d\tau^k} = (\vec{a}\cdot\vec{\partial})^kF(\vec{x})$. The Taylor series for G is:

$$
\begin{aligned}
&G(1) = G(0+1) = G(0) + \frac{dG(0)}{d\tau} + \frac{1}{2}\frac{d^2G(0)}{d\tau^2} + \ldots = \sum_{k=0}^{\infty}\frac{1}{k!}\frac{d^kG(0)}{d\tau^k}\\
\Rightarrow \quad & G(1) = F(\vec{x}+\vec{a}) = \sum_{k=0}^{\infty}\frac{1}{k!}(\vec{a}\cdot\vec{\partial})^kF(\vec{x}) = \exp(\vec{a}\cdot\vec{\partial})F(\vec{x}).
\end{aligned}
\tag{3.4.30}
$$

□

Definition 3.4.13. *Continuously differentiable, differential.*

F is continuously differentiable at $\vec{x}$, if for each fixed $\vec{a}$, $\vec{a}\cdot\vec{\partial}F(\vec{y})$ exists and is a continuous function of $\vec{y}$ for each $\vec{y}$ in a neighborhood of $\vec{x}$.

If F is defined and continuously differentiable at $\vec{x}$, then, for fixed $\vec{x}$, $\vec{a}\cdot\vec{\partial}F(\vec{x})$ is a linear function of $\vec{a}$, the (first) differential of F.

$$\underline{F}(\vec{a},\vec{x}) = F_{\vec{a}}(\vec{x}) \equiv \vec{a}\cdot\vec{\partial}F(\vec{x}). \tag{3.4.31}$$

([162], p. 107 uses $F' \equiv \underline{F}$.)

Suppressing $\vec{x}$, or for fixed $\vec{x}$:

$$\underline{F} = \underline{F}(\vec{a}) = F_{\vec{a}} \equiv \vec{a}\cdot\vec{\partial}F. \tag{3.4.32}$$

Proposition 3.4.14. *Linearity.*

$$\underline{F}(\vec{a}+\vec{b}) = \underline{F}(\vec{a}) + \underline{F}(\vec{b}), \qquad \lambda \text{ scalar } : \underline{F}(\lambda\vec{a}) = \lambda\underline{F}(\vec{a}). \tag{3.4.33}$$

Proof. Propositions 3.4.2 and 3.4.3. □

Proposition 3.4.15. *Linear approximation.*

For $|\vec{r}| = |\vec{x}-\vec{x}_0|$ sufficiently small:

$$F(\vec{x}) - F(\vec{x}_0) \approx \underline{F}(\vec{x}-\vec{x}_0) = \underline{F}(\vec{x}) - \underline{F}(\vec{x}_0). \tag{3.4.34}$$

Proof.

$$\begin{aligned} F(\vec{x}) = F(\vec{x}_0+\vec{r}) &\overset{\text{Prop. 3.4.12}}{=} F(\vec{x}_0) + \vec{r}\cdot\vec{\partial}F(\vec{x}_0) + \frac{1}{2}(\vec{r}\cdot\vec{\partial})^2F(\vec{x}_0) + \ldots \\ &= F(\vec{x}_0) + |\vec{r}|\hat{r}\cdot\vec{\partial}F(\vec{x}_0) + \frac{|\vec{r}|^2}{2}(\hat{r}\cdot\vec{\partial})^2F(\vec{x}_0) + \ldots \\ &\quad + \frac{|\vec{r}|^k}{k!}(\hat{r}\cdot\vec{\partial})^kF(\vec{x}_0) + \ldots \end{aligned} \tag{3.4.35}$$

with $\hat{r} \equiv \frac{\vec{r}}{|\vec{r}|}$. For sufficiently small $|\vec{r}|$:

$$\begin{aligned} F(\vec{x}) &= F(\vec{x}_0) + \vec{r}\cdot\vec{\partial}F(\vec{x}_0) = F(\vec{x}_0) + (\vec{x}-\vec{x}_0)\cdot\vec{\partial}F(\vec{x}_0) \\ &\overset{\text{Def. 3.4.13}}{=} F(\vec{x}_0) + \underline{F}(\vec{x}-\vec{x}_0,\vec{x}_0) \overset{\text{Def. 3.4.13}}{=} F(\vec{x}_0) + \underline{F}(\vec{x}-\vec{x}_0) \\ &\overset{\text{Prop. 3.4.14}}{=} F(\vec{x}_0) + \underline{F}(\vec{x}) - \underline{F}(\vec{x}_0) \end{aligned} \tag{3.4.36}$$

$$\Rightarrow \quad F(\vec{x}) - F(\vec{x}_0) \approx \underline{F}(\vec{x}-\vec{x}_0) = \underline{F}(\vec{x}) - \underline{F}(\vec{x}_0). \tag{3.4.37}$$

□

Proposition 3.4.16. *Chain rule.*

$$\frac{dF}{dt}(\vec{x}(t)) = \left(\frac{d}{dt}\vec{x}(t)\right)\cdot\vec{\partial}\,F(\vec{x})\Big|_{\vec{x}=\vec{x}(t)}. \tag{3.4.38}$$

Proof. Using the Taylor expansion

$$\vec{x}(t+\tau) = \vec{x}(t) + \tau\frac{d}{dt}\vec{x}(t) + \frac{\tau^2}{2}\frac{d^2}{dt^2}\vec{x}(t) + \ldots, \tag{3.4.39}$$

$$\begin{aligned}\frac{dF}{dt}(\vec{x}(t)) &= \lim_{\tau\to 0}\frac{F(\vec{x}(t+\tau)) - F(\vec{x}(t))}{\tau} \\ &\overset{\text{Taylor}}{=} \lim_{\tau\to 0}\frac{F(\vec{x}(t) + \tau\frac{d}{dt}\vec{x}(t)) - F(\vec{x}(t))}{\tau} \\ &\overset{\text{Def. 3.4.1}}{=} \left(\frac{d}{dt}\vec{x}(t)\right)\cdot\vec{\partial}F(\vec{x})\Big|_{\vec{x}=\vec{x}(t)}.\end{aligned} \tag{3.4.40}$$

□

Definition 3.4.17. *Vector derivative. Differentiation of F by its argument $\vec{x}$*

$$\vec{\partial}_{\vec{x}}F(\vec{x}) = \vec{\partial}F, \tag{3.4.41}$$

with the differential operator $\vec{\partial}$, assumed to

(i) *have the algebraic properties of a vector in $\mathcal{A}_n \equiv G^1(I)$, I unit pseudoscalar; and*

(ii) *that $\vec{a}\cdot\vec{\partial}_{\vec{x}}$ with $\vec{a}\in\mathcal{A}_n$ is $\vec{a}\cdot\vec{\partial}_{\vec{x}}F$ as in Def. 3.4.1.*

Proposition 3.4.18. *Algebraic properties of $\vec{\partial}_{\vec{x}}$.*

$$I\wedge\vec{\partial}_{\vec{x}} \overset{(3.3.31)}{=} 0. \tag{3.4.42}$$

$$I\vec{\partial}_{\vec{x}} \overset{(3.3.31)}{=} I\cdot\vec{\partial}_{\vec{x}} \tag{3.4.43}$$

$$\vec{\partial}_{\vec{x}} \overset{(3.3.32)}{=} \sum_{k=1}^{n}\vec{a}^k\vec{a}_k\cdot\vec{\partial}_{\vec{x}}, \tag{3.4.44}$$

where the $\vec{a}^k$ express the algebraic vector properties and the $\vec{a}_k\cdot\vec{\partial}_{\vec{x}}$ the scalar differential properties.

Definition 3.4.19. *Gradient. The vector field $\vec{f} = \vec{f}(\vec{x}) = \vec{\partial}_{\vec{x}}\Phi(\vec{x}) = \vec{\partial}\Phi$ for a scalar function $\Phi = \Phi(\vec{x})$ is called the gradient of Φ.*

Proposition 3.4.20. *3-dimensional cross product. For $\vec{b}$ independent of $\vec{x}\in\mathcal{A}_3\equiv\mathcal{G}^1(I_3)$:*

$$\vec{a}\cdot\vec{\partial}(\vec{x}\times\vec{b}) = \vec{a}\times\vec{b}. \tag{3.4.45}$$

Only here $\times$ means the 3-dimensional cross product (3.3.47)*, not the commutator product in* (3.3.45) *or in Proposition 3.4.81.*

Proof.

$$\begin{aligned}
\vec{a}\cdot\vec{\partial}(\vec{x}\times\vec{b}) &\overset{(3.3.47)}{=} -\vec{a}\cdot\vec{\partial}((\vec{x}\wedge\vec{b})I_3) \\
&\overset{\text{Prop. 3.4.5}}{=} [-\vec{a}\cdot\vec{\partial}(\vec{x}\wedge\vec{b})]I_3 - (\vec{x}\wedge\vec{b})\vec{a}\cdot\vec{\partial}I_3 \\
&\overset{I_3=\text{ const.,Prop. 3.4.9}}{=} [-\vec{a}\cdot\vec{\partial}(\vec{x}\wedge\vec{b})]I_3 \overset{(3.3.8)}{=} [-\vec{a}\cdot\vec{\partial}\left\langle\vec{x}\vec{b}\right\rangle_2]I_3 \\
&\overset{\text{Prop. 3.4.6}}{=} -\left\langle\vec{a}\cdot\vec{\partial}\vec{x}\vec{b}\right\rangle_2 I_3 \\
&\overset{\text{Prop. 3.4.5}}{=} -\left\langle(\vec{a}\cdot\vec{\partial}\vec{x})\vec{b}\right\rangle_2 I_3 - \left\langle\vec{x}(\vec{a}\cdot\vec{\partial}\vec{b})\right\rangle_2 I_3 \\
&\overset{\text{Props. 3.4.8, 3.4.9}}{=} -\left\langle\vec{a}\vec{b}\right\rangle_2 I_3 \overset{(3.3.8)}{=} -(\vec{a}\wedge\vec{b})I_3 \overset{(3.3.47)}{=} \vec{a}\times\vec{b}. \qquad (3.4.46)
\end{aligned}$$

□

Proposition 3.4.21.

$$\vec{a}\cdot\vec{\partial}\left(\vec{x}\cdot\langle A\rangle_r\right) = \vec{a}\cdot\langle A\rangle_r, \qquad (3.4.47)$$

$A\in\mathcal{G}(I)$ independent of $\vec{x}$. By linearity

$$\vec{a}\cdot\vec{\partial}\left(\vec{x}\cdot A\right) = \vec{a}\cdot A. \qquad (3.4.48)$$

Proof.

$$\begin{aligned}
\vec{a}\cdot\vec{\partial}\left(\vec{x}\cdot\langle A\rangle_r\right) &\overset{(3.3.16)}{=} \vec{a}\cdot\vec{\partial}\left\langle\vec{x}\langle A\rangle_r\right\rangle_{r-1} \overset{\text{Prop. 3.4.6}}{=} \left\langle\vec{a}\cdot\vec{\partial}(\vec{x}\langle A\rangle_r)\right\rangle_{r-1} \\
&\overset{\text{Prop. 3.4.5}}{=} \left\langle(\vec{a}\cdot\vec{\partial}\vec{x})\langle A\rangle_r + \vec{x}\,\vec{a}\cdot\vec{\partial}\langle A\rangle_r\right\rangle_{r-1} \\
&\overset{\text{Props. 3.4.8, 3.4.9}}{=} \left\langle\vec{a}\langle A\rangle_r\right\rangle_{r-1} \overset{(3.3.16)}{=} \vec{a}\cdot\langle A\rangle_r. \qquad (3.4.49)
\end{aligned}$$

□

Proposition 3.4.22.

$$\vec{a}\cdot\vec{\partial}[\vec{x}\cdot(\vec{x}\wedge\vec{b})] = \vec{a}\cdot(\vec{x}\wedge\vec{b}) + \vec{x}\cdot(\vec{a}\wedge\vec{b}). \qquad (3.4.50)$$

Proof.

$$\begin{aligned}
\vec{a}\cdot\vec{\partial}[\vec{x}\cdot(\vec{x}\wedge\vec{b})] &\overset{(3.3.41)}{=} \vec{a}\cdot\vec{\partial}[\vec{x}^2\vec{b} - \vec{x}\cdot\vec{b}\vec{x}] \\
&\overset{\text{Prop. 3.4.4,3.4.5}}{=} (\vec{a}\cdot\vec{\partial}\vec{x}^2)\vec{b} + \vec{x}^2\vec{a}\cdot\vec{\partial}\vec{b} - [\vec{a}\cdot\vec{\partial}(\vec{x}\cdot\vec{b})]\vec{x} - \vec{x}\cdot\vec{b}(\vec{a}\cdot\vec{\partial}\vec{x}) \\
&\overset{\text{Proof 3.4.10, Prop. 3.4.9,3.4.21,3.4.8}}{=} \vec{a}\cdot\vec{x}\vec{b} - \vec{a}\cdot\vec{b}\vec{x} + \vec{x}\cdot\vec{a}\vec{b} - \vec{x}\cdot\vec{b}\vec{a} \\
&\overset{(3.3.41)}{=} \vec{a}\cdot(\vec{x}\wedge\vec{b}) + \vec{x}\cdot(\vec{a}\wedge\vec{b}). \qquad (3.4.51)
\end{aligned}$$

□

Proposition 3.4.23. *For $\vec{x}'$ independent of $\vec{x}$ and $r\equiv|\vec{r}|=|\vec{x}-\vec{x}'|$:*

$$\vec{a}\cdot\vec{\partial}r = \vec{a}\cdot\frac{\vec{r}}{r} = \vec{a}\cdot\hat{r}, \qquad (3.4.52)$$

where $\hat{r}=\frac{\vec{r}}{r}$.

Proof. Compare [162], p. 681. $r^2 = (\vec{x} - \vec{x}')(\vec{x} - \vec{x}')$, then

$$\begin{aligned}\vec{a}\cdot\vec{\partial}r^2 &\overset{\text{Prop. 3.4.5}}{=} [\vec{a}\cdot\vec{\partial}(\vec{x}-\vec{x}')](\vec{x}-\vec{x}') + (\vec{x}-\vec{x}')[\vec{a}\cdot\vec{\partial}(\vec{x}-\vec{x}')] \\ &\overset{\text{Prop. 3.4.4,3.4.9}}{=} \vec{a}(\vec{x}-\vec{x}') + (\vec{x}-\vec{x}')\vec{a} = 2\vec{a}\cdot\vec{r},\end{aligned} \tag{3.4.53}$$

$$\begin{aligned}&\vec{a}\cdot\vec{\partial}r^2 \overset{\text{Prop. 3.4.5}}{=} 2r(\vec{a}\cdot\vec{\partial}r) \quad\Rightarrow\quad 2\vec{a}\cdot\vec{r} = 2r(\vec{a}\cdot\vec{\partial}r) \\ \Rightarrow\quad &\vec{a}\cdot\vec{\partial}r = \vec{a}\cdot\frac{\vec{r}}{r} \overset{(3.3.11)}{=} \vec{a}\cdot\hat{r}.\end{aligned} \tag{3.4.54}$$

□

Proposition 3.4.24.

$$\vec{a}\cdot\vec{\partial}\hat{r} = \frac{\hat{r}\hat{r}\wedge\vec{a}}{r}. \tag{3.4.55}$$

Proof. Compare [162], p. 681.

$$\begin{aligned}\vec{a}\cdot\vec{\partial}\hat{r} = \vec{a}\cdot\vec{\partial}\frac{\vec{r}}{r} &\overset{\text{Prop. 3.4.5}}{=} \frac{1}{r}\vec{a}\cdot\vec{\partial}\vec{r} + \vec{r}\vec{a}\cdot\vec{\partial}\frac{1}{r} \\ &\overset{\text{Prop. 3.4.7}}{=} \frac{1}{r}\vec{a}\cdot\vec{\partial}(\vec{x}-\vec{x}') - \vec{r}\frac{1}{r^2}\vec{a}\cdot\vec{\partial}r \overset{\text{Prop. 3.4.8, 3.4.23}}{=} \frac{1}{r}\vec{a} - \frac{\hat{r}}{r}\vec{a}\cdot\hat{r} \\ &\overset{\text{Prop. 3.4.8}}{=} \frac{\hat{r}\hat{r}\vec{a} - \hat{r}(\hat{r}\cdot\vec{a})}{r} \overset{(3.3.8)}{=} \frac{\hat{r}\hat{r}\wedge\vec{a}}{r}.\end{aligned} \tag{3.4.56}$$

□

Proposition 3.4.25.

$$\vec{a}\cdot\vec{\partial}(\hat{r}\cdot\vec{a}) = \frac{|\hat{r}\wedge\vec{a}|^2}{r}. \tag{3.4.57}$$

Proof.

$$\begin{aligned}\vec{a}\cdot\vec{\partial}(\hat{r}\cdot\vec{a}) = \vec{a}\cdot\vec{\partial}\frac{\vec{r}\cdot\vec{a}}{r} &\overset{\text{Prop. 3.4.5}}{=} \frac{1}{r}\vec{a}\cdot\vec{\partial}(\vec{r}\cdot\vec{a}) + \vec{r}\cdot\vec{a}(\vec{a}\cdot\vec{\partial})\frac{1}{r} \\ &\overset{\text{Prop. 3.4.7, 3.4.23}}{=} \frac{1}{r}\vec{a}\cdot\vec{\partial}(\vec{x}\cdot\vec{a} - \vec{x}'\cdot\vec{a}) - \frac{\vec{r}\cdot\vec{a}}{r^2}\vec{a}\cdot\hat{r}\hat{r}\wedge\vec{a}\frac{1}{r}\vec{a}\cdot\vec{a} - \frac{(\hat{r}\cdot\vec{a})^2}{r} \\ &\overset{(3.3.11)}{=} \frac{\hat{r}^2\vec{a}^2 - (\hat{r}\cdot\vec{a})^2}{r} \overset{(3.3.43)}{=} \frac{(\hat{r}\wedge\vec{a})\cdot(\hat{r}\wedge\vec{a})}{r} \overset{(3.3.43)}{=} \frac{|\hat{r}\wedge\vec{a}|^2}{r}.\end{aligned} \tag{3.4.58}$$

□

Proposition 3.4.26.

$$\vec{a}\cdot\vec{\partial}(\hat{r}\wedge\vec{a}) = \frac{\hat{r}\cdot\vec{a}\vec{a}\wedge\hat{r}}{r}. \tag{3.4.59}$$

Proof.

$$\begin{aligned}
\vec{a}\cdot\vec{\partial}(\hat{r}\wedge\vec{a}) &\overset{(3.3.8)}{=} \vec{a}\cdot\vec{\partial}(\hat{r}\vec{a}-\hat{r}\cdot\vec{a}) \overset{\text{Prop. 3.4.4}}{=} \vec{a}\cdot\vec{\partial}(\hat{r}\vec{a})-\vec{a}\cdot\vec{\partial}(\hat{r}\cdot\vec{a})\\
&\overset{\text{Prop. 3.4.5, 3.4.25}}{=} \vec{a}\cdot\vec{\partial}(\hat{r})\vec{a}+\hat{r}\vec{a}\cdot\vec{\partial}\vec{a}-\frac{|\hat{r}\wedge\vec{a}|^2}{r}\\
&\overset{\text{Prop. 3.4.9, 3.4.24, (3.3.43)}}{=} \frac{\hat{r}\hat{r}\wedge\vec{a}}{r}\vec{a}-\frac{(\vec{a}\wedge\hat{r})(\vec{a}\wedge\hat{r})}{r}\\
&\overset{(3.3.39),\,(3.3.40)}{=} \frac{\vec{a}\wedge\hat{r}\hat{r}\vec{a}-(\vec{a}\wedge\hat{r})(\hat{r}\wedge\vec{a})}{r}\\
&\overset{(3.3.3)}{=} \frac{\vec{a}\wedge\hat{r}(\hat{r}\vec{a}-(\hat{r}\wedge\vec{a}))}{r} \overset{(3.3.7)}{=} \frac{\vec{a}\wedge\hat{r}\hat{r}\cdot\vec{a}}{r}\\
&= \frac{\hat{r}\cdot\vec{a}\vec{a}\wedge\hat{r}}{r}.
\end{aligned}\tag{3.4.60}$$

□

Proposition 3.4.27.

$$\vec{a}\cdot\vec{\partial}|\hat{r}\wedge\vec{a}| = -\frac{\hat{r}\cdot\vec{a}|\hat{r}\wedge\vec{a}|}{r}.\tag{3.4.61}$$

Proof.

$$\vec{a}\cdot\vec{\partial}|\hat{r}\wedge\vec{a}|^2 \overset{\text{Prop. 3.4.5}}{=} 2|\hat{r}\wedge\vec{a}|\vec{a}\cdot\vec{\partial}|\hat{r}\wedge\vec{a}|,\tag{3.4.62}$$

$$\begin{aligned}
\vec{a}\cdot\vec{\partial}|\hat{r}\wedge\vec{a}|^2 &\overset{(3.3.43)}{=} \vec{a}\cdot\vec{\partial}[(\hat{r}\wedge\vec{a})\cdot(\vec{a}\wedge\hat{r})] \overset{(3.3.43)}{=} \vec{a}\cdot\vec{\partial}[\hat{r}^2\vec{a}^2-(\hat{r}\cdot\vec{a})]\\
&\overset{\text{Prop. 3.4.5}}{=} \vec{a}\cdot\vec{\partial}\vec{a}^2-2(\hat{r}\cdot\vec{a})\vec{a}\cdot\vec{\partial}(\hat{r}\cdot\vec{a})\\
&\overset{\text{Prop. 3.4.9, 3.4.25}}{=} -2(\hat{r}\cdot\vec{a})\frac{|\hat{r}\wedge\vec{a}|^2}{r}
\end{aligned}\tag{3.4.63}$$

$$\Rightarrow 2|\hat{r}\wedge\vec{a}|\vec{a}\cdot\vec{\partial}|\hat{r}\wedge\vec{a}| = -2\frac{(\hat{r}\cdot\vec{a})|\hat{r}\wedge\vec{a}|^2}{r}\tag{3.4.64}$$

$$\Rightarrow \vec{a}\cdot\vec{\partial}|\hat{r}\wedge\vec{a}| = -\frac{(\hat{r}\cdot\vec{a})|\hat{r}\wedge\vec{a}|}{r}.\tag{3.4.65}$$

□

Proposition 3.4.28.

$$\vec{a}\cdot\vec{\partial}\frac{1}{\vec{r}} = -\frac{1}{\vec{r}}\vec{a}\frac{1}{\vec{r}}.\tag{3.4.66}$$

Proof.

$$\begin{aligned}
\vec{a}\cdot\vec{\partial}\frac{1}{\vec{r}} &\overset{(3.3.12)}{=} \vec{a}\cdot\vec{\partial}\frac{\vec{r}}{r^2} \overset{\text{Prop. 3.4.5}}{=} \frac{1}{r^2}\vec{a}\cdot\vec{\partial}\vec{r}+\vec{r}\vec{a}\cdot\vec{\partial}\frac{1}{r^2}\\
&\overset{\text{Prop. 3.4.7}}{=} \frac{1}{r^2}\vec{a}+\vec{r}(-\frac{2}{r^3})\vec{a}\cdot\vec{\partial}r \overset{\text{Prop. 3.4.23}}{=} \frac{1}{r^2}\vec{a}-2\frac{\vec{r}}{r^3}\vec{a}\cdot\hat{r}\\
&\overset{(3.3.12),\,(3.3.11)}{=} \frac{1}{\vec{r}}\frac{1}{\vec{r}}\vec{a}-2\frac{1}{\vec{r}}\vec{a}\cdot\frac{1}{\vec{r}} \overset{(3.3.7)}{=} \frac{1}{\vec{r}}\frac{1}{\vec{r}}\vec{a}-\frac{1}{\vec{r}}\vec{a}\frac{1}{\vec{r}}-\frac{1}{\vec{r}}\frac{1}{\vec{r}}\vec{a}\\
&= -\frac{1}{\vec{r}}\vec{a}\frac{1}{\vec{r}}.
\end{aligned}\tag{3.4.67}$$

□

Proposition 3.4.29.

$$\vec{a} \cdot \vec{\partial} \frac{1}{r^2} = -2\frac{\vec{a} \cdot \hat{r}}{r^3}. \tag{3.4.68}$$

Proof.

$$\vec{a} \cdot \vec{\partial} \frac{1}{r^2} \overset{(3.3.7)}{=} -\frac{2}{r^3}\vec{a} \cdot \vec{\partial} r \overset{\text{Prop. 3.4.23}}{=} -\frac{2}{r^3}\vec{a} \cdot \hat{r}. \tag{3.4.69}$$

□

Proposition 3.4.30.

$$\frac{1}{2}(\vec{a} \cdot \vec{\partial})^2 \frac{1}{r^2} = \frac{3(\vec{a} \cdot \hat{r})^2 - |\hat{r} \wedge \vec{a}|^2}{r^4}. \tag{3.4.70}$$

Proof.

$$\begin{aligned}
\frac{1}{2}(\vec{a} \cdot \vec{\partial})^2 \frac{1}{r^2} &\overset{\text{Prop. 3.4.29}}{=} \frac{1}{2}\vec{a} \cdot \vec{\partial}(-2\frac{\vec{a} \cdot \hat{r}}{r^3}) \\
&\overset{\text{Prop. 3.4.5}}{=} -(\vec{a} \cdot \vec{\partial}\frac{1}{r^3})\vec{a} \cdot \hat{r} - \frac{1}{r^3}\vec{a} \cdot \vec{\partial}(\vec{a} \cdot \hat{r}) \\
&\overset{\text{Prop. 3.4.7,3.4.23,3.4.25}}{=} \frac{3\vec{a} \cdot \hat{r}}{r^4}\vec{a} \cdot \hat{r} - \frac{1}{r^3}\vec{a}\frac{|\hat{r} \cdot \vec{a}|^2}{r} \\
&= \frac{3(\vec{a} \cdot \hat{r})^2 - |\hat{r} \cdot \vec{a}|^2}{r^4}.
\end{aligned} \tag{3.4.71}$$

□

Proposition 3.4.31.

$$\frac{1}{6}(\vec{a} \cdot \vec{\partial})^3 \frac{1}{r^2} = \frac{-4(\vec{a} \cdot \hat{r})^3 + 4|\hat{r} \wedge \vec{a}|^2\vec{a} \cdot \hat{r}}{r^5}. \tag{3.4.72}$$

Proof.

$$\begin{aligned}
\frac{1}{6}(\vec{a} \cdot \vec{\partial})^3 \frac{1}{r^2} &= \frac{1}{3}(\vec{a} \cdot \vec{\partial})\frac{1}{2}(\vec{a} \cdot \vec{\partial})^2 \frac{1}{r^2} \overset{\text{Prop. 3.4.30}}{=} \frac{1}{3}(\vec{a} \cdot \vec{\partial})\frac{3(\vec{a} \cdot \hat{r})^2 - |\hat{r} \cdot \vec{a}|^2}{r^4} \\
&\overset{\text{Prop. 3.4.5,3.4.7}}{=} \frac{1}{3}\frac{3\,2(\vec{a} \cdot \hat{r})\vec{a} \cdot \vec{\partial}(\vec{a} \cdot \hat{r}) - 2|\hat{r} \wedge \vec{a}|\vec{a} \cdot \vec{\partial}|\hat{r} \wedge \vec{a}|}{r^4} \\
&\quad + \frac{1}{3}[3(\vec{a} \cdot \hat{r})^2 - |\hat{r} \wedge \vec{a}|^2](-4)\frac{1}{r^5}\vec{a} \cdot \vec{\partial} r \\
&\overset{\text{Prop. 3.4.23,3.4.25,3.4.27}}{=} \frac{2(\vec{a} \cdot \hat{r})\frac{|\hat{r} \wedge \vec{a}|^2}{r} + \frac{2}{3}|\hat{r} \wedge \vec{a}|\hat{r} \cdot \vec{a}\frac{|\hat{r} \wedge \vec{a}|}{r}}{r^4} \\
&\quad - \frac{4}{3}[3(\vec{a} \cdot \hat{r})^2 - |\hat{r} \wedge \vec{a}|^2]\frac{1}{r^5}\vec{a} \cdot \hat{r} \\
&= \frac{2(\vec{a} \cdot \hat{r})|\hat{r} \wedge \vec{a}|^2 + \frac{2}{3}|\hat{r} \wedge \vec{a}|^2\hat{r} \cdot \vec{a} - 4(\vec{a} \cdot \hat{r})^3 + \frac{4}{3}|\hat{r} \wedge \vec{a}|^2\hat{r} \cdot \vec{a}}{r^5} \\
&= \frac{-4(\vec{a} \cdot \hat{r})^3 + 4|\hat{r} \wedge \vec{a}|^2\vec{a} \cdot \hat{r}}{r^5}.
\end{aligned} \tag{3.4.73}$$

□

Proposition 3.4.32.

$$\vec{a} \cdot \vec{\partial} \log r = \frac{\vec{a} \cdot \vec{r}}{r^2} = \vec{a} \cdot \vec{r}^{-1}. \tag{3.4.74}$$

Proof.

$$\vec{a} \cdot \vec{\partial} \log r \overset{\text{Prop. 3.4.7}}{=} \frac{1}{r} \vec{a} \cdot \vec{\partial} r \overset{\text{Prop. 3.4.23}}{=} \frac{\vec{a} \cdot \hat{r}}{r} \overset{(3.3.11)}{=} \frac{\vec{a} \cdot \vec{r}}{r^2}. \tag{3.4.75}$$

□

Proposition 3.4.33. *For integer k and $\vec{r}$ if $k < 0$:*

$$\vec{a} \cdot \vec{\partial} \vec{r}^{2k} = 2k \vec{a} \cdot \vec{r} \vec{r}^{2(k-1)}. \tag{3.4.76}$$

Proof.

$$\begin{aligned} \vec{a} \cdot \vec{\partial} \vec{r}^{2k} &= \vec{a} \cdot \vec{\partial} r^{2k} \overset{\text{Prop. 3.4.7}}{=} 2k r^{2k-1} \vec{a} \cdot \vec{\partial} r \overset{\text{Prop. 3.4.23}}{=} 2k r^{2k-1} \vec{a} \cdot \hat{r} \\ &= 2k r^{2k-2} \vec{a} \cdot \vec{r} = 2k \vec{a} \cdot \vec{r} \vec{r}^{2(k-1)}. \end{aligned} \tag{3.4.77}$$

□

Proposition 3.4.34. *For integer k and $\vec{r} \neq 0$ if $2k + 1 < 0$:*

$$\vec{a} \cdot \vec{\partial} \vec{r}^{2k+1} = \vec{r}^{2k} (\vec{a} + 2k \vec{a} \cdot \hat{r} \hat{r}). \tag{3.4.78}$$

Proof.

$$\begin{aligned} \vec{a} \cdot \vec{\partial} \vec{r}^{2k+1} &= \vec{a} \cdot \vec{\partial} (\vec{r}^{2k} \vec{r}) \overset{\text{Prop. 3.4.5}}{=} \vec{a} \cdot \vec{\partial} (\vec{r}^{2k}) \vec{r} + \vec{r}^{2k} \vec{a} \cdot \vec{\partial} \vec{r} \\ &\overset{\text{Prop. 3.4.8,3.4.33}}{=} 2k \vec{a} \cdot \vec{r} \vec{r}^{2(k-)1} + \vec{r}^{2k} \vec{a} = \vec{r}^{2k} (\vec{a} + 2k \vec{a} \cdot \hat{r} \hat{r}). \end{aligned} \tag{3.4.79}$$

□

Proposition 3.4.35. *Taylor expansion of $\frac{1}{\vec{x} - \vec{a}}$.*

$$\frac{1}{\vec{x} - \vec{a}} = \frac{1}{\vec{x}} + \frac{1}{\vec{x}} \vec{a} \frac{1}{\vec{x}} + \frac{1}{\vec{x}} \vec{a} \frac{1}{\vec{x}} \vec{a} \frac{1}{\vec{x}} + \dots. \tag{3.4.80}$$

Proof.

$$\frac{1}{\vec{x}-\vec{a}} \overset{\text{Prop. 3.4.12}}{=} e^{-\vec{a}\cdot\vec{\partial}}\frac{1}{\vec{x}} = \sum_{k=0}^{\infty}\frac{(-\vec{a}\cdot\vec{\partial})^k}{k!}\frac{1}{\vec{x}}, \tag{3.4.81}$$

$$\begin{aligned}
(-\vec{a}\cdot\vec{\partial})^k\frac{1}{\vec{x}} &= (-\vec{a}\cdot\vec{\partial})^{k-1}(-\vec{a}\cdot\vec{\partial})\frac{1}{\vec{x}} \overset{\text{Prop. 3.4.28}}{=} (-\vec{a}\cdot\vec{\partial})^{k-1}\frac{1}{\vec{x}}\vec{a}\frac{1}{\vec{x}} \\
&= (-\vec{a}\cdot\vec{\partial})^{k-2}(-\vec{a}\cdot\vec{\partial})[\frac{1}{\vec{x}}\vec{a}\frac{1}{\vec{x}}] \\
&\overset{\text{Prop. 3.4.5}}{=} (-\vec{a}\cdot\vec{\partial})^{k-2}[(-\vec{a}\cdot\vec{\partial}\frac{1}{\vec{x}})\vec{a}\frac{1}{\vec{x}} + \frac{1}{\vec{x}}\vec{a}(-\vec{a}\cdot\vec{\partial}\frac{1}{\vec{x}})] \\
&\overset{\text{Prop. 3.4.28}}{=} 2(-\vec{a}\cdot\vec{\partial})^{k-2}\frac{1}{\vec{x}}\vec{a}\frac{1}{\vec{x}}\vec{a}\frac{1}{\vec{x}} \\
&\overset{\text{Props. 3.4.5,3.4.28}}{=} \ldots = 2\,3(-\vec{a}\cdot\vec{\partial})^{k-3}\frac{1}{\vec{x}}\vec{a}\frac{1}{\vec{x}}\vec{a}\frac{1}{\vec{x}}\vec{a}\frac{1}{\vec{x}} \\
&\overset{\text{Props. 3.4.5,3.4.28}}{=} \ldots = k!(-\vec{a}\cdot\vec{\partial})^{k-k}\frac{1}{\vec{x}}(\vec{a}\frac{1}{\vec{x}})^k \\
&= k!\frac{1}{\vec{x}}(\vec{a}\frac{1}{\vec{x}})^k
\end{aligned} \tag{3.4.82}$$

$$\Rightarrow \quad \frac{1}{k!}(-\vec{a}\cdot\vec{\partial})^k\frac{1}{\vec{x}} = \frac{1}{\vec{x}}(\vec{a}\frac{1}{\vec{x}})^k \tag{3.4.83}$$

$$\Rightarrow \quad \frac{1}{\vec{x}-\vec{a}} = \sum_{k=0}^{\infty}\frac{1}{\vec{x}}(\vec{a}\frac{1}{\vec{x}})^k. \tag{3.4.84}$$

□

Proposition 3.4.36. *Legendre Polynomials.*

The Legendre polynomials P_n (homogeneous and of grade n) are defined by:

$$\frac{1}{|\vec{x}-\vec{a}|} \equiv \sum_{n=0}^{\infty}\frac{P_n(\hat{x}\vec{a})}{|\vec{x}|^{n+1}} = \sum_{n=0}^{\infty}\frac{P_n(\vec{x}\vec{a})}{|\vec{x}|^{2n+1}}. \tag{3.4.85}$$

The explicit first four polynomials are:

$$P_0(\vec{x}\vec{a}) = 1, \tag{3.4.86}$$

$$P_1(\vec{x}\vec{a}) = \vec{x}\cdot\vec{a}, \tag{3.4.87}$$

$$P_2(\vec{x}\vec{a}) = \frac{1}{2}[3(\vec{x}\cdot\vec{a})^2 - \vec{a}^2\vec{x}^2] = (\vec{x}\cdot\vec{a})^2 + \frac{1}{2}(\vec{x}\wedge\vec{a})^2, \tag{3.4.88}$$

$$P_3(\vec{x}\vec{a}) = \frac{1}{2}[5(\vec{x}\cdot\vec{a})^3 - 3\vec{a}^2\vec{x}^2\vec{x}\cdot\vec{a}] = (\vec{x}\cdot\vec{a})^3 + \frac{3}{2}\vec{x}\cdot\vec{a}(\vec{x}\wedge\vec{a})^2, \tag{3.4.89}$$

$$P_n(\vec{x}\vec{a}) = |\vec{x}|^n P_n(\hat{x}\vec{a}) = |\vec{x}|^n|\vec{a}|^n P_n(\hat{x}\hat{a}). \tag{3.4.90}$$

Proof.

$$\begin{aligned}
F(\vec{x}-\vec{a}) &= \frac{1}{|\vec{x}-\vec{a}|} \overset{\text{Prop. 3.4.12}}{=} e^{-\vec{a}\cdot\vec{\partial}}F(\vec{x}) = \sum_{n=0}^{\infty}\frac{(-\vec{a}\cdot\vec{\partial})^n}{n!}\frac{1}{|\vec{x}|} \\
&= \frac{1}{|\vec{x}|} - \vec{a}\cdot\vec{\partial}\frac{1}{|\vec{x}|} + \frac{1}{2}(\vec{a}\cdot\vec{\partial})^2\frac{1}{|\vec{x}|} - \frac{1}{6}(\vec{a}\cdot\vec{\partial})^3\frac{1}{|\vec{x}|} + \ldots
\end{aligned} \tag{3.4.91}$$

$$\Rightarrow \quad P_0(\vec{x}\vec{a}) = 1, \tag{3.4.92}$$

$$-\vec{a}\cdot\vec{\partial}\frac{1}{|\vec{x}|} \overset{\text{Prop. 3.4.7}}{=} \frac{1}{|\vec{x}|^2}\vec{a}\cdot\vec{\partial}|\vec{x}| \overset{\text{Prop. 3.4.10}}{=} \frac{1}{|\vec{x}|^2}\vec{a}\cdot\hat{x}$$

$$\overset{\text{Prop. 3.4.11}}{=} \frac{\vec{a}\cdot\vec{x}}{|\vec{x}|^3}, \tag{3.4.93}$$

$$\Rightarrow \quad P_1(\vec{x}\vec{a}) = \vec{a}\cdot\vec{x}, \tag{3.4.94}$$

$$\frac{1}{2}(\vec{a}\cdot\vec{\partial})^2\frac{1}{|\vec{x}|} = -\frac{1}{2}(\vec{a}\cdot\vec{\partial})(-\vec{a}\cdot\vec{\partial})\frac{1}{|\vec{x}|} = -\frac{1}{2}(\vec{a}\cdot\vec{\partial})\frac{\vec{a}\cdot\vec{x}}{|\vec{x}|^3}$$

$$\overset{\text{Prop. 3.4.5}}{=} -\frac{1}{2}[\vec{a}\cdot\vec{\partial}\frac{1}{|\vec{x}|^3}]\vec{a}\cdot\vec{x} - \frac{1}{2}\frac{1}{|\vec{x}|^3}\vec{a}\cdot\vec{\partial}(\vec{a}\cdot\vec{x})$$

$$\overset{\text{Prop. 3.4.7,3.4.21}}{=} -\frac{1}{2}[\frac{-3}{|\vec{x}|^4}\vec{a}\cdot\vec{\partial}|\vec{x}|]\vec{a}\cdot\vec{x} - \frac{1}{2}\frac{1}{|\vec{x}|^3}\vec{a}\cdot\vec{a}$$

$$\overset{\text{Prop. 3.4.10}}{=} \frac{1}{2}\frac{3}{|\vec{x}|^4}\vec{a}\cdot\hat{x}(\vec{a}\cdot\vec{x}) - \frac{1}{2}\frac{1}{|\vec{x}|^3}\vec{a}^2$$

$$\overset{(3.3.11),(3.3.5)}{=} \frac{1}{2}\frac{3(\vec{a}\cdot\vec{x})^2 - \vec{x}^2\vec{a}^2}{|\vec{x}|^5} = \frac{(\vec{a}\cdot\vec{x})^2 + \frac{1}{2}\{(\vec{a}\cdot\vec{x})^2 - \vec{x}^2\vec{a}^2\}}{|\vec{x}|^5}$$

$$\overset{(3.3.43)}{=} \frac{(\vec{a}\cdot\vec{x})^2 + \frac{1}{2}\{\vec{a}\wedge\vec{x}\}^2}{|\vec{x}|^5} \tag{3.4.95}$$

$$\Rightarrow \quad P_2(\vec{x}\vec{a}) = \frac{1}{2}[3(\vec{a}\cdot\vec{x})^2 - \vec{x}^2\vec{a}^2]$$

$$= (\vec{a}\cdot\vec{x})^2 + \frac{1}{2}\{\vec{a}\wedge\vec{x}\}^2, \tag{3.4.96}$$

$$-\frac{1}{6}(\vec{a}\cdot\vec{\partial})^3\frac{1}{|\vec{x}|} = -\frac{1}{3}\vec{a}\cdot\vec{\partial}[\frac{1}{2}(\vec{a}\cdot\vec{\partial})^2\frac{1}{|\vec{x}|}] = -\frac{1}{3}\vec{a}\cdot\vec{\partial}[\frac{1}{2}\frac{3(\vec{a}\cdot\vec{x})^2 - \vec{x}^2\vec{a}^2}{|\vec{x}|^5}]$$

$$\overset{\text{Props. 3.4.4,3.4.5}}{=} -\frac{1}{6}(\vec{a}\cdot\vec{\partial}\frac{1}{|\vec{x}|^5})[3(\vec{a}\cdot\vec{x})^2 - \vec{x}^2\vec{a}^2]$$

$$- \frac{1}{6|\vec{x}|^5}[3\,2(\vec{a}\cdot\vec{x})\vec{a}\cdot\vec{\partial}(\vec{a}\cdot\vec{x}) - \vec{a}^2\vec{a}\cdot\vec{\partial}\vec{x}^2]$$

$$\overset{\text{Props. 3.4.5,3.4.7,3.4.21,3.4.23}}{=} -\frac{-5}{6|\vec{x}|^6}(\vec{a}\cdot\vec{\partial}|\vec{x}|)[3(\vec{a}\cdot\vec{x})^2 - \vec{x}^2\vec{a}^2]$$

$$- \frac{1}{6|\vec{x}|^5}[6(\vec{a}\cdot\vec{x})\vec{a}\cdot\vec{a} - \vec{a}^2 2(\vec{a}\cdot\vec{x})]$$

$$\overset{\text{Prop. 3.4.10}}{=} \frac{5}{6|\vec{x}|^6}\frac{\vec{a}\cdot\vec{x}}{|\vec{x}|}[3(\vec{a}\cdot\vec{x})^2 - \vec{x}^2\vec{a}^2] - \frac{4\vec{a}^2\vec{a}\cdot\vec{x}}{6|\vec{x}|^5}$$

$$= \frac{15(\vec{a}\cdot\vec{x})^3 - 5\vec{x}^2\vec{a}^2\vec{a}\cdot\vec{x} - 4\vec{x}^2\vec{a}^2\vec{a}\cdot\vec{x}}{6|\vec{x}|^7}$$

$$= \frac{1}{2|\vec{x}|^7}[5(\vec{a}\cdot\vec{x})^3 - 3\vec{x}^2\vec{a}^2\vec{a}\cdot\vec{x}]$$

$$= \frac{1}{|\vec{x}|^7}[(\vec{a}\cdot\vec{x})^3 + \frac{3}{2}\vec{a}\cdot\vec{x}((\vec{a}\cdot\vec{x})^2 - \vec{x}^2\vec{a}^2)]$$

$$\overset{(3.3.43)}{=} \frac{1}{|\vec{x}|^7}[(\vec{a}\cdot\vec{x})^3 + \frac{3}{2}\vec{a}\cdot\vec{x}(\vec{a}\wedge\vec{x})^2] \tag{3.4.97}$$

$$\Rightarrow \quad P_3(\vec{x}\vec{a}) = \frac{1}{2}[5(\vec{a}\cdot\vec{x})^3 - 3\vec{a}^2\vec{x}^2\vec{x}\cdot\vec{a}]$$
$$= (\vec{a}\cdot\vec{x})^3 + \frac{3}{2}\vec{x}\cdot\vec{a}(\vec{a}\wedge\vec{x})^2. \tag{3.4.98}$$

Homogeneity of degree n of the P_n :

$$\frac{(-\vec{a}\cdot\vec{\partial})^n}{n!}\frac{1}{|\vec{x}|} \overset{\text{def.}}{=} \frac{P_n(\vec{x}\vec{a})}{|\vec{x}|^{2n+1}} \quad \Rightarrow \quad (-\vec{a}\cdot\vec{\partial})^n\frac{1}{|\vec{x}|} = n!\frac{P_n(\vec{x}\vec{a})}{|\vec{x}|^{2n+1}} \tag{3.4.99}$$

and

$$(-\vec{a}\cdot\vec{\partial})^{n+1}\frac{1}{|\vec{x}|} = (n+1)!\frac{P_{n+1}(\vec{x}\vec{a})}{|\vec{x}|^{2(n+1)+1}}$$
$$\Rightarrow \quad \frac{(n+1)!}{n!}\frac{P_{n+1}(\vec{x}\vec{a})}{|\vec{x}|^{2(n+1)+1}} = (-\vec{a}\cdot\vec{\partial})\frac{P_n(\vec{x}\vec{a})}{|\vec{x}|^{2n+1}}$$
$$\overset{\text{Prop. 3.4.5}}{=} [-\vec{a}\cdot\vec{\partial}\frac{1}{|\vec{x}|^{2n+1}}]P_n(\vec{x}\vec{a}) - \frac{1}{|\vec{x}|^{2n+1}}\vec{a}\cdot\vec{\partial}P_n(\vec{x}\vec{a})$$
$$\overset{\text{Prop. 3.4.7}}{=} \frac{2n+1}{|\vec{x}|^{2n+2}}P_n(\vec{x}\vec{a})\vec{a}\cdot\vec{\partial}|\vec{x}| - \frac{1}{|\vec{x}|^{2n+1}}\vec{a}\cdot\vec{\partial}P_n(\vec{x}\vec{a})$$
$$\overset{\text{Prop. 3.4.10}}{=} \frac{2n+1}{|\vec{x}|^{2n+2}}P_n(\vec{x}\vec{a})\frac{\vec{a}\cdot\vec{x}}{|\vec{x}|} - \frac{1}{|\vec{x}|^{2n+1}}\vec{a}\cdot\vec{\partial}P_n(\vec{x}\vec{a})$$
$$= \frac{2n+1}{|\vec{x}|^{2n+3}}P_n(\vec{x}\vec{a})\vec{a}\cdot\vec{x} - \frac{1}{|\vec{x}|^{2n+1}}\vec{a}\cdot\vec{\partial}P_n(\vec{x}\vec{a}). \tag{3.4.100}$$

$P_n(\vec{x}\vec{a})\vec{a}\cdot\vec{x}$ is a homogeneous function of degree $n+1$, if we assume P_n to be homogeneous of degree n :

$$P_n(\vec{x}\vec{a}) \overset{\text{assume}}{=} \sum_{k=0}^{n} \alpha_k(\vec{x}\cdot\vec{a})^k|\vec{x}|^{n-k}\vec{a}^{n-k}, \tag{3.4.101}$$

$\alpha_k =$ const., which is especially true for $n = 0, 1, 2, 3$. The right term

$$\vec{a}\cdot\vec{\partial}P_n(\vec{x}\vec{a}) = \vec{a}\cdot\vec{\partial}\sum_{k=0}^{n}\alpha_k(\vec{x}\cdot\vec{a})^k|\vec{x}|^{n-k}\vec{a}^{n-k}$$
$$\overset{\text{Props. 3.4.4,3.4.5,3.4.9}}{=} \sum_{k=0}^{n}\Big\{\alpha_k[\vec{a}\cdot\vec{\partial}(\vec{x}\cdot\vec{a})^k]|\vec{x}|^{n-k}\vec{a}^{n-k}$$
$$+\alpha_k(\vec{x}\cdot\vec{a})^k[\vec{a}\cdot\vec{\partial}|\vec{x}|^{n-k}]\vec{a}^{n-k}\Big\}$$
$$\overset{\text{Props. 3.4.5,3.4.7}}{=} \sum_{k=0}^{n}\Big\{\alpha_k[k(\vec{x}\cdot\vec{a})^{k-1}\vec{a}\cdot\vec{\partial}(\vec{x}\cdot\vec{a})]|\vec{x}|^{n-k}\vec{a}^{n-k}$$
$$+\alpha_k(\vec{x}\cdot\vec{a})^k[(n-k)|\vec{x}|^{n-k-1}\vec{a}\cdot\vec{\partial}|\vec{x}|]\vec{a}^{n-k}\Big\}$$
$$\overset{\text{Props. 3.4.10,3.4.21}}{=} \sum_{k=0}^{n}\Big\{\alpha_k[k(\vec{x}\cdot\vec{a})^{k-1}\vec{a}\cdot\vec{a}]|\vec{x}|^{n-k}\vec{a}^{n-k}$$
$$+\alpha_k(\vec{x}\cdot\vec{a})^k[(n-k)|\vec{x}|^{n-k-1}\frac{\vec{a}\cdot\vec{x}}{|\vec{x}|}]\vec{a}^{n-k}\Big\}$$

$$= \sum_{k=0}^{n} \left\{ \alpha_k k(\vec{x}\cdot\vec{a})^{k-1} \frac{|\vec{x}|^2}{|\vec{x}|^2} |\vec{x}|^{n-k} \vec{a}^2 \vec{a}^{n-k} \right.$$
$$\left. + \alpha_k (n-k)(\vec{x}\cdot\vec{a})^{k+1} \frac{|\vec{x}|^{n-k-1}}{|\vec{x}|} \frac{|\vec{x}|}{|\vec{x}|} \vec{a}^{n-k} \right\}$$
$$= \frac{1}{|\vec{x}|^2} \sum_{k=0}^{n} \left\{ \alpha_k k(\vec{x}\cdot\vec{a})^{k-1} |\vec{x}|^{n-k+2} \vec{a}^{n-k+2} \right.$$
$$\left. + \alpha_k (n-k)(\vec{x}\cdot\vec{a})^{k+1} |\vec{x}|^{n-k} \vec{a}^{n-k} \right\} \tag{3.4.102}$$

yields $|\vec{x}|^2\, \vec{a}\cdot\vec{\partial} P_n(\vec{x}\vec{a})$, to be homogeneous of degree $n+1$. Hence

$$\frac{(n+1)!}{n!} \frac{P_{n+1}(\vec{x}\vec{a})}{|\vec{x}|^{2(n+1)+1}} = \frac{(\text{Polynomial homogeneous of degree } n+1)}{|\vec{x}|^{2(n+1)+1}}. \tag{3.4.103}$$

By induction every P_n will therefore be homogeneous of degree n. This and the explicit expressions above for $\frac{P_{n+1}(\vec{x}\vec{a})}{|\vec{x}|^{2(n+1)+1}}$ fully prove for all n :

$$P_n(\vec{x}\vec{a}) = |\vec{x}|^n P_n(\hat{x}\vec{a}) = |\vec{x}|^n |\vec{a}|^n P_n(\hat{x}\hat{a}). \tag{3.4.104}$$

□

Definition 3.4.37. *Redefinition of differential, over-dots.*

$$\underline{F}(\vec{a}) = \vec{a}\cdot\vec{\partial}F \overset{(3.3.7)}{=} \frac{1}{2}(\vec{a}\vec{\partial}F + \vec{\partial}\vec{a}\dot{F}), \tag{3.4.105}$$

where the over-dots indicate, that only F is to be differentiated and not $\vec{a}$.

Proposition 3.4.38. *For $\vec{a} \notin \mathcal{A}_n \equiv \mathcal{G}^1(I)$, $P = P_I$:*

$$\vec{a}\cdot\vec{\partial}_{\vec{x}} = \vec{a}\cdot P(\vec{\partial}_{\vec{x}}) = P(\vec{a})\cdot\vec{\partial}_{\vec{x}}, \quad P(\vec{a}) \in \mathcal{A}_n. \tag{3.4.106}$$

Proof.

$$\vec{a}\cdot\vec{\partial}_{\vec{x}} \overset{\text{Prop. 3.4.18}}{=} \vec{a}\cdot P(\vec{\partial}_{\vec{x}}) \overset{\text{Prop. 3.4.18}}{=} \sum_{k=1}^{n} \vec{a}\cdot\vec{a}^k(\vec{a}_k\cdot\vec{\partial}_{\vec{x}})$$
$$= \sum_{k=1}^{n} P(\vec{a})\cdot\vec{a}^k(\vec{a}_k\cdot\vec{\partial}_{\vec{x}}) = P(\vec{a})\cdot\vec{\partial}_{\vec{x}}. \tag{3.4.107}$$

□

Proposition 3.4.39.

$$\underline{F}(\vec{a}) = \underline{F}(P(\vec{a})) = P(\vec{a})\cdot\vec{\partial}F. \qquad \underline{F}(\vec{a}) = 0, \text{ if } P(\vec{a}) = 0. \tag{3.4.108}$$

Proof.

$$\underline{F}(\vec{a}) \overset{\text{Def. 3.4.13, Prop. 3.4.38}}{=} \underline{F}(P(\vec{a})) \overset{\text{Def. 3.4.13}}{=} P(\vec{a})\cdot\vec{\partial}F,$$
$$\Rightarrow \quad \underline{F}(\vec{a}) = 0, \text{ if } P(\vec{a}) = 0. \tag{3.4.109}$$

□

Proposition 3.4.40. *Differential of composite functions.*
For $F(\vec{x}) = G(f(\vec{x}))$ *and*

$$f : \vec{x} \in \mathcal{A}_n = \mathcal{G}^1(I) \to f(\vec{x}) \in \mathcal{A}'_n = \mathcal{G}^1(I'), \tag{3.4.110}$$

we have

$$\vec{a} \cdot \vec{\partial} F = \underline{f} \cdot \vec{\partial} G, \qquad \underline{F}(\vec{a}) = \underline{G}(f(\vec{x}), \underline{f}(\vec{x}, \vec{a})) \qquad (Def.\ 3.4.13). \tag{3.4.111}$$

The differential of composite functions is the composite of differentials :

$$\underline{F}(\vec{x}, \vec{a}) = \underline{G}(f(\vec{x}), \underline{f}(\vec{x}, \vec{a})) \quad (explicit). \tag{3.4.112}$$

Proof. Using Taylor expansion (Prop. 3.4.12):

$$f(\vec{x} + \tau\vec{a}) = f(\vec{x}) + \tau\vec{a} \cdot \vec{\partial} f(\vec{x}) + \frac{1}{2}\tau^2(\vec{a} \cdot \vec{\partial})^2 f(\vec{x}) + \ldots, \tag{3.4.113}$$

$$\begin{aligned}
\vec{a} \cdot \vec{\partial} G(f(\vec{x})) &\overset{\text{Def. 3.4.13}}{=} \frac{d}{d\tau} G(f(\vec{x} + \tau\vec{a}))\Big|_{\tau=0} \\
&\overset{\text{Taylor (Prop. 3.4.12), Def. 3.4.13}}{=} \frac{d}{d\tau} G(f(\vec{x}) + \tau \underline{f}(\vec{a}))\Big|_{\tau=0} \\
&\overset{\text{Def. 3.4.1}}{=} \underline{f}(\vec{a}) \cdot \vec{\partial}_{\vec{x}'} G(\vec{x}')\Big|_{\vec{x}'=f(\vec{x})} \\
&= \underline{f}(\vec{a}) \cdot \vec{\partial} G \quad \text{(evaluation at corresponding points.)}
\end{aligned} \tag{3.4.114}$$

□

Definition 3.4.41. *Second differential.*

$$F_{\vec{a}\vec{b}} = \vec{b} \cdot \dot{\vec{\partial}} \vec{a} \cdot \vec{\partial} \dot{F}(\vec{x}). \tag{3.4.115}$$

Suppressing $\vec{x}$: $F_{\vec{a}\vec{b}} = \vec{b} \cdot \dot{\vec{\partial}} \vec{a} \cdot \vec{\partial} \dot{F}$.

Proposition 3.4.42. *Integrability condition.*

$$F_{\vec{a}\vec{b}} = F_{\vec{b}\vec{a}}. \tag{3.4.116}$$

The second differential is a symmetric bilinear function of its differential arguments $\vec{a}, \vec{b}$.

Proof.

$$\begin{aligned}
F_{\vec{a}\vec{b}}(\vec{x}) \equiv \vec{b} \cdot \dot{\vec{\partial}} \vec{a} \cdot \vec{\partial} \dot{F}(\vec{x}) &\overset{\text{Def. 3.4.1}}{=} \vec{b} \cdot \dot{\vec{\partial}} \frac{d\dot{F}(\vec{x} + \tau\vec{a})}{d\tau}\Big|_{\tau=0} \\
&= \frac{d^2 \dot{F}(\vec{x} + \tau\vec{a} + \sigma\vec{b})}{d\sigma d\tau}\Big|_{\tau=0,\sigma=0} \\
&= \lim_{\sigma\to 0}\lim_{\tau \to 0} \frac{\frac{F(\vec{x}+\tau\vec{a}+\sigma\vec{b}) - F(\vec{x}+\sigma\vec{b})}{\tau} - \frac{F(\vec{x}+\tau\vec{a}) - F(\vec{x})}{\tau}}{\sigma},
\end{aligned} \tag{3.4.117}$$

which is symmetric under $(\vec{a}, \tau) \leftrightarrow (\vec{b}, \sigma)$. Hence

$$F_{\vec{a}\vec{b}}(\vec{x}) = \vec{b} \cdot \dot{\vec{\partial}} \vec{a} \cdot \vec{\partial} \dot{F}(\vec{x}) = \vec{a} \cdot \dot{\vec{\partial}} \vec{b} \cdot \vec{\partial} \dot{F}(\vec{x}) = F_{\vec{b}\vec{a}}(\vec{x}). \tag{3.4.118}$$

The bilinearity follows from the linearity in each argument (Props. 3.4.2, 3.4.3, 3.4.14). □

Proposition 3.4.43. *Differential of identity function.*

$$\vec{a} \cdot \vec{\partial}_{\vec{x}} \vec{x} = P(\vec{a}) = \vec{\partial}_{\vec{x}}(\vec{x} \cdot \vec{a}). \tag{3.4.119}$$

Proof. We first prove the first identity:

$$\vec{a} \cdot \vec{\partial}_{\vec{x}} \vec{x} \overset{\text{Prop. 3.4.38}}{=} P_I(\vec{a}) \cdot \vec{\partial}_{\vec{x}} \vec{x} \overset{\text{Prop. 3.4.8}}{=} P_I(\vec{a}). \tag{3.4.120}$$

Regarding the second identity, especially for base vectors $\vec{a}_k \in \mathcal{A}_n = \mathcal{G}^1(I)$:

$$\vec{a}_k \cdot \vec{\partial}_{\vec{x}} \vec{x} = P_I(\vec{a}_k) = \vec{a}_k, \tag{3.4.121}$$

$$\begin{aligned} \vec{\partial}_{\vec{x}}(\vec{x} \cdot \vec{a}) &\overset{\text{Prop. 3.4.18}}{=} \sum_k \vec{a}^k \vec{a}_k \cdot \vec{\partial}_{\vec{x}}(\vec{x} \cdot \vec{a}) \\ &\overset{\text{Prop. 3.4.21}}{=} \sum_k \vec{a}^k \vec{a}_k \cdot \vec{a} = P_I(\vec{a}). \end{aligned} \tag{3.4.122}$$

□

Proposition 3.4.44. *Operator identity.*

$$\vec{\partial}_{\vec{x}} = P_I(\vec{\partial}_{\vec{x}}) = \vec{\partial}_{\vec{a}}(\vec{a} \cdot \vec{\partial}_{\vec{x}}). \tag{3.4.123}$$

Proof. Proposition 3.4.18 and 3.4.43. □

Proposition 3.4.45. *Derivative from differential.*

$$\vec{\partial}_{\vec{x}} F(\vec{x}) = \vec{\partial}_{\vec{a}}(\vec{a} \cdot \vec{\partial}_{\vec{x}}) F(\vec{x}) = \vec{\partial}_{\vec{a}} \underline{F}(\vec{x}, \vec{a}) \tag{3.4.124}$$

Proof. Proposition 3.4.44 and Definition 3.4.13. □

Definition 3.4.46.

$$\vec{\partial} F = \vec{\partial}_{\vec{x}} F(\vec{x}) \overset{\textit{Prop. 3.4.45}}{=} \vec{\partial}_{\vec{a}} \underline{F}(\vec{x}, \vec{a}) = \underline{\vec{\partial}}\, \underline{F}, \tag{3.4.125}$$

where $\underline{\vec{\partial}}$ is the derivative with respect to the differential argument $\vec{a}$ of $\underline{F}(\vec{x}, \vec{a})$.

Proposition 3.4.47.

$$\vec{\partial} F = \vec{\partial} \cdot F + \vec{\partial} \wedge F. \tag{3.4.126}$$

Proof. Vector property (Prop. 3.4.18) of $\vec{\partial} = \vec{\partial}_{\vec{x}}$ and (3.3.18). □

Proposition 3.4.48. *Gradient. For scalar $F = \Phi(\vec{x})$:*

$$\vec{\partial} \cdot \Phi = 0, \qquad \vec{\partial} \Phi = \vec{\partial} \wedge \Phi = \dot{\Phi} \dot{\vec{\partial}}. \tag{3.4.127}$$

Proof. (3.3.21), Prop. 3.4.47 and (3.3.23). □

Remark 3.4.49. *In Prop. 3.4.48 the special definition of Hestenes and Sobczyk[161] in* (3.3.20) *and* (3.3.21) *for the inner product becomes important. It should be possible to make it more intuitive by replacing the inner product with the contraction [110].*

Definition 3.4.50. *Divergence and curl.*

- *Divergence of* $F : \vec{\partial} \cdot F$,
- *Curl of* $F : \vec{\partial} \wedge F$.

(Full vector derivative of F*:* $\vec{\partial} F$*.)*

Proposition 3.4.51. *Vector derivative of sums.*

$$\vec{\partial}(F+G) = \vec{\partial}F + \vec{\partial}G. \tag{3.4.128}$$

Proof.

$$\begin{aligned}
\vec{\partial}(F+G) &\overset{\text{Def. 3.4.46}}{=} \vec{\partial}(\underline{F+G}) \overset{\text{Prop. 3.4.4}}{=} \vec{\partial}(\underline{F}+\underline{G}) \\
&\overset{\text{Prop. 3.4.18}}{=} \sum_k \vec{a}^k \vec{a}_k \cdot \vec{\partial}_{\vec{a}}(\underline{F}+\underline{G}) \\
&\overset{\text{Prop. 3.4.4}}{=} \sum_k \vec{a}^k (\vec{a}_k \cdot \vec{\partial}_{\vec{a}}\underline{F} + \vec{a}_k \cdot \vec{\partial}_{\vec{a}}\underline{G}) \\
&\overset{\text{distributivity, (3.3.19)}}{=} \sum_k \vec{a}^k \vec{a}_k \cdot \vec{\partial}_{\vec{a}}\underline{F} + \sum_k \vec{a}^k \vec{a}_k \cdot \vec{\partial}_{\vec{a}}\underline{G} \\
&\overset{\text{Prop. 3.4.18, Def. 3.4.46}}{=} \vec{\partial}F + \vec{\partial}G.
\end{aligned} \tag{3.4.129}$$

Note that geometric multiplication is *distributive* with respect to addition. □

Proposition 3.4.52. *Vector derivative of products.*

$$\vec{\partial}(FG) = \dot{\vec{\partial}}\dot{F}G + \dot{\vec{\partial}}F\dot{G}. \tag{3.4.130}$$

Proof.

$$\begin{aligned}
\vec{\partial}(FG) &\overset{\text{Def. 3.4.46}}{=} \vec{\partial}(\underline{FG}) \overset{\text{Prop. 3.4.5}}{=} \vec{\partial}(\underline{F}G + F\underline{G}) \\
&\overset{\text{Def. 3.4.46, Prop. 3.4.51}}{=} \underline{\vec{\partial}}\,\underline{F}G + \underline{\vec{\partial}}F\underline{G} \\
&\overset{\text{Def. 3.4.46}}{=} \dot{\vec{\partial}}\dot{F}G + \dot{\vec{\partial}}F\dot{G}.
\end{aligned} \tag{3.4.131}$$

The third equality is a special case of Prop. 3.4.51, if we take the definition of $\underline{\vec{\partial}}$ in Def. 3.4.46 into account. The last term is to be interpreted as:

$$\dot{\vec{\partial}}F\dot{G} = \vec{\partial}_{\vec{y}}(F(\vec{x})G(\vec{y}))\Big|_{\vec{y}=\vec{x}}. \tag{3.4.132}$$

□

Proposition 3.4.53.

$$\vec{\partial}\vec{x}^2 = 2\vec{x}. \tag{3.4.133}$$

Proof.

$$\begin{aligned}\vec{\partial}\vec{x}^2 &= (\vec{\partial}\vec{x})\vec{x} + \dot{\vec{\partial}}\vec{x}\dot{\vec{x}} \overset{\text{Prop. 3.4.18}}{=} (\sum_k \vec{a}^k\vec{a}_k \cdot \vec{\partial}_{\vec{x}}\vec{x})\vec{x} + \sum_k \vec{a}^k\vec{x}\vec{a}_k \cdot \vec{\partial}_{\vec{x}}\vec{x} \\ &\overset{\text{Prop. 3.4.43}}{=} (\sum_k \vec{a}^k\vec{a}_k)\vec{x} + \sum_k \vec{a}^k\vec{x}\vec{a}_k \overset{(3.3.7)}{=} 2\sum_k \vec{a}^k\vec{x}\cdot\vec{a}_k \\ &\overset{(3.3.32)}{=} 2\vec{x}.\end{aligned} \tag{3.4.134}$$

□

Proposition 3.4.54.

$$\vec{\partial}|\vec{x}| = \hat{x}. \tag{3.4.135}$$

Proof.

$$\vec{\partial}|\vec{x}| \overset{\text{Prop. 3.4.18}}{=} \sum_k \vec{a}^k\vec{a}_k \cdot \vec{\partial}_{\vec{x}}|\vec{x}| \overset{\text{Prop. 3.4.10}}{=} \sum_k \vec{a}^k\vec{a}_k \cdot \hat{x} \overset{(3.3.32)}{=} \hat{x}. \tag{3.4.136}$$

□

Proposition 3.4.55. *For* $F = F(|\vec{x}|)$ *:*

$$\vec{\partial}F = \hat{x}\frac{dF}{d|\vec{x}|}. \tag{3.4.137}$$

Proof.

$$\vec{a}\cdot\vec{\partial}F \overset{\text{Prop. 3.4.7}}{=} \vec{a}\cdot\vec{\partial}|\vec{x}|\frac{dF}{d|\vec{x}|} \overset{\text{Prop. 3.4.54}}{=} \vec{a}\cdot\hat{x}\frac{dF}{d|\vec{x}|} \tag{3.4.138}$$

$$\Rightarrow \quad \vec{\partial}F \overset{\text{Def. 3.4.46}}{=} \vec{\partial}_{\vec{a}}(\vec{a}\cdot\vec{\partial})F = \vec{\partial}_{\vec{a}}\vec{a}\cdot\hat{x}\frac{dF}{d|\vec{x}|} \overset{\text{Prop. 3.4.43}}{=} \hat{x}\frac{dF}{d|\vec{x}|}. \tag{3.4.139}$$

□

Definition 3.4.56. *Sides of differentiation. Only right side differentiation:*

$$F\vec{\partial}G \overset{\text{Def. 3.4.46}}{=} F\underline{\vec{\partial}}\,\underline{G} = F\dot{\vec{\partial}}\dot{G}. \tag{3.4.140}$$

Left and right side differentiation (another form of the product rule Prop. 3.4.52):

$$\dot{F}\dot{\vec{\partial}}\dot{G} = \dot{F}\dot{\vec{\partial}}G + F\dot{\vec{\partial}}\dot{G}. \tag{3.4.141}$$

Definition 3.4.57. *Adjoint. For*

$$f : \vec{x} \in \mathcal{A}_n = \mathcal{G}^1(I) \to f(\vec{x}) \in \mathcal{A}'_n = \mathcal{G}^1(I'), \tag{3.4.142}$$

the entity

$$\overline{f}(\vec{a}') \equiv \underline{\vec{\partial}}(\underline{f}\cdot\vec{a}') \tag{3.4.143}$$

is the adjointof f or explicitly:

$$\overline{f}(\vec{x},\vec{a}') \equiv \vec{\partial}_{\vec{a}}\left[\Big((\vec{a}\cdot\vec{\partial}_{\vec{x}})f(\vec{x})\Big)\cdot\vec{a}'\right] \overset{\text{Def. 3.4.37}}{=} \vec{\partial}_{\vec{a}}[\underline{f}(\vec{x},\vec{a})\cdot\vec{a}']. \tag{3.4.144}$$

Proposition 3.4.58.

$$\overline{f}(\vec{a}') = \vec{\partial}(f \cdot \vec{a}'), \tag{3.4.145}$$

or explicitly:

$$\overline{f}(\vec{x}, \vec{a}') = \vec{\partial}_{\vec{x}}(f(\vec{x}) \cdot \vec{a}'). \tag{3.4.146}$$

Proof.

$$\begin{aligned}\overline{f}(\vec{x}, \vec{a}') &\overset{\text{Def. 3.4.57}}{=} \vec{\partial}_{\vec{a}} \left[\left((\vec{a} \cdot \vec{\partial}_{\vec{x}}) f(\vec{x}) \right) \cdot \vec{a}' \right] = \vec{\partial}_{\vec{a}} (\vec{a} \cdot \vec{\partial}_{\vec{x}}) \left[f(\vec{x}) \cdot \vec{a}' \right] \\ &\overset{\text{Prop. 3.4.45}}{=} \vec{\partial}_{\vec{x}} \left(f(\vec{x}) \cdot \vec{a}' \right). \end{aligned} \tag{3.4.147}$$

(Compare [161] p. 50; [107] p. 23 (1.109), p. 24 (1.118), and p. 104 (5.11).) □

Proposition 3.4.59. *Linearity of adjoint. For scalar* $\alpha \in \mathbb{R}$ *:*

$$\overline{f}(\vec{a}' + \vec{b}') = \overline{f}(\vec{a}') + \overline{f}(\vec{b}'), \qquad \overline{f}(\alpha \vec{a}') = \alpha \overline{f}(\vec{a}'). \tag{3.4.148}$$

Proof. Linearity of the inner product (3.3.9). □

Proposition 3.4.60.

$$P\left(\overline{f}(\vec{a}')\right) = \overline{f}(\vec{a}'). \tag{3.4.149}$$

Proof.

$$\begin{aligned} P\left(\overline{f}(\vec{a}')\right) &\overset{\text{[illegible]}}{=} P\left(\vec{\partial}_{\vec{x}}(f(\vec{x}) \cdot \vec{a}')\right) = P(\vec{\partial}_{\vec{x}}) f(\vec{x}) \cdot \vec{a}' \\ &\overset{\text{Prop. 3.4.44}}{=} \vec{\partial}_{\vec{x}}(f(\vec{x}) \cdot \vec{a}') \overset{\text{Prop. 3.4.58}}{=} \overline{f}(\vec{a}'). \end{aligned} \tag{3.4.150}$$

□

Proposition 3.4.61.

$$\overline{f}(\vec{a}') = \vec{\partial}_{\vec{a}}(\underline{f}(\vec{a}) \cdot \vec{a}') = \overline{f}(P'(\vec{a}')), \tag{3.4.151}$$

with P' *the projection into the range of* f *and* $\underline{f}$*, i.e. into* $\mathcal{A}_n' \equiv \mathcal{G}^1(I')$*.*

Proof.

$$\begin{aligned} \overline{f}(\vec{a}') &\overset{\text{Def. 3.4.57}}{=} \vec{\partial}_{\vec{a}}(\underline{f}(\vec{a}) \cdot \vec{a}') \overset{\text{Prop. 3.4.58}}{=} \vec{\partial}_{\vec{x}}(f(\vec{x}) \cdot \vec{a}') = \vec{\partial}_{\vec{x}}([P' f(\vec{x})] \cdot \vec{a}') \\ &= \vec{\partial}_{\vec{x}}(f(\vec{x}) \cdot P'(\vec{a}')) \overset{\text{Prop. 3.4.58}}{=} \overline{f}(P'(\vec{a}')). \end{aligned} \tag{3.4.152}$$

□

Proposition 3.4.62.

$$P \overline{f} P'(\vec{a}') = \overline{f}(\vec{a}'), \qquad P' \underline{f} P(\vec{a}) = \underline{f}(\vec{a}). \tag{3.4.153}$$

Proof. First identity: Definition 3.4.57 and Proposition 3.4.58.
Second identity: Proposition 3.4.39 and because the range of f and $\underline{f}$ is $\mathcal{A}_n' = \mathcal{G}^i(I')$.
□

Proposition 3.4.63. *Change of variables. For $F(\vec{x}) = G(f(\vec{x}))$, i.e. $\vec{x} \to f(\vec{x})$:*

$$\vec{\partial}_{\vec{x}} F(\vec{x}) = \overline{f}(\vec{\partial}_{\vec{x}'}) G(\vec{x}'), \tag{3.4.154}$$

i.e. $\vec{\partial}_{\vec{x}} = \overline{f}(\vec{\partial}_{\vec{x}'})$.

Proof.

$$\begin{aligned}\vec{\partial}_{\vec{x}} F(\vec{x}) &\overset{\text{Prop. 3.4.44}}{=} \vec{\partial}_{\vec{a}}[\vec{a}\cdot\vec{\partial}_{\vec{x}} F(\vec{x})] = \vec{\partial}_{\vec{a}}[\vec{a}\cdot\vec{\partial}_{\vec{x}} G(f(\vec{x}))]\\ &\overset{\text{Prop. 3.4.40}}{=} \vec{\partial}_{\vec{a}}[\underbrace{\underline{f}(\vec{a})\cdot\vec{\partial}_{\vec{x}'}}_{\text{scalar}} G(\vec{x}')]\Big|_{\vec{x}'=f(\vec{x})} = [\vec{\partial}_{\vec{a}}(\underline{f}(\vec{a})\cdot\vec{\partial}_{\vec{x}'})G(\vec{x}')]\Big|_{\vec{x}'=f(\vec{x})}\\ &\overset{\text{Def. 3.4.57}}{=} [\overline{f}(\vec{\partial}_{\vec{x}'})G(\vec{x}')]_{\vec{x}'=f(\vec{x})} = \overline{f}(\vec{\partial}_{\vec{x}'})G(\vec{x}'). \end{aligned} \tag{3.4.155}$$

□

Proposition 3.4.64. *Second derivative.*

$$\vec{\partial}^2_{\vec{x}} F(\vec{x}) = \vec{\partial}_{\vec{b}}\vec{\partial}_{\vec{a}} F_{\vec{a}\vec{b}} = (\vec{\partial}_{\vec{b}}\cdot\vec{\partial}_{\vec{a}} + \vec{\partial}_{\vec{b}}\wedge\vec{\partial}_{\vec{a}})F_{\vec{a}\vec{b}}. \tag{3.4.156}$$

Proof.

$$\vec{\partial}_{\vec{x}} F(\vec{x}) \overset{\text{Prop. 3.4.54}}{=} \vec{\partial}_{\vec{a}}(\vec{a}\cdot\vec{\partial}_{\vec{x}} F(\vec{x})) \overset{\text{Def. 3.4.13}}{=} \vec{\partial}_{\vec{a}} F_{\vec{a}} \tag{3.4.157}$$

$$\begin{aligned}\Rightarrow\quad \vec{\partial}^2_{\vec{x}} F(\vec{x}) &= \vec{\partial}_{\vec{x}}(\vec{\partial}_{\vec{x}} F(\vec{x})) = \vec{\partial}_{\vec{x}}(\vec{\partial}_{\vec{a}} F_{\vec{a}})\\ &\overset{\text{Prop. 3.4.54}}{=} \vec{\partial}_{\vec{b}}[\underbrace{\vec{b}\cdot\dot{\vec{\partial}}_{\vec{x}}}_{\text{scalar operator}}(\vec{\partial}_{\vec{a}}\dot{F}_{\vec{a}})] = \vec{\partial}_{\vec{b}}\vec{\partial}_{\vec{a}}[(\vec{b}\cdot\vec{\partial}_{\vec{x}})F_{\vec{a}}] \overset{\text{Def. 3.4.41}}{=} \vec{\partial}_{\vec{b}}\vec{\partial}_{\vec{a}} F_{\vec{a}\vec{b}}\\ &\overset{\text{Prop. 3.4.18}}{=} (\vec{\partial}_{\vec{b}}\cdot\vec{\partial}_{\vec{a}} + \vec{\partial}_{\vec{b}}\wedge\vec{\partial}_{\vec{a}})F_{\vec{a}\vec{b}}. \end{aligned} \tag{3.4.158}$$

□

Proposition 3.4.65. *Integrability condition for vector derivative.*

$$\vec{\partial}_{\vec{x}}\wedge\vec{\partial}_{\vec{x}} = 0 \quad\Leftrightarrow\quad F_{\vec{a}\vec{b}} = F_{\vec{b}\vec{a}}. \tag{3.4.159}$$

Proof. ($\Rightarrow$)

$$\vec{\partial}_{\vec{x}}\wedge\vec{\partial}_{\vec{x}} = 0 \tag{3.4.160}$$

$$\begin{aligned}\Rightarrow\quad 0 &= (\vec{a}\wedge\vec{b})\cdot(\vec{\partial}_{\vec{x}}\wedge\vec{\partial}_{\vec{x}})F \overset{(3.3.42)}{=} \vec{a}\cdot[\vec{b}\cdot(\vec{\partial}_{\vec{x}}\wedge\vec{\partial}_{\vec{x}})]F\\ &= \vec{a}\cdot[(\vec{b}\cdot\vec{\partial}_{\vec{x}})\vec{\partial}_{\vec{x}} - \vec{\partial}_{\vec{x}}(\vec{b}\cdot\vec{\partial}_{\vec{x}})]F = [(\vec{b}\cdot\vec{\partial}_{\vec{x}})(\vec{a}\cdot\vec{\partial}_{\vec{x}})F - (\vec{a}\cdot\vec{\partial}_{\vec{x}})(\vec{b}\cdot\vec{\partial}_{\vec{x}})F]\\ &\overset{\text{Def. 3.4.41}}{=} F_{\vec{a}\vec{b}} - F_{\vec{b}\vec{a}}, \end{aligned} \tag{3.4.161}$$

($\Leftarrow$) Integrability (Prop. 3.4.42) means: $F_{\vec{a}\vec{b}} = F_{\vec{b}\vec{a}}$. Obviously

$$\begin{aligned}0 &= \frac{1}{2}(\vec{\partial}_{\vec{b}}\vec{\partial}_{\vec{a}} F_{\vec{a}\vec{b}} - \vec{\partial}_{\vec{a}}\vec{\partial}_{\vec{b}} F_{\vec{b}\vec{a}}) \overset{\text{Prop. 3.4.42}}{=} \frac{1}{2}(\vec{\partial}_{\vec{b}}\vec{\partial}_{\vec{a}} - \vec{\partial}_{\vec{a}}\vec{\partial}_{\vec{b}})F_{\vec{a}\vec{b}} = \vec{\partial}_{\vec{b}}\wedge\vec{\partial}_{\vec{a}} F_{\vec{a}\vec{b}}\\ &\overset{\text{Prop. 3.4.64}}{=} \vec{\partial}_{\vec{x}}\wedge\vec{\partial}_{\vec{x}} F(\vec{x}) \end{aligned} \tag{3.4.162}$$

$$\Rightarrow\quad \vec{\partial}_{\vec{x}}\wedge\vec{\partial}_{\vec{x}} = 0. \tag{3.4.163}$$

□

Proposition 3.4.66. *Laplacian. Integrability of* $F \Leftrightarrow \vec{\partial}_{\vec{x}}^2 = \vec{\partial}_{\vec{x}} \cdot \vec{\partial}_{\vec{x}}$.

Proof. Integrability of $F \overset{\text{Prop. 3.4.65}}{\Leftrightarrow} \vec{\partial}_{\vec{x}} \wedge \vec{\partial}_{\vec{x}} = 0$

$$\Leftrightarrow \quad \vec{\partial}_{\vec{x}}^2 \overset{(3.3.6),\,(3.3.18)}{=} \vec{\partial}_{\vec{x}} \cdot \vec{\partial}_{\vec{x}} + \vec{\partial}_{\vec{x}} \wedge \vec{\partial}_{\vec{x}} = \vec{\partial}_{\vec{x}} \cdot \vec{\partial}_{\vec{x}}. \tag{3.4.164}$$

□

Proposition 3.4.67.

$$\vec{\partial}_{\vec{x}} \wedge \vec{x} = 0. \tag{3.4.165}$$

Proof.

$$\vec{\partial}_{\vec{x}} \wedge \vec{x} = \frac{1}{2}\vec{\partial}_{\vec{x}} \wedge (2\vec{x}) \overset{\text{Prop. 3.4.53}}{=} \frac{1}{2}(\vec{\partial}_{\vec{x}} \wedge \vec{\partial}_{\vec{x}})\vec{x}^2 \overset{\text{Prop. 3.4.65}}{=} 0, \tag{3.4.166}$$

because $F = \vec{x}^2$ is integrable. □

Proposition 3.4.68.

$$\vec{\partial}_{\vec{x}}\vec{x} = \vec{\partial}_{\vec{x}} \cdot \vec{x} \overset{(3.3.33)}{=} n. \tag{3.4.167}$$

Proof.

$$\begin{aligned}
\vec{\partial}_{\vec{x}}\vec{x} &\overset{\text{Prop. 3.4.47}}{=} \vec{\partial}_{\vec{x}} \cdot \vec{x} + \vec{\partial}_{\vec{x}} \wedge \vec{x} \overset{\text{Prop. 3.4.67}}{=} \vec{\partial}_{\vec{x}} \cdot \vec{x} \overset{\text{Prop. 3.4.18}}{=} \Big(\sum_{k=1}^{n} \vec{a}^k \vec{a}_k \cdot \vec{\partial}_{\vec{x}}\Big) \cdot \vec{x} \\
&= \sum_{k=1}^{n} \vec{a}^k \cdot (\vec{a}_k \cdot \vec{\partial}_{\vec{x}})\vec{x} \overset{\text{Prop. 3.4.43}}{=} \sum_{k=1}^{n} \vec{a}^k \cdot P(\vec{a}_k) \\
&= \sum_{k=1}^{n} \vec{a}^k \cdot \vec{a}_k = n.
\end{aligned} \tag{3.4.168}$$

□

Proposition 3.4.69.

$$\vec{\partial}_{\vec{x}}|\vec{x}|^k = k|\vec{x}|^{k-2}\vec{x}. \tag{3.4.169}$$

Proof.

$$\vec{\partial}_{\vec{x}}|\vec{x}|^k \overset{\text{Prop. 3.4.55}}{=} \hat{x}\frac{d|\vec{x}|^k}{d|\vec{x}|} = \hat{x}k|\vec{x}|^{k-1} = k|\vec{x}|^{k-2}|\vec{x}|\hat{x} = k|\vec{x}|^{k-2}\vec{x}. \tag{3.4.170}$$

□

Proposition 3.4.70.

$$\vec{\partial}_{\vec{x}}\Big(\frac{\vec{x}}{|\vec{x}|^k}\Big) = \frac{n-k}{|\vec{x}|^k}. \tag{3.4.171}$$

Proof.

$$\vec{\partial}_{\vec{x}}\Big(\frac{\vec{x}}{|\vec{x}|^k}\Big) \overset{\text{Prop. 3.4.52}}{=} \vec{\partial}_{\vec{x}}(\vec{x})\frac{1}{|\vec{x}|^k} + \vec{x}\vec{\partial}_{\vec{x}}|\vec{x}|^{-k}$$
$$\overset{\text{Prop. 3.4.68,3.4.69}}{=} n\frac{1}{|\vec{x}|^k} + \vec{x}(-k)|\vec{x}|^{-k-2}\vec{x} = n\frac{1}{|\vec{x}|^k} - k|\vec{x}|^{-k-2}\vec{x}^2$$
$$= \frac{n}{|\vec{x}|^k} - k|\vec{x}|^{-k} = \frac{n-k}{|\vec{x}|^k}. \tag{3.4.172}$$

□

Proposition 3.4.71.

$$\vec{\partial}_{\vec{x}} \log|\vec{x}| = \frac{\vec{x}}{|\vec{x}|^2} = \vec{x}^{-1}. \tag{3.4.173}$$

Proof.

$$\vec{\partial}_{\vec{x}} \log|\vec{x}| \overset{\text{Prop. 3.4.55}}{=} \dot{x}\frac{d\log|\vec{x}|}{|\vec{x}|} = \hat{x}\frac{1}{|\vec{x}|} \overset{(3.3.12)}{=} \frac{\vec{x}}{|\vec{x}|^2} \overset{(3.3.12)}{=} \vec{x}^{-1}. \tag{3.4.174}$$

□

Proposition 3.4.72. *For $A = P(A) = \langle A\rangle_r$:*

$$\dot{\vec{\partial}}_{\vec{x}}(\dot{\vec{x}} \cdot A) = A \cdot \vec{\partial}_{\vec{x}}\vec{x} = rA. \tag{3.4.175}$$

Proof. If A is a simple r-blade, then

$$\dot{\vec{\partial}}_{\vec{x}}(\dot{\vec{x}} \cdot A) \overset{(3.3.39)}{=} \dot{\vec{\partial}}_{\vec{x}}(A \cdot \dot{\vec{x}})(-1)^{r-1} = \frac{1}{2}[\dot{\vec{\partial}}_{\vec{x}}A\dot{\vec{x}} - (-1)^r \underbrace{(\vec{\partial}_{\vec{x}}\vec{x})}_{=n,\text{Prop. 3.4.68}} A](-1)^{r-1}$$
$$= \frac{1}{2}[\dot{\vec{\partial}}_{\vec{x}}A\dot{\vec{x}} - (-1)^r A(\vec{\partial}_{\vec{x}}\vec{x})](-1)^{r-1} = (\dot{\vec{\partial}}_{\vec{x}} \cdot A)\dot{\vec{x}}(-1)^{r-1}$$
$$\overset{(3.3.39)}{=} (A \cdot \vec{\partial}_{\vec{x}})\vec{x}, \tag{3.4.176}$$

$$\overset{(3.3.30)}{\Rightarrow} \quad A^{-1}\dot{\vec{\partial}}_{\vec{x}}(\dot{\vec{x}} \cdot A) = A^{-1}(A \cdot \vec{\partial}_{\vec{x}})\vec{x} \overset{(3.3.35)}{=} P_A(\vec{\partial}_{\vec{x}})\vec{x}$$
$$\overset{\text{Prop. 3.4.18}}{=} \Big(\sum_{k=1}^{r} \vec{a}^k\vec{a}_k \cdot \vec{\partial}_{\vec{x}}\Big)\vec{x} = \sum_{k=1}^{r} \vec{a}^k(\vec{a}_k \cdot \vec{\partial}_{\vec{x}})\vec{x}$$
$$\overset{\text{Prop. 3.4.43}}{=} \sum_{k=1}^{r} \vec{a}^k\vec{a}_k \overset{(3.3.33)}{=} r, \tag{3.4.177}$$

$$\Rightarrow \quad \dot{\vec{\partial}}_{\vec{x}}(\dot{\vec{x}} \cdot A) = (A \cdot \vec{\partial}_{\vec{x}})\vec{x} = rA. \tag{3.4.178}$$

Last step: Multiplication with A from the left. Distributivity (3.3.19), (3.3.25) gives the same result even for non-simple r-vectors A. □

Proposition 3.4.73. *For $A = P(A) = \langle A\rangle_r$:*

$$\dot{\vec{\partial}}_{\vec{x}}(\dot{\vec{x}} \wedge A) = A \wedge \vec{\partial}_{\vec{x}}\vec{x} = (n-r)A. \tag{3.4.179}$$

Proof. If A is a simple r-blade, then

$$
\begin{aligned}
\dot{\vec{\partial}}_{\vec{x}}(\dot{\vec{x}} \wedge A) &\overset{(3.3.40)}{=} \dot{\vec{\partial}}_{\vec{x}}(A \wedge \dot{\vec{x}})(-1)^r = \frac{1}{2}[\dot{\vec{\partial}}_{\vec{x}} A \dot{\vec{x}} + (-1)^r \underbrace{(\vec{\partial}_{\vec{x}} \vec{x})}_{=n,\text{Prop. 3.4.68}} A](-1)^r \\
&= \frac{1}{2}[\dot{\vec{\partial}}_{\vec{x}} A \dot{\vec{x}} + (-1)^r A(\vec{\partial}_{\vec{x}} \vec{x})](-1)^r = (\dot{\vec{\partial}}_{\vec{x}} \wedge A)\dot{\vec{x}}(-1)^r \\
&\overset{(3.3.40)}{=} (A \wedge \vec{\partial}_{\vec{x}})\vec{x},
\end{aligned} \tag{3.4.180}
$$

$$
\begin{aligned}
\overset{(3.3.30)}{\Rightarrow} \quad A^{-1}\dot{\vec{\partial}}_{\vec{x}}(\dot{\vec{x}} \wedge A) &= A^{-1}(A \wedge \vec{\partial}_{\vec{x}})\vec{x} = A^{-1}I^{-1}I(A \wedge \vec{\partial}_{\vec{x}})\vec{x} \\
&\overset{I\text{pseudoscalar, (3.3.40)}}{=} A^{-1}I^{-1}[\underbrace{I}_{n} \cdot \underbrace{(\vec{\partial}_{\vec{x}} \wedge A)}_{r+1 \text{ grades}}(-1)^r]\vec{x} \\
&\overset{(3.3.39)}{=} A^{-1}I^{-1}[(\vec{\partial}_{\vec{x}} \wedge A) \cdot I]\vec{x}(-1)^{r+(r+1)(n-r-1)} \\
&\overset{(3.3.38)}{=} A^{-1}I^{-1}[\vec{\partial}_{\vec{x}} \cdot (AI)]\vec{x}(-1)^{r+(r+1)(n-r)-r-1} \\
&\overset{(3.3.39)}{=} A^{-1}I^{-1}[\underbrace{(AI)}_{n-r} \cdot \vec{\partial}_{\vec{x}}]\vec{x}(-1)^{(r+1)(n-r)-1+n-r-1} \\
&\overset{(3.3.40)}{=} A^{-1}I^{-1}[\underbrace{(IA)}_{n-r} \cdot \vec{\partial}_{\vec{x}}]\vec{x}\underbrace{(-1)^{r(n-r)+r(n-r)}}_{=+1} \\
&\overset{(3.3.35),\,(3.3.36)}{=} P_{n-r}(\vec{\partial}_{\vec{x}})\vec{x} \overset{\text{Proof of Prop. 3.4.72}}{=} n - r.
\end{aligned} \tag{3.4.181}
$$

$$
\overset{AA^{-1}=1,(3.3.30)}{\Rightarrow} \quad \dot{\vec{\partial}}_{\vec{x}}(\dot{\vec{x}} \wedge A) = (A \wedge \vec{\partial}_{\vec{x}})\vec{x} = (n - r)A. \tag{3.4.182}
$$

Last step: Multiplication with A from the left. The distributive rule for the inner product gives the same result even for non-simple A. □

Proposition 3.4.74. *For* $A = P(A) = \langle A \rangle_r$ *:*

$$
\dot{\vec{\partial}}_{\vec{x}} A \dot{\vec{x}} = \sum_{k=1}^{n} \vec{a}^k A \vec{a}_k = (-1)^r(n - 2r)A. \tag{3.4.183}
$$

Proof.

$$
\begin{aligned}
\dot{\vec{\partial}}_{\vec{x}} A \dot{\vec{x}} &\overset{\text{Prop. 3.4.18}}{=} \sum_{k=1}^{n} \vec{a}^k(\vec{a}_k \cdot \dot{\vec{\partial}}_{\vec{x}})A\dot{\vec{x}} = \sum_{k=1}^{n} \vec{a}^k A(\vec{a}_k \cdot \dot{\vec{\partial}}_{\vec{x}})\dot{\vec{x}} \\
&\overset{\text{Prop. 3.4.43}}{=} \sum_{k=1}^{n} \vec{a}^k A \vec{a}_k,
\end{aligned} \tag{3.4.184}
$$

$$
\begin{aligned}
\dot{\vec{\partial}}_{\vec{x}} A \dot{\vec{x}} &= \dot{\vec{\partial}}_{\vec{x}}[A \cdot \dot{\vec{x}} + A \wedge \dot{\vec{x}}] \overset{(3.3.39),\,(3.3.40)}{=} \dot{\vec{\partial}}_{\vec{x}}[\dot{\vec{x}} \cdot A(-1)^{r-1} + \dot{\vec{x}} \wedge A(-1)^r] \\
&= (-1)^r[-\dot{\vec{\partial}}_{\vec{x}}(\dot{\vec{x}} \cdot A) + \dot{\vec{\partial}}_{\vec{x}}(\dot{\vec{x}} \wedge A)] \\
&\overset{\text{Props. 3.4.72, 3.4.73}}{=} (-1)^r[-rA + (n - r)A] \\
&= (-1)^r(n - 2r)A.
\end{aligned} \tag{3.4.185}
$$

□

Proposition 3.4.75. *For $\vec{a} = \vec{a}(\vec{x})$, $\vec{b} = \vec{b}(\vec{x})$:*

$$\vec{\partial}_{\vec{x}}(\vec{a} \cdot \vec{b}) = \vec{a} \cdot \vec{\partial}_{\vec{x}}\vec{b} + \vec{b} \cdot \vec{\partial}_{\vec{x}}\vec{a} - \vec{a} \cdot (\vec{\partial}_{\vec{x}} \wedge \vec{b}) - \vec{b} \cdot (\vec{\partial}_{\vec{x}} \wedge \vec{a}). \tag{3.4.186}$$

Proof.

$$\vec{a} \cdot (\vec{\partial}_{\vec{x}} \wedge \vec{b}) \overset{\text{Prop. 3.4.18, (3.3.41)}}{=} \vec{a} \cdot \dot{\vec{\partial}}_{\vec{x}}\dot{\vec{b}} - \dot{\vec{\partial}}_{\vec{x}}(\dot{\vec{b}} \cdot \vec{a}), \tag{3.4.187}$$

$$\vec{b} \cdot (\vec{\partial}_{\vec{x}} \wedge \vec{a}) \overset{\text{Prop. 3.4.18, (3.3.41)}}{=} \vec{b} \cdot \dot{\vec{\partial}}_{\vec{x}}\dot{\vec{a}} - \dot{\vec{\partial}}_{\vec{x}}(\dot{\vec{a}} \cdot \vec{b}), \tag{3.4.188}$$

$$\begin{aligned} \overset{\text{addition}}{\Rightarrow} \quad \vec{\partial}_{\vec{x}}(\vec{a} \cdot \vec{b}) &= \dot{\vec{\partial}}_{\vec{x}}(\dot{\vec{a}} \cdot \vec{b}) + \dot{\vec{\partial}}_{\vec{x}}(\dot{\vec{b}} \cdot \vec{a}) \\ &= \vec{a} \cdot \vec{\partial}_{\vec{x}}\vec{b} + \vec{b} \cdot \vec{\partial}_{\vec{x}}\vec{a} - \vec{a} \cdot (\vec{\partial}_{\vec{x}} \wedge \vec{b}) - \vec{b} \cdot (\vec{\partial}_{\vec{x}} \wedge \vec{a}). \end{aligned} \tag{3.4.189}$$

□

Definition 3.4.76. *Lie bracket. For $\vec{a} = \vec{a}(\vec{x})$, $\vec{b} = \vec{b}(\vec{x})$, the Lie bracket is defined as:*

$$[\vec{a}, \vec{b}] = \vec{a} \cdot \vec{\partial}_{\vec{x}}\,\vec{b} - \vec{b} \cdot \vec{\partial}_{\vec{x}}\,\vec{a}\,. \tag{3.4.190}$$

Proposition 3.4.77. *For $\vec{a} = \vec{a}(\vec{x})$, $\vec{b} = \vec{b}(\vec{x})$, we get the following expression for the Lie bracket :*

$$[\vec{a}, \vec{b}] = \vec{\partial}_{\vec{x}} \cdot (\vec{a} \wedge \vec{b}) - \vec{b}\vec{\partial}_{\vec{x}} \cdot \vec{a} + \vec{a}\vec{\partial}_{\vec{x}} \cdot \vec{b}. \tag{3.4.191}$$

Proof.

$$\begin{aligned} \vec{\partial}_{\vec{x}} \cdot (\vec{a} \wedge \vec{b}) &\overset{\text{Prop. 3.4.18, (3.3.41)}}{=} (\vec{\partial}_{\vec{x}} \cdot \vec{a})\vec{b} - (\vec{\partial}_{\vec{x}} \cdot \vec{b})\vec{a} \\ &= \vec{b}(\vec{\partial}_{\vec{x}} \cdot \vec{a}) + (\vec{a} \cdot \vec{\partial}_{\vec{x}})\vec{b} - \vec{a}(\vec{\partial}_{\vec{x}} \cdot \vec{b}) - (\vec{b} \cdot \vec{\partial}_{\vec{x}})\vec{a} \end{aligned} \tag{3.4.192}$$

$$\begin{aligned} \Rightarrow \quad [\vec{a}, \vec{b}] &\overset{\text{Def. 3.4.76}}{=} \vec{a} \cdot \vec{\partial}_{\vec{x}}\,\vec{b} - \vec{b} \cdot \vec{\partial}_{\vec{x}}\,\vec{a} \\ &= \vec{\partial}_{\vec{x}} \cdot (\vec{a} \wedge \vec{b}) - \vec{b}(\vec{\partial}_{\vec{x}} \cdot \vec{a}) + \vec{a}(\vec{\partial}_{\vec{x}} \cdot \vec{b}). \end{aligned} \tag{3.4.193}$$

□

Proposition 3.4.78. *For $\vec{a} = \vec{a}(\vec{x})$, $\vec{b} = \vec{b}(\vec{x})$, $\vec{c} = \vec{c}(\vec{x})$:*

$$\begin{aligned} (\vec{c} \wedge \vec{b}) \cdot (\vec{\partial}_{\vec{x}} \wedge \vec{a}) &= \vec{b} \cdot \dot{\vec{\partial}}_{\vec{x}}\dot{\vec{a}} \cdot \vec{c} - \vec{c} \cdot \dot{\vec{\partial}}_{\vec{x}}\dot{\vec{a}} \cdot \vec{b} \\ &= \vec{b} \cdot \vec{\partial}_{\vec{x}}(\vec{a} \cdot \vec{c}) - \vec{c} \cdot \vec{\partial}_{\vec{x}}(\vec{a} \cdot \vec{b}) + [\vec{c}, \vec{b}] \cdot \vec{a}, \end{aligned} \tag{3.4.194}$$

where the Lie bracket $[\vec{c}, \vec{b}]$ is defined according to Def. 3.4.76.

Proof.

$$\begin{aligned} (\vec{c} \wedge \vec{b}) \cdot (\vec{\partial}_{\vec{x}} \wedge \vec{a}) &\overset{(3.3.42)}{=} \vec{c} \cdot \Big(\vec{b} \cdot (\vec{\partial}_{\vec{x}} \wedge \vec{a})\Big) \overset{\text{Proof 3.4.75}}{=} \vec{c} \cdot \Big(\vec{b} \cdot \dot{\vec{\partial}}_{\vec{x}}\dot{\vec{a}} - \dot{\vec{\partial}}_{\vec{x}}(\dot{\vec{a}} \cdot \vec{b})\Big) \\ &= \vec{b} \cdot \dot{\vec{\partial}}_{\vec{x}}\dot{\vec{a}} \cdot \vec{c} - \vec{c} \cdot \dot{\vec{\partial}}_{\vec{x}}\dot{\vec{a}} \cdot \vec{b} \\ &= \vec{b} \cdot \vec{\partial}_{\vec{x}}(\vec{a} \cdot \vec{c}) - \vec{b} \cdot \dot{\vec{\partial}}_{\vec{x}}(\vec{a} \cdot \dot{\vec{c}}) - \vec{c} \cdot \vec{\partial}_{\vec{x}}(\vec{a} \cdot \vec{b}) + \vec{c} \cdot \dot{\vec{\partial}}_{\vec{x}}(\vec{a} \cdot \dot{\vec{b}}) \\ &= \vec{b} \cdot \vec{\partial}_{\vec{x}}(\vec{a} \cdot \vec{c}) - \vec{c} \cdot \vec{\partial}_{\vec{x}}(\vec{a} \cdot \vec{b}) + \vec{c} \cdot \dot{\vec{\partial}}_{\vec{x}}(\dot{\vec{b}} \cdot \vec{a}) - \vec{b} \cdot \dot{\vec{\partial}}_{\vec{x}}(\dot{\vec{c}} \cdot \vec{a}) \\ &\overset{\text{Def. 3.4.76}}{=} \vec{b} \cdot \vec{\partial}_{\vec{x}}(\vec{a} \cdot \vec{c}) - \vec{c} \cdot \vec{\partial}_{\vec{x}}(\vec{a} \cdot \vec{b}) + [\vec{c}, \vec{b}] \cdot \vec{a}. \end{aligned} \tag{3.4.195}$$

□

Proposition 3.4.79. *For* $\vec{a} = \vec{a}(\vec{x})$, $\vec{b} = \vec{b}(\vec{x})$ *:*

$$\begin{aligned}\vec{a} \cdot (\vec{\partial}_{\vec{x}} \wedge \vec{b}) &= \dot{\vec{b}} \cdot (\dot{\vec{\partial}}_{\vec{x}} \wedge \vec{a}) + \dot{\vec{\partial}}_{\vec{x}} \cdot (\vec{a} \wedge \dot{\vec{b}}) \\ &= (\vec{a} \wedge \vec{\partial}_{\vec{x}}) \cdot \vec{b} + \vec{a} \cdot \vec{\partial}_{\vec{x}} \vec{b} - \vec{a} \vec{\partial}_{\vec{x}} \cdot \vec{b}. \end{aligned} \tag{3.4.196}$$

Proof. It follows from the Jacobi identity (3.3.44) that

$$\vec{a} \cdot (\vec{\partial}_{\vec{x}} \wedge \vec{b}) + \dot{\vec{\partial}}_{\vec{x}} \cdot (\dot{\vec{b}} \wedge \vec{a}) + \dot{\vec{b}} \cdot (\vec{a} \wedge \dot{\vec{\partial}}_{\vec{x}}) = 0. \tag{3.4.197}$$

$$\begin{aligned}\Rightarrow \quad \vec{a} \cdot (\vec{\partial}_{\vec{x}} \wedge \vec{b}) &= \dot{\vec{b}} \cdot (\dot{\vec{\partial}}_{\vec{x}} \wedge \vec{a}) + \dot{\vec{\partial}}_{\vec{x}} \cdot (\vec{a} \wedge \dot{\vec{b}}) \\ &\overset{(3.3.39),(3.3.41)}{=} (\vec{a} \wedge \vec{\partial}_{\vec{x}}) \cdot \vec{b} + \vec{a} \cdot \vec{\partial}_{\vec{x}} \vec{b} - \vec{a} \, \vec{\partial}_{\vec{x}} \cdot \vec{b}. \end{aligned} \tag{3.4.198}$$

□

Proposition 3.4.80. *For* $\vec{a} = \vec{a}(\vec{x})$, $\vec{b} = \vec{b}(\vec{x})$ *:*

$$\begin{aligned}\dot{\vec{\partial}}_{\vec{x}} \cdot (\dot{\vec{a}} \wedge \dot{\vec{b}}) &= \dot{\vec{a}} \cdot (\dot{\vec{\partial}}_{\vec{x}} \wedge \dot{\vec{b}}) - \dot{\vec{b}} \cdot (\dot{\vec{\partial}}_{\vec{x}} \wedge \dot{\vec{a}}) \\ &= (\vec{b} \wedge \vec{\partial}_{\vec{x}}) \cdot \vec{a} + \vec{a} \cdot (\vec{\partial}_{\vec{x}} \wedge \vec{b}) \\ &\quad - (\vec{a} \wedge \vec{\partial}_{\vec{x}}) \cdot \vec{b} - \vec{b} \cdot (\vec{\partial}_{\vec{x}} \wedge \vec{a}). \end{aligned} \tag{3.4.199}$$

Proof. It follows from the Jacobi identity (3.3.44) that

$$\dot{\vec{\partial}}_{\vec{x}} \cdot (\dot{\vec{a}} \wedge \dot{\vec{b}}) + \dot{\vec{a}} \cdot (\dot{\vec{b}} \wedge \dot{\vec{\partial}}_{\vec{x}}) + \dot{\vec{b}} \cdot (\dot{\vec{\partial}}_{\vec{x}} \wedge \dot{\vec{a}}) = 0. \tag{3.4.200}$$

$$\begin{aligned}\Rightarrow \quad \dot{\vec{\partial}}_{\vec{x}} \cdot (\dot{\vec{a}} \wedge \dot{\vec{b}}) &= \dot{\vec{a}} \cdot (\dot{\vec{\partial}}_{\vec{x}} \wedge \dot{\vec{b}}) - \dot{\vec{b}} \cdot (\dot{\vec{\partial}}_{\vec{x}} \wedge \dot{\vec{a}}) \\ &= \dot{\vec{a}} \cdot (\dot{\vec{\partial}}_{\vec{x}} \wedge \vec{b}) + \vec{a} \cdot (\dot{\vec{\partial}}_{\vec{x}} \wedge \dot{\vec{b}}) - \dot{\vec{b}} \cdot (\dot{\vec{\partial}}_{\vec{x}} \wedge \vec{a}) - \vec{b} \cdot (\dot{\vec{\partial}}_{\vec{x}} \wedge \dot{\vec{a}}) \\ &\overset{(3.3.39)}{=} (\vec{b} \wedge \vec{\partial}_{\vec{x}}) \cdot \vec{a} + \vec{a} \cdot (\vec{\partial}_{\vec{x}} \wedge \vec{b}) \\ &\quad - (\vec{a} \wedge \vec{\partial}_{\vec{x}}) \cdot \vec{b} - \vec{b} \cdot (\vec{\partial}_{\vec{x}} \wedge \vec{a}). \end{aligned} \tag{3.4.201}$$

□

Proposition 3.4.81.

$$A \times (\vec{\partial}_{\vec{x}} \wedge \vec{b}) = A \cdot \vec{\partial}_{\vec{x}} \vec{b} - \dot{\vec{\partial}}_{\vec{x}} \dot{\vec{b}} \cdot A = A \wedge \vec{\partial}_{\vec{x}} \vec{b} - \dot{\vec{\partial}}_{\vec{x}} \dot{\vec{b}} \wedge A. \tag{3.4.202}$$

with the commutator product $A \times B$ *of multivectors* A, B*, introduced in* (3.3.45).

Proof.

$$\begin{aligned}A \times (\vec{\partial}_{\vec{x}} \wedge \vec{b}) &\overset{(3.3.45)}{=} -(\dot{\vec{\partial}}_{\vec{x}} \wedge \dot{\vec{b}}) \times A \overset{(3.3.46)}{=} -\dot{\vec{\partial}}_{\vec{x}} \dot{\vec{b}} \cdot A + A \cdot \vec{\partial}_{\vec{x}} \vec{b} \\ &\overset{(3.3.46)}{=} -\dot{\vec{\partial}}_{\vec{x}} \dot{\vec{b}} \wedge A + A \wedge \vec{\partial}_{\vec{x}} \vec{b}. \end{aligned} \tag{3.4.203}$$

□

Proposition 3.4.82. *For* $A = \langle A \rangle_r = A(\vec{x}),\ B = \langle B \rangle_s = B(\vec{x})$ *:*

$$\begin{aligned}\dot{A} \wedge \dot{\vec{\partial}}_{\vec{x}} \wedge \dot{B} &= (-1)^r \vec{\partial}_{\vec{x}} \wedge (A \wedge B) \\ &= A \wedge \vec{\partial}_{\vec{x}} \wedge B + (-1)^{r+s(r+1)} B \wedge \vec{\partial}_{\vec{x}} \wedge A. \end{aligned} \tag{3.4.204}$$

Proof.

$$\dot{A} \wedge \dot{\vec{\partial}}_{\vec{x}} \wedge \dot{B} \overset{(3.3.40)}{=} (-1)^r \vec{\partial}_{\vec{x}} \wedge (A \wedge B), \tag{3.4.205}$$

$$\begin{aligned}\dot{A} \wedge \dot{\vec{\partial}}_{\vec{x}} \wedge \dot{B} &= A \wedge \vec{\partial}_{\vec{x}} \wedge B + (\dot{A} \wedge \dot{\vec{\partial}}_{\vec{x}}) \wedge B \\ &\overset{(3.3.40)}{=} A \wedge \vec{\partial}_{\vec{x}} \wedge B + (-1)^r (\dot{\vec{\partial}}_{\vec{x}} \wedge \dot{A}) \wedge B \\ &\overset{(3.3.40)}{=} A \wedge \vec{\partial}_{\vec{x}} \wedge B + (-1)^{r+s(r+1)} B \wedge (\vec{\partial}_{\vec{x}} \wedge A). \end{aligned} \tag{3.4.206}$$

□

The following properties and definitions are needed to develop the subject of Clifford Fourier transforms.

Differentiating twice with the vector derivative, we get the differential Laplacian operator ∇^2. We can write $\nabla^2 = \nabla \cdot \nabla + \nabla \wedge \nabla$. But for integrable functions $\nabla \wedge \nabla = 0$. In this case we have $\nabla^2 = \nabla \cdot \nabla$.

Proposition 3.4.83. *(integration of parts)*

$$\int_{\mathbb{R}^3} g(\boldsymbol{x})[\boldsymbol{a} \cdot \nabla h(\boldsymbol{x})] d^3\boldsymbol{x} = \left[\int_{\mathbb{R}^2} g(\boldsymbol{x}) h(\boldsymbol{x}) d^2\boldsymbol{x}\right]_{a \cdot x = -\infty}^{a \cdot x = \infty} - \int_{\mathbb{R}^3} [\boldsymbol{a} \cdot \nabla g(\boldsymbol{x})] h(\boldsymbol{x}) d^3\boldsymbol{x}$$

We illustrate proposition 3.4.83 by inserting $\boldsymbol{a} = \boldsymbol{e}_3$, i.e.

$$\int_{\mathbb{R}^3} g(\boldsymbol{x})[\partial_3 h(\boldsymbol{x})] d^3\boldsymbol{x} = \left[\int_{\mathbb{R}^2} g(\boldsymbol{x}) h(\boldsymbol{x}) dx_1 dx_2\right]_{x_3 = -\infty}^{x_3 = \infty} - \int_{\mathbb{R}^3} [\partial_3 g(\boldsymbol{x})] h(\boldsymbol{x}) d^3\boldsymbol{x},$$

which is nothing but the usual integration of parts formula for the partial derivative $\partial_3 h(\boldsymbol{x})$.

3.5 SUMMARY ON VECTOR DIFFERENTIAL CALCULUS

Section 3 first summarized important geometric algebra relationships, which are necessary for the thorough and explicit development of the vector differential calculus part of universal geometric calculus.

It then showed how to differentiate multivector functions by a vector, including the results of standard vector analysis. The vector differential relationships are proven in a very explicit step by step way, enabling the reader, who is unfamiliar with the algebraic techniques to get complete comprehension. It may thus also serve as important reference material for studying and applying vector differential calculus in connection with Clifford Fourier transforms.

A corresponding thorough treatment of multivector differential calculus can be found in [175].

CHAPTER 4

Quaternion Fourier transforms

We study quaternion Fourier transforms (QFT). The two-sided QFT was introduced in [124] for the analysis of two-dimensional (2D) linear time-invariant partial-differential systems. In further theoretical investigations [184, 187] a special split of quaternions was introduced, then called ±split. We develop this split further, interpret it geometrically as *orthogonal 2D planes split* (OPS), and generalize it to a freely steerable split of $\mathbb{H}$ into two orthogonal 2D analysis planes. The new general form of the OPS split allows us to find new geometric interpretations for the action of the QFT on the signal. The second major result we present is a variety of *new steerable forms* of the QFT, their geometric interpretation, and for each form OPS split theorems, which allow fast and efficient numerical implementation with standard FFT software. We continue by studying properties of the QFT, like the uncertainty principle, convolution, and a Wiener Khinchine type theorem for the QFT. We also show how QFTs can be further specialized to a windowed QFT, a quaternionic Fourier-Mellin transform, and generalized to a quaternion domain FT, a volume-time FT, and a spacetime FT.

4.1 FUNDAMENTALS OF QUATERNION FOURIER TRANSFORMS (QFT)

4.1.1 Quaternion Fourier transform and the ±-split

We first treat the quaternionic Fourier transform (QFT) applied to quaternion fields and investigate QFT properties useful for applications. Different forms of the QFT lead us to different Plancherel theorems. We relate the QFT computation for quaternion fields to the QFT of real signals. We study the general linear (GL) transformation behavior of the QFT with matrices, Clifford geometric algebra and with examples. (Later, in Section 4.4 we reach wide-ranging non-commutative multivector FT generalizations of the QFT. Examples are the new volume-time and spacetime algebra Fourier transformations.) The content of this section is mainly based on [184].

DOI: 10.1201/9781003184478-4

4.1.1.1 Background

We strive to deepen the understanding of the quaternionic Fourier transform (QFT) applied to quaternion fields $f : \mathbb{R}^2 \to \mathbb{H}$, and not only to real signals $f : \mathbb{R}^2 \to \mathbb{R}$. We present QFT properties useful for applications to partial differential equations, image processing and optimized numerical implementations. We show how different forms of the QFT allow to establish different scalar and quaternion valued Plancherel theorems.

We explain how to reduce the computation for quaternion fields to the case of real signal computations, and on the other hand how results for real signals can be generalized to quaternion fields.

The third major focus of this section is on deriving the behavior of the QFT under $GL(\mathbb{R}^2)$ automorphisms. To do this we split the QFT appropriately, and work with invariant techniques of Clifford geometric algebra [161] to establish and understand the automorphism behavior. Details are brought to light by looking at the examples of stretches (dilations), reflections and rotations.

Together with isomorphisms (to Clifford subalgebras) we finally arrive at wide-ranging generalizations of the QFT. These new non-commutative multivector Fourier transforms operate on functions from domain spaces $\mathbb{R}^{m,n}$ (with $m, n \in \mathbb{N}_0$) to Clifford algebras $Cl(m, n)$ or subalgebras thereof. To practically demonstrate the method, we work out generalizations to volume-time and to spacetime algebra Fourier transformations, and provide some physical interpretation in Section 4.4.

4.1.1.1.1 Quaternions, split and module Quaternions were introduced in Section 2.8.1, part of Section 2.8 on the geometric meaning of quaternions. We remind the reader (compare Section 2.8.2) that in some applications it proves convenient to replace $\boldsymbol{k}$ with $\boldsymbol{k} = \boldsymbol{ij}$ and write a quaternion as

$$q = q_r + \boldsymbol{i}q_i + q_j\boldsymbol{j} + \boldsymbol{i}q_k\boldsymbol{j}, \tag{4.1.1}$$

neatly keeping all $\boldsymbol{i}$ to the left and all $\boldsymbol{j}$ to the right of each term. A second convenient form is the *split*

$$q = q_+ + q_-, \quad q_\pm = \frac{1}{2}(q \pm \boldsymbol{i}q\boldsymbol{j}). \tag{4.1.2}$$

Explicitly in real components $q_r, q_i, q_j, q_k \in \mathbb{R}$ using (2.8.1) the split (4.1.2) produces:

$$q_\pm = \{q_r \pm q_k + \boldsymbol{i}(q_i \mp q_j)\}\frac{1 \pm \boldsymbol{k}}{2} = \frac{1 \pm \boldsymbol{k}}{2}\{q_r \pm q_k + \boldsymbol{j}(q_j \mp q_i)\}. \tag{4.1.3}$$

For quaternion-valued functions $f, g : \mathbb{R}^2 \to \mathbb{H}$ we can define the quaternion-valued inner product

$$(f, g) = \int_{\mathbb{R}^2} f(\boldsymbol{x})\,\tilde{g}(\boldsymbol{x})\,d^2\boldsymbol{x}, \qquad \text{with} \quad d^2\boldsymbol{x} = dxdy, \tag{4.1.4}$$

with symmetric real scalar part [56]

$$\langle f, g\rangle = \frac{1}{2}[(f, g) + (g, f)] = \int_{\mathbb{R}^2} \langle f(\boldsymbol{x})\,\tilde{g}(\boldsymbol{x})\rangle_0 d^2\boldsymbol{x}. \tag{4.1.5}$$

Both (4.1.4) and (4.1.5) lead to the $L^2(\mathbb{R}^2;\mathbb{H})$-norm

$$\|f\| = \sqrt{(f,f)} = \sqrt{\langle f,f\rangle} = \int_{\mathbb{R}^2} |f(\boldsymbol{x})|^2 \, d^2\boldsymbol{x} \,. \tag{4.1.6}$$

A *quaternion module* $L^2(\mathbb{R}^2;\mathbb{H})$ is then defined as

$$L^2(\mathbb{R}^2;\mathbb{H}) = \{f | f : \mathbb{R}^2 \to \mathbb{H}, \|f\| < \infty\}. \tag{4.1.7}$$

4.1.1.2 The quaternion Fourier transform

Before defining the quaternion Fourier transform (QFT), we briefly outline its relationship with Clifford Fourier transformations, see also Chapter 1.

Brackx et al. [33] extended the Fourier transform to multivector valued function-distributions in $Cl_{0,n}$ with compact support. A related applied approach for hypercomplex Clifford Fourier transformations[1] in $Cl_{0,n}$ was followed by Bülow et. al. [58].

By extending the classical trigonometric exponential function $\exp(j\, \boldsymbol{x} * \boldsymbol{\xi})$ (where $*$ denotes the scalar product of $\boldsymbol{x} \in \mathbb{R}^m$ with $\boldsymbol{\xi} \in \mathbb{R}^m$, j the imaginary unit) in [243,259], McIntosh et. al. generalized the classical Fourier transform. Applied to a function of m real variables this generalized Fourier transform is holomorphic in m complex variables and its inverse is *monogenic* in $m + 1$ real variables, thereby effectively extending the function of m real variables to a monogenic function of $m+1$ real variables (with values in a *complex* Clifford algebra). This generalization has significant applications to harmonic analysis, especially to singular integrals on surfaces in $\mathbb{R}^{m+1}$. Based on this approach Kou and Qian obtained a Clifford Payley-Wiener theorem and derived Shannon interpolation of band-limited functions using the monogenic sinc function [286, and references therein]. The Clifford Payley-Wiener theorem also allows to derive left-entire (left-monogenic in the whole $\mathbb{R}^{m+1}$) functions from square integrable functions on $\mathbb{R}^m$ with compact support.

The real n-dimensional volume element $i_n = \boldsymbol{e}_1\boldsymbol{e}_2 \ldots \boldsymbol{e}_n$ of $Cl_{n,0}$ over the field of the reals $\mathbb{R}$ has been used in [117, 182, 185, 251] to construct and apply Clifford Fourier transformations for $n = 2, 3 \pmod 4$ with kernels $\exp(-i_n \boldsymbol{x} * \boldsymbol{u})$, $\boldsymbol{x}, \boldsymbol{u} \in \mathbb{R}^n$. This i_n has a clear geometric interpretation. Note that $i_n^2 = -1$ for $n = 2, 3 \pmod 4$.

Ell [124] defined the quaternion Fourier transform (QFT) for application to 2D linear time-invariant systems of PDEs. Ell's QFT belongs to the growing family of Clifford Fourier transformations because of (2.8.3). But the left and right placement of the exponential factors in Definition 4.1.1 distinguishes it. Later the QFT was applied extensively to 2D image processing, including color images [56, 58, 124]. This spurred research into optimized numerical implementations [132, 280]. Ell [124] and others [56,69] also investigated related *commutative* hypercomplex Fourier transforms like in the commutative subalgebra of $Cl_{4,0}$ with subalgebra basis $\{1, \boldsymbol{e}_{12}, \boldsymbol{e}_{34}, \boldsymbol{e}_{1234}\}$,

$$\boldsymbol{e}_{12}^2 = \boldsymbol{e}_{34}^2 = -1, \quad \boldsymbol{e}_{1234}^2 = +1. \tag{4.1.8}$$

[1]This is the kind of Clifford Fourier transform to which we will refer in Section 4.1.1.3.

Definition 4.1.1 (Quaternion Fourier transform (QFT)). *The quaternion Fourier transform*[2] $\hat{f}: \mathbb{R}^2 \to \mathbb{H}$ *of* $f \in L^2(\mathbb{R}^2;\mathbb{H})$, $\boldsymbol{x} = x\boldsymbol{e}_1 + y\boldsymbol{e}_2 \in \mathbb{R}^2$ *and* $\boldsymbol{u} = u\boldsymbol{e}_1 + v\boldsymbol{e}_2 \in \mathbb{R}^2$ *is defined*[3] *as*

$$\hat{f}(\boldsymbol{u}) = \int_{\mathbb{R}^2} e^{-\boldsymbol{i}xu} f(\boldsymbol{x})\, e^{-\boldsymbol{j}yv} d^2\boldsymbol{x}. \tag{4.1.9}$$

The QFT can be *inverted* by

$$f(\boldsymbol{x}) = \frac{1}{(2\pi)^2} \int_{\mathbb{R}^2} e^{\boldsymbol{i}xu} \hat{f}(\boldsymbol{u})\, e^{\boldsymbol{j}yv} d^2\boldsymbol{u}, \tag{4.1.10}$$

with $d^2\boldsymbol{u} = dudv$.

4.1.1.2.1 **Rewriting and splitting functions** Let $f: \mathbb{R}^2 \to \mathbb{H}$ (or $f \in L^2(\mathbb{R}^2;\mathbb{H})$). Using four $\mathbb{R}^2 \to \mathbb{R}$ (or $L^2(\mathbb{R}^2;\mathbb{R})$) real component functions f_r, f_i, f_j and f_k we can decompose and rewrite f with (4.1.1) as

$$f = f_r + f_i\boldsymbol{i} + f_j\boldsymbol{j} + f_k\boldsymbol{k} = f_r + \boldsymbol{i}f_i + f_j\boldsymbol{j} + \boldsymbol{i}f_k\boldsymbol{j}. \tag{4.1.11}$$

We can also split the functions f [similar to $q_\pm$ in (4.1.2)] into

$$f = f_+ + f_-, \quad f_+ = \frac{1}{2}(f + \boldsymbol{i}f\boldsymbol{j}), \quad f_- = \frac{1}{2}(f - \boldsymbol{i}f\boldsymbol{j}). \tag{4.1.12}$$

According to (4.1.3) the two components $f_\pm$ can also be rewritten as

$$f_\pm = \{f_r \pm f_k + \boldsymbol{i}(f_i \mp f_j)\}\frac{1 \pm \boldsymbol{k}}{2} = \frac{1 \pm \boldsymbol{k}}{2}\{f_r \pm f_k + \boldsymbol{j}(f_j \mp f_i)\}. \tag{4.1.13}$$

As an example, let us consider the split of the product of exponential functions under the QFT integral in (4.1.9). Using Euler's formula and trigonometric addition theorems the split leads to

$$\begin{aligned} K &= e^{-\boldsymbol{i}xu} e^{-\boldsymbol{j}yv} = K_+ + K_-, \\ K_\pm &= e^{-\boldsymbol{i}(xu \mp yv)}\frac{1 \pm \boldsymbol{k}}{2} = \frac{1 \pm \boldsymbol{k}}{2} e^{-\boldsymbol{j}(yv \mp xu)}. \end{aligned} \tag{4.1.14}$$

---

[2] We also assume always that $\int_{\mathbb{R}^2} |f(\boldsymbol{x})|\, d^2\boldsymbol{x}$ exists as well. But we do not explicitly write this condition again in the rest of the paper. Strictly speaking, the integral definition of Def. 4.1.1 only works for $f \in L^1(\mathbb{R}^2;\mathbb{H})$. But one can first define the QFT on the dense subset $L^1(\mathbb{R}^2;\mathbb{H}) \bigcap L^2(\mathbb{R}^2;\mathbb{H})$, and then use the continuity of the Fourier transform on $L^1(\mathbb{R}^2;\mathbb{H}) \bigcap L^2(\mathbb{R}^2;\mathbb{H})$, due to Plancherel's theorem Theorem 4.1.2, to define the QFT on $L^2(\mathbb{R}^2;\mathbb{H})$, see e.g. [135]. A similar argument applies for all quaternionic and Clifford Fourier transforms considered in this monograph.

[3] For real signals $f \in L^2(\mathbb{R}^2;\mathbb{R})$ the detailed relationship of the QFT of definition 4.1.1 with the conventional scalar FT, i.e. with the even cos-part and the odd sin-part are given on pp. 191 and 192 of [58]. With the help of (4.1.24) this can easily be extended to the full QFT of quaternion-valued $f \in L^2(\mathbb{R}^2;\mathbb{H})$.

Table 4.1 Properties of the quaternion Fourier transform (QFT) of quaternion functions (Quat. Funct.) $f, g \in L^2(\mathbb{R}^2;\mathbb{H})$, with $\boldsymbol{x}, \boldsymbol{u} \in \mathbb{R}^2$, constants $\alpha, \beta \in \{q| q = q_r + q_i\boldsymbol{i},\, q_r, q_i \in \mathbb{R}\}$, $\alpha', \beta' \in \{q| q = q_r + q_j\boldsymbol{j},\, q_r, q_j \in \mathbb{R}\}$, $a, b \in \mathbb{R}\setminus\{0\}$, $\boldsymbol{x}_0 = x_0\boldsymbol{e}_1 + y_0\boldsymbol{e}_2$, $\boldsymbol{u}_0 = u_0\boldsymbol{e}_1 + v_0\boldsymbol{e}_2 \in \mathbb{R}^2$ and $m, n \in \mathbb{N}_0$. Source: [184, Tab. 1].

Property	**Quat. Funct.**	**QFT**
Left linearity	$\alpha f(\boldsymbol{x})+\beta\, g(\boldsymbol{x})$	$\alpha\hat{f}(\boldsymbol{u})+\,\beta\hat{g}(\boldsymbol{u})$
Right linearity	$f(\boldsymbol{x})\alpha'+\,g(\boldsymbol{x})\beta'$	$\hat{f}(\boldsymbol{u})\alpha'+\,\hat{g}(\boldsymbol{u})\beta'$
$\boldsymbol{x}$-Shift	$f(\boldsymbol{x}-\boldsymbol{x}_0)$	$e^{-\boldsymbol{i}x_0u}\hat{f}(\boldsymbol{u})\,e^{-\boldsymbol{j}y_0v}$
Modulation	$e^{\boldsymbol{i}xu_0}f(\boldsymbol{x})\,e^{\boldsymbol{j}yv_0}$	$\hat{f}(\boldsymbol{u}-\boldsymbol{u}_0)$
Dilation[4]	$f(a\,x\boldsymbol{e}_1+b\,y\boldsymbol{e}_2)$	$\frac{1}{\lvert ab\rvert}\hat{f}(\frac{u}{a}\boldsymbol{e}_1+\frac{v}{b}\boldsymbol{e}_2)$
Part. deriv.	$\frac{\partial^{m+n}}{\partial x^m\partial y^n}f(\boldsymbol{x})$	$(\boldsymbol{i}u)^m\hat{f}(\boldsymbol{u})(\boldsymbol{j}v)^n$
Powers[5] of x, y	$x^m y^n f(\boldsymbol{x})$	$\boldsymbol{i}^m\frac{\partial^{m+n}}{\partial u^m\partial v^n}\hat{f}(\boldsymbol{u})\,\boldsymbol{j}^n$
Powers[5] of $\boldsymbol{i}, \boldsymbol{j}$	$\boldsymbol{i}^m f(\boldsymbol{x})\,\boldsymbol{j}^n$	$\boldsymbol{i}^m\hat{f}(\boldsymbol{u})\,\boldsymbol{j}^n$
Plancherel[5]	$\langle f, g\rangle =$	$\frac{1}{(2\pi)^2}\langle\hat{f},\hat{g}\rangle$
Parseval[6]	$\lVert f\rVert =$	$\frac{1}{2\pi}\lVert\hat{f}\rVert$

4.1.1.2.2 Useful properties of the QFT We first show a *new* Plancherel theorem with respect to the scalar product (4.1.5).

Theorem 4.1.2 (QFT Plancherel). *The scalar product (4.1.5) of two quaternion module functions $f, g \in L^2(\mathbb{R}^2;\mathbb{H})$ is given by the scalar product of the of the corresponding QFTs $\hat{f}$ and $\hat{g}$*

$$\langle f, g\rangle = \frac{1}{(2\pi)^2}\langle\hat{f},\hat{g}\rangle. \tag{4.1.15}$$

Proof. For $f, g \in L^2(\mathbb{R}^2;\mathbb{H})$ we calculate the scalar product (4.1.5)

$$\begin{aligned}
\langle f, g\rangle &= \int_{\mathbb{R}^2}\langle f(\boldsymbol{x})\tilde{g}(\boldsymbol{x})\rangle_0 d^2\boldsymbol{x}\\
&= \frac{1}{(2\pi)^2}\int_{\mathbb{R}^2}\langle\int_{\mathbb{R}^2} e^{\boldsymbol{i}ux}\hat{f}(\boldsymbol{u})e^{\boldsymbol{j}vy}d^2\boldsymbol{u}\,\tilde{g}(\boldsymbol{x})\rangle_0 d^2\boldsymbol{x}\\
&= \frac{1}{(2\pi)^2}\int_{\mathbb{R}^2}\langle\hat{f}(\boldsymbol{u})\int_{\mathbb{R}^2} e^{\boldsymbol{j}vy}\tilde{g}(\boldsymbol{x})e^{\boldsymbol{i}ux}d^2\boldsymbol{x}\rangle_0 d^2\boldsymbol{u}\\
&= \frac{1}{(2\pi)^2}\int_{\mathbb{R}^2}\langle\hat{f}(\boldsymbol{u})[\int_{\mathbb{R}^2} e^{-\boldsymbol{i}ux}g(\boldsymbol{x})e^{-\boldsymbol{j}vy}d^2\boldsymbol{x}]^{\sim}\rangle_0 d^2\boldsymbol{u}\\
&= \frac{1}{(2\pi)^2}\int_{\mathbb{R}^2}\langle\hat{f}(\boldsymbol{u})\tilde{\hat{g}}(\boldsymbol{u})\rangle_0 d^2\boldsymbol{u} = \frac{1}{(2\pi)^2}\langle\hat{f},\hat{g}\rangle.
\end{aligned} \tag{4.1.16}$$

[4] Bülow [56] omits the absolute value signs for the determinant of the transformation.
[5] Theorems 4.1.4, 4.1.5 and 4.1.2.
[6] Corollary 4.1.3.

In the second equality of (4.1.16) we replaced f with its inverse QFT expression (4.1.10). In the third equality we exchanged the order of integration and we used the cyclic symmetry (2.8.14). For the fourth equality we simply pulled the reversion outside the square brackets $[\dots]$ and obtained the QFT $\hat{g}(\boldsymbol{u})$, which proves (4.1.15) according to (4.1.5). $\square$

For $g = f$ the Plancherel theorem 4.1.2 has a QFT Parseval theorem (also called Rayleigh's theorem) as a direct corollary.

Corollary 4.1.3 (QFT Parseval). *The $L^2(\mathbb{R}^2;\mathbb{H})$-norm of a quaternion module function $f \in L^2(\mathbb{R}^2;\mathbb{H})$ is given by the $L^2(\mathbb{R}^2;\mathbb{H})$-norm of its QFT multiplied by $1/(2\pi)$*

$$\|f\| = \frac{1}{2\pi}\|\hat{f}\|. \tag{4.1.17}$$

This leads to the following observations:

- The way we obtained the Parseval theorem of cor. 4.1.3 is much simpler than the proofs in [56, 124].
- For two-dimensional linear time-invariant partial differential systems the Parseval theorem provides an appropriate method to measure controller performance.
- In signal processing it states that the signal energy is preserved by the QFT.

For solving PDEs with quaternionic (or real) coefficient polynomials in $x, y \in \mathbb{R}^2$ we show the following two theorems. In this context we note again that every quaternionic (or real) coefficient polynomial in the variables $x, y \in \mathbb{R}^2$ can be brought into a form having factors of $\boldsymbol{i} \in \mathbb{H}$ to the left side of each term and factors of $\boldsymbol{j} \in \mathbb{H}$ to the right side of each term (compare (4.1.11)).

Theorem 4.1.4 (Powers of x, y). *The QFT of a quaternion module function $x^m y^n f(\boldsymbol{x}) \in L^2(\mathbb{R}^2;\mathbb{H})$, $\boldsymbol{x} = x\boldsymbol{e}_1 + y\boldsymbol{e}_2 \in \mathbb{R}^2$, $f \in L^2(\mathbb{R}^2;\mathbb{H})$, $m, n \in \mathbb{N}_0$ is given by*

$$\widehat{x^m y^n f}(\boldsymbol{u}) = \boldsymbol{i}^m \frac{\partial^{m+n}}{\partial u^m \partial v^n}\hat{f}(\boldsymbol{u})\,\boldsymbol{j}^n. \tag{4.1.18}$$

Proof. The proof is done by induction. It is trivial for $m = n = 0$.
For $m = 1, n = 0$ we calculate the QFT of $\widehat{xf}$ according to (4.1.9)

$$\begin{aligned}\widehat{xf}(\boldsymbol{u}) &= \int_{\mathbb{R}^2} e^{-\boldsymbol{i}xu} x f(\boldsymbol{x})\, e^{-\boldsymbol{j}yv} d^2\boldsymbol{x} = \int_{\mathbb{R}^2} \boldsymbol{i}\frac{\partial}{\partial u} e^{-\boldsymbol{i}xu} f(\boldsymbol{x})\, e^{-\boldsymbol{j}yv} d^2\boldsymbol{x} \\ &= \boldsymbol{i}\frac{\partial}{\partial u}\int_{\mathbb{R}^2} e^{-\boldsymbol{i}xu} f(\boldsymbol{x})\, e^{-\boldsymbol{j}yv} d^2\boldsymbol{x} = \boldsymbol{i}\frac{\partial}{\partial u}\hat{f}(\boldsymbol{u}).\end{aligned} \tag{4.1.19}$$

In second equality we used $\frac{\partial}{\partial u}e^{-\boldsymbol{i}xu} = -\boldsymbol{i}xe^{-\boldsymbol{i}xu}$ and $\boldsymbol{i}(-\boldsymbol{i}) = 1$.
Completely analogous for $m = 0, n = 1$ we find

$$\widehat{yf}(\boldsymbol{u}) = \frac{\partial}{\partial v}\int_{\mathbb{R}^2} e^{-\boldsymbol{i}xu} f(\boldsymbol{x})\, e^{-\boldsymbol{j}yv} d^2\boldsymbol{x}\,\boldsymbol{j} = \frac{\partial}{\partial v}\hat{f}(\boldsymbol{u})\boldsymbol{j}\,. \tag{4.1.20}$$

Because of non-commutativity $\boldsymbol{j}$ appears to the right of $\hat{f}$. Induction over $m, n \in \mathbb{N}$ completes the proof. $\square$

Theorem 4.1.5 (Powers of $\boldsymbol{i}, \boldsymbol{j}$). *The QFT of a quaternion module function* $\boldsymbol{i}^m f(\boldsymbol{x})\boldsymbol{j}^n \in L^2(\mathbb{R}^2;\mathbb{H})$, $f \in L^2(\mathbb{R}^2;\mathbb{H})$, $m, n \in \mathbb{N}_0$ *is given by*

$$\widehat{\boldsymbol{i}^m f \boldsymbol{j}^n}(\boldsymbol{u}) = \boldsymbol{i}^m \hat{f}(\boldsymbol{u})\, \boldsymbol{j}^n. \tag{4.1.21}$$

Proof. Similar to the left and right linearities of Table 4.1, theorem 4.1.5 follows directly from the definition 4.1.1 of the QFT, using the commutation relationships

$$\exp(-\boldsymbol{i}xu)\boldsymbol{i}^m = \boldsymbol{i}^m \exp(-\boldsymbol{i}xu) \quad \text{and} \quad \exp(-\boldsymbol{j}yv)\boldsymbol{j}^n = \boldsymbol{j}^n \exp(-\boldsymbol{j}yv). \tag{4.1.22}$$

□

For every $f \in L^2(\mathbb{R}^2;\mathbb{H})$ we can always rewrite $f = f_r + f_i\boldsymbol{i} + f_j\boldsymbol{j} + f_k\boldsymbol{k}$ as in (4.1.11) to the form

$$f = f_r + \boldsymbol{i}f_i + f_j\boldsymbol{j} + \boldsymbol{i}f_k\boldsymbol{j}. \tag{4.1.23}$$

Accordingly we now can make the following two important observations:

- Theorem 4.1.5 reduces the computation of the QFT of any $f \in L^2(\mathbb{R}^2;\,\mathbb{H})$ to the computation of four QFTs of the real functions $f_r, f_i, f_j, f_k \in L^2(\mathbb{R}^2;\mathbb{R})$ as in
$$\hat{f} = \hat{f}_r + \boldsymbol{i}\hat{f}_i + \hat{f}_j\boldsymbol{j} + \boldsymbol{i}\hat{f}_k\boldsymbol{j}. \tag{4.1.24}$$
- On the other hand theorem 4.1.5 reveals that every *theorem for the QFT of real functions* $g \in L^2(\mathbb{R}^2;\mathbb{R})$ immediately results via (4.1.24) in a corresponding *theorem for quaternion module functions* $f \in L^2(\mathbb{R}^2;\,\mathbb{H})$. We simply need to apply the theorem for the QFT of real functions to each of the four real component functions $f_r, f_i, f_j, f_k \in L^2(\mathbb{R}^2;\mathbb{R})$. This fact is rather useful, because often in image processing theorems are only established for real image signals [56].

4.1.1.2.3 Example: $GL(\mathbb{R}^2)$ transformation properties of the QFT To give an example for the second observation at the end of Section 4.1.1.2.2 we use it to generalize the general linear real non-singular transformation property of the QFT of real 2D functions $f \in L^2(\mathbb{R}^2;\mathbb{R})$ of [56] to quaternion module functions $f \in L^2(\mathbb{R}^2;\mathbb{H})$. This property of real 2D signals states that for

$$\boldsymbol{x}' = A\boldsymbol{x} = (ax + by)\boldsymbol{e}_1 + (cx + dy)\boldsymbol{e}_2 \tag{4.1.25}$$

with non-singular real transformation matrix

$$A = \begin{pmatrix} a & b \\ c & d \end{pmatrix} \tag{4.1.26}$$

the QFT of a *real* signal $f : \mathbb{R}^2 \to \mathbb{R}$ is[7]

$$\widehat{f(A\boldsymbol{x})}(\boldsymbol{u}) = \frac{|\det \mathcal{B}|}{2}\left(\hat{f}(\mathcal{B}_+\,\boldsymbol{u}) + \hat{f}(\mathcal{B}_-\,\boldsymbol{u}) + \boldsymbol{i}\left\{\hat{f}(\mathcal{B}_+\,\boldsymbol{u}) - \hat{f}(\mathcal{B}_-\,\boldsymbol{u})\right\}\boldsymbol{j}\right). \tag{4.1.27}$$

[7]Bülow [56] omits the absolute value signs for the determinant of the transformation.

In (4.1.27) the two linear real non-singular transformations $\mathcal{B}_+$ and $\mathcal{B}_-$ have corresponding matrices and the (same) determinant

$$B_+ = A^{-1T}, \quad B_- = \frac{1}{\det A}\begin{pmatrix} d & c \\ b & a \end{pmatrix},$$
$$\det \mathcal{B} = \det B_+ = \det B_- = (\det A)^{-1}. \tag{4.1.28}$$

We can now establish the generalization from $f \in L^2(\mathbb{R}^2;\mathbb{R})$ to $f \in L^2(\mathbb{R}^2;\mathbb{H})$ functions.

Theorem 4.1.6. *The QFT of a quaternion-module function $f \in L^2(\mathbb{R}^2;\mathbb{H})$ with a $GL(\mathbb{R}^2)$ transformation $\mathcal{A}$ of its vector arguments (4.1.25) is also given by (4.1.27).*

Proof. We only sketch the proof, because writing out all expressions explicitly would consume too much space:

- Applying (4.1.27) and (4.1.28) to each component of (4.1.24) and
- Rearranging the sum (of 16 terms) yields the validity of (4.1.27) together with (4.1.28) also for quaternion-valued $f \in L^2(\mathbb{R}^2;\mathbb{H})$.
- It is again crucial that in each term all factors $\boldsymbol{i}$ are always kept to the left and all factors $\boldsymbol{j}$ are always kept to the right.

□

We remark that resorting to matrices and matrix manipulations is geometrically not very intuitive, so in Section 4.1.1.4 an alternative more geometric approach is taken to derive the transformation properties of general $f \in L^2(\mathbb{R}^2;\mathbb{H})$. This geometric approach has far reaching consequences for the generalization of the QFT, exploited in later sections.

But before geometrically reanalyzing QFT transformation properties we look at the following variant of the QFT with some desirable properties not valid for the QFT of Definition 4.1.1.

4.1.1.3 The right-side quaternion Fourier transform (QFTr)

We observe that it is not possible to establish a general Plancherel theorem for the QFT of the inner product (f,g) of (4.1.4), because the product (4.1.4) lacks the cyclic symmetry (2.8.14) applied in the proof of theorem 4.1.2. To obtain a Plancherel theorem it is therefore either necessary to modify the symmetry properties of the inner product as in (4.1.5) or to modify the QFT itself. In this section we explore the second possibility.

Definition 4.1.7 (Right-side QFT (QFTr)). *The right-side quaternion Fourier transform*[8] $\overset{\triangleright}{f} : \mathbb{R}^2 \to \mathbb{H}$ *of* $f \in L^2(\mathbb{R}^2;\mathbb{H})$, $\boldsymbol{x} = x\boldsymbol{e}_1 + y\boldsymbol{e}_2 \in \mathbb{R}^2$ *and* $\boldsymbol{u} = u\boldsymbol{e}_1 + v\boldsymbol{e}_2 \in \mathbb{R}^2$ *is defined as*

$$\overset{\triangleright}{f}(\boldsymbol{u}) = \int_{\mathbb{R}^2} f(\boldsymbol{x})\, e^{-\boldsymbol{i}xu} e^{-\boldsymbol{j}yv} d^2\boldsymbol{x} \qquad \textit{with} \quad d^2\boldsymbol{x} = dxdy. \tag{4.1.29}$$

[8]For the right-side QFT applied to L^1 and L^2 functions, see Footnote 2 on page 144.

Table 4.2 Properties of the *right sided* quaternion Fourier transform (QFTr) of quaternion functions (Quat. Funct.) $f, g \in L^2(\mathbb{R}^2; \mathbb{H})$, with $\boldsymbol{x}, \boldsymbol{u} \in \mathbb{R}^2$, constants $\alpha, \beta \in \mathbb{R}$, $\alpha', \beta' \in \mathbb{H}$, $a, b \in \mathbb{R} \setminus \{0\}$, $\boldsymbol{x}_0 = x_0\boldsymbol{e}_1 + y_0\boldsymbol{e}_2$, $\boldsymbol{u}_0 = u_0\boldsymbol{e}_1 + v_0\boldsymbol{e}_2 \in \mathbb{R}^2$ and $m, n \in \mathbb{N}$. Source: [184, Tab. 2].

Property	Quat. Funct.	QFTr
Linearity[9]	$\alpha f(\boldsymbol{x}) + \beta\, g(\boldsymbol{x})$	$\alpha \overset{\triangleright}{f}(\boldsymbol{u}) + \beta \overset{\triangleright}{g}(\boldsymbol{u})$
Left linearity	$\alpha' f(\boldsymbol{x}) + \beta' g(\boldsymbol{x})$	$\alpha' \overset{\triangleright}{f}(\boldsymbol{u}) + \beta' \overset{\triangleright}{g}(\boldsymbol{u})$
$\boldsymbol{x}$-Shift[10]	$f(\boldsymbol{x} - \boldsymbol{x}_0)$	$\mathcal{F}_{\triangleright}\{f e^{-\boldsymbol{i}x_0 u}\}(\boldsymbol{u})\, e^{-\boldsymbol{j}y_0 v}$
Dilation	$f(a\,x\boldsymbol{e}_1 + b\,y\boldsymbol{e}_2)$	$\frac{1}{\lvert ab\rvert} \overset{\triangleright}{f}(\frac{u}{a}\boldsymbol{e}_1 + \frac{v}{b}\boldsymbol{e}_2)$
Part. deriv.[11]	$\frac{\partial^{m+n}}{\partial x^m \partial y^n} f(\boldsymbol{x}) \boldsymbol{i}^{-m}$	$u^m \overset{\triangleright}{f}(\boldsymbol{u}) (\boldsymbol{j}v)^n$
Powers[12] of x, y	$x^m y^n f(\boldsymbol{x}) \boldsymbol{i}^{-m}$	$\frac{\partial^{m+n}}{\partial u^m \partial v^n} \overset{\triangleright}{f}(\boldsymbol{u})\, \boldsymbol{j}^n$
Powers[13] of $\boldsymbol{i}, \boldsymbol{j}$	$\boldsymbol{i}^m \boldsymbol{j}^n f(\boldsymbol{x})$	$\boldsymbol{i}^m \boldsymbol{j}^n \overset{\triangleright}{f}(\boldsymbol{u})$
Plancherel[14]	$(f, g) =$	$\frac{1}{(2\pi)^2} (\overset{\triangleright}{f}, \overset{\triangleright}{g})$
Plancherel[15]	$\langle f, g \rangle =$	$\frac{1}{(2\pi)^2} \langle \overset{\triangleright}{f}, \overset{\triangleright}{g} \rangle$
Parseval	$\lVert f \rVert =$	$\frac{1}{2\pi} \lVert \overset{\triangleright}{f} \rVert$

The QFTr is known as Clifford Fourier transform [33, 58], because of the isomorphism $\mathbb{H} \cong Cl_{0,2}$. Further freedoms in alternative definitions would be to exchange the order of the exponentials in (4.1.29) or to wholly shift both exponential factors to the left side instead. The former would simply exchange the roles of $\boldsymbol{i}$ and $\boldsymbol{j}$, but the latter would not serve our purpose as will soon become clear. The QFTr can be inverted [33, 58] using

$$f(\boldsymbol{x}) = \frac{1}{(2\pi)^2} \int_{\mathbb{R}^2} \overset{\triangleright}{f}(\boldsymbol{u})\, e^{\boldsymbol{j}yv} e^{\boldsymbol{i}xu} d^2\boldsymbol{u}, \tag{4.1.30}$$

with $d^2\boldsymbol{u} = dudv$. Attention needs to be paid to the reversed order of the exponential factors in (4.1.30) compared to (4.1.29).

4.1.1.3.1 Properties of the QFTr

For general $f, g \in L^2(\mathbb{R}^2; \mathbb{H})$ *left linearity* and *dilation* properties of Table 4.1 hold. The left linearity coefficients can now be fully quaternionic constants $\alpha', \beta' \in \mathbb{H}$.

---

[9] The positions of the real scalars α, β before or after the functions f, g do not matter.

[10] Only for quaternion module functions $f \in L^2(\mathbb{R}^2; \mathbb{H})$ with $\boldsymbol{i}f = f\boldsymbol{i}$, i.e. $f = f_r + \boldsymbol{i}f_i$ with $f_r, f_i \in L^2(\mathbb{R}^2; \mathbb{R})$ do we get $\mathcal{F}_{\triangleright}\{f(\boldsymbol{x} - \boldsymbol{x}_0)\}(\boldsymbol{u}) = e^{-\boldsymbol{i}x_0 u} \overset{\triangleright}{f}(\boldsymbol{u})\, e^{-\boldsymbol{j}y_0 v}$.

[11] Only for $\boldsymbol{i}f = f\boldsymbol{i}$ do we get $\mathcal{F}_{\triangleright}\{\frac{\partial^{m+n}}{\partial x^m \partial y^n} f\}(\boldsymbol{u}) = (\boldsymbol{i}u)^m \overset{\triangleright}{f}(\boldsymbol{u}) (\boldsymbol{j}v)^n$.

[12] Only for $\boldsymbol{i}f = f\boldsymbol{i}$ do we get $\mathcal{F}_{\triangleright}\{x^m y^n f\}(\boldsymbol{u}) = \boldsymbol{i}^m \frac{\partial^{m+n}}{\partial u^m \partial v^n} \overset{\triangleright}{f}(\boldsymbol{u})\, \boldsymbol{j}^n$.

[13] Here the powers of $\boldsymbol{i}$, $\boldsymbol{j}$ law is a direct consequence of the left linearity.

[14] Compare theorem 4.1.8.

[15] A direct consequence of symmetrizing theorem 4.1.8.

But $\boldsymbol{x}$-shift, partial derivative, and powers of $x^m y^n$ properties need to be modified as in Table 4.2. Regarding (2.8.1) it is clear that $\boldsymbol{i}f = f\boldsymbol{i}$ holds iff $f = f_r + f_i\,\boldsymbol{i}$, $f_r, f_i \in \mathbb{R}$, which is slightly more general than the restriction of [56] to $f = f_r \in \mathbb{R}$. A modulation property analogous to the one in Table 4.1 does not hold. It is obstructed by the non-commutativity of the exponential factors

$$\exp(\boldsymbol{j}yv_0)\,\exp(\boldsymbol{i}xu) \neq \exp(\boldsymbol{i}xu)\,\exp(\boldsymbol{j}yv_0). \tag{4.1.31}$$

For a powers of $\boldsymbol{i}, \boldsymbol{j}$ property to hold for the QFTr, we need to shift the factors $\boldsymbol{j}^n$ also to the left of the quaternion function $f(\boldsymbol{x})$.

For fully general quaternion-valued $f, g \in L^2(\mathbb{R}^2; \mathbb{H})$ we can establish for the QFTr the following quaternion-valued Plancherel theorem based on the inner product (4.1.4).

Theorem 4.1.8 (QFTr Plancherel). *The (quaternion-valued) inner product (4.1.4) of two quaternion module functions $f, g \in L^2(\mathbb{R}^2; \mathbb{H})$ is given by the inner product of the corresponding QFTrs $\overset{\triangleright}{f}$ and $\overset{\triangleright}{g}$*

$$(f, g) = \frac{1}{(2\pi)^2}(\overset{\triangleright}{f}, \overset{\triangleright}{g}). \tag{4.1.32}$$

Proof. For $f, g \in L^2(\mathbb{R}^2; \mathbb{H})$ we calculate the inner product (4.1.4)

$$\begin{aligned}
(f, g) &= \int_{\mathbb{R}^2} f(\boldsymbol{x})\tilde{g}(\boldsymbol{x})d^2\boldsymbol{x} \\
&= \frac{1}{(2\pi)^2}\int_{\mathbb{R}^2}\int_{\mathbb{R}^2} \overset{\triangleright}{f}(\boldsymbol{u})e^{\boldsymbol{j}vy}e^{\boldsymbol{i}ux}d^2\boldsymbol{u}\,\tilde{g}(\boldsymbol{x})d^2\boldsymbol{x} \\
&= \frac{1}{(2\pi)^2}\int_{\mathbb{R}^2} \overset{\triangleright}{f}(\boldsymbol{u})\int_{\mathbb{R}^2} e^{\boldsymbol{j}vy}e^{\boldsymbol{i}ux}\tilde{g}(\boldsymbol{x})d^2\boldsymbol{x}d^2\boldsymbol{u} \\
&= \frac{1}{(2\pi)^2}\int_{\mathbb{R}^2} \overset{\triangleright}{f}(\boldsymbol{u})[\int_{\mathbb{R}^2} g(\boldsymbol{x})e^{-\boldsymbol{i}ux}e^{-\boldsymbol{j}vy}d^2\boldsymbol{x}]^{\sim}\,d^2\boldsymbol{u} \\
&= \frac{1}{(2\pi)^2}\int_{\mathbb{R}^2} \overset{\triangleright}{f}(\boldsymbol{u})\tilde{\overset{\triangleright}{g}}(\boldsymbol{u})d^2\boldsymbol{u} = \frac{1}{(2\pi)^2}(\overset{\triangleright}{f}, \overset{\triangleright}{g}).
\end{aligned} \tag{4.1.33}$$

In the second equality of (4.1.33) we replaced f with its inverse QFTr expression (4.1.30). In the third equality we exchanged the order of integration. For the fourth equality we simply pulled the reversion outside the square brackets [...] and obtained the QFTr $\overset{\triangleright}{g}(\boldsymbol{u})$, which proves (4.1.32) according to (4.1.4). □

For $g = f$ theorem 4.1.8 has a corresponding QFTr Parseval theorem as a direct corollary.

Corollary 4.1.9 (QFTr Parseval). *The $L^2(\mathbb{R}^2;\mathbb{H})$-norm of a quaternion module function $f \in L^2(\mathbb{R}^2;\mathbb{H})$ is given by the $L^2(\mathbb{R}^2;\mathbb{H})$-norm of its QFTr $\overset{\triangleright}{f}$ multiplied by $1/(2\pi)$*

$$\|f\| = \frac{1}{2\pi}\|\overset{\triangleright}{f}\| = \frac{1}{2\pi}\|\hat{f}\|. \tag{4.1.34}$$

Proof. The first identity follows from setting $g = f$ in theorem 4.1.8 (QFTr Plancherel). The second identity follows from comparing with corollary 4.1.3 (QFT Parseval). □

To facilitate the use of the QFTr and comparison with the QFT (Table 4.1) we list the main QFTr properties in Table 4.2.

4.1.1.4 Understanding the $GL(\mathbb{R}^2)$ transformation properties of the QFT

We begin with noting that the matrix transformation law (4.1.27), derived by Bülow [56] for real signals $f \in L^2(\mathbb{R}^2;\mathbb{R})$, and generalized in Theorem 4.1.6 of Section 4.1.1.2.3 to quaternion-valued signals[16] $f \in L^2(\mathbb{R}^2;\mathbb{H})$, with four terms on the right side, allows no straightforward geometric interpretation. Yet a clear geometric interpretation is not only needed in many applications, such an interpretation is also very instructive in order to successfully generalize the QFT to higher dimensions.

Toward this aim we observe that the split (4.1.14) of the exponentials K under the QFT integral results in two (single exponential) complex kernels $K_\pm$ with complex units $\boldsymbol{i}$ (or $\boldsymbol{j}$) apart from the right (or left) factor $(1 \pm \boldsymbol{k})/2$.

This and the known elegant monomial transformation properties of complex Fourier transforms (also preserved in the Clifford FT of [251]) motivates us to geometrically re-analyze the $GL(\mathbb{R}^2)$ transformation properties of the QFT of $f \in L^2(\mathbb{R}^2;\mathbb{H})$ in terms of its two components $f_\pm$ as given in (4.1.12).

Theorem 4.1.10 (QFT of $f_\pm$). *The QFT of the $f_\pm$ split parts of a quaternion module function $f \in L^2(\mathbb{R}^2,\mathbb{H})$ have the complex forms*

$$\hat{f}_\pm = \int_{\mathbb{R}^2} f_\pm e^{-\boldsymbol{j}(yv \mp xu)} d^2x = \int_{\mathbb{R}^2} e^{-\boldsymbol{i}(xu \mp yv)} f_\pm d^2x \,. \tag{4.1.35}$$

---

[16] Remember that Bülow [56] proved his transformation law only for real signals. But in Theorem 4.1.6 of Section 4.1.1.2.3, we used (4.1.24) and Theorem 4.1.5 to generalize from real signals $f \in L^2(\mathbb{R}^2,\mathbb{R})$ to quaternion valued signals $f \in L^2(\mathbb{R}^2,\mathbb{H})$.

Proof.

$$\begin{aligned}
\hat{f}_{\pm} &= \int_{\mathbb{R}^2} e^{-\boldsymbol{i}xu}\{f_r \pm f_k + \boldsymbol{i}(f_i \mp f_j)\}\frac{1\pm\boldsymbol{k}}{2}e^{-\boldsymbol{j}yv}\,d^2x \\
&= \int_{\mathbb{R}^2} \{f_r \pm f_k + \boldsymbol{i}(f_i \mp f_j)\}\, e^{-\boldsymbol{i}xu}\frac{1\pm\boldsymbol{k}}{2}e^{-\boldsymbol{j}yv}\,d^2x \\
&= \int_{\mathbb{R}^2} \{f_r \pm f_k + \boldsymbol{i}(f_i \mp f_j)\}\, \underbrace{\frac{1\pm\boldsymbol{k}}{2}e^{-\boldsymbol{j}(yv\mp xu)}}_{=K_{\pm}}\; d^2x \\
&\overset{(4.1.13)}{=} \int_{\mathbb{R}^2} f_{\pm}e^{-\boldsymbol{j}(yv\mp xu)}d^2x = \int_{\mathbb{R}^2} e^{-\boldsymbol{i}(xu\mp yv)}f_{\pm}d^2x\;,
\end{aligned} \tag{4.1.36}$$

where for the third equality we did a number of quaternion algebra manipulations, involving Euler's formula and trigonometric addition theorems. The last equality of (4.1.36) follows analogously by replacing $f_{\pm}$ with the third expression in (4.1.13), etc. □

We learn from the third line of (4.1.36) that the behavior of the two parts (4.1.14) under automorphisms $\mathcal{A} \in GL(\mathbb{R}^2)$ also determines the automorphism properties of the QFTs $\hat{f}_{\pm}$, where due to theorem 4.1.5 the QFT operation and the split operation (4.1.12) commute.

4.1.1.4.1 Geometric interpretation and coordinate independent formulation of $GL(\mathbb{R}^2)$ transformations of the QFT We begin with noting that according to the *polar decomposition theorem* [162] every automorphism $\mathcal{A} \in GL(\mathbb{R}^2)$ has a unique decomposition $\mathcal{A} = \mathcal{T}\mathcal{R} = \mathcal{R}\mathcal{S}$, where $\mathcal{R}$ is a rotation and $\mathcal{T}$ and $\mathcal{S}$ are symmetric with positive and negative eigenvalues.

Positive eigenvalues correspond to stretches by the eigenvalue in the direction of the eigenvector. Negative eigenvalues correspond to reflections at the line (hyperplane) normal to the eigenvector, composed with stretches by the absolute value of the eigenvalue in the direction of the eigenvector.

Stretches (positive eigenvalues) $\mathcal{D} \in GL(\mathbb{R}^2)$ were already fully treated in [56] (compare also Table 4.1).

Rotations correspond to two reflections [68, 85] at lines subtending half the angle of the resulting rotation $\mathcal{R}_{\boldsymbol{ab}} = \mathcal{U}_{\boldsymbol{a}}\mathcal{U}_{\boldsymbol{b}}$. The elementary transformations that compose all automorphisms $\mathcal{A} \in GL(\mathbb{R}^2)$ are therefore stretches and reflections.

In geometric algebra reflections $\mathcal{U}_{\boldsymbol{n}}$ at a hyperplane (line in 2D) through the origin can be characterized by normal vectors $\boldsymbol{n}$

$$\mathcal{U}_{\boldsymbol{n}}\boldsymbol{x} = -\boldsymbol{n}^{-1}\boldsymbol{x}\boldsymbol{n}. \tag{4.1.37}$$

The length of $\boldsymbol{n}$ does not matter. $\mathcal{U}_{\boldsymbol{n}}$ preserves (reverses) the component parallel (perpendicular) to the hyperplane of reflection.

With the vectors $\boldsymbol{x} = x\boldsymbol{e}_1 + y\boldsymbol{e}_2$, $\boldsymbol{u} = u\boldsymbol{e}_1 + v\boldsymbol{e}_2$ we now rewrite coordinate free[17]

[17]The fact that the reflection $\mathcal{U}_{\boldsymbol{e}_1}$ with the special hyperplane normal to vector $\boldsymbol{e}_1$ is needed stems from the arbitrary initial association of the $\boldsymbol{e}_1$-coordinate product xu with $\boldsymbol{i}$ and of the $\boldsymbol{e}_2$-coordinate product yv with $\boldsymbol{j}$.

the *angles in the exponentials* of $\hat{f}_\pm$ as

$$-xu + yv = \boldsymbol{x} \cdot (\mathcal{U}_{\boldsymbol{e}_1}\boldsymbol{u}), \quad xu + yv = \boldsymbol{x} \cdot \boldsymbol{u}. \tag{4.1.38}$$

Hence we get for the QFTs of $f_\pm$

$$\hat{f}_+ = \int_{\mathbb{R}^2} f_+ e^{-\boldsymbol{j}\,\boldsymbol{x}\cdot(\mathcal{U}_{\boldsymbol{e}_1}\boldsymbol{u})} d^2x, \quad \hat{f}_- = \int_{\mathbb{R}^2} f_- e^{-\boldsymbol{j}\,\boldsymbol{x}\cdot\boldsymbol{u}} d^2x. \tag{4.1.39}$$

The QFT of f_- is analogous to a complex 2D Fourier transform, only in general f_- and the exponential factor do not commute. The QFT of f_+ is similar except for the reflection $\mathcal{U}_{\boldsymbol{e}_1}$.

We are now in a position to apply any automorphism $\mathcal{A} \in GL(\mathbb{R}^2)$ to the spatial argument of the $f_\pm$ components of any $f \in L^2(\mathbb{R}^2, \mathbb{H})$. We begin with

$$\begin{aligned}\widehat{f_-(\mathcal{A}\boldsymbol{x})}(\boldsymbol{u}) &= \int_{\mathbb{R}^2} f_-(\mathcal{A}\boldsymbol{x}) e^{-\boldsymbol{j}\,\boldsymbol{x}\cdot\boldsymbol{u}} d^2x \overset{\boldsymbol{z}=\mathcal{A}\boldsymbol{x}}{=} \int_{\mathbb{R}^2} f_-(\boldsymbol{z}) e^{-\boldsymbol{j}(\mathcal{A}^{-1}\boldsymbol{z})\cdot\boldsymbol{u}} |\det \mathcal{A}^{-1}| d^2z \\ &= |\det \mathcal{A}^{-1}| \int_{\mathbb{R}^2} f_-(\boldsymbol{z}) e^{-\boldsymbol{j}\,\boldsymbol{z}\cdot(\overline{\mathcal{A}^{-1}}\boldsymbol{u})}\, d^2z = |\det \mathcal{A}^{-1}|\, \hat{f}_-(\overline{\mathcal{A}^{-1}}\boldsymbol{u}),\end{aligned} \tag{4.1.40}$$

where $\overline{\mathcal{A}^{-1}}$ indicates the adjoint automorphism of $\mathcal{A}^{-1}$. The absolute value of the determinant $\det \mathcal{A}^{-1}$ needs to be used, because of the interchange of integration boundaries for a negative determinant. We continue with

$$\begin{aligned}\widehat{f_+(\mathcal{A}\boldsymbol{x})}(\boldsymbol{u}) &= \int_{\mathbb{R}^2} f_+(\mathcal{A}\boldsymbol{x}) e^{-\boldsymbol{j}\,\boldsymbol{x}\cdot(\mathcal{U}_{\boldsymbol{e}_1}\boldsymbol{u})} d^2x \\ &\overset{\boldsymbol{z}=\mathcal{A}\boldsymbol{x}}{=} \int_{\mathbb{R}^2} f_+(\boldsymbol{z}) e^{-\boldsymbol{j}(\mathcal{A}^{-1}\boldsymbol{z})\cdot(\mathcal{U}_{\boldsymbol{e}_1}\boldsymbol{u})} |\det \mathcal{A}^{-1}| d^2z \\ &= |\det \mathcal{A}^{-1}| \int_{\mathbb{R}^2} f_+(\boldsymbol{z}) e^{-\boldsymbol{j}\,\boldsymbol{z}\cdot(\overline{\mathcal{A}^{-1}}\mathcal{U}_{\boldsymbol{e}_1}\boldsymbol{u})}\, d^2z \\ &= |\det \mathcal{A}^{-1}| \int_{\mathbb{R}^2} f_+(\boldsymbol{z}) e^{-\boldsymbol{j}\,\boldsymbol{z}\cdot(\mathcal{U}_{\boldsymbol{e}_1}\mathcal{U}_{\boldsymbol{e}_1}\overline{\mathcal{A}^{-1}}\mathcal{U}_{\boldsymbol{e}_1}\boldsymbol{u})}\, d^2z \\ &= |\det \mathcal{A}^{-1}|\, \hat{f}_+(\mathcal{U}_{\boldsymbol{e}_1}\overline{\mathcal{A}^{-1}}\mathcal{U}_{\boldsymbol{e}_1}\boldsymbol{u}),\end{aligned} \tag{4.1.41}$$

which is very similar to the previous calculation for $\hat{f}_-$. The only difference is that in line 4 we insert $1 = \mathcal{U}_{\boldsymbol{e}_1}\mathcal{U}_{\boldsymbol{e}_1}$ before $\overline{\mathcal{A}^{-1}}$, and that the argument of the transformed $\hat{f}_+$ now has the *reflected* version $\mathcal{U}_{\boldsymbol{e}_1}\overline{\mathcal{A}^{-1}}\mathcal{U}_{\boldsymbol{e}_1}$ of the adjoint inverse transformation $\overline{\mathcal{A}^{-1}}$. Recombining $\hat{f}_+$ and $\hat{f}_-$ we get from (4.1.40) and (4.1.41)

Theorem 4.1.11 ($GL(\mathbb{R}^2)$ transformation properties of the QFT). *The QFT of a quaternion module function $f \in L^2(\mathbb{R}^2; \mathbb{H})$ with a $GL(\mathbb{R}^2)$ transformation $\mathcal{A}$ of its vector argument is given by*

$$\widehat{f(\mathcal{A}\boldsymbol{x})}(\boldsymbol{u}) = |\det \mathcal{A}^{-1}| \{ \hat{f}_-(\overline{\mathcal{A}^{-1}}\boldsymbol{u}) + \hat{f}_+(\mathcal{U}_{\boldsymbol{e}_1}\overline{\mathcal{A}^{-1}}\mathcal{U}_{\boldsymbol{e}_1}\boldsymbol{u}) \}\, . \tag{4.1.42}$$

Theorem 4.1.11 corresponds exactly to equation (4.1.27) with (4.1.28), if the matrix expression (4.1.26) is used for the automorphism $\mathcal{A}$ and if the $f_\pm$ split formulas (4.1.12) are used. The four terms of (4.1.27) together with all the matrices involved therefore get in theorem 4.1.11 a clear geometric interpretation. In order to be even more explicit we specify below the full geometric algebra expressions for stretches, reflections and rotations.

4.1.1.4.2 **Explicit examples: stretches, reflections & rotations** To deepen our geometrical understanding we now look at stretches, reflections (and rotations) which *compose* every general automorphism $\mathcal{A} \in GL(\mathbb{R}^2)$.

Stretches expressed by $\mathcal{A}_s \boldsymbol{x} = ax\mathbf{e}_1 + by\mathbf{e}_2$, with $a, b \in \mathbb{R} \setminus \{0\}$, result because of $\mathcal{U}_{\mathbf{e}_1}\mathcal{A}_s\mathcal{U}_{\mathbf{e}_1} = \mathcal{A}_s$ in

$$\widehat{f(\mathcal{A}_s \boldsymbol{x})}(\boldsymbol{u}) = |\det \mathcal{A}_s^{-1}|\, \hat{f}(\mathcal{A}_s^{-1}\boldsymbol{u}) = \frac{1}{|ab|}\hat{f}(\frac{u}{a}\mathbf{e}_1 + \frac{v}{b}\mathbf{e}_2). \tag{4.1.43}$$

Reflections in hyperplanes normal to $\boldsymbol{a}$ expressed by $\mathcal{U}_{\boldsymbol{a}}\boldsymbol{x} = -\boldsymbol{a}^{-1}\boldsymbol{x}\boldsymbol{a}$, with $|\det \mathcal{U}_{\boldsymbol{a}}| = 1$, $\overline{\mathcal{U}_{\boldsymbol{a}}} = \mathcal{U}_{\boldsymbol{a}}$, $\mathcal{U}_{\mathbf{e}_1}\mathcal{U}_{\boldsymbol{a}}\mathcal{U}_{\mathbf{e}_1} = \mathcal{U}_{\boldsymbol{a}'}$ and $\boldsymbol{a}' = \mathcal{U}_{\mathbf{e}_1}\boldsymbol{a}$ result in

$$\widehat{f(\mathcal{U}_{\boldsymbol{a}}\boldsymbol{x})}(\boldsymbol{u}) = \hat{f}_-(\mathcal{U}_{\boldsymbol{a}}\boldsymbol{u}) + \hat{f}_+(\mathcal{U}_{\boldsymbol{a}'}\boldsymbol{u}). \tag{4.1.44}$$

Finally rotations (equivalent to two reflections at lines subtending half the rotation angle) expressed by $\mathcal{R}_{\boldsymbol{ab}}\boldsymbol{x} = \mathcal{U}_{\boldsymbol{b}}\mathcal{U}_{\boldsymbol{a}}\boldsymbol{x}$, with $|\det \mathcal{R}_{\boldsymbol{ab}}| = 1$, $\mathcal{R}_{\boldsymbol{ab}}^{-1} = \mathcal{R}_{\boldsymbol{ba}}$, and $\mathcal{U}_{\boldsymbol{a}'}\mathcal{U}_{\boldsymbol{b}'} = \mathcal{U}_{\mathbf{e}_1}\mathcal{R}_{\boldsymbol{ab}}^{-1}\mathcal{U}_{\mathbf{e}_1}$, result in

$$\begin{aligned}\widehat{f(\mathcal{R}_{\boldsymbol{ab}}\boldsymbol{x})}(\boldsymbol{u}) &= \hat{f}(\mathcal{U}_{\boldsymbol{b}}\mathcal{U}_{\boldsymbol{a}}\boldsymbol{x})(\boldsymbol{u}) = \hat{f}_-(\mathcal{U}_{\boldsymbol{a}}\mathcal{U}_{\boldsymbol{b}}\boldsymbol{u}) + \hat{f}_+(\mathcal{U}_{\boldsymbol{a}'}\mathcal{U}_{\boldsymbol{b}'}\boldsymbol{u}) \\ &= \hat{f}_-(\mathcal{R}_{\boldsymbol{ab}}^{-1}\boldsymbol{u}) + \hat{f}_+(\mathcal{U}_{\mathbf{e}_1}\mathcal{R}_{\boldsymbol{ab}}^{-1}\mathcal{U}_{\mathbf{e}_1}\boldsymbol{u})\end{aligned} \tag{4.1.45}$$

In two dimensions[18] the formula for rotations of the spatial argument of a quaternion module function f subject to the QFT can be further simplified to

$$\widehat{f(\mathcal{R}_{\boldsymbol{ab}}\boldsymbol{x})}(\boldsymbol{u}) \stackrel{\text{in 2D}}{=} \hat{f}_-(\mathcal{R}_{\boldsymbol{ab}}^{-1}\boldsymbol{u}) + \hat{f}_+(\mathcal{R}_{\boldsymbol{ab}}\boldsymbol{u}), \tag{4.1.46}$$

because in two dimensions we have $\mathcal{U}_{\mathbf{e}_1}\mathcal{R}_{\boldsymbol{ab}}^{-1}\mathcal{U}_{\mathbf{e}_1} = \mathcal{R}_{\boldsymbol{ab}}$.

Theorems 4.1.10 and 4.1.11 together with their clear geometric interpretation with the help of geometric algebra pave the way for wide-ranging generalizations of the QFT of Definition 4.1.1. In this book we cannot fully treat all possible generalizations.

But in order to demonstrate the method, we show in the Section 4.4 how to generalize the QFT to a new general non-commutative Fourier transformation of functions from spacetime $\mathbb{R}^{3,1}$ to the spacetime algebra [158] of $\mathbb{R}^{3,1}$, i.e. to the Clifford geometric algebra $Cl(3,1)$. An intermediate step will be the generalization to a new Fourier transform of functions from spacetime $\mathbb{R}^{3,1}$ to a volume-time subalgebra of the spacetime algebra.

4.1.1.5 Summary on QFT and the ±-split

We employed a convenient rewriting of quaternions only in terms of $\boldsymbol{i}$ and $\boldsymbol{j}$, keeping one to the left and the other to the right; and a quaternion split, which in spacetime applications (Section 4.4) is closely related the choice of the time direction. This allowed us to investigate a range of properties of the QFT, last but not least the behavior of the QFT under general linear automorphisms.

---

[18]In Section 4.4, we generalize theorem 4.1.11 to higher dimensions, but for rotations the expression for $\hat{f}_+$ on the right hand side of (4.1.46) will in general not be valid for higher dimensions.

General coordinate free formulation in combination with quaternion to Clifford subalgebra isomorphisms opens the door to a wide range of QFT generalizations to these Clifford geometric algebras, studied further in Section 4.4.

4.1.2 QFT and orthogonal 2D planes split of quaternions

The current treatment is based on [194, 205].

4.1.2.1 Background

The two-sided quaternionic Fourier transformation (QFT) was introduced in [124] for the analysis of 2D linear time-invariant partial-differential systems. Subsequently it has been applied in many fields, including colour image processing [291]. This led to further theoretical investigations [184, 187], where a special split of quaternions was introduced, then called ±split, see Sections 2.8.2, and 4.1.1. An interesting physical consequence was that this split resulted a in left and right traveling multivector wave packet analysis, when generalizing the QFT to a full spacetime Fourier transform (SFT), see Section 4.4. In the current section we investigate this split further, interpret it geometrically and generalize it to a *freely steerable* split of $\mathbb{H}$ into two orthogonal 2D analysis planes. For reasons to become obvious we prefer to call it from now on the *orthogonal 2D planes split* (OPS).

The general form of the OPS split allows us to find new geometric interpretations for the action of the QFT on the signal. The second major result of this work are a variety of new forms of the QFT, their detailed geometric interpretation, and for each form, OPS split theorems, which allow fast and efficient numerical implementation with standard FFT software. A preliminary formal investigation of these new OPS-QFTs can also be found in [194].

This section is organized as follows. We first introduce in Section 4.1.2.2 several properties of quaternions together with a brief review of the *±-split* of [184, 187]. In Section 4.1.2.3, we generalize this split to a *freely steerable orthogonal 2D planes split* (OPS) of quaternions $\mathbb{H}$. In Section 4.1.2.4, we use the general OPS of Section 4.1.2.3 to generalize the two sided QFT to a *new two sided QFT* with freely *steerable analysis planes*, complete with a detailed *local geometric transformation interpretation*. The geometric interpretation of the OPS in Section 4.1.2.3 further allows the construction of a new type of *steerable QFT with a direct phase angle interpretation*. In Section 4.1.2.5, we finally investigate *new steerable QFTs involving quaternion conjugation*. Their local geometric interpretation crucially relies on the notion of *4D rotation reflections*.

4.1.2.2 Orthogonal planes split of quaternions with two orthonormal pure unit quaternions

For an introduction to quaternions we refer to Section 2.8.1.

We remind the reader of the *orthogonal 2D planes split* (OPS) of quaternions with respect to the orthonormal pure unit quaternions $\boldsymbol{i}, \boldsymbol{j}$ [184, 187], see also Section

2.8.2, defined by

$$q = q_+ + q_-, \quad q_\pm = \frac{1}{2}(q \pm \boldsymbol{i} q \boldsymbol{j}). \tag{4.1.47}$$

Explicitly in real components $q_r, q_i, q_j, q_k \in \mathbb{R}$ using (2.8.1) we get

$$q_\pm = \{q_r \pm q_k + \boldsymbol{i}(q_i \mp q_j)\}\frac{1 \pm \boldsymbol{k}}{2} = \frac{1 \pm \boldsymbol{k}}{2}\{q_r \pm q_k + \boldsymbol{j}(q_j \mp q_i)\}. \tag{4.1.48}$$

This leads to the following new Pythagorean *modulus identity* [187]

Lemma 4.1.12 (Modulus identity). *For $q \in \mathbb{H}$*

$$|q|^2 = |q_-|^2 + |q_+|^2. \tag{4.1.49}$$

Lemma 4.1.13 (Orthogonality of OPS split parts). *Given any two quaternions $p, q \in \mathbb{H}$ and applying the OPS of (4.1.47) the resulting parts are orthogonal*

$$\mathrm{S}(p_+\overline{q_-}) = 0, \qquad \mathrm{S}(p_-\overline{q_+}) = 0, \tag{4.1.50}$$

i.e. $p_+ \perp q_-$ and $p_- \perp q_+$.

In Lemma 4.1.13 (proved in [187]) the second identity follows from the first by $\mathrm{S}(\overline{x}) = \mathrm{S}(x)$, $\forall x \in \mathbb{H}$, and $\overline{p_-\overline{q_+}} = q_+\overline{p_-}$.

It is evident, that instead of $\boldsymbol{i}, \boldsymbol{j}$ any pair of orthonormal pure quaternions can be used to produce an analogous split. This is a first indication, that the OPS of (4.1.47) is in fact *steerable*. We observe, that $\boldsymbol{i}q\boldsymbol{j} = q_+ - q_-$, i.e. under the map $\boldsymbol{i}(\,)\boldsymbol{j}$ the q_+ part is invariant, the q_- part changes sign. Both parts are according to (4.1.48) two-dimensional, and by Lemma 4.1.13 they span two completely orthogonal planes. The q_+ plane is spanned by the orthogonal quaternions $\{\boldsymbol{i} - \boldsymbol{j}, 1 + \boldsymbol{i}\boldsymbol{j} = 1 + \boldsymbol{k}\}$, whereas the q_- plane is e.g. spanned by $\{\boldsymbol{i} + \boldsymbol{j}, 1 - \boldsymbol{i}\boldsymbol{j} = 1 - \boldsymbol{k}\}$, i.e. we have the two 2D subspace basis

$$q_+\text{-basis: } \{\boldsymbol{i} - \boldsymbol{j}, 1 + \boldsymbol{i}\boldsymbol{j} = 1 + \boldsymbol{k}\}, \qquad q_-\text{-basis: } \{\boldsymbol{i} + \boldsymbol{j}, 1 - \boldsymbol{i}\boldsymbol{j} = 1 - \boldsymbol{k}\}. \tag{4.1.51}$$

Note that all basis vectors of (4.1.51)

$$\{\boldsymbol{i} - \boldsymbol{j}, 1 + \boldsymbol{i}\boldsymbol{j}, \boldsymbol{i} + \boldsymbol{j}, 1 - \boldsymbol{i}\boldsymbol{j}\} \tag{4.1.52}$$

together form an orthogonal basis of $\mathbb{H}$ interpreted as $\mathbb{R}^4$.

The map $\boldsymbol{i}(\,)\boldsymbol{j}$ rotates the q_- plane by 180° around the 2D q_+ axis plane. Note that in agreement with its geometric interpretation, the map $\boldsymbol{i}(\,)\boldsymbol{j}$ is an *involution*, because applying it twice leads to identity

$$\boldsymbol{i}(\boldsymbol{i}q\boldsymbol{j})\boldsymbol{j} = \boldsymbol{i}^2 q \boldsymbol{j}^2 = (-1)^2 q = q. \tag{4.1.53}$$

4.1.2.3 General orthogonal 2D planes split

We will study generalizations of the OPS split by replacing $\boldsymbol{i}, \boldsymbol{j}$ by arbitrary unit quaternions f, g. Even with this generalization, the map $f(\,)g$ continues to be an involution, because $f^2 q g^2 = (-1)^2 q = q$. For clarity we study the cases $f \neq \pm g$, and $f = g$ separately, though they have a lot in common, and do not always need to be distinguished in concrete applications. In the following we further extend the treatment given in Section 2.8.3.

4.1.2.3.1 Orthogonal 2D planes split using two linearly independent pure unit quaternions

Our result is now, that all these properties hold, even if in the above considerations the pair $\boldsymbol{i}, \boldsymbol{j}$ is replaced by an arbitrary pair of linearly independent nonorthogonal pure quaternions f, g, $f^2 = g^2 = -1, f \neq \pm g$. The OPS is then *re-defined* with respect to the linearly independent pure unit quaternions f, g as

$$q_{\pm} = \frac{1}{2}(q \pm fqg). \tag{4.1.54}$$

Equation (4.1.47) is a special case with $f = \boldsymbol{i}, g = \boldsymbol{j}$. We observe from (4.1.54), that $fqg = q_+ - q_-$, i.e. under the map $f(\,)g$ the q_+ part is invariant, but the q_- part changes sign

$$fq_{\pm}g = \frac{1}{2}(fqg \pm f^2qg^2) = \frac{1}{2}(fqg \pm q) = \pm\frac{1}{2}(q \pm fqg) = \pm q_{\pm}. \tag{4.1.55}$$

We now show that even for (4.1.54) both parts are two-dimensional, and span two completely orthogonal planes. The q_+ plane is spanned by the *orthogonal* pair of quaternions $\{f - g, 1 + fg\}$:

$$\begin{aligned} \mathrm{S}\Big((f-g)\overline{(1+fg)}\Big) &= \mathrm{S}((f-g)(1+(-g)(-f))) \\ &= \mathrm{S}\big(f + fgf - g - g^2f\big) \overset{(2.8.13)}{=} \mathrm{S}\big(f + f^2g - g + f\big) \\ &= 2\,\mathrm{S}(f-g) = 0, \end{aligned} \tag{4.1.56}$$

whereas the q_- plane is e.g. spanned by $\{f+g, 1-fg\}$. The quaternions $f+g, 1-fg$ can be proved to be mutually *orthogonal* by simply replacing $g \to -g$ in (4.1.56). Note that we have

$$\begin{aligned} f(f-g)g &= f^2g - fg^2 = -g + f = f - g, \\ f(1+fg)g &= fg + f^2g^2 = fg + 1 = 1 + fg, \end{aligned} \tag{4.1.57}$$

as well as

$$\begin{aligned} f(f+g)g &= f^2g + fg^2 = -g - f = -(f+g), \\ f(1-fg)g &= fg - f^2g^2 = fg - 1 = -(1-fg). \end{aligned} \tag{4.1.58}$$

We now want to generalize Lemma 4.1.13 to

Lemma 4.1.14 (Orthogonality of two OPS planes). *Given any two quaternions $q, p \in \mathbb{H}$ and applying the OPS (4.1.54) with respect to two linearly independent pure unit quaternions f, g we get zero for the scalar part of the mixed products*

$$\mathrm{S}(p_+\overline{q_-}) = 0, \qquad \mathrm{S}(p_-\overline{q_+}) = 0. \tag{4.1.59}$$

We prove the first identity, the second follows from $\mathrm{S}(x) = \mathrm{S}(\overline{x})$.

$$\begin{aligned} \mathrm{S}(p_+\overline{q_-}) &= \frac{1}{4}\mathrm{S}((p + fpg)(\overline{q} - g\overline{q}f)) = \frac{1}{4}\mathrm{S}(p\overline{q} - fpgg\overline{q}f + fpg\overline{q} - pg\overline{q}f) \\ &\overset{(2.8.15),(2.8.13)}{=} \frac{1}{4}\mathrm{S}(p\overline{q} - p\overline{q} + pg\overline{q}f - pg\overline{q}f) = 0. \end{aligned} \tag{4.1.60}$$

Thus the set

$$\{f-g, 1+fg, f+g, 1-fg\} \tag{4.1.61}$$

forms a 4D orthogonal basis of $\mathbb{H}$ interpreted by (2.8.16) as $\mathbb{R}^4$, where we have for the orthogonal 2D planes the subspace basis:

$$q_+\text{-basis: } \{f-g, 1+fg\}, \qquad q_-\text{-basis: } \{f+g, 1-fg\}. \tag{4.1.62}$$

We can therefore use the following representation for every $q \in \mathbb{H}$ by means of four real coefficients $q_1, q_2, q_3, q_4 \in \mathbb{R}$

$$q = q_1(1+fg) + q_2(f-g) + q_3(1-fg) + q_4(f+g), \tag{4.1.63}$$

where

$$\begin{aligned} q_1 &= \mathrm{S}\Big(q(1+fg)^{-1}\Big), & q_2 &= \mathrm{S}\Big(q(f-g)^{-1}\Big), \\ q_3 &= \mathrm{S}\Big(q(1-fg)^{-1}\Big), & q_4 &= \mathrm{S}\Big(q(f+g)^{-1}\Big). \end{aligned}$$

As an example we have for $f = \boldsymbol{i}, g = \boldsymbol{j}$ according to (4.1.48)

$$q_1 = \frac{1}{2}(q_r + q_k), \quad q_2 = \frac{1}{2}(q_i - q_j), \quad q_3 = \frac{1}{2}(q_r - q_k), \quad q_4 = \frac{1}{2}(q_i + q_j). \tag{4.1.64}$$

Moreover, using

$$f - g = f(1+fg) = (1+fg)(-g), \quad f + g = f(1-fg) = (1-fg)g, \tag{4.1.65}$$

we have the following left and right factoring properties

$$q_+ = q_1(1+fg) + q_2(f-g) = (q_1 + q_2 f)(1+fg) = (1+fg)(q_1 - q_2 g), \tag{4.1.66}$$

$$q_- = q_3(1-fg) + q_4(f+g) = (q_3 + q_4 f)(1-fg) = (1-fg)(q_3 + q_4 g). \tag{4.1.67}$$

Equations (4.1.57) and (4.1.58) further show that the map $f()g$ rotates the q_- plane by 180° around the q_+ axis plane. We found that our interpretation of the map $f()g$ is in perfect agreement with Coxeter's notion of *half-turn* in [84]. This opens the way for new types of QFTs, where the pair of square roots of -1 involved does not necessarily need to be orthogonal.

Before suggesting a generalization of the QFT, we will establish a new set of very useful algebraic identities. Based on (4.1.66) and (4.1.67) we get for $\alpha, \beta \in \mathbb{R}$

$$\begin{aligned} e^{\alpha f} q e^{\beta g} &= e^{\alpha f} q_+ e^{\beta g} + e^{\alpha f} q_- e^{\beta g}, \\ e^{\alpha f} q_+ e^{\beta g} &= (q_1 + q_2 f) e^{\alpha f}(1+fg) e^{\beta g} = e^{\alpha f}(1+fg) e^{\beta g}(q_1 - q_2 g), \\ e^{\alpha f} q_- e^{\beta g} &= (q_3 + q_4 f) e^{\alpha f}(1-fg) e^{\beta g} = e^{\alpha f}(1-fg) e^{\beta g}(q_3 + q_4 g). \end{aligned} \tag{4.1.68}$$

Using (4.1.66) again we obtain

$$\begin{aligned} e^{\alpha f}(1+fg) &= (\cos\alpha + f\sin\alpha)(1+fg) \\ &\overset{(4.1.66)}{=} (1+fg)(\cos\alpha - g\sin\alpha) = (1+fg)e^{-\alpha g}, \end{aligned} \tag{4.1.69}$$

where we set $q_1 = \cos\alpha$, $q_2 = \sin\alpha$ for applying (4.1.66). Replacing in (4.1.69) $-\alpha \to \beta$ we get

$$e^{-\beta f}(1+fg) = (1+fg)e^{\beta g}, \tag{4.1.70}$$

Furthermore, replacing in (4.1.69) $g \to -g$ and subsequently $\alpha \to \beta$ we get

$$e^{\alpha f}(1-fg) = (1-fg)e^{\alpha g}, \qquad e^{\beta f}(1-fg) = (1-fg)e^{\beta g}, \tag{4.1.71}$$

Applying (4.1.66), (4.1.68), (4.1.69) and (4.1.70) we can rewrite

$$\begin{aligned} e^{\alpha f}q_+e^{\beta g} &\overset{(4.1.68)}{=} (q_1+q_2 f)e^{\alpha f}(1+fg)e^{\beta g} \overset{(4.1.69)}{=} (q_1+q_2 f)(1+fg)e^{(\beta-\alpha)g} \\ &\overset{(4.1.66)}{=} q_+e^{(\beta-\alpha)g}, \end{aligned} \tag{4.1.72}$$

or equivalently as

$$\begin{aligned} e^{\alpha f}q_+e^{\beta g} &\overset{(4.1.68)}{=} e^{\alpha f}(1+fg)e^{\beta g}(q_1-q_2 g) \overset{(4.1.70)}{=} e^{(\alpha-\beta)f}(1+fg)(q_1-q_2 g) \\ &\overset{(4.1.66)}{=} e^{(\alpha-\beta)f}q_+. \end{aligned} \tag{4.1.73}$$

In the same way by changing $g \to -g, \beta \to -\beta$ in (4.1.72) and (4.1.73) we can rewrite

$$e^{\alpha f}q_-e^{\beta g} = e^{(\alpha+\beta)f}q_- = q_-e^{(\alpha+\beta)g}. \tag{4.1.74}$$

The result is therefore

$$e^{\alpha f}q_\pm e^{\beta g} = q_\pm e^{(\beta\mp\alpha)g} = e^{(\alpha\mp\beta)f}q_\pm. \tag{4.1.75}$$

4.1.2.3.2 Orthogonal 2D planes split using one pure unit quaternion We now treat the case for $g = f, f^2 = -1$. We then have the map $f(\,)f$, and the OPS split with respect to $f \in \mathbb{H}, f^2 = -1$,

$$q_\pm = \frac{1}{2}(q \pm fqf). \tag{4.1.76}$$

The pure quaternion $\boldsymbol{i}$ can be rotated by $R = \boldsymbol{i}(\boldsymbol{i}+f)$, see (4.1.79), into the quaternion unit f and back. Therefore studying the map $\boldsymbol{i}(\,)\boldsymbol{i}$ is up to the constant rotation between $\boldsymbol{i}$ and f the same as studying $f(\,)f$. This gives

$$\boldsymbol{i}q\boldsymbol{i} = \boldsymbol{i}(q_r + q_i\boldsymbol{i} + q_j\boldsymbol{j} + q_k\boldsymbol{k})\boldsymbol{i} = -q_r - q_i\boldsymbol{i} + q_j\boldsymbol{j} + q_k\boldsymbol{k}. \tag{4.1.77}$$

The OPS with respect to $f = g = \boldsymbol{i}$ gives

$$q_\pm = \frac{1}{2}(q \pm \boldsymbol{i}q\boldsymbol{i}), \quad q_+ = q_j\boldsymbol{j} + q_k\boldsymbol{k} = (q_j + q_k\boldsymbol{i})\boldsymbol{j}, \quad q_- = q_r + q_i\boldsymbol{i}, \tag{4.1.78}$$

where the q_+ plane is two-dimensional and manifestly orthogonal to the 2D q_- plane. This form (4.1.78) of the OPS is therefore identical to the quaternionic simplex/perplex split of [127], see an application in Fig. 4.1 from [127].

For $g = f$ the q_- plane is always spanned by $\{1, f\}$. The rotation operator

Figure 4.1 Simplex/perplex split with gray line $f = g = (\boldsymbol{i} + \boldsymbol{j} + \boldsymbol{k})/\sqrt{3}$. Photo: E. Hitzer. Image processing: S.J. Sangwine. *Top left*: Original. *Top right*: q_--part (luminance). *Bottom left*: q_+-part ($q_i \leftrightarrow q_j$) (chrominance). *Bottom right*: Sum.

$R = \boldsymbol{i}(\boldsymbol{i} + f)$, with norm square $|R|^2 = |\boldsymbol{i}(\boldsymbol{i} + f)|^2 = |(\boldsymbol{i} + f)|^2 = -(\boldsymbol{i} + f)^2$, rotates $\boldsymbol{i}$ into f according to

$$R^{-1}\boldsymbol{i}R = \frac{\overline{R}}{|R|^2}\boldsymbol{i}R = \frac{(\boldsymbol{i}+f)\boldsymbol{i}\boldsymbol{i}\boldsymbol{i}(\boldsymbol{i}+f)}{-(\boldsymbol{i}+f)^2} = \frac{(\boldsymbol{i}+f)\boldsymbol{i}(\boldsymbol{i}(-f)+1)f}{(\boldsymbol{i}+f)^2}$$
$$= \frac{(\boldsymbol{i}+f)(f+\boldsymbol{i})f}{(\boldsymbol{i}+f)^2} = f. \qquad (4.1.79)$$

The rotation R leaves 1 invariant and thus rotates the whole $\{1, \boldsymbol{i}\}$ plane into the q_- plane spanned by $\{1, f\}$. Consequently R also rotates the $\{\boldsymbol{j}, \boldsymbol{k}\}$ plane into the q_+ plane spanned by $\{\boldsymbol{j}' = R^{-1}\boldsymbol{j}R,\ \boldsymbol{k}' = R^{-1}\boldsymbol{k}R\}$. We thus constructively obtain the fully *orthonormal* 4D basis of $\mathbb{H}$ as

$$\{1, f, \boldsymbol{j}', \boldsymbol{k}'\} = R^{-1}\{1, \boldsymbol{i}, \boldsymbol{j}, \boldsymbol{k}\}R, \qquad R = \boldsymbol{i}(\boldsymbol{i} + f), \qquad (4.1.80)$$

for any chosen pure unit quaternion f.

Remark 4.1.15. *Note, that there is a gauge freedom in this split by changing* $R \to R\exp(f\varphi/2)$, $\varphi \in [0, 2\pi)$, *i.e. a rotation freedom in the* q_+*-plane. The units* $\{1, f, \boldsymbol{j}', \boldsymbol{k}'\}$ *form another equivalent representation of quaternions* $\mathbb{H}$.

We further have for the orthogonal 2D planes created in (4.1.76) the subspace basis:

$$q_+\text{-basis: } \{\boldsymbol{j}', \boldsymbol{k}'\}, \qquad q_-\text{-basis: } \{1, f\}. \qquad (4.1.81)$$

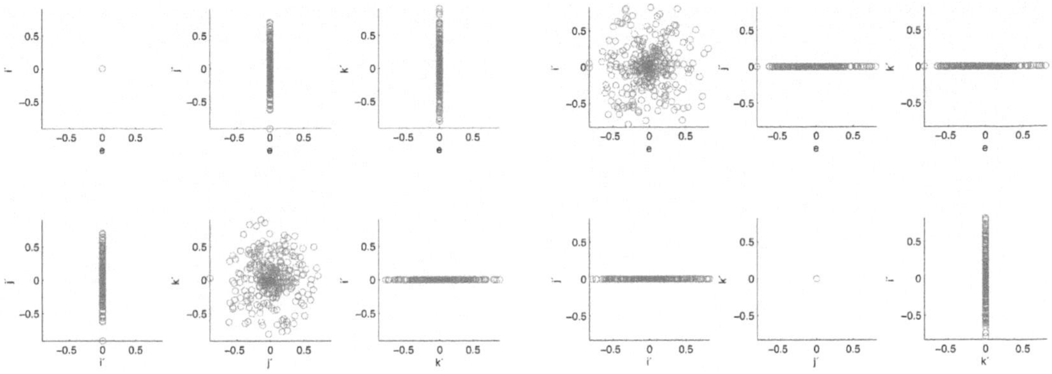

Figure 4.2 4D scatter plot of quaternions decomposed using the orthogonal planes split of (4.1.76) with one unit pure quaternion $f = \boldsymbol{i}' = \frac{1}{\sqrt{3}}(\boldsymbol{i}+\boldsymbol{j}+\boldsymbol{k}) = g$. Left: q_+ component; right: q_- component. Source: [205, Fig. 1].

The rotation R (an orthogonal transformation!) of (4.1.79) preserves the fundamental quaternionic orthonormality and the anticommutation relations

$$f\boldsymbol{j}' = \boldsymbol{k}' = -\boldsymbol{j}'f, \qquad \boldsymbol{k}'f = \boldsymbol{j}' = -f\boldsymbol{k}' \qquad \boldsymbol{j}'\boldsymbol{k}' = f = -\boldsymbol{k}'\boldsymbol{j}'. \tag{4.1.82}$$

Hence

$$fqf = f(q_+ + q_-)f = q_+ - q_-, \quad \text{i.e.} \quad fq_\pm f = \pm q_\pm, \tag{4.1.83}$$

which represents again a half turn by 180° in the 2D q_- plane around the 2D q_+ plane (as axis). Equivalently, the q_- part *commutes* with f, whereas the q_+ part *anticommutes*

$$q_- f = fq_-, \qquad q_+ f = -fq_+. \tag{4.1.84}$$

Figures 4.2 and 4.3 illustrate this decomposition for the case where f is the unit pure quaternion $\frac{1}{\sqrt{3}}(\boldsymbol{i}+\boldsymbol{j}+\boldsymbol{k})$. This decomposition corresponds (for pure quaternions) to the classical *luminance-chrominance* decomposition used in colour image processing, as illustrated, for example, in [127, Figure 2]. Three hundred unit quaternions randomly oriented in 4-space were decomposed. Figure 4.2 shows the three hundred points in 4-space, projected onto the six orthogonal planes $\{e,\boldsymbol{i}'\},\{e,\boldsymbol{j}'\},\{e,\boldsymbol{k}'\},\{\boldsymbol{i}',\boldsymbol{j}'\},\{\boldsymbol{j}',\boldsymbol{k}'\},\{\boldsymbol{k}',\boldsymbol{i}'\}$ where $e = 1$ and $\boldsymbol{i}' = f$, as given in (4.1.80). The six views on the left show the q_+ plane, and the six on the right show the q_- plane.

Figure 4.3 shows the vector parts of the decomposed quaternions. The basis for the plot is $\{\boldsymbol{i}',\boldsymbol{j}',\boldsymbol{k}'\}$, where $\boldsymbol{i}' = f$ as given in (4.1.80). The green circles show the components in the $\{1, f\}$ plane, which intersects the 3-space of the vector part only along the line f (which is the *luminance* or *grey line* of colour image pixels). The red line on the figure corresponds to f. The blue circles show the components in

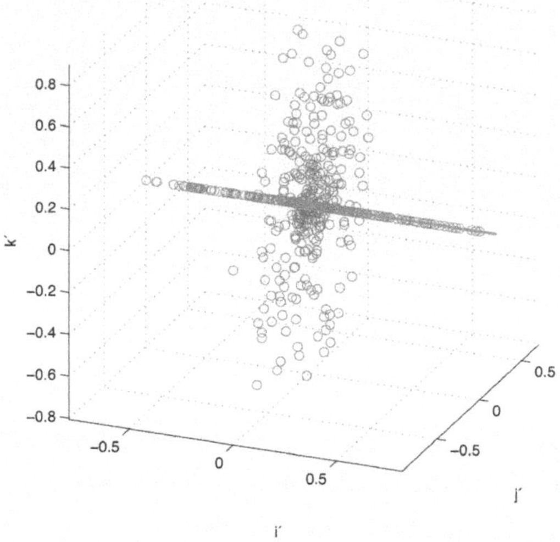

Figure 4.3 Scatter plot of vector parts of quaternions decomposed using the orthogonal planes split of (4.1.76) with one pure unit quaternion $f = \boldsymbol{i}' = \frac{1}{\sqrt{3}}(\boldsymbol{i} + \boldsymbol{j} + \boldsymbol{k}) = g$. The red line corresponds to the direction of f. Source: [205, Fig. 2].

the $[\boldsymbol{j}', \boldsymbol{k}']$ plane, which is entirely within the 3-space. It is orthogonal to f and corresponds to the *chrominance plane* of colour image processing.

The next question is the influence the current OPS (4.1.76) has for left and right exponential factors of the form

$$e^{\alpha f} q_\pm e^{\beta f}. \tag{4.1.85}$$

We learn from (4.1.82) that

$$e^{\alpha f} q_\pm e^{\beta f} = e^{(\alpha \mp \beta) f} q_\pm = q_\pm e^{(\beta \mp \alpha) f}, \tag{4.1.86}$$

which is identical to (4.1.75), if we insert $g = f$ in (4.1.75).

Next, we consider $g = -f, f^2 = -1$. We then have the map $f(\,)(-f)$, and the OPS split with respect to $f, -f \in \mathbb{H}, f^2 = -1$,

$$q_\pm = \frac{1}{2}(q \pm f q(-f)) = \frac{1}{2}(q \mp f q f). \tag{4.1.87}$$

Again we can study $f = \boldsymbol{i}$ first, because for general pure unit quaternions f the unit quaternion $\boldsymbol{i}$ can be rotated by (4.1.79) into the quaternion unit f and back. Therefore studying the map $\boldsymbol{i}(\,)(-\boldsymbol{i})$ is up to the constant rotation R of (4.1.79) the same as studying $f(\,)(-f)$. This gives the map

$$\boldsymbol{i}q(-\boldsymbol{i}) = \boldsymbol{i}(q_r + q_i\boldsymbol{i} + q_j\boldsymbol{j} + q_k\boldsymbol{k})(-\boldsymbol{i}) = q_r + q_i\boldsymbol{i} - q_j\boldsymbol{j} - q_k\boldsymbol{k}. \tag{4.1.88}$$

The OPS with respect to $f = \boldsymbol{i}, g = -\boldsymbol{i}$ gives

$$q_\pm = \frac{1}{2}(q \pm \boldsymbol{i}q(-\boldsymbol{i})), \quad q_- = q_j\boldsymbol{j} + q_k\boldsymbol{k} = (q_j + q_k\boldsymbol{i})\boldsymbol{j}, \quad q_+ = q_r + q_i\boldsymbol{i}, \tag{4.1.89}$$

where, compared to $f = g = \boldsymbol{i}$, the 2D q_+ plane and the 2D q_- plane appear interchanged. The form (4.1.89) of the OPS is again identical to the quaternionic simplex/perplex split of [127], but the simplex and perplex parts appear interchanged.

For $g = -f$ the q_+ plane is always spanned by $\{1, f\}$. The rotation R of (4.1.79) rotates $\boldsymbol{i}$ into f and leaves 1 invariant and thus rotates the whole $\{1, \boldsymbol{i}\}$ plane into the q_+ plane spanned by $\{1, f\}$. Consequently, R of (4.1.79) also rotates the $\{\boldsymbol{j}, \boldsymbol{k}\}$ plane into the q_- plane spanned by $\{\boldsymbol{j}' = R^{-1}\boldsymbol{j}R,\ \boldsymbol{k}' = R^{-1}\boldsymbol{k}R\}$.

We therefore have for the orthogonal 2D planes created in (4.1.87) the subspace basis:

$$q_+\text{-basis: } \{1, f\}, \qquad q_-\text{-basis: } \{\boldsymbol{j}', \boldsymbol{k}'\}. \tag{4.1.90}$$

We again obtain the fully *orthonormal* 4D basis (4.1.80) of $\mathbb{H}$, preserving the fundamental quaternionic orthonormality and the anticommutation relations (4.1.82).

Hence for (4.1.87)

$$fq(-f) = f(q_+ + q_-)(-f) = q_+ - q_-, \quad \text{i.e.} \quad fq_\pm(-f) = \pm q_\pm, \tag{4.1.91}$$

which represents again a half turn by 180° in the 2D q_- plane around the 2D q_+ plane (as axis).

The remaining question is the influence the current OPS (4.1.87) has for left and right exponential factors of the form

$$e^{\alpha f} q_\pm e^{-\beta f}. \tag{4.1.92}$$

We learn from (4.1.82) that

$$e^{\alpha f} q_\pm e^{-\beta f} = e^{(\alpha \mp \beta) f} q_\pm = q_\pm e^{-(\beta \mp \alpha) f}, \tag{4.1.93}$$

which is identical to (4.1.75), if we insert $g = -f$ in (4.1.75).

For (4.1.75) therefore, we do not any longer need to distinguish the cases $f \neq \pm g$ and $f = \pm g$. This motivates us to a general OPS definition for any pair of pure quaternions f, g, and we get a general lemma.

Definition 4.1.16 (General orthogonal 2D planes split). *Let $f, g \in \mathbb{H}$ be an arbitrary pair of pure quaternions f, g, $f^2 = g^2 = -1$, including the cases $f = \pm g$. The general OPS is then defined with respect to the two pure unit quaternions f, g as*

$$q_\pm = \frac{1}{2}(q \pm fqg). \tag{4.1.94}$$

Remark 4.1.17. *The three generalized OPS (4.1.54), (4.1.76) and (4.1.87) are formally identical and are now subsumed in (4.1.94) of Definition 4.1.16, where the values $g = \pm f$ are explicitly included, i.e. any pair of pure unit quaternions $f, g \in \mathbb{H}$, $f^2 = g^2 = -1$, is admissible.*

Lemma 4.1.18. *With respect to the general OPS of Definition 4.1.16 we have for left and right exponential factors the identity*

$$e^{\alpha f} q_\pm e^{\beta g} = q_\pm e^{(\beta \mp \alpha) g} = e^{(\alpha \mp \beta) f} q_\pm. \tag{4.1.95}$$

4.1.2.3.3 Geometric interpretation of left and right exponential factors in f, g We obtain the following general *geometric interpretation*. The map $f(\,)g$ always represents a rotation by angle π in the q_- plane (around the q_+ plane), the map $f^t(\,)g^t$, $t \in \mathbb{R}$, similarly represents a rotation by angle $t\pi$ in the q_- plane (around the q_+ plane as axis). Replacing $g \to -g$ in the map $f(\,)g$ we further find that

$$fq_{\pm}(-g) = \mp q_{\pm}. \tag{4.1.96}$$

Therefore the map $f(\,)(-g) = f(\,)g^{-1}$, because $g^{-1} = -g$, represents a rotation by angle π in the q_+ plane (around the q_- plane), exchanging the roles of 2D rotation plane and 2D rotation axis. Similarly, the map $f^s(\,)g^{-s}$, $s \in \mathbb{R}$, represents a rotation by angle $s\pi$ in the q_+ plane (around the q_- plane as axis).

The product of these two rotations gives

$$f^{t+s}qg^{t-s} = e^{(t+s)\frac{\pi}{2}f}qe^{(t-s)\frac{\pi}{2}g} = e^{\alpha f}qe^{\beta g}, \quad \alpha = (t+s)\frac{\pi}{2}, \quad \beta = (t-s)\frac{\pi}{2}, \tag{4.1.97}$$

where based on (2.8.10) we used the identities $f = e^{\frac{\pi}{2}f}$ and $g = e^{\frac{\pi}{2}g}$.

The *geometric interpretation* of (4.1.97) is a rotation by angle $\alpha+\beta$ in the q_- plane (around the q_+ plane), and a second rotation by angle $\alpha - \beta$ in the q_+ plane (around the q_- plane). For $\alpha = \beta = \pi/2$ we recover the map $f(\,)g$, and for $\alpha = -\beta = \pi/2$ we recover the map $f(\,)g^{-1}$.

4.1.2.3.4 Determination of f, g for given steerable pair of orthogonal 2D planes Equations (4.1.62), (4.1.81), and (4.1.90) tell us how the pair of pure unit quaternions $f, g \in \mathbb{H}$ used in the general OPS of Definition 4.1.16, leads to an explicit basis for the resulting two orthogonal 2D planes, the q_+ plane and the q_- plane. We now ask the *opposite* question: How can we determine from a given steerable pair of orthogonal 2D planes in $\mathbb{H}$ the pair of pure unit quaternions $f, g \in \mathbb{H}$, which splits $\mathbb{H}$ exactly into this given pair of orthogonal 2D planes?

To answer this question, we first observe that in a 4D space it is sufficient to know only one 2D plane explicitly, specified e.g. by a pair of orthogonal unit quaternions $a, b \in \mathbb{H}$, $|a| = |b| = 1$, and without restriction of generality $b^2 = -1$, i.e. b can be a pure unit quaternion $b = \boldsymbol{\mu}(b)$. But for $a = \cos\alpha + \boldsymbol{\mu}(a)\sin\alpha$, compare (2.8.10), we must distinguish $\mathrm{S}(a) = \cos\alpha \neq 0$ and $\mathrm{S}(a) = \cos\alpha = 0$, i.e. of a also being a pure quaternion with $a^2 = -1$. The second orthogonal 2D plane is then simply the *orthogonal complement* in $\mathbb{H}$ to the a, b plane.

Let us first treat the case $\mathrm{S}(a) = \cos\alpha \neq 0$. We set

$$f := ab, \qquad g := \overline{a}b. \tag{4.1.98}$$

With this setting we get for the basis of the q_- plane

$$\begin{aligned} f + g &= ab + \overline{a}b = 2\,\mathrm{S}(a)\,b, \\ 1 - fg &= 1 - ab\overline{a}b = 1 - a^2b^2 = 1 + a^2 \\ &= 1 + \cos^2\alpha - \sin^2\alpha + 2\boldsymbol{\mu}(a)\cos\alpha\sin\alpha \\ &= 2\cos\alpha(\cos\alpha + \boldsymbol{\mu}(a)\sin\alpha) = 2\,\mathrm{S}(a)\,a. \end{aligned} \tag{4.1.99}$$

For the equality $ab\overline{a}b = a^2b^2$ we used the orthogonality of a, b, which means that the vector part of a must be orthogonal to the pure unit quaternion b, i.e. it must anticommute with b

$$ab = b\overline{a}, \qquad ba = \overline{a}b. \tag{4.1.100}$$

Comparing (4.1.62) and (4.1.99), the plane spanned by the two orthogonal unit quaternions $a, b \in \mathbb{H}$ is indeed the q_- plane for $S(a) = \cos\alpha \neq 0$. The orthogonal q_+ plane is simply given by its basis vectors (4.1.62), inserting (4.1.98). This leads to the pair of orthogonal unit quaternions c, d for the q_+ plane as

$$c = \frac{f-g}{|f-g|} = \frac{ab-\overline{a}b}{|(a-\overline{a})b|} = \frac{a-\overline{a}}{|a-\overline{a}|}b = \boldsymbol{\mu}(a)b, \tag{4.1.101}$$

$$d = \frac{1+fg}{|1+fg|} = \frac{f-g}{|f-g|}g = cg = \boldsymbol{\mu}(a)bg = \boldsymbol{\mu}(a)b\overline{a}b = \boldsymbol{\mu}(a)ab^2 = -\boldsymbol{\mu}(a)a, \tag{4.1.102}$$

where we have used (4.1.65) for the second, and (4.1.100) for the sixth equality in (4.1.102).

Let us also verify that f, g of (4.1.98) are both pure unit quaternions using (4.1.100)

$$f^2 = abab = (a\overline{a})bb = -1, \qquad g^2 = \overline{a}b\overline{a}b = (\overline{a}a)bb = -1. \tag{4.1.103}$$

Note, that if we would set in (4.1.98) for $g := -\overline{a}b$, then the a, b plane would have become the q_+ plane instead of the q_- plane. We can therefore determine by the sign in the definition of g, which of the two OPS planes the a, b plane is to represent.

For both a and b being two pure orthogonal quaternions, we can again set

$$f := ab \Rightarrow f^2 = abab = -a^2b^2 = -1, \qquad g := \overline{a}b = -ab = -f, \tag{4.1.104}$$

where due to the orthogonality of the pure unit quaternions a, b we were able to use $ba = -ab$. In this case $f = ab$ is thus also shown to be a pure unit quaternion. Now the q_- plane of the corresponding OPS (4.1.87) is spanned by $\{a, b\}$, whereas the q_+ plane is spanned by $\{1, f\}$. Setting instead $g := -\overline{a}b = ab = f$, the q_- plane of the corresponding OPS (4.1.76) is spanned by $\{1, f\}$, whereas the q_+ plane is spanned by $\{a, b\}$.

We summarize our results in

Theorem 4.1.19. ***(Determination of f, g from given steerable 2D plane)*** *Given any 2D plane in $\mathbb{H}$ in terms of two unit quaternions a, b, where b is without restriction of generality pure, i.e. $b^2 = -1$, we can make the given plane the q_- plane of the OPS $q_\pm = \frac{1}{2}(q \pm fqg)$, by setting*

$$f := ab, \qquad g := \overline{a}b. \tag{4.1.105}$$

For $S(a) \neq 0$ the orthogonal q_+ plane is fully determined by the orthogonal unit quaternions

$$c = \boldsymbol{\mu}(a)b, \qquad d = -\boldsymbol{\mu}(a)a. \tag{4.1.106}$$

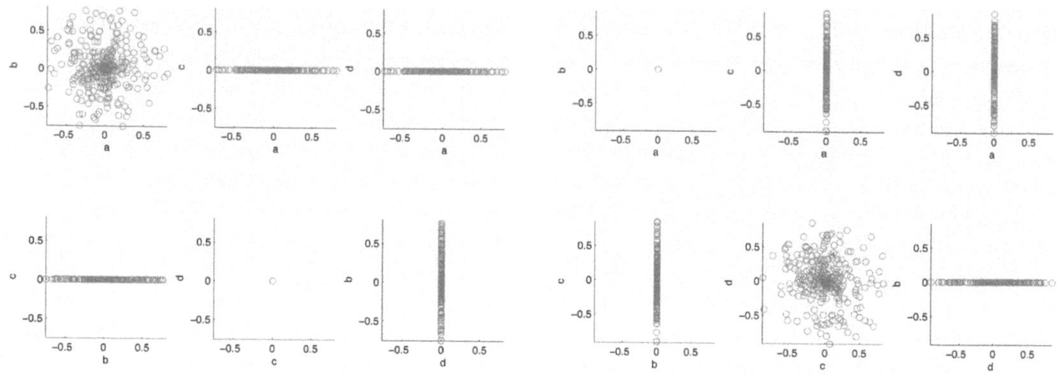

Figure 4.4 4D scatter plot of quaternions decomposed using the orthogonal planes split of Definition 4.1.16. Left: q_+ component; right: q_- component. Source: [205, Fig. 3].

where $\boldsymbol{\mu}(a)$ is as defined in (2.8.10). For $\mathrm{S}(a) = 0$ the orthogonal q_+ plane with basis $\{1, f\}$ is instead fully determined by $f = -g = ab$ and (4.1.90).

Setting alternatively

$$f := ab, \qquad g := -\bar{a}b \tag{4.1.107}$$

makes the given a, b plane the q_+ plane instead. For $\mathrm{S}(a) \neq 0$ the orthogonal q_- plane is then fully determined by (4.1.107) *and* (4.1.62)*, with the same orthogonal unit quaternions $c = \boldsymbol{\mu}(a)b$, $d = -\boldsymbol{\mu}(a)a$ as in (4.1.106). For $\mathrm{S}(a) = 0$ the orthogonal q_- plane with basis $\{1, f\}$ is then instead fully determined by $f = g = ab$ and (4.1.81).*

An illustration of the decomposition is given in Figures 4.4 and 4.5. Again, three hundred unit pure quaternions randomly oriented in 4-space have been decomposed into two sets using the decomposition of Definition 4.1.16 and two unit pure quaternions f and g computed as in Theorem 4.1.19. b was the pure unit quaternion $\frac{1}{\sqrt{3}}(\boldsymbol{i} + \boldsymbol{j} + \boldsymbol{k})$ and a was the full unit quaternion $\frac{1}{\sqrt{2}} + \frac{1}{2}(\boldsymbol{i} - \boldsymbol{j})$. c and d were computed by (4.1.106) as $c = \boldsymbol{\mu}(a)b$ and $d = -\boldsymbol{\mu}(a)a$.

Figure 4.4 shows the three hundred points in 4-space, projected onto the six orthogonal planes $\{c, d\}, \{c, b\}, \{c, a\}, \{d, b\}, \{b, a\}, \{a, d\}$ where the orthonormal 4-space basis $\{c, d, b, a\} = \{(f - g)/|f - g|, (1 + fg)/|1 + fg|, (f + g)/|f + g|, (1 - fg)/|1 - fg|\}$. The six views on the left show the q_+ plane, and the six on the right show the q_- plane. Figure 4.5 shows the vector parts of the decomposed quaternions.

4.1.2.4 New QFT forms: OPS-QFTs with two pure unit quaternions f, g

We note that the versions of the QFT defined here on L^1 can also be defined on L^2 as explained in Footnote 2 on page 144.

4.1.2.4.1 Generalized OPS leads to new steerable type of QFT We begin with a straightforward generalization of the (double-sided form of the) QFT [184, 187] in $\mathbb{H}$ by replacing $\boldsymbol{i} \to f$ and $\boldsymbol{j} \to g$defined as

Definition 4.1.20. ***(QFT with respect to two pure unit quaternions f, g)***

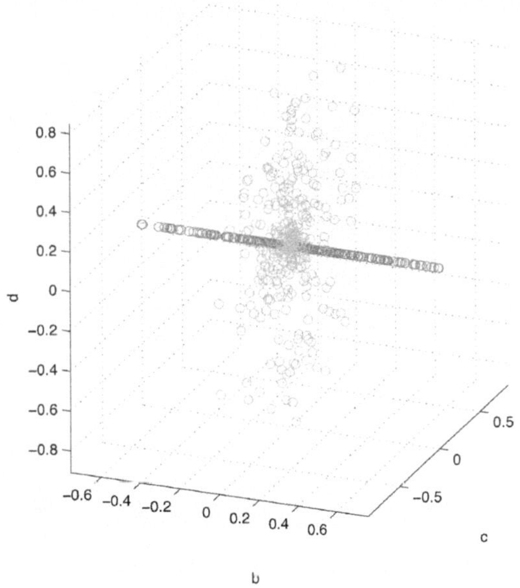

Figure 4.5 Scatter plot of vector parts of quaternions decomposed using the orthogonal planes split of Definition 4.1.16. Source: [205, Fig. 4].

Let $f, g \in \mathbb{H}$, $f^2 = g^2 = -1$, be any two pure unit quaternions. The quaternion Fourier transform with respect to f, g is

$$\mathcal{F}^{f,g}\{h\}(\boldsymbol{\omega}) = \int_{\mathbb{R}^2} e^{-fx_1\omega_1} h(\boldsymbol{x})\, e^{-gx_2\omega_2} d^2\boldsymbol{x}, \tag{4.1.108}$$

where $h \in L^1(\mathbb{R}^2, \mathbb{H})$, $d^2\boldsymbol{x} = dx_1 dx_2$ and $\boldsymbol{x}, \boldsymbol{\omega} \in \mathbb{R}^2$.

Note, that the pure unit quaternions f, g in Definition 4.1.20 do not need to be orthogonal, and that the cases $f = \pm g$ are fully included.

Linearity of the integral (4.1.108) allows us to use the OPS split $h = h_- + h_+$

$$\mathcal{F}^{f,g}\{h\}(\boldsymbol{\omega}) = \mathcal{F}^{f,g}\{h_-\}(\boldsymbol{\omega}) + \mathcal{F}^{f,g}\{h_+\}(\boldsymbol{\omega}) = \mathcal{F}^{f,g}_-\{h\}(\boldsymbol{\omega}) + \mathcal{F}^{f,g}_+\{h\}(\boldsymbol{\omega}), \tag{4.1.109}$$

since by their construction the operators of the Fourier transformation $\mathcal{F}^{f,g}$, and of the OPS with respect to f, g commute. From Lemma 4.1.18 follows

Theorem 4.1.21 (QFT of $h_\pm$). *The QFT of the $h_\pm$ OPS split parts, with respect to two unit quaternions f, g, of a quaternion module function $h \in L^1(\mathbb{R}^2, \mathbb{H})$ have the quasi-complex forms*

$$\mathcal{F}^{f,g}_\pm\{h\} = \mathcal{F}^{f,g}\{h_\pm\}$$
$$= \int_{\mathbb{R}^2} h_\pm e^{-g(x_2\omega_2 \mp x_1\omega_1)} d^2x = \int_{\mathbb{R}^2} e^{-f(x_1\omega_1 \mp x_2\omega_2)} h_\pm d^2x\, . \tag{4.1.110}$$

Remark 4.1.22. *The quasi-complex forms in Theorem 4.1.21 allow to establish discretized and fast versions of the QFT of Definition 4.1.20 as sums of two complex discretized and fast Fourier transformations (FFT), respectively.*

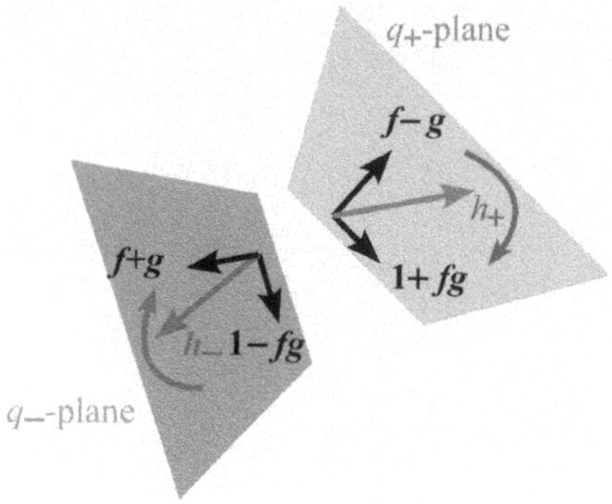

Figure 4.6 Geometric interpretation of integrand of $\mathrm{QFT}^{f,g}$ in Definition 4.1.20 in terms of local phase rotations in $q_\pm$-planes. Source: [205, Fig. 5].

We can now give a *geometric interpretation* of the integrand the $\mathrm{QFT}^{f,g}$ in Definition 4.1.20 in terms of local phase rotations, compare Section 4.1.2.3.3. The integrand product

$$e^{-fx_1\omega_1}h(\boldsymbol{x})\,e^{-gx_2\omega_2} \tag{4.1.111}$$

means to *locally rotate* by the phase angle $-(x_1\omega_1 + x_2\omega_2)$ in the q_- plane, and by phase angle $-(x_1\omega_1 - x_2\omega_2) = x_2\omega_2 - x_1\omega_1$ in the orthogonal q_+ plane, compare Figure 4.6, which depicts two completely orthogonal planes in four dimensions.

Based on Theorem 4.1.19 the two phase rotation planes (analysis planes) can be freely *steered* by defining the two pure unit quaternions f, g used in Definition 4.1.20 according to (4.1.105) or (4.1.107).

4.1.2.4.2 Two phase angle version of QFT The above newly gained geometric understanding motivates us to propose a further new version of the $\mathrm{QFT}^{f,g}$, with a straightforward two *phase angle* interpretation.

Definition 4.1.23. ***(Phase angle QFT with respect to f,g)***
Let $f, g \in \mathbb{H}$, $f^2 = g^2 = -1$, be any two pure unit quaternions. The phase angle quaternion Fourier transform with respect to f, g is

$$\mathcal{F}_D^{f,g}\{h\}(\boldsymbol{\omega}) = \int_{\mathbb{R}^2} e^{-f\frac{1}{2}(x_1\omega_1+x_2\omega_2)}h(\boldsymbol{x})\,e^{-g\frac{1}{2}(x_1\omega_1-x_2\omega_2)}d^2\boldsymbol{x}. \tag{4.1.112}$$

where again $h \in L^1(\mathbb{R}^2, \mathbb{H})$, $d^2\boldsymbol{x} = dx_1dx_2$ and $\boldsymbol{x}, \boldsymbol{\omega} \in \mathbb{R}^2$.

The *geometric interpretation* of the integrand of (4.1.112) is a local phase rotation by angle $-(x_1\omega_1+x_2\omega_2)/2-(x_1\omega_1-x_2\omega_2)/2 = -x_1\omega_1$ in the q_- plane, and a second local phase rotation by angle $-(x_1\omega_1 + x_2\omega_2)/2 + (x_1\omega_1 - x_2\omega_2)/2 = -x_2\omega_2$ in the q_+ plane, compare Section 4.1.2.3.3.

If we apply the $\mathrm{OPS}^{f,g}$ split to (4.1.112) we obtain the following theorem.

Theorem 4.1.24 (Phase angle QFT of $h_\pm$). *The phase angle QFT of Definition*

4.1.23 applied to the $h_\pm$ OPS split parts, with respect to two pure unit quaternions f, g, of a quaternion module function $h \in L^1(\mathbb{R}^2, \mathbb{H})$ leads to the quasi-complex expressions

$$\mathcal{F}^{f,g}_{D+}\{h\} = \mathcal{F}^{f,g}_{D}\{h_+\} = \int_{\mathbb{R}^2} h_+ e^{+gx_2\omega_2} d^2x = \int_{\mathbb{R}^2} e^{-fx_2\omega_2} h_+ d^2x, \tag{4.1.113}$$

$$\mathcal{F}^{f,g}_{D-}\{h\} = \mathcal{F}^{f,g}_{D}\{h_-\} = \int_{\mathbb{R}^2} h_- e^{-gx_1\omega_1} d^2x = \int_{\mathbb{R}^2} e^{-fx_1\omega_1} h_- d^2x, \tag{4.1.114}$$

Note that based on Theorem 4.1.19 the two phase rotation planes (analysis planes) are again freely steerable.

Theorem 4.1.24 allows to establish *discretized* and *fast* versions of the phase angle QFT of Definition 4.1.23 as sums of two complex discretized and fast Fourier transformations (FFT), respectively.

The maps $f(\,)g$ considered so far did not involve *quaternion conjugation* $q \to \overline{q}$. In the following we investigate maps which additionally quaternion conjugate the argument, i.e. of type $f\overline{(\,)}g$, which are also *involutions*.

4.1.2.5 *Involutions and QFTs involving quaternion conjugation*

4.1.2.5.1 Involutions involving quaternion conjugations The simplest case is quaternion conjugation itself

$$q \to \overline{q} = q_r - q_i\boldsymbol{i} - q_j\boldsymbol{j} - q_k\boldsymbol{k}, \tag{4.1.115}$$

which can be interpreted as a *reflection at the real line* q_r. The real line through the origin remains pointwise invariant, while every other point in the 3D hyperplane of pure quaternions is reflected to the opposite side of the real line. The related involution

$$q \to -\overline{q} = -q_r + q_i\boldsymbol{i} + q_j\boldsymbol{j} + q_k\boldsymbol{k}, \tag{4.1.116}$$

is the *reflection at the 3D hyperplane* of pure quaternions (which stay invariant), i.e. only the real line is changed into its negative $q_r \to -q_r$.

Similarly any pure unit quaternion factor like $\boldsymbol{i}$ in the map

$$q \to \boldsymbol{i}\,\overline{q}\boldsymbol{i} = -q_r + q_i\boldsymbol{i} - q_j\boldsymbol{j} - q_k\boldsymbol{k}, \tag{4.1.117}$$

leads to a reflection at the (pointwise invariant) line through the origin with direction $\boldsymbol{i}$, while the map

$$q \to -\boldsymbol{i}\,\overline{q}\boldsymbol{i} = q_r - q_i\boldsymbol{i} + q_j\boldsymbol{j} + q_k\boldsymbol{k}, \tag{4.1.118}$$

leads to a reflection at the invariant 3D hyperplane orthogonal to the line through the origin with direction $\boldsymbol{i}$.

$$q \to f\,\overline{q}f, \tag{4.1.119}$$

leads to a reflection at the (pointwise invariant) line with direction f through the origin, while the map

$$q \to -f\,\overline{q}f, \tag{4.1.120}$$

leads to a reflection at the invariant 3D hyperplane orthogonal to the line with direction f through the origin.

Next we turn to a map of the type

$$q \to -e^{\alpha f}\bar{q}e^{\alpha f}. \tag{4.1.121}$$

Its set of pointwise invariants is given by

$$\begin{aligned} q = -e^{\alpha f}\bar{q}e^{\alpha f} \Leftrightarrow\ & e^{-\alpha f}q = -\bar{q}e^{\alpha f} \ \Leftrightarrow\ e^{-\alpha f}q + \bar{q}e^{\alpha f} = 0 \\ \Leftrightarrow\ & \mathrm{S}\left(\bar{q}e^{\alpha f}\right) = 0 \ \Leftrightarrow\ q \perp e^{\alpha f}. \end{aligned} \tag{4.1.122}$$

We further observe that

$$e^{\alpha f} \to -e^{\alpha f}e^{-\alpha f}e^{\alpha f} = -e^{\alpha f}. \tag{4.1.123}$$

The map $-a\overline{(\ )}a$, with unit quaternion $a = e^{\alpha f}$, therefore represents a reflection at the invariant 3D hyperplane orthogonal to the line through the origin with direction a.

Similarly, the map $a\overline{(\ )}a$, with unit quaternion $a = e^{\alpha f}$, then represents a reflection at the (pointwise invariant) line with direction a through the origin.

The combination of two such reflections (both at 3D hyperplanes, or both at lines), given by unit quaternions a, b, leads to a rotation

$$\begin{aligned} &-b\overline{-a\bar{q}a}b = b\overline{a\bar{q}a}b = b\bar{a}q\bar{a}b = rqs, \\ r = b\bar{a}, \qquad & s = \bar{a}b, \qquad |r| = |b|\,|a| = 1 = |s|, \end{aligned} \tag{4.1.124}$$

in two orthogonal planes, exactly as studied in Section 4.1.2.3.3.

The combination of three reflections at 3D hyperplanes, given by unit quaternions a, b, c , leads to

$$\begin{aligned} -c\overline{[-b\overline{-a\bar{q}a}b]}c = d\,\bar{q}t, \quad d = -c\bar{b}a,\ \ t = a\bar{b}c, \\ |d| = |c|\,|b|\,|a| = |t| = 1. \end{aligned} \tag{4.1.125}$$

The product of the reflection map $-\bar{q}$ of (4.1.116) with $d\,\bar{q}t$ leads to $-dqt$, a *double rotation* as studied in Sections 4.1.2.4. Therefore $d\,\overline{(\,)}\,t$ represents a *rotary reflection*. The three reflections $-a\bar{q}a$, $-b\bar{q}b$, $-c\bar{q}c$ have the intersection of the three 3D hyperplanes as a resulting common pointwise invariant line, which is $d+t$, because

$$d\,\overline{(d+t)}\,t = d\,\bar{t}t + d\,\bar{d}t = d + t. \tag{4.1.126}$$

In the remaining 3D hyperplane, orthogonal to the pointwise invariant line through the origin in direction $d+t$, the axis of the rotary reflection is

$$d\,\overline{(d-t)}\,t = -d\,\bar{t}t + d\,\bar{d}t = -d + t = -(d-t). \tag{4.1.127}$$

We now also understand that a sign change of $d \to -d$ (compare three reflections at three 3D hyperplanes $-c\overline{[-b\overline{(-a\bar{q}a)}b]}c$ with three reflections at three lines $+c\overline{[+b\overline{(+a\bar{q}a)}b]}c$) simply exchanges the roles of pointwise invariant line $d+t$ and rotary reflection axis $d-t$.

Next, we seek for an explicit description of the rotation plane of the rotary reflection $d\overline{()}t$. We find that for the unit quaternions $d = e^{\alpha g}, t = e^{\beta f}$ the commutator

$$[d,t] = dt - td = e^{\alpha g}e^{\beta f} - e^{\beta f}e^{\alpha g} = (gf - fg)\sin\alpha\sin\beta, \qquad (4.1.128)$$

is a pure quaternion, because

$$\overline{gf - fg} = fg - gf = -(gf - fg). \qquad (4.1.129)$$

Moreover, $[d,t]$ is orthogonal to d and t, and therefore orthogonal to the plane spanned by the pointwise invariant line $d+t$ and the rotary reflection axis $d-t$, because

$$S\big([d,t]\overline{d}\big) = S\big(dt\overline{d} - td\overline{d}\big) = 0, \qquad S([d,t]\overline{t}) = 0. \qquad (4.1.130)$$

We obtain a second quaternion in the plane of $d+t, d-t$ by applying the rotation reflection to $[d,t]$

$$d\,\overline{[d,t]}\,t = -d[d,t]t = -[d,t]\overline{d}t, \qquad (4.1.131)$$

because d is orthogonal to the pure quaternion $[d,t]$. We can construct an *orthogonal basis of the plane of the rotation reflection* $d\overline{()}t$ by computing the pair of orthogonal quaternions

$$v_{1,2} = [d,t] \mp d\,\overline{[d,t]}\,t = [d,t] \pm [d,t]\overline{d}t = [d,t](1 \pm \overline{d}t). \qquad (4.1.132)$$

For finally computing the rotation angle, we need to know the relative length of the two orthogonal quaternions v_1, v_2 of (4.1.132). For this it helps to represent the unit quaternion $\overline{d}t$ as

$$\overline{d}t = e^{\gamma u}, \qquad \gamma \in \mathbb{R},\ u \in \mathbb{H},\ u^2 = -1. \qquad (4.1.133)$$

We then obtain for the length ratio

$$\begin{aligned} r^2 = \frac{|v_1|^2}{|v_2|^2} = \frac{|1+\overline{d}t|^2}{|1-\overline{d}t|^2} = \frac{(1+\overline{d}t)(1+\overline{t}d)}{(1-\overline{d}t)(1-\overline{t}d)} = \frac{1+\overline{d}t\overline{t}d + \overline{d}t + \overline{t}d}{1+\overline{d}t\overline{t}d - \overline{d}t - \overline{t}d} \\ = \frac{2+2\cos\gamma}{2-2\cos\gamma} = \frac{1+\cos\gamma}{1-\cos\gamma}. \end{aligned} \qquad (4.1.134)$$

By applying the rotation reflection $d\overline{()}t$ to v_1 and decomposing the result with respect to the pair of orthogonal quaternions in the rotation reflection plane (4.1.132) we can compute the rotation angle. Applying the rotation reflection to v_1 gives

$$\begin{aligned} d\,\overline{v_1}\,t = d\,\overline{[d,t] - d\overline{[d,t]}t}\,t = d\overline{[d,t](1+\overline{d}t)}t = d(1+\overline{t}d)\overline{[d,t]}t \\ = d(1+\overline{t}d)(-[d,t])t = -[d,t](\overline{d}t + \overline{d}t\overline{d}t). \end{aligned} \qquad (4.1.135)$$

The square of $\overline{d}t$ is

$$\begin{aligned} (\overline{d}t)^2 = (\cos\gamma + u\sin\gamma)^2 = -1 + 2\cos\gamma\,[\cos\gamma + u\sin\gamma] \\ = -1 + 2\cos\gamma\,\overline{d}t. \end{aligned} \qquad (4.1.136)$$

We therefore get

$$\begin{aligned} d\,\overline{v_1}\,t &= -[d,t](\overline{d}t - 1 + 2\cos\gamma\overline{d}t) = [d,t](1-(1+2\cos\gamma)\overline{d}t) \\ &= a_1 v_1 + a_2 r v_2, \end{aligned} \tag{4.1.137}$$

and need to solve

$$1-(1+2\cos\gamma)\overline{d}t = a_1(1+\overline{d}t) + a_2 r(1-\overline{d}t), \tag{4.1.138}$$

which leads to

$$d\,\overline{v_1}\,t = -\cos\gamma v_1 + \sin\gamma r v_2 = \cos(\pi-\gamma)v_1 + \sin(\pi-\gamma) r v_2. \tag{4.1.139}$$

The rotation angle of the rotation reflection $d\,\overline{(\,)}\,t$ in its rotation plane v_1, v_2 is therefore

$$\Gamma = \pi - \gamma, \qquad \gamma = \arccos \mathrm{S}\left(\overline{d}t\right). \tag{4.1.140}$$

In terms of $d = e^{\alpha g}, t = e^{\beta f}$ we get

$$\overline{d}t = \cos\alpha\cos\beta - g\sin\alpha\cos\beta + f\cos\alpha\sin\beta - gf\sin\alpha\sin\beta. \tag{4.1.141}$$

And with the angle ω between g and f

$$\begin{aligned} gf &= \frac{1}{2}(gf+fg) + \frac{1}{2}(gf-fg) = \mathrm{S}(gf) + \frac{1}{2}[g,f] \\ &= -\cos\omega + \sin\omega\frac{[g,f]}{|[g,f]|}, \end{aligned} \tag{4.1.142}$$

we finally obtain for γ the scalar part $\mathrm{S}\left(\overline{d}t\right)$ as

$$\begin{aligned} \mathrm{S}\left(\overline{d}t\right) &= \cos\gamma = \cos\alpha\cos\beta + \cos\omega\sin\alpha\sin\beta \\ &= \cos\alpha\cos\beta - \mathrm{S}(gf)\sin\alpha\sin\beta. \end{aligned} \tag{4.1.143}$$

In the special case of $g = \pm f$, $\mathrm{S}(gf) = \mp 1$, i.e. for $\omega = 0, \pi$, we get from (4.1.143) that

$$\begin{aligned} \mathrm{S}\left(\overline{d}t\right) &= \cos\alpha\cos\beta \pm \sin\alpha\sin\beta = \cos\alpha\cos\beta + \sin(\pm\alpha)\sin\beta \\ &= \cos(\pm\alpha - \beta), \end{aligned} \tag{4.1.144}$$

and thus using (4.1.140) the rotation angle would become

$$\Gamma = \pi - (\pm\alpha - \beta) = \pi \mp \alpha + \beta. \tag{4.1.145}$$

Yet (4.1.140) was derived assuming $[d,t] \neq 0$. But direct inspection shows that (4.1.145) is indeed correct: For $g = \pm f$ the plane $d+t, d-t$ is identical to the $1, f$ plane. The rotation plane is thus a plane of pure quaternions orthogonal to the $1, f$ plane. The quaternion conjugation in $q \mapsto d\,\overline{q}\,t$ leads to a rotation by π and the left and right factors lead to further rotations by $\mp\alpha$ and β, respectively. Thus (4.1.145) is verified as a special case of (4.1.140) for $g = \pm f$.

By substituting in Lemma 4.1.18 $(\alpha, \beta) \to (-\beta, -\alpha)$, and by taking the quaternion conjugate we obtain the following Lemma.

Lemma 4.1.25. *Let $q_\pm = \frac{1}{2}(q \pm fqg)$ be the OPS of Definition 4.1.16. For left and right exponential factors we have the identity*

$$e^{\alpha g}\,\overline{q_\pm} e^{\beta f} = \overline{q_\pm} e^{(\beta\mp\alpha)f} = e^{(\alpha\mp\beta)g}\,\overline{q_\pm}. \tag{4.1.146}$$

4.1.2.5.2 New steerable QFTs with quaternion conjugation and two pure unit quaternions f, g We therefore consider now the following new variant of the (double-sided form of the) QFT [184, 187] in $\mathbb{H}$ (replacing both $\boldsymbol{i} \to g$ and $\boldsymbol{j} \to f$, and using quaternion conjugation). It is essentially the quaternion conjugate of the new QFT of Definition 4.1.20, but because of its distinct local transformation geometry deserves separate treatment.

Definition 4.1.26. ***(QFT with respect to f, g, including quaternion conjugation)***
Let $f, g \in \mathbb{H}$, $f^2 = g^2 = -1$, be any two pure unit quaternions. The quaternion Fourier transform with respect to f, g, involving quaternion conjugation, is

$$\mathcal{F}_c^{g,f}\{h\}(\boldsymbol{\omega}) = \overline{\mathcal{F}^{f,g}\{h\}(-\boldsymbol{\omega})} = \int_{\mathbb{R}^2} e^{-gx_1\omega_1}\,\overline{h(\boldsymbol{x})}\,e^{-fx_2\omega_2} d^2\boldsymbol{x}, \tag{4.1.147}$$

where $h \in L^1(\mathbb{R}^2, \mathbb{H})$, $d^2\boldsymbol{x} = dx_1 dx_2$ and $\boldsymbol{x}, \boldsymbol{\omega} \in \mathbb{R}^2$.

Linearity of the integral in (4.1.147) of Definition 4.1.26 leads to the following corollary to Theorem 4.1.21.

Corollary 4.1.27 (QFT $\mathcal{F}_c^{g,f}$ of $h_\pm$). *The QFT $\mathcal{F}_c^{g,f}$ (4.1.147) of the $h_\pm = \frac{1}{2}(h \pm fhg)$ OPS split parts, with respect to any two unit quaternions f, g, of a quaternion module function $h \in L^1(\mathbb{R}^2, \mathbb{H})$ have the quasi-complex forms*

$$\begin{aligned}\mathcal{F}_c^{g,f}\{h_\pm\}(\boldsymbol{\omega}) &= \overline{\mathcal{F}^{f,g}\{h_\pm\}(-\boldsymbol{\omega})}\\ &= \int_{\mathbb{R}^2} \overline{h_\pm} e^{-f(x_2\omega_2\mp x_1\omega_1)} d^2x = \int_{\mathbb{R}^2} e^{-g(x_1\omega_1\mp x_2\omega_2)}\,\overline{h_\pm} d^2x\,.\end{aligned} \tag{4.1.148}$$

Note, that the pure unit quaternions f, g in Definition 4.1.26 and Corollary 4.1.27 do not need to be orthogonal, and that the cases $f = \pm g$ are fully included. Corollary 4.1.27 leads to discretized and fast versions of the QFT with quaternion conjugation of Definition 4.1.26.

It is important to note that the roles (sides) of f, g appear exchanged in (4.1.147) of Definition 4.1.26 and in Corollary 4.1.27, although the same OPS of Definition 4.1.16 is applied to the signal h as in Sections 4.1.2.3 and 4.1.2.4. This role change is due to the presence of quaternion conjugation in Definition 4.1.26. Note that it is possible to first apply (4.1.147) to h, and subsequently split the integral with the $\text{OPS}^{g,f}$ $\mathcal{F}_{c,\pm} = \frac{1}{2}(\mathcal{F}_c \pm g\mathcal{F}_c f)$, where the particular order of g from the left and f from the right is due to the application of conjugation in (4.1.148) to $h_\pm$ *after* h is split with (4.1.94) into h_+ and h_-.

4.1.2.5.3 Local geometric interpretation of the QFT with quaternion conjugation Regarding the *local geometric interpretation* of the QFT with quaternion conjugation of

Definition 4.1.26 we need to distinguish the following cases, depending on $[d,t]$ and on whether the left and right phase factors

$$d = e^{-gx_1\omega_1}, \qquad t = e^{-fx_2\omega_2}, \tag{4.1.149}$$

attain scalar values ± 1 or not.

Let us first assume that $[d,t] \neq 0$, which by (4.1.128) is equivalent to $g \neq \pm f$, and $\sin(x_1\omega_1) \neq 0$, and $\sin(x_2\omega_2) \neq 0$. Then we have the generic case of a local rotation reflection with pointwise invariant line of direction

$$d + t = e^{-gx_1\omega_1} + e^{-fx_2\omega_2}, \tag{4.1.150}$$

rotation axis in direction

$$d - t = e^{-gx_1\omega_1} - e^{-fx_2\omega_2}, \tag{4.1.151}$$

rotation plane with basis (4.1.132), and by (4.1.140) and (4.1.143) the general rotation angle

$$\Gamma = \pi - \arccos \mathrm{S}\left(\overline{d}t\right),$$
$$\mathrm{S}\left(\overline{d}t\right) = \cos(x_1\omega_1)\cos(x_2\omega_2) - \mathrm{S}(gf)\sin(x_1\omega_1)\sin(x_2\omega_2). \tag{4.1.152}$$

Whenever $g = \pm f$, or when $\sin(x_1\omega_1) = 0$ ($x_1\omega_1 = 0, \pi[\mathrm{mod}\, 2\pi]$, i.e. $d = \pm 1$), we get for the pointwise invariant line in direction $d + t$ the simpler unit quaternion direction expression $e^{-\frac{1}{2}(\pm x_1\omega_1 + x_2\omega_2)f}$, because we can apply

$$e^{\alpha f} + e^{\beta f} = e^{\frac{1}{2}(\alpha+\beta)f}\left(e^{\frac{1}{2}(\alpha-\beta)f} + e^{\frac{1}{2}(\beta-\alpha)f}\right) = e^{\frac{1}{2}(\alpha+\beta)f} 2\cos\frac{\alpha-\beta}{2}, \tag{4.1.153}$$

and similarly for the rotation axis $d - t$ we obtain the direction $e^{-\frac{1}{2}(\pm x_1\omega_1 + x_2\omega_2 + \pi)f}$, whereas the rotation angle is by (4.1.145) simply

$$\Gamma = \pi \pm x_1\omega_1 - x_2\omega_2. \tag{4.1.154}$$

For $\sin(x_2\omega_2) = 0$ ($x_2\omega_2 = 0, \pi[\mathrm{mod}\, 2\pi]$, i.e. $t = \pm 1$), the pointwise invariant line in direction $d + t$ simplifies by (4.1.153) to $e^{-\frac{1}{2}(x_1\omega_1 + x_2\omega_2)g}$, and the rotation axis with direction $d - t$ simplifies to $e^{-\frac{1}{2}(x_1\omega_1 + x_2\omega_2 + \pi)g}$, whereas the angle of rotation is by (4.1.145) simply

$$\Gamma = \pi + x_1\omega_1 - x_2\omega_2. \tag{4.1.155}$$

4.1.2.5.4 Phase angle QFT with respect to f, g, including quaternion conjugation Even when quaternion conjugation is applied to the signal h we can propose a further new version of the $\mathrm{QFT}_c^{g,f}$, with a straight forward two *phase angle* interpretation. The following definition to some degree ignores the resulting local rotation reflection effect of combining quaternion conjugation and left and right phase factors of Section 4.1.2.5.3, but depending on the application context, it may nevertheless be of interest in its own right.

Definition 4.1.28. ***(Phase angle QFT with respect to f, g, including quaternion conjugation)***
Let $f, g \in \mathbb{H}$, $f^2 = g^2 = -1$, be any two pure unit quaternions. The phase angle quaternion Fourier transform with respect to f, g, involving quaternion conjugation, is

$$\begin{aligned}\mathcal{F}^{g,f}_{cD}\{h\}(\boldsymbol{\omega}) &= \overline{\mathcal{F}^{f,g}_{D}\{h\}(-\omega_1,\omega_2)} \\ &= \int_{\mathbb{R}^2} e^{-g\frac{1}{2}(x_1\omega_1+x_2\omega_2)}\overline{h(\boldsymbol{x})}\, e^{-f\frac{1}{2}(x_1\omega_1-x_2\omega_2)} d^2\boldsymbol{x}. \end{aligned} \tag{4.1.156}$$

where again $h \in L^1(\mathbb{R}^2, \mathbb{H})$, $d^2\boldsymbol{x} = dx_1dx_2$ and $\boldsymbol{x}, \boldsymbol{\omega} \in \mathbb{R}^2$.

Based on Lemma 4.1.25, one possible *geometric interpretation* of the integrand of (4.1.156) is a local phase rotation of $\overline{h_+}$ by angle $-(x_1\omega_1 - x_2\omega_2)/2 + (x_1\omega_1 + x_2\omega_2)/2 = +x_2\omega_2$ in the $\overline{q_+}$ plane, and a second local phase rotation of $\overline{h_-}$ by angle $-(x_1\omega_1 - x_2\omega_2)/2 - (x_1\omega_1 + x_2\omega_2)/2 = -x_1\omega_1$ in the $\overline{q_-}$ plane. This is expressed in the following corollary to Theorem 4.1.24.

Corollary 4.1.29 (Phase angle QFT of $h_\pm$, involving quaternion conjugation). *The phase angle QFT with quaternion conjugation of Definition 4.1.28 applied to the $h_\pm$ OPS split parts, with respect to any two pure unit quaternions f, g, of a quaternion module function $h \in L^1(\mathbb{R}^2, \mathbb{H})$ leads to the quasi-complex expressions*

$$\begin{aligned}\mathcal{F}^{g,f}_{cD}\{h_+\}(\boldsymbol{\omega}) &= \overline{\mathcal{F}^{g,f}_{D}\{h_+\}(-\omega_1,\omega_2)} \\ &= \int_{\mathbb{R}^2} \overline{h_+} e^{+fx_2\omega_2} d^2x = \int_{\mathbb{R}^2} e^{-gx_2\omega_2}\overline{h_+} d^2x\ , \end{aligned} \tag{4.1.157}$$

$$\begin{aligned}\mathcal{F}^{g,f}_{cD}\{h_-\}(\boldsymbol{\omega}) &= \overline{\mathcal{F}^{g,f}_{D}\{h_-\}(-\omega_1,\omega_2)} \\ &= \int_{\mathbb{R}^2} \overline{h_-} e^{-fx_1\omega_1} d^2x = \int_{\mathbb{R}^2} e^{-gx_1\omega_1}\overline{h_-} d^2x\ . \end{aligned} \tag{4.1.158}$$

Note that based on Theorem 4.1.19 the two phase rotation planes (analysis planes) are again freely steerable. Corollary 4.1.29 leads to discretized and fast versions of the phase angle QFT with quaternion conjugation of Definition 4.1.28.

4.1.2.6 Summary on QFT and OPS of quaternions

The involution maps $\boldsymbol{i}(\,)\boldsymbol{j}$ and more general $f(\,)g$ have led us to explore a range of similar quaternionic maps $q \mapsto aqb$ and $q \mapsto a\bar{q}b$, where a, b are taken to be unit quaternions. Geometric interpretations of these maps as reflections, rotations and rotary reflections in 4D can mostly be found in [84], see also Section 2.8. We have further developed these geometric interpretations to gain a *complete local transformation geometric understanding* of the integrands of the proposed new quaternion Fourier transformations (QFTs) applied to general quaternionic signals $h \in L^1(\mathbb{R}^2, \mathbb{H})$. This new geometric understanding is also valid for the special cases of the hitherto well known left sided, right sided and left and right sided (two-sided) QFTs of [124,127,184,291] and numerous other references.

Our newly gained geometric understanding itself motivated us to study *new types* of QFTs with specific geometric properties. The investigation of these new types of QFTs with the generalized form of the orthogonal 2D planes split of Definition 4.1.16 lead to important QFT split theorems , which allow the use of discrete and (complex) Fourier transform software for efficient discretized and fast numerical implementations.

Finally, our geometric interpretation of old and new QFTs paves the way for new applications, e.g., regarding *steerable filter design* for specific tasks in image, color image and signal processing, etc.

4.2 PROPERTIES OF QUATERNION FOURIER TRANSFORM

4.2.1 Uncertainty principle for quaternion Fourier transform

This section derives a directional uncertainty principle for quaternion valued functions subject to the quaternion Fourier transformation. This can be generalized to establish directional uncertainty principles in Clifford geometric algebras with quaternion subalgebras. We demonstrate this later in Section 4.4 with the example of a directional spacetime algebra function uncertainty principle related to multivector wave packets. Our treatment is based on [187]. See [255] for a strictly coordinate wise approach.

The Heisenberg uncertainty principle and quaternions are both fundamental for quantum mechanics, including the spin of elementary particles. The quaternion Fourier transform (QFT) [131, 184] is used in image and signal processing. It allows to formulate a component wise uncertainty principle [56, 255, 256].

In Clifford geometric algebras, which generalize real and complex numbers and quaternions to higher dimensions, the vector differential allows to formulate a more general directional uncertainty principle [185, 251]. The present text formulates a directional uncertainty principle for quaternion functions. As prerequisite for its proof we further investigate the split of quaternions introduced in [184, 265].

4.2.1.1 Uncertainty principle

4.2.1.1.1 The Heisenberg uncertainty principle in physics Photons (quanta of light) scattering off particles to be detected cause minimal position momentum uncertainty during detection. The uncertainty principle has played a fundamental role in the development and understanding of quantum physics. It is also central for information processing [287].

In quantum physics it states, e.g., that particle momentum and position cannot be simultaneously measured with arbitrary precision. The hypercomplex (quaternion or more general multivector) function $f(\boldsymbol{x})$ would represent the spatial part of a separable wave function and its Fourier transform (QFT, CFT, SFT, ...) $\mathcal{F}\{f\}(\boldsymbol{\omega})$ the same wave function in momentum space (compare [108, 299, 319]). The variance in space $\mathbb{R}^n$ (in physics often $n = 3$) would then be calculated as $(1 \leq k \leq n)$

$$(\Delta x_k)^2 = \int_{\mathbb{R}^n} \langle f(\boldsymbol{x})(\boldsymbol{e}_k \cdot \boldsymbol{x})^2 \tilde{f}(\boldsymbol{x}) \rangle \, d^n\boldsymbol{x} = \int_{\mathbb{R}^n} (\boldsymbol{e}_k \cdot \boldsymbol{x})^2 |f(\boldsymbol{x})|^2 \, d^n\boldsymbol{x},$$

where it is customary to set without loss of generality the mean value of $\boldsymbol{e}_k \cdot \boldsymbol{x}$ to zero [319]. The variance in momentum space would be calculated as $(1 \leq l \leq n)$

$$\begin{aligned}(\Delta\omega_l)^2 &= \frac{1}{(2\pi)^n}\int_{\mathbb{R}^n}\langle \mathcal{F}\{f\}(\boldsymbol{\omega})(\boldsymbol{e}_l\cdot\boldsymbol{\omega})^2\widetilde{\mathcal{F}}\{f\}(\boldsymbol{\omega})\rangle\, d^n\boldsymbol{\omega} \\ &= \frac{1}{(2\pi)^n}\int_{\mathbb{R}^n}(\boldsymbol{e}_l\cdot\boldsymbol{\omega})^2\, |\mathcal{F}\{f\}(\boldsymbol{\omega})|^2 d^n\boldsymbol{\omega}.\end{aligned}$$

Again the mean value of $\boldsymbol{e}_l \cdot \boldsymbol{\omega}$ is customarily set to zero, it merely corresponds to a phase shift [319]. Using our mathematical units, the position-momentum uncertainty relation of quantum mechanics is then expressed by (compare e.g. with (4.9) of [299, page 86])

$$\Delta x_k \Delta\omega_l = \frac{1}{2}\,\delta_{k,l} F, \tag{4.2.1}$$

where $\delta_{k,l}$ is the usual Kronecker symbol. Note that we have not normalized the squares of the variances by division with $F = \int_{\mathbb{R}^n} |f(\boldsymbol{x})|^2\, d^n\boldsymbol{x}$, therefore the extra factor F on the right side of (4.2.1). Further explicit examples from image processing can be found in [132].

In general in Fourier analysis such conjugate entities correspond to the variances of a function and its Fourier transform which cannot both be simultaneously sharply localized (e.g. [74, 287]). Material on the classical uncertainty principle for the general case of $L^2(\mathbb{R}^n)$ without the additional condition $\lim_{|x|\to\infty} |x|^2|f(x)| = 0$ can be found in [250] and [147]. Felsberg [132] even notes for two dimensions: *In 2D however, the uncertainty relation is still an open problem. In [145] it is stated that there is no straightforward formulation for the 2D uncertainty relation.*

Let us now briefly define the notion of *directional* uncertainty principle in Clifford geometric algebra.

4.2.1.1.2 Directional uncertainty principle in Clifford geometric algebra From the view point of Clifford geometric algebra an uncertainty principle gives us information about how the variance of a multivector valued function and the variance of its Clifford Fourier transform are related. We can shed the restriction to the parallel $(k = l)$ and orthogonal $(k \neq l)$ cases of (4.2.1) by looking at the $\boldsymbol{x} \in \mathbb{R}^n$ variance in an arbitrary but fixed direction $\boldsymbol{a} \in \mathbb{R}^n$ and at the $\boldsymbol{\omega} \in \mathbb{R}^n$ variance in an arbitrary but fixed direction $\boldsymbol{b} \in \mathbb{R}^n$. We are now concerned with Clifford geometric algebra multivector functions $f : \mathbb{R}^n \to \mathcal{G}_n$, $n = 2, 3 \pmod 4$ with GA FT $\mathcal{F}\{f\}(\boldsymbol{\omega})$ such that[19]

$$\int_{\mathbb{R}^n} |f(\boldsymbol{x})|^2\, d^n\boldsymbol{x} = F < \infty. \tag{4.2.2}$$

The directional uncertainty principle [185, 251] with arbitrary constant vectors $\boldsymbol{a}$, $\boldsymbol{b} \in \mathbb{R}^n$ means that

$$\frac{1}{F}\int_{\mathbb{R}^n}(\boldsymbol{a}\cdot\boldsymbol{x})^2|f(\boldsymbol{x})|^2\, d^n\boldsymbol{x}\, \frac{1}{(2\pi)^n F}\int_{\mathbb{R}^n}(\boldsymbol{b}\cdot\boldsymbol{\omega})^2\, |\mathcal{F}\{f\}(\boldsymbol{\omega})|^2 d^n\boldsymbol{\omega} \geq \frac{(\boldsymbol{a}\cdot\boldsymbol{b})^2}{4}. \tag{4.2.3}$$

[19] Regarding the definition of the GA FT (CFT) on L^2, see Footnote 2 on page 144.

Equality (minimal bound) in (4.2.3) is achieved for optimal *Gaussian* multivector functions

$$f(\boldsymbol{x}) = C_0\, e^{-\alpha\, \boldsymbol{x}^2}, \tag{4.2.4}$$

$C_0 \in \mathcal{G}_n$ arbitrary const. multivector, $0 < \alpha \in \mathbb{R}$.

So far there seems to be no directional uncertainty principle for quaternion functions over $\mathbb{R}^2$ subject to the QFT. But there already exists a component wise uncertainty principle.

4.2.1.1.3 Component-wise uncertainty principle for right-sided QFT

For the right sided QFT[20] $\mathcal{F}_r\{f\}\colon \mathbb{R}^2 \to \mathbb{H}$ of $f \in L^1(\mathbb{R}^2;\mathbb{H})$ given by [56, 255]

$$\mathcal{F}_r\{f\}(\boldsymbol{\omega}) = \int_{\mathbb{R}^2} f(\boldsymbol{x}) e^{-\boldsymbol{i}\omega_1 x_1} e^{-\boldsymbol{j}\omega_2 x_2}\, d^2\boldsymbol{x}, \tag{4.2.5}$$

where $\boldsymbol{x} = x_1\boldsymbol{e}_1 + x_2\boldsymbol{e}_2$, $\boldsymbol{\omega} = \omega_1\boldsymbol{e}_1 + \omega_2\boldsymbol{e}_2$, and the quaternion exponential product $e^{-\boldsymbol{i}\omega_1 x_1} e^{-\boldsymbol{j}\omega_2 x_2}$ is the quaternion Fourier *kernel*, it is possible to establish a component uncertainty principle ($k = 1, 2$)

$$\int_{\mathbb{R}^2} x_k^2 |f(\boldsymbol{x})|^2 d^2\boldsymbol{x} \int_{\mathbb{R}^2} \omega_k^2 |\mathcal{F}_r\{f\}(\boldsymbol{\omega})|^2 d^2\boldsymbol{\omega} \geq \frac{(2\pi)^2}{4} \left\{ \int_{\mathbb{R}^2} |f(\boldsymbol{x})|^2 d^2\boldsymbol{x} \right\}^2, \tag{4.2.6}$$

where $f \in L^2(\mathbb{R}^2;\mathbb{H})$ is now a quaternion-valued signal such that both $(1+|x_k|)f(\boldsymbol{x}) \in L^2(\mathbb{R}^2;\mathbb{H})$ and $\frac{\partial}{\partial x_k} f(\boldsymbol{x}) \in L^2(\mathbb{R}^2;\mathbb{H})$. It is further possible to prove [255] that equality holds if and only if f is a Gaussian quaternion function

$$f(\boldsymbol{x}) = C_0 e^{-\alpha_1 x_1^2 - \alpha_2 x_2^2}, \qquad \text{const. } C_0 \in \mathbb{H}, \qquad 0 < \alpha_1, \alpha_2 \in \mathbb{R}. \tag{4.2.7}$$

Remark 4.2.1. *Now we want to address the very important question: Is it also possible to establish a full directional uncertainty principle for the QFT?*

In the following we will try to answer this question after investigating some relevant properties of the *double-sided* QFT, which is slightly different from (4.2.5).

4.2.1.2 *Quaternion Fourier transform (QFT)*

From now on we only consider the double-sided QFT $\mathcal{F}\{f\}\colon \mathbb{R}^2 \to \mathbb{H}$ of $f \in L^1(\mathbb{R}^2;\mathbb{H})$ in the form[21],

$$\mathcal{F}\{f\}(\boldsymbol{\omega}) = \hat{f}(\boldsymbol{\omega}) = \int_{\mathbb{R}^2} e^{-\boldsymbol{i}x_1\omega_1} f(\boldsymbol{x})\, e^{-\boldsymbol{j}x_2\omega_2} d^2\boldsymbol{x}. \tag{4.2.8}$$

Linearity allows to also split up the QFT itself as

$$\mathcal{F}\{f\}(\boldsymbol{\omega}) = \mathcal{F}\{f_- + f_+\}(\boldsymbol{\omega}) = \mathcal{F}\{f_-\}(\boldsymbol{\omega}) + \mathcal{F}\{f_+\}(\boldsymbol{\omega}). \tag{4.2.9}$$

[20] For the QFT on L^2, see Footnote 2 on page 144.
[21] For the QFT on L^2, see Footnote 2 on page 144.

4.2.1.2.1 Simple complex forms for QFT of $f_\pm$ The QFT of the $f_\pm$ split parts of a quaternion function $f \in L^2(\mathbb{R}^2, \mathbb{H})$ have simple complex forms [184]

$$\hat{f}_\pm = \int_{\mathbb{R}^2} f_\pm e^{-\boldsymbol{j}(x_2\omega_2 \mp x_1\omega_1)} d^2x = \int_{\mathbb{R}^2} e^{-\boldsymbol{i}(x_1\omega_1 \mp x_2\omega_2)} f_\pm d^2x. \tag{4.2.10}$$

We can rewrite this free of coordinates as

$$\hat{f}_- = \int_{\mathbb{R}^2} f_- e^{-\boldsymbol{j}\,\boldsymbol{x}\cdot\boldsymbol{\omega}} d^2x, \quad \hat{f}_+ = \int_{\mathbb{R}^2} f_+ e^{-\boldsymbol{j}\,\boldsymbol{x}\cdot(\mathcal{U}_1\boldsymbol{\omega})} d^2x, \tag{4.2.11}$$

where the reflection $\mathcal{U}_1\boldsymbol{\omega}$ changes component $\omega_1 \to -\omega_1$. That this applies to ω_1 is due to the choice of the kernel in (4.2.8).

4.2.1.2.2 Preparations for the full directional QFT uncertainty principle Now we continue to lay the ground work for the full directional QFT uncertainty principle. We begin with the following lemma.

Lemma 4.2.2 (Integration of parts). *With the vector differential $\boldsymbol{a}\cdot\nabla = a_1\partial_1 + a_2\partial_2$, with arbitrary constant $\boldsymbol{a} \in \mathbb{R}^2$, $g, h \in L^2(\mathbb{R}^2, \mathbb{H})$*

$$\begin{aligned}&\int_{\mathbb{R}^2} g(\boldsymbol{x})[\boldsymbol{a}\cdot\nabla h(\boldsymbol{x})]d^2\boldsymbol{x}\\ &\quad= \left[\int_{\mathbb{R}} g(\boldsymbol{x})h(\boldsymbol{x})d\boldsymbol{x}\right]_{a\cdot x=-\infty}^{a\cdot x=\infty} - \int_{\mathbb{R}^2}[\boldsymbol{a}\cdot\nabla g(\boldsymbol{x})]h(\boldsymbol{x})d^2\boldsymbol{x}.\end{aligned} \tag{4.2.12}$$

Remark 4.2.3. *The proof of Lemma 4.2.2 works very similar to the proof of integration of parts in [251]. We therefore don't repeat it here.*

For a quaternion function f and its QFT $\mathcal{F}\{f\}$ itself we also get important modulus identities.

Lemma 4.2.4 (Modulus identities). *Due to $|q|^2 = |q_-|^2 + |q_+|^2$ of Lemma 4.1.12 we get for $f : \mathbb{R}^2 \to \mathbb{H}$ the following identities*

$$|f(\boldsymbol{x})|^2 = |f_-(\boldsymbol{x})|^2 + |f_+(\boldsymbol{x})|^2, \tag{4.2.13}$$

$$|\mathcal{F}\{f\}(\boldsymbol{\omega})|^2 = |\mathcal{F}\{f_-\}(\boldsymbol{\omega})|^2 + |\mathcal{F}\{f_+\}(\boldsymbol{\omega})|^2. \tag{4.2.14}$$

We further establish formulas for the vector differentials of the QFTs of the split function parts f_- and f_+.

Lemma 4.2.5 (QFT of vector differentials). *Using the split $f = f_- + f_+$ we get the QFTs of the split parts. Let $\boldsymbol{b} \in \mathbb{R}^2$ be an arbitrary constant vector.*

$$\mathcal{F}\{\boldsymbol{b}\cdot\nabla f_-\}(\boldsymbol{\omega}) = \boldsymbol{b}\cdot\boldsymbol{\omega}\mathcal{F}\{f_-\}(\boldsymbol{\omega})\,\boldsymbol{j}, \tag{4.2.15}$$

$$\mathcal{F}\{\boldsymbol{b}\cdot\nabla f_+\}(\boldsymbol{\omega}) = \boldsymbol{b}\cdot(\mathcal{U}_1\boldsymbol{\omega})\,\mathcal{F}\{f_+\}(\boldsymbol{\omega})\,\boldsymbol{j}, \tag{4.2.16}$$

$$\mathcal{F}\{(\mathcal{U}_1\boldsymbol{b})\cdot\nabla f_+\}(\boldsymbol{\omega}) = \boldsymbol{b}\cdot\boldsymbol{\omega}\,\mathcal{F}\{f_+\}(\boldsymbol{\omega})\,\boldsymbol{j}. \tag{4.2.17}$$

Remark 4.2.6. *The previously explained complex forms of the QFTs make the proof of Lemma 4.2.5 very similar to the case $n = 2$ of [185]. Noncommutativity must be duly taken into account!*

Lemma 4.2.7 (Schwarz inequality). *Two quaternion functions $g, h \in L^2(\mathbb{R}^2, \mathbb{H})$ obey the following Schwarz inequality*

$$\int_{\mathbb{R}^2} |g(\boldsymbol{x})|^2 d^2\boldsymbol{x} \int_{\mathbb{R}^2} |h(\boldsymbol{x})|^2 d^2\boldsymbol{x} \geq \frac{1}{4}\left[\int_{\mathbb{R}^2} g(\boldsymbol{x})\tilde{h}(\boldsymbol{x}) + h(\boldsymbol{x})\tilde{g}(\boldsymbol{x}) d^2\boldsymbol{x}\right]^2$$
$$= \left[\int_{\mathbb{R}^2} Sc(g(\boldsymbol{x})\tilde{h}(\boldsymbol{x})) d^2\boldsymbol{x}\right]^2. \tag{4.2.18}$$

Remark 4.2.8. *The proof of Lemma 4.2.7 can be based on the following inequality ($0 < \epsilon \in \mathbb{R}$)*

$$\int_{\mathbb{R}^2} [g(\boldsymbol{x}) + \epsilon h(\boldsymbol{x})][g(\boldsymbol{x}) + \epsilon h(\boldsymbol{x})]^{\sim} d^2\boldsymbol{x} \geq 0. \tag{4.2.19}$$

4.2.1.3 Directional QFT uncertainty principle

In this section we state the directional QFT uncertainty principle and prove it step by step.

Theorem 4.2.9 (Directional QFT UP). *For two arbitrary constant vectors $\boldsymbol{a} = a_1\boldsymbol{e}_1 + a_2\boldsymbol{e}_2 \in \mathbb{R}^2$, $\boldsymbol{b} = b_1\boldsymbol{e}_1 + b_2\boldsymbol{e}_2 \in \mathbb{R}^2$ (selecting two directions), and $f \in L^2(\mathbb{R}^2, \mathbb{H})$, $|\boldsymbol{x}|^{1/2} f \in L^2(\mathbb{R}^2, \mathbb{H})$ we obtain*

$$\int_{\mathbb{R}^2} (\boldsymbol{a}\cdot\boldsymbol{x})^2 |f(\boldsymbol{x})|^2 d^2\boldsymbol{x} \int_{\mathbb{R}^2} (\boldsymbol{b}\cdot\boldsymbol{\omega})^2 |\mathcal{F}\{f\}(\boldsymbol{\omega})|^2 d^2\boldsymbol{\omega}$$
$$\geq \frac{(2\pi)^2}{4}\left[(\boldsymbol{a}\cdot\boldsymbol{b})^2 F_-^2 + (\boldsymbol{a}\cdot\boldsymbol{b}')^2 F_+^2\right], \tag{4.2.20}$$

with the energies

$$F_\pm = \int_{\mathbb{R}^2} |f_\pm(\boldsymbol{x})|^2 d^2\boldsymbol{x}, \qquad \boldsymbol{b}' = \mathcal{U}_1\boldsymbol{b} = -b_1\boldsymbol{e}_1 + b_2\boldsymbol{e}_2 \in \mathbb{R}^2. \tag{4.2.21}$$

Proof. We now prove the directional QFT uncertainty principle by the following direct calculation.

$$\int_{\mathbb{R}^2} (\boldsymbol{a}\cdot\boldsymbol{x})^2 |f(\boldsymbol{x})|^2 d^2\boldsymbol{x} \int_{\mathbb{R}^2} (\boldsymbol{b}\cdot\boldsymbol{\omega})^2 |\mathcal{F}\{f\}(\boldsymbol{\omega})|^2 d^2\boldsymbol{\omega}$$
$$\overset{\pm\text{split}}{=} \left[\int_{\mathbb{R}^2} (\boldsymbol{a}\cdot\boldsymbol{x})^2 |f_-(\boldsymbol{x})|^2 d^2\boldsymbol{x} + \int_{\mathbb{R}^2} (\boldsymbol{a}\cdot\boldsymbol{x})^2 |f_+(\boldsymbol{x})|^2 d^2\boldsymbol{x}\right]$$
$$\left[\int_{\mathbb{R}^2} (\boldsymbol{b}\cdot\boldsymbol{\omega})^2 |\mathcal{F}\{f_-\}(\boldsymbol{\omega})|^2 d^2\boldsymbol{\omega} + \int_{\mathbb{R}^2} (\boldsymbol{b}\cdot\boldsymbol{\omega})^2 |\mathcal{F}\{f_+\}(\boldsymbol{\omega})|^2 d^2\boldsymbol{\omega}\right]$$
$$\overset{\text{Lem.4.2.5}}{=} [\ldots]\left[\int_{\mathbb{R}^2} |\mathcal{F}\{(\boldsymbol{b}\cdot\nabla)f_-\}(\boldsymbol{\omega})(-\boldsymbol{j})|^2 d^2\boldsymbol{\omega}\right.$$
$$\left. + \int_{\mathbb{R}^2} |\mathcal{F}\{(\boldsymbol{b}'\cdot\nabla)f_+\}(\boldsymbol{\omega})(-\boldsymbol{j})|^2 d^2\boldsymbol{\omega}\right]$$

$$\overset{\text{Parsev.}}{=} \left[\int_{\mathbb{R}^2} |(\boldsymbol{a}\cdot\boldsymbol{x})f_-(\boldsymbol{x})|^2 d^2\boldsymbol{x} + \int_{\mathbb{R}^2} |(\boldsymbol{a}\cdot\boldsymbol{x})f_+(\boldsymbol{x})|^2 d^2\boldsymbol{x}\right]$$
$$(2\pi)^2 \left[\int_{\mathbb{R}^2} |(\boldsymbol{b}\cdot\nabla)f_-(\boldsymbol{x})|^2 d^2\boldsymbol{x} + \int_{\mathbb{R}^2} |(\boldsymbol{b}'\cdot\nabla)f_+(\boldsymbol{x})|^2 d^2\boldsymbol{x}\right]$$
$$= (2\pi)^2 \left[\int_{\mathbb{R}^2} |(\boldsymbol{a}\cdot\boldsymbol{x})f_-(\boldsymbol{x})|^2 d^2\boldsymbol{x} \int_{\mathbb{R}^2} |(\boldsymbol{b}\cdot\nabla)f_-(\boldsymbol{x})|^2 d^2\boldsymbol{x}\right.$$
$$+ \int_{\mathbb{R}^2} |(\boldsymbol{a}\cdot\boldsymbol{x})f_+(\boldsymbol{x})|^2 d^2\boldsymbol{x} \int_{\mathbb{R}^2} |(\boldsymbol{b}'\cdot\nabla)f_+(\boldsymbol{x})|^2 d^2\boldsymbol{x}$$
$$+ \int_{\mathbb{R}^2} |(\boldsymbol{a}\cdot\boldsymbol{x})f_-(\boldsymbol{x})|^2 d^2\boldsymbol{x} \int_{\mathbb{R}^2} |(\boldsymbol{b}'\cdot\nabla)f_+(\boldsymbol{x})|^2 d^2\boldsymbol{x}$$
$$\left. + \int_{\mathbb{R}^2} |(\boldsymbol{a}\cdot\boldsymbol{x})f_+(\boldsymbol{x})|^2 d^2\boldsymbol{x} \int_{\mathbb{R}^2} |(\boldsymbol{b}\cdot\nabla)f_-(\boldsymbol{x})|^2 d^2\boldsymbol{x}\right]$$

$$\overset{\text{Schwarz}}{\leq} (2\pi)^2 \left[\left\{\int_{\mathbb{R}^2} Sc\left(\boldsymbol{a}\cdot\boldsymbol{x}f_-\,\boldsymbol{b}\cdot\nabla\widetilde{f_-}\right) d^2\boldsymbol{x}\right\}^2\right.$$
$$+ \left\{\int_{\mathbb{R}^2} Sc\left(\boldsymbol{a}\cdot\boldsymbol{x}f_+\,\boldsymbol{b}'\cdot\nabla\widetilde{f_+}\right) d^2\boldsymbol{x}\right\}^2$$
$$+ \left\{\int_{\mathbb{R}^2} \underbrace{Sc\left(\boldsymbol{a}\cdot\boldsymbol{x}f_-\,\boldsymbol{b}'\cdot\nabla\widetilde{f_+}\right)}_{=0\ \text{(Lem.4.1.13)}} d^2\boldsymbol{x}\right\}^2$$
$$\left. + \left\{\int_{\mathbb{R}^2} \underbrace{Sc\left(\boldsymbol{a}\cdot\boldsymbol{x}f_+\,\boldsymbol{b}\cdot\nabla\widetilde{f_-}\right)}_{=0\ \text{(Lem.4.1.13)}} d^2\boldsymbol{x}\right\}^2\right]$$
$$= (2\pi)^2 \left[\left\{\int_{\mathbb{R}^2} \boldsymbol{a}\cdot\boldsymbol{x}\frac{1}{2}\boldsymbol{b}\cdot\nabla(f_-\widetilde{f_-})d^2\boldsymbol{x}\right\}^2\right.$$
$$\left. + \left\{\int_{\mathbb{R}^2} \boldsymbol{a}\cdot\boldsymbol{x}\frac{1}{2}\boldsymbol{b}'\cdot\nabla(f_+\widetilde{f_+})d^2\boldsymbol{x}\right\}^2\right]$$
$$\overset{\text{Int. by p.}}{=} \frac{(2\pi)^2}{4}\left[\left\{\int_{\mathbb{R}^2} \underbrace{\boldsymbol{b}\cdot\nabla(\boldsymbol{a}\cdot\boldsymbol{x})}_{=\boldsymbol{a}\cdot\boldsymbol{b}} |f_-|^2 d^2\boldsymbol{x}\right\}^2\right.$$
$$\left. + \left\{\int_{\mathbb{R}^2} \underbrace{\boldsymbol{b}'\cdot\nabla(\boldsymbol{a}\cdot\boldsymbol{x})}_{=\boldsymbol{a}\cdot\boldsymbol{b}'} |f_+|^2 d^2\boldsymbol{x}\right\}^2\right]$$
$$= \frac{(2\pi)^2}{4}\left[(\boldsymbol{a}\cdot\boldsymbol{b})^2\ F_-^2 + (\boldsymbol{a}\cdot\boldsymbol{b}')^2\ F_+^2\right]. \tag{4.2.22}$$

After using the Schwarz inequality of Lemma 4.2.7 we further used

$$Sc\left(f_-\,\boldsymbol{b}\cdot\nabla\widetilde{f_-}\right) = \frac{1}{2}\boldsymbol{b}\cdot\nabla\left(f_-\,\widetilde{f_-}\right), Sc\left(f_+\,\boldsymbol{b}\cdot\nabla\widetilde{f_+}\right) = \frac{1}{2}\boldsymbol{b}\cdot\nabla\left(f_+\,\widetilde{f_+}\right). \tag{4.2.23}$$

□

4.2.1.4 *Summary on directional QFT uncertainty principle*

We first studied the $\pm$ split of quaternions and its effects on the double-sided QFT. Based on this we have established a new *directional* QFT uncertainty principle.

Finally, see [255] for a strictly coordinate wise approach.

4.2.2 Quaternion Fourier transform and convolution

4.2.2.1 *Background*

In this treatment we use the general two-sided quaternion Fourier transform (QFT), and relate the classical convolution of quaternion-valued signals over $\mathbb{R}^2$ with the Mustard convolution. A Mustard convolution can be expressed in the spectral domain as the point wise product of the QFTs of the factor functions. In full generality do we express the classical convolution of quaternion signals in terms of finite linear combinations of Mustard convolutions, and vice versa the Mustard convolution of quaternion signals in terms of finite linear combinations of classical convolutions. The treatment in this section is based on [213].

The two-sided quaternionic Fourier transformation (QFT) of Section 4.1.1 was introduced in [124] for the analysis of 2D linear time-invariant partial-differential systems. Subsequently it has been further studied in [56] and applied in many fields, including color image processing [291], edge detection and image filtering [104, 293], watermarking [17], pattern recognition [268, 308], quaternionic multiresolution analysis (a generalization of discrete wavelets) [24], speech recognition [25], noise removal from video images [225], and efficient and robust image feature detection [153].

This led to further theoretical investigations [184, 187], where a special split of quaternions was introduced, then called $\pm$split. An interesting physical consequence was that this split resulted in a left and right traveling multivector wave packet analysis, when generalizing the QFT to a full spacetime Fourier transform (SFT), compare Section 4.4. Later [194, 205, 212] this split has been analyzed further, interpreted geometrically and generalized to a freely steerable split of $\mathbb{H}$ into two orthogonal 2D analysis planes, then appropriately named *orthogonal 2D planes split* (OPS), see Section 4.1.2.

A key strength of the classical complex Fourier transform is its easy and fast application to filtering problems. The convolution of a signal with its filter function becomes in the spectral domain a point wise product of the respective Fourier transformations. This is not the case for the convolution of two quaternion-valued signals over $\mathbb{R}^2$, due to non-commutativity. Yet it is possible to define from the point wise product of the QFTs of two quaternion signals a new type of convolution, called Mustard convolution [270]. This Mustard convolution can be expressed in terms of sums of classical convolutions and vice versa. For the left-sided QFT this has recently been carried out in [97]. We expand this work in full generality to the two-sided QFT, making significant and efficient use of the two-sided orthogonal planes split of quaternions.

This Section is organized as follows. Section 4.2.2.1.1 introduces quaternion commutators, notation and briefly reviews the general orthogonal two-dimensional planes split of quaternions. Section 4.2.2.2 introduces to the general form of the two-sided version of the quaternion Fourier transform, and a related mixed exponential-sine transform. Section 4.2.2.3 introduces the Mustard type convolutions based on the QFT and contains the main results of this paper. That is the formulation of the QFT of the classical convolution of quaternion signals in Theorem 4.2.17, and specialized to the simpler case of only one transform axis in Corollary 4.2.18. Moreover, the expression of the classical convolution of quaternion signals in terms of the Mustard type convolutions is given in Theorem 4.2.19, for only one transform axis in Corollary 4.2.21, and using only standard Mustard convolutions fully general and explicit in Theorem 4.2.22. Finally the Mustard convolution is fully expanded in terms of classical convolutions in Theorem 4.2.23.

4.2.2.1.1 Quaternions, commutators, notation orthogonal planes split The quaternion algebra and its properties are described in Section 2.8.1.

The *commutator* of any two quaternions $p, q \in \mathbb{H}$ is a pure quaternion (because $S(pq) = S(qp)$) defined as

$$[p, q] = pq - qp. \tag{4.2.24}$$

For example,

$$[\boldsymbol{i}, \boldsymbol{j}] = \boldsymbol{ij} - \boldsymbol{ji} = 2\boldsymbol{k}. \tag{4.2.25}$$

The commutator $[f, g]$ of any two pure unit quaternions f, g gives therefore another pure quaternion (or zero for $f = \alpha g$, $\alpha \in \mathbb{R}$). It appears in the commutation of exponentials, e.g.

$$e^{\alpha f} e^{\beta g} = e^{\beta g} e^{\alpha f} + [f, g] \sin(\alpha) \sin(\beta), \tag{4.2.26}$$

which (in the same form for general multivector square roots of -1) has been used in [209] in order to derive a general convolution theorem for Clifford Fourier transformations. We furthermore note the useful anticommutation relationships

$$g[f, g] = -[f, g]g, \qquad f[f, g] = -[f, g]f, \tag{4.2.27}$$

and therefore

$$e^{\alpha f}[f, g] = [f, g]e^{-\alpha f}, \qquad e^{\beta g}[f, g] = [f, g]e^{-\beta g}. \tag{4.2.28}$$

We will subsequently adopt the following notation for reflecting the argument of functions ([97], Notation 2.4, page 583)

Notation 4.2.10 (Argument reflection). *For a function $h : \mathbb{R}^2 \to \mathbb{H}$ and a multi-index $\phi = (\phi_1, \phi_2)$ with $\phi_1, \phi_2 \in \{0, 1\}$ we set*

$$h^{\phi} = h^{(\phi_1, \phi_2)}(\boldsymbol{x}) := h((-1)^{\phi_1} x_1, (-1)^{\phi_2} x_2). \tag{4.2.29}$$

4.2.2.1.2 Review of general orthogonal two-dimensional planes split (OPS) This section provides only a brief review and clarifies notation used in the rest of Section 4.2.2. For details we refer to Section 4.1.2.

Assume in the following an arbitrary pair of pure unit quaternions f, g, $f^2 =$

$g^2 = -1$. The *orthogonal 2D planes split (OPS)* is then defined with respect to *any two pure unit quaternions* f, g as

Definition 4.2.11 (General orthogonal 2D planes split [205]). *Let $f, g \in \mathbb{H}$ be an arbitrary pair of pure quaternions f, g, $f^2 = g^2 = -1$, including the cases $f = \pm g$. The general OPS is then defined with respect to the two pure unit quaternions f, g as*

$$q_{\pm} = \frac{1}{2}(q \pm fqg). \tag{4.2.30}$$

Note that

$$fqg = q_+ - q_-, \tag{4.2.31}$$

i.e. under the map $f()g$ the q_+ part is invariant, but the q_- part changes sign.

Both parts are two-dimensional, and span two completely orthogonal planes. For $f \neq \pm g$ the q_+ plane is spanned by two orthogonal quaternions $\{f - g, 1 + fg = -f(f-g)\}$, the q_- plane is e.g. spanned by $\{f+g, 1-fg = -f(f+g)\}$. For $g = f$ a fully *orthonormal* four-dimensional basis of $\mathbb{H}$ is (R acts as rotation operator (rotor))

$$\{1, f, \boldsymbol{j}', \boldsymbol{k}'\} = R^{-1}\{1, \boldsymbol{i}, \boldsymbol{j}, \boldsymbol{k}\}R, \qquad R = \boldsymbol{i}(\boldsymbol{i} + f), \tag{4.2.32}$$

and the two orthogonal two-dimensional planes basis:

$$q_+\text{-basis: } \{\boldsymbol{j}', \boldsymbol{k}' = f\boldsymbol{j}'\}, \qquad q_-\text{-basis: } \{1, f\}. \tag{4.2.33}$$

Note the notation for normed vectors in [261] $\{q_1, q_2, q_3, q_4\}$ for the resulting total *orthonormal basis of* $\mathbb{H}$. In (4.2.33), the q_- part commutes with f and is also known as simplex part of q, whereas the q_+ anticommutes with f, and $-q_+\boldsymbol{j}'$ is known as perplex part of q, see [127].

Lemma 4.2.12 (Orthogonality of two OPS planes [205]). *Given any two quaternions q, p and applying the OPS with respect to any two pure unit quaternions f, g we get zero for the scalar part of the mixed products*

$$Sc(p_+\overline{q}_-) = 0, \qquad Sc(p_-\overline{q}_+) = 0. \tag{4.2.34}$$

Next we mention the possibility to perform a split along any given set of two (two-dimensional) analysis planes. It has been found, that any two-dimensional plane in $\mathbb{R}^4$ determines in an elementary way an OPS split and vice versa, compare Theorem 3.5 of [205].

Let us turn to the geometric interpretation of the map $f()g$. It *rotates* the q_- plane by 180° around the q_+ axis plane. This is in perfect agreement with Coxeter's notion of *half-turn* [84], see the right side of Figure 2.6. The following *identities* hold

$$e^{\alpha f} q_{\pm} e^{\beta g} = q_{\pm} e^{(\beta \mp \alpha) g} = e^{(\alpha \mp \beta) f} q_{\pm}, \tag{4.2.35}$$

as well as their quaternion conjugate (with $\alpha \to -\alpha$, $\beta \to -\beta$)

$$e^{\beta g} \overline{q_{\pm}} e^{\alpha f} = \overline{q_{\pm}} e^{(\alpha \mp \beta) f} = e^{(\beta \mp \alpha) g} \overline{q_{\pm}}. \tag{4.2.36}$$

This leads to a straightforward geometric interpretation of the integrands of the quaternion Fourier transform (QFT or OPS-QFT) with two pure quaternions f, g [205]. Particularly useful cases of (4.2.35) are $(\alpha, \beta) = (\pi/2, 0)$ and $(0, \pi/2)$:

$$f q_\pm = \mp q_\pm g, \qquad q_\pm g = \mp f q_\pm. \tag{4.2.37}$$

We further note, that with respect to any pure unit quaternion $f \in \mathbb{H}$, $f^2 = -1$, every quaternion $A \in \mathbb{H}$ can be similarly split into *commuting* and *anticommuting* parts [203].

Lemma 4.2.13 (Commuting and anticommuting with pure unit quaternion [203]). *Every quaternion $A \in \mathbb{H}$ has, with respect to any pure unit quaternion $f \in \mathbb{H}$, $f^2 = -1$, i.e. $f^{-1} = -f$, the unique decomposition denoted by*

$$A_{+f} = \frac{1}{2}(A + f^{-1}Af), \qquad A_{-f} = \frac{1}{2}(A - f^{-1}Af)$$
$$A = A_{+f} + A_{-f}, \qquad A_{+f}\, f = f A_{+f}, \qquad A_{-f}\, f = -f A_{-f}. \tag{4.2.38}$$

Note, that in Lemma 4.2.13 the commuting part A_{+f} is also known as simplex part of A, and the anticommuting part is up to a pure quaternion factor the perplex part of A, see [127].

4.2.2.2 The general two-sided QFT

Definition 4.2.14. ***(QFT with respect to two pure unit quaternions** f, g [205])*
Let $f, g \in \mathbb{H}$, $f^2 = g^2 = -1$, be any two pure unit quaternions. The quaternion Fourier transform (QFT) with respect to f, g is

$$\mathcal{F}\{h\}(\boldsymbol{\omega}) = \mathcal{F}^{f,g}\{h\}(\boldsymbol{\omega}) = \int_{\mathbb{R}^2} e^{-fx_1\omega_1} h(\boldsymbol{x})\, e^{-gx_2\omega_2} d^2\boldsymbol{x}, \tag{4.2.39}$$

where $h \in L^1(\mathbb{R}^2, \mathbb{H})$, $d^2\boldsymbol{x} = dx_1 dx_2$ and $\boldsymbol{x}, \boldsymbol{\omega} \in \mathbb{R}^2$.

The QFT can be inverted with

$$h(\boldsymbol{x}) = \mathcal{F}^{-1}\{h\}(\boldsymbol{x}) = \frac{1}{(2\pi)^2}\int_{\mathbb{R}^2} e^{fx_1\omega_1} \mathcal{F}\{h\}(\boldsymbol{\omega})\, e^{gx_2\omega_2} d^2\boldsymbol{\omega}, \tag{4.2.40}$$

with $d^2\boldsymbol{\omega} = d\omega_1 d\omega_2$.

Remark 4.2.15. *Note, that the general pair of pure unit quaternions f, g in Definition 4.2.14 includes orthogonal pairs $f \perp g$, non-orthogonal pairs $f \not\perp g$, and parallel pairs $f = \pm g$ (only one transform axis). In the rest of this paper the theorems will be valid for fully general pairs f, g, if not otherwise specified. To avoid clutter we often omit the indication of the pair f, g as in $\mathcal{F} = \mathcal{F}^{f,g}$.*

Linearity of the integral (4.2.39) allows us to use the OPS split $h = h_- + h_+$

$$\mathcal{F}^{f,g}\{h\}(\boldsymbol{\omega}) = \mathcal{F}^{f,g}\{h_-\}(\boldsymbol{\omega}) + \mathcal{F}^{f,g}\{h_+\}(\boldsymbol{\omega}) = \mathcal{F}_-^{f,g}\{h\}(\boldsymbol{\omega}) + \mathcal{F}_+^{f,g}\{h\}(\boldsymbol{\omega}), \tag{4.2.41}$$

since by its construction the operators of the QFT $\mathcal{F}^{f,g}$, and of the OPS with respect to f, g commute. From (4.2.35) follows

Theorem 4.2.16 (QFT of $h_\pm$ [205]). *The QFT of the $h_\pm$ OPS split parts, with respect to two linearly independent unit quaternions f, g, of a quaternion module function $h \in L^1(\mathbb{R}^2, \mathbb{H})$ have the quasi-complex forms*

$$\mathcal{F}_\pm^{f,g}\{h\} = \mathcal{F}^{f,g}\{h_\pm\}$$

$$= \int_{\mathbb{R}^2} h_\pm e^{-g(x_2\omega_2 \mp x_1\omega_1)} d^2x = \int_{\mathbb{R}^2} e^{-f(x_1\omega_1 \mp x_2\omega_2)} h_\pm d^2x \ . \tag{4.2.42}$$

We further define for later use the following two mixed *exponential-sine* Fourier transforms

$$\mathcal{F}^{f,\pm s}\{h\}(\boldsymbol{\omega}) = \int_{\mathbb{R}^2} e^{-fx_1\omega_1} h(\boldsymbol{x})(\pm 1)\sin(-x_2\omega_2) d^2\boldsymbol{x}, \tag{4.2.43}$$

$$\mathcal{F}^{\pm s,g}\{h\}(\boldsymbol{\omega}) = \int_{\mathbb{R}^2} (\pm 1)\sin(-x_1\omega_1) h(\boldsymbol{x}) e^{-gx_2\omega_2} d^2\boldsymbol{x}. \tag{4.2.44}$$

With the help of

$$\sin(-x_1\omega_1) = \frac{f}{2}(e^{-fx_1\omega_1} - e^{fx_1\omega_1}), \ \sin(-x_2\omega_2) = \frac{g}{2}(e^{-gx_2\omega_2} - e^{gx_2\omega_2}), \tag{4.2.45}$$

we can rewrite the above mixed exponential-sine Fourier transforms in terms of the QFTs of Definition 4.2.14 as

$$\mathcal{F}^{f,\pm s}\{h\} = \pm(\mathcal{F}^{f,g}\{h\frac{g}{2}\} - \mathcal{F}^{f,-g}\{h\frac{g}{2}\}), \tag{4.2.46}$$

$$\mathcal{F}^{\pm s,g}\{h\} = \pm(\mathcal{F}^{f,g}\{\frac{f}{2}h\} - \mathcal{F}^{-f,g}\{\frac{f}{2}h\}). \tag{4.2.47}$$

We further note the following useful relationships using the argument reflection of Notation 4.2.10

$$\mathcal{F}^{-f,g}\{h\} = \mathcal{F}^{f,g}\{h^{(1,0)}\},$$
$$\mathcal{F}^{f,-g}\{h\} = \mathcal{F}^{f,g}\{h^{(0,1)}\},$$
$$\mathcal{F}^{-f,-g}\{h\} = \mathcal{F}^{f,g}\{h^{(1,1)}\}, \tag{4.2.48}$$

and similarly

$$\mathcal{F}^{f,-s}\{h\} = \mathcal{F}^{f,s}\{h^{(0,1)}\}, \quad \mathcal{F}^{-s,g}\{h\} = \mathcal{F}^{s,g}\{h^{(1,0)}\}. \tag{4.2.49}$$

4.2.2.3 Convolution and Mustard convolution

We define the *convolution* of two quaternion signals $a, b \in L^1(\mathbb{R}^2; \mathbb{H})$ as

$$(a \star b)(\boldsymbol{x}) = \int_{\mathbb{R}^2} a(\boldsymbol{y}) b(\boldsymbol{x} - \boldsymbol{y}) d^2\boldsymbol{y}, \tag{4.2.50}$$

provided that the integral exists.

The *Mustard* convolution [62, 270] of two quaternion signals $a, b \in L^1(\mathbb{R}^2; \mathbb{H})$ is defined as

$$(a \star_M b)(\boldsymbol{x}) = (\mathcal{F}^{f,g})^{-1}(\mathcal{F}^{f,g}\{a\}\mathcal{F}^{f,g}\{b\}) \tag{4.2.51}$$

provided that the integral exists. The Mustard convolution has the conceptual and computational advantage to simply yield as spectrum in the QFT Fourier domain the point wise product of the QFTs of the two signals, just as the classical complex Fourier transform.

We additionally define a further type of *exponential-sine* Mustard convolution as

$$(a \star_{Ms} b)(\boldsymbol{x}) = (\mathcal{F}^{f,g})^{-1}(\mathcal{F}^{f,s}\{a\}\mathcal{F}^{s,g}\{b\}). \tag{4.2.52}$$

In the following two subsections we will first express the convolution (4.2.50) in terms of the Mustard convolution (4.2.51) and vice versa.

4.2.2.3.1 Expressing the convolution in terms of the Mustard convolution

In [97] Theorem 4.1 on page 584 expresses the classical convolution of two quaternion functions with the help of the general left-sided QFT as a sum of 40 Mustard convolutions. In our approach we use Theorem 5.12 on page 327 of [209], which expresses the convolution of two Clifford signal functions (higher dimensional generalizations of quaternion functions) in the Clifford Fourier domain with the help of the general two-sided Clifford Fourier transform (CFT), the latter is in turn a generalization of the QFT. We restate this theorem here again, specialized for quaternion functions and the QFT of Definition 4.2.14.

Theorem 4.2.17 (QFT of convolution). *Assuming a general pair of unit pure quaternions f, g, the general two-sided QFT of the convolution (4.2.50) of two functions $a, b \in L^1(\mathbb{R}^2; \mathbb{H})$ can then be expressed as*

$$\begin{aligned}
&\mathcal{F}^{f,g}\{a \star b\} = \\
&\quad \mathcal{F}^{f,g}\{a_{+f}\}\mathcal{F}^{f,g}\{b_{+g}\} + \mathcal{F}^{f,-g}\{a_{+f}\}\mathcal{F}^{f,g}\{b_{-g}\} \\
&\quad + \mathcal{F}^{f,g}\{a_{-f}\}\mathcal{F}^{-f,g}\{b_{+g}\} + \mathcal{F}^{f,-g}\{a_{-f}\}\mathcal{F}^{-f,g}\{b_{-g}\} \\
&\quad + \mathcal{F}^{f,s}\{a_{+f}\}[f,g]\mathcal{F}^{s,g}\{b_{+g}\} + \mathcal{F}^{f,-s}\{a_{+f}\}[f,g]\mathcal{F}^{s,g}\{b_{-g}\} \\
&\quad + \mathcal{F}^{f,s}\{a_{-f}\}[f,g]\mathcal{F}^{-s,g}\{b_{+g}\} + \mathcal{F}^{f,-s}\{a_{-f}\}[f,g]\mathcal{F}^{-s,g}\{b_{-g}\}.
\end{aligned} \tag{4.2.53}$$

Note that due to the commutation properties of (4.2.43) and (4.2.44) we can place the commutator $[f, g]$ also inside the exponential-sine transform terms as e.g. in

$$\begin{aligned}
\mathcal{F}^{f,s}\{a_{+f}\}[f,g]\mathcal{F}^{s,g}\{b_{+g}\} &= \mathcal{F}^{f,s}\{a_{+f}[f,g]\}\mathcal{F}^{s,g}\{b_{+g}\} \\
&= \mathcal{F}^{f,s}\{a_{+f}\}\mathcal{F}^{s,g}\{[f,g]b_{+g}\}.
\end{aligned} \tag{4.2.54}$$

For the special case of parallel unit pure quaternions $f = \pm g$, the commutator vanishes $[f, g] = 0$, and we get the following corollary. Note that in this case $b_{\pm g} = b_{\pm f}$.

Corollary 4.2.18 (QFT of convolution with $f \parallel g$). *Assuming a parallel pair of unit pure quaternions $f = \pm g$, the general two-sided QFT of the convolution (4.2.50) of two functions $a, b \in L^1(\mathbb{R}^2; \mathbb{H})$ can be expressed as*

$$\begin{aligned}
&\mathcal{F}^{f,g}\{a \star b\} = \\
&\quad \mathcal{F}^{f,g}\{a_{+f}\}\mathcal{F}^{f,g}\{b_{+f}\} + \mathcal{F}^{f,-g}\{a_{+f}\}\mathcal{F}^{f,g}\{b_{-f}\} \\
&\quad + \mathcal{F}^{f,g}\{a_{-f}\}\mathcal{F}^{-f,g}\{b_{+f}\} + \mathcal{F}^{f,-g}\{a_{-f}\}\mathcal{F}^{-f,g}\{b_{-f}\}.
\end{aligned} \tag{4.2.55}$$

By applying the inverse QFT, we can now easily express the convolution of two quaternion signals $a \star b$, in terms of only eight Mustard convolutions (4.2.51) and (4.2.52).

Theorem 4.2.19 (Convolution in terms of two types of Mustard convolution)**.** *Assuming a general pair of unit pure quaternions f, g, the convolution* (4.2.50) *of two quaternion functions $a, b \in L^1(\mathbb{R}^2; \mathbb{H})$ can be expressed in terms of four Mustard convolutions* (4.2.51) *and four exponential-sine Mustard convolutions* (4.2.52) *as*

$$\begin{aligned} a \star b = {} & a_{+f} \star_M b_{+g} + a_{+f}^{(0,1)} \star_M b_{-g} + a_{-f} \star_M b_{+g}^{(1,0)} + a_{-f}^{(0,1)} \star_M b_{-g}^{(1,0)} \\ & + a_{+f} \star_{Ms} [f, g] b_{+g} + a_{+f}^{(0,1)} \star_{Ms} [f, g] b_{-g} \\ & + a_{-f} \star_{Ms} [f, g] b_{+g}^{(1,0)} + a_{-f}^{(0,1)} \star_{Ms} [f, g] b_{-g}^{(1,0)} . \end{aligned} \tag{4.2.56}$$

Remark 4.2.20. *We use the convention, that terms such as $a_{+f} \star_{Ms} [f, g] b_{+g}$, should be understood with brackets $a_{+f} \star_{Ms} ([f, g] b_{+g})$, which are omitted to avoid clutter.*

Assuming $f \parallel g$, the standard Mustard convolution is sufficient to express the classical convolution.

Corollary 4.2.21 (Convolution in terms of Mustard convolution with parallel axes)**.** *Assuming a parallel pair of unit pure quaternions $f \parallel g$, the convolution* (4.2.50) *of two quaternion functions $a, b \in L^1(\mathbb{R}^2; \mathbb{H})$ can be expressed in terms of four Mustard convolutions* (4.2.51) *as*

$$a \star b = a_{+f} \star_M b_{+f} + a_{+f}^{(0,1)} \star_M b_{-f} + a_{-f} \star_M b_{+f}^{(1,0)} + a_{-f}^{(0,1)} \star_M b_{-f}^{(1,0)} \tag{4.2.57}$$

Furthermore, applying (4.2.46) and (4.2.47), we can expand the terms in (4.2.53) with exponential-sine transforms into sums of products of QFTs. For example, the first term gives

$$\begin{aligned} & \mathcal{F}^{f,s}\{a_{+f}\}[f, g]\mathcal{F}^{s,g}\{b_{+g}\} \\ & = \frac{1}{4}\left(\mathcal{F}^{f,g}\{a_{+f}g\} - \mathcal{F}^{f,-g}\{a_{+f}g\}\right)\left(\mathcal{F}^{f,g}\{f[f, g]b_{+g}\} - \mathcal{F}^{-f,g}\{f[f, g]b_{+g}\}\right) \\ & = \frac{1}{4}\left(\mathcal{F}\{a_{+f}g\}\mathcal{F}\{f[f, g]b_{+g}\} - \mathcal{F}\{a_{+f}g\}\mathcal{F}\{f[f, g]b_{+g}^{(1,0)}\}\right. \\ & \qquad \left. - \mathcal{F}\{a_{+f}^{(0,1)}g\}\mathcal{F}\{f[f, g]b_{+g}\} + \mathcal{F}\{a_{+f}^{(0,1)}g\}\mathcal{F}\{f[f, g]b_{+g}^{(1,0)}\}\right) . \end{aligned} \tag{4.2.58}$$

This now allows us in turn to express the quaternion signal convolution purely in terms of standard Mustard convolutions

Theorem 4.2.22 (Convolution in terms of Mustard convolution)**.** *Assuming a general pair of unit pure quaternions f, g, the convolution* (4.2.50) *of two quaternion functions $a, b \in L^1(\mathbb{R}^2; \mathbb{H})$ can be expressed in terms of twenty standard Mustard*

convolutions (4.2.51) *as*

$$
\begin{aligned}
a \star b = {} & a_{+f} \star_M b_{+g} + a_{+f}^{(0,1)} \star_M b_{-g} + a_{-f} \star_M b_{+g}^{(1,0)} + a_{-f}^{(0,1)} \star_M b_{-g}^{(1,0)} \\
& + \tfrac{1}{4}\big(a_{+f}g \star_M fcb_{+g} - a_{+f}g \star_M fcb_{+g}^{(1,0)} - a_{+f}^{(0,1)}g \star_M fcb_{+g} + a_{+f}^{(0,1)}g \star_M fcb_{+g}^{(1,0)} \\
& + a_{+f}^{(0,1)}g \star_M fcb_{-g} - a_{+f}^{(0,1)}g \star_M fcb_{-g}^{(1,0)} - a_{+f}g \star_M fcb_{-g} + a_{+f}g \star_M fcb_{-g}^{(1,0)} \\
& + a_{-f}g \star_M fcb_{+g}^{(1,0)} - a_{-f}g \star_M fcb_{+g} - a_{-f}^{(0,1)}g \star_M fcb_{+g}^{(1,0)} + a_{-f}^{(0,1)}g \star_M fcb_{+g} \\
& + a_{-f}^{(0,1)}g \star_M fcb_{-g}^{(1,0)} - a_{-f}^{(0,1)}g \star_M fcb_{-g} - a_{-f}g \star_M fcb_{-g}^{(1,0)} + a_{-f}g \star_M fcb_{-g} \big),
\end{aligned}
\tag{4.2.59}
$$

with the abbreviation $c = [f, g]$.

4.2.2.3.2 Expressing the Mustard convolution in terms of the convolution Now we will simply write out the Mustard convolution (4.2.51) and simplify it until only standard convolutions (4.2.50) remain. In this subsection, we will use the general OPS split of Definition 4.2.11. Our result should be compared with the Theorem 2.5 on page 584 of [97] for the left-sided QFT with 32 classical convolutions for expressing the Mustard convolution of quaternion functions (and for the two-sided QFT in [62], Section 4.4.2, with 16 classical convolutions, stated in a different but apparently equivalent form to Theorem 4.2.23).

We begin by writing the Mustard convolution (4.2.51) of two quaternion functions $a, b \in L^2(\mathbb{R}^2, \mathbb{H})$

$$
\begin{aligned}
a \star_M b(\boldsymbol{x}) &= \frac{1}{(2\pi)^2} \int_{\mathbb{R}^2} e^{f x_1 \omega_1} \mathcal{F}\{a\}(\boldsymbol{\omega}) \mathcal{F}\{b\}(\boldsymbol{\omega}) e^{g x_2 \omega_2} d^2\boldsymbol{\omega} \\
&= \frac{1}{(2\pi)^2} \int_{\mathbb{R}^2} e^{-f x_1 \omega_1} \int_{\mathbb{R}^2} e^{-f y_1 \omega_1} a(\boldsymbol{y}) e^{-g y_2 \omega_2} d^2\boldsymbol{y} \\
&\qquad \int_{\mathbb{R}^2} e^{-f z_1 \omega_1} b(\boldsymbol{z}) e^{-g z_2 \omega_2} d^2\boldsymbol{z} e^{g x_2 \omega_2} d^2\boldsymbol{\omega} \\
&= \frac{1}{(2\pi)^2} \int_{\mathbb{R}^2} \int_{\mathbb{R}^2} \int_{\mathbb{R}^2} e^{f(x_1 - y_1)\omega_1} (a_+(\boldsymbol{y}) + a_-(\boldsymbol{y})) e^{-g y_2 \omega_2} \\
&\qquad e^{-f z_1 \omega_1} (b_+(\boldsymbol{z}) + b_-(\boldsymbol{z})) e^{g(x_2 - z_2)\omega_2} d^2\boldsymbol{y} d^2\boldsymbol{z} d^2\boldsymbol{\omega}.
\end{aligned}
\tag{4.2.60}
$$

Next, we use the identities (4.2.35) in order to shift the inner factor $e^{-g y_2 \omega_2}$ to the left and $e^{-f z_1 \omega_1}$ to the right, respectively. We abbreviate $\int_{\mathbb{R}^2} \int_{\mathbb{R}^2} \int_{\mathbb{R}^2}$ to $\iiint$.

$$
\begin{aligned}
& a \star_M b(\boldsymbol{x}) = \\
&= \frac{1}{(2\pi)^2} \iiint e^{f(x_1 - y_1)\omega_1} e^{f y_2 \omega_2} a_+(\boldsymbol{y}) b_+(\boldsymbol{z}) e^{g z_1 \omega_1} e^{g(x_2 - z_2)\omega_2} d^2\boldsymbol{y} d^2\boldsymbol{z} d^2\boldsymbol{\omega} \\
&+ \frac{1}{(2\pi)^2} \iiint e^{f(x_1 - y_1)\omega_1} e^{f y_2 \omega_2} a_+(\boldsymbol{y}) b_-(\boldsymbol{z}) e^{-g z_1 \omega_1} e^{g(x_2 - z_2)\omega_2} d^2\boldsymbol{y} d^2\boldsymbol{z} d^2\boldsymbol{\omega} \\
&+ \frac{1}{(2\pi)^2} \iiint e^{f(x_1 - y_1)\omega_1} e^{-f y_2 \omega_2} a_-(\boldsymbol{y}) b_+(\boldsymbol{z}) e^{g z_1 \omega_1} e^{g(x_2 - z_2)\omega_2} d^2\boldsymbol{y} d^2\boldsymbol{z} d^2\boldsymbol{\omega} \\
&+ \frac{1}{(2\pi)^2} \iiint e^{f(x_1 - y_1)\omega_1} e^{-f y_2 \omega_2} a_-(\boldsymbol{y}) b_-(\boldsymbol{z}) e^{-g z_1 \omega_1} e^{g(x_2 - z_2)\omega_2} d^2\boldsymbol{y} d^2\boldsymbol{z} d^2\boldsymbol{\omega}
\end{aligned}
\tag{4.2.61}
$$

Furthermore, we abbreviate the inner function products as $ab_{\pm\pm}(\boldsymbol{y}, \boldsymbol{z}) := a_\pm(\boldsymbol{y}) b_\pm(\boldsymbol{z})$, and apply the general OPS split of Definition 4.2.11 once again to obtain

$ab_{\pm\pm}(\boldsymbol{y},\boldsymbol{z}) = [ab_{\pm\pm}(\boldsymbol{y},\boldsymbol{z})]_+ + [ab_{\pm\pm}(\boldsymbol{y},\boldsymbol{z})]_- = ab_{\pm\pm}(\boldsymbol{y},\boldsymbol{z})_+ + ab_{\pm\pm}(\boldsymbol{y},\boldsymbol{z})_-$. We omit the square brackets and use the convention that the final OPS split indicated by the final $\pm$ index should be performed last. This allows to further apply (4.2.35) again in order to shift the factors $e^{\pm g z_1 \omega_1}$ $e^{g(x_2-z_2)\omega_2}$ to the left. We end up with the following eight terms

$$
\begin{aligned}
a \star_M b(\boldsymbol{x}) &= \\
&= \frac{1}{(2\pi)^2}\iiint e^{f(x_1-y_1-z_1)\omega_1} e^{f(y_2-(x_2-z_2))\omega_2} ab_{++}(\boldsymbol{y},\boldsymbol{z})_+ d^2\boldsymbol{y}d^2\boldsymbol{z}d^2\boldsymbol{\omega} \\
&+\frac{1}{(2\pi)^2}\iiint e^{f(x_1-y_1+z_1)\omega_1} e^{f(y_2+(x_2-z_2))\omega_2} ab_{++}(\boldsymbol{y},\boldsymbol{z})_- d^2\boldsymbol{y}d^2\boldsymbol{z}d^2\boldsymbol{\omega} \\
&+\frac{1}{(2\pi)^2}\iiint e^{f(x_1-y_1+z_1)\omega_1} e^{f(y_2-(x_2-z_2))\omega_2} ab_{+-}(\boldsymbol{y},\boldsymbol{z})_+ d^2\boldsymbol{y}d^2\boldsymbol{z}d^2\boldsymbol{\omega} \\
&+\frac{1}{(2\pi)^2}\iiint e^{f(x_1-y_1-z_1)\omega_1} e^{f(y_2+(x_2-z_2))\omega_2} ab_{+-}(\boldsymbol{y},\boldsymbol{z})_- d^2\boldsymbol{y}d^2\boldsymbol{z}d^2\boldsymbol{\omega} \\
&+\frac{1}{(2\pi)^2}\iiint e^{f(x_1-y_1-z_1)\omega_1} e^{f(-y_2-(x_2-z_2))\omega_2} ab_{-+}(\boldsymbol{y},\boldsymbol{z})_+ d^2\boldsymbol{y}d^2\boldsymbol{z}d^2\boldsymbol{\omega} \\
&+\frac{1}{(2\pi)^2}\iiint e^{f(x_1-y_1+z_1)\omega_1} e^{f(-y_2+(x_2-z_2))\omega_2} ab_{-+}(\boldsymbol{y},\boldsymbol{z})_- d^2\boldsymbol{y}d^2\boldsymbol{z}d^2\boldsymbol{\omega} \\
&+\frac{1}{(2\pi)^2}\iiint e^{f(x_1-y_1+z_1)\omega_1} e^{f(-y_2-(x_2-z_2))\omega_2} ab_{--}(\boldsymbol{y},\boldsymbol{z})_+ d^2\boldsymbol{y}d^2\boldsymbol{z}d^2\boldsymbol{\omega} \\
&+\frac{1}{(2\pi)^2}\iiint e^{f(x_1-y_1-z_1)\omega_1} e^{f(-y_2+(x_2-z_2))\omega_2} ab_{--}(\boldsymbol{y},\boldsymbol{z})_- d^2\boldsymbol{y}d^2\boldsymbol{z}d^2\boldsymbol{\omega}. \qquad (4.2.62)
\end{aligned}
$$

We now only show explicitly how to simplify the second triple integral, the others follow the same pattern.

$$
\begin{aligned}
&\frac{1}{(2\pi)^2}\iiint e^{f(x_1-y_1+z_1)\omega_1} e^{f(y_2+(x_2-z_2))\omega_2} [a_+(\boldsymbol{y})b_+(\boldsymbol{z})]_- d^2\boldsymbol{y}d^2\boldsymbol{z}d^2\boldsymbol{\omega} \\
&= \frac{1}{(2\pi)^2}\iint \int_{\mathbb{R}} e^{f(x_1-y_1+z_1)\omega_1} d\omega_1 \int_{\mathbb{R}} e^{f(y_2+(x_2-z_2))\omega_2} d\omega_2 [a_+(\boldsymbol{y})b_+(\boldsymbol{z})]_- d^2\boldsymbol{y}d^2\boldsymbol{z} \\
&= \iint \delta(x_1-y_1+z_1)\delta(y_2+(x_2-z_2))[a_+(\boldsymbol{y})b_+(z_1,z_2)]_- d^2\boldsymbol{y}d^2\boldsymbol{z} \\
&= \int_{\mathbb{R}^2} [a_+(\boldsymbol{y})b_+(-(x_1-y_1), x_2+y_2)]_- d^2\boldsymbol{y} \\
&= \int_{\mathbb{R}^2} [a_+(\boldsymbol{y})b_+(-(x_1-y_1), -(-x_2-y_2))]_- d^2\boldsymbol{y} \\
&= \int_{\mathbb{R}^2} [a_+(\boldsymbol{y})b_+^{(1,1)}(x_1-y_1, -x_2-y_2)]_- d^2\boldsymbol{y} \\
&= [a_+ \star b_+^{(1,1)}(x_1, -x_2)]_-. \qquad (4.2.63)
\end{aligned}
$$

Note that $a_+ \star b_+^{(1,1)}(x_1, -x_2)$ means to first apply the convolution to the pair of functions a_+ and $b_+^{(1,1)}$, and only then to evaluate them with the argument $(x_1, -x_2)$. So in general $a_+ \star b_+^{(1,1)}(x_1, -x_2) \neq a_+ \star b_+(-x_1, x_2)$. Simplifying the other seven triple integrals similarly we finally obtain the desired decomposition of the Mustard convolution (4.2.51) in terms of the classical convolution.

Theorem 4.2.23 (Mustard convolution in terms of standard convolution). *The Mustard convolution* (4.2.51) *of two quaternion functions* $a, b \in L^1(\mathbb{R}^2; \mathbb{H})$ *can be expressed in terms of eight standard convolutions* (4.2.50) *as*

$$\begin{aligned} a \star_M b(\boldsymbol{x}) &= \\ &= [a_+ \star b_+(\boldsymbol{x})]_+ + [a_+ \star b_+^{(1,1)}(x_1, -x_2)]_- \\ &\quad + [a_+ \star b_-^{(1,0)}(\boldsymbol{x})]_+ + [a_+ \star b_-^{(0,1)}(x_1, -x_2)]_- \\ &\quad + [a_- \star b_+^{(0,1)}(x_1, -x_2)]_+ + [a_- \star b_+^{(1,0)}(\boldsymbol{x})]_- \\ &\quad + [a_- \star b_-^{(1,1)}(x_1, -x_2)]_+ + [a_- \star b_-(\boldsymbol{x})]_-. \end{aligned} \tag{4.2.64}$$

Remark 4.2.24 (Levels of computation). *Equation* (4.2.64) *involves five levels of operation: primary (inner) and secondary (outer) OPS, argument reflections before (pre) and after (post) the actual convolution, and the convolution itself in each of the eight terms.*

Remark 4.2.25 (Efficiency of notation and interpretation). *If we would explicitly insert according Definition 4.2.11* $a_\pm = \frac{1}{2}(a \pm f a g)$ *and* $b_\pm = \frac{1}{2}(b \pm f b g)$, *and similarly explicitly insert the second level OPS split* $[\ldots]_\pm$, *we would (potentially) obtain up to a maximum of 64 terms*[22]. *It is therefore obvious how straightforward, significant and efficient the use of the general OPS split is in this context. Efficiency first of all with respect to concise (compact) notation, which in turn may assist geometric (or physical) interpretation in concrete applications.*

Regarding *derivation efficiency*, we needed two pages to derive (4.2.64), but in order to strike the balance with sufficient detail for reasonably selfcontained comprehension, this level of detail may be justified. Whether the compact eight term form of (4.2.64) is advantageous for actual *numerical computations*, is an open question, which requires application to concrete representatives problems, e.g. in the area of quaternionic image processing (see e.g. [17,25,56,104,127,153,225,268,291,293,308]). And it may depend on the concrete hardware architecture, e.g. how many parallel channels of computation (e.g. number of parallel GPUs) are available, whether quaternion operations are hard coded or need breaking down in elementary real (or matrix) computations; and whether optimized software packages like the precompiler GAALOP [167] or the MATLAB package QFMT [297] would be used, etc.

4.2.2.4 Summary on QFT and convolution

In Section 4.2.2, we have briefly reviewed the general orthogonal two-dimensional planes split of quaternions, the general two-sided quaternion Fourier transform, and introduced a related mixed quaternionic exponential-sine Fourier transform. We defined the notions of convolution of two quaternion valued functions over $\mathbb{R}^2$, the Mustard convolution (with its QFT as the point wise product of the QFTs of the factor functions), and a special Mustard convolution involving the point wise products

[22]Note that in [62], Section 4.4.2, a fully explicit form is given with only 16 terms.

of mixed exponential-sine QFTs. The main results are: an efficient decomposition of the classical convolution of quaternion signals in terms of eight Mustard type convolutions. For the special case of parallel pure unit quaternion axis in the QFT, only four terms of the standard Mustard convolution are sufficient. Even in the case of two general pure unit quaternion axis in the QFT, the classical convolution of two quaternion signals can always be fully expanded in terms of standard Mustard convolutions. Finally we showed how to fully generally expand the Mustard convolution of two quaternion signals in terms of eight classical convolutions.

In view of the many applications of the QFT explained in the introduction, we expect the results described to be of great interest, especially for filter design and feature extraction in signal and color image processing.

4.2.3 Quaternionic Wiener Khintchine type theorems and convolution

4.2.3.1 Background

In this part we use the general two-sided quaternion Fourier transform (QFT) of Section 4.1.1 as generalized in Section 4.1.2, in order to derive Wiener-Khintchine type theorems for the cross-correlation and for the auto-correlation of quaternion signals. Furthermore, we show how to derive a new four term spectral representation for the convolution of quaternion signals. The treatment in this section is based on [214].

The two-sided quaternionic Fourier transformation (QFT) was introduced in [124] for the analysis of 2D linear time-invariant partial-differential systems. Subsequently it has been further studied in [56] and applied in many fields, including color image processing [291], edge detection and image filtering [104, 293], watermarking [17], pattern recognition [268, 308], quaternionic multiresolution analysis (a generalization of discrete wavelets) [24], speech recognition [25], noise removal from video images [225], and efficient and robust image feature detection [153].

This led to further theoretical investigations [184, 187], where a special split of quaternions was introduced, then called ±split. An interesting physical consequence was that this split resulted in a left and right traveling multivector wave packet analysis, when generalizing the QFT to a full spacetime Fourier transform (SFT). Later [194, 205, 212] this split has been analyzed further, interpreted geometrically and generalized to a freely steerable split of $\mathbb{H}$ into two orthogonal 2D analysis planes, then appropriately named *orthogonal 2D planes split* (OPS).

A hypercomplex (quaternionic) Wiener-Khintchine theorem for a pair of discrete quaternionic signals applied to color image correlation can be found in [126, 267], using a right-sided two-dimensional quaternionic Fourier transform. In our contribution we generalize the Wiener-Khintchine theorem to continuous quaternion signals, using the general steerable two-sided QFT (Theorem 4.2.27). Furthermore a corollary (Corollary 4.2.28) presents the auto-correlation of a quaternion signal. Finally we use a similar approach to derive a new four term spectral representation of the convolution of two quaternion signals (Theorem 4.2.29).

This Section is organized as follows. The following paragraphs introduce quaternions and the two-dimensional orthogonal planes split of quaternions. Then

Section 4.2.2.2 explains the general steerable two-sided quaternion Fourier transform. Section 4.2.3.2 defines the notions of convolution, cross-correlation and auto-correlation applied to quaternion signal functions. Section 4.2.3.3 derives the Wiener-Khintchine theorem for quaternion signals based on the steerable two-sided QFT, which gives quaternionic spectral representations for the cross-correlation and the auto-correlation of quaternion signals. Finally, Section 4.2.3.4 derives a new spectral representation of the convolution of two quaternion signals.

For extensive treatment of quaternions, we refer to Section 2.8.1. Here we only provide some extra notation relevant to Section 4.2.3. Regarding quaternion commutators, we refer to Section 4.2.2.1.1. Also the argument reflection Notation 4.2.10 should be noted.

We further recall, that with respect to any pure unit quaternion $f \in \mathbb{H}$, $f^2 = -1$, every quaternion $A \in \mathbb{H}$ can be similarly split into *commuting* and *anticommuting* parts [203], see Lemma 4.2.13.

Remark 4.2.26. *Note, that the general pair of pure unit quaternions f, g in the general Definition 4.2.14 of a quaternionic FT includes orthogonal pairs $f \perp g$, non-orthogonal pairs $f \not\perp g$, and parallel pairs $f = \pm g$ (only one transform axis). In the rest of this section the theorems will be valid for fully general pairs f, g, if not otherwise specified. To avoid clutter we often omit the indication of the pair f, g as in $\mathcal{F} = \mathcal{F}^{f,g}$.*

We further note the following useful relationships using the argument reflection of Notation 4.2.10

$$\begin{aligned}\mathcal{F}^{-f,g}\{h\} &= \mathcal{F}^{f,g}\{h^{(1,0)}\},\\ \mathcal{F}^{f,-g}\{h\} &= \mathcal{F}^{f,g}\{h^{(0,1)}\},\\ \mathcal{F}^{-f,-g}\{h\} &= \mathcal{F}^{f,g}\{h^{(1,1)}\}.\end{aligned} \tag{4.2.65}$$

4.2.3.2 Convolution, cross-correlation and auto-correlation of quaternion signals

We define the *convolution* of two quaternion signals $a, b \in L^1(\mathbb{R}^2; \mathbb{H})$ as

$$(a \star b)(\boldsymbol{x}) = \int_{\mathbb{R}^2} a(\boldsymbol{y})b(\boldsymbol{x}-\boldsymbol{y})d^2\boldsymbol{y} = \int_{\mathbb{R}^2} a(\boldsymbol{x}-\boldsymbol{y})b(\boldsymbol{y})d^2\boldsymbol{y}, \tag{4.2.66}$$

provided that the integral exists.

We define the *cross-correlation* of two quaternion signals $a, b \in L^1(\mathbb{R}^2; \mathbb{H})$ as

$$(a \star_c b)(\boldsymbol{x}) = \int_{\mathbb{R}^2} a(\boldsymbol{y})\overline{b(\boldsymbol{y}-\boldsymbol{x})}d^2\boldsymbol{y} = \int_{\mathbb{R}^2} a(\boldsymbol{y}+\boldsymbol{x})\overline{b(\boldsymbol{y})}d^2\boldsymbol{y}, \tag{4.2.67}$$

provided that the integral exists.

We define the *auto-correlation* of a quaternion signal $a \in L^1(\mathbb{R}^2; \mathbb{H})$ as

$$(a \star_a a)(\boldsymbol{x}) = \int_{\mathbb{R}^2} a(\boldsymbol{y})\overline{a(\boldsymbol{y}-\boldsymbol{x})}d^2\boldsymbol{y} = \int_{\mathbb{R}^2} a(\boldsymbol{y}+\boldsymbol{x})\overline{a(\boldsymbol{y})}d^2\boldsymbol{y}, \tag{4.2.68}$$

provided that the integral exists.

4.2.3.3 The Wiener-Khintchine theorem for quaternion signals using the steerable two-sided QFT

For establishing the Wiener-Khintchine theorem for quaternion signals using the steerable two-sided QFT we begin with the cross correlation (4.2.67) and apply the inverse two-sided QFT of (4.1.10), where we set $\widehat{b} = \mathcal{F}^{f,g}\{b\}$:

$$
\begin{aligned}
(a \star_c b)(\boldsymbol{x}) &= \int_{\mathbb{R}^2} a(\boldsymbol{y})\overline{b(\boldsymbol{y}-\boldsymbol{x})}d^2\boldsymbol{y} \\
&= \int_{\mathbb{R}^2} a(\boldsymbol{y})\overline{\mathcal{F}^{-1}\{\widehat{b}\}(\boldsymbol{y}-\boldsymbol{x})}d^2\boldsymbol{y} \\
&= \frac{1}{(2\pi)^2}\int_{\mathbb{R}^2} a(\boldsymbol{y})\overline{\int_{\mathbb{R}^2} e^{f(y_1-x_1)\omega_1}\widehat{b}(\boldsymbol{\omega})e^{g(y_2-x_2)\omega_2}d^2\boldsymbol{\omega}}d^2\boldsymbol{y} \\
&= \frac{1}{(2\pi)^2}\int_{\mathbb{R}^2} a(\boldsymbol{y})\int_{\mathbb{R}^2} e^{-g(y_2-x_2)\omega_2}\overline{\widehat{b}(\boldsymbol{\omega})}e^{-f(y_1-x_1)\omega_1}d^2\boldsymbol{y}d^2\boldsymbol{\omega} \\
&= \frac{1}{(2\pi)^2}\int_{\mathbb{R}^2}\int_{\mathbb{R}^2} a(\boldsymbol{y})e^{-gy_2\omega_2}e^{gx_2\omega_2}\overline{\widehat{b}(\boldsymbol{\omega})}e^{-fy_1\omega_1}e^{fx_1\omega_1}d^2\boldsymbol{y}d^2\boldsymbol{\omega} \\
&= \frac{1}{(2\pi)^2}\int_{\mathbb{R}^2}\int_{\mathbb{R}^2}[a_-(\boldsymbol{y})+a_+(\boldsymbol{y})] \\
&\qquad e^{-gy_2\omega_2}e^{gx_2\omega_2}[\overline{\widehat{b}_-(\boldsymbol{\omega})}+\overline{\widehat{b}_+(\boldsymbol{\omega})}]e^{-fy_1\omega_1}e^{fx_1\omega_1}d^2\boldsymbol{y}d^2\boldsymbol{\omega} \\
&= \frac{1}{(2\pi)^2}\int_{\mathbb{R}^2}\int_{\mathbb{R}^2} a_-(\boldsymbol{y})e^{-gy_2\omega_2}e^{gx_2\omega_2}\overline{\widehat{b}_-(\boldsymbol{\omega})}e^{-fy_1\omega_1}e^{fx_1\omega_1}d^2\boldsymbol{y}d^2\boldsymbol{\omega} \\
&\quad + \frac{1}{(2\pi)^2}\int_{\mathbb{R}^2}\int_{\mathbb{R}^2} a_-(\boldsymbol{y})e^{-gy_2\omega_2}e^{gx_2\omega_2}\overline{\widehat{b}_+(\boldsymbol{\omega})}e^{-fy_1\omega_1}e^{fx_1\omega_1}d^2\boldsymbol{y}d^2\boldsymbol{\omega} \\
&\quad + \frac{1}{(2\pi)^2}\int_{\mathbb{R}^2}\int_{\mathbb{R}^2} a_+(\boldsymbol{y})e^{-gy_2\omega_2}e^{gx_2\omega_2}\overline{\widehat{b}_-(\boldsymbol{\omega})}e^{-fy_1\omega_1}e^{fx_1\omega_1}d^2\boldsymbol{y}d^2\boldsymbol{\omega} \\
&\quad + \frac{1}{(2\pi)^2}\int_{\mathbb{R}^2}\int_{\mathbb{R}^2} a_+(\boldsymbol{y})e^{-gy_2\omega_2}e^{gx_2\omega_2}\overline{\widehat{b}_+(\boldsymbol{\omega})}e^{-fy_1\omega_1}e^{fx_1\omega_1}d^2\boldsymbol{y}d^2\boldsymbol{\omega}.
\end{aligned}
\tag{4.2.69}
$$

In the following we transform the last four integrals I_1, I_2, I_3, I_4 of (4.2.69) one by one, applying (4.2.35) and (4.2.36) repeatedly, and setting $\widehat{a_\pm} = \mathcal{F}^{f,g}\{a_\pm\}$.

$$
\begin{aligned}
I_1 &= \frac{1}{(2\pi)^2}\int_{\mathbb{R}^2}\int_{\mathbb{R}^2} a_-(\boldsymbol{y})e^{-gy_2\omega_2}e^{gx_2\omega_2}\overline{\widehat{b}_-(\boldsymbol{\omega})}e^{-fy_1\omega_1}e^{fx_1\omega_1}d^2\boldsymbol{y}d^2\boldsymbol{\omega} \\
&= \frac{1}{(2\pi)^2}\int_{\mathbb{R}^2}\int_{\mathbb{R}^2} a_-(\boldsymbol{y})e^{-gy_2\omega_2}e^{gx_2\omega_2}e^{-gy_1\omega_1}\overline{\widehat{b}_-(\boldsymbol{\omega})}e^{fx_1\omega_1}d^2\boldsymbol{y}d^2\boldsymbol{\omega} \\
&= \frac{1}{(2\pi)^2}\int_{\mathbb{R}^2}\int_{\mathbb{R}^2} a_-(\boldsymbol{y})e^{-gy_1\omega_1}e^{-gy_2\omega_2}d^2\boldsymbol{y}e^{gx_2\omega_2}\overline{\widehat{b}_-(\boldsymbol{\omega})}e^{fx_1\omega_1}d^2\boldsymbol{\omega} \\
&= \frac{1}{(2\pi)^2}\int_{\mathbb{R}^2}\int_{\mathbb{R}^2} e^{-fy_1\omega_1}a_-(\boldsymbol{y})e^{-gy_2\omega_2}d^2\boldsymbol{y}\overline{\widehat{b}_-(\boldsymbol{\omega})}e^{f(x_1\omega_1+x_2\omega_2)}d^2\boldsymbol{\omega} \\
&= \frac{1}{(2\pi)^2}\int_{\mathbb{R}^2} \widehat{a_-}(\boldsymbol{\omega})\,\overline{\widehat{b}_-(\boldsymbol{\omega})}e^{f\boldsymbol{x}\cdot\boldsymbol{\omega}}d^2\boldsymbol{\omega}.
\end{aligned}
\tag{4.2.70}
$$

Computing the second integral we further apply Notation 4.2.10 and similarly get

$$\begin{aligned}
I_2 &= \frac{1}{(2\pi)^2}\int_{\mathbb{R}^2}\int_{\mathbb{R}^2} a_-(\boldsymbol{y})e^{-gy_2\omega_2}e^{gx_2\omega_2}\overline{\widehat{b_+}(\boldsymbol{\omega})}e^{-fy_1\omega_1}e^{fx_1\omega_1}\,d^2\boldsymbol{y}\,d^2\boldsymbol{\omega}\\
&= \frac{1}{(2\pi)^2}\int_{\mathbb{R}^2}\int_{\mathbb{R}^2} a_-(\boldsymbol{y})e^{-gy_2\omega_2}e^{gx_2\omega_2}e^{+gy_1\omega_1}\overline{\widehat{b_+}(\boldsymbol{\omega})}e^{fx_1\omega_1}\,d^2\boldsymbol{y}\,d^2\boldsymbol{\omega}\\
&= \frac{1}{(2\pi)^2}\int_{\mathbb{R}^2}\int_{\mathbb{R}^2} a_-(\boldsymbol{y})e^{+gy_1\omega_1}e^{-gy_2\omega_2}\,d^2\boldsymbol{y}\,e^{gx_2\omega_2}\overline{\widehat{b_+}(\boldsymbol{\omega})}e^{fx_1\omega_1}\,d^2\boldsymbol{\omega}\\
&= \frac{1}{(2\pi)^2}\int_{\mathbb{R}^2}\int_{\mathbb{R}^2} e^{+fy_1\omega_1}a_-(\boldsymbol{y})e^{-gy_2\omega_2}\,d^2\boldsymbol{y}\,\overline{\widehat{b_+}(\boldsymbol{\omega})}e^{f(x_1\omega_1-x_2\omega_2)}\,d^2\boldsymbol{\omega}\\
&= \frac{1}{(2\pi)^2}\int_{\mathbb{R}^2} \widehat{a_-}(-\omega_1,\omega_2)\,\overline{\widehat{b_+}(\boldsymbol{\omega})}e^{f(x_1\omega_1-x_2\omega_2)}\,d^2\boldsymbol{\omega}\\
&= \frac{1}{(2\pi)^2}\int_{\mathbb{R}^2} \widehat{a_-}^{(1,0)}(\boldsymbol{\omega})\,\overline{\widehat{b_+}(\boldsymbol{\omega})}e^{f(x_1\omega_1-x_2\omega_2)}\,d^2\boldsymbol{\omega}.
\end{aligned} \tag{4.2.71}$$

Computing the third integral we get

$$\begin{aligned}
I_3 &= \frac{1}{(2\pi)^2}\int_{\mathbb{R}^2}\int_{\mathbb{R}^2} a_+(\boldsymbol{y})e^{-gy_2\omega_2}e^{gx_2\omega_2}\overline{\widehat{b_-}(\boldsymbol{\omega})}e^{-fy_1\omega_1}e^{fx_1\omega_1}\,d^2\boldsymbol{y}\,d^2\boldsymbol{\omega}\\
&= \frac{1}{(2\pi)^2}\int_{\mathbb{R}^2}\int_{\mathbb{R}^2} a_+(\boldsymbol{y})e^{-gy_2\omega_2}e^{gx_2\omega_2}e^{-gy_1\omega_1}\overline{\widehat{b_-}(\boldsymbol{\omega})}e^{fx_1\omega_1}\,d^2\boldsymbol{y}\,d^2\boldsymbol{\omega}\\
&= \frac{1}{(2\pi)^2}\int_{\mathbb{R}^2}\int_{\mathbb{R}^2} a_+(\boldsymbol{y})e^{-gy_1\omega_1}e^{-gy_2\omega_2}\,d^2\boldsymbol{y}\,e^{gx_2\omega_2}\overline{\widehat{b_-}(\boldsymbol{\omega})}e^{fx_1\omega_1}\,d^2\boldsymbol{\omega}\\
&= \frac{1}{(2\pi)^2}\int_{\mathbb{R}^2}\int_{\mathbb{R}^2} e^{+fy_1\omega_1}a_+(\boldsymbol{y})e^{-gy_2\omega_2}\,d^2\boldsymbol{y}\,\overline{\widehat{b_-}(\boldsymbol{\omega})}e^{f(x_1\omega_1+x_2\omega_2)}\,d^2\boldsymbol{\omega}\\
&= \frac{1}{(2\pi)^2}\int_{\mathbb{R}^2} \widehat{a_+}(-\omega_1,\omega_2)\,\overline{\widehat{b_-}(\boldsymbol{\omega})}e^{f\boldsymbol{x}\cdot\boldsymbol{\omega}}\,d^2\boldsymbol{\omega}\\
&= \frac{1}{(2\pi)^2}\int_{\mathbb{R}^2} \widehat{a_+}^{(1,0)}(\boldsymbol{\omega})\,\overline{\widehat{b_-}(\boldsymbol{\omega})}e^{f\boldsymbol{x}\cdot\boldsymbol{\omega}}\,d^2\boldsymbol{\omega}.
\end{aligned} \tag{4.2.72}$$

Computing the fourth integral we get

$$\begin{aligned}
I_4 &= \frac{1}{(2\pi)^2}\int_{\mathbb{R}^2}\int_{\mathbb{R}^2} a_+(\boldsymbol{y})e^{-gy_2\omega_2}e^{gx_2\omega_2}\overline{\widehat{b_+}(\boldsymbol{\omega})}e^{-fy_1\omega_1}e^{fx_1\omega_1}\,d^2\boldsymbol{y}\,d^2\boldsymbol{\omega}\\
&= \frac{1}{(2\pi)^2}\int_{\mathbb{R}^2}\int_{\mathbb{R}^2} a_+(\boldsymbol{y})e^{-gy_2\omega_2}e^{gx_2\omega_2}e^{+gy_1\omega_1}\overline{\widehat{b_+}(\boldsymbol{\omega})}e^{fx_1\omega_1}\,d^2\boldsymbol{y}\,d^2\boldsymbol{\omega}
\end{aligned}$$

$$\begin{aligned}
&= \frac{1}{(2\pi)^2}\int_{\mathbb{R}^2}\int_{\mathbb{R}^2} a_+(\boldsymbol{y})e^{+gy_1\omega_1}e^{-gy_2\omega_2}\,d^2\boldsymbol{y}\,e^{gx_2\omega_2}\overline{\widehat{b_+}(\boldsymbol{\omega})}e^{fx_1\omega_1}\,d^2\boldsymbol{\omega}\\
&= \frac{1}{(2\pi)^2}\int_{\mathbb{R}^2}\int_{\mathbb{R}^2} e^{-fy_1\omega_1}a_+(\boldsymbol{y})e^{-gy_2\omega_2}\,d^2\boldsymbol{y}\,\overline{\widehat{b_+}(\boldsymbol{\omega})}e^{f(x_1\omega_1-x_2\omega_2)}\,d^2\boldsymbol{\omega}\\
&= \frac{1}{(2\pi)^2}\int_{\mathbb{R}^2} \widehat{a_+}(\boldsymbol{\omega})\,\overline{\widehat{b_+}(\boldsymbol{\omega})}e^{f(x_1\omega_1-x_2\omega_2)}\,d^2\boldsymbol{\omega}.
\end{aligned} \tag{4.2.73}$$

Collecting the results of the four integrals I_1, I_2, I_3, I_4, we obtain two relevant forms of the cross-correlation

$$\begin{aligned}
&(a \star_c b)(\boldsymbol{x}) \\
&= \frac{1}{(2\pi)^2} \int_{\mathbb{R}^2} [\widehat{a_-}(\boldsymbol{\omega})\, e^{gx_2\omega_2}\overline{\widehat{b}_-(\boldsymbol{\omega})e^{fx_1\omega_1}} + \widehat{a_-}^{(1,0)}(\boldsymbol{\omega})\, e^{gx_2\omega_2}\overline{\widehat{b}_+(\boldsymbol{\omega})e^{fx_1\omega_1}} \\
&\quad + \widehat{a_+}^{(1,0)}(\boldsymbol{\omega})\, e^{gx_2\omega_2}\overline{\widehat{b}_-(\boldsymbol{\omega})e^{fx_1\omega_1}} + \widehat{a_+}(\boldsymbol{\omega})\, e^{gx_2\omega_2}\overline{\widehat{b}_+(\boldsymbol{\omega})e^{fx_1\omega_1}}]d^2\boldsymbol{\omega} \\
&= \frac{1}{(2\pi)^2} \int_{\mathbb{R}^2} \Big([\widehat{a_-}(\boldsymbol{\omega}) + \widehat{a_+}^{(1,0)}(\boldsymbol{\omega})]e^{gx_2\omega_2}\overline{\widehat{b}_-(\boldsymbol{\omega})e^{fx_1\omega_1}} + \\
&\quad + [\widehat{a_-}^{(1,0)}(\boldsymbol{\omega}) + \widehat{a_+}(\boldsymbol{\omega})]e^{gx_2\omega_2}\overline{\widehat{b}_+(\boldsymbol{\omega})e^{fx_1\omega_1}}\Big)\, d^2\boldsymbol{\omega} \\
&= \frac{1}{(2\pi)^2} \int_{\mathbb{R}^2} \Big([\widehat{a_-}(\boldsymbol{\omega}) + \widehat{a_+}^{(1,0)}(\boldsymbol{\omega})]\overline{e^{-fx_1\omega_1}\widehat{b}_-(\boldsymbol{\omega})e^{-gx_2\omega_2}} + \\
&\quad + [\widehat{a_-}^{(1,0)}(\boldsymbol{\omega}) + \widehat{a_+}(\boldsymbol{\omega})]\overline{e^{-fx_1\omega_1}\widehat{b}_+(\boldsymbol{\omega})e^{-gx_2\omega_2}}\Big)\, d^2\boldsymbol{\omega},
\end{aligned} \tag{4.2.74}$$

and alternatively

$$\begin{aligned}
(a \star_c b)(\boldsymbol{x}) &= \frac{1}{(2\pi)^2} \int_{\mathbb{R}^2} \Big([\widehat{a_-}(\boldsymbol{\omega}) + \widehat{a_+}^{(1,0)}(\boldsymbol{\omega})]\overline{\widehat{b}_-(\boldsymbol{\omega})}e^{f\boldsymbol{x}\cdot\boldsymbol{\omega}} \\
&\quad + [\widehat{a_-}^{(1,0)}(\boldsymbol{\omega}) + \widehat{a_+}(\boldsymbol{\omega})]\overline{\widehat{b}_+(\boldsymbol{\omega})}e^{f(x_1\omega_1 - x_2\omega_2)}\Big)\, d^2\boldsymbol{\omega}.
\end{aligned} \tag{4.2.75}$$

We note, that the last form appears most similar to the conventional complex Wiener-Khintchine theorem. Formal identity with the conventional complex case is obtained for $a_+ = b_+ = 0$. Setting $b = a$, and noting that with Theorem 4.2.16 $\widehat{a}_\pm = \widehat{a_\pm}$, we get important identities for the auto-correlation

$$\begin{aligned}
&(a \star_a a)(\boldsymbol{x}) \\
&= \frac{1}{(2\pi)^2} \int_{\mathbb{R}^2} \Big([\widehat{a_-}(\boldsymbol{\omega}) + \widehat{a_+}^{(1,0)}(\boldsymbol{\omega})]\overline{e^{-fx_1\omega_1}\widehat{a}_-(\boldsymbol{\omega})e^{-gx_2\omega_2}} + \\
&\quad + [\widehat{a_-}^{(1,0)}(\boldsymbol{\omega}) + \widehat{a_+}(\boldsymbol{\omega})]\overline{e^{-fx_1\omega_1}\widehat{a}_+(\boldsymbol{\omega})e^{-gx_2\omega_2}}\Big)\, d^2\boldsymbol{\omega} \\
&= \frac{1}{(2\pi)^2} \int_{\mathbb{R}^2} \Big([\widehat{a_-}(\boldsymbol{\omega}) + \widehat{a_+}^{(1,0)}(\boldsymbol{\omega})]\overline{\widehat{a}_-(\boldsymbol{\omega})}e^{f\boldsymbol{x}\cdot\boldsymbol{\omega}} \\
&\quad + [\widehat{a_-}^{(1,0)}(\boldsymbol{\omega}) + \widehat{a_+}(\boldsymbol{\omega})]\overline{\widehat{a}_+(\boldsymbol{\omega})}e^{f(x_1\omega_1 - x_2\omega_2)}\Big)\, d^2\boldsymbol{\omega} \\
&= \frac{1}{(2\pi)^2} \int_{\mathbb{R}^2} \Big([|\widehat{a_-}(\boldsymbol{\omega})|^2 + \widehat{a_+}^{(1,0)}(\boldsymbol{\omega})\overline{\widehat{a_-}(\boldsymbol{\omega})}]e^{f\boldsymbol{x}\cdot\boldsymbol{\omega}} \\
&\quad + [\widehat{a_-}^{(1,0)}(\boldsymbol{\omega})\overline{\widehat{a_+}(\boldsymbol{\omega})} + |\widehat{a_+}(\boldsymbol{\omega})|^2]e^{f(x_1\omega_1 - x_2\omega_2)}\Big)\, d^2\boldsymbol{\omega}.
\end{aligned} \tag{4.2.76}$$

Formal identity with the conventional complex auto-correlation is obtained for $a_+ = 0$. We summarize our results in one theorem and one corollary.

Theorem 4.2.27 (Wiener-Khintchine theorem for cross-correlation of quaternionic signals). *The cross-correlation of two quaternionic signals $a, b \in L^1(\mathbb{R}^2; \mathbb{H})$ can be expressed as*

$$
\begin{aligned}
&(a \star_c b)(\boldsymbol{x})\\
&= \frac{1}{(2\pi)^2}\int_{\mathbb{R}^2}\Big([\widehat{a_-}(\boldsymbol{\omega}) + \widehat{a_+}^{(1,0)}(\boldsymbol{\omega})]e^{gx_2\omega_2}\overline{\widehat{b}_-(\boldsymbol{\omega})e^{fx_1\omega_1}}+\\
&\quad + [\widehat{a_-}^{(1,0)}(\boldsymbol{\omega}) + \widehat{a_+}(\boldsymbol{\omega})]e^{gx_2\omega_2}\overline{\widehat{b}_+(\boldsymbol{\omega})e^{fx_1\omega_1}}\Big)\, d^2\boldsymbol{\omega}\\
&= \frac{1}{(2\pi)^2}\int_{\mathbb{R}^2}\Big([\widehat{a_-}(\boldsymbol{\omega}) + \widehat{a_+}^{(1,0)}(\boldsymbol{\omega})]\overline{e^{-fx_1\omega_1}\widehat{b}_-(\boldsymbol{\omega})e^{-gx_2\omega_2}}+\\
&\quad + [\widehat{a_-}^{(1,0)}(\boldsymbol{\omega}) + \widehat{a_+}(\boldsymbol{\omega})]\overline{e^{-fx_1\omega_1}\widehat{b}_+(\boldsymbol{\omega})e^{-gx_2\omega_2}}\Big)\, d^2\boldsymbol{\omega}\\
&= \frac{1}{(2\pi)^2}\int_{\mathbb{R}^2}\Big([\widehat{a_-}(\boldsymbol{\omega}) + \widehat{a_+}^{(1,0)}(\boldsymbol{\omega})]\overline{\widehat{b}_-(\boldsymbol{\omega})}e^{f\boldsymbol{x}\cdot\boldsymbol{\omega}}\\
&\quad + [\widehat{a_-}^{(1,0)}(\boldsymbol{\omega}) + \widehat{a_+}(\boldsymbol{\omega})]\overline{\widehat{b}_+(\boldsymbol{\omega})}e^{f(x_1\omega_1 - x_2\omega_2)}\Big)\, d^2\boldsymbol{\omega}.
\end{aligned}
\tag{4.2.77}
$$

Corollary 4.2.28 (Wiener-Khintchine theorem for auto-correlation of quaternionic signals). *The cross-correlation of a quaternionic signal* $a \in L^1(\mathbb{R}^2;\mathbb{H})$ *can be expressed as*

$$
\begin{aligned}
&(a \star_a a)(\boldsymbol{x})\\
&= \frac{1}{(2\pi)^2}\int_{\mathbb{R}^2}\Big([\widehat{a_-}(\boldsymbol{\omega}) + \widehat{a_+}^{(1,0)}(\boldsymbol{\omega})]\overline{e^{-fx_1\omega_1}\widehat{a}_-(\boldsymbol{\omega})e^{-gx_2\omega_2}}+\\
&\quad + [\widehat{a_-}^{(1,0)}(\boldsymbol{\omega}) + \widehat{a_+}(\boldsymbol{\omega})]\overline{e^{-fx_1\omega_1}\widehat{a}_+(\boldsymbol{\omega})e^{-gx_2\omega_2}}\Big)\, d^2\boldsymbol{\omega}\\
&= \frac{1}{(2\pi)^2}\int_{\mathbb{R}^2}\Big([|\widehat{a_-}(\boldsymbol{\omega})|^2 + \widehat{a_+}^{(1,0)}(\boldsymbol{\omega})\overline{\widehat{a_-}(\boldsymbol{\omega})}]e^{f\boldsymbol{x}\cdot\boldsymbol{\omega}}\\
&\quad + [\widehat{a_-}^{(1,0)}(\boldsymbol{\omega})\overline{\widehat{a_+}(\boldsymbol{\omega})} + |\widehat{a_+}(\boldsymbol{\omega})|^2]e^{f(x_1\omega_1 - x_2\omega_2)}\Big)\, d^2\boldsymbol{\omega}.
\end{aligned}
\tag{4.2.78}
$$

4.2.3.4 *New spectral representation for convolution of quaternion signals*

It is known, that due to the non-commutativity of quaternion signals and the quaternionic kernels of the QFT, the spectral representation of the convolution may include many terms. See e.g. Theorem 4.6 in [213] for a representation involving as many as twenty terms. But the expressions in (4.2.66) and (4.2.67) for convolution and cross-correlation of two quaternion signals are related by

$$
(a \star b)(\boldsymbol{x}) = (a^{(1,1)} \star_c \overline{b})(\boldsymbol{x}). \tag{4.2.79}
$$

This indeed raises the expectation of finding similar to Theorem 4.2.27 a new short and compact spectral representation for the convolution by reflecting the argument of the first signal $a \to a^{(1,1)}$ and inserting the quaternion conjugate of the second signal $b \to \overline{b}$. But in doing this all operations in the derivation of Theorem 4.2.27 would have to be carefully checked for their validity, which is not as trivial as it may seem at first. Therefore starting all over with the very definition of the convolution of two quaternion signals (4.2.66) seems to be better and clearly reveals why the result is both similar, yet also characteristically different from the above cross-correlation result.

In the following and throughout the current section we will apply two different splits to the signal functions a and b, respectively defined as

$$a_{\pm} = \tfrac{1}{2}(a \pm gaf), \qquad b_{\pm} = \tfrac{1}{2}(b \pm fbg), \tag{4.2.80}$$

note in particular the different positions of f and g. Two different splits are needed, because in the convolution the second signal is not quaternion conjugated. To avoid clutter we will use the notation $\widehat{b} = \mathcal{F}^{f,g}\{b\}$, and for the inverse transform we use a lower index: $\mathcal{F}^{f,g}_{-1}$.

$$\begin{aligned}
(a \star b)(\boldsymbol{x}) &= \int_{\mathbb{R}^2} a(\boldsymbol{y}) b(\boldsymbol{x}-\boldsymbol{y}) d^2\boldsymbol{y} \\
&= \int_{\mathbb{R}^2} a(\boldsymbol{y}) \mathcal{F}^{f,g}_{-1}\{\widehat{b}\}(\boldsymbol{x}-\boldsymbol{y}) d^2\boldsymbol{y} \\
&= \int_{\mathbb{R}^2} a(\boldsymbol{y}) \int_{\mathbb{R}^2} e^{f(x_1-y_1)\omega_1} \widehat{b}(\boldsymbol{\omega}) e^{g(x_2-y_2)\omega_2} d^2\boldsymbol{\omega} d^2\boldsymbol{y} \\
&= \int_{\mathbb{R}^2}\int_{\mathbb{R}^2} a(\boldsymbol{y}) e^{-fy_1\omega_1} e^{fx_1\omega_1} \widehat{b}(\boldsymbol{\omega}) e^{-gy_2\omega_2} e^{gx_2\omega_2} d^2\boldsymbol{\omega} d^2\boldsymbol{y} \\
&= \int_{\mathbb{R}^2}\int_{\mathbb{R}^2} [a_-(\boldsymbol{y}) + a_+(\boldsymbol{y})] e^{-fy_1\omega_1} e^{fx_1\omega_1} [\widehat{b}_-(\boldsymbol{\omega}) + \widehat{b}_+(\boldsymbol{\omega})] \\
&\quad e^{-gy_2\omega_2} e^{gx_2\omega_2} d^2\boldsymbol{\omega} d^2\boldsymbol{y} \\
&= \int_{\mathbb{R}^2}\int_{\mathbb{R}^2} a_-(\boldsymbol{y}) e^{-fy_1\omega_1} e^{fx_1\omega_1} \widehat{b}_-(\boldsymbol{\omega}) e^{-gy_2\omega_2} e^{gx_2\omega_2} d^2\boldsymbol{\omega} d^2\boldsymbol{y} \\
&\quad + \int_{\mathbb{R}^2}\int_{\mathbb{R}^2} a_-(\boldsymbol{y}) e^{-fy_1\omega_1} e^{fx_1\omega_1} \widehat{b}_+(\boldsymbol{\omega}) e^{-gy_2\omega_2} e^{gx_2\omega_2} d^2\boldsymbol{\omega} d^2\boldsymbol{y} \\
&\quad + \int_{\mathbb{R}^2}\int_{\mathbb{R}^2} a_+(\boldsymbol{y}) e^{-fy_1\omega_1} e^{fx_1\omega_1} \widehat{b}_-(\boldsymbol{\omega}) e^{-gy_2\omega_2} e^{gx_2\omega_2} d^2\boldsymbol{\omega} d^2\boldsymbol{y} \\
&\quad + \int_{\mathbb{R}^2}\int_{\mathbb{R}^2} a_+(\boldsymbol{y}) e^{-fy_1\omega_1} e^{fx_1\omega_1} \widehat{b}_+(\boldsymbol{\omega}) e^{-gy_2\omega_2} e^{gx_2\omega_2} d^2\boldsymbol{\omega} d^2\boldsymbol{y} \\
&= \int_{\mathbb{R}^2}\int_{\mathbb{R}^2} e^{-gy_1\omega_1} a_-(\boldsymbol{y}) e^{-fy_2\omega_2} d^2\boldsymbol{y} e^{fx_1\omega_1} \widehat{b}_-(\boldsymbol{\omega}) e^{gx_2\omega_2} d^2\boldsymbol{\omega} \\
&\quad + \int_{\mathbb{R}^2}\int_{\mathbb{R}^2} e^{-gy_1\omega_1} a_-(\boldsymbol{y}) e^{fy_2\omega_2} d^2\boldsymbol{y} e^{fx_1\omega_1} \widehat{b}_+(\boldsymbol{\omega}) e^{gx_2\omega_2} d^2\boldsymbol{\omega} \\
&\quad + \int_{\mathbb{R}^2}\int_{\mathbb{R}^2} e^{gy_1\omega_1} a_+(\boldsymbol{y}) e^{-fy_2\omega_2} d^2\boldsymbol{y} e^{fx_1\omega_1} \widehat{b}_-(\boldsymbol{\omega}) e^{gx_2\omega_2} d^2\boldsymbol{\omega} \\
&\quad + \int_{\mathbb{R}^2}\int_{\mathbb{R}^2} e^{gy_1\omega_1} a_+(\boldsymbol{y}) e^{fy_2\omega_2} d^2\boldsymbol{y} e^{fx_1\omega_1} \widehat{b}_+(\boldsymbol{\omega}) e^{gx_2\omega_2} d^2\boldsymbol{\omega}
\end{aligned}$$

$$
\begin{aligned}
&= \int_{\mathbb{R}^2} \mathcal{F}^{g,f}\{a_-\}(\boldsymbol{\omega}) e^{f x_1 \omega_1} \widehat{b}_-(\boldsymbol{\omega}) e^{g x_2 \omega_2} d^2\boldsymbol{\omega} \\
&\quad + \int_{\mathbb{R}^2} \mathcal{F}^{g,f}\{a_-^{(0,1)}\}(\boldsymbol{\omega}) e^{f x_1 \omega_1} \widehat{b}_+(\boldsymbol{\omega}) e^{g x_2 \omega_2} d^2\boldsymbol{\omega} \\
&\quad + \int_{\mathbb{R}^2} \mathcal{F}^{g,f}\{a_+^{(1,0)}\}(\boldsymbol{\omega}) e^{f x_1 \omega_1} \widehat{b}_-(\boldsymbol{\omega}) e^{g x_2 \omega_2} d^2\boldsymbol{\omega} \\
&\quad + \int_{\mathbb{R}^2} \mathcal{F}^{g,f}\{a_+^{(1,1)}\}(\boldsymbol{\omega}) e^{f x_1 \omega_1} \widehat{b}_+(\boldsymbol{\omega}) e^{g x_2 \omega_2} d^2\boldsymbol{\omega} \\
&= \int_{\mathbb{R}^2} \mathcal{F}^{g,f}\{a_- + a_+^{(1,0)}\}(\boldsymbol{\omega}) e^{f x_1 \omega_1} \widehat{b}_-(\boldsymbol{\omega}) e^{g x_2 \omega_2} d^2\boldsymbol{\omega} \\
&\quad + \int_{\mathbb{R}^2} \mathcal{F}^{g,f}\{a_-^{(0,1)} + a_+^{(1,1)}\}(\boldsymbol{\omega}) e^{f x_1 \omega_1} \widehat{b}_+(\boldsymbol{\omega}) e^{g x_2 \omega_2} d^2\boldsymbol{\omega},
\end{aligned} \tag{4.2.81}
$$

where we repeatedly applied that $\widehat{b}_\pm \exp(\beta g) = \exp(\mp\beta f)\widehat{b}_\pm$, and $a_\pm \exp(\alpha f) = \exp(\mp\alpha g)a_\pm$, and note the use of $\mathcal{F}^{g,f}\{a_\pm\}$. We therefore obtain the following theorem.

Theorem 4.2.29 (Spectral representation of convolution of quaternion signals). *The convolution of two quaternionic signals $a, b \in L^1(\mathbb{R}^2;\mathbb{H})$ can be expressed as*

$$
\begin{aligned}
(a \star b)(\boldsymbol{x}) &= \int_{\mathbb{R}^2} \mathcal{F}^{g,f}\{a_- + a_+^{(1,0)}\}(\boldsymbol{\omega}) e^{f x_1 \omega_1} \widehat{b}_-(\boldsymbol{\omega}) e^{g x_2 \omega_2} d^2\boldsymbol{\omega} \\
&\quad + \int_{\mathbb{R}^2} \mathcal{F}^{g,f}\{a_-^{(0,1)} + a_+^{(1,1)}\}(\boldsymbol{\omega}) e^{f x_1 \omega_1} \widehat{b}_+(\boldsymbol{\omega}) e^{g x_2 \omega_2} d^2\boldsymbol{\omega},
\end{aligned} \tag{4.2.82}
$$

where $\widehat{b} = \mathcal{F}^{f,g}\{b\}$, $a_\pm = \frac{1}{2}(a \pm gaf)$, and $b_\pm = \frac{1}{2}(b \pm fbg)$.

4.2.3.5 *Summary on QFT Wiener-Khintchine theorems*

In Section 4.2.3, we defined the notions of convolution and cross-correlation of two quaternion valued functions over $\mathbb{R}^2$, and the auto-correlation of a quaternion valued function over $\mathbb{R}^2$.

The main results are a Wiener-Khintchine theorem for the cross-correlation of two quaternion signals, based on the steerable two-dimensional QFT, and as a corollary for the auto-correlation. Furthermore we obtain a new four term long spectral representation of the convolution of two quaternion signals.

In view of the many applications of the QFT referred to in Section 4.2.3.1, we expect these results to be of great interest, especially for filter design and feature extraction in signal and color image processing.

4.3 SPECIAL QUATERNION FOURIER TRANSFORMS

4.3.1 Windowed quaternion Fourier transform (QWFT)

4.3.1.1 *Background*

In this section, we generalize the classical windowed Fourier transform (WFT) to quaternion-valued signals, calling it the *quaternionic windowed Fourier transform* (QWFT). Using the spectral representation of the quaternionic Fourier transform

(QFT), we derive several important properties such as reconstruction formula, reproducing kernel, isometry and orthogonality relation. Taking the Gaussian function as window function we obtain quaternionic Gabor filters which play the role of coefficient functions when decomposing the signal in the quaternionic Gabor basis. We apply the QWFT properties and the (right-sided) QFT to establish a Heisenberg type uncertainty principle for the QWFT. Finally, we briefly introduce an application of the QWFT to a linear time-varying system. This section is based on [256], but for lack of space and because most proofs are fairly elementary, we omit proofs. The interested reader is invited to try the proofs himself, or simply look them up in [256].

One of the basic problems encountered in signal representations using the conventional Fourier transform (FT) is the ineffectiveness of the Fourier kernel to represent and compute location information. One method to overcome such a problem is the windowed Fourier transform (WFT). Some authors [147, 148, 318] have extensively studied the WFT and its properties from a mathematical point of view. In [234, 337] the WFT has been successfully applied as a tool of spatial-frequency analysis which is able to characterize the local frequency at any location in a fringe pattern.

On the other hand, the quaternionic Fourier transform (QFT), which, as we have seen, is a nontrivial generalization of the real and complex Fourier transform (FT) using the quaternion algebra [156], has been of interest to researchers for some years (see, for example, [25, 56, 58, 124, 184, 187, 255, 280]). They found that many FT properties still hold and others have to be modified. Based on the (right-sided) QFT, one may extend the WFT to quaternion algebra while enjoying similar properties as in the classical case.

The idea of extending the WFT to the quaternion algebra setting has already been studied by Bülow and Sommer [56, 58]. They introduced a special case of the QWFT known as quaternionic Gabor filters. They applied these filters to obtain a local two-dimensional quaternionic phase. Their generalization is obtained using the inverse (two-sided) quaternion Fourier kernel. Hahn [154] constructed a Fourier-Wigner distribution of 2-D quaternionic signals which is in fact closely related to the QWFT. In [254], the extension of the WFT to Clifford (geometric) algebra was discussed. This extension used the kernel of the Clifford Fourier transform (CFT) [185]. In general a CFT replaces the complex imaginary unit $i \in \mathbb{C}$ by a geometric root [193, 294] of -1, i.e. any element of a Clifford (geometric) algebra squaring to -1.

The main goal of this section is to thoroughly study the generalization of the classical WFT to quaternion algebra, which we call the *quaternionic windowed Fourier transform* (QWFT), and investigate important properties of the QWFT such as (specific) shift, reconstruction formula, reproducing kernel, isometry and orthogonality relation. We emphasize that the QWFT proposed in the present work is significantly different from [254] in the definition of the exponential kernel. In the present approach, we use the kernel of the (right-sided) QFT. We present several examples to show the differences between the QWFT and the WFT. Using the (right-sided) QFT properties and its uncertainty principle [255] we establish a generalized QWFT uncertainty principle. We will also study an application of the QWFT to a linear time-varying system.

The organization of the treatment is as follows. In the remainder of this background section we briefly describe notation and review the (right-sided) QFT. In Section 4.3.1.2, we discuss the basic ideas for the construction of the QWFT and derive several important properties of the QWFT using the (right-sided) QFT, including the Heisenberg uncertainty principle in Section 4.3.1.3 We also give some examples of the QWFT. In Section 4.3.1.4, an application of the QWFT to a linear time varying system is presented.

We will make use of an inner product for two functions $f, g : \mathbb{R}^2 \longrightarrow \mathbb{H}$ as follows:

$$\langle f, g\rangle_{L^2(\mathbb{R}^2;\mathbb{H})} = \int_{\mathbb{R}^2} f(\boldsymbol{x})\overline{g(\boldsymbol{x})}\, d^2\boldsymbol{x}, \tag{4.3.1}$$

where the overline indicates the quaternion conjugation of the function. In particular, if $f = g$, we obtain the associated norm

$$\|f\|_{L^2(\mathbb{R}^2;\mathbb{H})} = \langle f, f\rangle^{1/2}_{L^2(\mathbb{R}^2;\mathbb{H})} = \left(\int_{\mathbb{R}^2} |f(\boldsymbol{x})|^2\, d^2\boldsymbol{x}\right)^{1/2}. \tag{4.3.2}$$

As a consequence of the inner product (4.3.1) we obtain the *quaternion Cauchy-Schwarz* inequality

$$|Sc\langle f, g\rangle| \leq \|f\|_{L^2(\mathbb{R}^2;\mathbb{H})}\|g\|_{L^2(\mathbb{R}^2;\mathbb{H})}, \qquad \forall f, g \in L^2(\mathbb{R}^2;\mathbb{H}). \tag{4.3.3}$$

Now we recall the (right-sided) QFT needed later in section 4.3.1.2 to establish the QWFT.

Definition 4.3.1 (Right-sided QFT). *The (right sided) quaternion Fourier transform (QFT) of $f \in L^1(\mathbb{R}^2;\mathbb{H})$ is the function*[23] *$\mathcal{F}_q\{f\}\colon \mathbb{R}^2 \to \mathbb{H}$ given by*

$$\mathcal{F}_q\{f\}(\boldsymbol{\omega}) = \int_{\mathbb{R}^2} f(\boldsymbol{x}) e^{-\boldsymbol{i}\omega_1 x_1} e^{-\boldsymbol{j}\omega_2 x_2}\, d^2\boldsymbol{x}, \tag{4.3.4}$$

where $\boldsymbol{x} = x_1\boldsymbol{e}_1 + x_2\boldsymbol{e}_2$, $\boldsymbol{\omega} = \omega_1\boldsymbol{e}_1 + \omega_2\boldsymbol{e}_2$, and the quaternion exponential product $e^{-\boldsymbol{i}\omega_1 x_1} e^{-\boldsymbol{j}\omega_2 x_2}$ is the quaternion Fourier kernel.

Theorem 4.3.2 (Inverse QFT). *Suppose that $f \in L^2(\mathbb{R}^2;\mathbb{H})$ and $\mathcal{F}_q\{f\} \in L^1(\mathbb{R}^2;\mathbb{H})$. Then the QFT of f is an invertible transform and its inverse is given by*

$$\mathcal{F}_q^{-1}[\mathcal{F}_q\{f\}](\boldsymbol{x}) = f(\boldsymbol{x}) = \frac{1}{(2\pi)^2}\int_{\mathbb{R}^2} \mathcal{F}_q\{f\}(\boldsymbol{\omega}) e^{\boldsymbol{j}\omega_2 x_2} e^{\boldsymbol{i}\omega_1 x_1}\, d^2\boldsymbol{\omega}, \tag{4.3.5}$$

where the quaternion exponential product $e^{\boldsymbol{j}\omega_2 x_2} e^{\boldsymbol{i}\omega_1 x_1}$ is called the inverse (right-sided) quaternion Fourier kernel.

Detailed information about the QFT and its properties can be found in [56, 58, 124, 184, 255].

[23] For the QFT on L^2, see Footnote 2 on page 144.

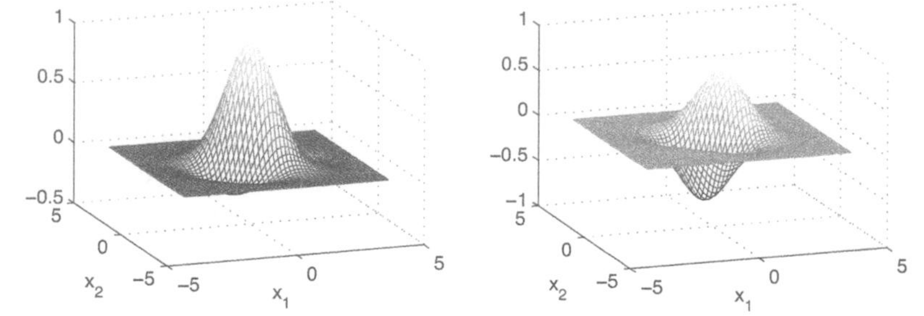

Figure 4.7 Representation of complex Gabor filter for $\sigma_1 = \sigma_2 = 1, u_0 = v_0 = 1$ in the spatial domain with its real part (*left*) and imaginary part (*right*). Source: [256, Fig. 1].

4.3.1.2 Quaternionic windowed Fourier transform

This section generalizes the classical WFT to quaternion algebra. Using the definition of the (right-sided) QFT described before, we extend the WFT to the QWFT. We shall later see how some properties of the WFT are extended in the new definition. For this purpose we briefly review the two-dimensional WFT.

4.3.1.2.1 Two-dimensional WFT The FT is a powerful tool for the analysis of stationary signals but it is not well suited for the analysis of non-stationary signals because it is a global transformation with poor spatial localization [007]. However, in practice, most natural signals are non-stationary. In order to characterize a non-stationary signal properly, the WFT is commonly used.

Definition 4.3.3 (WFT). *The WFT of a two-dimensional real signal $f \in L^2(\mathbb{R}^2)$ with respect to the window function $g \in L^2(\mathbb{R}^2) \setminus \{0\}$ is given by*

$$\mathcal{G}_g f(\boldsymbol{\omega}, \boldsymbol{b}) = \frac{1}{(2\pi)^2} \int_{\mathbb{R}^2} f(\boldsymbol{x}) \, \overline{g_{\boldsymbol{\omega},\boldsymbol{b}}(\boldsymbol{x})} \, d^2\boldsymbol{x}, \tag{4.3.6}$$

where the window daughter function $g_{\boldsymbol{\omega},\boldsymbol{b}}$ is called the windowed Fourier kernel defined by

$$g_{\boldsymbol{\omega},\boldsymbol{b}}(\boldsymbol{x}) = g(\boldsymbol{x} - \boldsymbol{b}) e^{\sqrt{-1}\,\boldsymbol{\omega}\cdot\boldsymbol{x}}. \tag{4.3.7}$$

Equation (4.3.6) shows that the image of a WFT is a complex 4-D coefficient function.

Most applications make use of the Gaussian window function g which is non-negative and well localized around the origin in both spatial and frequency domains. The Gaussian window function can be expressed as

$$g(\boldsymbol{x}, \sigma_1, \sigma_2) = e^{-\left[(x_1/\sigma_1)^2 + (x_2/\sigma_2)^2\right]/2}, \tag{4.3.8}$$

where σ_1 and σ_2 are the standard deviations of the Gaussian function and determine the width of the window. We call (4.3.7), for fixed $\boldsymbol{\omega} = \boldsymbol{\omega}_0 = u_0\boldsymbol{e}_1 + v_0\boldsymbol{e}_2$, and

$b_1 = b_2 = 0$, a complex Gabor filter as shown in Figure 4.7 if g is the Gaussian function (4.3.8), i.e.

$$g_{c,\boldsymbol{\omega}_0}(\boldsymbol{x}, \sigma_1, \sigma_2) = e^{\sqrt{-1}\,(u_0 x_1 + v_0 x_2)} g(\boldsymbol{x}, \sigma_1, \sigma_2). \tag{4.3.9}$$

In general, when the Gaussian function (4.3.8) is chosen as the window function, the WFT in (4.3.6) is called *Gabor transform*. We observe that the WFT localizes the signal f in the neighborhood of $\boldsymbol{x} = \boldsymbol{b}$. For this reason, the WFT is often called *short time Fourier transform*.

4.3.1.2.2 Definition of the QWFT Bülow [56] extended the complex Gabor filter (4.3.9) to quaternion algebra by replacing the complex kernel $e^{\sqrt{-1}(u_0 x_1 + v_0 x_2)}$ with the inverse (two-sided) quaternion Fourier kernel $e^{\boldsymbol{i} u_0 x_1} e^{\boldsymbol{j} v_0 x_2}$. His extension then takes the form

$$g_q(\boldsymbol{x}, \sigma_1, \sigma_2) = e^{\boldsymbol{i} u_0 x_1} e^{\boldsymbol{j} v_0 x_2} e^{-\left[(x_1/\sigma_1)^2 + (x_2/\sigma_2)^2\right]/2}, \tag{4.3.10}$$

which he called *quaternionic Gabor filter*[24] as shown in Figure 4.8 and applied it to get the local quaternionic phase of a two-dimensional real signal. Bayro-Corrochano et al. [25] also used quaternionic Gabor filters for the preprocessing of two-dimensional speech representations.

The extension of the WFT to quaternion algebra using the (two-sided) QFT is rather complicated, due to the non-commutativity of quaternion functions. Alternatively, we use the (right-sided) QFT to define the QWFT. We therefore introduce the following general QWFT of a two-dimensional quaternion signal $f \in L^2(\mathbb{R}^2; \mathbb{H})$ in Def. 4.3.5.

Definition 4.3.4. *A quaternion window function is a function* $\phi \in L^2(\mathbb{R}^2; \mathbb{H}) \setminus \{0\}$ *such that* $|\boldsymbol{x}|^{1/2}\phi(\boldsymbol{x}) \in L^2(\mathbb{R}^2; \mathbb{H})$ *too. We call*

$$\phi_{\boldsymbol{\omega},\boldsymbol{b}}(\boldsymbol{x}) = \frac{1}{(2\pi)^2} e^{\boldsymbol{j}\omega_2 x_2} e^{\boldsymbol{i}\omega_1 x_1} \phi(\boldsymbol{x} - \boldsymbol{b}), \tag{4.3.11}$$

a quaternionic window daughter function.

If we fix $\boldsymbol{\omega} = \boldsymbol{\omega}_0$, and $b_1 = b_2 = 0$, and take the Gaussian function as the window function of (4.3.11), then we get the quaternionic Gabor filter shown in Figure 4.9,

$$g_q(\boldsymbol{x}, \sigma_1, \sigma_2) = \frac{1}{(2\pi)^2} e^{\boldsymbol{j} v_0 x_2} e^{\boldsymbol{i} u_0 x_1} e^{-\left[(x_1/\sigma_1)^2 + (x_2/\sigma_2)^2\right]/2}. \tag{4.3.12}$$

---

[24] If we would have interchanged the order of the two exponentials in Definition 4.3.1, which we are always free to do, then (4.3.10) and (4.3.12) would agree fully, except for the factor $(2\pi)^{-2}$. Figures 4.8 and 4.9 illustrate the two different kinds of quaternionic Gabor filters that arise. The differences can be made obvious by decomposition of the two exponential products $e^{\boldsymbol{i} u_0 x_1} e^{\boldsymbol{j} v_0 x_2}$ and $e^{\boldsymbol{j} v_0 x_2} e^{\boldsymbol{i} u_0 x_1}$. The imaginary $\boldsymbol{i}$-part of Figure 4.8 is the imaginary $\boldsymbol{j}$-part of Figure 4.9 and vice versa. Note also that the imaginary $\boldsymbol{k}$-parts of Figures 4.8 and 4.9 are essentially the same, because they only have different signs.

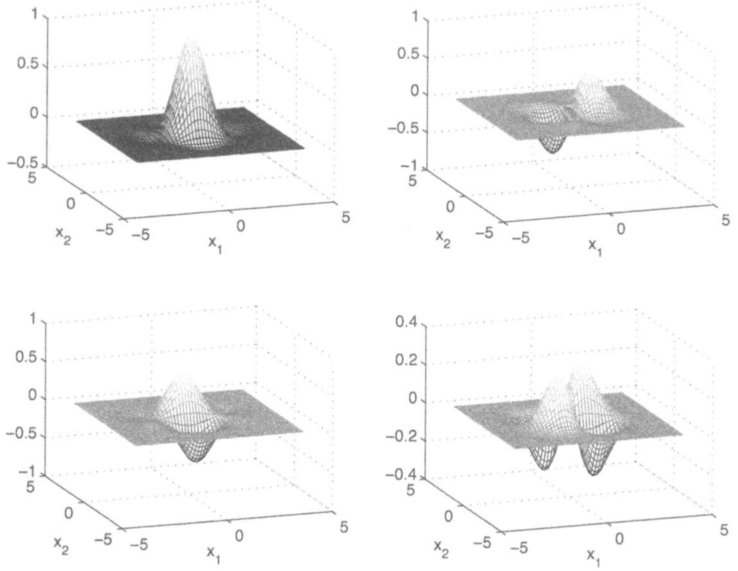

Figure 4.8 Bülow's quaternionic Gabor filter (4.3.10) ($\sigma_1 = \sigma_2 = 1, u_0 = v_0 = 1$) in the spatial domain with real part (*top left*) and imaginary $\boldsymbol{i}$-part (*top right*), $\boldsymbol{j}$-part (*bottom left*), and $\boldsymbol{k}$-part (*bottom right*). Source: [256, Fig. 2].

Definition 4.3.5 (QWFT). *Denote the QWFT on $L^2(\mathbb{R}^2;\mathbb{H})$ by G_ϕ. Then the QWFT of $f \in L^2(\mathbb{R}^2;\mathbb{H})$ is defined by*

$$\begin{aligned} f(\boldsymbol{x}) \quad \longrightarrow \quad & G_\phi f(\boldsymbol{\omega}, \boldsymbol{b}) = \langle f, \phi_{\boldsymbol{\omega},\boldsymbol{b}} \rangle_{L^2(\mathbb{R}^2;\mathbb{H})} \\ = \quad & \int_{\mathbb{R}^2} f(\boldsymbol{x})\, \overline{\phi_{\boldsymbol{\omega},\boldsymbol{b}}(\boldsymbol{x})}\, d^2\boldsymbol{x} \\ = \quad & \frac{1}{(2\pi)^2} \int_{\mathbb{R}^2} f(\boldsymbol{x})\, \overline{e^{\boldsymbol{j}\omega_2 x_2} e^{\boldsymbol{i}\omega_1 x_1} \phi(\boldsymbol{x} - \boldsymbol{b})} d^2\boldsymbol{x} \\ = \quad & \frac{1}{(2\pi)^2} \int_{\mathbb{R}^2} f(\boldsymbol{x})\, \overline{\phi(\boldsymbol{x} - \boldsymbol{b})} e^{-\boldsymbol{i}\omega_1 x_1} e^{-\boldsymbol{j}\omega_2 x_2} d^2\boldsymbol{x}. \end{aligned} \tag{4.3.13}$$

Please note that the order of the exponentials in (4.3.13) is fixed because of the non-commutativity of the product of quaternions. Changing the order yields another quaternion valued function which differs by the signs of the terms. Equation (4.3.13) clearly shows that the QWFT can be regarded as the (right-sided) QFT (compare (5.5.35)) of the product of a quaternion-valued signal f and a shifted and quaternion conjugate version of the quaternion window function or as an inner product (4.3.1) of f and the quaternionic window daughter function. In contrast to the QFT basis $e^{-\boldsymbol{i}\omega_1 x_1} e^{-\boldsymbol{j}\omega_1 x_2}$ which has an infinite spatial extension, the QWFT basis $\phi(\boldsymbol{x} - \boldsymbol{b})\, e^{-\boldsymbol{i}\omega_1 x_1} e^{-\boldsymbol{j}\omega_1 x_2}$ has a limited spatial extension due to the local quaternion window function $\phi(\boldsymbol{x} - \boldsymbol{b})$.

The energy density is defined as the modulus square of the QWFT (4.3.13) given

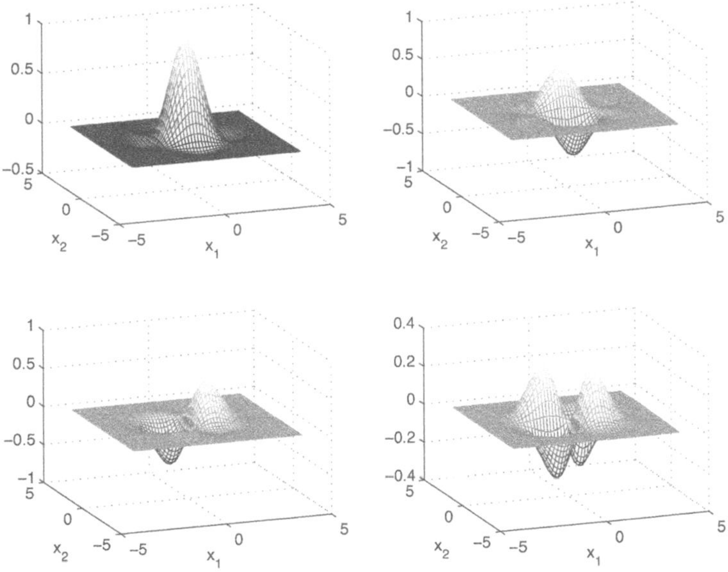

Figure 4.9 The real part (*top left*) and imaginary $\boldsymbol{i}$-part (*top right*), $\boldsymbol{j}$-part (*bottom left*), and $\boldsymbol{k}$-part (*bottom right*) of a quaternionic Gabor filter ($\sigma_1 = \sigma_2 = 1, u_0 = v_0 = 1$) in the spatial domain. Source: [256, Fig. 3].

by

$$|G_\phi f(\boldsymbol{\omega}, \boldsymbol{b})|^2 = \frac{1}{(2\pi)^4} \left| \int_{\mathbb{R}^2} f(\boldsymbol{x}) \, \overline{\phi(\boldsymbol{x} - \boldsymbol{b})} e^{-\boldsymbol{i}\omega_1 x_1} e^{-\boldsymbol{j}\omega_2 x_2} \, d^2\boldsymbol{x} \right|^2 . \tag{4.3.14}$$

Equation (4.3.14) is often called a spectrogram which measures the energy of a quaternion-valued function f in the position-frequency neighborhood of $(\boldsymbol{b}, \boldsymbol{\omega})$.

A good choice for the window function ϕ is the Gaussian quaternion function because, according to Heisenberg's uncertainty principle, the Gaussian quaternion signal can simultaneously minimize the spread in both spatial and quaternionic frequency domains, and it is smooth in both domains. The uncertainty principle can be written in the following form [255]

$$\triangle g_{x_1} \triangle g_{x_2} \triangle g_{\omega_1} \triangle g_{\omega_2} \geq \frac{1}{4}, \tag{4.3.15}$$

where $\triangle g_{x_k}$, $k = 1, 2$, are the effective spatial widths of the quaternion function g and $\triangle g_{\omega_k}$, $k = 1, 2$, are its effective bandwidths.

4.3.1.2.3 Examples of the QWFT

For illustrative purposes, we shall discuss examples of the QWFT. We begin with a straightforward example.

Example 4.3.6. *Consider the two-dimensional first order B-spline window function (see [313]) defined by*

$$\phi(\boldsymbol{x}) = \begin{cases} 1, & \text{if } -1 \leq x_1 \leq 1 \text{ and } -1 \leq x_2 \leq 1, \\ 0, & \text{otherwise.} \end{cases} \tag{4.3.16}$$

Obtain the QWFT of the function defined as follows:

$$f(\boldsymbol{x}) = \begin{cases} e^{x_1+x_2}, & \text{if } -\infty < x_1 < 0 \text{ and } -\infty < x_2 < 0, \\ 0, & \text{otherwise.} \end{cases} \tag{4.3.17}$$

By applying the definition of the QWFT we have

$$G_\phi f(\boldsymbol{\omega}, \boldsymbol{b}) = \frac{1}{(2\pi)^2} \int_{-1+b_1}^{m_1} \int_{-1+b_2}^{m_2} e^{x_1+x_2} e^{-\boldsymbol{i}\omega_1 x_1} e^{-\boldsymbol{j}\omega_2 x_2} dx_1 dx_2,$$
$$m_1 = \min(0, 1+b_1), \quad m_2 = \min(0, 1+b_2). \tag{4.3.18}$$

Simplifying (4.3.18) yields

$$\begin{aligned} G_\phi f(\boldsymbol{\omega}, \boldsymbol{b}) &= \frac{1}{(2\pi)^2} \int_{-1+b_1}^{m_1} \int_{-1+b_2}^{m_2} e^{x_1(1-\boldsymbol{i}\omega_1)} e^{x_2(1-\boldsymbol{j}\omega_2)}\, d^2\boldsymbol{x} \qquad (4.3.19) \\ &= \frac{1}{(2\pi)^2} \int_{-1+b_1}^{m_1} e^{x_1(1-\boldsymbol{i}\omega_1)} dx_1 \int_{-1+b_2}^{m_2} e^{x_2(1-\boldsymbol{j}\omega_2)}\, dx_2 \\ &= \frac{1}{(2\pi)^2} e^{x_1(1-\boldsymbol{i}\omega_1)} (1-\boldsymbol{i}\omega_1) \Big|_{-1+b_1}^{m_1} \frac{e^{x_2(1-\boldsymbol{j}\omega_2)}}{(1-\boldsymbol{j}\omega_2)} \Big|_{-1+b_2}^{m_2} \\ &= \frac{(e^{m_1(1-\boldsymbol{i}\omega_1)} - e^{(-1+b_1)(1-\boldsymbol{i}\omega_1)})(e^{m_2(1-\boldsymbol{j}\omega_2)} - e^{(-1+b_2)(1-\boldsymbol{j}\omega_2)})}{(2\pi)^2(1-\boldsymbol{i}\omega_1 - \boldsymbol{j}\omega_2 + \boldsymbol{k}\omega_1\omega_2)}. \end{aligned}$$

Using the properties of quaternions we obtain

$$\begin{aligned} G_\phi f(\boldsymbol{\omega}, \boldsymbol{b}) = &\frac{(e^{m_1(1-\boldsymbol{i}\omega_1)} - e^{(-1+b_1)(1-\boldsymbol{i}\omega_1)})(e^{m_2(1-\boldsymbol{j}\omega_2)} - e^{(-1+b_2)(1-\boldsymbol{j}\omega_2)})}{(2\pi)^2} \times \\ &\times \frac{(1+\boldsymbol{i}\omega_1 + \boldsymbol{j}\omega_2 - \boldsymbol{k}\omega_1\omega_2)}{(1+\omega_1^2+\omega_2^2+\omega_1^2\omega_2^2)}. \end{aligned} \tag{4.3.20}$$

Example 4.3.7. *Given the window function of the two-dimensional Haar function defined by*

$$\phi(\boldsymbol{x}) = \begin{cases} 1, & \text{for } 0 \le x_1 < 1/2 \text{ and } 0 \le x_2 < 1/2, \\ -1, & \text{for } 1/2 \le x_1 < 1 \text{ and } 1/2 \le x_2 < 1, \\ 0, & \text{otherwise,} \end{cases} \tag{4.3.21}$$

find the QWFT of the Gaussian function $f(\boldsymbol{x}) = e^{-(x_1^2+x_2^2)}$.

From Definition 4.3.5 we obtain

$$\begin{aligned} G_\phi f(\boldsymbol{\omega}, \boldsymbol{b}) &= \frac{1}{(2\pi)^2} \int_{\mathbb{R}^2} f(\boldsymbol{x}) \overline{\phi(\boldsymbol{x}-\boldsymbol{b})} e^{-\boldsymbol{i}\omega_1 x_1} e^{-\boldsymbol{j}\omega_2 x_2} d^2\boldsymbol{x} \\ &= \frac{1}{(2\pi)^2} \int_{b_1}^{1/2+b_1} e^{-x_1^2} e^{-\boldsymbol{i}\omega_1 x_1}\, dx_1 \int_{b_2}^{1/2+b_2} e^{-x_2^2} e^{-\boldsymbol{j}\omega_2 x_2}\, dx_2 \\ &\quad - \frac{1}{(2\pi)^2} \int_{1/2+b_1}^{1+b_1} e^{-x_1^2} e^{-\boldsymbol{i}\omega_1 x_1}\, dx_1 \int_{1/2+b_2}^{1+b_2} e^{-x_2^2} e^{-\boldsymbol{j}\omega_2 x_2}\, dx_2. \end{aligned} \tag{4.3.22}$$

By completing squares, we have

$$G_\phi f(\boldsymbol{\omega},\boldsymbol{b}) = \frac{1}{(2\pi)^2}\int_{b_1}^{1/2+b_1} e^{-(x_1+\boldsymbol{i}\omega_1/2)^2-\omega_1^2/4}dx_1 \int_{b_2}^{1/2+b_2} e^{-(x_2+\boldsymbol{j}\omega_2/2)^2-\omega_2^2/4}dx_2 - \frac{1}{(2\pi)^2}\int_{1/2+b_1}^{1+b_1} e^{-(x_1+\boldsymbol{i}\omega_1/2)^2-\omega_1^2/4}dx_1 \int_{1/2+b_2}^{1+b_2} e^{-(x_2+\boldsymbol{j}\omega_2/2)^2-\omega_2^2/4}dx_2. \tag{4.3.23}$$

Making the substitutions $y_1 = x_1 + \boldsymbol{i}\,\frac{\omega_1}{2}$ and $y_2 = x_2 + \boldsymbol{j}\,\frac{\omega_2}{2}$ in the above expression we immediately obtain

$$\begin{aligned} G_\phi f(\boldsymbol{\omega},\boldsymbol{b}) &= \frac{e^{-(\omega_1^2+\omega_2^2)/4}}{(2\pi)^2}\int_{b_1+\boldsymbol{i}\omega_1/2}^{1/2+b_1+\boldsymbol{i}\omega_1/2} e^{-y_1^2}\,dy_1 \int_{b_2+\boldsymbol{j}\omega_2/2}^{1/2+b_2+\boldsymbol{j}\omega_2/2} e^{-y_2^2}\,dy_2 \\ &\quad - \frac{e^{-(\omega_1^2+\omega_2^2)/4}}{(2\pi)^2}\int_{1/2+b_1+\boldsymbol{i}\omega_1/2}^{1+b_1+\boldsymbol{i}\omega_1/2} e^{-y_1^2}\,dy_1 \int_{1/2+b_2+\boldsymbol{j}\omega_2/2}^{1+b_2+\boldsymbol{j}\omega_2/2} e^{-y_2^2}\,dy_2 \\ &= \frac{e^{-(\omega_1^2+\omega_2^2)/4}}{(2\pi)^2}\Bigg[\left(\int_0^{b_1+\boldsymbol{i}\omega_1/2}(-e^{-y_1^2})\,dy_1 + \int_0^{1/2+b_1+\boldsymbol{i}\omega_1/2} e^{-y_1^2}\,dy_1\right) \\ &\quad \times \left(\int_0^{b_2+\boldsymbol{j}\omega_2/2}(-e^{-y_2^2})\,dy_2 + \int_0^{1/2+b_2+\boldsymbol{j}\omega_2/2} e^{-y_2^2}\,dy_2\right) \\ &\quad - \left(\int_0^{1/2+b_1+\boldsymbol{i}\omega_1/2}(-e^{-y_1^2})\,dy_1 + \int_0^{1+b_1+\boldsymbol{i}\omega_1/2} e^{-y_1^2}\,dy_1\right) \\ &\quad \times \left(\int_0^{1/2+b_2+\boldsymbol{j}\omega_2/2}(-e^{-y_2^2})\,dy_2 + \int_0^{1+b_2+\boldsymbol{j}\omega_2/2} e^{-y_2^2}\,dy_2\right)\Bigg]. \end{aligned} \tag{4.3.24}$$

Equation (4.3.24) can be written in the form

$$\begin{aligned} G_\phi f(\boldsymbol{\omega},\boldsymbol{b}) = \frac{e^{-(\omega_1^2+\omega_2^2)/4}}{(2\sqrt{\pi})^3}\Bigg\{&\left[-\mathrm{erf}\left(b_1+\frac{\boldsymbol{i}}{2}\omega_1\right) + \mathrm{erf}\left(\frac{1}{2}+b_1+\frac{\boldsymbol{i}}{2}\omega_1\right)\right] \\ &\times\left[-\mathrm{erf}\left(b_2+\frac{\boldsymbol{j}}{2}\omega_2\right) + \mathrm{erf}\left(\frac{1}{2}+b_2+\frac{\boldsymbol{j}}{2}\omega_2\right)\right] \\ &-\left[-\mathrm{erf}\left(\frac{1}{2}+b_1+\frac{\boldsymbol{i}}{2}\omega_1\right) + \mathrm{erf}\left(1+b_1+\frac{\boldsymbol{i}}{2}\omega_1\right)\right] \\ &\times\left[-\mathrm{erf}\left(\frac{1}{2}+b_2+\frac{\boldsymbol{j}}{2}\omega_2\right) + \mathrm{erf}\left(1+b_2+\frac{\boldsymbol{j}}{2}\omega_2\right)\right]\Bigg\}, \end{aligned} \tag{4.3.25}$$

where $\mathrm{erf}(x) = \frac{2}{\sqrt{\pi}}\int_0^x e^{-t^2}\,dt$.

4.3.1.2.4 Properties of the QWFT In this subsection, we describe the properties of the QWFT. We must exercise care in extending the properties of the WFT to the QWFT because of the general non-commutativity of quaternion multiplication. We will find most of the properties of the WFT are still valid for the QWFT, however with some modifications.

Theorem 4.3.8 (Left linearity). *Let $\phi \in L^2(\mathbb{R}^2;\mathbb{H})$ be a quaternion window function. The QWFT of $f, g \in L^2(\mathbb{R}^2;\mathbb{H})$ is a left linear operator, which means*

$$[G_\phi(\lambda f + \boldsymbol{\mu} g)](\boldsymbol{\omega}, \boldsymbol{b}) = \lambda G_\phi f(\boldsymbol{\omega}, \boldsymbol{b}) + \boldsymbol{\mu} G_\phi g(\boldsymbol{\omega}, \boldsymbol{b}), \tag{4.3.26}$$

for arbitrary quaternion constants $\lambda, \boldsymbol{\mu} \in \mathbb{H}$.

Remark 4.3.9. *Restricting the constants in Theorem 4.3.8 to $\lambda, \boldsymbol{\mu} \in \mathbb{R}$ we get both left and right linearity of the QWFT.*

Theorem 4.3.10 (Parity). *Let $\phi \in L^2(\mathbb{R}^2;\mathbb{H})$ be a quaternion window function. Then we have*

$$G_{P\phi}\{Pf\}(\boldsymbol{\omega}, \boldsymbol{b}) = G_\phi f(-\boldsymbol{\omega}, -\boldsymbol{b}), \tag{4.3.27}$$

where $P\phi(\boldsymbol{x}) = \phi(-\boldsymbol{x}), \forall \phi \in L^2(\mathbb{R}^2;\mathbb{H})$.

Theorem 4.3.11 (Specific shift). *Let ϕ be a quaternion window function. Assume that*

$$f = f_0 + \boldsymbol{i} f_1 \quad \text{and} \quad \phi = \phi_0 + \boldsymbol{i}\phi_1. \tag{4.3.28}$$

Then we obtain

$$G_\phi T_{\boldsymbol{x}_0} f(\boldsymbol{\omega}, \boldsymbol{b}) = e^{-\boldsymbol{i}\omega_1 x_0} \left(G_\phi f(\boldsymbol{\omega}, \boldsymbol{b} - \boldsymbol{x}_0)\right) e^{-\boldsymbol{j}\omega_2 y_0}, \tag{4.3.29}$$

where $T_{\boldsymbol{x}_0}$ denotes the translation operator by $\boldsymbol{x}_0 = x_0\boldsymbol{e}_1 + y_0\boldsymbol{e}_2$, i.e. $T_{\boldsymbol{x}_0} f = f(\boldsymbol{x} - \boldsymbol{x}_0)$.

Equation (4.3.29) describes that if the original function $f(\boldsymbol{x})$ is shifted by $\boldsymbol{x}_0$, its window function will be shifted by $\boldsymbol{x}_0$, the frequency will remain unchanged, and the phase will be changed by the left and right phase factors $e^{-\boldsymbol{i}\omega_1 x_0}$ and $e^{-\boldsymbol{j}\omega_2 y_0}$.

Remark 4.3.12. *Like for the (right-sided) QFT, the usual form of the modulation property of the QWFT does not hold [184, 255]. It is obstructed by the non-commutativity of the quaternion exponential product factors*

$$e^{-\boldsymbol{i}\omega_1 x_1} e^{-\boldsymbol{j}\omega_2 x_2} \neq e^{-\boldsymbol{j}\omega_2 x_2} e^{-\boldsymbol{i}\omega_1 x_1}. \tag{4.3.30}$$

The following theorem tells us that the QWFT is invertible, that is, the original quaternion signal f can be recovered simply by taking the inverse QWFT.

Theorem 4.3.13 (Reconstruction formula). *Let ϕ be a quaternion window function. Then every 2-D quaternion signal $f \in L^2(\mathbb{R}^2;\mathbb{H})$ can be fully reconstructed by*

$$f(\boldsymbol{x}) = \frac{(2\pi)^2}{\|\phi\|^2_{L^2(\mathbb{R}^2;\mathbb{H})}} \int_{\mathbb{R}^2}\int_{\mathbb{R}^2} G_\phi f(\boldsymbol{\omega}, \boldsymbol{b}) \phi_{\boldsymbol{\omega},\boldsymbol{b}}(\boldsymbol{x})\, d^2\boldsymbol{b}\, d^2\boldsymbol{\omega}. \tag{4.3.31}$$

Set $C_\phi = \|\phi\|^2_{L^2(\mathbb{R}^2;\mathbb{H})}$ and assume that $0 < C_\phi < \infty$. Then, the reconstruction formula (4.3.31) can also be written as

$$\begin{aligned} f(\boldsymbol{x}) &= \frac{(2\pi)^2}{C_\phi} \int_{\mathbb{R}^2}\int_{\mathbb{R}^2} G_\phi f(\boldsymbol{\omega}, \boldsymbol{b})\, \phi_{\boldsymbol{\omega},\boldsymbol{b}}\, d^2\boldsymbol{b}\, d^2\boldsymbol{\omega} \\ &= \frac{(2\pi)^2}{C_\phi} \int_{\mathbb{R}^2}\int_{\mathbb{R}^2} \langle f, \phi_{\boldsymbol{\omega},\boldsymbol{b}}\rangle_{L^2(\mathbb{R}^2;\mathbb{H})} \phi_{\boldsymbol{\omega},\boldsymbol{b}}\, d^2\boldsymbol{b}\, d^2\boldsymbol{\omega}. \end{aligned} \tag{4.3.32}$$

More properties of the QWFT are given in the following theorems.

Theorem 4.3.14 (Orthogonality relation). *Let ϕ be a quaternion window function and $f, g \in L^2(\mathbb{R}^2;\mathbb{H})$ arbitrary. Then we have*

$$\int_{\mathbb{R}^2}\int_{\mathbb{R}^2} \langle f, \phi_{\boldsymbol{\omega},\boldsymbol{b}}\rangle_{L^2(\mathbb{R}^2;\mathbb{H})} \overline{\langle g, \phi_{\boldsymbol{\omega},\boldsymbol{b}}\rangle_{L^2(\mathbb{R}^2;\mathbb{H})}} d^2\boldsymbol{\omega}\, d^2\boldsymbol{b} = \frac{C_\phi}{(2\pi)^2}\langle f, g\rangle_{L^2(\mathbb{R}^2;\mathbb{H})}. \tag{4.3.33}$$

As an easy consequence of the previous theorem, we immediately obtain the following corollary.

Corollary 4.3.15. *If $f, \phi \in L^2(\mathbb{R}^2;\mathbb{H})$ are two quaternion-valued signals, then*

$$\int_{\mathbb{R}^2}\int_{\mathbb{R}^2} |G_\phi f(\boldsymbol{\omega}, \boldsymbol{b})|^2\, d^2\boldsymbol{b}\, d^2\boldsymbol{\omega} = \frac{1}{(2\pi)^2}\|f\|^2_{L^2(\mathbb{R}^2;\mathbb{H})}\|\phi\|^2_{L^2(\mathbb{R}^2;\mathbb{H})}. \tag{4.3.34}$$

In particular, if the quaternion window function is normalized to $\|\phi\|_{L^2(\mathbb{R}^2;\mathbb{H})} = 1$, then (4.3.34) becomes

$$\int_{\mathbb{R}^2}\int_{\mathbb{R}^2} |G_\phi f(\boldsymbol{\omega}, \boldsymbol{b})|^2\, d^2\boldsymbol{b}\, d^2\boldsymbol{\omega} = \frac{1}{(2\pi)^2}\|f\|^2_{L^2(\mathbb{R}^2;\mathbb{H})}. \tag{4.3.35}$$

Equation (4.3.35) shows that the QWFT is an *isometry* from $L^2(\mathbb{R}^2;\mathbb{H})$ into $L^2(\mathbb{R}^2;\mathbb{H})$. In other words, up to a factor of $\frac{1}{(2\pi)^2}$ the *total energy* of a quaternion-valued signal computed in the spatial domain is equal to the total energy computed in the quaternionic windowed Fourier domain, compare (4.3.14) for the corresponding energy density.

Theorem 4.3.16 (Reproducing kernel). *Let be $\phi \in L^2(\mathbb{R}^2;\mathbb{H})$ be a quaternion window function. If*

$$\mathbb{K}_\phi(\boldsymbol{\omega}, \boldsymbol{b}; \boldsymbol{\omega}', \boldsymbol{b}') = \frac{(2\pi)^2}{C_\phi}\langle \phi_{\boldsymbol{\omega},\boldsymbol{b}}, \phi_{\boldsymbol{\omega}',\boldsymbol{b}'}\rangle_{L^2(\mathbb{R}^2;\mathbb{H})}, \tag{4.3.36}$$

then $\mathbb{K}_\phi(\boldsymbol{\omega}, \boldsymbol{b}; \boldsymbol{\omega}', \boldsymbol{b}')$ is a reproducing kernel, i.e.

$$G_\phi f(\boldsymbol{\omega}', \boldsymbol{b}') = \int_{\mathbb{R}^2}\int_{\mathbb{R}^2} G_\phi f(\boldsymbol{\omega}, \boldsymbol{b})\mathbb{K}_\phi(\boldsymbol{\omega}, \boldsymbol{b}; \boldsymbol{\omega}', \boldsymbol{b}')\, d^2\boldsymbol{\omega}\, d^2\boldsymbol{b} \tag{4.3.37}$$

The above properties of the QWFT are summarized in Table 4.3.

4.3.1.3 Heisenberg's uncertainty principle for the QWFT

The classical uncertainty principle of harmonic analysis states that a non-trivial function and its Fourier transform can not both be simultaneously sharply localized [76, 319]. In quantum mechanics an uncertainty principle asserts one can not at the same time be certain of the position and of the velocity of an electron (or any particle). That is, increasing the knowledge of the position decreases the knowledge of the velocity or momentum of an electron. This section extends the uncertainty

Table 4.3 Properties of the QWFT of $f, g \in L^2(\mathbb{R}^2;\mathbb{H})$, where $\lambda, \boldsymbol{\mu} \in \mathbb{H}$ are constants and $\boldsymbol{x}_0 = x_0\boldsymbol{e}_1 + y_0\boldsymbol{e}_2 \in \mathbb{R}^2$. Source: [256, Tab. 1].

Property	Quat. Function	QWFT
Left linearity	$\lambda f(\boldsymbol{x}) + \boldsymbol{\mu} g(\boldsymbol{x})$	$\lambda G_\phi f(\boldsymbol{\omega}, \boldsymbol{b}) + \boldsymbol{\mu} G_\phi g(\boldsymbol{\omega}, \boldsymbol{b})$
Parity	$G_{P\phi}\{Pf\}(\boldsymbol{\omega}, \boldsymbol{b})$	$G_\phi f(-\boldsymbol{\omega}, -\boldsymbol{b})$
Specific shift	$f(\boldsymbol{x} - \boldsymbol{x}_0)$	$e^{-\boldsymbol{i}\omega_1 x_0}\, G_\phi f(\boldsymbol{\omega}, \boldsymbol{b} - \boldsymbol{x}_0)\, e^{-\boldsymbol{j}\omega_2 y_0}$, if $f = f_0 + \boldsymbol{i} f_1$ and $\phi = \phi_0 + \boldsymbol{i}\phi_1$
Formula		
Orthogonality	$\frac{\Vert\phi\Vert^2_{L^2(\mathbb{R}^2;\mathbb{H})}}{(2\pi)^2} \langle f, g\rangle_{L^2(\mathbb{R}^2;\mathbb{H})} =$	$\int_{\mathbb{R}^2}\int_{\mathbb{R}^2} \langle f, \phi_{\boldsymbol{\omega},\boldsymbol{b}}\rangle_{L^2(\mathbb{R}^2;\mathbb{H})} \overline{\langle g, \phi_{\boldsymbol{\omega},\boldsymbol{b}}\rangle_{L^2(\mathbb{R}^2;\mathbb{H})}}\, d^2\boldsymbol{\omega}\, d^2\boldsymbol{b}$
Reconstruction	$f(\boldsymbol{x}) =$	$\frac{(2\pi)^2}{\Vert\phi\Vert^2_{L^2(\mathbb{R}^2;\mathbb{H})}} \int_{\mathbb{R}^2}\int_{\mathbb{R}^2} G_\phi f(\boldsymbol{\omega}, \boldsymbol{b})\, \phi_{\boldsymbol{\omega},\boldsymbol{b}}(\boldsymbol{x})\, d^2\boldsymbol{b}\, d^2\boldsymbol{\omega}$
Isometry	$\frac{1}{(2\pi)^2}\Vert f\Vert^2_{L^2(\mathbb{R}^2;\mathbb{H})} =$	$\int_{\mathbb{R}^2}\int_{\mathbb{R}^2} \vert G_\phi f(\boldsymbol{\omega}, \boldsymbol{b})\vert^2\, d^2\boldsymbol{b}\, d^2\boldsymbol{\omega}$, if $\Vert\phi\Vert_{L^2(\mathbb{R}^2;\mathbb{H})} = 1$
Reproducing Kernel	$G_\phi f(\boldsymbol{\omega}', \boldsymbol{b}') =$	$\int_{\mathbb{R}^2}\int_{\mathbb{R}^2} G_\phi f(\boldsymbol{\omega}, \boldsymbol{b})\, \mathbb{K}_\phi(\boldsymbol{\omega}, \boldsymbol{b}; \boldsymbol{\omega}', \boldsymbol{b}')\, d^2\boldsymbol{\omega}\, d^2\boldsymbol{b}$, $\mathbb{K}_\phi(\boldsymbol{\omega}, \boldsymbol{b}; \boldsymbol{\omega}', \boldsymbol{b}') = \frac{(2\pi)^2}{\Vert\phi\Vert^2_{L^2(\mathbb{R}^2;\mathbb{H})}} \langle \phi_{\boldsymbol{\omega},\boldsymbol{b}}, \phi_{\boldsymbol{\omega}',\boldsymbol{b}'}\rangle_{L^2(\mathbb{R}^2;\mathbb{H})}$

principle which is valid for the (right-sided) QFT [255] to the setting of the QWFT. A directional QFT uncertainty principle has been studied in [187].

In [255] a component-wise uncertainty principle for the QFT establishes a lower bound on the product of the effective widths of quaternion-valued signals in the spatial and frequency domains. This uncertainty can be written in the following form.

Theorem 4.3.17 (QFT uncertainty principle). *Let $f \in L^2(\mathbb{R}^2;\mathbb{H})$ be a quaternion-valued function. If $\mathcal{F}_q\{f\}(\boldsymbol{\omega}) \in L^2(\mathbb{R}^2;\mathbb{H})$ too, then we have the inequality (no summation over k, $k = 1, 2$)*

$$\int_{\mathbb{R}^2} x_k^2 |f(\boldsymbol{x})|^2\, d^2\boldsymbol{x} \int_{\mathbb{R}^2} \omega_k^2 |\mathcal{F}_q\{f\}(\boldsymbol{\omega})|^2 d^2\boldsymbol{\omega} \geq \frac{(2\pi)^2}{4}\left(\int_{\mathbb{R}^2} |f(\boldsymbol{x})|^2 d^2\boldsymbol{x}\right)^2. \tag{4.3.38}$$

Equality holds if and only if f is the Gaussian quaternion function, i.e.

$$f(\boldsymbol{x}) = C_0\, e^{-(a_1 x_1^2 + a_2 x_2^2)}, \tag{4.3.39}$$

where C_0 is a quaternion constant and a_1, a_2 are positive real constants.

Applying the Parseval theorem for the QFT [184] to the right-hand side of (4.3.38) we get the following corollary.

Corollary 4.3.18. *Under the above assumptions, we have*

$$\int_{\mathbb{R}^2} x_k^2 |\mathcal{F}_q^{-1}[\mathcal{F}_q\{f\}](\boldsymbol{x})|^2 \, d^2\boldsymbol{x} \int_{\mathbb{R}^2} \omega_k^2 |\mathcal{F}_q\{f\}(\boldsymbol{\omega})|^2 d^2\boldsymbol{\omega} \geq \left(\frac{1}{4\pi} \int_{\mathbb{R}^2} |\mathcal{F}_q\{f\}(\boldsymbol{\omega})|^2 \, d^2\boldsymbol{\omega} \right)^2. \tag{4.3.40}$$

Let us now establish a generalization of the Heisenberg type uncertainty principle for the QWFT. From a mathematical point of view this principle describes how the spatial extension of a two-dimensional quaternion function relates to the bandwidth of its QWFT.

Theorem 4.3.19 (QWFT uncertainty principle)**.** *Let $\phi \in L^2(\mathbb{R}^2;\mathbb{H})$ be a quaternion window function and let $G_\phi f \in L^2(\mathbb{R}^2;\mathbb{H})$ be the QWFT of f such that $\omega_k G_\phi f \in L^2(\mathbb{R}^2;\mathbb{H})$, $k = 1, 2$. Then for every $f \in L^2(\mathbb{R}^2;\mathbb{H})$ we have the following inequality:*

$$\left(\int_{\mathbb{R}^2} \int_{\mathbb{R}^2} \omega_k^2 |G_\phi f(\boldsymbol{\omega}, \boldsymbol{b})|^2 \, d^2\boldsymbol{\omega} \, d^2\boldsymbol{b} \right)^{1/2} \left(\int_{\mathbb{R}^2} x_k^2 |f(\boldsymbol{x})|^2 \, d^2\boldsymbol{x} \right)^{1/2} \geq \frac{1}{4\pi} \|f\|^2_{L^2(\mathbb{R}^2;\mathbb{H})} \|\phi\|_{L^2(\mathbb{R}^2;\mathbb{H})}. \tag{4.3.41}$$

In order to prove this theorem, we need to introduce the following lemma.

Lemma 4.3.20. *Under the assumptions of Theorem 4.3.19, we have for $k = 1, 2$ that*

$$\frac{\|\phi\|^2_{L^2(\mathbb{R}^2;\mathbb{H})}}{(2\pi)^4} \int_{\mathbb{R}^2} x_k^2 |f(\boldsymbol{x})|^2 \, d^2\boldsymbol{x} = \int_{\mathbb{R}^2} \int_{\mathbb{R}^2} x_k^2 |\mathcal{F}_q^{-1}\{G_\phi f(\boldsymbol{\omega}, \boldsymbol{b})\}(\boldsymbol{x})|^2 \, d^2\boldsymbol{x} d^2\boldsymbol{b}. \tag{4.3.42}$$

Remark 4.3.21. *According to the properties of the QFT and its uncertainty principle, Theorem 4.3.19 does not hold for summation over k. If we introduce summation, we would have to replace the factor $\frac{1}{4\pi}$ on the right hand side of (4.3.41) by $\frac{1}{2\pi}$.*

4.3.1.4 Application of the QWFT

The WFT plays a fundamental role in the analysis of signals and linear time-varying (TV) systems [140,147,313]. The effectiveness of the WFT is a result of its providing a unique representation for the signals in terms of the windowed Fourier kernel. It is natural to ask whether the QWFT can also be applied to such problems. This section briefly discusses the application of the QWFT to study two-dimensional linear TV systems (see Figure 4.10). We may regard the QWFT as a linear TV band-pass filter element of a filter-bank spectrum analyzer and, therefore, the TV spectrum obtained by the QWFT can also be interpreted as the output of such a linear TV band-pass filter element. For this purpose let us introduce the following definition.

Definition 4.3.22. *Consider a two-dimensional linear TV system with $h(\cdot,\cdot,\cdot)$ denoting the quaternion impulse response of the filter. The output $r(\cdot,\cdot)$ of the linear TV system is defined by*

$$r(\boldsymbol{\omega}, \boldsymbol{b}) = \int_{\mathbb{R}^2} f(\boldsymbol{x}) h(\boldsymbol{\omega}, \boldsymbol{b}, \boldsymbol{b} - \boldsymbol{x}) \, d^2\boldsymbol{x} = \int_{\mathbb{R}^2} f(\boldsymbol{b} - \boldsymbol{x}) h(\boldsymbol{\omega}, \boldsymbol{b}, \boldsymbol{x}) \, d^2\boldsymbol{x}, \tag{4.3.43}$$

where $f(\cdot)$ is a two-dimensional quaternion valued input signal.

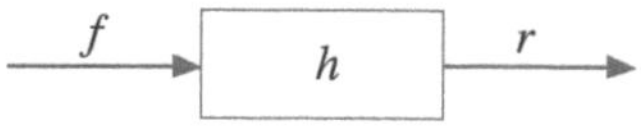

Figure 4.10 Block diagram of a two-dimensional linear time-varying system. Source: [256, Fig. 4].

We then obtain the transfer function $R(\cdot,\cdot)$ of the quaternion impulse response $h(\cdot,\cdot,\cdot)$ of the TV filter as

$$R(\boldsymbol{\omega},\boldsymbol{b}) = \int_{\mathbb{R}^2} h(\boldsymbol{\omega},\boldsymbol{b},\boldsymbol{\alpha})\, e^{-\boldsymbol{i}\omega_1\alpha_1} e^{-\boldsymbol{j}\omega_2\alpha_2} d^2\boldsymbol{\alpha}, \quad \boldsymbol{\alpha} = \alpha_1\boldsymbol{e}_1 + \alpha_2\boldsymbol{e}_2 \in \mathbb{R}^2. \tag{4.3.44}$$

The following simple theorem (compare to Ghosh and Sreenivas [140]) relates the QWFT to the output of a linear TV band-pass filter.

Theorem 4.3.23. *Consider a linear TV band-pass filter. Let the TV quaternion impulse response $h_1(\cdot,\cdot,\cdot)$ of the filter be defined by*

$$h_1(\boldsymbol{\omega},\boldsymbol{b},\boldsymbol{\alpha}) = \frac{1}{(2\pi)^2}\,\overline{\phi(-\boldsymbol{\alpha})}\, e^{-\boldsymbol{i}\omega_1(b_1-\alpha_1)} e^{-\boldsymbol{j}\omega_2(b_2-\alpha_2)}, \tag{4.3.45}$$

where $\phi(\cdot)$ is the quaternion window function. The output $r_1(\cdot,\cdot)$ of the TV system is equal to the QWFT of the quaternion input signal $f(\boldsymbol{x})$.

This shows that the choice of the quaternion impulse response of the filter will determine a characteristic output of the linear TV systems. For example, if we translate the TV quaternion impulse response $h_1(\cdot,\cdot,\cdot)$ by $\boldsymbol{b}_0 = b_{01}\boldsymbol{e}_1 + b_{02}\boldsymbol{e}_2$, i.e.

$$\begin{aligned} h_1(\boldsymbol{\omega},\boldsymbol{b},\boldsymbol{\alpha}) \to h_1(\boldsymbol{\omega},\boldsymbol{b},\boldsymbol{\alpha}-\boldsymbol{b}_0) &= \frac{1}{(2\pi)^2}\,\overline{\phi(-(\boldsymbol{\alpha}-\boldsymbol{b}_0))}\, e^{-\boldsymbol{i}\omega_1(b_1-(\alpha_1-b_{01}))} \\ &\quad \times e^{-\boldsymbol{j}\omega_2(b_2-(\alpha_2-b_{02}))}, \end{aligned} \tag{4.3.46}$$

then the output is according to Theorem 4.3.11

$$r_{1,\boldsymbol{b}_0}(\boldsymbol{\omega},\boldsymbol{b}) = e^{-\boldsymbol{i}\omega_1 b_{01}}\, G_\phi f(\boldsymbol{\omega},\boldsymbol{b}-\boldsymbol{b}_0)\, e^{-\boldsymbol{j}\omega_2 b_{02}}. \tag{4.3.47}$$

In this case, we assumed that the input $f\boldsymbol{i} = \boldsymbol{i}f$ and the window function $\phi\boldsymbol{i} = \boldsymbol{i}\phi$.

Theorem 4.3.24. *Consider a linear TV band-pass filter with the TV quaternion impulse response $h_2(\cdot,\cdot,\cdot)$ defined by*

$$h_2(\boldsymbol{\omega},\boldsymbol{b},\boldsymbol{\alpha}) = e^{-\boldsymbol{i}\omega_1(b_1-\alpha_1)} e^{-\boldsymbol{j}\omega_2(b_2-\alpha_2)}, \tag{4.3.48}$$

If the input to this system is the quaternion signal $f(\boldsymbol{x})$, its output $r_2(\boldsymbol{\omega}) = r_2(\boldsymbol{\omega},\cdot)$ is, independent of the $\boldsymbol{b}$-argument, equal to the QFT of f:

$$r_2(\boldsymbol{\omega}) = \mathcal{F}_q\{f\}(\boldsymbol{\omega}). \tag{4.3.49}$$

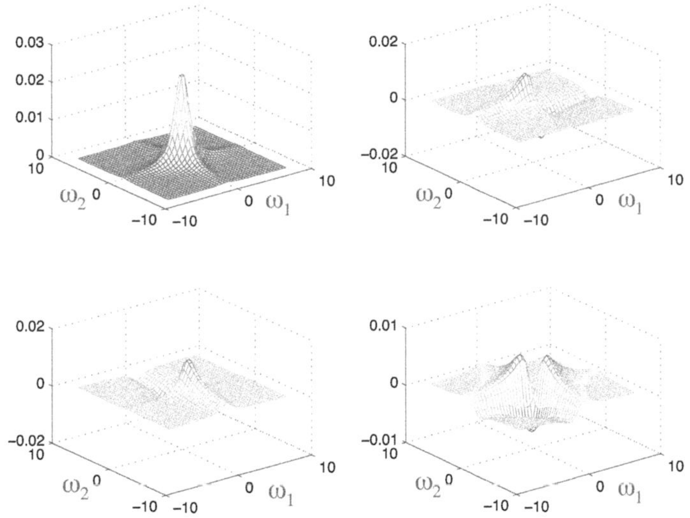

Figure 4.11 The real part (*top left*) and imaginary $\boldsymbol{i}$-part (*top right*), $\boldsymbol{j}$-part (*bottom left*), and $\boldsymbol{k}$-part (*bottom right*) of the output $r_2(\cdot)$ in Example 4.3.25, i.e. the QFT (4.3.51) of (4.3.50). Source: [256, Fig. 5].

Example 4.3.25. *Given the TV quaternion impulse response defined by (4.3.48). Find the output $r_2(\cdot)$ (see Figure 4.11) of the following input*

$$f(\boldsymbol{x}) = \begin{cases} e^{-(x_1+x_2)}, & \text{if } x_1 \geq 0 \text{ and } x_2 \geq 0, \\ 0, & \text{otherwise.} \end{cases} \tag{4.3.50}$$

From Theorem 4.3.24, we obtain the QFT of f

$$\begin{aligned} r_2(\boldsymbol{\omega}) &= \frac{1}{(2\pi)^2} \int_0^\infty \int_0^\infty e^{-x_1(1+\boldsymbol{i}\omega_1)} e^{-x_2(1+\boldsymbol{j}\omega_2)}\, d^2\boldsymbol{x} \\ &= \frac{1}{(2\pi)^2} \frac{-1}{1+\boldsymbol{i}\omega_1} e^{-\boldsymbol{i}\omega_1 x_1} e^{-x_1}\Big|_0^\infty \frac{-1}{(1+\boldsymbol{j}\omega_2)} e^{-\boldsymbol{j}\omega_2 x_2} e^{-x_2}\Big|_0^\infty \\ &= \frac{1}{(2\pi)^2} \frac{1}{1+\boldsymbol{i}\omega_1+\boldsymbol{j}\omega_2+\boldsymbol{k}\omega_1\omega_2} \\ &= \frac{1}{(2\pi)^2} \frac{1-\boldsymbol{i}\omega_1-\boldsymbol{j}\omega_2-\boldsymbol{k}\omega_1\omega_2}{1+\omega_1^2+\omega_2^2+\omega_1^2\omega_2^2}. \end{aligned} \tag{4.3.51}$$

Example 4.3.26. *Consider the TV quaternion impulse response defined by (4.3.45) with respect to the first order two-dimensional B-spline window function* (4.3.16) *in Example 4.3.6. Find the output $r_1(\cdot,\cdot)$ (see Figure 4.12) of the input* (4.3.50) *defined in Example 4.3.25.*

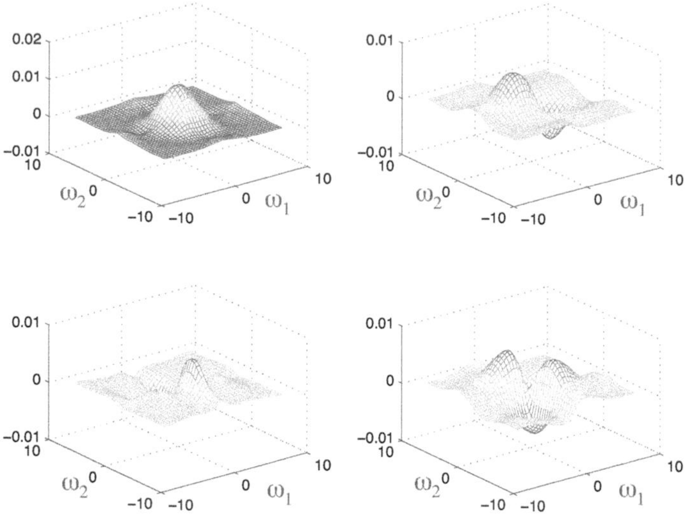

Figure 4.12 The real part (*top left*) and imaginary $\boldsymbol{i}$-part (*top right*), $\boldsymbol{j}$-part (*bottom left*), and $\boldsymbol{k}$-part (*bottom right*) of the output $r_1(\boldsymbol{\omega}, \boldsymbol{b} = 0)$ in Example 4.3.26, i.e. the QWFT (4.3.52) of (4.3.45) with $b_1 = b_2 = 0$. Source: [256, Fig. 6].

With $m_1 = \max(0, -1 + b_1), m_2 = \max(0, -1 + b_2)$, Theorem 4.3.23 gives

$$\begin{aligned} r_1(\boldsymbol{\omega}, \boldsymbol{b}) &= \mathcal{G}_\phi f(\boldsymbol{\omega}, \boldsymbol{b}) \qquad (4.3.52) \\ &= \frac{1}{(2\pi)^2} \int_{m_1}^{1+b_1} e^{-x_1} e^{-\boldsymbol{i} x_1 \omega_1} dx_1 \int_{m_2}^{1+b_2} e^{-x_2} e^{-\boldsymbol{j} x_2 \omega_2} dx_2 \\ &= \frac{-1}{(2\pi)^2 (1 + \boldsymbol{i}\omega_1)} e^{-x_1(1+\boldsymbol{i}\omega_1)} \Big|_{m_1}^{1+b_1} \frac{-1}{(1 + \boldsymbol{j}\omega_2)} e^{-x_2(1+\boldsymbol{j}\omega_1)} \Big|_{m_2}^{1+b_2} \\ &= \frac{\left(e^{-m_1(1+\boldsymbol{i}\omega_1)} - e^{-(1+b_1)(1+\boldsymbol{i}\omega_1)}\right)\left(e^{-m_2(1+\boldsymbol{j}\omega_2)} - e^{-(1+b_2)(1+\boldsymbol{j}\omega_2)}\right)}{(2\pi)^2 (1 + \boldsymbol{i}\omega_1 + \boldsymbol{j}\omega_2 + \boldsymbol{k}\omega_1\omega_2)}. \end{aligned}$$

For the sake of simplicity, we take the parameters $b_1 = b_2 = 0 \Rightarrow m_1 = m_2 = 0$, to obtain

$$\begin{aligned} r_1(\boldsymbol{\omega}, \boldsymbol{b} = 0) &= \frac{(1 - e^{-(1+\boldsymbol{i}\omega_1)})(1 - e^{-(1+\boldsymbol{j}\omega_2)})}{(2\pi)^2 (1 + \boldsymbol{i}\omega_1 + \boldsymbol{j}\omega_2 + \boldsymbol{k}\omega_1\omega_2)} \\ &= \frac{1 - e^{-1}\cos\omega_1 - e^{-1}\cos\omega_2 + e^{-2}\cos\omega_1\cos\omega_2}{(2\pi)^2 (1 + \boldsymbol{i}\omega_1 + \boldsymbol{j}\omega_2 + \boldsymbol{k}\omega_1\omega_2)} \\ &\quad + \frac{\boldsymbol{i}(e^{-1}\sin\omega_1 - e^{-2}\sin\omega_1\cos\omega_2) + \boldsymbol{j}(e^{-1}\sin\omega_2 - e^{-2}\cos\omega_1\sin\omega_2)}{(2\pi)^2 (1 + \boldsymbol{i}\omega_1 + \boldsymbol{j}\omega_2 + \boldsymbol{k}\omega_1\omega_2)} \\ &\quad + \frac{\boldsymbol{k}\, e^{-2}\sin\omega_1\sin\omega_2}{(2\pi)^2 (1 + \boldsymbol{i}\omega_1 + \boldsymbol{j}\omega_2 + \boldsymbol{k}\omega_1\omega_2)}. \qquad (4.3.53) \end{aligned}$$

We may regard the QFT (4.3.51) as the QWFT with an infinite window function. Since the integration domain of the QWFT (4.3.52) is smaller than that of the QFT (4.3.51), the QWFT output (4.3.52) is more localized in the base space than the QFT output of (4.3.51). In addition, according to the Paley-Wiener theorem the QWFT output of (4.3.52) is very smooth. This means that it provides accurate information on the output $r(\cdot,\cdot)$ due to the local window function ϕ.

4.3.1.5 Summary on windowed QFT

Using the basic concepts of quaternion algebra and the (right-sided) QFT we introduced the windowed QFT (QWFT). Important properties of the QWFT, such as left linearity, parity, reconstruction formula, reproducing kernel, isometry and orthogonality relation, were demonstrated. Because of the non-commutativity of multiplication in the quaternion algebra $\mathbb{H}$, not all properties of the classical WFT can be established for the QWFT, such as general shift and modulation properties. This generalization also enables us to construct quaternionic Gabor filters (compare to Bülow [56, 58]), which can extend the applications of the 2D complex Gabor filters to texture segmentation and disparity estimation [56].

We have established a new uncertainty principle for the QWFT. This principle is founded on the QWFT properties and the uncertainty principle for the (right-sided) QFT. We also applied the QWFT to a linear time-varying (TV) system. We showed that the output of a linear TV system can result in a QFT or a QWFT of the quaternion input signal, depending on the choice of the quaternion impulse response of the filter.

4.3.2 Quaternionic Fourier Mellin transform

4.3.2.1 Background

In this part we generalize the classical Fourier Mellin transform [109], which transforms functions f representing, e.g., a gray level image defined over a compact set of $\mathbb{R}^2$. The resulting quaternionic Fourier Mellin transform (QFMT) applies to functions $f : \mathbb{R}^2 \to \mathbb{H}$, for which $|f|$ is summable over $\mathbb{R}_+^* \times \mathbb{S}^1$ under the measure $d\theta \frac{dr}{r}$. $\mathbb{R}_+^*$ is the multiplicative group of positive and non-zero real numbers. We investigate the properties of the QFMT similar to the investigation of the quaternionic Fourier Transform (QFT) in [184, 187]. This section is based on [198].

4.3.2.2 The quaternionic Fourier Mellin transformations (QFMT)

4.3.2.2.1 Robert Hjalmar Mellin (1854–1933) Robert Hjalmar Mellin (1854–1933) [260], Figure 4.13, was a Finnish mathematician, a student of G. Mittag-Leffler and K. Weierstrass. He became the director of the Polytechnic Institute in Helsinki, and in 1908 first professor of mathematics at Technical University of Finland. He was a fervent fennoman with fiery temperament, and co-founder of the Finnish Academy of Sciences. He became known for the *Mellin transform* with major applications to the evaluation of integrals, see [284], which lists 1624 references. During his last 10 years, he tried to refute Einstein's theory of relativity as logically untenable.

Definition 4.3.27 (Classical Fourier Mellin transform (FMT)).

$$\forall (v,k) \in \mathbb{R} \times \mathbb{Z}, \quad \mathcal{M}\{h\}(v,k) = \frac{1}{2\pi} \int_0^{\infty} \int_0^{2\pi} h(r,\theta) r^{-iv} e^{-ik\theta} d\theta \frac{dr}{r}, \tag{4.3.54}$$

where $h : \mathbb{R}^2 \to \mathbb{R}$ *denotes a function representing, e.g., a gray level image defined over a compact set of* $\mathbb{R}^2$.

Well known applications are to shape recognition (independent of rotation and scale), image registration and similarity.

We now define the generalization of the FMT to quaternionic signals.

Definition 4.3.28 (Quaternionic Fourier Mellin transform (QFMT)). *Let* $f, g \in \mathbb{H} : f^2 = g^2 = -1$ *be any pair of pure unit quaternions. The quaternionic Fourier Mellin transform (QFMT) is given by*

$$\forall (v,k) \in \mathbb{R} \times \mathbb{Z},$$

$$\hat{h}(v,k) = \mathcal{M}\{h\}(v,k) = \frac{1}{2\pi} \int_0^{\infty} \int_0^{2\pi} r^{-fv} h(r,\theta) e^{-gk\theta} d\theta \frac{dr}{r}, \tag{4.3.55}$$

where $h : \mathbb{R}^2 \to \mathbb{H}$ *denotes a function from* $\mathbb{R}^2$ *into the algebra of quaternions* $\mathbb{H}$*, such that* $|h|$ *is summable over* $\mathbb{R}_+^* \times \mathbb{S}^1$ *under the measure* $d\theta \frac{dr}{r}$. $\mathbb{R}_+^*$ *is the multiplicative group of positive and non-zero real numbers.*

For $f = i$, $g = j$ we have the special case

$$\forall (v,k) \in \mathbb{Z} \times \mathbb{R},$$

$$\hat{h}(v,k) = \mathcal{M}\{h\}(v,k) = \frac{1}{2\pi} \int_0^{\infty} \int_0^{2\pi} r^{-iv} h(r,\theta) e^{-jk\theta} d\theta \frac{dr}{r}. \tag{4.3.56}$$

Note, that the $\pm$ split and the QFMT commute:

$$\mathcal{M}\{h_{\pm}\} = \mathcal{M}\{h\}_{\pm}. \tag{4.3.57}$$

Theorem 4.3.29 (Inverse QFMT). *The QFMT can be inverted by*

$$h(r,\theta) = \mathcal{M}^{-1}\{h\}(r,\theta) = \frac{1}{2\pi} \int_{-\infty}^{\infty} \sum_{k \in \mathbb{Z}} r^{fv}\, \hat{h}(v,k)\, e^{gk\theta} dv. \tag{4.3.58}$$

Figure 4.13 Robert Hjalmar Mellin (1854–1933). Image: Wikipedia.

The proof uses

$$\frac{1}{2\pi}\sum_{k\in\mathbb{Z}} e^{gk(\theta-\theta')} = \delta(\theta-\theta'), \qquad r^{fv} = e^{fv\ln r},$$
$$\frac{1}{2\pi}\int_0^{2\pi} e^{fv(\ln(r)-s)}dv = \delta(\ln(r)-s). \tag{4.3.59}$$

We now investigate the basic properties of the QFMT. First, left linearity: For $\alpha, \beta \in \{q \mid q = q_r + q_f f,\ q_r, q_f \in \mathbb{R}\}$,

$$m(r,\theta) = \alpha h_1(r,\theta) + \beta h_2(r,\theta) \implies \hat{m}(v,k) = \alpha\hat{h}_1(v,k) + \beta\hat{h}_2(v,k). \tag{4.3.60}$$

Second, right linearity: For $\alpha', \beta' \in \{q \mid q = q_r + q_g g,\ q_r, q_g \in \mathbb{R}\}$,

$$m(r,\theta) = h_1(r,\theta)\alpha' + h_2(r,\theta)\beta' \implies \hat{m}(v,k) = \hat{h}_1(v,k)\alpha' + \hat{h}_2(v,k)\beta'. \tag{4.3.61}$$

The linearity of the QFMT leads to

$$\mathcal{M}\{h\}(v,k) = \mathcal{M}\{h_- + h_+\}(v,k) = \mathcal{M}\{h_-\}(v,k) + \mathcal{M}\{h_+\}(v,k), \tag{4.3.62}$$

which gives rise to the following theorem.

Theorem 4.3.30 (Quasi-complex FMT like forms for QFMT of $h_\pm$). *The QFMT of $h_\pm$ parts of $h \in L^2(\mathbb{R}^2, \mathbb{H})$ have simple quasi-complex forms*[25]

$$\begin{aligned}\mathcal{M}\{h_\pm\} &= \frac{1}{2\pi}\int_0^\infty\int_0^{2\pi} h_\pm r^{\pm gv} e^{-gk\theta} d\theta \frac{dr}{r} \\ &= \frac{1}{2\pi}\int_0^\infty\int_0^{2\pi} r^{-fv} e^{\pm fk\theta} h_\pm d\theta \frac{dr}{r}.\end{aligned} \tag{4.3.63}$$

Theorem 4.3.30 allows to use discrete and fast software to compute the QFMT based on a pair of complex FMT transformations.

For the two split parts of the QFMT, we have the following lemma.

Lemma 4.3.31 (Modulus identities). *Due to $|q|^2 = |q_-|^2 + |q_+|^2$ we get for $f : \mathbb{R}^2 \to \mathbb{H}$ the following identities*

$$\begin{aligned}|h(r,\theta)|^2 &= |h_-(r,\theta)|^2 + |h_+(r,\theta)|^2, \\ |\mathcal{M}\{h\}(v,k)|^2 &= |\mathcal{M}\{h_-\}(v,k)|^2 + |\mathcal{M}\{h_+\}(v,k)|^2.\end{aligned} \tag{4.3.64}$$

Further properties are *scaling* and *rotation*: For $m(r,\theta) = h(ar, \theta+\phi),\ a > 0,\ 0 \leq \phi \leq 2\pi$,

$$\widehat{m}(v,k) = a^{fv}\hat{h}(v,k)e^{gk\phi}. \tag{4.3.65}$$

Moreover, we have the following magnitude identity:

$$|\widehat{m}(v,k)| = |\hat{h}(v,k)|, \tag{4.3.66}$$

[25]For quaternionic Fourier transforms on L^1 and L^2, see Footnote 2 on page 144.

i.e. the magnitude of the QFMT of a scaled and rotated quaternion signal $m(r,\theta) = h(ar,\theta+\phi)$ is identical to the magnitude of the QFMT of h. Equation (4.3.66) forms the basis for applications to rotation and scale invariant shape recognition and image registration. This may now be extended to color images, since quaternions can encode colors RGB in their $\boldsymbol{i},\boldsymbol{j},\boldsymbol{k}$ components.

The reflection at the unit circle ($r \to \frac{1}{r}$) leads to

$$m(r,\theta) = h(\frac{1}{r},\theta) \qquad \Longrightarrow \qquad \widehat{m}(v,k) = \hat{h}(-v,k). \tag{4.3.67}$$

Reversing the sense of rotation ($\theta \to -\theta$) yields

$$m(r,\theta) = h(r,-\theta) \qquad \Longrightarrow \qquad \widehat{m}(v,k) = \hat{h}(v,-k). \tag{4.3.68}$$

Regarding radial and rotary modulation we assume

$$m(r,\theta) = r^{fv_0}\, h(r,\theta)\, e^{gk_0\theta}, \qquad v_0 \in \mathbb{R},\ k_0 \in \mathbb{Z}. \tag{4.3.69}$$

Then we get

$$\widehat{m}(v,k) = \hat{h}(v-v_0, k-k_0). \tag{4.3.70}$$

4.3.2.2.2 QFMT derivatives and power scaling We note for the logarithmic derivative that $\frac{d}{d\ln r} = r\frac{d}{dr} = r\partial_r$,

$$\mathcal{M}\{(r\partial_r)^n h\}(v,k) = (fv)^n \hat{h}(v,k), \qquad n \in \mathbb{N}. \tag{4.3.71}$$

Applying the angular derivative with respect to θ we obtain

$$\mathcal{M}\{\partial_\theta^n h\}(v,k) = \hat{h}(v,k)(gk)^n, \qquad n \in \mathbb{N}. \tag{4.3.72}$$

Finally, power scaling with $\ln r$ and θ leads to

$$\mathcal{M}\{(\ln r)^m \theta^n h\}(v,k) = f^m\, \partial_v^m \partial_k^n \hat{h}(v,k)\, g^n, \qquad m,n \in \mathbb{N}. \tag{4.3.73}$$

4.3.2.2.3 QFMT Plancherel and Parseval theorems For the QFMT we have the following two theorems.

Theorem 4.3.32 (QFMT Plancherel theorem). *The scalar part of the inner product of two functions $h,m : \mathbb{R}^2 \to \mathbb{H}$ is*

$$\langle h,m\rangle = \langle \hat{h},\widehat{m}\rangle. \tag{4.3.74}$$

Theorem 4.3.33 (QFMT Parseval theorem). *Let $h : \mathbb{R}^2 \to \mathbb{H}$. Then*

$$\|h\| = \|\hat{h}\|, \qquad \|h\|^2 = \|\hat{h}\|^2 = \|\hat{h}_+\|^2 + \|\hat{h}_-\|^2. \tag{4.3.75}$$

4.3.2.3 Symmetry and kernel structures of 2D FMT, FT, QFT, QFMT

The QFMT of real signals analyzes symmetry. The following notation will be used.[26] The function h_{ee} is *even* with respect to (w.r.t.) $r \to \frac{1}{r} \iff \ln r \to -\ln r$, i.e. w.r.t. the reflection at the unit circle, and *even* w.r.t. $\theta \to -\theta$, i.e. w.r.t. reversing the sense of rotation (reflection at the $\theta = 0$ line of polar coordinates in the (r, θ)-plane). Similarly we denote by h_{eo} even-odd symmetry, by h_{oe} odd-even symmetry, and by h_{oo} odd-odd symmetry.

Let h be a real valued function $\mathbb{R}^2 \to \mathbb{R}$. The QFMT of h results in

$$\hat{h}(v,k) = \underbrace{\hat{h}_{ee}(v,k)}_{\text{real part}} + \underbrace{\hat{h}_{eo}(v,k)}_{f\text{-part}} + \underbrace{\hat{h}_{oe}(v,k)}_{g\text{-part}} + \underbrace{\hat{h}_{oo}(v,k)}_{fg\text{-part}} . \tag{4.3.76}$$

The QFMT of a real signal therefore automatically separates components with different combinations of symmetry w.r.t. reflection at the unit circle and reversal of the sense of rotation. The four components of the QFMT kernel differ by radial and angular phase shifts, see the left side of Figure 4.17. The symmetries of $r \to 1/r$ (reflection at *yellow unit circle*), and $\theta \to -\theta$ (reflection at *green line*) can be clearly seen on the right of Figure 4.17.

Figure 4.18 shows real the component of the QFMT kernel for various values of v, k, demonstrating various angular and radial resolutions. Figure 4.19 shows the real component of the QFMT kernel for $v = k = 4$ at three different scales. Similar patterns appear at all scales. Figure 4.14 shows the kernels of complex 2D Fourier transform (FT) $e^{-i(ux+vy)}$, $i \in \mathbb{C}$, and the QFT $e^{-\boldsymbol{i}ux}e^{-\boldsymbol{j}vy}$, $\boldsymbol{i}, \boldsymbol{j} \in \mathbb{H}$, taken from [56], which treats applications to 2D gray scale images. Corresponding applications to color images can be found in [127]. The 2D FT is intrinsically 1D, the QFT is intrinsically 2D, which makes it superior in disparity estimation and 2D texture segmentation, etc.

Figure 4.15 compares the kernels (real parts) of 2D complex FMT and the QFMT. Obviously the 2D QFMT can analyze genuine 2D textures better than the 2D complex FMT. Finally, Figure 4.16 compares the kernels of the QFT (left) and the QFMT (right). The scale invariant feature of the QFMT is obvious. Compared with the left side of Figure 4.15, the QFMT is obviously the linear superposition of two quasi-complex FMTs with opposite winding sense, as shown in Theorem 4.3.30.

4.3.2.4 Summary on quaternionic Fourier Mellin transformation

The algebra of quaternions allows to construct a variety of quaternionic Fourier-Mellin transformations (QFMT), dependent on the choice of $f, g \in \mathbb{H}$, $f^2 = g^2 = -1$. Further variations would be to place both kernel factors initially at the left or right of the signal $h(r, \theta)$. The whole QFMT concept can easily be generalized to Clifford algebras $Cl(p, q)$, based on the general theory of square roots of -1 in $Cl(p, q)$, see Section 2.7.

The modulus of the transform is scale and rotation invariant. Preceded by 2D FT or by QFT, this allows translation, scale and rotation invariant object description.

[26] In this section we assume $g \neq \pm f$, but a similar study is possible for $g = \pm f$.

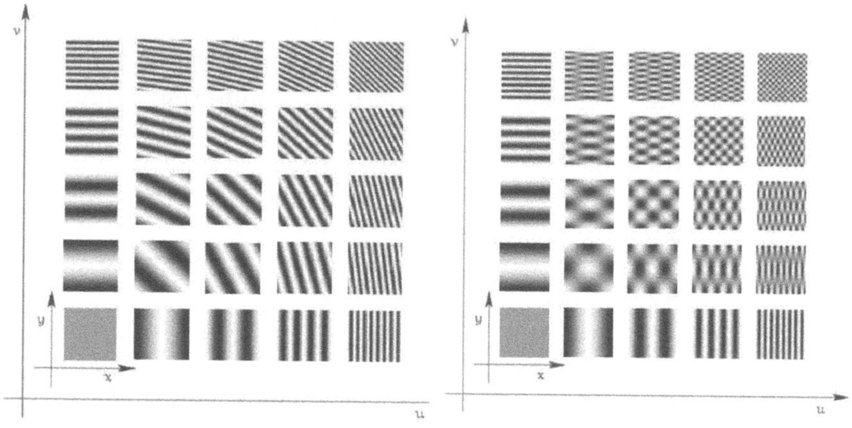

Figure 4.14 Left: 2D FT is intrinsically 1D. Right: QFT is intrinsically 2D. Source: [56].

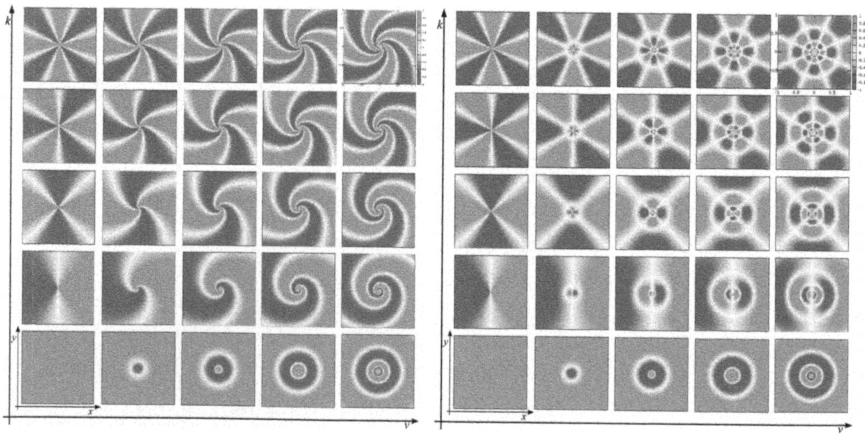

Figure 4.15 *Left*: Kernel of FMT. *Right*: Kernel of QFMT. $k, v \in \{0, 1, 2, 3, 4\}$. Source: [198, Fig. 5].

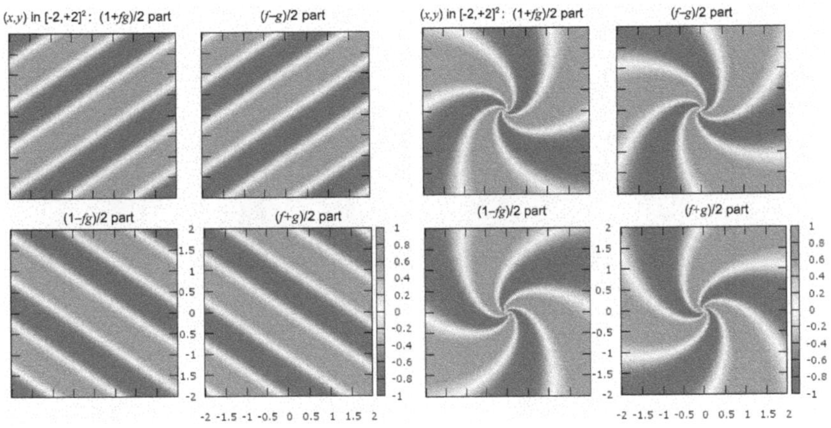

Figure 4.16 Left: QFT kernel, right: QFMT kernel. *Top row*: q_+ parts: $1 + fg$ and $f - g$ components. *Bottom row*: q_- parts: $1 - fg$ and $f + g$ components. Source: [198, Fig. 6].

A diverse range of applications can therefore be imagined: Color object shape recognition, color image registration, application to evaluation of hypercomplex integrals, etc.

We will encounter an extension of the QFMT to Clifford algebras in Section 5.5.2. The future may bring extensions for the QFMT to windowed and wavelet transforms, discretization and numerical implementations.

4.3.3 The Quaternion domain Fourier transform

4.3.3.1 Background

So far quaternion Fourier transforms have been mainly defined over $\mathbb{R}^2$ as signal domain space, see e.g. Sections 4.1 and 4.2. But it seems natural to define a quaternion Fourier transform for quaternion valued signals over *quaternion domains.* This *quaternion domain Fourier transform* (QDFT) transforms quaternion valued signals (e.g. electromagnetic scalar-vector potentials, color data, space-time data, etc.) defined over a quaternion domain (space-time or other 4D domains) from a quaternion "position" space to a quaternion "frequency" space. The QDFT uses the full potential provided by hypercomplex algebra in higher dimensions and may moreover be useful for solving quaternion partial differential equations or functional equations, and in crystallographic texture analysis. We define the QDFT and analyze its *main properties*, including quaternion dilation, modulation and shift properties, Plancherel and Parseval identities, covariance under orthogonal transformations, transformations of coordinate polynomials and differential operator polynomials, transformations of derivative and Dirac derivative operators, as well as signal width related to band width uncertainty relationships. This section is based on [211]. See also [208, 210].

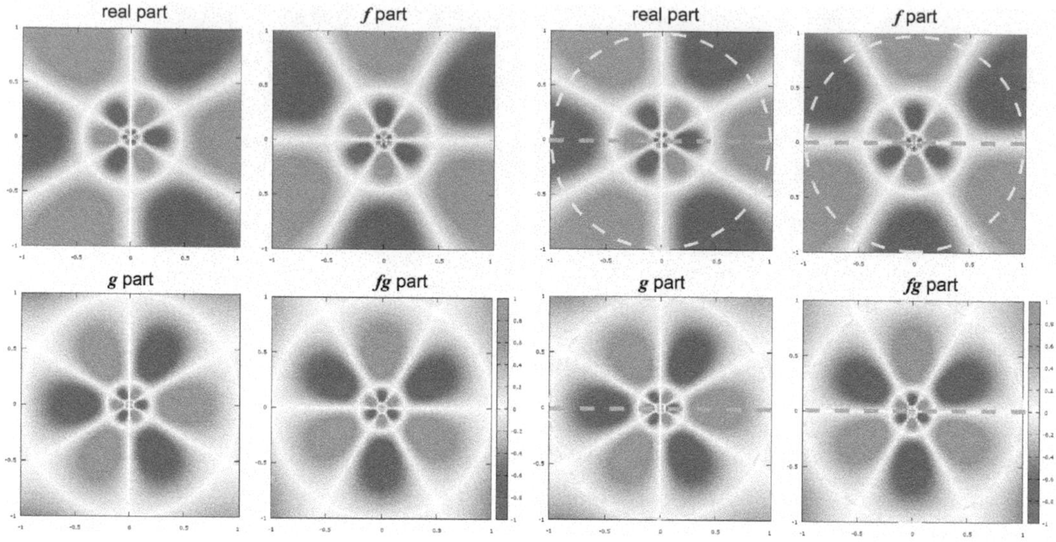

Figure 4.17 *Left:* Four components of the QFMT kernel ($v = 2, k = 3$). *Right:* Symmetries of four components of the QFMT kernel ($v = 2, k = 3$). Source: [198, Fig. 7].

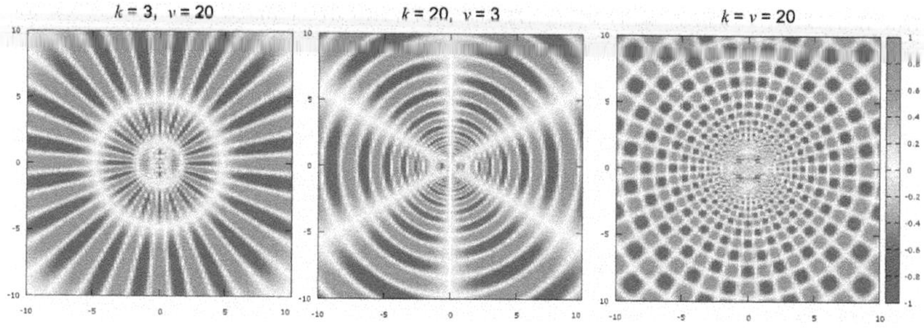

Figure 4.18 *Left*: High angular resolution. *Center*: High radial resolution. *Right*: High radial and angular resolution. Source: [198, Fig. 8].

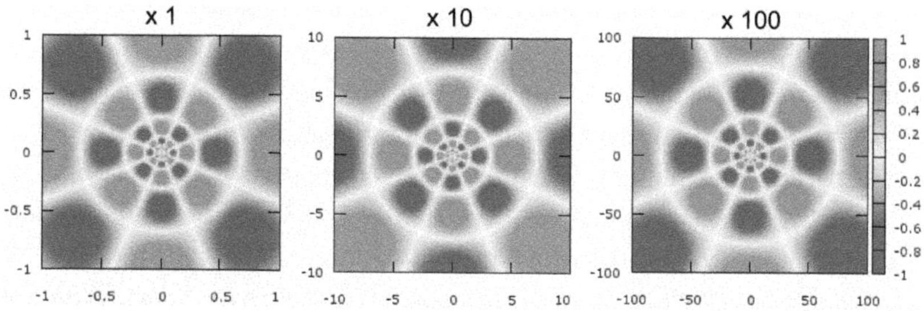

Figure 4.19 Illustration of QFMT scaling. Source: [198, Fig. 9].

Generally speaking, quaternion Fourier transforms (QFT) are since over 20 years a mathematically well researched and frequently applied subject [51]. Yet interesting enough most publications on QFTs concentrate on transformations for signals with domain $\mathbb{R}^2$. First motivated by private communication with T.L. Saaty related to *quaternion valued functions over the domain of quaternions*, we establish here a genuine Fourier transform with a quaternionic kernel operating on such functions.

This section begins by introducing quaternions and their relevant properties, including quaternion domain functions in Section 4.3.3.1.1. For more details on quaternions, we refer the reader to Section 2.8.1. The quaternion domain Fourier transform (QDFT) is defined in Section 4.3.3.2. Many fundamental properties of the QDFT are investigated in Section 4.3.3.3. The conclusions in Section 4.3.3.4 give an outlook into the wide area of possible applications and the rich possibilities of studying related transforms for quaternion domain signals.

4.3.3.1.1 About quaternions $\mathbb{H}$

Quaternion multiplication pq can be alternatively represented by the following *matrix vector multiplication* [326]

$$\begin{pmatrix} Sc(pq) \\ (pq)_i \\ (pq)_j \\ (pq)_k \end{pmatrix} = \begin{pmatrix} p_r & -p_i & -p_j & -p_k \\ p_i & p_r & -p_k & p_j \\ p_j & p_k & p_r & -p_i \\ p_k & -p_j & p_i & p_r \end{pmatrix} \begin{pmatrix} q_r \\ q_i \\ q_j \\ q_k \end{pmatrix}. \tag{4.3.77}$$

The *determinant* of the above matrix $P(p)$ is simply

$$\det P(p) = |p|^4. \tag{4.3.78}$$

If we interpret the four real coefficients of $x \in \mathbb{H}$, $x_r, x_i, x_j, x_k \in \mathbb{R}$ as *coordinates* in $\mathbb{R}^4$, with *infinitesimal volume element* $d^4x = dx_r dx_i dx_j dx_k$, then the substitution $z = ax$, $a \in \mathbb{H}$, yields

$$z = ax \quad \Rightarrow \quad d^4z = |a|^4 d^4x, \qquad d^4x = |a|^{-4} d^4z, \tag{4.3.79}$$

assuming $a \neq 0$ for the last identity.

For the transformation $z = axb$, $a, b, x \in \mathbb{H}$, we set $y = xb$ and then

$$d^4z = |a|^4 d^4y. \tag{4.3.80}$$

Quaternion *conjugation* leads to

$$\tilde{y} = \tilde{b}\tilde{x}, \tag{4.3.81}$$

such that

$$-d^4y = d^4\tilde{y} = |\tilde{b}|^4 d^4\tilde{x} = -|b|^4 d^4x, \tag{4.3.82}$$

because $|\tilde{b}| = |b|$, $d^4\tilde{x} = dx_r(-dx_i)(-dx_j)(-dx_k) = -d^4x$ and similarly $d^4\tilde{y} = -d^4y$. Hence

$$d^4z = |a|^4 d^4y = |a|^4|b|^4 d^4x = |ab|^4 d^4x. \tag{4.3.83}$$

As expected the *rotation* (4.3.88) does not change the infinitesimal volume element

$$z = axa^{-1} \quad \Rightarrow \quad d^4x = |aa^{-1}|^4 d^4x = d^4x. \tag{4.3.84}$$

We follow [151] in defining the following *derivative operators*

$$\widetilde{\partial} = \partial_{x_r} + \partial_{x_i}\boldsymbol{i} + \partial_{x_j}\boldsymbol{j} + \partial_{x_k}\boldsymbol{k}, \tag{4.3.85}$$

$$\partial = \partial_{x_r} - \partial_{x_i}\boldsymbol{i} - \partial_{x_j}\boldsymbol{j} - \partial_{x_k}\boldsymbol{k}, \tag{4.3.86}$$

where $\partial_{x_r} = \partial/\partial x_r$, etc. We further define the three-dimensional *Dirac operator*

$$D = \widetilde{\partial} - \partial_{x_r} = \partial_{x_i}\boldsymbol{i} + \partial_{x_j}\boldsymbol{j} + \partial_{x_k}\boldsymbol{k}, \qquad \widetilde{\partial} = \partial_{x_r} + D. \tag{4.3.87}$$

4.3.3.1.2 Quaternions and reflections and rotations in three and four dimensions The *geometry of reflections and rotations* in three and four dimensions, expressed in the language of quaternions is discussed in [84, 205, 212, 261], see also Section 2.8. We now briefly recall how important *orthogonal transformations* in three-dimensional and four-dimensional Euclidean space can be expressed by means of quaternions.

A three-dimensional *rotation* of the vector part $\mathbf{x}$ of the quaternion $x \in \mathbb{H}$ by the angle 2α around the axis $\hat{\mathbf{a}}$, leaving the scalar part x_r invariant, is given by

$$x' = axa^{-1}, \qquad a = e^{\alpha\hat{\mathbf{a}}}, \qquad \hat{\mathbf{a}}^2 = -1. \tag{4.3.88}$$

For example $a = \cos\alpha + \boldsymbol{k}\sin\alpha = \exp(\boldsymbol{k}\alpha)$ rotates $x = \boldsymbol{i}$ to

$$\begin{aligned} x' = a\boldsymbol{i}a^{-1} &= e^{\boldsymbol{k}\alpha}\boldsymbol{i}e^{-\boldsymbol{k}\alpha} = e^{2\boldsymbol{k}\alpha}\boldsymbol{i} = (\cos 2\alpha + \boldsymbol{k}\sin 2\alpha)\boldsymbol{i} \\ &= \cos(2\alpha)\boldsymbol{i} + \sin(2\alpha)\boldsymbol{j}. \end{aligned} \tag{4.3.89}$$

We further note, that the transformation

$$x' = axb, \qquad a = e^{\alpha\hat{\mathbf{a}}}, \qquad b = e^{\beta\hat{\mathbf{a}}}, \tag{4.3.90}$$

rotates the x_--part by the angle $\alpha + \beta$ in the q_--plane (determined according to Section 4.1.2.3.2, setting $f = \hat{\mathbf{a}}$), and rotates the x_+-part by $\alpha - \beta$ in the q_+-plane.

The *4D reflection* at the *real line* is given by quaternion conjugation $x \to \tilde{x}$, leaving the real line pointwise invariant.

The *4D reflection* at the *3D hyperplane* of pure quaternions is therefore given by $x \to -\tilde{x}$, leaving the 3D hyperplane of pure quaternions pointwise invariant.

A *reflection* at a (pointwise invariant) *general line* in $\mathbb{R}^4$ in the direction of the unit quaternion $a \in \mathbb{H}$, $|a| = 1$, is given by $x \to a\tilde{x}a$.

A *reflection* at the (pointwise invariant) *three-dimensional hyperplane* orthogonal to the direction in four dimensions given by the unit quaternion a, $|a| = 1$, is given by $x \to -a\tilde{x}a$.

A *general rotation* in $\mathbb{R}^4$ is given by

$$x \to axb, \qquad a, b \in \mathbb{H}, \qquad |a| = |b| = 1. \tag{4.3.91}$$

To understand the *geometry of this rotation* [205], we rewrite the unit quaternions a, b as

$$a = e^{\alpha\hat{\mathbf{a}}}, \qquad b = e^{\beta\hat{\mathbf{b}}}. \tag{4.3.92}$$

The pure unit quaternions $\hat{\mathbf{a}}$ and $\hat{\mathbf{b}}$ define two orthogonal two-dimensional rotation planes in $\mathbb{R}^4$, where without restriction of generality we assume $\hat{\mathbf{a}} \neq \hat{\mathbf{b}}$, because the case $\hat{\mathbf{a}} = \hat{\mathbf{b}}$ has already been discussed in (4.3.90). The $q_+^{a,b}$ plane with orthogonal basis and projection

$$q_+^{a,b}\text{basis} : \{\hat{\mathbf{a}} - \hat{\mathbf{b}}, 1 + \hat{\mathbf{a}}\hat{\mathbf{b}}\}, \qquad q_+^{a,b} = \frac{1}{2}(q + \hat{\mathbf{a}}q\hat{\mathbf{b}}), \tag{4.3.93}$$

and the orthogonal $q_-^{a,b}$ plane orthogonal basis and projection

$$q_-^{a,b}\text{basis} : \{\hat{\mathbf{a}} + \hat{\mathbf{b}}, 1 - \hat{\mathbf{a}}\hat{\mathbf{b}}\}, \qquad q_-^{a,b} = \frac{1}{2}(q - \hat{\mathbf{a}}q\hat{\mathbf{b}}), \tag{4.3.94}$$

such that $q = q_+^{a,b} + q_-^{a,b}$, for all $q \in \mathbb{H}$. The transformation $x \to axb$ of (4.3.91) then means geometrically a rotation by the angle $\alpha - \beta$ in the $q_+^{a,b}$ plane (around the $q_-^{a,b}$ plane as axis) and a rotation by the angle $\alpha + \beta$ in the $q_-^{a,b}$ plane (around the $q_+^{a,b}$ plane as axis). This also tells us, that for $\alpha = \beta$ the rotation degenerates to a single two-dimensional rotation by 2α in the $q_-^{a,b}$ plane, and for $\alpha = -\beta$ it degenerates to a single two-dimensional rotation by 2α in the $q_+^{a,b}$ plane.

A *general rotary reflection* (rotation reflection) in $\mathbb{R}^4$ is given by

$$x \to a\tilde{x}b, \qquad a, b \in \mathbb{H}, \qquad |a| = |b| = 1. \tag{4.3.95}$$

This rotary reflection has the pointwise invariant line through $a+b$. In the remaining three-dimensional hyperplane, orthogonal to the $a + b$ line, the axis of the rotary reflection is the line in the direction $a - b$, because $a\widetilde{(a-b)}b = -(a-b)$. The rotation plane of the rotary reflection is spanned by the two orthogonal quaternions $v_{1,2} = [a, b](1 \pm \tilde{a}b)$, $[a, b] = ab - ba$, and the angle of rotation is $\Gamma = \pi - \text{acrccos}(Sc(\tilde{a}b))$, [205].

4.3.3.1.3 Quaternion domain functions Every real valued quaternion domain function f maps $\mathbb{H} \to \mathbb{R}$:

$$f : x \mapsto f(x) \in \mathbb{R}, \quad \forall x \in \mathbb{H}. \tag{4.3.96}$$

Every *quaternion valued quaternion domain function* f maps $\mathbb{H} \to \mathbb{H}$, its four coefficient functions f_r, f_i, f_j, f_k, are in turn real valued quaternion domain functions:

$$f : x \mapsto f(x) = f_r(x) + f_i(x)\boldsymbol{i} + f_j(x)\boldsymbol{j} + f_k(x)\boldsymbol{k} \in \mathbb{H}. \tag{4.3.97}$$

Quaternion valued quaternion domain functions have been historically studied in [137, 275, 298, 311], and applications are described in [151].

We define for two functions $f, g : \mathbb{H} \to \mathbb{H}$ the following *quaternion valued inner product*

$$(f, g) = \int_{\mathbb{H}} f(x)\tilde{g}(x)d^4x \tag{4.3.98}$$

with $d^4x = dx_r dx_i dx_j dx_k \in \mathbb{R}$. Note that quaternion conjugation yields

$$\widetilde{(f, g)} = (g, f). \tag{4.3.99}$$

This means that the real scalar part of the *inner product* (f,g) is *symmetric*

$$\langle f,g\rangle = \frac{1}{2}[(f,g)+(g,f)] = \int_{\mathbb{H}} \langle f(x)\tilde{g}(x)\rangle_0 d^4x \in \mathbb{R}, \qquad \langle f,g\rangle = \langle g,f\rangle. \quad (4.3.100)$$

We further define the $L^2(\mathbb{H};\mathbb{H})$-*norm*[27] as

$$\|f\| = \sqrt{(f,f)} = \sqrt{\langle f,f\rangle} = \sqrt{\int_{\mathbb{H}} |f(x)|^2 d^4x} \geq 0. \quad (4.3.101)$$

The *quaternion domain module* $L^2(\mathbb{H};\mathbb{H})$ is the set of all finite $L^2(\mathbb{H};\mathbb{H})$-norm functions

$$L^2(\mathbb{H};\mathbb{H}) = \{f | f: \mathbb{H} \to \mathbb{H}, \|f\| \leq \infty\}. \quad (4.3.102)$$

The *convolution* of two functions $f,g \in L^2(\mathbb{H};\mathbb{H})$ is defined as

$$(f * g)(x) = \int_{\mathbb{H}} f(y)g(x-y)d^4y. \quad (4.3.103)$$

For unit norm signals $f \in L^2(\mathbb{H};\mathbb{H})$, $\|f\| = 1$, we define the *effective spatial width* or spatial uncertainty (or signal width) of f *in the direction* of the unit quaternion $a \in \mathbb{H}$, $|a| = 1$, as the square root of the variance of the energy distribution of f along the a-axis

$$\Delta x_a = \|(x \cdot a)f\| = \sqrt{\int_{\mathbb{H}} (x \cdot a)^2 |f(x)|^2 d^4x}. \quad (4.3.104)$$

Also for unit norm signals f, we define the *effective spatial width* (spatial uncertainty) as the square root of the variance of the energy distribution of f

$$\Delta x = \|xf\| = \sqrt{\int_{\mathbb{H}} |x|^2 |f(x)|^2 d^4x}. \quad (4.3.105)$$

4.3.3.2 The quaternion domain Fourier transform

Since the traditional quaternion Fourier transform (QFT) [56, 124, 184] is only defined for real or quaternion valued signals over the domain $\mathbb{R}^2$, we newly define the *quaternion domain Fourier transform* (QDFT) for $h \in L^1(\mathbb{H};\mathbb{H})$ as[28]

$$\mathcal{F}\{h\}(\omega) = \hat{h}(\omega) = \frac{1}{(2\pi)^2}\int_{\mathbb{H}} h(x)e^{-Ix\cdot\omega}d^4x, \quad (4.3.106)$$

with $x,\omega \in \mathbb{H}$, $d^4x = dx_r dx_i dx_j dx_k \in \mathbb{R}$, and some constant $I \in \mathbb{H}$, $I^2 = -1$. The constant unit pure quaternion I can be chosen specific for each problem.

Note that the QDFT of (4.3.106) is *steerable* due to the free choice of the unit pure quaternion unit $I \in S^2$.

[27] Note that in equation (13) of [184] the square root is missing over the integral in the definition of the $L^2(\mathbb{R}^2;\mathbb{H})$-norm.

[28] For quaternionic Fourier transforms on L^2, see Footnote 2 on page 144.

This QDFT definition is *left linear*

$$\mathcal{F}\{\alpha h + \beta g\}(\omega) = \alpha \hat{h}(\omega) + \beta \hat{g}(\omega), \qquad (4.3.107)$$

for $g, h \in L^1(\mathbb{H}; \mathbb{H})$ and constants $\alpha, \beta \in \mathbb{H}$. See (4.3.126) for linear combinations of signals with constant quaternion coefficients multiplied from the right.

Applying the orthogonal planes split of Section 4.1.2.3.2 to the signal function $h = h_+ + h_-$ and to the QDFT $\hat{h}$ we find

$$\hat{h}(\omega) = \hat{h}_+(\omega) + \hat{h}_-(\omega), \qquad (4.3.108)$$

$$\hat{h}_+(\omega) = \frac{1}{(2\pi)^2}\int_{\mathbb{H}} h_+(x)e^{-Ix\cdot\omega}d^4x = \frac{1}{(2\pi)^2}\int_{\mathbb{H}} e^{+Ix\cdot\omega}h_+(x)d^4x, \qquad (4.3.109)$$

$$\hat{h}_-(\omega) = \frac{1}{(2\pi)^2}\int_{\mathbb{H}} h_-(x)e^{-Ix\cdot\omega}d^4x = \frac{1}{(2\pi)^2}\int_{\mathbb{H}} e^{-Ix\cdot\omega}h_-(x)d^4x. \qquad (4.3.110)$$

Example 4.3.34. *Following the suggestion of T. L. Saaty, we QDFT transform the functional quaternion equation*[29]

$$h(ax) = bh(x), \quad h: \mathbb{H} \to \mathbb{H}, \qquad (4.3.111)$$

with quaternion constants $a, b \in \mathbb{H}$. We define the auxiliary function $h_a(x) = h(ax)$ and compute

$$\hat{h}_a(\omega) = \frac{1}{(2\pi)^2}\int_{\mathbb{H}} h(ax)e^{-Ix\cdot\omega}d^4x = \frac{1}{(2\pi)^2}\int_{\mathbb{H}} h(ax)e^{-ISc((ax)^{\sim}\tilde{a}^{-1}\omega)}d^4x$$

$$\overset{z=ax}{=} \frac{1}{(2\pi)^2}\int_{\mathbb{H}} h(z)e^{-ISc(\tilde{z}\tilde{a}^{-1}\omega)}|a|^{-4}d^4z = |a|^{-4}\hat{h}(\tilde{a}^{-1}\omega), \qquad (4.3.112)$$

where we used that

$$x\cdot\omega = Sc(\tilde{x}\omega) = Sc(\tilde{x}\tilde{a}\tilde{a}^{-1}\omega) = Sc((ax)^{\sim}\tilde{a}^{-1}\omega) = (ax)\cdot(\tilde{a}^{-1}\omega). \qquad (4.3.113)$$

For $a = \alpha \in \mathbb{R}$ we get

$$\hat{h}_\alpha(\omega) = \frac{1}{|\alpha|^4}\hat{h}(\frac{\omega}{\alpha}), \qquad (4.3.114)$$

and for $a \in \mathbb{H}$, $|a| = 1$, we get

$$\hat{h}_a(\omega) = \hat{h}(a\omega), \qquad (4.3.115)$$

because for $a \in \mathbb{H}$, $|a| = 1$, we have $\tilde{a}^{-1} = a$. Using relationship (4.3.112) *and left linearity we arrive at the QDFT of* (4.3.111)

$$|a|^{-4}\hat{h}(\tilde{a}^{-1}\omega) = b\hat{h}(\omega), \qquad (4.3.116)$$

or equivalently

$$\hat{h}(\tilde{a}^{-1}\omega) = |a|^4 b\hat{h}(\omega), \qquad (4.3.117)$$

[29] The simplest solutions of this equation take the form $h(x) = cxd$, $bc = ca$, with quaternion constants $c, d \in \mathbb{H}$. (I thank the anonymous reviewer for this hint.) In the complex domain T. L. Saaty has developed interesting solutions [289].

which seems neither less nor more complicated to solve than the original equation (4.3.111)*. But note, that for* $a \in \mathbb{H}$, $|a| = 1$, *equations* (4.3.111) *and* (4.3.117) *become identical, because* (4.3.117) *then reads*

$$\hat{h}(a\omega) = b\hat{h}(\omega), \tag{4.3.118}$$

i.e. then (4.3.111) *is invariant under the QDFT operator.*

An application of (4.3.112) is the four-dimensional inversion at the origin $x \to -x$ results in

$$\hat{h}_{-1}(\omega) = \hat{h}(-\omega). \tag{4.3.119}$$

The QDFT can *separate* the two components of a "complex" signal $f : \mathbb{H} \to \mathbb{R} + \boldsymbol{i}\mathbb{R}$, $f(x) = f_r(x) + \boldsymbol{i} f_i(x)$, into *even* and *odd components* with respect to the inversion $x \to -x$. Let

$$f(x) = f_r(x) + \boldsymbol{i} f_i(x) = f_r^e(x) + f_r^o(x) + \boldsymbol{i}(f_i^e(x) + f_i^o(x)), \tag{4.3.120}$$

with

$$\begin{aligned} f_r^e(-x) &= f_r^e(x) = \frac{1}{2}(f_r(x) + f_r(-x)), \\ f_r^o(-x) &= -f_r^o(x) = \frac{1}{2}(f_r(x) - f_r(-x)), \\ f_i^e(x) &= f_i^e(-x), \qquad f_i^o(-x) = -f_i^o(x). \end{aligned} \tag{4.3.121}$$

Using the steerability of the QDFT (4.3.106) we select for $I = \boldsymbol{j}$ (we could also set $I = \boldsymbol{k}$ or any other pure quaternion $\perp \boldsymbol{i}$) and have by linearity

$$\begin{aligned} \hat{f}(\omega) = &\ \hat{f}_r^e(\omega) + \hat{f}_r^o(\omega) + \boldsymbol{i}(\hat{f}_i^e(\omega) + \hat{f}_i^o(\omega)) \\ = &\int_{\mathbb{H}} f_r^e(x) \cos(x \cdot \omega) d^4\omega + \boldsymbol{j} \int_{\mathbb{H}} f_r^o(x) \sin(x \cdot \omega) d^4\omega \\ &+ \boldsymbol{i} \int_{\mathbb{H}} f_i^e(x) \cos(x \cdot \omega) d^4\omega + \boldsymbol{k} \int_{\mathbb{H}} f_i^o(x) \sin(x \cdot \omega) d^4\omega. \end{aligned} \tag{4.3.122}$$

Compare [240] for a similar approach to the symmetry analysis of signals $f : \mathbb{R} \to \mathbb{C}$.

4.3.3.3 Properties of the QDFT

Properties of the QDFT that can easily be established are *inversion*

$$h(x) = \frac{1}{(2\pi)^2} \int_{\mathbb{H}} \hat{h}(x) e^{+Ix\cdot\omega} d^4\omega, \tag{4.3.123}$$

a *shift theorem* for $g(x) = h(x - a)$, constant $a \in \mathbb{H}$,

$$\hat{g}(\omega) = \hat{h}(\omega) e^{-Ia\cdot\omega}, \tag{4.3.124}$$

and a *modulation theorem* for $m(x) = h(x)e^{Ix\cdot\omega_0}$, constant $\omega_0 \in \mathbb{H}$,

$$\hat{m}(\omega) = \hat{h}(\omega - \omega_0). \tag{4.3.125}$$

Linear combinations with constant quaternion coefficients $\alpha, \beta \in \mathbb{H}$ from the *right* lead due to (4.1.84) to

$$\mathcal{F}\{h\alpha + g\beta\} = \hat{h}(\omega)\alpha_+ + \hat{h}(-\omega)\alpha_- + \hat{g}(\omega)\beta_+ + \hat{g}(-\omega)\beta_-. \tag{4.3.126}$$

We define $g_l(x) = \partial_{x_l} h(x)$, $l \in \{r, i, j, k\}$ for the *partial derivative* of the signal function h and obtain its QDFT as

$$\widehat{g_l}(\omega) = \hat{h}(\omega) I \omega_l. \tag{4.3.127}$$

For example for $l = r$ we obtain

$$\widehat{\partial_{x_r} h}(\omega) = \omega_r \hat{h}(\omega) I. \tag{4.3.128}$$

This leads to the QDFT of the *derivative operators* (4.3.85) and (4.3.86)

$$\widehat{\tilde{\partial}^m h}(\omega) = \omega^m \hat{h}(\omega) I^m, \qquad \widehat{\partial^m h}(\omega) = \widetilde{\omega}^m \hat{h}(\omega) I^m, \qquad m \in \mathbb{N}. \tag{4.3.129}$$

Applying the *derivative operators from the right* to the signal function h we further obtain

$$\widehat{h\tilde{\partial}^m}(\omega) = \widehat{h\omega^m}(\omega) I^m, \qquad \widehat{h\partial^m}(\omega) = \widehat{h\widetilde{\omega}^m}(\omega) I^m, \qquad m \in \mathbb{N}. \tag{4.3.130}$$

QDFT transformations of the *Dirac operator* D applied from the left and right, respectively, give

$$\widehat{D^m h}(\omega) = \boldsymbol{\omega}^m \hat{h}(\omega) I^m, \qquad \widehat{hD^m}(\omega) = \widehat{h\boldsymbol{\omega}^m}(\omega) I^m, \qquad m \in \mathbb{N}. \tag{4.3.131}$$

where the pure quaternion part of the quaternion frequency ω is $\boldsymbol{\omega} = \omega - \omega_r$.

The QDFT of m-fold *powers of coordinates* x_l, $l \in \{r, i, j, k\}$, $m \in \mathbb{N}$, times the signal function h leads to (dual to (4.3.127))

$$\widehat{x_l^m h}(\omega) = \partial_{\omega_l}^m \hat{h}(\omega) I^m. \tag{4.3.132}$$

For example for $l = r$ we obtain

$$\widehat{x_r h}(\omega) = \partial_{\omega_r} \hat{h}(\omega) I. \tag{4.3.133}$$

If $P(x_r, x_i, x_j, x_k) = \sum_{m_r, m_i, m_j, m_k} \alpha_{m_r, m_i, m_j, m_k} x_r^{m_r} x_i^{m_i} x_j^{m_j} x_k^{m_k}$, with quaternion coefficients $\alpha_{m_r, m_i, m_j, m_k} \in \mathbb{H}$, is a *polynomial* of the four *coordinates* $\{x_r, x_i, x_j, x_k\}$, then the QDFT yields

$$\mathcal{F}\{P(x_r, x_i, x_j, x_k) h\}(\omega) = \sum_{m_r, m_i, m_j, m_k} \alpha_{m_r, m_i, m_j, m_k} \partial_{\omega_r}^{m_r} \partial_{\omega_i}^{m_i} \partial_{\omega_j}^{m_j} \partial_{\omega_k}^{m_k} \hat{h}(\omega) I^{m_r + m_i + m_j + m_k}. \tag{4.3.134}$$

For example for $P(x) = a \cdot x = a_r x_r + a_i x_i + a_j x_j + a_k x_k$ we obtain

$$\mathcal{F}\{(a \cdot x) h\}(\omega) = (a \cdot \tilde{\partial}_\omega) \hat{h}(\omega) I, \tag{4.3.135}$$

with $\tilde{\partial}_\omega = \partial_{\omega_r} + \partial_{\omega_i}\boldsymbol{i} + \partial_{\omega_j}\boldsymbol{j} + \partial_{\omega_k}\boldsymbol{k}$ and $a \cdot \tilde{\partial}_\omega = a_r\partial_{\omega_r} + a_i\partial_{\omega_i} + a_j\partial_{\omega_j} + a_k\partial_{\omega_k}$. We have the *dual* to (4.3.134) result that

$$\mathcal{F}\{P(\partial_{x_r}, \partial_{x_i}, \partial_{x_j}, \partial_{x_k})h\}(\omega) = \sum_{m_r,m_i,m_j,m_k} \alpha_{m_r,m_i,m_j,m_k}{\omega_r}^{m_r}{\omega_i}^{m_i}{\omega_j}^{m_j}{\omega_k}^{m_k}\hat{h}(\omega)I^{m_r+m_i+m_j+m_k}, \qquad (4.3.136)$$

with the special case (dual to (4.3.135))

$$\mathcal{F}\{(a \cdot \tilde{\partial})h\}(\omega) = (a \cdot \omega)\hat{h}(\omega)I. \qquad (4.3.137)$$

Note that (4.3.136) shows how the QDFT (with $t = x_0$, $x_1 = x_i$, $x_2 = x_j$, $x_3 = x_k$) can be used to *treat* important *partial differential equations in physics*, e.g. the heat equation, wave equation, Klein-Gordon equation, the Maxwell equations in vacuum, free particle Schrödinger and Dirac equations [320, 324, 325, 327].

Equation (4.3.134) leads further (dual to left side of (4.3.129)) to,

$$\widehat{xh}(\omega) = \tilde{\partial}\hat{h}(\omega)I, \qquad \widehat{x^m h}(\omega) = \tilde{\partial}^m\hat{h}(\omega)I^m, \qquad m \in \mathbb{N}. \qquad (4.3.138)$$

Multiplying instead with the *quaternion conjugate* $\tilde{x}$ we obtain (*dual* to right side of (4.3.129))

$$\widehat{\tilde{x}h}(\omega) = \partial\hat{h}(\omega)I, \quad \widehat{\tilde{x}^m h}(\omega) = \partial^m\hat{h}(\omega)I^m, \qquad m \in \mathbb{N}. \qquad (4.3.139)$$

Taking only the *pure vector part* of x, $\mathbf{x} = x - x_r$ we obtain (*dual* to (4.3.131))

$$\widehat{\mathbf{x}h}(\omega) = D_\omega\hat{h}(\omega)I, \quad \widehat{\mathbf{x}^m h}(\omega) = D_\omega^m\hat{h}(\omega)I^m, \qquad m \in \mathbb{N}, \qquad (4.3.140)$$

where $D_\omega = \partial_{\omega_i}\boldsymbol{i} + \partial_{\omega_j}\boldsymbol{j} + \partial_{\omega_k}\boldsymbol{k}$.

We further obtain the following *QDFT Plancherel identity*, which expresses, that the quaternion valued *inner product* (4.3.98) of two quaternion domain module functions $f, g \in L^2(\mathbb{H};\mathbb{H})$ is given by the quaternion valued inner product of the corresponding QDFTs $\hat{f}$ and $\hat{g}$

$$(f, g) = (\hat{f}, \hat{g}). \qquad (4.3.141)$$

As corollaries we get the corresponding *QDFT Plancherel identity* for the *scalar inner product* of equation (4.3.100)

$$\langle f, g\rangle = \langle \hat{f}, \hat{g}\rangle, \qquad (4.3.142)$$

as well as the *QDFT Parseval identity*

$$\|f\| = \|\hat{f}\|. \qquad (4.3.143)$$

The QDFT Parseval identity means, that the QDFT preserves the signal energy when applied in signal processing.

We now define analogous to (4.3.104) for unit norm signals $f \in L^2(\mathbb{H};\mathbb{H})$, $\|f\| = 1$, the *effective spectral width* (or band width) of f *in the direction* of the unit quaternion

$a \in \mathbb{H}$, $|a| = 1$, as the square root of the variance of the frequency spectrum of f along the a-axis

$$\Delta\omega_a = \|(\omega \cdot a)\hat{f}\| = \sqrt{\int_{\mathbb{H}} (\omega \cdot a)^2 |\hat{f}(\omega)|^2 d^4\omega}. \tag{4.3.144}$$

We further define the *effective spectral width* (frequency uncertainty) as the square root of the variance of the energy distribution of $\hat{f}$

$$\Delta\omega = \|\omega\hat{f}\| = \sqrt{\int_{\mathbb{H}} |\omega|^2 |\hat{f}(\omega)|^2 d^4\omega}. \tag{4.3.145}$$

We can now state the *directional uncertainty principle* for the QDFT of unit norm signals $f \in L^2(\mathbb{H};\mathbb{H})$, $\|f\| = 1$ as

$$\Delta x_a \Delta\omega_b \geq \frac{|a \cdot b|}{2}. \tag{4.3.146}$$

The *uncertainty principle* takes the form

$$\Delta x \Delta\omega \geq 1. \tag{4.3.147}$$

Equality holds in (4.3.146) and (4.3.147) for *Gaussian signals*

$$f(x) = Ce^{-c|x|^2}, \qquad 0 < c \in \mathbb{R}, \tag{4.3.148}$$

with constant factor $C \in \mathbb{H}$.

The QDFT of the *convolution* (4.3.103) of two functions $f, g \in L^2(\mathbb{H};\mathbb{H})$ results in

$$\widehat{(f * g)}(\omega) = (2\pi)^2[\hat{f}(\omega)\hat{g}_-(\omega) + \hat{f}(-\omega)\hat{g}_+(\omega)]. \tag{4.3.149}$$

Note that for $\hat{g}_+(\omega) = 0$ or if $\hat{f}(\omega) = \hat{f}(-\omega)$ we obtain

$$\widehat{(f * g)}(\omega) = (2\pi)^2 \hat{f}(\omega)\hat{g}(\omega). \tag{4.3.150}$$

The QDFT of the convolution of the two functions $f, g \in L^2(\mathbb{H};\mathbb{H})$ in opposite order results in

$$\widehat{(g * f)}(\omega) = (2\pi)^2[\hat{g}(\omega)\hat{f}_-(\omega) + \hat{g}(-\omega)\hat{f}_+(\omega)], \tag{4.3.151}$$

and is usually different from (4.3.149), because of the general non-commutativity of $f, g \in L^2(\mathbb{H};\mathbb{H})$.

An application of the QDFT convolution (4.3.149) is, e.g., the *fast convolution* (via simple multiplication of the QDFTs in the Fourier domain) of a *quaternion domain signal* $f : \mathbb{H} \to \mathbb{H}$ with a *pair of complex filters* $g_1(x) = g_{1,r}(x) + g_{1,i}(x)\boldsymbol{i} = g_-(x)$, $g_2(x) = g_{2,r}(x) + g_{2,i}(x)\boldsymbol{i} = g_+(x)(-\boldsymbol{j})$, choosing $I = \boldsymbol{i}$ in (4.3.106).

Next, we study the *covariance properties* of the QDFT under *orthogonal transformations.* We find that a *three-dimensional rotation* (4.3.88) of the argument $g(x) = h(a^{-1}xa)$ leads to

$$\hat{g}(\omega) = \hat{h}(a^{-1}\omega a). \tag{4.3.152}$$

The *reflection at* the pointwise invariant *real scalar line* $x \to \tilde{x}$, $g(x) = h(\tilde{x})$ gives

$$\hat{g}(\omega) = -\hat{h}(\tilde{\omega}). \tag{4.3.153}$$

The *reflection at* the three-dimensional *hyperplane of pure quaternions* $x \to -\tilde{x}$, $g(x) = h(-\tilde{x})$ results in

$$\hat{g}(\omega) = -\hat{h}(-\tilde{\omega}). \tag{4.3.154}$$

The *reflection at* the pointwise invariant *line through* $a \in \mathbb{H}$, $|a| = 1$, $x \to a\tilde{x}a$, $g(x) = h(a\tilde{x}a)$ gives

$$\hat{g}(\omega) = -\hat{h}(\tilde{a}^{-1}\tilde{\omega}\tilde{a}^{-1}) = -\hat{h}(a\tilde{\omega}a), \tag{4.3.155}$$

because $\tilde{a}^{-1} = a$ for $|a| = 1$. The *reflection at* the three-dimensional *hyperplane orthogonal to* the line through $a \in \mathbb{H}$, $|a| = 1$, $x \to -a\tilde{x}a$, $g(x) = h(-a\tilde{x}a)$ results in

$$\hat{g}(\omega) = -\hat{h}(-a\tilde{\omega}a). \tag{4.3.156}$$

A general *four-dimensional rotation* in $\mathbb{R}^4$, $x \to axb$, $a, b \in \mathbb{H}$, $|a| = |b| = 1$, $g(x) = h(axb)$ leads to

$$\hat{g}(\omega) = \hat{h}(a\omega b). \tag{4.3.157}$$

We have thus studied the behavior of the QDFT under *all point group transformations* in three and four dimensions (reflections, rotations, rotary reflections, inversions), which are of importance in crystallography. We note, that quaternions have already been employed for the description of crystallographic symmetry in [12] and for the description of root systems of finite groups in three and four dimensions in [99].

4.3.3.4 Summary on QDFT

We first briefly *reviewed quaternion algebra*, its expression in terms of matrices and vectors, quaternion derivatives and the Dirac derivative, the orthogonal planes split of quaternions with respect to a single unit pure quaternion, and the description of three-dimensional and four-dimensional orthogonal transformations. Then we defined *quaternion domain functions*, the quaternion module $L^2(\mathbb{H};\mathbb{H})$, convolution of quaternion domain functions, and their effective spatial width (uncertainty).

We established the *steerable quaternion domain Fourier transform* (QDFT) with a free choice a single constant pure unit quaternion in the kernel. We examined the *properties* of left and right linearity, orthogonal plane split property, and gave an example of the QDFT applied to a functional equation. Further properties studied are the inverse QDFT, shift and modulation theorems, the QDFT of quaternion coordinate polynomials[30] multiplied with quaternion domain signals, as well as products with powers of the signal argument x, and the corresponding dual properties (polynomials of partial differential operators, quaternion derivatives and Dirac derivatives). We found that the QDFT can separate the symmetry components of complex signals,

[30]Note that real and complex polynomial generated moment invariants have been successfully used for translation, rotation and scale invariant normalized moment description of vector field features, including flow fields [63–65].

and can be applied to many partial differential equations in physics. Quaternion *non-commutativity* means, that multiplication from the right and left need to be distinguished carefully. Next we established Parseval and Plancherel identities, uncertainty principles and convolution properties for the QDFT. The convolution allows, e.g. fast filtering with pairs of complex filters. Finally we studied the covariance properties of the QDFT under orthogonal transformations of the signal arguments, which may a.o. be of importance for applications in crystallography.

We expect that this relatively new quaternionic Fourier transformation may find rich *applications* in mathematics (e.g. higher dimensional holomorphic functions [151]) and physics, including relativity and spacetime physics, in three-dimensional color field processing, neural signal processing, space color video processing, crystallography, quaternion analysis, and for the solution of many types of quaternionic differential equations. We further expect that the QDFT can be successfully *extended* to *localized transforms*, e.g. quaternion domain window Fourier transforms, and continuous quaternionic wavelets and quaternionic ridgelets [67][31]. Further research should be done into *operator versions* of the QDFT, and its related *linear canonical transforms*, which may open up many further areas of interesting applications, including quantum physics. Especially for applications, *discretization* and *fast implementation* with pairs of complex fast Fourier transforms will be of great interest.

4.4 GENERALIZING THE QFT TO A VOLUME-TIME FT AND A SPACETIME FT

We begin by recalling isomorphisms of quaternion algebra to subalgebras of higher order Clifford algebras [247] such as $\mathbb{H} \cong Cl(0,2) \cong Cl^+(3,0)$. Such isomorphisms together with the generalized $GL(\mathbb{R}^{n,m})$ transformation laws for $\{\hat{f}_{\pm}(\mathcal{A}\boldsymbol{x})\}(\boldsymbol{u})$ allow us now to generalize the QFT to higher dimensions.

This indeed opens up a vast new field of related *multivector Fourier transforms*, which are in general non-commutative.

4.4.1 QFT generalization to volume-time functions

One of these quaternion algebra to Clifford sub-algebra isomorphisms that is of particular relevance in physics exists with a subalgebra of the *spacetime algebra* [158] $Cl_{3,1}$, which is of prime importance in physics, and in applications where time matters as well (motion in time, video sequences, flow fields, global navigation satellite systems (GNSS) ...). The quaternion subalgebra allows to introduce generalizations of the QFT to functions in these higher order Clifford geometric algebras. For example, it allows to generalize the QFT to a spacetime FT [184].

The spacetime algebra is the geometric algebra of $\mathbb{R}^{3,1}$. We express this isomorphism by introducing an orthonormal (grade 1) vector basis for $\mathbb{R}^{3,1}$

$$\{\boldsymbol{e}_0, \boldsymbol{e}_1, \boldsymbol{e}_2, \boldsymbol{e}_3\}, \quad -\boldsymbol{e}_0^2 = \boldsymbol{e}_1^2 = \boldsymbol{e}_2^2 = \boldsymbol{e}_3^2 = 1. \tag{4.4.1}$$

[31]Note that in [188] real Clifford Fourier transforms served as spectral representation for the construction and study of Clifford wavelets. Analogously we expect that the QDFT can serve as suitable spectral representation in the construction of quaternionic wavelets and ridgelets.

Using this vector basis of $\mathbb{R}^{3,1}$, the spatial unit volume trivector i_3 and total four-dimensional (hyper volume) pseudoscalar i_4 can be expressed by

$$i_3 = \boldsymbol{e}_1\boldsymbol{e}_2\boldsymbol{e}_3, \quad i_3^2 = -1, \quad i_4 = \boldsymbol{e}_0\boldsymbol{e}_1\boldsymbol{e}_2\boldsymbol{e}_3, \quad i_4^2 = -1. \tag{4.4.2}$$

We emphasize the fact that the vector $\boldsymbol{e}_0$, the 3D volume trivector i_3, and the 4D pseudoscalar i_4, all square to minus one and anti-commute mutually. Examining the geometric algebra multiplication laws of $\boldsymbol{e}_0$, i_3 and i_4 shows indeed that the arising volume-time subalgebra V_t of the spacetime algebra is isomorphic (see Sections 4.1 and 4.2 of [142]) to the quaternion algebra $\mathbb{H}$

$$V_t = \text{span}\{1, \boldsymbol{e}_0, i_3, i_4\} \cong \mathbb{H} = \text{span}\{1, \boldsymbol{i}, \boldsymbol{j}, \boldsymbol{k}\}, \tag{4.4.3}$$

Note especially that

$$i_3 = -\boldsymbol{e}_0\boldsymbol{e}_0 i_3 = \boldsymbol{e}_0(-i_4) = \boldsymbol{e}_0\, i_4^{-1} = \boldsymbol{e}_0^*, \tag{4.4.4}$$

which shows that i_3 is *dual* to $\boldsymbol{e}_0$ in $Cl_{3,1}$. The time direction $\boldsymbol{e}_0$ determines therefore the complimentary 3D space with pseudoscalar i_3 as well!

Based on the isomorphism (4.4.3) we now define a Fourier transform for volume-time module functions $f \in L^2(\mathbb{R}^{3,1}; V_t)$.

Definition 4.4.1 (volume-time Fourier transform (VtFT)). *The volume-time Fourier transform $\mathring{f} : \mathbb{R}^{3,1} \to V_t$ of volume-time module functions*[32] $f \in L^2(\mathbb{R}^{3,1}; V_t)$, *with spacetime vectors $\boldsymbol{x} = t\boldsymbol{e}_0 + \vec{x} \in \mathbb{R}^{3,1}$, $\vec{x} = x\boldsymbol{e}_1 + y\boldsymbol{e}_2 + z\boldsymbol{e}_3 \in \mathbb{R}^3$, and spacetime frequency vectors $\boldsymbol{u} = s\boldsymbol{e}_0 + \vec{u} \in \mathbb{R}^{3,1}$, $\vec{u} = u\boldsymbol{e}_1 + v\boldsymbol{e}_2 + w\boldsymbol{e}_3 \in \mathbb{R}^3$ is defined as*

$$\mathring{f}(\boldsymbol{u}) = \int_{\mathbb{R}^{3,1}} e^{-\boldsymbol{e}_0 ts} f(\boldsymbol{x})\, e^{-i_3\vec{x}\cdot\vec{u}} d^4\boldsymbol{x}, \tag{4.4.5}$$

with the differential spacetime integration volume $d^4\boldsymbol{x} = dtdxdydz$.

Remark 4.4.2. *The 3D integration part*

$$\int f(\boldsymbol{x})\, e^{-i_3\vec{x}\cdot\vec{\omega}} d^3\vec{x} \tag{4.4.6}$$

in (4.4.5) *and* (4.4.17) *fully corresponds to the GA FT in $Cl(3,0)$, as defined in* (5.1.7), *compare [185, 251].*

The VtFT can be *inverted* in close analogy to (4.1.10) by using

$$f(\boldsymbol{x}) = \frac{1}{(2\pi)^4} \int_{\mathbb{R}^{3,1}} e^{\boldsymbol{e}_0 ts} \mathring{f}(\boldsymbol{u})\, e^{i_3\vec{x}\cdot\vec{u}} d^4\boldsymbol{u}, \tag{4.4.7}$$

with $d^4\boldsymbol{u} = dsdudvdw$.

[32]For the Fourier transform on L^1 and L^2, see Footnote 2 on page 144.

The $f_\pm$ split (4.1.12) of the QFT combined with the isomorphism (4.4.3) now yields for volume-time module functions $f \in L^2(\mathbb{R}^{3,1}; V_t)$

$$f = f_+ + f_-, \quad f_+ = \frac{1}{2}(f + \boldsymbol{e}_0 f i_3), \quad f_- = \frac{1}{2}(f - \boldsymbol{e}_0 f i_3). \tag{4.4.8}$$

Rewriting the split (4.4.8) with the duality relation (4.4.4) to

$$f_\pm = \frac{1}{2}(f \pm \boldsymbol{e}_0 f \boldsymbol{e}_0^*) \tag{4.4.9}$$

shows that it naturally only depends on the physical spacetime split, i.e. on the choice of the time direction $\boldsymbol{e}_0$. Applying our new VtFT of Definition 4.4.1 to the split functions $f_\pm$ of (4.4.9) results in a VtFT formula which corresponds to Theorem 4.3.30

$$\mathring{f} = \mathring{f}_+ + \mathring{f}_-,$$
$$\mathring{f}_\pm = \int_{\mathbb{R}^{3,1}} f_\pm \, e^{-i_3(\vec{x}\cdot\vec{u} \mp ts)} d^4x = \int_{\mathbb{R}^{3,1}} e^{-\boldsymbol{e}_0(ts \mp \vec{x}\cdot\vec{u})} f_\pm \, d^4x \,. \tag{4.4.10}$$

Note especially that the $\mathring{f}_+$ part in (4.4.10) has the kernel with the flat Minkowski metric $ts - \vec{x}\cdot\vec{u}$ in the exponent. (Compare Section 4.4.3 for further interpretation.)

Definition 4.4.1 preserves the form of the GL transformation properties of Section 4.1.1.4. We get the $GL(\mathbb{R}^{3,1})$ transformation properties of (4.4.5) simply by inserting in Theorem 4.1.11 transformations $\mathcal{A} \in GL(\mathbb{R}^{3,1})$ and replacing $\mathcal{U}_{\boldsymbol{e}_1}$ by $\mathcal{U}_{\boldsymbol{e}_0}$.

Theorem 4.4.3 ($GL(\mathbb{R}^{3,1})$ transformation properties of the VtFT)**.** *The VtFT of a V_t module function $f \in L^2(\mathbb{R}^2; V_t)$ with a $GL(\mathbb{R}^{3,1})$ transformation $\mathcal{A}$ of its vector argument is given by*

$$\{f(\mathcal{A}\boldsymbol{x})\}^\circ(\boldsymbol{u}) = |\det \mathcal{A}^{-1}| \, \{ \mathring{f}_-(\overline{\mathcal{A}^{-1}}\boldsymbol{u}) + \mathring{f}_+(\mathcal{U}_{\boldsymbol{e}_0}\overline{\mathcal{A}^{-1}}\mathcal{U}_{\boldsymbol{e}_0}\boldsymbol{u}) \}. \tag{4.4.11}$$

In physical applications proper Lorentz transformations with $|\det \mathcal{A}| = 1$ are most relevant, so the $|\det \mathcal{A}^{-1}|$ factor in (4.4.11) can then naturally be omitted.

For all kinds of applications it is of interest to know whether we can push the QFT generalization established by the VtFT for volume-time module functions $f \in L^2(\mathbb{R}^{3,1}; V_t)$ even further, i.e. if even more general spacetime algebra functions can be treated meaningfully with the VtFT. That this is indeed the case will be shown in the next subsection.

4.4.2 Generalization to full spacetime algebra functions

We now explain how we can drop in the VtFT Definition 4.4.1 the restriction to volume-time functions $f \in L^2(\mathbb{R}^{3,1}; V_t)$. The key to this is found in the commutativity of the unit volume trivector i_3 of the right side exponential factor in (4.4.5) with all spatial vectors $\{\boldsymbol{e}_1, \boldsymbol{e}_2, \boldsymbol{e}_3\}$

$$i_3 \, \boldsymbol{e}_k = \boldsymbol{e}_k \, i_3 \,, \quad 1 \le k \le 3. \tag{4.4.12}$$

This directly leads us to the *right linearity* of the VtFT

$$\{f\alpha\}^{\circ}(\boldsymbol{u}) = \int_{\mathbb{R}^{3,1}} e^{-\boldsymbol{e}_0\, ts} f(\boldsymbol{x})\, \alpha\, e^{-i_3 \vec{x}\cdot\vec{u}} d^4\boldsymbol{x}$$
$$= \int_{\mathbb{R}^{3,1}} e^{-\boldsymbol{e}_0\, ts} f(\boldsymbol{x})\, e^{-i_3 \vec{x}\cdot\vec{u}} d^4\boldsymbol{x}\ \alpha = \mathring{f}(\boldsymbol{u})\, \alpha, \quad \forall \text{ const. } \alpha \in Cl_{3,0}, \tag{4.4.13}$$

where $Cl(3,0)$ is the eight-dimensional Clifford geometric algebra of $\mathbb{R}^{3,0}$, i.e. the 3D space subalgebra of $Cl(3,1)$ spanned by

$$\{1, \boldsymbol{e}_1, \boldsymbol{e}_2, \boldsymbol{e}_3, \boldsymbol{e}_2\boldsymbol{e}_3, \boldsymbol{e}_3\boldsymbol{e}_1, \boldsymbol{e}_1\boldsymbol{e}_2, i_3\}. \tag{4.4.14}$$

Naturally this right linearity also holds for the inverse transformation

$$f(\boldsymbol{x})\, \alpha = \frac{1}{(2\pi)^4} \int_{\mathbb{R}^{3,1}} e^{\boldsymbol{e}_0\, ts}\, \mathring{f}(\boldsymbol{u})\alpha\, e^{i_3 \vec{x}\cdot\vec{u}} d^4\boldsymbol{u} \quad \forall \text{ const. } \alpha \in Cl_{3,0}. \tag{4.4.15}$$

Now all 16 basis multivectors of $Cl_{3,1}$ can be obtained by successive geometric multiplications of 1 and $\boldsymbol{e}_0$ (or alternatively of i_3 and i_4, etc.) with the three spatial vectors $\{\boldsymbol{e}_1, \boldsymbol{e}_2, \boldsymbol{e}_3\}$ from the right[33]

$$\{1, \boldsymbol{e}_1, \boldsymbol{e}_2, \boldsymbol{e}_3, \boldsymbol{e}_2\boldsymbol{e}_3, \boldsymbol{e}_3\boldsymbol{e}_1, \boldsymbol{e}_1\boldsymbol{e}_2, i_3,$$
$$\boldsymbol{e}_0, \boldsymbol{e}_0\boldsymbol{e}_1, \boldsymbol{e}_0\boldsymbol{e}_2, \boldsymbol{e}_0\boldsymbol{e}_3, \boldsymbol{e}_0\boldsymbol{e}_2\boldsymbol{e}_3, \boldsymbol{e}_0\boldsymbol{e}_3\boldsymbol{e}_1, \boldsymbol{e}_0\boldsymbol{e}_1\boldsymbol{e}_2, \boldsymbol{e}_0 i_3\}. \tag{4.4.16}$$

We now have laid all the groundwork for the full spacetime algebra generalization of the VtFT of Definition 4.4.1:

Definition 4.4.4 (Spacetime Fourier transform (SFT)) *The SFT* $\hat{f}: \mathbb{R}^{3,1} \to Cl(3,1)$ *of a (16 dimensional) spacetime algebra* $Cl(3,1)$ *module function*[34] $f \in L^2(\mathbb{R}^{3,1}; Cl(3,1))$ *with spacetime vectors* $\boldsymbol{x} = t\boldsymbol{e}_0 + \vec{x} \in \mathbb{R}^{3,1}$, $\vec{x} = x\boldsymbol{e}_1 + y\boldsymbol{e}_2 + z\boldsymbol{e}_3 \in \mathbb{R}^3$, *and spacetime frequency vectors* $\boldsymbol{u} = s\boldsymbol{e}_0 + \vec{u} \in \mathbb{R}^{3,1}$, $\vec{u} = u\boldsymbol{e}_1 + v\boldsymbol{e}_2 + w\boldsymbol{e}_3 \in \mathbb{R}^3$ *is defined by*

$$\mathcal{F}_{SFT}\{f\}(\boldsymbol{u}) = \overset{\diamond}{f}(\boldsymbol{u}) = \int_{\mathbb{R}^{3,1}} e^{-\boldsymbol{e}_0\, ts} f(\boldsymbol{x})\, e^{-i_3 \vec{x}\cdot\vec{u}} d^4\boldsymbol{x}, \tag{4.4.17}$$

with $d^4\boldsymbol{x} = dtdxdydz$.

Because of (4.4.13) Definition 4.4.4 is fully compatible with Definition 4.4.1, since (4.4.17) is nothing but a (right) linear combination of (4.4.5). To show this, we can use $Cl(3,1) \cong V_t \otimes Cl(3,0)^+ \cong V_t \otimes \{1, \boldsymbol{e}_1, \boldsymbol{e}_2, \boldsymbol{e}_3\}$, or we can e.g. rewrite a general spacetime algebra module function $f \in L^2(\mathbb{R}^{3,1}; Cl_{3,1})$ as a (right) linear combination of four volume-time subalgebra module $L^2(\mathbb{R}^{3,1}; V_t)$ functions

$$\begin{aligned} f &= f_s + f_1\boldsymbol{e}_1 + f_2\boldsymbol{e}_2 + f_3\boldsymbol{e}_3 + f_{23}\boldsymbol{e}_2\boldsymbol{e}_3 + f_{31}\boldsymbol{e}_3\boldsymbol{e}_1 + f_{12}\boldsymbol{e}_1\boldsymbol{e}_2 + f_{123} i_3 \\ &\quad + f_0\boldsymbol{e}_0 + f_{01}\boldsymbol{e}_0\boldsymbol{e}_1 + f_{02}\boldsymbol{e}_0\boldsymbol{e}_2 + f_{03}\boldsymbol{e}_0\boldsymbol{e}_3 \\ &\quad + f_{023}\boldsymbol{e}_0\boldsymbol{e}_2\boldsymbol{e}_3 + f_{031}\boldsymbol{e}_0\boldsymbol{e}_3\boldsymbol{e}_1 + f_{012}\boldsymbol{e}_0\boldsymbol{e}_1\boldsymbol{e}_2 + f_4 i_4 \\ &= f_s + f_0\boldsymbol{e}_0 + f_{123} i_3 + f_4 i_4 + \{f_1 + f_{01}\boldsymbol{e}_0 + f_{23} i_3 + f_{023} i_4\}\, \boldsymbol{e}_1 + \\ &\quad + \{f_2 + f_{02}\boldsymbol{e}_0 + f_{31} i_3 + f_{031} i_4\}\, \boldsymbol{e}_2 \\ &\quad + \{f_3 + f_{03}\boldsymbol{e}_0 + f_{12} i_3 + f_{012} i_4\}\, \boldsymbol{e}_3. \end{aligned} \tag{4.4.18}$$

---

[33] $Cl_{3,1}$ is also isomorphic to the tensor product $V_t \otimes Cl^+_{3,0}$, with V_t defined as in Section 4.4.1 and $Cl^+_{3,0}$ defined as in Section 4.1.1.1.1. (See [142], Sections 4.1 and 4.2.)

[34] For the Fourier transform on L^1 and L^2, see Footnote 2 on page 144.

The four $L^2(\mathbb{R}^{3,1}; V_t)$ functions of (4.4.18) are $\{f_s + f_0 \boldsymbol{e}_0 + f_{123} i_3 + f_4 i_4,\ f_1 + f_{01}\boldsymbol{e}_0 + f_{23} i_3 + f_{023} i_4,\ f_2 + f_{02}\boldsymbol{e}_0 + f_{31} i_3 + f_{031} i_4,\ f_3 + f_{03}\boldsymbol{e}_0 + f_{12} i_3 + f_{012} i_4\}$, where all 16 coefficient functions $\{f_s, f_0, f_1, \cdots, f_4\}$ belong to $L^2(\mathbb{R}^{3,1}, \mathbb{R})$.

Because of (4.4.15) the general SFT of Clifford module $L^2(\mathbb{R}^{3,1}; Cl_{3,1})$ functions of Definition 4.4.4 is also invertible

$$f(\boldsymbol{x}) = \frac{1}{(2\pi)^4} \int_{\mathbb{R}^{3,1}} e^{\boldsymbol{e}_0\, ts}\, \overset{\diamond}{f}(\boldsymbol{u})\, e^{i_3 \vec{x}\cdot\vec{u}} d^4\boldsymbol{u}. \tag{4.4.19}$$

4.4.3 Spacetime Fourier transform of $f_\pm$ split parts and physical interpretation

Further application of analogous (right) linearity arguments also yield that the split (4.4.8) and (4.4.9) can also be applied to general multivector Clifford module functions $f \in L^2(\mathbb{R}^{3,1}; Cl_{3,1})$. In (4.4.8) and (4.4.9), we can thus simply replace the $L^2(\mathbb{R}^{3,1}; V_t)$ functions by $L^2(\mathbb{R}^{3,1}; Cl_{3,1})$ functions[35]. This carries on to the general SFTs of the split functions $f_\pm$, which are formally identical to (4.4.10) if we again replace the $L^2(\mathbb{R}^{3,1}; V_t)$ functions by $L^2(\mathbb{R}^{3,1}; Cl_{3,1})$ functions.

We can therefore rewrite the SFT (4.4.17) for $f \in L^2(\mathbb{R}^{3,1}; Cl_{3,1})$ as

$$\begin{aligned} \overset{\diamond}{f} = \overset{\diamond}{f}_+ + \overset{\diamond}{f}_- &= \int_{\mathbb{R}^{3,1}} f_+\, e^{-i_3(\vec{x}\cdot\vec{u} - ts)} d^4x + \int_{\mathbb{R}^{3,1}} f_-\, e^{-i_3(\vec{x}\cdot\vec{u} + ts)} d^4x \\ &= \int_{\mathbb{R}^{3,1}} e^{-\boldsymbol{e}_0(ts - \vec{x}\cdot\vec{u})} f_+\, d^4x + \int_{\mathbb{R}^{3,1}} e^{-\boldsymbol{e}_0(ts + \vec{x}\cdot\vec{u})} f_-\, d^4x. \end{aligned} \tag{4.4.20}$$

Complex spacetime Fourier transformations, with $\exp\{-i(\vec{x}\cdot\vec{u} - ts)\}$ (where $i \in \mathbb{C}$) as the related complex kernel, are e.g. used for electromagnetic fields in spatially dispersive media [8] or in electromagnetic wavelet theory [231].

In physics f_+ can be interpreted as (time dependent) multivector amplitude of a *rightward* (forward) moving wave packetixmultivector wave packet!left, right and f_- as that of a *leftward* (backward) moving wave packet. We therefore see that these physically important wave packets arise absolutely naturally from elementary purely algebraic considerations. But we emphasize that both the non-commutative multivector structure and the geometric interpretation (e.g. of i_3 as oriented 3D spatial volume trivector) go beyond conventional treatment.

We get the consequent generalization of Theorem 4.1.11, i.e. the $GL(\mathbb{R}^{3,1})$ transformation properties of the SFT in the form of

Theorem 4.4.5 ($GL(\mathbb{R}^{3,1})$ transformation properties of the SFT)**.** *The SFT of a $Cl_{3,1}$ module function $f \in L^2(\mathbb{R}^2; Cl_{3,1})$ with a $GL(\mathbb{R}^{3,1})$ transformation $\mathcal{A}$ of its vector argument is given by*

$$\{f(\mathcal{A}\boldsymbol{x})\}^\diamond(\boldsymbol{u}) = |\det \mathcal{A}^{-1}|\, \{\, \overset{\diamond}{f}_-(\overline{\mathcal{A}^{-1}}\boldsymbol{u}) + \overset{\diamond}{f}_+(\mathcal{U}_{\boldsymbol{e}_0}\overline{\mathcal{A}^{-1}}\,\mathcal{U}_{\boldsymbol{e}_0}\boldsymbol{u})\,\}. \tag{4.4.21}$$

This concludes our brief example of a higher dimensional multivector generalization of the QFT for $L^2(\mathbb{R}^2; \mathbb{H})$ functions to a SFT for $L^2(\mathbb{R}^{3,1}; Cl_{3,1})$ functions. We again emphasize that mathematically many other generalizations are in fact possible and we expect a number of them to be of great utility in applications.

[35] Again the $f_\pm$ split (4.4.9) solely depends on the choice of time direction $\boldsymbol{e}_0$.

4.4.4 Directional 4D spacetime uncertainty principle

We now generalize the directional QFT uncertainty principle of Section 4.2.1 to spacetime algebra multivector functions.

Theorem 4.4.6 (Directional 4D spacetime uncertainty principle). *For two arbitrary constant spacetime vectors* $\boldsymbol{a}, \boldsymbol{b} \in \mathbb{R}^{3,1}$ *(selecting two directions), and* $f \in L^2(\mathbb{R}^{3,1}, \mathcal{G}_{3,1})$, $|\boldsymbol{x}|^{1/2} f \in L^2(\mathbb{R}^{3,1}, \mathcal{G}_{3,1})$ *we obtain*

$$\int_{\mathbb{R}^{3,1}} (a_t t - \vec{a}\cdot\vec{x})^2 |f(\boldsymbol{x})|^2 d^4\boldsymbol{x} \int_{\mathbb{R}^{3,1}} (b_t\omega_t - \vec{b}\cdot\vec{\omega})^2 |\mathcal{F}\{f\}(\boldsymbol{\omega})|^2 d^4\boldsymbol{\omega}$$
$$\geq \frac{(2\pi)^4}{4}\left[(a_t b_t - \vec{a}\cdot\vec{b})^2 F_-^2 + (a_t b_t + \vec{a}\cdot\vec{b})^2 F_+^2\right],$$

with energies of the left and right traveling wave packets:

$$F_{\pm} = \int_{\mathbb{R}^2} |f_{\pm}(\boldsymbol{x})|^2 d^2\boldsymbol{x}. \tag{4.4.22}$$

Remark 4.4.7. *The proof is strictly analogous to the proof of the directional QFT uncertainty principle in Section 4.2.1. In that proof we simply need to replace the integral* $\int_{\mathbb{R}^2}$ *by the integral* $\int_{\mathbb{R}^{3,1}}$*, the infinitesimal 2D area element* $d^2\boldsymbol{x}$ *by the infinitesimal 4D volume element* $d^4\boldsymbol{x}$*, etc. Applying the Parseval theorem in four dimensions leads to the factor of* $(2\pi)^4$ *instead of* $(2\pi)^2$.

4.4.5 Summary of QFT generalization to spacetime FT

As already pointed out at the end of Section 4.1.1, general coordinate free formulation in combination with quaternion to Clifford subalgebra isomorphisms opens the door to a wide range of QFT generalizations to these Clifford geometric algebras.

These non-commutative multivector Fourier transforms act on functions from $\mathbb{R}^{m,n}$, $m, n \in \mathbb{N}_0$ to Clifford geometric algebras $Cl_{m,n}$ (or appropriate subalgebras). We demonstrated this by establishing two multivector Fourier transforms: the volume-time and the spacetime Fourier transforms.

There the quaternion $\pm$ split corresponds to the relativistic split of spacetime into time and space, e.g. the rest frame of an observer. The split of the SFT corresponds then to analyzing a spacetime multivector function in terms of left and right traveling multivector wave packets [184].

This generalization to higher dimensional algebras is also possible for the new directional QFT uncertainty principle. We demonstrated this with the physically most relevant generalization to spacetime algebra functions resulting in a (left- and right traveling) multivector wave packet analysis. The energies of these wave packets determine together with two arbitrary spacetime directions (one spacetime vector and one relativistic wave vector) the resulting uncertainty threshold.

Regarding the proliferation of higher dimensional theories in theoretical physics, and the model of Euclidean space in the so-called conformal geometric algebra $Cl(4,1)$, which has recently become of interest for applications in computer graphics, computer vision and geometric reasoning [111, 244], even higher dimensional generalizations of the directional QFT uncertainty principle may be of interest in the future.

They await application, e.g. in the fields of dynamic fluid and gas flows, seismic analysis, to electromagnetic phenomena, in short wherever spatial data are recorded with time. We expect other generalizations of the QFT obtained by the same methods to be of great potential use as well.

The SFT has been further studied in [218] and [122].

CHAPTER 5

Clifford Fourier transforms

In the field of applied mathematics the Fourier transform has developed into an important tool. It is a powerful method for solving partial differential equations. The Fourier transform provides also a technique for signal analysis where the signal from the original domain is transformed to the spectral or frequency domain. In the frequency domain many characteristics of the signal are revealed. But how to extend the Fourier transform to Clifford's geometric algebras[1]?

This central chapter begins with an introduction to the Clifford Fourier transforms (CFT) in $Cl(3,0)$, the geometric algebra of three-dimensional Euclidean space $\mathbb{R}^3$, one of the most studied and applied versions of the CFT. Next is the generalization to one-sided CFTs in n-dimensional Euclidean space Clifford algebras $Cl(n,0)$ and quadratic space Clifford algebras $Cl(p,q)$. After that an alternative operator exponential approach to one-sided CFTs is presented. Then we turn our attention to two-sided CFTs in Clifford algebras $Cl(p,q)$, which makes sense due to the non-commutativity of Clifford algebras. Distinctions are made based on the number of square roots of -1 (and therefore kernel factors) involved. Of further interest are the uncertainty and convolution properties of CFTs. Finally, special CFTs (windowed, Fourier-Mellin and in conformal geometric algebra) are investigated.

Note that a thorough understanding of the CFT is also essential for constructing Clifford wavelets.

5.1 OVERVIEW OF CLIFFORD FOURIER TRANSFORMS

In this overview we mainly follow the tutorial [183]. We concentrate on introducing Clifford Fourier transforms (CFT) taking multivector functions with values in $Cl(3,0)$ as example. In this overview we omit proofs, because as it turns out the CFT in $Cl(3,0)$ is a special case of the CFT in $Cl(n,0)$ treated in full detail in Section 5.2.1. We proceed as follows.

[1]In the geometric algebra literature [161] instead of the mathematical notation $Cl(p,q) = Cl_{p,q}$ the notation $\mathcal{G}_{p,q}$ is widely in use. It is convention to abbreviate $\mathcal{G}_{n,0}$ to $\mathcal{G}_n$. We use the words Clifford algebra and geometric algebra interchangeably, similarly the notions of geometric algebra FT and Clifford FT, and we use both notations $Cl_{p,q}$, $Cl(p,q)$ and $\mathcal{G}_{p,q}$. Nowadays geometric algebra is often understood as Clifford algebra together with geometric interpretation based on the underlying vector space and its subspaces.

DOI: 10.1201/9781003184478-5

First, the basic concept of multivector functions and their vector derivative in geometric algebra (GA) is briefly reviewed. Second, beginning with the Fourier transform on a scalar function we generalize to a real Fourier transform on GA multivector-valued functions ($f : \mathbb{R}^3 \rightarrow Cl(3,0)$). Third, we show a set of important properties of the Clifford Fourier transform (CFT) on $Cl(3,0)$ multivector-valued functions such as differentiation properties, and the Plancherel theorem. We round off the treatment of the CFT (at the end of this overview) by applying the Clifford Fourier transform properties for showing an uncertainty principle for $Cl(3,0)$ multivector functions.

Historically, Brackx et al. [33] extended the Fourier transform to multivector-valued function-distributions in $Cl(0,n)$ with compact support. A related applied approach for hypercomplex Clifford Fourier transformations in $Cl(0,n)$ was followed by Bülow et. al. [58].

By extending the classical trigonometric exponential function $\exp(j\,\boldsymbol{x}*\boldsymbol{\xi})$ (where $*$ denotes the scalar product of $\boldsymbol{x} \in \mathbb{R}^m$ with $\boldsymbol{\xi} \in \mathbb{R}^m$, j the imaginary unit) in [243,259], McIntosh et. al. generalized the classical Fourier transform. Applied to a function of m real variables this generalized Fourier transform is holomorphic in m complex variables and its inverse is *monogenic* in $m+1$ real variables, thereby effectively extending the function of m real variables to a monogenic function of $m+1$ real variables (with values in a *complex* Clifford algebra). This generalization showed significant applications to harmonic analysis, especially to singular integrals on surfaces in $\mathbb{R}^{m+1}$. Based on this approach Kou and Qian obtained a Clifford Payley-Wiener theorem and derived Shannon interpolation of band-limited functions using the monogenic sinc function [286, and references therein]. The Clifford Payley-Wiener theorem also allows to derive left-entire (left-monogenic in the whole $\mathbb{R}^{m+1}$) functions from square integrable functions on $\mathbb{R}^m$ with compact support.

In this introductory overview we adopt and expand the generalization of the Fourier transform in Clifford geometric algebra $\mathcal{G}_3 = Cl(3,0)$ as suggested by Ebling and Scheuermann [117]. We explicitly introduce properties of the real[2] Clifford geometric algebra Fourier transform (CFT), which are useful to establish the uncertainty principle for $\mathcal{G}_3$ multivector functions.

5.1.1 Clifford Fourier transform (CFT)

We first present the classical Fourier transform in $\mathbb{R}$ and generalize it to Clifford's geometric algebra $\mathcal{G}_3 = Cl(3,0)$. Generalizations to other dimensions can be found in [182, 184, 185].

5.1.1.1 Fourier transform in $\mathbb{R}$

Popoulis [277] defined the Fourier transform and its inverse as follows:

[2]The meaning of *real* in this context is that we use the oriented three dimensional volume element $i_3 = \boldsymbol{e}_{123}$ of the geometric algebra $\mathcal{G}_3 = Cl(3,0)$ over the field of the reals $\mathbb{R}$ to construct the kernel of the Clifford Fourier transformation of Definition 5.1.3. This i_3 has a clear geometric interpretation.

Table 5.1 Properties of the traditional Fourier transform. Source: [251, Tab. 2].

Property	Function	Fourier Transform
Linearity	$\alpha f(x)+\beta g(x)$	$\alpha\mathcal{F}\{f\}(\omega)+\beta\mathcal{F}\{g\}(\omega)$
Delay	$f(x-a)$	$e^{-i\omega a}\mathcal{F}\{f\}(\omega)$
Shift	$e^{i\omega_0 x}f(x)$	$\mathcal{F}\{f\}(\omega-\omega_0)$
Scaling	$f(ax)$	$\frac{1}{\lvert a\rvert}\mathcal{F}\{f\}(\frac{\omega}{a})$
Convolution	$(f\star g)(x)$	$\mathcal{F}\{f\}(\omega)\,\mathcal{F}\{g\}(\omega)$
Derivative	$f^{(n)}(x)$	$(i\omega)^n\mathcal{F}\{f\}(\omega)$
Parseval theorem	$\int_{\mathbb{R}}\lvert f(x)\rvert^2\,dx$	$\frac{1}{2\pi}\int_{\mathbb{R}}\lvert\mathcal{F}\{f\}(\omega)\rvert^2\,d\omega$

Definition 5.1.1. *For an integrable function*[3] $f\in L^2(\mathbb{R})$, *the Fourier transform of* f *is the function* $\mathcal{F}\{f\}\colon\mathbb{R}\to\mathbb{C}$ *given by*

$$\mathcal{F}\{f\}(\omega)=\int_{\mathbb{R}}f(x)\,e^{-i\omega x}\,dx, \tag{5.1.1}$$

where i, $i^2=-1$, *is the unit imaginary, and* $\exp(-i\omega x)=\cos(\omega x)+i\sin(\omega x)$.

The function $\mathcal{F}\{f\}(\omega)$ has the general form

$$\mathcal{F}\{f\}(\omega)=A(\omega)+iB(\omega)=C(\omega)e^{i\phi(\omega)}. \tag{5.1.2}$$

$C(\omega)$ is called the Fourier spectrum of $f(t)$, $C^2(\omega)$ its energy spectrum, and $\phi(\omega)$ its phase angle.

Definition 5.1.2. *If* $\mathcal{F}\{f\}(\omega)\in L^2(\mathbb{R})$ *and* $f\in L^2(\mathbb{R})$, *the inverse Fourier transform is given by*

$$\mathcal{F}^{-1}[\mathcal{F}\{f\}(\omega)]=f(x)=\frac{1}{2\pi}\int_{\mathbb{R}}\mathcal{F}\{f\}(\omega)\,e^{i\omega x}\,d\omega. \tag{5.1.3}$$

The following Table 5.1 summarizes some basic properties of the Fourier transform.

5.1.1.2 Clifford geometric algebra Fourier transform in $\mathcal{G}_3=Cl(3,0)$

Here we only give a brief overview.

Consider a multivector valued function $f(\boldsymbol{x})$ in $\mathcal{G}_3$, i.e. $f\colon\mathbb{R}^3\to\mathcal{G}_3=Cl(3,0)$ where $\boldsymbol{x}$ is a vector variable. $f(\boldsymbol{x})$ can be decomposed as

$$\begin{aligned} f(\boldsymbol{x})=\sum_A f_A(\boldsymbol{x})\boldsymbol{e}_A &= f_0(\boldsymbol{x})+f_1(\boldsymbol{x})\boldsymbol{e}_1+f_2(\boldsymbol{x})\boldsymbol{e}_2+f_3(\boldsymbol{x})\boldsymbol{e}_3\\ &+f_{12}(\boldsymbol{x})\,\boldsymbol{e}_{12}+f_{31}(\boldsymbol{x})\boldsymbol{e}_{31}+f_{23}(\boldsymbol{x})\boldsymbol{e}_{23}+f_{123}(\boldsymbol{x})\boldsymbol{e}_{123}, \end{aligned} \tag{5.1.4}$$

[3]For the Fourier transform on L^1 and L^2, see Footnote 2 on page 144.

where the f_A are eight real-valued functions. Equation (5.1.4) can also be written as (compare Table 2.2)

$$\begin{aligned} f(\boldsymbol{x}) &= [f_0(\boldsymbol{x}) + f_{123}(\boldsymbol{x})i_3] + [f_1(\boldsymbol{x}) + f_{23}(\boldsymbol{x})i_3]\,\boldsymbol{e}_1 \\ &\quad + [f_2(\boldsymbol{x}) + f_{31}(\boldsymbol{x})i_3]\,\boldsymbol{e}_2 + [f_3(\boldsymbol{x}) + f_{12}(\boldsymbol{x})i_3]\,\boldsymbol{e}_3. \end{aligned} \tag{5.1.5}$$

Equation (5.1.5) can be regarded as a set of four complex functions. This motivates the extension of the Fourier transform to $\mathcal{G}_3$ multivector functions f. We will call this the Clifford Fourier transform (CFT).

Alternatively to (5.1.7), Bülow et. al. [58] extended the real Fourier transform to the n-dimensional geometric algebra $\mathcal{G}_{0,n} = Cl(0,n)$. This variant of the Clifford Fourier transform of a multivector valued function in $\mathcal{G}_{0,n}$ is given by

$$\mathcal{F}\{f\}(\boldsymbol{\omega}) = \int_{\mathbb{R}^n} f(\boldsymbol{x}) \prod_{k=1}^{n} e^{-\boldsymbol{e}_k 2\pi\omega_k x_k}\, d^n\boldsymbol{x}, \tag{5.1.6}$$

where

$$\boldsymbol{x} = \sum_{k=1}^{k=n} x_k \boldsymbol{e}_k, \quad \boldsymbol{\omega} = \sum_{k=1}^{k=n} \omega_k \boldsymbol{e}_k, \quad \text{and} \quad \boldsymbol{e}_i \cdot \boldsymbol{e}_j = -\delta_{ij} \quad i,j = 1,2,\dots,n.$$

Yet in the following we will adopt (compare [117])

Definition 5.1.3. *The Clifford Fourier transform*[4] *of $f(\boldsymbol{x})$ is the function $\mathcal{F}\{f\}$: $\mathbb{R}^3 \to \mathcal{G}_3$ given by*

$$\mathcal{F}\{f\}(\boldsymbol{\omega}) = \int_{\mathbb{R}^3} f(\boldsymbol{x})\, e^{-i_3 \boldsymbol{\omega}\cdot\boldsymbol{x}}\, d^3\boldsymbol{x}, \tag{5.1.7}$$

Note that we can write $\boldsymbol{\omega} = \omega_1\boldsymbol{e}_1 + \omega_2\boldsymbol{e}_2 + \omega_3\boldsymbol{e}_3$, $\boldsymbol{x} = x_1\boldsymbol{e}_1 + x_2\boldsymbol{e}_2 + x_3\boldsymbol{e}_3$ with $\boldsymbol{e}_1, \boldsymbol{e}_2, \boldsymbol{e}_3$ the basis vectors of $\mathbb{R}^3$ and

$$d^3\boldsymbol{x} = \frac{d\boldsymbol{x}_1 \wedge d\boldsymbol{x}_2 \wedge d\boldsymbol{x}_3}{i_3} \tag{5.1.8}$$

Theorem 5.1.4. *The Clifford Fourier transform $\mathcal{F}\{f\}$ of $f \in L^2(\mathbb{R}^3, \mathcal{G}_3)$, $\int_{\mathbb{R}^3} \|f\|^2 d^3\boldsymbol{x} < \infty$ is invertible and its inverse is calculated by*

$$\mathcal{F}^{-1}[\mathcal{F}\{f\}](\boldsymbol{x}) = f(\boldsymbol{x}) = \frac{1}{(2\pi)^3} \int_{\mathbb{R}^3} \mathcal{F}\{f\}(\boldsymbol{\omega})\, e^{i_3 \boldsymbol{\omega}\cdot\boldsymbol{x}}\, d^3\boldsymbol{\omega}. \tag{5.1.9}$$

Equation (5.1.9) is called the Clifford Fourier integral theorem. It describes how to get from the transform back to the original function f.

[4]For the Fourier transform on L^1 and L^2, see Footnote 2 on page 144.

5.1.2 Overview of basic CFT properties

We summarize some important properties of the Clifford Fourier transform which are similar to the traditional scalar Fourier transform properties. Most can be proven via substitution of variables. Details of proofs are given in Section 5.2.1 for the more general case of $Cl(n,0)$.

Linearity. If $f(\boldsymbol{x}) = \alpha f_1(\boldsymbol{x}) + \beta f_2(\boldsymbol{x})$ for constants α and β, $f_1(\boldsymbol{x})$, $f_2(\boldsymbol{x}) \in \mathcal{G}_3$ then by construction

$$\mathcal{F}\{f\}(\boldsymbol{\omega}) = \alpha\mathcal{F}\{f_1\}(\boldsymbol{\omega}) + \beta\mathcal{F}\{f_2\}(\boldsymbol{\omega}). \tag{5.1.10}$$

Delay property. If the argument of $f(\boldsymbol{x})$ is offset by a constant vector $\boldsymbol{a}$, i.e. $f_d(\boldsymbol{x}) = f(\boldsymbol{x} - \boldsymbol{a})$, then

$$\mathcal{F}\{f_d\}(\boldsymbol{\omega}) = e^{-i_3\boldsymbol{\omega}\cdot\boldsymbol{a}}\mathcal{F}\{f\}(\boldsymbol{\omega}). \tag{5.1.11}$$

Scaling property. Let a be a positive scalar constant, then the Clifford Fourier transform of the function $f_a(\boldsymbol{x}) = f(a\boldsymbol{x})$ becomes

$$\mathcal{F}\{f_a\}(\boldsymbol{\omega}) = \frac{1}{a^3}\mathcal{F}\{f\}(\frac{\boldsymbol{\omega}}{a}). \tag{5.1.12}$$

Shift property. If $\boldsymbol{\omega}_0 \in \mathbb{R}^3$ and $f_0(\boldsymbol{x}) = f(\boldsymbol{x})e^{i_3\boldsymbol{\omega}_0\cdot\boldsymbol{x}}$, then

$$\mathcal{F}\{f_0\}(\boldsymbol{\omega}) = \mathcal{F}\{f\}(\boldsymbol{\omega} - \boldsymbol{\omega}_0) \tag{5.1.13}$$

The shift property shows that the multiplication by $e^{i_3\boldsymbol{\omega}_0\cdot\boldsymbol{x}}$ shifts the CFT of the multivector function $f(\boldsymbol{x})$ so that it becomes centered on the point $\boldsymbol{\omega} = \boldsymbol{\omega}_0$ in the frequency domain.

5.1.3 Clifford Fourier transform and differentiation

The CFT differentiation properties also resemble that of the traditional scalar Fourier transform of Table 5.1. For details of proofs we refer to Section 5.2.1 on the more general case of $Cl(n,0)$.

The Clifford Fourier transform of the *vector differential* of $f(\boldsymbol{x})$ is

$$\mathcal{F}\{\boldsymbol{a}\cdot\nabla f\}(\boldsymbol{\omega}) = i_3\,\boldsymbol{a}\cdot\boldsymbol{\omega}\,\mathcal{F}\{f\}(\boldsymbol{\omega}). \tag{5.1.14}$$

The vector differential in the $\boldsymbol{a}$ direction is

$$\boldsymbol{a}\cdot\nabla f(\boldsymbol{x}) \quad = \quad i_3\,\mathcal{F}^{-1}[\boldsymbol{a}\cdot\boldsymbol{\omega}\mathcal{F}\{f\}](\boldsymbol{x}).$$

This proves (5.1.14). Setting $\boldsymbol{a} = \boldsymbol{e}_k$ we get for a *partial derivative* of $f(\boldsymbol{x})$

$$\mathcal{F}\{\partial_k f\}(\boldsymbol{\omega}) = i_3\omega_k\mathcal{F}\{f\}(\boldsymbol{\omega}), \qquad k = 1,2,3. \tag{5.1.15}$$

Similarly, we can find the derivatives of second order, i.e.

$$\mathcal{F}\{\boldsymbol{a}\cdot\nabla\;\boldsymbol{b}\cdot\nabla f\}(\boldsymbol{\omega}) = -\boldsymbol{a}\cdot\boldsymbol{\omega}\;\boldsymbol{b}\cdot\boldsymbol{\omega}\;\mathcal{F}\{f\}(\boldsymbol{\omega}). \tag{5.1.16}$$

For $\boldsymbol{a} = \boldsymbol{e}_k, \boldsymbol{b} = \boldsymbol{e}_l$ we therefore get

$$\mathcal{F}\{\partial_k\partial_l\}(\boldsymbol{\omega}) = -\omega_k\omega_l\mathcal{F}\{f\}(\boldsymbol{\omega}) \qquad k, l = 1, 2, 3. \tag{5.1.17}$$

If $\boldsymbol{x}$ is a vector variable, then

$$\mathcal{F}\{\boldsymbol{x}f(\boldsymbol{x})\}(\boldsymbol{\omega}) = i_3\nabla_{\boldsymbol{\omega}}\mathcal{F}\{f\}(\boldsymbol{\omega}) \tag{5.1.18}$$

The Clifford Fourier transform of $\boldsymbol{a} \cdot \boldsymbol{x}\ f(\boldsymbol{x})$ gives

$$\mathcal{F}\{\boldsymbol{a} \cdot \boldsymbol{x}\ f(\boldsymbol{x})\}(\boldsymbol{\omega}) \quad = \quad i_3\, \boldsymbol{a} \cdot \nabla_{\boldsymbol{\omega}}\mathcal{F}\{f\}(\boldsymbol{\omega}). \tag{5.1.19}$$

For $\boldsymbol{a} = e_k$ $(k = 1,\ 2,\ 3)$ we get

$$\mathcal{F}\{x_k\ f(\boldsymbol{x})\}(\boldsymbol{\omega}) = i_3\ \frac{\partial}{\partial\omega_k}\mathcal{F}\{f\}(\boldsymbol{\omega}). \tag{5.1.20}$$

The Clifford Fourier transform of the *vector derivative* is

$$\mathcal{F}\{\nabla f\}(\boldsymbol{\omega}) = i_3\boldsymbol{\omega}\mathcal{F}\{f\}(\boldsymbol{\omega}) \tag{5.1.21}$$

and of the *Laplace operator*

$$\mathcal{F}\{\nabla^2 f\}(\boldsymbol{\omega}) = -\boldsymbol{\omega}^2\mathcal{F}\{f\}(\boldsymbol{\omega}) \tag{5.1.22}$$

In general[5] we get

$$\mathcal{F}\{\nabla^m f\} = (i_3\boldsymbol{\omega})^m\mathcal{F}\{f\}(\boldsymbol{\omega}), \qquad m \in \mathbb{N}. \tag{5.1.23}$$

5.1.4 Convolution and CFT

The most important property of the Clifford Fourier transform for signal processing applications is the convolution theorem. For detailed proofs we refer to Section 5.2.1 on the more general case of $Cl(n, 0)$.

Because of the non-Abelian geometric product we have the following definition:

Definition 5.1.5. *Let f and g be multivector valued functions and both have Clifford Fourier transforms, then the convolution of f and g is denoted $f\star g$, and defined by*

$$(f\star g)(\boldsymbol{x}) = \int_{\mathbb{R}^3} f(\boldsymbol{y})g(\boldsymbol{x} - \boldsymbol{y})\, d^3\boldsymbol{y}, \tag{5.1.24}$$

Theorem 5.1.6. *The Clifford Fourier transform of the convolution of $f(\boldsymbol{x})$ and $g(\boldsymbol{x})$ is equal to the product of the Clifford Fourier transforms of $f(\boldsymbol{x})$ and $g(\boldsymbol{x})$, i.e*

$$\mathcal{F}\{f\star g\}(\boldsymbol{\omega}) = \mathcal{F}\{f\}(\boldsymbol{\omega})\mathcal{F}\{g\}(\boldsymbol{\omega}), \tag{5.1.25}$$

---

[5]This general formula should prove very useful for transforming partial differential equations (more precisely: vector derivative equations) into algebraic equations.

5.1.5 Plancherel and Parseval theorems

Just as in the case of the traditional scalar Fourier transform, the **Plancherel theorem** in the geometric algebra $\mathcal{G}_3 = Cl(3,0)$ relates two multivector functions with their Clifford Fourier transforms. For detailed proofs we refer to Section 5.2.1 on the more general case of $Cl(n,0)$, $\widetilde{f}$ means the reverse of f, see Section 2.1.1.6.

Theorem 5.1.7. *Assume that* $f_1(\boldsymbol{x}), f_2(\boldsymbol{x}) \in \mathcal{G}_3$ *with Clifford Fourier transform* $\mathcal{F}\{f_1\}(\boldsymbol{\omega})$ *and* $\mathcal{F}\{f_2\}(\boldsymbol{\omega})$ *respectively, then*

$$\langle f_1(\boldsymbol{x})\, \widetilde{f_2(\boldsymbol{x})}\rangle_V = \frac{1}{(2\pi)^3}\langle \mathcal{F}\{f_1\}(\boldsymbol{\omega})\widetilde{\mathcal{F}\{f_2\}(\boldsymbol{\omega})}\rangle_V, \tag{5.1.26}$$

where we define the volume integral

$$\langle f_1(\boldsymbol{x})\, \widetilde{f_2(\boldsymbol{x})}\rangle_V = \int_{\mathbb{R}^3} f_1(\boldsymbol{x})\widetilde{f_2(\boldsymbol{x})} d^3\boldsymbol{x} \tag{5.1.27}$$

In particular, with $f_1(\boldsymbol{x}) = f_2(\boldsymbol{x}) = f(\boldsymbol{x})$, we get the (multivector) **Parseval theorem**, i.e.

$$\langle f(\boldsymbol{x})\, \widetilde{f(\boldsymbol{x})}\rangle_V = \frac{1}{(2\pi)^3}\langle \mathcal{F}\{f\}(\boldsymbol{\omega})\, \widetilde{\mathcal{F}\{f\}(\boldsymbol{\omega})}\rangle_V. \tag{5.1.28}$$

Note that equation (5.1.26) is multivector valued. This theorem holds for each grade k of the multivectors on both sides of equation (5.1.26)

$$\langle\langle f_1(\boldsymbol{x})\, \widetilde{f_2(\boldsymbol{x})}\rangle_V\rangle_k = \frac{1}{(2\pi)^3}\langle\langle \mathcal{F}\{f_1\}(\boldsymbol{\omega})\, \widetilde{\mathcal{F}\{f_2\}(\boldsymbol{\omega})}\rangle_V\rangle_k, \quad k = 0,1,2,3. \tag{5.1.29}$$

For $k = 0$ and according to equations (5.1.27) and (5.2.147), the (scalar) Parseval theorem becomes

$$\int_{\mathbb{R}^3} \|f(\boldsymbol{x})\|^2\, d^3\boldsymbol{x} = \frac{1}{(2\pi)^3}\int_{\mathbb{R}^3} \|\mathcal{F}\{f\}(\boldsymbol{\omega})\|^2\, d^3\boldsymbol{\omega}. \tag{5.1.30}$$

Because of the similarity with equation (5.1.2) we call $\int_{\mathbb{R}^3} \|f(\boldsymbol{x})\|^2\, d^3\boldsymbol{x}$ the energy of f. A summary of the properties of the Clifford Fourier transform (CFT) for $Cl(3,0)$ can be found in Table 5.2.

5.1.6 Precision limits – minimal variance products for geometric algebra Fourier transforms

The uncertainty principle plays an important role in the development and understanding of quantum physics. It is also central for information processing [287]. In quantum physics it states, e.g., that particle momentum and position cannot be simultaneously known. In Fourier analysis such conjugate entities correspond to a function and its Fourier transform which cannot both be simultaneously sharply localized. Furthermore, much work (e.g. [74, 287]) has been devoted to extending the uncertainty principle to a function and its Fourier transform. From the view point of geometric algebra an uncertainty principle gives us information about how a multivector valued function and its Clifford Fourier transform are related. For detailed proofs we refer to Section 5.2.1 on the more general case of $Cl(n,0)$.

Table 5.2 Properties of the Clifford Fourier transform (CFT). Source: [251, Tab. 3].

Property	Multivector Function	CFT
Linearity	$\alpha f(\boldsymbol{x})+\beta\, g(\boldsymbol{x})$	$\alpha\mathcal{F}\{f\}(\boldsymbol{\omega})+\ \beta\mathcal{F}\{g\}(\boldsymbol{\omega})$
Delay	$f(\boldsymbol{x}-\boldsymbol{a})$	$e^{-i_3\boldsymbol{\omega}\cdot\boldsymbol{a}}\mathcal{F}\{f\}(\boldsymbol{\omega})$
Shift	$e^{i_3\boldsymbol{\omega}_0\cdot\boldsymbol{x}}f(\boldsymbol{x})$	$\mathcal{F}\{f\}(\boldsymbol{\omega}-\boldsymbol{\omega}_0)$
Scaling	$f(a\boldsymbol{x})$	$\frac{1}{a^3}\mathcal{F}\{f\}(\frac{\boldsymbol{\omega}}{a})$
Convolution	$(f\star g)(\boldsymbol{x})$	$\mathcal{F}\{f\}(\boldsymbol{\omega})\,\mathcal{F}\{g\}(\boldsymbol{\omega})$
Vec. diff.	$\boldsymbol{a}\cdot\nabla f(\boldsymbol{x})$	$i_3\ \boldsymbol{a}\cdot\boldsymbol{\omega}\mathcal{F}\{f\}(\boldsymbol{\omega})$
	$\boldsymbol{a}\cdot\boldsymbol{x}\ f(\boldsymbol{x})$	$i_3\ \boldsymbol{a}\cdot\nabla_{\boldsymbol{\omega}}\ \mathcal{F}\{f\}(\boldsymbol{\omega})$
	$\boldsymbol{x}f(\boldsymbol{x})$	$i_3\ \nabla_{\boldsymbol{\omega}}\ \mathcal{F}\{f\}(\boldsymbol{\omega})$
Vec. deriv.	$\nabla^m f(\boldsymbol{x})$	$(i_3\ \boldsymbol{\omega})^m\mathcal{F}\{f\}(\boldsymbol{\omega})$
Plancherel T.	$\langle f_1(\boldsymbol{x})\widetilde{f_2(\boldsymbol{x})}\rangle_V =$	$\frac{1}{(2\pi)^3}\langle\mathcal{F}\{f_1\}(\boldsymbol{\omega})\widetilde{\mathcal{F}\{f_2\}(\boldsymbol{\omega})}\rangle_V$
sc. Parseval T.	$\int_{\mathbb{R}^3}\|f(\boldsymbol{x})\|^2\,d^3\boldsymbol{x} =$	$\frac{1}{(2\pi)^3}\int_{\mathbb{R}^3}\|\mathcal{F}\{f\}(\boldsymbol{\omega})\|^2\,d^3\boldsymbol{\omega}$

Theorem 5.1.8. *Let f be a multivector valued function in $\mathcal{G}_3$ which has the Clifford Fourier transform $\mathcal{F}\{f\}(\boldsymbol{\omega})$. Assume $\int_{\mathbb{R}^3}\|f(\boldsymbol{x})\|^2\,d^3\boldsymbol{x} = F < \infty$, then the following inequality holds for arbitrary constant vectors $\boldsymbol{a}$, $\boldsymbol{b}$:*

$$\int_{\mathbb{R}^3}(\boldsymbol{a}\cdot\boldsymbol{x})^2\|f(\boldsymbol{x})\|^2\,d^3\boldsymbol{x}\int_{\mathbb{R}^3}(\boldsymbol{b}\cdot\boldsymbol{\omega})^2\,\|\mathcal{F}\{f\}(\boldsymbol{\omega})\|^2 d^3\boldsymbol{\omega} \geq (\boldsymbol{a}\cdot\boldsymbol{b})^2\,\frac{(2\pi)^3}{4}\,F^2 \qquad (5.1.31)$$

Choosing $\boldsymbol{b}=\pm\boldsymbol{a}$, with $\boldsymbol{a}^2=1$ we get the following **uncertainty principle**, i.e.

$$\int_{\mathbb{R}^3}(\boldsymbol{a}\cdot\boldsymbol{x})^2\,\|f(\boldsymbol{x})\|^2\,d^3\boldsymbol{x}\int_{\mathbb{R}^3}(\boldsymbol{a}\cdot\boldsymbol{\omega})^2\,\|\mathcal{F}\{f\}(\boldsymbol{\omega})\|^2 d^3\boldsymbol{\omega} \geq \frac{(2\pi)^3}{4}\,F^2. \qquad (5.1.32)$$

In (5.1.32) equality holds for Gaussian multivector valued functions (See appendix A.3)

$$f(\boldsymbol{x}) = C_0\,e^{-k\,\boldsymbol{x}^2} \qquad (5.1.33)$$

where $C_0\in\mathcal{G}_3$ is a constant multivector, $0<k\in\mathbb{R}$.

Theorem 5.1.9. *For $\boldsymbol{a}\cdot\boldsymbol{b}$ =0, we get*

$$\int_{\mathbb{R}^3}(\boldsymbol{a}\cdot\boldsymbol{x})^2\,\|f(\boldsymbol{x})\|^2\,d^3\boldsymbol{x}\int_{\mathbb{R}^3}(\boldsymbol{b}\cdot\boldsymbol{\omega})^2\,\|\mathcal{F}\{f\}(\boldsymbol{\omega})\|^2 d^3\boldsymbol{\omega}\ \geq 0. \qquad (5.1.34)$$

Note that with

$$\boldsymbol{x}^2=\sum_{k=1}^{3}x_k^2=\sum_{k=1}^{3}(\boldsymbol{e}_k\cdot\boldsymbol{x})^2, \qquad \boldsymbol{\omega}^2=\sum_{l=1}^{3}\omega_l^2=\sum_{l=1}^{3}(\boldsymbol{e}_l\cdot\boldsymbol{\omega})^2 \qquad (5.1.35)$$

we can extend the formula of the uncertainty principle to

Theorem 5.1.10. *Under the same assumptions as in theorem 5.1.8, we obtain*

$$\int_{\mathbb{R}^3}\boldsymbol{x}^2\,\|f(\boldsymbol{x})\|^2\,d^3\boldsymbol{x}\int_{\mathbb{R}^3}\boldsymbol{\omega}^2\,\|\mathcal{F}\{f\}(\boldsymbol{\omega})\|^2 d^3\boldsymbol{\omega} \geq 3\,\frac{(2\pi)^3}{4}F^2. \qquad (5.1.36)$$

5.1.7 Summary of CFT overview

We introduced how the (real) Clifford geometric algebra Fourier transform (CFT) extends the traditional Fourier transform on scalar functions to multivector functions. Basic properties and rules for differentiation, convolution, the Plancherel and Parseval theorems were given for this new CFT.

It is known that the Fourier transform is successfully applied to solving physical equations such as the heat equation, wave equations, etc. Therefore geometric algebra and the Clifford Fourier transform can now be applied to solve such problems involving scalar, vector, bivector and pseudoscalar fields. A close relative of the geometric algebra CFT is the quaternion Fourier transform (QFT) of Chapter 4, since quaternions are nothing but the bivector subalgebra of the geometric algebra of three-dimensional Euclidean space $\mathbb{R}^3$. Based on the QFT far reaching generalizations to high dimensional geometric algebra are possible, including the geometric algebra of spacetime [184], see also Section 4.4.

We also gave uncertainty principles in the geometric algebra $\mathcal{G}_3 = Cl(3,0)$ which describe how a multivector-valued function and its Clifford geometric algebra Fourier transform relate.

In the field of information theory and image processing these geometric algebra uncertainty principles establish bounds or limits for the effective width times frequency extension of processed signals or images.

5.2 ONE-SIDED CLIFFORD FOURIER TRANSFORMS

We now generalize the CFT for functions valued in $Cl(3,0)$ of Section 5.1 to higher dimensions.

5.2.1 One-sided Clifford Fourier transforms in $Cl(n,0)$, $n = 2\,(\mathrm{mod}\,4)$ and $n = 3\,(\mathrm{mod}\,4)$

5.2.1.1 *Preliminaries*

We define a generalized real Fourier transform on Clifford multivector-valued functions ($f : \mathbb{R}^n \to Cl(n,0)$, $n = 2,3\,(\mathrm{mod}\,4)$). Next, we show a set of important properties of the Clifford Fourier transform on $Cl(n,0)$, $n = 2,3\,(\mathrm{mod}\,4)$ such as differentiation properties, and the Plancherel theorem, independent of special commutation properties. Then, we develop and utilize commutation properties for giving explicit formulas for $f\boldsymbol{x}^m$, $f\,\nabla^m$ and for the Clifford convolution. Finally, we apply Clifford Fourier transform properties for proving an uncertainty principle for $Cl(n,0)$, $n = 2,3\,(\mathrm{mod}\,4)$ multivector functions. The treatment in this section is based on [185].

Historically, Brackx et al. [33] extended the Fourier transform to multivector valued function-distributions in $Cl_{0,n}$ with compact support. They also showed some properties of this generalized Fourier transform. A related applied approach for hypercomplex Clifford Fourier transformations in $Cl_{0,n}$ was followed by Bülow et. al. [58]. In [243], Li et. al. extended the Fourier transform holomorphically to a function of m complex variables. For more background and context, we refer to Chapter 1.

In this section, we adopt and expand[6] to $\mathcal{G}_n$, $n = 2, 3 \pmod 4$ the generalization of the Fourier transform in Clifford geometric algebra $\mathcal{G}_3$ previously suggested by Ebling and Scheuermann [117], based on [132]. To avoid ambiguities we recall that

$$\begin{aligned} n = 2 \pmod 4 \Leftrightarrow n = 2 + 4k, \; k \in \mathbb{N}, \\ n = 3 \pmod 4 \Leftrightarrow n = 3 + 4l, \; l \in \mathbb{N}. \end{aligned} \tag{5.2.1}$$

We explicitly show detailed properties of the real[7] Clifford geometric algebra Fourier transform (CFT). As a first application we subsequently use some of these properties to define and prove the uncertainty principle for $\mathcal{G}_n$ multivector functions.

5.2.1.1.1 Clifford's geometric algebra $\mathcal{G}_n = Cl(n, 0)$ of $\mathbb{R}^n$ Readers not familiar with Clifford's geometric algebras are referred to Chapter 2.

Let us consider now and in the following an orthonormal vector basis $\{\boldsymbol{e}_1, \boldsymbol{e}_2, \ldots, \boldsymbol{e}_n\}$ of the real n-dimensional Euclidean vector space $\mathbb{R}^n$ with $n = 2, 3 \pmod 4$. The geometric algebra over $\mathbb{R}^n$ denoted by $\mathcal{G}_n = Cl(n, 0)$ then has the graded 2^n-dimensional basis

$$\{1, \boldsymbol{e}_1, \boldsymbol{e}_2, \ldots, \boldsymbol{e}_n, \boldsymbol{e}_{12}, \boldsymbol{e}_{31}, \boldsymbol{e}_{23}, \ldots, i_n = \boldsymbol{e}_1\boldsymbol{e}_2 \ldots \boldsymbol{e}_n\}. \tag{5.2.2}$$

Remark 5.2.1. *The fact that we begin by introducing orthonormal bases for both the vector space $\mathbb{R}^n$ and for its associated geometric algebra $\mathcal{G}_n$ is only because we assume readers to be familiar with these concepts. As is well-known, the definitions of vector spaces and geometric algebras are generically basis independent [161]. The definition of the vector derivative of Section 5.2.1.1.2 is basis independent, too. Only when we introduce the infinitesimal scalar volume element for integration over $\mathbb{R}^n$ in Section 5.2.1.2 and in the proof of the last Theorem 5.2.59 in [251] do we use a basis explicitly. In the latter case it may well be possible to formulate a basis independent proof. All results derived in this section are therefore manifestly invariant (independent of the use of coordinate systems), apart from the proof of Theorem 5.2.59.*

The squares of vectors are positive definite scalars (Euclidean metric) and so are all even powers of vectors

$$\boldsymbol{x}^2 \geq 0, \quad \boldsymbol{x}^m \geq 0 \;\; \text{for} \;\; m = 2m', m' \in \mathbb{N}. \tag{5.2.3}$$

Therefore given a multivector $M \in \mathcal{G}_n$

$$\boldsymbol{x}^m M = M\, \boldsymbol{x}^m, \;\; m = 2m', m' \in \mathbb{N}. \tag{5.2.4}$$

[6]In the following we mean with $n = 2, 3 \pmod 4$ that $n = 2 \pmod 4$ or $n = 3 \pmod 4$, i.e. with (5.2.1) that $n \in \{2, 3, 6, 7, 10, 11, \ldots\}$. For further details in the case of $n = 3$ compare Section 5.1. In the geometric algebra literature [161] instead of the mathematical notation $Cl_{p,q}$ the notation $\mathcal{G}_{p,q}$ is widely in use. It is convention to abbreviate $\mathcal{G}_{n,0}$ to $\mathcal{G}_n$.

[7]The meaning of *real* in this context is, that we use the oriented n-dimensional unit volume element i_n of the geometric algebra $\mathcal{G}_n$ over the field of the reals $\mathbb{R}$ to construct the kernel of the Clifford Fourier transformation of Definition 5.2.21. This i_n has a clear geometric interpretation, e.g. as n-dimensional hypercube with side length one, and in an orthonormal basis of $\mathbb{R}^n$ it can be factorized as $i_n = \boldsymbol{e}_1\boldsymbol{e}_2 \ldots \boldsymbol{e}_n$.

Note that for $n = 2, 3 \pmod 4$)

$$i_n^2 = -1, \quad i_n^{-1} = -i_n, \quad i_n^m = (-1)^{\frac{m}{2}} \text{ for } m = 2m', m' \in \mathbb{Z}, \tag{5.2.5}$$

similar to the complex imaginary unit.

As a consequence of the *square norm* of M, $\|M\|^2 = \langle M\widetilde{M}\rangle$, we obtain the *multivector Cauchy-Schwarz inequality*

$$|\langle M\widetilde{N}\rangle|^2 \leq \|M\|^2\,\|N\|^2 \quad \forall\, M, N \in \mathcal{G}_n. \tag{5.2.6}$$

5.2.1.1.2 Multivector functions, vector differentiation and convolution in the context of one-sided CFTs in $Cl(n, 0)$ We only include here a very brief summary of vector differential calculus. For details and proofs we refer to Chapter 3. This section deals throughout with multivector valued functions $f : \mathbb{R}^n \to Cl(n, 0)$.

Remark 5.2.2. *The vector differential $\boldsymbol{a} \cdot \nabla$ is a scalar operator, therefore the left and right vector differentials*[8] *agree, i.e.*

$$\boldsymbol{a} \cdot \dot{\nabla} \dot{f}(\boldsymbol{x}) = \dot{f}(\boldsymbol{x})\, \boldsymbol{a} \cdot \dot{\nabla}. \tag{5.2.7}$$

The basis independent *vector derivative* ∇ is defined in [161, 174] to have the algebraic properties of a grade one vector in $\mathbb{R}^n$ and to obey equation (3.2.9) for all vectors $\boldsymbol{a} \in \mathbb{R}^n$.

Exercise 5.2.3. *Here we give a set of multivector functions $f : \mathbb{R}^6 \to \mathcal{G}_6$, their vector differentials and vector derivatives [174]. We assume that $\boldsymbol{e}_{1256} = \boldsymbol{e}_1\boldsymbol{e}_2\boldsymbol{e}_5\boldsymbol{e}_6$, constant $\boldsymbol{x}_0 \in \mathbb{R}^6$, $\boldsymbol{r} = \boldsymbol{x} - \boldsymbol{x}_0$, $r = \|\boldsymbol{r}\|$, and $\boldsymbol{r}^{-1} = \frac{\boldsymbol{r}}{\|\boldsymbol{r}\|^2}$.*

$$f_1 = \boldsymbol{x}, \quad \boldsymbol{a} \cdot \nabla f_1 = \boldsymbol{a}, \quad \nabla f_1 = 6, \tag{5.2.8}$$

$$f_2 = \boldsymbol{x}^2, \quad \boldsymbol{a} \cdot \nabla f_2 = 2\boldsymbol{a} \cdot \boldsymbol{x}, \quad \nabla f_2 = 2\boldsymbol{x}, \tag{5.2.9}$$

$$f_3 = \|\boldsymbol{x}\|, \quad \boldsymbol{a} \cdot \nabla f_3 = \boldsymbol{a} \cdot \boldsymbol{x}/\|\boldsymbol{x}\|, \quad \nabla f_3 = \boldsymbol{x}/\|\boldsymbol{x}\|, \tag{5.2.10}$$

$$f_4 = \boldsymbol{x} \cdot \boldsymbol{e}_{1256}, \quad \boldsymbol{a} \cdot \nabla f_4 = \boldsymbol{a} \cdot \boldsymbol{e}_{1256}, \quad \nabla f_4 = 4\boldsymbol{e}_{1256}, \tag{5.2.11}$$

$$f_5 = \log r, \quad \boldsymbol{a} \cdot \nabla f_5 = \boldsymbol{a} \cdot \boldsymbol{r}^{-1}, \quad \nabla f_5 = \boldsymbol{r}^{-1}. \tag{5.2.12}$$

For convenience, we now summarize some properties of the vector derivative, including its application from the left and the right.

Proposition 5.2.4 (Left and right linearity).

$$\nabla(f + g) = \nabla f + \nabla g, \quad (f + g)\nabla = f\nabla + g\nabla\,. \tag{5.2.13}$$

Proposition 5.2.5. *For $f(\boldsymbol{x}) = g(\lambda(\boldsymbol{x}))$, $\lambda(\boldsymbol{x}) \in \mathbb{R}$,*

$$\boldsymbol{a} \cdot \nabla f = f\, \boldsymbol{a} \cdot \nabla = \{\boldsymbol{a} \cdot \nabla\lambda(\boldsymbol{x})\}\frac{\partial g}{\partial \lambda}\,. \tag{5.2.14}$$

[8]The point symbols specify on which function the vector derivative is supposed to act.

Proposition 5.2.6 (Left and right derivative from differential).

$$\nabla f = \nabla_{\boldsymbol{a}} (\boldsymbol{a} \cdot \nabla f), \quad f\nabla = (\boldsymbol{a} \cdot \nabla f)\nabla_{\boldsymbol{a}} . \tag{5.2.15}$$

Proposition 5.2.7 (Left and right product rules).

$$\nabla(fg) = (\dot{\nabla}\dot{f})g + \dot{\nabla} f \dot{g} = (\dot{\nabla}\dot{f})g + \nabla_{\boldsymbol{a}} f(\boldsymbol{a} \cdot \nabla g) . \tag{5.2.16}$$

$$(fg)\nabla = f(\dot{g}\dot{\nabla}) + \dot{f} g \dot{\nabla} = f(\dot{g}\dot{\nabla}) + (\boldsymbol{a} \cdot \nabla f)\, g \nabla_{\boldsymbol{a}} . \tag{5.2.17}$$

Note that the multivector functions f and g in (5.2.16) and (5.2.17) do not necessarily commute.

Exercise 5.2.8. *For two functions $f, g : \mathbb{R}^2 \to \mathcal{G}_2$, $f = \boldsymbol{x}, g = \boldsymbol{x} \cdot \boldsymbol{e}_{12}$ we calculate [174]*

$$\begin{aligned}
(fg)\nabla &= f(\dot{g}\dot{\nabla}) + (\boldsymbol{a} \cdot \nabla f)\, g \nabla_{\boldsymbol{a}} \\
= \boldsymbol{x}[(\boldsymbol{x} \cdot \boldsymbol{e}_{12})\nabla] + (\boldsymbol{a} \cdot \nabla \boldsymbol{x})(\boldsymbol{x} \cdot \boldsymbol{e}_{12})\nabla_{\boldsymbol{a}} &= \boldsymbol{x}(-2\boldsymbol{e}_{12}) + \boldsymbol{a}(\boldsymbol{x} \cdot \boldsymbol{e}_{12})\nabla_{\boldsymbol{a}} \\
&= -2\boldsymbol{x}\boldsymbol{e}_{12} + \boldsymbol{e}_1(\boldsymbol{x} \cdot \boldsymbol{e}_{12})\boldsymbol{e}_1 + \boldsymbol{e}_2(\boldsymbol{x} \cdot \boldsymbol{e}_{12})\boldsymbol{e}_2 = -2\boldsymbol{x}\boldsymbol{e}_{12}.
\end{aligned} \tag{5.2.18}$$

Because ∇^2 is a scalar operator, the left and right Laplace derivatives agree, i.e. $\nabla^2 f = f \nabla^2$. More generally all even powers of the left and right vector derivative agree

$$\nabla^m f = f \nabla^m \quad \text{for} \quad m = 2m', m' \in \mathbb{N}. \tag{5.2.19}$$

Proposition 5.2.9 (Integration of parts).

$$\int_{\mathbb{R}^n} f(\boldsymbol{x})[\boldsymbol{a} \cdot \nabla g(\boldsymbol{x})]\, d^n\boldsymbol{x} =$$

$$\left[\int_{\mathbb{R}^{n-1}} f(\boldsymbol{x})g(\boldsymbol{x})\, d^{n-1}\boldsymbol{x}\right]_{a\cdot x=-\infty}^{a\cdot x=\infty} - \int_{\mathbb{R}^n} [\boldsymbol{a} \cdot \nabla f(\boldsymbol{x})]g(\boldsymbol{x})\, d^n\boldsymbol{x}. \tag{5.2.20}$$

Remark 5.2.10. *Proposition 5.2.9 reduces to the familiar coordinate form, if we insert for $\boldsymbol{a}$ the grade 1 basis vectors $\boldsymbol{e}_k, 1 \leq k \leq n$ of (5.2.2), because*

$$\boldsymbol{e}_k \cdot \nabla = \partial_k \quad \textit{and} \quad \boldsymbol{e}_k \cdot \boldsymbol{x} = x_k. \tag{5.2.21}$$

We also note that because of (5.2.4) even powers of vectors commute with multivector valued functions $f \in L^2(\mathbb{R}^n, \mathcal{G}_n)$

$$\boldsymbol{x}^m f = f\, \boldsymbol{x}^m \quad \text{for} \quad m = 2m', m' \in \mathbb{N}. \tag{5.2.22}$$

Theorem 5.2.11. *For all geometric algebras $\mathcal{G}_n$, $n \in \mathbb{N}$ we have for $f : (\mathbb{R}^n \to \mathcal{G}_n)$, $\boldsymbol{a}, \boldsymbol{x}, \boldsymbol{\omega} \in \mathbb{R}^n$ and $\lambda = \pm\boldsymbol{\omega} \cdot \boldsymbol{x}$*

$$\boldsymbol{a} \cdot \nabla_{\boldsymbol{\omega}} f(\boldsymbol{x})(\pm i_n^{-1})e^{\pm i_n \boldsymbol{\omega}\cdot\boldsymbol{x}} = \boldsymbol{a} \cdot \boldsymbol{x} f(\boldsymbol{x})e^{\pm i_n \boldsymbol{\omega}\cdot\boldsymbol{x}}. \tag{5.2.23}$$

For $f(\boldsymbol{x}) = 1$ we get

$$\boldsymbol{a} \cdot \nabla_{\boldsymbol{\omega}}(\pm i_n^{-1})\, e^{\pm i_n \boldsymbol{\omega}\cdot\boldsymbol{x}} = \boldsymbol{a} \cdot \boldsymbol{x}\, e^{\pm i_n \boldsymbol{\omega}\cdot\boldsymbol{x}}. \tag{5.2.24}$$

Proof.

$$\boldsymbol{a}\cdot\nabla_{\boldsymbol{\omega}} f(\boldsymbol{x})(\pm i_n^{-1})e^{\pm i_n\boldsymbol{\omega}\cdot\boldsymbol{x}} = f(\boldsymbol{x})(\pm i_n^{-1})\,\boldsymbol{a}\cdot\nabla_{\boldsymbol{\omega}} e^{\pm i_n\boldsymbol{\omega}\cdot\boldsymbol{x}}$$
$$= f(\boldsymbol{x})(\pm i_n^{-1})\,\boldsymbol{a}\cdot\nabla_{\boldsymbol{\omega}}(\pm\boldsymbol{\omega}\cdot\boldsymbol{x})\,\frac{\partial e^{i_n\lambda}}{\partial\lambda} = f(\boldsymbol{x})\,i_n^{-1}\,\boldsymbol{a}\cdot\boldsymbol{x}\,i_n\,e^{i_n\lambda}$$
$$= \boldsymbol{a}\cdot\boldsymbol{x}f(\boldsymbol{x})\,i_n^{-1}\,i_n\,e^{i_n\lambda} = \boldsymbol{a}\cdot\boldsymbol{x}f(\boldsymbol{x})\,e^{\pm i_n\boldsymbol{\omega}\cdot\boldsymbol{x}}. \qquad (5.2.25)$$

For the second equality we used proposition 5.2.5, and for the third equality proposition 21 of [174]. □

Exercise 5.2.12. *Functions $f : \mathbb{R}^6 \to \mathcal{G}_6$, like $\sin(\boldsymbol{x})$ and $e^{i_6\boldsymbol{\omega}\cdot\boldsymbol{x}}$ with $i_6 = \boldsymbol{e}_1\boldsymbol{e}_2\boldsymbol{e}_3\boldsymbol{e}_4\boldsymbol{e}_5\boldsymbol{e}_6$, $i_6^{-1} = \tilde{i}_6 = -i_6$ can be defined by power series. An example for (5.2.23) is therefore*

$$\boldsymbol{a}\cdot\nabla_{\boldsymbol{\omega}}\sin(\boldsymbol{x})(-i_6)e^{i_6\boldsymbol{\omega}\cdot\boldsymbol{x}} = \boldsymbol{a}\cdot\boldsymbol{x}\sin(\boldsymbol{x})e^{i_6\boldsymbol{\omega}\cdot\boldsymbol{x}}. \qquad (5.2.26)$$

By exchanging the roles of $\boldsymbol{x}$ and $\boldsymbol{\omega}$ in (5.2.24) we obtain

Corollary 5.2.13.

$$\boldsymbol{a}\cdot\nabla(\pm i_n^{-1})\,e^{\pm i_n\boldsymbol{\omega}\cdot\boldsymbol{x}} = \boldsymbol{a}\cdot\boldsymbol{\omega}\,e^{\pm i_n\boldsymbol{\omega}\cdot\boldsymbol{x}}. \qquad (5.2.27)$$

Applying proposition 5.2.6 to corollary 5.2.13 and multiplying both sides with $\pm i_n$ from the right we get

Corollary 5.2.14.

$$\nabla e^{\pm i_n\boldsymbol{\omega}\cdot\boldsymbol{x}} = \boldsymbol{\omega}\,(\pm i_n)\,e^{\pm i_n\boldsymbol{\omega}\cdot\boldsymbol{x}}, \qquad (5.2.28)$$

and its reverse

Corollary 5.2.15.

$$e^{\pm i_n\boldsymbol{\omega}\cdot\boldsymbol{x}}\,\nabla = e^{\pm i_n\boldsymbol{\omega}\cdot\boldsymbol{x}}\,(\pm i_n)\,\boldsymbol{\omega}. \qquad (5.2.29)$$

Theorem 5.2.16. *For all geometric algebras $\mathcal{G}_n$, $n \in \mathbb{N}$ we have for $f : (\mathbb{R}^n \to \mathcal{G}_n)$, $\boldsymbol{a}, \boldsymbol{x}, \boldsymbol{\omega} \in \mathbb{R}^n$, and $\lambda = \pm\boldsymbol{\omega}\cdot\boldsymbol{x}$*

$$\nabla_{\boldsymbol{\omega}} f(\boldsymbol{x})(\pm i_n^{-1})e^{\pm i_n\boldsymbol{\omega}\cdot\boldsymbol{x}} = \boldsymbol{x}f(\boldsymbol{x})e^{\pm i_n\boldsymbol{\omega}\cdot\boldsymbol{x}}. \qquad (5.2.30)$$

For $f(\boldsymbol{x}) = 1$ we get

$$\nabla_{\boldsymbol{\omega}}(\pm i_n^{-1})e^{\pm i_n\boldsymbol{\omega}\cdot\boldsymbol{x}} = \boldsymbol{x}\,e^{\pm i_n\boldsymbol{\omega}\cdot\boldsymbol{x}}. \qquad (5.2.31)$$

Proof. We first proof (5.2.30).

$$\nabla_{\boldsymbol{\omega}} f(\boldsymbol{x})(\pm i_n^{-1})e^{\pm i_n\boldsymbol{\omega}\cdot\boldsymbol{x}} = \nabla_{\boldsymbol{a}}[\boldsymbol{a}\cdot\nabla_{\boldsymbol{\omega}} f(\boldsymbol{x})(\pm i_n^{-1})e^{\pm i_n\boldsymbol{\omega}\cdot\boldsymbol{x}}]$$
$$= \nabla_{\boldsymbol{a}}(\boldsymbol{a}\cdot\boldsymbol{x})f(\boldsymbol{x})\,e^{\pm i_n\boldsymbol{\omega}\cdot\boldsymbol{x}} = \boldsymbol{x}f(\boldsymbol{x})\,e^{\pm i_n\boldsymbol{\omega}\cdot\boldsymbol{x}}. \qquad (5.2.32)$$

For the first equality we used proposition 5.2.6, for the second equality Theorem 5.2.11, and for the third equality proposition 72 of [174]. □

Exercise 5.2.17. *Similar to example 5.2.12 functions* $f : \mathbb{R}^7 \to \mathcal{G}_7$, *like* $\cos(\boldsymbol{x})$ *and* $e^{i_7\boldsymbol{\omega}\cdot\boldsymbol{x}}$ *with* $i_7 = \boldsymbol{e}_1\boldsymbol{e}_2\boldsymbol{e}_3\boldsymbol{e}_4\boldsymbol{e}_5\boldsymbol{e}_6\boldsymbol{e}_7$, $i_7^{-1} = \tilde{i}_7 = -i_7$ *can be defined by power series. An example for (5.2.30) is therefore*

$$\nabla_{\boldsymbol{\omega}} \cos(\boldsymbol{x})\, i_7\, e^{-i_7\boldsymbol{\omega}\cdot\boldsymbol{x}} = \boldsymbol{x}\cos(\boldsymbol{x})e^{i_7\boldsymbol{\omega}\cdot\boldsymbol{x}}. \tag{5.2.33}$$

The reverse of (5.2.31) gives

Corollary 5.2.18.

$$e^{\pm i_n\boldsymbol{\omega}\cdot\boldsymbol{x}}\,(\pm i_n^{-1})\,\nabla_{\boldsymbol{\omega}} = e^{\pm i_n\boldsymbol{\omega}\cdot\boldsymbol{x}}\,\boldsymbol{x}. \tag{5.2.34}$$

Convolution is an important operation for smoothing images and for edge detection in image processing. The *Clifford Convolution* of multivector valued functions is defined for arbitrary n.

Definition 5.2.19 (Clifford Convolution). *The Clifford Convolution of* $f, g \in L^2(\mathbb{R}^n, \mathcal{G}_n)$ *is defined as*

$$(f \star g)(\boldsymbol{x}) = \int_{\mathbb{R}^n} f(\boldsymbol{y})g(\boldsymbol{x}-\boldsymbol{y})d^n\boldsymbol{y}. \tag{5.2.35}$$

Exercise 5.2.20. *As an example let us compute the convolution of two exponential functions* $f, g : \mathbb{R}^2 \to \mathcal{G}_2$, $f(\boldsymbol{x}) = \boldsymbol{e}_2\exp(-i_2\boldsymbol{\omega}\cdot\boldsymbol{x})$, $g(\boldsymbol{x}) = 3\boldsymbol{e}_1\exp(i_2\boldsymbol{\omega}'\cdot\boldsymbol{x})$ *with* $\boldsymbol{\omega}, \boldsymbol{\omega}' \in \mathbb{R}^2$, $i_2 = \boldsymbol{e}_1\boldsymbol{e}_2$, $i_2\boldsymbol{e}_{1,2} = -\boldsymbol{e}_{1,2}i_2$,

$$(f \star g)(\boldsymbol{x}) = \int_{\mathbb{R}^2} \boldsymbol{e}_2 e^{-i_2\boldsymbol{\omega}\cdot\boldsymbol{y}} 3\boldsymbol{e}_1 e^{i_2\boldsymbol{\omega}'\cdot(\boldsymbol{x}-\boldsymbol{y})} d^2\boldsymbol{y}$$

$$= 3\boldsymbol{e}_2\boldsymbol{e}_1 e^{i_2\boldsymbol{\omega}'\cdot\boldsymbol{x}} \int_{\mathbb{R}^2} e^{i_2(\boldsymbol{\omega}-\boldsymbol{\omega}')\cdot\boldsymbol{y}} d^2\boldsymbol{y} = -3(2\pi)^2 i_2 e^{i_2\boldsymbol{\omega}'\cdot\boldsymbol{x}}\delta(\boldsymbol{\omega}-\boldsymbol{\omega}'). \tag{5.2.36}$$

Exchanging the order of the functions we get

$$(g \star f)(\boldsymbol{x}) = 3(2\pi)^2 i_2 e^{-i_2\boldsymbol{\omega}'\cdot\boldsymbol{x}}\delta(\boldsymbol{\omega}-\boldsymbol{\omega}'), \tag{5.2.37}$$

illustrating the general non-commutativity $(f \star g) \neq (g \star f)$ *due to the geometric product.*

Note that the following identity, which follows from the substitution of variables ($\boldsymbol{z} = \boldsymbol{x} - \boldsymbol{y}$), is valid for all dimensions n. Let $f, g \in L^2(\mathbb{R}^n, \mathcal{G}_n)$ then

$$\int_{\mathbb{R}^n} f(\boldsymbol{x}-\boldsymbol{y})g(\boldsymbol{y})d^n\boldsymbol{y} = \int_{\mathbb{R}^n} f(\boldsymbol{z})g(\boldsymbol{x}-\boldsymbol{z})d^n\boldsymbol{z}. \tag{5.2.38}$$

Ebling and Scheuermann [117] distinguish between right and left convolution. They are right that products of multivector valued functions do not commute (compare example 5.2.20), so after e.g. a linear and shift-invariant (LSI) multivector filter is chosen it matters if a multivector image function is multiplied with the filter from the right or from the left. But because of (5.2.38) we only define one kind of convolution and leave it up to particular applications which factor is taken as multivector filter and which for the multivector image function, etc. The CFT formulas of the convolution which we derive for $n = 2, 3 \pmod 4$ are valid for whatever choice is made in applications.

5.2.1.2 One-sided Clifford Fourier transform (CFT)

Definition 5.2.21. *The Clifford Fourier transform*[9] *of* $f(\boldsymbol{x})$ *is the function* $\mathcal{F}\{f\}$: $\mathbb{R}^n \to \mathcal{G}_n$, $n = 2, 3 \pmod 4$ *given by*

$$\mathcal{F}\{f\}(\boldsymbol{\omega}) = \int_{\mathbb{R}^n} f(\boldsymbol{x})\, e^{-i_n \boldsymbol{\omega}\cdot\boldsymbol{x}}\, d^n\boldsymbol{x}, \tag{5.2.39}$$

with $\boldsymbol{x}, \boldsymbol{\omega} \in \mathbb{R}^n$.

Note that

$$d^n\boldsymbol{x} = d\boldsymbol{x_1} \wedge d\boldsymbol{x_2} \wedge \ldots \wedge d\boldsymbol{x_n}\, i_n^{-1} \tag{5.2.40}$$

is scalar valued ($d\boldsymbol{x_k} = dx_k\boldsymbol{e_k} \in \mathbb{R}^n$, $k = 1, 2, \ldots, n$, no summation). Because for $n = 3 \pmod 4$ i_n commutes with every element of $\mathcal{G}_n$, the Clifford Fourier kernel $e^{-i_n\boldsymbol{\omega}\cdot\boldsymbol{x}}$ will also commute with every element of $\mathcal{G}_n$. This is not the case for $n = 2 \pmod 4$.

Exercise 5.2.22. *We give an example for an integrable function* $f : \mathbb{R}^n \to \mathcal{G}_n$, *the* n*-dimensional* $\mathrm{rect}(\boldsymbol{x})$ *function, which can be given in terms of the real scalar* $\mathrm{rect}(x)$ *function with* $x \in \mathbb{R}$ *as*

$$\mathrm{rect}(\boldsymbol{x}) = \prod_{k=1}^{n} \mathrm{rect}(x_k)\boldsymbol{e}_k = i_n \prod_{k=1}^{n} \mathrm{rect}(x_k). \tag{5.2.41}$$

The CFT of $\mathrm{rect}(\boldsymbol{x})$ *gives*

$$\begin{aligned}\mathcal{F}\{\mathrm{rect}\}(\boldsymbol{\omega}) &= \int_{\mathbb{R}^n} \mathrm{rect}(\boldsymbol{x}) e^{-i_n\boldsymbol{\omega}\cdot\boldsymbol{x}}\, d^n\boldsymbol{x} \\ &= i_n \int_{\mathbb{R}^n} \prod_{k=1}^{n} \mathrm{rect}(x_k) e^{-i_n\boldsymbol{\omega}\cdot\boldsymbol{x}}\, d^n\boldsymbol{x} = i_n \prod_{k=1}^{n} \mathrm{sinc}(\frac{\omega_k}{2\pi}).\end{aligned} \tag{5.2.42}$$

Theorem 5.2.23. *The Clifford Fourier transform* $\mathcal{F}\{f\}$ *of* $f \in L^2(\mathbb{R}^n, \mathcal{G}_n)$, $n = 2, 3 \pmod 4$, $\int_{\mathbb{R}^n} \|f\|^2 d^n\boldsymbol{x} < \infty$ *is invertible and its inverse is calculated by*

$$\mathcal{F}^{-1}[\mathcal{F}\{f\}](\boldsymbol{x}) = f(\boldsymbol{x}) = \frac{1}{(2\pi)^n} \int_{\mathbb{R}^n} \mathcal{F}\{f\}(\boldsymbol{\omega})\, e^{i_n\boldsymbol{\omega}\cdot\boldsymbol{x}}\, d^n\boldsymbol{\omega}. \tag{5.2.43}$$

For a full proof of Theorem 5.2.23 in dimension $n = 3$ that can be generalized straight forwardly to dimensions $n = 2, 3 \pmod 4$ see [251]. Though Definition 5.2.21 and Theorem 5.2.23 are the same for the dimensions $n = 2 \pmod 4$ and $n = 3 \pmod 4$, care has to be taken of the general non-commutativity of i_n for $n = 2 \pmod 4$. However it turns out, that many properties of the CFT for $n = 2, 3 \pmod 4$ can be expressed independent of the commutation properties of i_n, if sufficient care is taken to avoid commuting i_n with other multivectors (except scalars and powers of i_n itself). Exceptions are the CFT of the Clifford convolution, $f\boldsymbol{x}^m$ and $f\,\nabla^m$, which need to be studied dimension dependent.

[9] Compare e.g. [221], article 160 for the precise conditions on the existence of Fourier integrals. For the Fourier transform on L^1 and L^2, see Footnote 2 on page 144.

We therefore continue with a general section on investigating properties of the CFT for $n = 2\,(\mathrm{mod}\,4)$ and $n = 3\,(\mathrm{mod}\,4)$ aiming at expressions that do not depend on the commutation properties of i_n. This will be followed by a section on the properties of the CFT of the Clifford convolution, $f\boldsymbol{x}^m$ and $f\,\nabla^m$. This second section will also include one table each for $n = 2\,(\mathrm{mod}\,4)$ and $n = 3\,(\mathrm{mod}\,4)$ that summarize all the properties of the CFT studied in this text and fully utilize the commutation properties of i_n.

5.2.1.2.1 **Properties of the CFT for $n = 2, 3\,(\mathrm{mod}\,4)$ expressed independent of i_n commutations** The properties of the CFT we will treat now are linearity, scaling, delay, shift, transformations of powers of the vector differential, of left and right powers of the vector derivative, of the vector variable $\boldsymbol{x} \in \mathbb{R}^n$, and finally the Plancherel and Parseval theorems. The unique feature of our study is the independence of theorem formulations and proofs on the commutation properties of the pseudoscalars i_n in dimensions $n = 2, 3\,(\mathrm{mod}\,4)$. If not otherwise stated, n is assumed to be $n = 2, 3\,(\mathrm{mod}\,4)$ in the remainder of this section.

Theorem 5.2.24 (Left linearity). *For $f(\boldsymbol{x}) = \alpha f_1(\boldsymbol{x}) + \beta f_2(\boldsymbol{x})$ with constants $\alpha, \beta \in \mathcal{G}_n$, and functions $f_1(\boldsymbol{x}),\ f_2(\boldsymbol{x}) \in \mathcal{G}_n$ we have*

$$\mathcal{F}\{f\}(\boldsymbol{\omega}) = \alpha\mathcal{F}\{f_1\}(\boldsymbol{\omega}) + \beta\mathcal{F}\{f_2\}(\boldsymbol{\omega}). \tag{5.2.44}$$

Proof. Follows from the linearity of the geometric product and the integration involved in the Definition 5.2.21 of the CFT. □

Remark 5.2.25. *Restricting the constants in Theorem 5.2.24 to $\alpha, \beta \in \mathbb{R}$ we get both left and right linearity of the CFT.*

Theorem 5.2.26 (Scaling). *Let $a \in \mathbb{R}, a \neq 0$ be a scalar constant, then the Clifford Fourier transform of the function $f_a(\boldsymbol{x}) = f(a\boldsymbol{x})$ becomes*

$$\mathcal{F}\{f_a\}(\boldsymbol{\omega}) = \frac{1}{|a|^n}\mathcal{F}\{f\}(\frac{\boldsymbol{\omega}}{a}). \tag{5.2.45}$$

Proof. Follows from variable substitution $\boldsymbol{u} = a\boldsymbol{x}$. □

Theorem 5.2.27 (Shift in space domain, delay). *If the argument of $f(\boldsymbol{x}) \in \mathcal{G}_n$ is offset by a constant vector $\boldsymbol{a} \in \mathbb{R}^n$, i.e. $f_d(\boldsymbol{x}) = f(\boldsymbol{x} - \boldsymbol{a})$, then*

$$\mathcal{F}\{f_d\}(\boldsymbol{\omega}) = \mathcal{F}\{f\}(\boldsymbol{\omega})e^{-i_n\boldsymbol{\omega}\cdot\boldsymbol{a}}. \tag{5.2.46}$$

Proof. Definition 5.2.21 gives

$$\mathcal{F}\{f_d\}(\boldsymbol{\omega}) = \int_{\mathbb{R}^n} f(\boldsymbol{x} - \boldsymbol{a})e^{-i_n\boldsymbol{\omega}\cdot\boldsymbol{x}}\,d^n\boldsymbol{x}.$$

We substitute $\boldsymbol{t}$ for $\boldsymbol{x} - \boldsymbol{a}$ in the above expression, and get with $d^n\boldsymbol{x} = d^n\boldsymbol{t}$

$$\mathcal{F}\{f_d\}(\boldsymbol{\omega}) = \int_{\mathbb{R}^n} f(\boldsymbol{t})\,e^{-i_n\boldsymbol{\omega}\cdot\boldsymbol{t}}e^{-i_n\boldsymbol{\omega}\cdot\boldsymbol{a}}\,d^n\boldsymbol{t} = \mathcal{F}\{f\}(\boldsymbol{\omega})e^{-i_n\boldsymbol{\omega}\cdot\boldsymbol{a}}. \tag{5.2.47}$$

This proves (5.2.46). □

Exercise 5.2.28. *Using example 5.2.22 we can calculate the CFT of a shifted n-dimensional* $\text{rect}(\boldsymbol{x})$ *function with center at* $\boldsymbol{a} = 3\boldsymbol{e}_2$, $\omega_2 = \boldsymbol{\omega} \cdot \boldsymbol{e}_2$ *as*

$$\mathcal{F}\{\text{rect}_d\}(\boldsymbol{\omega}) = \int_{\mathbb{R}^n} \text{rect}(\boldsymbol{x} - 3\boldsymbol{e}_2) e^{-i_n \boldsymbol{\omega}\cdot\boldsymbol{x}}\, d^n\boldsymbol{x} = i_n\, e^{-3i_n\omega_2} \prod_{k=1}^{n} \text{sinc}(\frac{\omega_k}{2\pi}). \quad (5.2.48)$$

Theorem 5.2.29 (Shift in frequency domain). *If* $\boldsymbol{\omega}_0 \in \mathbb{R}^n$ *and* $f_0(\boldsymbol{x}) = f(\boldsymbol{x})\, e^{i_n \boldsymbol{\omega}_0 \cdot \boldsymbol{x}}$, *then*

$$\mathcal{F}\{f_0\}(\boldsymbol{\omega}) = \mathcal{F}\{f\}(\boldsymbol{\omega} - \boldsymbol{\omega}_0) \quad (5.2.49)$$

Proof. Using Definition 5.2.21 and simplifying it we obtain

$$\mathcal{F}\{f_0\}(\boldsymbol{\omega}) = \int_{\mathbb{R}^n} f(\boldsymbol{x}) e^{-i_n(\boldsymbol{\omega}-\boldsymbol{\omega}_0)\cdot\boldsymbol{x}}\, d^n\boldsymbol{x} = \mathcal{F}\{f\}(\boldsymbol{\omega} - \boldsymbol{\omega}_0). \quad (5.2.50)$$

□

The CFT $\mathcal{F}\{f\}(\boldsymbol{\omega} - \boldsymbol{\omega}_0)$ is centered on the point $\boldsymbol{\omega} = \boldsymbol{\omega}_0$ in the frequency domain.

Theorem 5.2.30 (Powers of $\boldsymbol{x} \in \mathbb{R}^n$ from left).

$$\mathcal{F}\{\boldsymbol{x}^m f(\boldsymbol{x})\}(\boldsymbol{\omega}) = \nabla^m_{\boldsymbol{\omega}} \mathcal{F}\{f\}(\boldsymbol{\omega})\, i_n^m, \quad m \in \mathbb{N}. \quad (5.2.51)$$

Proof. We first proof Theorem 5.2.30 for $m = 1$. Direct calculation leads to

$$\begin{aligned} \mathcal{F}\{\boldsymbol{x} f(\boldsymbol{x})\}(\boldsymbol{\omega}) &= \int_{\mathbb{R}^n} \boldsymbol{x} f(\boldsymbol{x})\, e^{-i_n \boldsymbol{\omega}\cdot\boldsymbol{x}}\, d^n\boldsymbol{x} = \int_{\mathbb{R}^n} \nabla_{\boldsymbol{\omega}} f(\boldsymbol{x})\, i_n\, e^{-i_n \boldsymbol{\omega}\cdot\boldsymbol{x}}\, d^n\boldsymbol{x} \\ &= \nabla_{\boldsymbol{\omega}} \int_{\mathbb{R}^n} f(\boldsymbol{x})\, e^{-i_n \boldsymbol{\omega}\cdot\boldsymbol{x}}\, d^n\boldsymbol{x}\; i_n = \nabla_{\boldsymbol{\omega}} \mathcal{F}\{f\}(\boldsymbol{\omega})\, i_n, \end{aligned} \quad (5.2.52)$$

where we have used Definition 5.2.21 and (5.2.30) of Theorem 5.2.16. We therefore have

$$\mathcal{F}\{\boldsymbol{x} f(\boldsymbol{x})\}(\boldsymbol{\omega}) = \nabla_{\boldsymbol{\omega}} \mathcal{F}\{f\}(\boldsymbol{\omega})\, i_n. \quad (5.2.53)$$

Repeating this process $m - 1$ times we get

$$\mathcal{F}\{\boldsymbol{x}^m f(\boldsymbol{x})\}(\boldsymbol{\omega}) = \nabla^m_{\boldsymbol{\omega}} \mathcal{F}\{f\}(\boldsymbol{\omega})\, i_n^m, \quad m \in \mathbb{N}. \quad (5.2.54)$$

□

Exercise 5.2.31. *The CFT of a Gaussian function* $f(\boldsymbol{x}) = \exp(-\boldsymbol{x}^2), \boldsymbol{x} \in \mathbb{R}^n$ *is again a Gaussian function*

$$\mathcal{F}\{f\}(\boldsymbol{\omega}) = \int_{\mathbb{R}^n} e^{-\boldsymbol{x}^2} e^{-i_n \boldsymbol{\omega}\cdot\boldsymbol{x}}\, d^n\boldsymbol{x} = \pi^{\frac{n}{2}} e^{-\frac{\boldsymbol{\omega}^2}{4}} \quad (5.2.55)$$

The CFT of its first moment is therefore according to Theorem 5.2.30 and propositions 5.2.5 and 5.2.6

$$\mathcal{F}\{\boldsymbol{x} f\}(\boldsymbol{\omega}) = \int_{\mathbb{R}^n} \boldsymbol{x} e^{-\boldsymbol{x}^2} e^{-i_n \boldsymbol{\omega}\cdot\boldsymbol{x}}\, d^n\boldsymbol{x} = \pi^{\frac{n}{2}} \nabla_{\boldsymbol{\omega}} e^{-\frac{\boldsymbol{\omega}^2}{4}}\, i_n = -\frac{\pi^{\frac{n}{2}}}{2}\, \boldsymbol{\omega}\, i_n\, e^{-\frac{\boldsymbol{\omega}^2}{4}}. \quad (5.2.56)$$

Theorem 5.2.32 ($\boldsymbol{x}^m$ from right).

$$\mathcal{F}\{f(\boldsymbol{x})\,\boldsymbol{x}^m\}(\boldsymbol{\omega}) = \int_{\mathbb{R}^n} f(\boldsymbol{x})\nabla_{\boldsymbol{\omega}}^m\, e^{-i_n\boldsymbol{\omega}\cdot\boldsymbol{x}}\, d^n\boldsymbol{x}\; i_n^m, \quad m \in \mathbb{N}. \tag{5.2.57}$$

Proof. Direct calculation leads to

$$\begin{aligned}\mathcal{F}\{f(\boldsymbol{x})\,\boldsymbol{x}\}(\boldsymbol{\omega}) &= \int_{\mathbb{R}^n} f(\boldsymbol{x})\,\boldsymbol{x}\, e^{-i_n\boldsymbol{\omega}\cdot\boldsymbol{x}}\, d^n\boldsymbol{x} \\ = \int_{\mathbb{R}^n} f(\boldsymbol{x})\,\nabla_{\boldsymbol{\omega}} i_n\, e^{-i_n\boldsymbol{\omega}\cdot\boldsymbol{x}}\, d^n\boldsymbol{x} &= \int_{\mathbb{R}^n} f(\boldsymbol{x})\nabla_{\boldsymbol{\omega}}\, e^{-i_n\boldsymbol{\omega}\cdot\boldsymbol{x}}\, d^n\boldsymbol{x}\; i_n\end{aligned} \tag{5.2.58}$$

where we have used Definition 5.2.21 and (5.2.31) of Theorem 5.2.16. Replacing f $m-1$ times by $f\boldsymbol{x}$ and converting the additional right factor $\boldsymbol{x}$ each time into a derivative of $e^{-i_n\boldsymbol{\omega}\cdot\boldsymbol{x}}$ leads to the full proof of the theorem. □

Remark 5.2.33. *In the next section, we will use Theorem 5.2.32 and the dimension dependent commutation properties of i_n to derive final formulas for the CFTs of $f(\boldsymbol{x})\,\boldsymbol{x}^m$, $m \in \mathbb{N}$.*

Theorem 5.2.34.

$$\mathcal{F}\{(\boldsymbol{a}\cdot\boldsymbol{x})^m f(\boldsymbol{x})\}(\boldsymbol{\omega}) = (\boldsymbol{a}\cdot\nabla_{\boldsymbol{\omega}})^m\, \mathcal{F}\{f\}(\boldsymbol{\omega})\, i_n^m, \quad m \in \mathbb{N}. \tag{5.2.59}$$

Proof. We first proof Theorem 5.2.34 for $m = 1$.

$$\begin{aligned}\mathcal{F}\{\boldsymbol{a}\cdot\boldsymbol{x}\, f(\boldsymbol{x})\}(\boldsymbol{\omega}) &= \int_{\mathbb{R}^n} \boldsymbol{a}\cdot\boldsymbol{x}\, f(\boldsymbol{x})\, e^{-i_n\boldsymbol{\omega}\cdot\boldsymbol{x}}\, d^n\boldsymbol{x} \\ &= \int_{\mathbb{R}^n} f(\boldsymbol{x})\, \boldsymbol{a}\cdot\boldsymbol{x}\, e^{-i_n\boldsymbol{\omega}\cdot\boldsymbol{x}}\, d^n\boldsymbol{x} \\ &\overset{\text{Theor. 5.2.11}}{=} \int_{\mathbb{R}^n} f(\boldsymbol{x})\, \boldsymbol{a}\cdot\nabla_{\boldsymbol{\omega}}\, i_n\, e^{-i_n\boldsymbol{\omega}\cdot\boldsymbol{x}}\, d^n\boldsymbol{x} \\ &= \boldsymbol{a}\cdot\nabla_{\boldsymbol{\omega}} \int_{\mathbb{R}^n} f(\boldsymbol{x})\, e^{-i_n\boldsymbol{\omega}\cdot\boldsymbol{x}}\, d^n\boldsymbol{x}\; i_n \\ &= \boldsymbol{a}\cdot\nabla_{\boldsymbol{\omega}}\, \mathcal{F}\{f\}(\boldsymbol{\omega})\, i_n.\end{aligned} \tag{5.2.60}$$

Repeatedly inserting $\boldsymbol{a}\cdot\boldsymbol{x}f$ for f in (5.2.60) we obtain Theorem 5.2.34 for every $m \in \mathbb{N}$. □

Inserting $\boldsymbol{b}\cdot\boldsymbol{x}f$ with $\boldsymbol{b} \in \mathbb{R}^n$ for f in (5.2.60) we obtain the following corollary.

Corollary 5.2.35.

$$\mathcal{F}\{\boldsymbol{a}\cdot\boldsymbol{x}\;\boldsymbol{b}\cdot\boldsymbol{x}f(\boldsymbol{x})\}(\boldsymbol{\omega}) = -\,\boldsymbol{a}\cdot\nabla_{\boldsymbol{\omega}}\;\boldsymbol{b}\cdot\nabla_{\boldsymbol{\omega}}\,\mathcal{F}\{f\}(\boldsymbol{\omega}). \tag{5.2.61}$$

Theorem 5.2.36 (Vector differential). *The Clifford Fourier transform of the mth power vector differential of $f(\boldsymbol{x})$ is*

$$\mathcal{F}\{(\boldsymbol{a}\cdot\nabla)^m f\}(\boldsymbol{\omega}) = (\boldsymbol{a}\cdot\boldsymbol{\omega})^m\, \mathcal{F}\{f\}(\boldsymbol{\omega})\, i_n^m, \quad m \in \mathbb{N}. \tag{5.2.62}$$

Proof. We first proof Theorem 5.2.36 for $m = 1$.

$$\begin{aligned}
\boldsymbol{a}\cdot\nabla f(\boldsymbol{x}) &= \boldsymbol{a}\cdot\nabla\frac{1}{(2\pi)^n}\int_{\mathbb{R}^n}\mathcal{F}\{f\}(\boldsymbol{\omega})\,e^{i_n\boldsymbol{\omega}\cdot\boldsymbol{x}}\,d^n\boldsymbol{\omega}\\
&= \frac{1}{(2\pi)^n}\int_{\mathbb{R}^n}\mathcal{F}\{f\}(\boldsymbol{\omega})\,\boldsymbol{a}\cdot\nabla\, e^{i_n\boldsymbol{\omega}\cdot\boldsymbol{x}}\,d^n\boldsymbol{\omega}\\
&\overset{\text{Cor. 5.2.13}}{=} \frac{1}{(2\pi)^n}\int_{\mathbb{R}^n}\mathcal{F}\{f\}(\boldsymbol{\omega})\,\boldsymbol{a}\cdot\boldsymbol{\omega}i_n\, e^{i_n\boldsymbol{\omega}\cdot\boldsymbol{x}}\,d^n\boldsymbol{\omega}\\
&= \mathcal{F}^{-1}[\boldsymbol{a}\cdot\boldsymbol{\omega}\,\mathcal{F}\{f\}\,i_n](\boldsymbol{x}). \qquad (5.2.63)
\end{aligned}$$

Application of the inverse CFT Theorem 5.2.23 proves Theorem 5.2.36 for $m = 1$

$$\mathcal{F}\{\boldsymbol{a}\cdot\nabla f(\boldsymbol{x})\} = \boldsymbol{a}\cdot\boldsymbol{\omega}\,\mathcal{F}\{f\}\,i_n. \qquad (5.2.64)$$

By repeatedly replacing f with $\boldsymbol{a}\cdot\nabla f$ in (5.2.64) we obtain Theorem 5.2.36 for all $m \in \mathbb{N}$. □

Theorem 5.2.37 (Left vector derivative). *The Clifford Fourier transform of the mth power vector derivative of $f(\boldsymbol{x})$ is*

$$\mathcal{F}\{\nabla^m f\}(\boldsymbol{\omega}) = \boldsymbol{\omega}^m\,\mathcal{F}\{f\}(\boldsymbol{\omega})\,i_n^m, \quad m \in \mathbb{N}. \qquad (5.2.65)$$

Proof. We first proof Theorem 5.2.37 for $m = 1$. According to proposition 5.2.6 we can calculate the derivative from the differential of Theorem 5.2.36

$$\begin{aligned}
\mathcal{F}\{\nabla f(\boldsymbol{x})\} &= \mathcal{F}\{\nabla_{\boldsymbol{a}}[\boldsymbol{a}\cdot\nabla f(\boldsymbol{x})]\} = \nabla_{\boldsymbol{a}}\mathcal{F}\{\boldsymbol{a}\cdot\nabla f(\boldsymbol{x})\}\\
&= \nabla_{\boldsymbol{a}}(\boldsymbol{a}\cdot\boldsymbol{\omega})\,\mathcal{F}\{f\}\,i_n = \boldsymbol{\omega}\,\mathcal{F}\{f\}\,i_n \qquad (5.2.66)
\end{aligned}$$

By repeatedly replacing f with ∇f in (5.2.66) we obtain Theorem 5.2.37 for all $m \in \mathbb{N}$. □

Exercise 5.2.38. *Using the CFT of a Gaussian function* $f(\boldsymbol{x}) = \exp(-\boldsymbol{x}^2)$ *of (5.2.55) we can calculate the CFT of its third vector derivative with (5.2.65) as*

$$\mathcal{F}\{\nabla^3 f\}(\boldsymbol{\omega}) = \int_{\mathbb{R}^n}\nabla^3 e^{-\boldsymbol{x}^2}e^{-i_n\boldsymbol{\omega}\cdot\boldsymbol{x}}d^n\boldsymbol{x} = -\pi^{\frac{n}{2}}\,\boldsymbol{\omega}^3\,i_n\,e^{-\frac{\boldsymbol{\omega}^2}{4}}. \qquad (5.2.67)$$

We now prove a theorem, which we will use in the next section together with the dimension dependent commutation properties of i_n to derive final formulas for the CFT of powers of the right vector derivative $f(\boldsymbol{x})\,\nabla^m$, $m \in \mathbb{N}$.

Theorem 5.2.39 (∇^m from right).

$$f(\boldsymbol{x})\nabla^m = \frac{1}{(2\pi)^n}\int_{\mathbb{R}^n}\mathcal{F}\{f\}(\boldsymbol{\omega})\,i_n^m\,e^{i_n\boldsymbol{\omega}\cdot\boldsymbol{x}}\,\boldsymbol{\omega}^m\,d^n\boldsymbol{\omega}. \qquad (5.2.68)$$

Proof. We first proof Theorem 5.2.39 for $m = 1$.

$$\begin{aligned} f(\boldsymbol{x})\nabla &= \frac{1}{(2\pi)^3}\int_{\mathbb{R}^n} \mathcal{F}\{f\}(\boldsymbol{\omega})\, e^{i_n \boldsymbol{\omega}\cdot\boldsymbol{x}} d^n\boldsymbol{\omega}\nabla \\ &= \frac{1}{(2\pi)^3}\int_{\mathbb{R}^n} \mathcal{F}\{f\}(\boldsymbol{\omega})\, e^{i_n \boldsymbol{\omega}\cdot\boldsymbol{x}}\, \nabla\, d^n\boldsymbol{\omega} \\ &= \frac{1}{(2\pi)^3}\int_{\mathbb{R}^n} \mathcal{F}\{f\}(\boldsymbol{\omega})\, e^{i_n \boldsymbol{\omega}\cdot\boldsymbol{x}}\, i_n\, \boldsymbol{\omega}\, d^n\boldsymbol{\omega} \\ &= \frac{1}{(2\pi)^3}\int_{\mathbb{R}^n} \mathcal{F}\{f\}(\boldsymbol{\omega})\, i_n\, e^{i_n \boldsymbol{\omega}\cdot\boldsymbol{x}}\, \boldsymbol{\omega}\, d^n\boldsymbol{\omega}, \end{aligned} \tag{5.2.69}$$

where we used Corollary 5.2.15 for the third equality. Repeating the application of the vector derivative ∇ from the right $m-1$ times to both sides of (5.2.69) completes the proof. □

Next we will prove a Plancherel theorem and deduce a scalar Parseval theorem, which we need in the last section on the uncertainty principles.

Theorem 5.2.40 (Plancherel). *Assume that $f_1(\boldsymbol{x}), f_2(\boldsymbol{x}) \in \mathcal{G}_n$ with Clifford Fourier transform $\mathcal{F}\{f_1\}(\boldsymbol{\omega})$ and $\mathcal{F}\{f_2\}(\boldsymbol{\omega})$ respectively, then*

$$\int_{\mathbb{R}^n} f_1(\boldsymbol{x})\, \widetilde{f_2(\boldsymbol{x})}\, d^n\boldsymbol{x} = \frac{1}{(2\pi)^n}\int_{\mathbb{R}^n} \mathcal{F}\{f_1\}(\boldsymbol{\omega})\widetilde{\mathcal{F}\{f_2\}(\boldsymbol{\omega})}\, d^n\boldsymbol{x}. \tag{5.2.70}$$

Proof. Direct calculation yields

$$\begin{aligned} &\int_{\mathbb{R}^n} f_1(\boldsymbol{x})\, \widetilde{f_2(\boldsymbol{x})}\, d^n\boldsymbol{x} \\ &\quad= \frac{1}{(2\pi)^n}\int_{\mathbb{R}^n}\Big[\int_{\mathbb{R}^n} \mathcal{F}\{f_1\}(\boldsymbol{\omega})\, e^{i_n\boldsymbol{\omega}\cdot\boldsymbol{x}} d^n\boldsymbol{\omega}\Big]\, \widetilde{f_2(\boldsymbol{x})} d^n\boldsymbol{x} \\ &\quad= \frac{1}{(2\pi)^n}\int_{\mathbb{R}^n} \mathcal{F}\{f_1\}(\boldsymbol{\omega})\Big[\int_{\mathbb{R}^n} f_2(\boldsymbol{x})\, e^{-i_n\boldsymbol{\omega}\cdot\boldsymbol{x}}\, d^n\boldsymbol{x}\Big]^{\sim} d^n\boldsymbol{\omega} \\ &\quad= \frac{1}{(2\pi)^n}\int_{\mathbb{R}^n} \mathcal{F}\{f_1\}(\boldsymbol{\omega})\widetilde{\mathcal{F}\{f_2\}(\boldsymbol{\omega})} d^n\boldsymbol{\omega}. \end{aligned} \tag{5.2.71}$$

□

Note that Theorem 5.2.40 is multivector valued. It holds for each grade k, $0 \leq k \leq n$ of the multivectors on both sides of equation (5.2.70). We therefore have

Corollary 5.2.41.

$$\langle\int_{\mathbb{R}^n} f_1(\boldsymbol{x})\, \widetilde{f_2(\boldsymbol{x})}\, d^n\boldsymbol{x}\rangle_k = \frac{1}{(2\pi)^n}\langle\int_{\mathbb{R}^n} \mathcal{F}\{f_1\}(\boldsymbol{\omega})\widetilde{\mathcal{F}\{f_2\}(\boldsymbol{\omega})}\, d^n\boldsymbol{x}\rangle_k. \tag{5.2.72}$$

Note further, that with $f_1(\boldsymbol{x}) = f_2(\boldsymbol{x}) = f(\boldsymbol{x})$, we get the following multivector version of the *Parseval theorem*, i.e.

Theorem 5.2.42 (Multivector Parseval).

$$\int_{\mathbb{R}^n} f(\boldsymbol{x})\, \widetilde{f(\boldsymbol{x})}\, d^n\boldsymbol{x} = \frac{1}{(2\pi)^n}\int_{\mathbb{R}^n} \mathcal{F}\{f\}(\boldsymbol{\omega})\, \widetilde{\mathcal{F}\{f\}(\boldsymbol{\omega})}\, d^n\boldsymbol{x}. \tag{5.2.73}$$

Exercise 5.2.43. *According to (5.2.55) and left linearity of Theorem 5.2.24 the CFT of the function* $f(\boldsymbol{x}) = (1+\boldsymbol{e}_1)\exp(-\boldsymbol{x}^2), \boldsymbol{x} \in \mathbb{R}^2$ *is*

$$\mathcal{F}\{f\}(\boldsymbol{\omega}) = \int_{\mathbb{R}^2} (1+\boldsymbol{e}_1) e^{-\boldsymbol{x}^2} e^{-i_2\boldsymbol{\omega}\cdot\boldsymbol{x}} d^2\boldsymbol{x} = (1+\boldsymbol{e}_1)\pi e^{-\frac{\boldsymbol{\omega}^2}{4}} \tag{5.2.74}$$

Inserting this f *into (5.2.73) gives on the left side*

$$\int_{\mathbb{R}^2} f(\boldsymbol{x})\, \widetilde{f(\boldsymbol{x})}\, d^2\boldsymbol{x} = (1+\boldsymbol{e}_1)^2 \int_{\mathbb{R}^2} e^{-2\boldsymbol{x}^2}\, d^2\boldsymbol{x} = (1+\boldsymbol{e}_1)\pi. \tag{5.2.75}$$

We can check (5.2.73) by inserting (5.2.74) on the right side. The scalar part of the result is π*, the vector part is* $\pi\boldsymbol{e}_1$*.*

The scalar part of Theorem 5.2.42 together with (5.2.147), gives us the scalar Parseval theorem

Theorem 5.2.44 (Scalar Parseval).

$$\int_{\mathbb{R}^n} \|f(\boldsymbol{x})\|^2\, d^n\boldsymbol{x} = \frac{1}{(2\pi)^n}\int_{\mathbb{R}^n} \|\mathcal{F}\{f\}(\boldsymbol{\omega})\|^2\, d^n\boldsymbol{\omega}. \tag{5.2.76}$$

5.2.1.2.2 Properties of the CFT for $n = 2, 3 \,(\mathrm{mod}\, 4)$ dependent on i_n commutations

Now we concentrate on properties of the CFT, which need to make use of the commutation properties of the unit oriented pseudoscalar $i_n \in \mathcal{G}_n$ in order to be fully developed. In this category, we have for $m \in \mathbb{N}$ the CFTs of $f\boldsymbol{x}^m$, $f\nabla^m$ and the CFT of the Clifford convolution with distinct expressions for the dimensions of $n = 2\,(\mathrm{mod}\,4)$ and $n = 3\,(\mathrm{mod}\,4)$. Before we proceed, we note that for $n = 3\,(\mathrm{mod}\,4)$ the pseudoscalar i_n commutes with all elements of the algebra. For the case of $n = 2\,(\mathrm{mod}\,4)$ we first establish a theorem and two corollaries.

Theorem 5.2.45. *Any odd grade multivector* $A_r \in \mathcal{G}_n, r = 2s+1, s \in \mathbb{N}, s < \frac{n}{2}$ *anti-commutes with* i_n *for* $n = 2\,(\mathrm{mod}\,4)$

$$A_r i_n = -i_n A_r. \tag{5.2.77}$$

Any even grade multivector $A_r \in \mathcal{G}_n, r = 2s, s \in \mathbb{N}, s \leq n/2$ *commutes with* i_n *for* $n = 2\,(\mathrm{mod}\,4)$

$$A_r i_n = +i_n A_r. \tag{5.2.78}$$

Proof. For the case of $n = 2$ we have [58] for a vector $\boldsymbol{a} \in \mathbb{R}^2$

$$i_2\boldsymbol{a} = -\boldsymbol{a}i_2. \tag{5.2.79}$$

For the general case of $n = 2\,(\mathrm{mod}\,4)$ we can factorize i_n for any vector $\boldsymbol{a} \in \mathbb{R}^n$ such that

$$i_n = i_{n-2}\hat{\boldsymbol{b}}\,\hat{\boldsymbol{a}},\;\; \hat{\boldsymbol{b}} * \hat{\boldsymbol{a}} = \hat{\boldsymbol{b}}\rfloor i_{n-2} = \hat{\boldsymbol{a}}\rfloor i_{n-2} = 0,\;\; \hat{\boldsymbol{a}} = \frac{\boldsymbol{a}}{a},\;\; \hat{\boldsymbol{b}}^2 = 1. \tag{5.2.80}$$

i_{n-2} will therefore be of grade 0 (mod 4), represent a subspace[10] of $\mathbb{R}^n$ perpendicular to $\boldsymbol{a}$ and therefore commute with $\boldsymbol{a}$. According to (5.2.79) and (5.2.80) the two-blade $\hat{\boldsymbol{b}}\,\hat{\boldsymbol{a}}$ anti-commutes with $\boldsymbol{a}$ and hence

$$i_n\boldsymbol{a} = i_{n-2}\hat{\boldsymbol{b}}\,\hat{\boldsymbol{a}}\boldsymbol{a} = -i_{n-2}\boldsymbol{a}\,\hat{\boldsymbol{b}}\,\hat{\boldsymbol{a}} = -\boldsymbol{a}i_{n-2}\hat{\boldsymbol{b}}\,\hat{\boldsymbol{a}} = -\boldsymbol{a}i_n. \tag{5.2.81}$$

Any odd grade multivector A_{odd} can be written as a sum over homogeneous odd grade parts. These parts can in turn be decomposed into sums of odd grade blades, which can be factorized into products of an odd number of vectors [58, 179]. Since a single vector anti commutes with i_n, a geometric product of an odd number of vectors will also anti-commute with i_n and hence by linearity any odd grade multivector will anti-commute with i_n. This proves (5.2.77). Similarly any even grade multivector A_{even} can be written as a sum over homogeneous even grade parts. These parts can in turn be decomposed into sums of even grade blades, which can be factorized into geometric products of an even number of vectors [58, 179]. The even number of commutations with an even number of vector factors leads via linearity to the total commutation relationship (5.2.78). □

Based on Theorem 5.2.45 we derive two useful corollaries.

Corollary 5.2.46. *For $n = 2\,(\text{mod}\,4)$ and $\boldsymbol{a} \in \mathbb{R}$ we have for even $m \in \mathbb{N}$*

$$(\boldsymbol{a}\,i_n)^m = \boldsymbol{a}^m \tag{5.2.82}$$

and for odd $m \in \mathbb{N}$

$$(\boldsymbol{a}\,i_n)^m = \boldsymbol{a}^m\,i_n \tag{5.2.83}$$

Proof. We have

$$(\boldsymbol{a}\,i_n)^2 = \boldsymbol{a}\,i_n\boldsymbol{a}\,i_n = \boldsymbol{a}\boldsymbol{a}(-i_ni_n) = \boldsymbol{a}^2, \tag{5.2.84}$$

where we used (5.2.77) for the second equality. Using (5.2.84) $m/2$ times [$(m-1)/2$ times] we arrive at equations (5.2.82) [and (5.2.83)]. □

Corollary 5.2.47. *Let the odd grade part of a general multivector $A \in \mathcal{G}_n$ be $A_{odd} = \langle A\rangle_{odd}$ and the even grade part be $A_{even} = \langle A\rangle_{even}$. Then for $n = 2\,(\text{mod}\,4)$ we have*

$$A\,i_n = A_{odd}\,i_n + A_{even}\,i_n = -i_n\,A_{odd} + i_n\,A_{even}, \tag{5.2.85}$$

and for $\lambda \in \mathbb{R}$

$$Ae^{i_n\lambda} = e^{-i_n\lambda}A_{odd} + e^{+i_n\lambda}A_{even}, \tag{5.2.86}$$

and

$$e^{i_n\lambda}A = A_{odd}\,e^{-i_n\lambda} + A_{even}\,e^{+i_n\lambda}. \tag{5.2.87}$$

Proof. Corollary 5.2.47 follows from Theorem 5.2.45 and the fact that $e^{i_n\lambda}$ is a power series of i_n. □

---

[10] A subspace in the sense of outer product null space.

Theorem 5.2.48 (Powers of $\boldsymbol{x} \in \mathbb{R}^n$ from right). *For $n = 2 \pmod 4$ we have*

$$\mathcal{F}\{f(\boldsymbol{x})\, \boldsymbol{x}^m\}(\boldsymbol{\omega}) = \mathcal{F}\{f\}((-1)^m\boldsymbol{\omega})\, \nabla_{\boldsymbol{\omega}}^m\, i_n^m\,, \quad m \in \mathbb{N}. \tag{5.2.88}$$

For $n = 3 \pmod 4$ we have

$$\mathcal{F}\{f(\boldsymbol{x})\, \boldsymbol{x}^m\}(\boldsymbol{\omega}) = i_n^m \mathcal{F}\{f\}(\boldsymbol{\omega})\, \nabla_{\boldsymbol{\omega}}^m\,, \quad m \in \mathbb{N}. \tag{5.2.89}$$

Proof. We first proof Theorem 5.2.48 for $n = 2 \pmod 4$. We start with Theorem 5.2.32 and apply corollary 5.2.47 to commute the vector derivative $\nabla_{\boldsymbol{\omega}}$ to the right of the integral

$$\mathcal{F}\{f(\boldsymbol{x})\, \boldsymbol{x}^m\}(\boldsymbol{\omega}) = \int_{\mathbb{R}^n} f(\boldsymbol{x}) \nabla_{\boldsymbol{\omega}}^m\, e^{-i_n \boldsymbol{\omega}\cdot\boldsymbol{x}}\, d^n\boldsymbol{x}\; i_n^m$$
$$= \int_{\mathbb{R}^n} f(\boldsymbol{x}) e^{-i_n(-1)^m \boldsymbol{\omega}\cdot\boldsymbol{x}}\, d^n\boldsymbol{x}\, \nabla_{\boldsymbol{\omega}}^m\, i_n^m = \mathcal{F}\{f\}((-1)^m\boldsymbol{\omega})\, \nabla_{\boldsymbol{\omega}}^m\, i_n^m. \tag{5.2.90}$$

The proof of (5.2.89) with $n = 3 \pmod 4$ is the same, except that the sign of $\boldsymbol{\omega}$ in the exponent does not change and that we can freely commute i_n to the left. □

Theorem 5.2.49 (Right vector derivative). *The Clifford Fourier transform of the mth power vector derivative (applied from the right) of $f(\boldsymbol{x})$ is for $n = 2 \pmod 4$*

$$\mathcal{F}\{f\, \nabla^m\}(\boldsymbol{\omega}) = \mathcal{F}\{f\}((-1)^m\boldsymbol{\omega})\, \boldsymbol{\omega}^m\, i_n^m, \quad m \in \mathbb{N}, \tag{5.2.91}$$

and for $n = 3 \pmod 4$

$$\mathcal{F}\{f\, \nabla^m\}(\boldsymbol{\omega}) = i_n^m \mathcal{F}\{f\}(\boldsymbol{\omega})\, \boldsymbol{\omega}^m, \quad m \in \mathbb{N}. \tag{5.2.92}$$

Proof. We first proof Theorem 5.2.49 for $n = 2 \pmod 4$. We start with Theorem 5.2.39 and apply corollary 5.2.47 to commute the vector $\boldsymbol{\omega}^m$ with $e^{i_n \boldsymbol{\omega}\cdot\boldsymbol{x}}$

$$\begin{aligned}
f(\boldsymbol{x})\nabla^m &= \frac{1}{(2\pi)^n} \int_{\mathbb{R}^n} \mathcal{F}\{f\}(\boldsymbol{\omega})\, i_n^m\, e^{i_n \boldsymbol{\omega}\cdot\boldsymbol{x}}\, \boldsymbol{\omega}^m\, d^n\boldsymbol{\omega} \\
&= \frac{1}{(2\pi)^n} \int_{\mathbb{R}^n} \mathcal{F}\{f\}(\boldsymbol{\omega})\, i_n^m\, \boldsymbol{\omega}^m e^{i_n(-1)^m \boldsymbol{\omega}\cdot\boldsymbol{x}}\, d^n\boldsymbol{\omega} \\
&= \frac{1}{(2\pi)^n} \int_{\mathbb{R}^n} \mathcal{F}\{f\}((-1)^m\boldsymbol{\omega})\, i_n^m\, (-\boldsymbol{\omega})^m e^{i_n \boldsymbol{\omega}\cdot\boldsymbol{x}}\, d^n\boldsymbol{\omega} \\
&= \mathcal{F}^{-1}[\mathcal{F}\{f\}((-1)^m\boldsymbol{\omega})\, \boldsymbol{\omega}^m\, i_n^m](\boldsymbol{x}),
\end{aligned} \tag{5.2.93}$$

where for odd m we substituted $-\boldsymbol{\omega} \to \boldsymbol{\omega}$ for the third equality. For the fourth equality we applied apply corollary 5.2.47 once more to commute $\boldsymbol{\omega}^m$ and i_n^m. Equation (5.2.91) is obtained by applying the inverse CFT Theorem 5.2.23 to both sides of (5.2.93).

Once again the proof of (5.2.92) with $n = 3 \pmod 4$ is the same, except that the sign of $\boldsymbol{\omega}$ in the exponent does not change and that we can freely commute i_n to the left. □

Remark 5.2.50. *Theorem 5.2.37 and (5.2.91) show that all signs of the right hand sides of all five lines in the derivative Theorem 5.7 in [117] are wrong. Line three there contains further errors.*

For even m we get from Theorems 5.2.30 and 5.2.48, and from (5.2.5)

Corollary 5.2.51.

$$\mathcal{F}\{\boldsymbol{x}^m f(\boldsymbol{x})\}(\boldsymbol{\omega}) = \mathcal{F}\{f(\boldsymbol{x})\,\boldsymbol{x}^m\}(\boldsymbol{\omega}) = (-1)^{\frac{m}{2}} \nabla_{\boldsymbol{\omega}}^m \mathcal{F}\{f\}(\boldsymbol{\omega}). \tag{5.2.94}$$

We further get for even m from Theorems 5.2.37 and 5.2.49, and from (5.2.5)

Corollary 5.2.52.

$$\mathcal{F}\{\nabla^m f(\boldsymbol{x})\} = \mathcal{F}\{f(\boldsymbol{x})\,\nabla^m\} = (-1)^{\frac{m}{2}} \boldsymbol{\omega}^m\, \mathcal{F}\{f\}(\boldsymbol{\omega}). \tag{5.2.95}$$

Theorem 5.2.53 (CFT of Clifford Convolution). *For $n = 2\,(\mathrm{mod}\,4)$, $f, g \in L^2(\mathbb{R}^n, \mathcal{G}_n)$, and g_{odd} (g_{even}) the odd (even) grade part of g we have*

$$\mathcal{F}\{f{\star}g\}(\boldsymbol{\omega}) = \mathcal{F}\{f\}(-\boldsymbol{\omega})\,\mathcal{F}\{g_{odd}\}(\boldsymbol{\omega}) + \mathcal{F}\{f\}(\boldsymbol{\omega})\,\mathcal{F}\{g_{even}\}(\boldsymbol{\omega}). \tag{5.2.96}$$

For $n = 3\,(\mathrm{mod}\,4)$ we have

$$\mathcal{F}\{f{\star}g\}(\boldsymbol{\omega}) = \mathcal{F}\{f\}(\boldsymbol{\omega})\,\mathcal{F}\{g\}(\boldsymbol{\omega}). \tag{5.2.97}$$

Proof. For $n = 2\,(\mathrm{mod}\,4)$ we have

$$\begin{aligned}
\mathcal{F}\{f{\star}g\}(\boldsymbol{\omega}) &= \int_{\mathbb{R}^n}[\int_{\mathbb{R}^n} f(\boldsymbol{y})g(\boldsymbol{x}-\boldsymbol{y})d^n\boldsymbol{y}]e^{-i_n\boldsymbol{\omega}\cdot\boldsymbol{x}}d^n\boldsymbol{x} \\
&= \int_{\mathbb{R}^n} f(\boldsymbol{y})[\int_{\mathbb{R}^n} g(\boldsymbol{x}-\boldsymbol{y})e^{-i_n\boldsymbol{\omega}\cdot\boldsymbol{x}}d^n\boldsymbol{x}]d^n\boldsymbol{y} \\
&= \int_{\mathbb{R}^n} f(\boldsymbol{y})[\int_{\mathbb{R}^n} g(\boldsymbol{z})e^{-i_n\boldsymbol{\omega}\cdot(\boldsymbol{y}+\boldsymbol{z})}d^n\boldsymbol{z}]d^n\boldsymbol{y} \\
&= \int_{\mathbb{R}^n} f(\boldsymbol{y}) \int_{\mathbb{R}^n}[e^{+i_n\boldsymbol{\omega}\cdot\boldsymbol{y}}g_{odd}(\boldsymbol{z}) + e^{-i_n\boldsymbol{\omega}\cdot\boldsymbol{y}}g_{even}(\boldsymbol{z})]e^{-i_n\boldsymbol{\omega}\cdot\boldsymbol{z}}d^n\boldsymbol{z}d^n\boldsymbol{y} \\
&= \int_{\mathbb{R}^n} f(\boldsymbol{y})e^{-i_n(-\boldsymbol{\omega})\cdot\boldsymbol{y}}d^n\boldsymbol{y}\mathcal{F}\{g_{odd}\}(\boldsymbol{\omega}) + \int_{\mathbb{R}^n} f(\boldsymbol{y})e^{-i_n\boldsymbol{\omega}\cdot\boldsymbol{y}}d^n\boldsymbol{y}\mathcal{F}\{g_{even}\}(\boldsymbol{\omega}) \\
&= \mathcal{F}\{f\}(-\boldsymbol{\omega})\,\mathcal{F}\{g_{odd}\}(\boldsymbol{\omega}) + \mathcal{F}\{f\}(\boldsymbol{\omega})\,\mathcal{F}\{g_{even}\}(\boldsymbol{\omega}).
\end{aligned} \tag{5.2.98}$$

For the third equality we used the variable substitution $\boldsymbol{z} = \boldsymbol{x} - \boldsymbol{y}$ and for the fourth equality we used Corollary 5.2.47.

The corresponding proof in [252] for $n = 3$, can be generalized straight forwardly to $n = 3\,(\mathrm{mod}\,4)$. This part of the proof is also strictly invariant. □

Remark 5.2.54. *The above proof of Theorem 5.2.53 for $n = 2\,(\mathrm{mod}\,4)$ depends on the i_n commutation relationships. But on the other hand, it has the advantage of being manifestly invariant, since no coordinate system needed to be introduced.*

Table 5.3 Properties of the Clifford Fourier transform (CFT) with $n = 3 \pmod 4$. Multivector functions (Multiv. Funct.) $f, g, f_1, f_2 \in L^2(\mathbb{R}^n, Cl(n,0))$, constants $\alpha, \beta \in Cl(n,0)$, $0 \neq a \in \mathbb{R}$, $\boldsymbol{a}, \boldsymbol{\omega}_0 \in \mathbb{R}^n$ and $m \in \mathbb{N}$. Source: [185, Tab. 1.1].

Property	**Multiv. Funct.**	**CFT**
Left lin.	$\alpha f(\boldsymbol{x}) + \beta\, g(\boldsymbol{x})$	$\alpha \mathcal{F}\{f\}(\boldsymbol{\omega}) + \beta \mathcal{F}\{g\}(\boldsymbol{\omega})$
$\boldsymbol{x}$-Shift	$f(\boldsymbol{x} - \boldsymbol{a})$	$e^{-i_n \boldsymbol{\omega}\cdot\boldsymbol{a}} \mathcal{F}\{f\}(\boldsymbol{\omega})$
$\boldsymbol{\omega}$-Shift	$e^{i_n \boldsymbol{\omega}_0\cdot\boldsymbol{x}} f(\boldsymbol{x})$	$\mathcal{F}\{f\}(\boldsymbol{\omega} - \boldsymbol{\omega}_0)$
Scaling	$f(a\boldsymbol{x}),\ a \in \mathbb{R}\setminus\{0\}$	$\frac{1}{\lvert a\rvert^n} \mathcal{F}\{f\}(\frac{\boldsymbol{\omega}}{a})$
Convolution	$(f \star g)(\boldsymbol{x})$	$\mathcal{F}\{f\}(\boldsymbol{\omega})\, \mathcal{F}\{g\}(\boldsymbol{\omega})$
Vec. diff.	$(\boldsymbol{a}\cdot\nabla)^m f(\boldsymbol{x})$	$i_n^m\, (\boldsymbol{a}\cdot\boldsymbol{\omega})^m \mathcal{F}\{f\}(\boldsymbol{\omega})$
	$(\boldsymbol{a}\cdot\boldsymbol{x})^m\, f(\boldsymbol{x})$	$i_n^m\, (\boldsymbol{a}\cdot\nabla_{\boldsymbol{\omega}})^m\, \mathcal{F}\{f\}(\boldsymbol{\omega})$
Powers of $\boldsymbol{x}$	$\boldsymbol{x}^m f(\boldsymbol{x})$	$i_n^m\, \nabla_{\boldsymbol{\omega}}^m\, \mathcal{F}\{f\}(\boldsymbol{\omega})$
	$f(\boldsymbol{x})\boldsymbol{x}^m$	$i_n^m\, \mathcal{F}\{f\}(\boldsymbol{\omega})\, \nabla_{\boldsymbol{\omega}}^m$
Vec. deriv.	$\nabla^m f(\boldsymbol{x})$	$i_n^m\, \boldsymbol{\omega}^m \mathcal{F}\{f\}(\boldsymbol{\omega})$
	$f(\boldsymbol{x})\nabla^m$	$i_n^m\, \mathcal{F}\{f\}(\boldsymbol{\omega})\, \boldsymbol{\omega}^m$
Plancherel	$\int_{\mathbb{R}^n} f_1(\boldsymbol{x})\widetilde{f_2(\boldsymbol{x})}\, d^n\boldsymbol{x}$	$\frac{1}{(2\pi)^n}\int_{\mathbb{R}^n} \mathcal{F}\{f_1\}(\boldsymbol{\omega})\widetilde{\mathcal{F}\{f_2\}(\boldsymbol{\omega})}\, d^n\boldsymbol{\omega}$
sc. Parseval	$\int_{\mathbb{R}^n} \lVert f(\boldsymbol{x})\rVert^2\, d^n\boldsymbol{x}$	$\frac{1}{(2\pi)^n}\int_{\mathbb{R}^n} \lVert \mathcal{F}\{f\}(\boldsymbol{\omega})\rVert^2\, d^n\boldsymbol{\omega}$

Theorem 5.2.53 correctly generalizes the results for $n = 2$ in [117] to $n = 2 \pmod 4$. Comparing Theorem 5.2.53 with the convolution Theorem 5.6 of [117] in two dimensions, we see that the fourth line of convolution Theorem 5.6 in [117] must be wrong. On the right hand side of this formula the dot over the vector filter function $\mathbf{h}$ under the CFT indicating $\dot{\mathbf{h}}(\boldsymbol{x}) = \mathbf{h}(-\boldsymbol{x})$ is incorrect. Because of (5.2.38) $\mathbf{h}$ should also have no dot, in agreement with the correct dot-free vector filter function $\mathbf{f}$ on the right hand side of line two of Theorem 5.6 in [117].

In order to give an overview of the properties of the CFT we list its properties for $n = 3 \pmod 4$ in Table 5.3 and for $n = 2 \pmod 4$ in Table 5.4. Comparing the tables, the differences caused by the different commutation rules for the pseudoscalars i_n in $n = 3 \pmod 4$ and $n = 2 \pmod 4$ dimensions are obvious. In Table 5.3 the positions of i_n and of exponentials $e^{i_n\lambda}$, $\lambda \in \mathbb{R}$ are arbitrary. In Table 5.4 the pseudoscalar i_n and its exponentials $e^{i_n\lambda}$, $\lambda \in \mathbb{R}$ cannot be freely commuted.

5.2.1.3 Uncertainty principle

The uncertainty principle plays an important role in the development and understanding of quantum physics. It is also central for information processing [287]. In quantum physics it states e.g. that conjugate properties like particle momentum and position cannot be be simultaneously measured accurately. In Fourier analysis such conjugate entities correspond to a function and its Fourier transform which cannot both be simultaneously sharply localized. Furthermore much work (e.g. [74, 287]) has been devoted to extending the uncertainty principle to a function and its Fourier transform. Yet Felsberg [132] notes even for two dimensions: *In two dimensions*

Table 5.4 Properties of the Clifford Fourier transform (CFT) with $n = 2 \pmod 4$. Multivector functions (Multiv. Funct.) $f, g, f_1, f_2 \in L^2(\mathbb{R}^n, Cl(n,0))$, constants $\alpha, \beta \in Cl(n,0)$, $0 \neq a \in \mathbb{R}$, $\boldsymbol{a}, \boldsymbol{\omega}_0 \in \mathbb{R}^n$ and $m \in \mathbb{N}$. Source: [185, Tab. 1.2].

Property	Multiv. Funct.	CFT
Left lin.	$\alpha f(\boldsymbol{x}) + \beta\, g(\boldsymbol{x})$	$\alpha \mathcal{F}\{f\}(\boldsymbol{\omega}) + \beta \mathcal{F}\{g\}(\boldsymbol{\omega})$
$\boldsymbol{x}$-Shift	$f(\boldsymbol{x} - \boldsymbol{a})$	$\mathcal{F}\{f\}(\boldsymbol{\omega})\, e^{-i_n \boldsymbol{\omega}\cdot\boldsymbol{a}}$
$\boldsymbol{\omega}$-Shift	$f(\boldsymbol{x})\, e^{i_n \boldsymbol{\omega}_0\cdot\boldsymbol{x}}$	$\mathcal{F}\{f\}(\boldsymbol{\omega} - \boldsymbol{\omega}_0)$
Scaling	$f(a\boldsymbol{x}),\ a \in \mathbb{R}\setminus\{0\}$	$\frac{1}{\lvert a\rvert^n}\mathcal{F}\{f\}(\frac{\boldsymbol{\omega}}{a})$
Convolution	$(f \star g)(\boldsymbol{x})$	$\mathcal{F}\{f\}(-\boldsymbol{\omega})\, \mathcal{F}\{g_{odd}\}(\boldsymbol{\omega})$ $+\mathcal{F}\{f\}(\boldsymbol{\omega})\, \mathcal{F}\{g_{even}\}(\boldsymbol{\omega})$
Vec. diff.	$(\boldsymbol{a}\cdot\nabla)^m f(\boldsymbol{x})$	$(\boldsymbol{a}\cdot\boldsymbol{\omega})^m \mathcal{F}\{f\}(\boldsymbol{\omega})\, i_n^m$
	$(\boldsymbol{a}\cdot\boldsymbol{x})^m\, f(\boldsymbol{x})$	$(\boldsymbol{a}\cdot\nabla_{\boldsymbol{\omega}})^m\, \mathcal{F}\{f\}(\boldsymbol{\omega})\, i_n^m$
Powers of $\boldsymbol{x}$	$\boldsymbol{x}^m f(\boldsymbol{x})$	$\nabla_{\boldsymbol{\omega}}^m\, \mathcal{F}\{f\}(\boldsymbol{\omega})\, i_n^m$
	$f(\boldsymbol{x})\, \boldsymbol{x}^m$	$\mathcal{F}\{f\}((-1)^m\boldsymbol{\omega})\, \nabla_{\boldsymbol{\omega}}^m\, i_n^m$
Vec. deriv.	$\nabla^m f(\boldsymbol{x})$	$\boldsymbol{\omega}^m\, \mathcal{F}\{f\}(\boldsymbol{\omega})\, i_n^m$
	$f(\boldsymbol{x})\nabla^m$	$\mathcal{F}\{f\}((-1)^m\boldsymbol{\omega})\, \boldsymbol{\omega}^m\, i_n^m$
Plancherel	$\int_{\mathbb{R}^n} f_1(\boldsymbol{x})\widetilde{f_2(\boldsymbol{x})}\, d^n\boldsymbol{x}$	$\frac{1}{(2\pi)^n}\int_{\mathbb{R}^n} \mathcal{F}\{f_1\}(\boldsymbol{\omega})\widetilde{\mathcal{F}\{f_2\}(\boldsymbol{\omega})}\, d^n\boldsymbol{\omega}$
sc. Parseval	$\int_{\mathbb{R}^n} \lVert f(\boldsymbol{x})\rVert^2\, d^n\boldsymbol{x}$	$\frac{1}{(2\pi)^n}\int_{\mathbb{R}^n} \lVert \mathcal{F}\{f\}(\boldsymbol{\omega})\rVert^2\, d^n\boldsymbol{\omega}$

however, the uncertainty relation is still an open problem. In [145] it is stated that there is no straightforward formulation for the two-dimensional uncertainty relation.

From the view point of geometric algebra an uncertainty principle gives us information about how the variations of a multivector valued function and its Clifford Fourier transform are related.

The theorems and the corollary below have all been proven with great detail for the case of $n = 3$ in [251]. The key steps of the proofs there involve the CFT of the vector differential of Theorem 5.2.36 and the (scalar) Parseval Theorem 5.2.44, the Cauchy Schwarz inequality (5.2.6) for multivectors, and finally the coordinate free integration of parts formula of Proposition 5.2.9. Otherwise the proofs are very analogous, and do not involve dimension dependent i_n commutations. Therefore we don't repeat them here, we only list the resulting formulas.

Theorem 5.2.55 (Directional uncertainty principle). *Let f be a multivector valued function in $\mathcal{G}_n$, $n = 2, 3 \pmod 4$, which has the Clifford Fourier transform $\mathcal{F}\{f\}(\boldsymbol{\omega})$. Assume $\int_{\mathbb{R}^n} \lVert f(\boldsymbol{x})\rVert^2\, d^n\boldsymbol{x} = F < \infty$, then the following inequality holds for arbitrary constant vectors $\boldsymbol{a}$, $\boldsymbol{b}$:*

$$\int_{\mathbb{R}^n} (\boldsymbol{a}\cdot\boldsymbol{x})^2 \lVert f(\boldsymbol{x})\rVert^2\, d^n\boldsymbol{x}\, \frac{1}{(2\pi)^n}\int_{\mathbb{R}^n} (\boldsymbol{b}\cdot\boldsymbol{\omega})^2\, \lVert \mathcal{F}\{f\}(\boldsymbol{\omega})\rVert^2 d^n\boldsymbol{\omega}$$
$$\geq (\boldsymbol{a}\cdot\boldsymbol{b})^2\, \frac{1}{4}\, F^2. \tag{5.2.99}$$

Proof. Applying the results stated so far we have[11]

$$\begin{aligned}
&\int_{\mathbb{R}^n} (\boldsymbol{a}\cdot\boldsymbol{x})^2\, \|f(\boldsymbol{x})\|^2\, d^n\boldsymbol{x}\, \frac{1}{(2\pi)^n} \int_{\mathbb{R}^n} (\boldsymbol{b}\cdot\boldsymbol{\omega})^2\, \|\mathcal{F}\{f\}(\boldsymbol{\omega})\|^2 d^n\boldsymbol{\omega} \\
&\overset{\text{tab. 5.3,5.4, line 6}}{=} \int_{\mathbb{R}^n} (\boldsymbol{a}\cdot\boldsymbol{x})^2\, \|f(\boldsymbol{x})\|^2\, d^n\boldsymbol{x}\, \frac{1}{(2\pi)^n} \int_{\mathbb{R}^n} \|\mathcal{F}\{\boldsymbol{b}\cdot\nabla f\}(\boldsymbol{\omega})\|^2\, d^n\boldsymbol{\omega} \\
&\overset{\text{sc. Parseval}}{=} \int_{\mathbb{R}^n} (\boldsymbol{a}\cdot\boldsymbol{x})^2\, \|f(\boldsymbol{x})\|^2\, d^n\boldsymbol{x} \int_{\mathbb{R}^n} \|\boldsymbol{b}\cdot\nabla f(\boldsymbol{x})\|^2\, d^n\boldsymbol{x} \\
&\overset{\text{footnote 11}}{\geq} \left(\int_{\mathbb{R}^n} \boldsymbol{a}\cdot\boldsymbol{x}\, \|f(\boldsymbol{x})\|\, \|\boldsymbol{b}\cdot\nabla f(\boldsymbol{x})\|\, d^n\boldsymbol{x} \right)^2 \\
&\overset{(5.2.6)}{\geq} \left(\int_{\mathbb{R}^n} \boldsymbol{a}\cdot\boldsymbol{x} |\langle \widetilde{f(\boldsymbol{x})}\, \boldsymbol{b}\cdot\nabla f(\boldsymbol{x})\rangle|\, d^n\boldsymbol{x} \right)^2 \\
&\geq \left(\int_{\mathbb{R}^n} \boldsymbol{a}\cdot\boldsymbol{x} \langle \widetilde{f(\boldsymbol{x})}\, \boldsymbol{b}\cdot\nabla f(\boldsymbol{x})\rangle\, d^n\boldsymbol{x} \right)^2 . \qquad (5.2.100)
\end{aligned}$$

Because of (5.2.147) and (5.2.148)

$$(\boldsymbol{b}\cdot\nabla)\|f\|^2 = 2\langle \tilde{f}\, (\boldsymbol{b}\cdot\nabla) f\rangle, \qquad (5.2.101)$$

we furthermore obtain

$$\begin{aligned}
&\int_{\mathbb{R}^n} (\boldsymbol{a}\cdot\boldsymbol{x})^2\, \|f(\boldsymbol{x})\|^2\, d^n\boldsymbol{x}\, \frac{1}{(2\pi)^n} \int_{\mathbb{R}^n} (\boldsymbol{b}\cdot\boldsymbol{\omega})^2\, \|\mathcal{F}\{f\}(\boldsymbol{\omega})\|^2 d^n\boldsymbol{\omega} \\
&\geq \left(\int_{\mathbb{R}^n} \boldsymbol{a}\cdot\boldsymbol{x}\, \frac{1}{2}\, (\boldsymbol{b}\cdot\nabla\|f\|^2)\, d^n\boldsymbol{x} \right)^2 \\
&\overset{\text{Prop. 5.2.9}}{=} \frac{1}{4} \left(\left[\int_{\mathbb{R}^{n-1}} \boldsymbol{a}\cdot\boldsymbol{x} \|f(\boldsymbol{x})\|^2 d^{n-1}\boldsymbol{x} \right]_{b\cdot x=-\infty}^{b\cdot x=\infty} \right. \\
&\qquad \left. - \int_{\mathbb{R}^n} [(\boldsymbol{b}\cdot\nabla)(\boldsymbol{a}\cdot\boldsymbol{x})]\, \|f(\boldsymbol{x})\|^2\, d^n\boldsymbol{x} \right)^2 \\
&= \frac{1}{4} \left(0 - \boldsymbol{a}\cdot\boldsymbol{b} \int_{\mathbb{R}^n} \|f(\boldsymbol{x})\|^2\, d^n\boldsymbol{x} \right)^2 \\
&= (\boldsymbol{a}\cdot\boldsymbol{b})^2\, \frac{1}{4}\, F^2. \qquad (5.2.102)
\end{aligned}$$

□

Choosing $\boldsymbol{b} = \pm\boldsymbol{a}$, i.e. $\boldsymbol{b} \parallel \boldsymbol{a}$, with $\boldsymbol{a}^2 = 1$ we get from Theorem 5.2.55 the following

Corollary 5.2.56 (Uncertainty principle).

$$\int_{\mathbb{R}^n} (\boldsymbol{a}\cdot\boldsymbol{x})^2 \|f(\boldsymbol{x})\|^2\, d^n\boldsymbol{x}\, \frac{1}{(2\pi)^n} \int_{\mathbb{R}^n} (\boldsymbol{a}\cdot\boldsymbol{\omega})^2\, \|\mathcal{F}\{f\}(\boldsymbol{\omega})\|^2 d^n\boldsymbol{\omega} \geq \frac{1}{4}\, F^2. \qquad (5.2.103)$$

[11] $\phi, \psi : \mathbb{R}^n \to \mathbb{C}, \quad \int_{\mathbb{R}^n} |\phi(x)|^2 d^n x \int_{\mathbb{R}^n} |\psi(x)|^2 d^n x \geq (\int_{\mathbb{R}^n} \phi(x)\overline{\psi(x)}\, d^n x)^2$

Remark 5.2.57. *In (5.2.103) equality holds for Gaussian multivector valued functions*

$$f(\boldsymbol{x}) = C_0\, e^{-k\, \boldsymbol{x}^2}, \tag{5.2.104}$$

where $C_0 \in \mathcal{G}_n$ is an arbitrary but constant multivector, $0 < k \in \mathbb{R}$. This follows from the observation that we have for the f of (5.2.104)

$$-2k\, \boldsymbol{a} \cdot \boldsymbol{x}\, f = \boldsymbol{a} \cdot \nabla f. \tag{5.2.105}$$

Theorem 5.2.58. *For $\boldsymbol{a} \cdot \boldsymbol{b} = 0$, i.e. $\boldsymbol{b} \perp \boldsymbol{a}$, we get*

$$\int_{\mathbb{R}^n} (\boldsymbol{a} \cdot \boldsymbol{x})^2 \|f(\boldsymbol{x})\|^2\, d^n\boldsymbol{x}\, \frac{1}{(2\pi)^n} \int_{\mathbb{R}^n} (\boldsymbol{b} \cdot \boldsymbol{\omega})^2\, \|\mathcal{F}\{f\}(\boldsymbol{\omega})\|^2 d^n\boldsymbol{\omega} \geq 0. \tag{5.2.106}$$

Theorem 5.2.59. *Under the same assumptions as in Theorem 5.2.55, we obtain*

$$\int_{\mathbb{R}^n} \boldsymbol{x}^2\, \|f(\boldsymbol{x})\|^2\, d^n\boldsymbol{x}\, \frac{1}{(2\pi)^n} \int_{\mathbb{R}^n} \boldsymbol{\omega}^2\, \|\mathcal{F}\{f\}(\boldsymbol{\omega})\|^2 d^n\boldsymbol{\omega} \geq n\, \frac{1}{4} F^2. \tag{5.2.107}$$

Remark 5.2.60. *To prove Theorem 5.2.59 we first insert $\boldsymbol{x}^2 = \sum_{k=1}^n (\boldsymbol{e}_k \cdot \boldsymbol{x})^2$, $\boldsymbol{\omega}^2 = \sum_{l=1}^n (\boldsymbol{e}_l \cdot \boldsymbol{\omega})^2$. After that we apply (5.2.103) and (5.2.106) depending on the relative directions of the vectors $\boldsymbol{e}_k$ and $\boldsymbol{e}_l$.*

5.2.1.4 *Summary on CFTs in $Cl(n,0)$, $n = 2\,(\mathrm{mod}\,4)$ and $n = 3\,(\mathrm{mod}\,4)$*

The (real) Clifford Fourier transform extends the traditional Fourier transform on scalar functions to $\mathcal{G}_n = Cl(n,0)$ multivector functions with $n = 2, 3\,(\mathrm{mod}\,4)$ over the vector space domain $\mathbb{R}^n$. Basic properties and rules for differentiation, convolution, the Plancherel and Parseval theorems have been demonstrated in a manifestly invariant fashion. We then presented an uncertainty principle in the geometric algebra $\mathcal{G}_n$, which describes how a multivector-valued function and its Clifford Fourier transform relate.

In many fields the Fourier transform is successfully applied to solving physical equations such as heat and wave equations, in optics, in signal and image processing, etc. Therefore, we can apply geometric algebra and the Clifford Fourier transform to solve such problems involving the whole range of k-vector fields ($k = 0, 1, 2, \ldots, n$) in geometric algebras $\mathcal{G}_n = Cl(n,0)$ with $n = 2, 3\,(\mathrm{mod}\,4)$. The calculations will be real, have clear geometric interpretations and manifestly invariant. The use of coordinate bases is optional.

5.2.2 One-sided Clifford Fourier transforms in $Cl(p,q)$

We now generalize the one-sided CFTs to multivector valued functions in $Cl(p,q)$.

5.2.2.1 *Background*

We will now use the comprehensive research [193, 203] on the *manifolds of square roots of* -1 in real Clifford's geometric algebras $Cl(p,q)$ in order to construct the *Clifford Fourier transform.* Basically in the kernel of the complex Fourier transform the imaginary unit $j \in \mathbb{C}$ is replaced by a square root of -1 in $Cl(p,q)$. The Clifford Fourier transform (CFT) thus obtained generalizes previously known and applied CFTs [117, 132, 185], which replaced $j \in \mathbb{C}$ only by blades (usually pseudoscalars) squaring to -1. A major advantage of real Clifford algebra CFTs is their completely real geometric interpretation. We study (left and right) linearity of the CFT for constant multivector coefficients $\in Cl(p,q)$, translation ($\boldsymbol{x}$-shift) and modulation ($\boldsymbol{\omega}$-shift) properties and signal dilations. We show an inversion theorem. We establish the CFT of vector differentials, partial derivatives, vector derivatives and spatial moments of the signal. We also derive Plancherel and Parseval identities as well as a general convolution theorem. The treatment in this section is based on [201].

We note that quaternion, Clifford and geometric algebra Fourier transforms (QFT, CFT, GAFT) [184, 185, 194, 202] have already proven *very useful* tools for applications in non-marginal color image processing, image diffusion, electromagnetism, multi-channel processing, vector field processing, shape representation, linear scale invariant filtering, fast vector pattern matching, phase correlation, analysis of non-stationary improper complex signals, flow analysis, partial differential systems, disparity estimation, texture segmentation, as spectral representations for Clifford wavelet analysis, etc.

All these Fourier transforms essentially analyze scalar, vector and multivector signals in terms of sine and cosine waves with multivector coefficients. For this purpose the imaginary unit $j \in \mathbb{C}$ in $e^{j\phi} = \cos\phi + j\sin\phi$ can be replaced by any *square root of* -1 *in a real Clifford algebra* $Cl(p,q)$. The replacement by pure quaternions and blades with negative square [61, 184] has already yielded a wide variety of results with a clear geometric interpretation. It is well-known that there are elements other than blades, squaring to -1. Motivated by their special relevance for new types of CFTs, they have by now been studied thoroughly [193, 203, 294].

We therefore tap into these new results on square roots of -1 in Clifford algebras and fully general construct CFTs, with one general square root of -1 in $Cl(p,q)$. Our new CFTs form therefore a more general class of CFTs, subsuming and generalizing previous results[12]. A further benefit is, that these new CFTs become *fully steerable* within the continuous Clifford algebra submanifolds of square roots of -1. We thus obtain a comprehensive *new mathematical framework* for the investigation and application of Clifford Fourier transforms together with *new properties* (full steerability). Regarding the question of the *most suitable* CFT for a certain application, we are only just beginning to leave the terra cognita of familiar transforms to map out the vast array of possible CFTs in $Cl(p,q)$.

This part is organized as follows. We first recall in Section 5.2.2.1.1 some key notions of Clifford algebra, *multivector signal functions* and the recent results on *square roots of* -1 in Clifford algebras. Next, in Section 5.2.2.2 we define the central notion of *Clifford Fourier transforms* with respect to any square root of -1 in Clifford algebras $Cl(p,q)$. Then we study in Section 5.2.2.3 (left and right) linearity of the

[12]This is only the first step towards generalization. The non-commutativity of the geometric product of multivectors makes it meaningful to investigate CFTs with several kernel factors to both sides of the signal function. Each kernel factor may use a different square root of -1, compare Section 5.3.

CFT for constant multivector coefficients $\in Cl(p,q)$, translation ($\boldsymbol{x}$-shift) and modulation ($\boldsymbol{\omega}$-shift) properties, and signal dilations, followed by an inversion theorem. We establish the CFT of vector differentials, partial derivatives, vector derivatives and spatial moments of the signal. We also show Plancherel and Parseval identities as well as a general convolution theorem.

For an introduction to Clifford algebras we refer to Chapter 2, for vector differential calculus to Chapter 3.

5.2.2.1.1 Principal reverse and modulus in $Cl(p,q)$ The principal reverse[13], compare Section 2.1.1.6, of every basis element $e_A \in Cl(p,q)$, $1 \leq A \leq 2^n$, has the property

$$\widetilde{e_A} * e_B = \delta_{AB}, \qquad 1 \leq A, B \leq 2^n, \tag{5.2.108}$$

where the Kronecker delta $\delta_{AB} = 1$ if $A = B$, and $\delta_{AB} = 0$ if $A \neq B$. For the vector space $\mathbb{R}^{p,q}$ this leads to a reciprocal basis e^l, $1 \leq l, k \leq n$

$$e^l := \widetilde{e_l} = \varepsilon_l e_l, \quad e^l * e_k = e^l \cdot e_k = \begin{cases} 1, & \text{for } l = k \\ 0, & \text{for } l \neq k \end{cases}. \tag{5.2.109}$$

For $M, N \in Cl(p,q)$ we get $M * \widetilde{N} = \sum_A M_A N_A$. Two multivectors $M, N \in Cl(p,q)$ are *orthogonal* if and only if $M * \widetilde{N} = 0$. The modulus $|M|$ of a multivector $M \in Cl(p,q)$ is defined as

$$|M|^2 = M * \widetilde{M} = \sum_A M_A^2. \tag{5.2.110}$$

5.2.2.1.2 Square roots of -1 in Clifford algebras Square roots of -1 in Clifford algebras are extensively studied in Section 2.7 of this book.

With respect to any square root $i \in Cl(p,q)$ of -1, $i^2 = -1$, every multivector $A \in Cl(p,q)$ can be split into *commuting* A_{+i} and *anticommuting* A_{-i} parts, see Lemma 2.7.5 and [203].

5.2.2.2 The one-sided Clifford Fourier transform in $Cl(p,q)$

The *general Clifford Fourier transform* (CFT), to be introduced now, can be understood as a generalization of known CFTs [185] to a general real Clifford algebra setting. Most previously known CFTs use in their kernels specific square roots of -1, like bivectors, pseudoscalars, unit pure quaternions or blades [61]. For an introduction to known CFTs see [51], and for their various applications see [202]. We will *remove all these restrictions* on the square root of -1 used in a CFT.[14]

[13] Note that in the current section, we use the principal reverse throughout. But depending on the context another involution or anti-involution of Clifford algebra may be more appropriate for specific Clifford algebras, or for the purpose of a specific geometric interpretation.

[14] For example, the use of the square root $i = f_1$ of Table 2.7 would lead to a new type of CFT, which has so far not been studies anywhere in the literature, apart from [201].

Definition 5.2.61 (CFT with respect to one square root of -1). *Let $i \in Cl(p,q)$, $i^2 = -1$, be any square root of -1. The general Clifford Fourier transform*[15] *(CFT) of $f \in L^1(\mathbb{R}^{p,q}; Cl(p,q))$, with respect to i is*

$$\mathcal{F}^i\{f\}(\boldsymbol{\omega}) = \int_{\mathbb{R}^{p,q}} f(\boldsymbol{x})\, e^{-iu(\boldsymbol{x},\boldsymbol{\omega})} d^n\boldsymbol{x}, \tag{5.2.111}$$

where $d^n\boldsymbol{x} = dx_1 \dots dx_n$, $\boldsymbol{x}, \boldsymbol{\omega} \in \mathbb{R}^{p,q}$, and $u : \mathbb{R}^{p,q} \times \mathbb{R}^{p,q} \to \mathbb{R}$.

Since square roots of -1 in $Cl(p,q)$ populate *continuous submanifolds* in $Cl(p,q)$, the CFT of Definition 5.2.61 is generically *steerable* within these manifolds, see (5.2.127). In Definition 5.2.61, the square roots $i \in Cl(p,q)$ of -1 may be from any component of any conjugacy class. The choice of the geometric product in the integrand of (5.2.111) is very important. Because only this choice allowed, e.g. in [117], to define and apply a holistic vector field convolution, without loss of information.

5.2.2.3 *Properties of the one-sided CFT in $Cl(p,q)$*

We now study important properties of the general CFT of Definition 5.2.61. The proofs in this section may seem deceptively similar to standard proofs of properties of the classical complex Fourier transform. But the inherent non-commutativity of the geometric product of multivectors, makes it necessary to carefully respect the order of factors. Already for the first property of left and right linearity in (5.2.112) and (5.2.113), respectively, the order of factors leads to crucial differences. We therefore give detailed proofs of all properties.

5.2.2.3.1 Linearity, shift, modulation, dilation and powers of f, g, steerability Regarding *left and right linearity* of the general CFT of Definition 5.2.61 we can establish with the help of Lemma 2.7.5 that for $h_1, h_2 \in L^1(\mathbb{R}^{p,q}; Cl(p,q))$, and constants $\alpha, \beta \in Cl(p,q)$

$$\mathcal{F}^i\{\alpha h_1 + \beta h_2\}(\boldsymbol{\omega}) = \alpha \mathcal{F}^i\{h_1\}(\boldsymbol{\omega}) + \beta \mathcal{F}^i\{h_2\}(\boldsymbol{\omega}), \tag{5.2.112}$$

$$\begin{aligned}\mathcal{F}^i\{h_1\alpha + h_2\beta\}(\boldsymbol{\omega}) &= \mathcal{F}^i\{h_1\}(\boldsymbol{\omega})\alpha_{+i} + \mathcal{F}^{-i}\{h_1\}(\boldsymbol{\omega})\alpha_{-i} \\ &\quad + \mathcal{F}^i\{h_2\}(\boldsymbol{\omega})\beta_{+i} + \mathcal{F}^{-i}\{h_2\}(\boldsymbol{\omega})\beta_{-i}\,.\end{aligned} \tag{5.2.113}$$

Proof. Based on Lemma 2.7.5 we have

$$\begin{aligned}\alpha = \alpha_{+i} + \alpha_{-i}, \qquad \alpha_{+i}i &= i\alpha_{+i}, \qquad \alpha_{-i}i = -i\alpha_{-i} \\ \Rightarrow \quad \alpha e^{-iu} &= (\alpha_{+i} + \alpha_{-i})e^{-iu} = \alpha_{+i}e^{-iu} + \alpha_{-i}e^{-iu} \\ &= e^{-iu}\alpha_{+i} + e^{-(-i)u}\alpha_{-i},\end{aligned} \tag{5.2.114}$$

and similarly

$$\beta = \beta_{+i} + \beta_{-i}, \qquad \beta e^{-iu} = e^{-iu}\beta_{+i} + e^{-(-i)u}\beta_{-i}. \tag{5.2.115}$$

[15]For the Fourier transform on L^1 and on L^2, see Footnote 2 on page 144.

We apply Definition 5.2.61 and get

$$\mathcal{F}^i\{\alpha h_1 + \beta h_2\}(\boldsymbol{\omega}) = \int_{\mathbb{R}^{p,q}} \{\alpha h_1 + \beta h_2\}\, e^{-iu} d^n\boldsymbol{x}$$
$$= \alpha \mathcal{F}^i\{h_1\}(\boldsymbol{\omega}) + \beta \mathcal{F}^i\{h_2\}(\boldsymbol{\omega}). \tag{5.2.116}$$

By inserting (5.2.114) and (5.2.115) into Definition 5.2.61 we can further derive

$$\mathcal{F}^i\{h_1\alpha + h_2\beta\}(\boldsymbol{\omega}) = \mathcal{F}^i\{h_1\}(\boldsymbol{\omega})\alpha_{+i} + \mathcal{F}^{-i}\{h_1\}(\boldsymbol{\omega})\alpha_{-i}$$
$$+ \mathcal{F}^i\{h_2\}(\boldsymbol{\omega})\beta_{+i} + \mathcal{F}^{-i}\{h_2\}(\boldsymbol{\omega})\beta_{-i}\,. \tag{5.2.117}$$

□

For i *power factors* in $h_{a,b}(\boldsymbol{x}) = i^a h(\boldsymbol{x}) i^b$, $a, b \in \mathbb{Z}$, we obtain as an application of linearity

$$\mathcal{F}^i\{h_{a,b}\}(\boldsymbol{\omega}) = i^a \mathcal{F}^i\{h\}(\boldsymbol{\omega}) i^b. \tag{5.2.118}$$

Regarding the $\boldsymbol{x}$-*shifted* function $h_0(\boldsymbol{x}) = h(\boldsymbol{x} - \boldsymbol{x}_0)$ we obtain with constant $\boldsymbol{x}_0 \in \mathbb{R}^{p,q}$, assuming linearity of $u(\boldsymbol{x}, \boldsymbol{\omega})$ in its vector space argument $\boldsymbol{x}$,

$$\mathcal{F}^i\{h_0\}(\boldsymbol{\omega}) = \mathcal{F}^i\{h\}(\boldsymbol{\omega})\, e^{-iu(\boldsymbol{x}_0, \boldsymbol{\omega})}. \tag{5.2.119}$$

Proof. We assume linearity of $u(\boldsymbol{x}, \boldsymbol{\omega})$ in its vector space argument $\boldsymbol{x}$. Inserting $h_0(\boldsymbol{x}) = h(\boldsymbol{x} - \boldsymbol{x}_0)$ in Definition 5.2.61 we obtain

$$\mathcal{F}^i\{h_0\}(\boldsymbol{\omega}) = \int_{\mathbb{R}^{p,q}} h(\boldsymbol{x} - \boldsymbol{x}_0)\, e^{-iu(\boldsymbol{x}, \boldsymbol{\omega})} d^n\boldsymbol{x}$$
$$= \int_{\mathbb{R}^{p,q}} h(\boldsymbol{y})\, e^{-iu(\boldsymbol{y} + \boldsymbol{x}_0, \boldsymbol{\omega})} d^n\boldsymbol{y}$$
$$= \int_{\mathbb{R}^{p,q}} h(\boldsymbol{y})\, e^{-iu(\boldsymbol{y}, \boldsymbol{\omega})} e^{-iu(\boldsymbol{x}_0, \boldsymbol{\omega})} d^n\boldsymbol{y}$$
$$= \int_{\mathbb{R}^{p,q}} h(\boldsymbol{y})\, e^{-iu(\boldsymbol{y}, \boldsymbol{\omega})} d^n\boldsymbol{y}\, e^{-iu(\boldsymbol{x}_0, \boldsymbol{\omega})}$$
$$= \mathcal{F}^i\{h\}(\boldsymbol{\omega})\, e^{-iu(\boldsymbol{x}_0, \boldsymbol{\omega})}, \tag{5.2.120}$$

where we have substituted $\boldsymbol{y} = \boldsymbol{x} - \boldsymbol{x}_0$ for the second equality, we used the linearity of $u(\boldsymbol{x}, \boldsymbol{\omega})$ in its vector space argument $\boldsymbol{x}$ for the third equality, and that $e^{-iu(\boldsymbol{x}_0, \boldsymbol{\omega})}$ is independent of $\boldsymbol{y}$ for the fourth equality.

□

For the purpose of *modulation* we make the special assumption, that the function $u(\boldsymbol{x}, \boldsymbol{\omega})$ is linear in its frequency argument $\boldsymbol{\omega}$. Then we obtain for $h_m(\boldsymbol{x}) = h(x)\, e^{-iu(\boldsymbol{x}, \boldsymbol{\omega}_0)}$, and constant $\boldsymbol{\omega}_0 \in \mathbb{R}^{p,q}$ the modulation formula

$$\mathcal{F}^i\{h_m\}(\boldsymbol{\omega}) = \mathcal{F}^i\{h\}(\boldsymbol{\omega} + \boldsymbol{\omega}_0). \tag{5.2.121}$$

Proof. We assume, that the function $u(\boldsymbol{x},\boldsymbol{\omega})$ is linear in its frequency argument $\boldsymbol{\omega}$. Inserting $h_m(\boldsymbol{x}) = h(x)\, e^{-iu(\boldsymbol{x},\boldsymbol{\omega}_0)}$ in Definition 5.2.61 we obtain

$$\begin{aligned}
\mathcal{F}^i\{h_m\}(\boldsymbol{\omega}) &= \int_{\mathbb{R}^{p,q}} h_m(\boldsymbol{x})\, e^{-iu(\boldsymbol{x},\boldsymbol{\omega})} d^n\boldsymbol{x} \\
&= \int_{\mathbb{R}^{p,q}} h(x)\, e^{-iu(\boldsymbol{x},\boldsymbol{\omega}_0)}\, e^{-iu(\boldsymbol{x},\boldsymbol{\omega})} d^n\boldsymbol{x} \\
&= \int_{\mathbb{R}^{p,q}} h(x)\, e^{-iu(\boldsymbol{x},\boldsymbol{\omega}+\boldsymbol{\omega}_0)} d^n\boldsymbol{x} \\
&= \mathcal{F}^i\{h\}(\boldsymbol{\omega}+\boldsymbol{\omega}_0), \qquad (5.2.122)
\end{aligned}$$

where we used the linearity of $u(\boldsymbol{x},\boldsymbol{\omega})$ in its frequency argument $\boldsymbol{\omega}$ for the third equality.

□ Regarding *dilations*, we make the special assumption, that for constants $a_1,\ldots,a_n \in \mathbb{R}\setminus\{0\}$, and $\boldsymbol{x}' = \sum_{k=1}^n a_k x^k \boldsymbol{e}_k$, we have $u(\boldsymbol{x}',\boldsymbol{\omega}) = u(\boldsymbol{x},\boldsymbol{\omega}')$, with $\boldsymbol{\omega}' = \sum_{k=1}^n a_k\omega^k \boldsymbol{e}_k$. We then obtain for $h_d(\boldsymbol{x}) = h(\boldsymbol{x}')$ that

$$\mathcal{F}^i\{h_d\}(\boldsymbol{\omega}) = \frac{1}{|a_1\ldots a_n|}\mathcal{F}^i\{h\}(\boldsymbol{\omega}_d), \qquad \boldsymbol{\omega}_d = \sum_{k=1}^n \frac{1}{a_k}\omega^k \boldsymbol{e}_k. \qquad (5.2.123)$$

For $a_1 = \ldots = a_n = a \in \mathbb{R}\setminus\{0\}$ this simplifies under the same special assumption to

$$\mathcal{F}^i\{h_d\}(\boldsymbol{\omega}) = \frac{1}{|a|^n}\mathcal{F}^i\{h\}(\frac{1}{a}\boldsymbol{\omega}). \qquad (5.2.124)$$

Note, that the above assumption would, e.g., be fulfilled for $u(\boldsymbol{x},\boldsymbol{\omega}) = \boldsymbol{x} * \widetilde{\boldsymbol{\omega}} = \sum_{k=1}^n x^k\omega^k = \sum_{k=1}^n x_k\omega_k$.

Proof. We assume for constants $a_1,\ldots,a_n \in \mathbb{R}\setminus\{0\}$, and $\boldsymbol{x}' = \sum_{k=1}^n a_k x^k \boldsymbol{e}_k$, that we have $u(\boldsymbol{x}',\boldsymbol{\omega}) = u(\boldsymbol{x},\boldsymbol{\omega}')$, with $\boldsymbol{\omega}' = \sum_{k=1}^n a_k\omega^k \boldsymbol{e}_k$. Inserting $h_d(\boldsymbol{x}) = h(\boldsymbol{x}')$ in Definition 5.2.61 we obtain

$$\begin{aligned}
\mathcal{F}^i\{h_d\}(\boldsymbol{\omega}) &= \int_{\mathbb{R}^{p,q}} h_d(\boldsymbol{x})\, e^{-iu(\boldsymbol{x},\boldsymbol{\omega})} d^n\boldsymbol{x} \\
&= \int_{\mathbb{R}^{p,q}} h(\boldsymbol{x}')\, e^{-iu(\boldsymbol{x},\boldsymbol{\omega})} d^n\boldsymbol{x} \\
&= \frac{1}{|a_1\ldots a_n|}\int_{\mathbb{R}^{p,q}} h(\boldsymbol{y})\, e^{-iu(\boldsymbol{y}',\boldsymbol{\omega})} d^n\boldsymbol{y} \\
&= \frac{1}{|a_1\ldots a_n|}\int_{\mathbb{R}^{p,q}} h(\boldsymbol{y})\, e^{-iu(\boldsymbol{y},\boldsymbol{\omega}_d)} d^n\boldsymbol{y} \\
&= \frac{1}{|a_1\ldots a_n|}\mathcal{F}^i\{h\}(\boldsymbol{\omega}_d), \qquad (5.2.125)
\end{aligned}$$

where we substituted $\boldsymbol{y} = \boldsymbol{x}' = \sum_{k=1}^n a_k x^k \boldsymbol{e}_k$ and $\boldsymbol{x} = \sum_{k=1}^n \frac{1}{a_k} y^k \boldsymbol{e}_k = \boldsymbol{y}'$ for the third equality. Note that in this step each negative $a_k < 0, 1 \le k \le n$, leads to a factor $\frac{1}{|a_k|}$, because the negative sign is absorbed by interchanging the resulting integration boundaries $\{+\infty,-\infty\}$ to $\{-\infty,+\infty\}$. For the fourth equality we applied the assumption $u(\boldsymbol{y}',\boldsymbol{\omega}) = u(\boldsymbol{y},\boldsymbol{\omega}')$, and defined $\boldsymbol{\omega}_d = \boldsymbol{\omega}' = \sum_{k=1}^n a_k\omega^k \boldsymbol{e}_k$.

□

Within the same conjugacy class of square roots of -1 the CFTs of Definition 5.2.61 are related by the following equation, and therefore steerable. Let $i, i' \in Cl(p,q)$ be any two square roots of -1 in the same conjugacy class, i.e. $i' = a^{-1}ia$, $a \in Cl(p,q)$, a being invertible. As a consequence of this relationship we also have

$$e^{-i'u} = a^{-1}e^{-iu}a, \quad \forall u \in \mathbb{R}. \tag{5.2.126}$$

This in turn leads to the following *steerability relationship* of all CFTs with square roots of -1 from the same conjugacy class:

$$\mathcal{F}^{i'}\{h\}(\boldsymbol{\omega}) = \mathcal{F}^{i}\{ha^{-1}\}(\boldsymbol{\omega})a, \tag{5.2.127}$$

where ha^{-1} means to multiply the signal function h by the constant multivector $a^{-1} \in Cl(p,q)$.

5.2.2.3.2 CFT inversion, moments, derivatives, Plancherel, Parseval For establishing an inversion formula, moment and derivative properties, Plancherel and Parseval identities, certain *assumptions* about the phase function $u(\boldsymbol{x},\boldsymbol{\omega})$ need to be made. In principle these assumptions could be made based on the desired properties of the resulting CFT. One possibility is, e.g., to assume

$$u(\boldsymbol{x},\boldsymbol{\omega}) = \boldsymbol{x} * \widetilde{\boldsymbol{\omega}} = \sum_{l=1}^{n} x^l \omega^l = \sum_{l=1}^{n} x_l \omega_l, \tag{5.2.128}$$

which will be assumed for the current subsection.

We then get the following *inversion* formula[16]

$$h(\boldsymbol{x}) = \mathcal{F}^{i}_{-1}\{\mathcal{F}^{i}\{h\}\}(\boldsymbol{x}) = \frac{1}{(2\pi)^n}\int_{\mathbb{R}^{p,q}} \mathcal{F}^{i}\{h\}(\boldsymbol{\omega})\, e^{iu(\boldsymbol{x},\boldsymbol{\omega})} d^n\boldsymbol{\omega}, \tag{5.2.129}$$

where $d^n\boldsymbol{\omega} = d\omega_1 \ldots d\omega_n$, $\boldsymbol{x},\boldsymbol{\omega} \in \mathbb{R}^{p,q}$. For the existence of (5.2.129) we need $\mathcal{F}^{i}\{h\} \in L^1(\mathbb{R}^{p,q};\ Cl(p,q))$.

Proof. By direct computation we find

$$\begin{aligned}
&\frac{1}{(2\pi)^n}\int_{\mathbb{R}^{p,q}} \mathcal{F}^{i}\{h\}(\boldsymbol{\omega})\, e^{iu(\boldsymbol{x},\boldsymbol{\omega})} d^n\boldsymbol{\omega} \\
&= \frac{1}{(2\pi)^n}\int_{\mathbb{R}^{p,q}}\int_{\mathbb{R}^{p,q}} h(\boldsymbol{y})\, e^{-iu(\boldsymbol{y},\boldsymbol{\omega})} e^{iu(\boldsymbol{x},\boldsymbol{\omega})} d^n\boldsymbol{y}\, d^n\boldsymbol{\omega} \\
&= \frac{1}{(2\pi)^n}\int_{\mathbb{R}^{p,q}}\int_{\mathbb{R}^{p,q}} h(\boldsymbol{y})\, e^{iu(\boldsymbol{x}-\boldsymbol{y},\boldsymbol{\omega})} d^n\boldsymbol{\omega}\, d^n\boldsymbol{y} \\
&= \frac{1}{(2\pi)^n}\int_{\mathbb{R}^{p,q}}\int_{\mathbb{R}^{p,q}} h(\boldsymbol{y})\, e^{i\sum_{m=1}^{n}(x_m-y_m)\omega_m} d^n\boldsymbol{\omega}\, d^n\boldsymbol{y}
\end{aligned}$$

[16]Note, that we show the inversion symbol -1 as lower index in $\mathcal{F}^{i}_{-1}$, in order to avoid a possible confusion by using two upper indice. The inversion could also be written with the help of the CFT itself as $\mathcal{F}^{i}_{-1} = \frac{1}{(2\pi)^n}\mathcal{F}^{-i}$.

$$
\begin{aligned}
&= \frac{1}{(2\pi)^n} \int_{\mathbb{R}^{p,q}} \int_{\mathbb{R}^{p,q}} h(\boldsymbol{y}) \prod_{m=1}^{n} e^{i(x_m - y_m)\omega_m} d^n\boldsymbol{\omega} d^n\boldsymbol{y} \\
&= \int_{\mathbb{R}^{p,q}} h(\boldsymbol{y}) \prod_{m=1}^{n} \delta(x_m - y_m) d^n\boldsymbol{y} \\
&= h(\boldsymbol{x}),
\end{aligned}
\tag{5.2.130}
$$

where we have inserted Definition 5.2.61 for the first equality, used the linearity of u according to (5.2.128) for the second equality, as well as inserted (5.2.128) for the third equality, and that $\frac{1}{2\pi}\int_{\mathbb{R}} e^{i(x_m - y_m)\omega_m} d\omega_m = \delta(x_m - y_m)$, $1 \leq m \leq n$, for the fifth equality.

□

Additionally, we get the transformation law for *partial derivatives* $h_l'(\boldsymbol{x}) = \partial_{x_l} h(\boldsymbol{x})$, $1 \leq l \leq n$, for h piecewise smooth and integrable, and $h, h_l' \in L^1(\mathbb{R}^{p,q};\ Cl(p,q))$ as

$$
\mathcal{F}^i\{h_l'\}(\boldsymbol{\omega}) = \omega_l\, \mathcal{F}^i\{h\}(\boldsymbol{\omega}) i, \quad \text{for } 1 \leq l \leq n. \tag{5.2.131}
$$

Proof. We have

$$
\begin{aligned}
\mathcal{F}^i\{h_l'\}(\boldsymbol{\omega}) &= \int_{\mathbb{R}^{p,q}} h_l'(\boldsymbol{x})\, e^{-iu(\boldsymbol{y},\boldsymbol{\omega})} d^n\boldsymbol{x} \\
&= \int_{\mathbb{R}^{p,q}} \partial_{x_l} h(\boldsymbol{x})\, e^{-iu(\boldsymbol{y},\boldsymbol{\omega})} d^n\boldsymbol{x} \\
&= \int_{\mathbb{R}^{p,q}} \partial_{x_l} h(\boldsymbol{x})\, e^{-i\sum_{l=1}^{n} x_l\omega_l} d^n\boldsymbol{x} \\
&= -\int_{\mathbb{R}^{p,q}} h(\boldsymbol{x})\, \partial_{x_l} \left(e^{-i\sum_{l=1}^{n} x_l\omega_l}\right) d^n\boldsymbol{x} \\
&= -\int_{\mathbb{R}^{p,q}} h(\boldsymbol{x})\, e^{-i\sum_{l=1}^{n} x_l\omega_l} d^n\boldsymbol{x} (-i\omega_l) \\
&= \omega_l \mathcal{F}^i\{h\}(\boldsymbol{\omega}) i,
\end{aligned}
\tag{5.2.132}
$$

where we inserted u of (5.2.128) for the third equality and performed integration by parts for the fourth equality.

□

The *vector derivative* of $h \in L^1(\mathbb{R}^{p,q};\ Cl(p,q))$ with $h_l' \in L^1(\mathbb{R}^{p,q};\ Cl(p,q))$ gives therefore due to the linearity (5.2.112) of the CFT integral

$$
\mathcal{F}^i\{\nabla h\}(\boldsymbol{\omega}) = \mathcal{F}^i\{\sum_{l=1}^{n} e^l h_l'\}(\boldsymbol{\omega}) = \boldsymbol{\omega}\mathcal{F}^i\{h\}(\boldsymbol{\omega}) i. \tag{5.2.133}
$$

For the transformation of the *spatial moments* with $h_l(\boldsymbol{x}) = x_l h(\boldsymbol{x})$, $1 \leq l \leq n$, $h, h_l \in L^1(\mathbb{R}^{p,q};\ Cl(p,q))$, we obtain

$$
\mathcal{F}^i\{h_l\}(\boldsymbol{\omega}) = \partial_{\omega_l} \mathcal{F}^i\{h\}(\boldsymbol{\omega}) i. \tag{5.2.134}
$$

Proof. We compute

$$
\begin{aligned}
-h_l(\boldsymbol{x})i &= h(\boldsymbol{x})(-ix_l) = \mathcal{F}^i_{-1}\{\mathcal{F}^i\{h\}\}(\boldsymbol{x})(-ix_l) \\
&= \frac{1}{(2\pi)^n}\int_{\mathbb{R}^{p,q}} \mathcal{F}^i\{h\}(\boldsymbol{\omega})\, e^{iu(\boldsymbol{x},\boldsymbol{\omega})} d^n\boldsymbol{\omega}(-ix_l) \\
&= \frac{1}{(2\pi)^n}\int_{\mathbb{R}^{p,q}} \mathcal{F}^i\{h\}(\boldsymbol{\omega})\, e^{i\sum_{l=1}^n x_l\omega_l}(-ix_l) d^n\boldsymbol{\omega} \\
&= -\frac{1}{(2\pi)^n}\int_{\mathbb{R}^{p,q}} \mathcal{F}^i\{h\}(\boldsymbol{\omega})\, \partial_{\omega_l}\left(e^{i\sum_{l=1}^n x_l\omega_l}\right) d^n\boldsymbol{\omega} \\
&= \frac{1}{(2\pi)^n}\int_{\mathbb{R}^{p,q}} \left[\partial_{\omega_l}\mathcal{F}^i\{h\}(\boldsymbol{\omega})\right]\, e^{i\sum_{l=1}^n x_l\omega_l} d^n\boldsymbol{\omega} \\
&= \mathcal{F}^i_{-1}\left[\partial_{\omega_l}\mathcal{F}^i\{h\}\right](\boldsymbol{x}),
\end{aligned} \tag{5.2.135}
$$

where we used the inversion formula (5.2.129) for the second equality, integration by parts for the sixth equality and (5.2.129) again for the seventh equality. Moreover, by applying the CFT $\mathcal{F}^i$ to both sides of (5.2.135) we finally obtain

$$
\mathcal{F}^i\{h_l(-i)\}(\boldsymbol{\omega}) = \partial_{\omega_l}\mathcal{F}^i\{h\}(\boldsymbol{\omega}) \quad \Leftrightarrow \quad \mathcal{F}^i\{h_l\}(\boldsymbol{\omega}) = \partial_{\omega_l}\mathcal{F}^i\{h\}(\boldsymbol{\omega})i, \tag{5.2.136}
$$

because $\mathcal{F}^i\{h_l(-i)\} = \mathcal{F}^i\{h_l\}(-i)$. Note that in (5.2.136) the notation $(-i)$ indicates a constant right side multivector factor and not an argument of the function h_l.
□

For the *spatial vector moment* we obtain due to the linearity (5.2.112) of the CFT integral

$$
\mathcal{F}^i\{\boldsymbol{x}h\}(\boldsymbol{\omega}) = \mathcal{F}^i\{\sum_{l=1}^n e^l x_l h\}(\boldsymbol{\omega}) = \nabla_{\boldsymbol{\omega}}\mathcal{F}^i\{h\}(\boldsymbol{\omega})i, \tag{5.2.137}
$$

Note that for $Cl(p,q) \cong \mathcal{M}(2d,\mathbb{C})$ or $\mathcal{M}(d,\mathbb{H})$ or $\mathcal{M}(d,\mathbb{H}^2)$, or for i being a blade in $Cl(p,q) \cong \mathcal{M}(2d,\mathbb{R})$ or $\mathcal{M}(2d,\mathbb{R}^2)$, we have $\widetilde{i} = -i$. We assume this for the CFT $\mathcal{F}^i$ in the following Plancherel and Parseval identities.

For the functions $h_1, h_2, h \in L^2(\mathbb{R}^{p,q};\ Cl(p,q))$ we obtain the *Plancherel* identity

$$
\langle h_1, h_2\rangle = \frac{1}{(2\pi)^n}\langle \mathcal{F}^i\{h_1\}, \mathcal{F}^i\{h_2\}\rangle, \tag{5.2.138}
$$

as well as the *Parseval* identity

$$
\|h\| = \frac{1}{(2\pi)^{n/2}}\left\|\mathcal{F}^i\{h\}\right\|. \tag{5.2.139}
$$

Proof. We only need to proof the Plancherel identity, because the Parseval identity follows from it by setting $h_1 = h_2 = h$ and by taking the square root on both sides. Assume that $\widetilde{i} = -i$. We abbreviate $\int = \int_{\mathbb{R}^{p,q}}$, and compute

$$
\begin{aligned}
&\langle \mathcal{F}^i\{h_1\}, \mathcal{F}^i\{h_2\}\rangle \\
&= \int \langle \mathcal{F}^i\{h_1\}(\boldsymbol{\omega})[\mathcal{F}^i\{h_2\}(\boldsymbol{\omega})]^{\sim}\rangle d^n\boldsymbol{\omega} \\
&= \int\int\int \langle h_1(\boldsymbol{x})\, e^{-iu(\boldsymbol{x},\boldsymbol{\omega})} d^n\boldsymbol{x}[h_2(\boldsymbol{y})\, e^{-iu(\boldsymbol{y},\boldsymbol{\omega})} d^n\boldsymbol{y}]^{\sim}\rangle d^n\boldsymbol{\omega}
\end{aligned}
$$

$$
\begin{aligned}
&= \int\int\int \langle h_1(\boldsymbol{x})\, e^{-iu(\boldsymbol{x},\boldsymbol{\omega})} \widetilde{e^{-iu(\boldsymbol{y},\boldsymbol{\omega})} h_2(\boldsymbol{y})}\, d^n\boldsymbol{y}\rangle d^n\boldsymbol{x} d^n\boldsymbol{\omega} \\
&= \int\int\int \langle h_1(\boldsymbol{x})\, e^{-iu(\boldsymbol{x},\boldsymbol{\omega})} e^{iu(\boldsymbol{y},\boldsymbol{\omega})} \widetilde{h_2(\boldsymbol{y})}\, d^n\boldsymbol{\omega}\, d^n\boldsymbol{y}\rangle d^n\boldsymbol{x} \\
&= \int\int\int \langle h_1(\boldsymbol{x})\, e^{-iu(\boldsymbol{x}-\boldsymbol{y},\boldsymbol{\omega})} \widetilde{h_2(\boldsymbol{y})}\, d^n\boldsymbol{\omega}\, d^n\boldsymbol{y}\rangle d^n\boldsymbol{x} \\
&= (2\pi)^n \int\int\int \langle h_1(\boldsymbol{x}) \frac{e^{-i\sum_{m=1}^n (x_m - y_m)\omega_m}}{(2\pi)^n} \widetilde{h_2(\boldsymbol{y})}\, d^n\boldsymbol{\omega}\, d^n\boldsymbol{y}\rangle d^n\boldsymbol{x} \\
&= (2\pi)^n \int\int \langle h_1(\boldsymbol{x}) \prod_{m=1}^{n} \delta(x_m - y_m)\, \widetilde{h_2(\boldsymbol{y})}\, d^n\boldsymbol{y}\rangle d^n\boldsymbol{x} \\
&= (2\pi)^n \int \langle h_1(\boldsymbol{x}) \widetilde{h_2(\boldsymbol{x})}\rangle d^n\boldsymbol{x} \\
&= (2\pi)^n \langle h_1, h_2\rangle,
\end{aligned}
\tag{5.2.140}
$$

where we inserted (3.2.4) for the first equality, the Definition 5.2.61 of the CFT $\mathcal{F}^i$ for the second equality, applied the principal reverse for the third equality, and the symmetry of the scalar product and that $\widetilde{i} = -i$ for the fourth equality, the linearity of u according to (5.2.128) for the fifth equality, inserted the explicit forms of u of (5.2.128) for the sixth equality, and that $\frac{1}{2\pi}\int_{\mathbb{R}} e^{i(x_m - y_m)\omega_m} d\omega_m = \delta(x_m - y_m)$, $1 \leq m \leq n$, for the seventh equality, and again (3.2.4) for the last equality. Division of both sides with $(2\pi)^n$ finally gives the Plancherel identity (5.2.138).
□

5.2.2.3.3 Convolution We define the *convolution* of two multivector signals $a, b \in L^1(R^{p,q}; Cl(p,q))$ as

$$
(a \star b)(\boldsymbol{x}) = \int_{\mathbb{R}^{p,q}} a(\boldsymbol{y}) b(\boldsymbol{x} - \boldsymbol{y}) d^n\boldsymbol{y}. \tag{5.2.141}
$$

We assume that the function u is linear with respect to its first argument. The *CFT of the convolution* (5.2.141) can then be expressed as

$$
\mathcal{F}^i\{a \star b\}(\boldsymbol{\omega}) = \mathcal{F}^{-i}\{a\}(\boldsymbol{\omega})\mathcal{F}^i\{b_{-i}\}(\boldsymbol{\omega}) + \mathcal{F}^i\{a\}(\boldsymbol{\omega})\mathcal{F}^i\{b_{+i}\}(\boldsymbol{\omega}). \tag{5.2.142}
$$

Proof. We now proof (5.2.142).

$$
\begin{aligned}
&\mathcal{F}^i\{a \star b\}(\boldsymbol{\omega}) \\
&= \int_{\mathbb{R}^{p,q}} (a \star b)(\boldsymbol{x})\, e^{-iu(\boldsymbol{x},\boldsymbol{\omega})} d^n\boldsymbol{x} \\
&= \int_{\mathbb{R}^{p,q}} \int_{\mathbb{R}^{p,q}} a(\boldsymbol{y}) b(\boldsymbol{x} - \boldsymbol{y}) d^n\boldsymbol{y}\, e^{-iu(\boldsymbol{x},\boldsymbol{\omega})} d^n\boldsymbol{x} \\
&= \int_{\mathbb{R}^{p,q}} \int_{\mathbb{R}^{p,q}} a(\boldsymbol{y}) b(\boldsymbol{z}) d^n\boldsymbol{y}\, e^{-iu(\boldsymbol{y}+\boldsymbol{z},\boldsymbol{\omega})} d^n\boldsymbol{z} \\
&= \int_{\mathbb{R}^{p,q}} \int_{\mathbb{R}^{p,q}} a(\boldsymbol{y}) b(\boldsymbol{z}) d^n\boldsymbol{y}\, e^{-iu(\boldsymbol{y},\boldsymbol{\omega})} e^{-iu(\boldsymbol{z},\boldsymbol{\omega})} d^n\boldsymbol{z} \\
&= \int_{\mathbb{R}^{p,q}} \int_{\mathbb{R}^{p,q}} a(\boldsymbol{y}) [b_{+i}(\boldsymbol{z}) + b_{-i}(\boldsymbol{z})] d^n\boldsymbol{y}\, e^{-iu(\boldsymbol{y},\boldsymbol{\omega})} e^{-iu(\boldsymbol{z},\boldsymbol{\omega})} d^n\boldsymbol{z},
\end{aligned}
\tag{5.2.143}
$$

where we used the substitution $\boldsymbol{z} = \boldsymbol{x} - \boldsymbol{y}$, $\boldsymbol{x} = \boldsymbol{y} + \boldsymbol{z}$. To simplify (5.2.143) we expand the inner expression of the integrand to obtain

$$\begin{aligned} & a(\boldsymbol{y})\left[b_{+i}(\boldsymbol{z}) + b_{-i}(\boldsymbol{z})\right] e^{-iu(\boldsymbol{y},\boldsymbol{\omega})} \\ &= a(\boldsymbol{y})[e^{-iu(\boldsymbol{y},\boldsymbol{\omega})} b_{+i}(\boldsymbol{z}) + e^{+iu(\boldsymbol{y},\boldsymbol{\omega})} b_{-i}(\boldsymbol{z})] \\ &= a(\boldsymbol{y}) e^{-iu(\boldsymbol{y},\boldsymbol{\omega})} b_{+i}(\boldsymbol{z}) + a(\boldsymbol{y}) e^{+iu(\boldsymbol{y},\boldsymbol{\omega})} b_{-i}(\boldsymbol{z}). \end{aligned} \tag{5.2.144}$$

Reinserting (5.2.144) into (5.2.143) we get

$$\begin{aligned} & \mathcal{F}^{i}\{a \star b\}(\boldsymbol{\omega}) \\ &= \int_{\mathbb{R}^{p,q}} a(\boldsymbol{y}) e^{-iu(\boldsymbol{y},\boldsymbol{\omega})} d^n\boldsymbol{y} \int_{\mathbb{R}^{p,q}} b_{+i}(\boldsymbol{z}) e^{-iu(\boldsymbol{z},\boldsymbol{\omega})} d^n\boldsymbol{z} \\ &\quad + \int_{\mathbb{R}^{p,q}} a(\boldsymbol{y}) e^{+iu(\boldsymbol{y},\boldsymbol{\omega})} d^n\boldsymbol{y} \int_{\mathbb{R}^{p,q}} b_{-i}(\boldsymbol{z}) e^{-iu(\boldsymbol{z},\boldsymbol{\omega})} d^n\boldsymbol{z} \\ &= \mathcal{F}^{i}\{a\}(\boldsymbol{\omega}) \mathcal{F}^{i}\{b_{+i}\}(\boldsymbol{\omega}) + \mathcal{F}^{-i}\{a\}(\boldsymbol{\omega}) \mathcal{F}^{i}\{b_{-i}\}(\boldsymbol{\omega}). \end{aligned} \tag{5.2.145}$$

□

We point out that the above convolution theorem of equation (5.2.142) is a special case of a more general convolution theorem recently derived in [59].

5.2.2.4 *Summary on one-sided Clifford Fourier transforms in $Cl(p,q)$*

We have established a comprehensive *new mathematical framework* for the investigation and application of Clifford Fourier transforms (CFTs) together with *new properties*. Our new CFTs form a more general class of CFTs, subsuming and generalizing previous results. We have applied new results (see Section 2.7) on square roots of -1 in Clifford algebras to fully general construct CFTs, with a general square root of -1 in real Clifford algebras $Cl(p,q)$. The new CFTs are *fully steerable* within the continuous Clifford algebra submanifolds of square roots of -1. We have thus left the terra cognita of familiar transforms to outline the vast array of possible one-sided CFTs in $Cl(p,q)$.

We defined the central notion of the *Clifford Fourier transform* with respect to any square root of -1 in real Clifford algebras. Finally, we investigated important *properties* of these new CFTs: linearity, shift, modulation, dilation, moments, inversion, partial and vector derivatives, Plancherel and Parseval formulas, as well as a convolution theorem.

Regarding numerical implementations, usually 2^n complex Fourier transformations (FTs) are sufficient. In some cases this can be reduced to $2^{(n-1)}$ complex FTs, e.g. when the square root of -1 is a pseudoscalar. Further algebraic studies may widen the class of CFTs, where $2^{(n-1)}$ complex FTs are sufficient. Numerical implementation is then possible with 2^n (or $2^{(n-1)}$) discrete complex FTs, which can also be fast Fourier transforms (FFTs), leading to fast CFT implementations.

A well-known example of a CFT is the quaternion FT (QFT) [56,58,124,127,184,194,291] of Chapter 4, which is particularly used in applications to partial differential systems, color image processing, filtering, disparity estimation (two images differ by local translations) and texture segmentation. Another example is the spacetime FT, see Section 4.4, which leads to a multivector wave packet analysis of spacetime signals (e.g. electro-magnetic signals), applicable even to relativistic signals [184, 187].

Depending on the choice of the phase functions $u(\boldsymbol{x}, \boldsymbol{\omega})$ the multivector basis coefficient functions of the CFT result carry information on the symmetry of the signal, similar to the special case of the QFT [56].

The convolution theorem allows to design and apply multivector valued filters to multivector valued signals.

Research on the application of CFTs with general square roots of -1 is ongoing. Further results, including special choices of square roots of -1 for certain applications are to be expected.

5.2.3 Operator exponential approach to one-sided Clifford Fourier transform

5.2.3.1 Motivation and background

In this section, we study Clifford Fourier transforms (CFT) of multivector functions taking values in Clifford's geometric algebra, hereby using techniques coming from Clifford analysis (the multivariate function theory for the Dirac operator). In these CFTs on multivector signals, the complex unit $i \in \mathbb{C}$ is replaced by a multivector square root of -1, see Section 2.7, which may be a pseudoscalar in the simplest case. For these integral transforms we derive an operator representation expressed as the Hamilton operator of a harmonic oscillator. This section is based on [121]. See also [120].

Historically,[17] the Clifford Fourier transform (CFT) referred to above was originally introduced by B. Jancewicz [222] for electro-magnetic field computations in Clifford's geometric algebra $\mathcal{G}_3 = Cl(3,0)$ of $\mathbb{R}^3$, replacing the imaginary complex unit $i \in \mathbb{C}$ by the central unit pseudoscalar $i_3 \in \mathcal{G}_3$, which squares to minus one. This type of CFT was subsequently expanded to $\mathcal{G}_2$, instead using $i_2 \in \mathcal{G}_2$, and applied to image structure computations by M. Felsberg, see e.g. [132]. Subsequently J. Ebling and G. Scheuermann [117,118] applied both CFTs to the study of vector fields, as they occur in physical flows in dimensions 2 and 3. E. Hitzer and B. Mawardi [185] extended these CFTs to general higher dimensions $n = 2, 3 (\text{mod } 4)$ and studied their properties, including the physically uncertainty principle for multivector fields.

Independently De Bie et al [90] showed how Fourier transforms can be generalized to Clifford algebras in arbitrary dimensions, by introducing an operator form for the complex Fourier transform and generalizing it in the framework of Clifford analysis. This approach relies on a particular realization for the Lie algebra $\mathfrak{sl}(2)$, closely connected to the Hamilton operator of a harmonic oscillator, which makes it possible to introduce a fractional Clifford Fourier transform (see [92]) and to study radial deformations (see [95, 96]).

In the present section, we show how the CFT of [185] on multivector fields in $\mathcal{S}(\mathbb{R}^n, \mathcal{G}_n)$ can be written in the form of an *exponential of the Hamilton operator of the harmonic oscillator* in n dimensions and a constant phase factor depending

[17] For more historical details, we refer to Chapter 1.

on the underlying dimension. The proof of this result relies on the properties of monogenic functions of degree k and (in particular) of Clifford-Hermite functions. The computation of an integral transformation is thus replaced by the application of a differential operator, which has profound physical meaning. We therefore expect the content of this section to be valuable not only as another way to represent and compute the CFT, but beyond its mathematical aspect to shed new light on the closely related roles of the harmonic oscillator and the CFT in nature, in particular in physics, and in its wide field of technical applications.

Note that in quantum physics the Fourier transform of a wave function is called the momentum representation, and that the multiplication of the wave function with the exponential of the Hamilton operator multiplied with time represents the transition between the Schrödinger- and the Heisenberg representations [299]. The fact that the exponential of the Hamilton operator can also produce the change from position to momentum representation in quantum mechanics adds a very interesting facet to the picture of quantum mechanics.

The section is structured as follows. Section 5.2.3.2 introduces the field of Clifford analysis (a function theory for the Dirac operator). Section 5.2.3.3 finally reviews the notion of CFT and derives the operator representation of the CFTs we are looking for by exponentiating the Hamiltonian of a harmonic oscillator, hereby using a multivector function representation in terms of monogenic polynomials.

For a systematic introduction to Clifford's geometric algebras, we refer to Chapter 2.

We remind the reader, that the *reverse* of $M \in \mathcal{G}_n = Cl(n,0)$ is defined by the anti-automorphism

$$\widetilde{M} = \sum_{k=0}^{n} (-1)^{\frac{k(k-1)}{2}} \langle M \rangle_k. \tag{5.2.146}$$

The *square norm* of M is defined by

$$\|M\|^2 = \langle M\widetilde{M} \rangle \,, \tag{5.2.147}$$

which can also be expressed as $M * \widetilde{M}$, in terms of the real-valued (scalar) inner product

$$M * \widetilde{N} := \langle M\widetilde{N} \rangle \,. \tag{5.2.148}$$

Remark 5.2.62. *For vectors $\boldsymbol{a}, \boldsymbol{b} \in \mathbb{R}^n \subset \mathcal{G}_n$ the inner product is identical with the scalar product* (5.2.148)*,*

$$\boldsymbol{a} \cdot \boldsymbol{b} = \boldsymbol{a} * \widetilde{\boldsymbol{b}} = \boldsymbol{a} * \boldsymbol{b} \,.$$

As a consequence we obtain the *multivector Cauchy-Schwarz inequality*

$$|\langle M\widetilde{N} \rangle|^2 \;\leq\; \|M\|^2 \; \|N\|^2 \quad \forall\, M, N \in \mathcal{G}_n. \tag{5.2.149}$$

5.2.3.2 Clifford analysis

We add a brief excursion into Clifford analysis. For vector differential calculus, we refer to Chapter 3. The function theory for the operator ∇, often denoted by means of $\underline{\partial}_x$ in the literature (and referred to as the Dirac operator), is known as Clifford analysis. This is a multivariate function theory, which can be described as a higher-dimensional version of complex analysis or a refinement of harmonic analysis on $\mathbb{R}^n$. The latter is a simple consequence of the fact that ∇^2 yields the Laplace operator, the former refers to the fact that the functions on which ∇ acts take their values in the geometric algebra $\mathcal{G}_n$, which then generalizes the algebra $\mathbb{C}$ of complex numbers.

Remark 5.2.63. *Note that in classical Clifford analysis, for which we refer to e.g. [33, 100, 141, 149], one usually works with the geometric algebra (also known as a Clifford algebra) of signature $(0, n)$, which obviously lies closer to the idea of having complex units (n non-commuting complex units $e_j^2 = -1$, to be precise). However, in this paper we have chosen to work with the geometric algebra $\mathcal{G}_n$ associated to the Euclidean signature $(n, 0)$ to stay closer to the situation as it is used in physics. It is important to add that this has little influence on the final conclusions, as most of the results in Clifford analysis (in particular the ones we need in this paper) can be formulated independent of the signature.*

Clifford analysis can then essentially be described as the function theory for the Dirac operator, in which properties of functions $f(\underline{x}) \in \ker \nabla$ are studied. In this section, we will list a few properties. For the main part, we refer to the aforementioned references, or the excellent overview paper [101]. An important definition, in which the analogues of holomorphic powers z^k are introduced, is the following:

Definition 5.2.64. *For all integers $k \geq 0$, the vector space $\mathcal{M}_k(\mathbb{R}^n, \mathcal{G}_n)$ is defined by means of*

$$\mathcal{M}_k(\mathbb{R}^n, \mathcal{G}_n) := \text{Pol}_k(\mathbb{R}^n, \mathcal{G}_n) \cap \ker \nabla.$$

This is the vector space of k-homogeneous monogenics on $\mathbb{R}^n$, containing polynomial null solutions for the Dirac operator.

As the Dirac operator ∇ is surjective on polynomials, see e.g. [100], one easily finds for all $k \geq 1$ that

$$d_k := \dim \mathcal{M}_k(\mathbb{R}^n, \mathcal{G}_n) = 2^n \binom{n+k-2}{k} \qquad (5.2.150)$$

It is trivial to see that $d_0 = \dim \mathcal{G}_n$. In view of the fact that the operator ∇ factorizes the Laplace operator, null solutions of which are called harmonics, it is obvious that each monogenic polynomial is also harmonic. Indeed, denoting the space of k-homogeneous ($\mathcal{G}_n$-valued) harmonic polynomials by means of $\mathcal{H}_k(\mathbb{R}^n, \mathcal{G}_n)$,

$$\mathcal{M}_k(\mathbb{R}^n, \mathcal{G}_n) \subset \mathcal{H}_k(\mathbb{R}^n, \mathcal{G}_n). \qquad (5.2.151)$$

As ∇^2 is a scalar operator, one can even decompose monogenic polynomials into harmonic ones (using the basis for $\mathcal{G}_n$):

$$M_k(\underline{x}) = \sum_A e_A M_k^{(A)}(\underline{x}) \Rightarrow \nabla^2 M_k^{(A)}(\underline{x}) = 0. \qquad (5.2.152)$$

This will turn out to be an important property, since we will use a crucial algebraic characterization of the space of real-valued harmonic polynomials in the next section. Another crucial decomposition, which will then be used in the next section, is the so-called Fischer decomposition for $\mathcal{G}_n$-valued polynomials on $\mathbb{R}^n$ (see e.g. [100] for a proof):

Theorem 5.2.65. *For any $k \in \mathbb{N}$, the space $\mathcal{P}_k(\mathbb{R}^n, \mathcal{G}_n)$ decomposes into a direct sum of monogenic polynomials:*

$$\mathcal{P}_k(\mathbb{R}^n, \mathcal{G}_n) = \bigoplus_{j=0}^{k} \underline{x}^j \mathcal{M}_{k-j}(\mathbb{R}^n, \mathcal{G}_n). \tag{5.2.153}$$

We will often rely on the fact that the operator ∇ and the one-vector $\underline{x} \in \mathcal{G}_n$ (considered as a multiplication operator, acting on $\mathcal{G}_n$-valued functions) span the Lie superalgebra $\mathfrak{osp}(1,2)$, which appears as a Howe dual partner for the spin group (see [48] for more details). This translates itself into a collection of operator identities, and we hereby list the most crucial ones for what follows. Note that $\mathbb{E}_x := \sum_j x_j \partial_{x_j}$ denotes the Euler operator on $\mathbb{R}^m$, and $\{A, B\} = AB+BA$ (resp. $[A, B] = AB-BA$) denotes the anti-commutator (resp. the commutator) of two operators A and B:

$$\begin{aligned} \{\underline{x}, \nabla\} &= 2\left(\mathbb{E}_x + \tfrac{n}{2}\right) & [\nabla^2, ||\underline{x}||^2] &= 4\left(\mathbb{E}_x + \tfrac{n}{2}\right) \\ \{\underline{x}, \underline{x}\} &= 2||\underline{x}||^2 & [\nabla, ||\underline{x}||^2] &= 2\underline{x} \\ \{\nabla, \nabla\} &= 2\nabla^2 & [\nabla^2, \underline{x}] &= 2\nabla \end{aligned}$$

In particular, there also exists an operator which anti-commutes with the generators $\underline{x}$ and ∇ of $\mathfrak{osp}(1,2)$, see [13]. This operator, which is known as the Scasimir operator in abstract representation theory (it factorizes the Casimir operator), is related to the Gamma operator from Clifford analysis. To define this latter operator, we need a polar decomposition of the Dirac operator. Due to the change of signature, mentioned in the remark above, we will find a Gamma operator which differs from the one obtained in e.g. [100] by an overall minus sign. In order to retain the most important properties of this operator, we will therefore introduce a new notation Γ_∇, for the operator defined below[18]:

$$\underline{x}\nabla = \mathbb{E}_x + \Gamma_\nabla := \sum_{j=1}^{n} x_j \partial_{x_j} + \sum_{i<j} e_{ij}(x_i \partial_{x_j} - x_j \partial_{x_i}). \tag{5.2.154}$$

Remark 5.2.66. *Note that the operators $dH_{ij}^x := x_i \partial_{x_j} - x_j \partial_{x_i}$ appearing in the definition for the Gamma operator are known as the angular momentum operators. For $n = 3$, they are denoted by means of (L_x, L_y, L_z) and appear in quantum mechanics as the generators of the Lie algebra $\mathfrak{so}(3)$. In the next section, we will need the higher-dimensional analogue.*

Definition 5.2.67. *The Scasimir operator is defined by means of*

$$Sc := \frac{1}{2}[\underline{x}, \nabla] + \frac{1}{2} = \Gamma_\nabla - \frac{n-1}{2}\ . \tag{5.2.155}$$

[18]One simply has that $\Gamma_\nabla = -\Gamma$.

This operator satisfies $\{Sc, \underline{x}\} = 0 = \{Sc, \nabla\} = 0$, which can easily be verified (although this is superfluous, as the operator is specifically designed to satisfy these two conditions, see [13]). For example, one has that

$$\{Sc, \underline{x}\} = \frac{1}{2}\{[\underline{x}, \nabla] + 1, \underline{x}\} = \frac{1}{2}[\underline{x}, \nabla]\underline{x} + \frac{1}{2}\underline{x}[\underline{x}, \nabla] + \underline{x} = 0, \tag{5.2.156}$$

hereby using that $[\underline{x}^2, \nabla] = -2\underline{x}$ (see the relations above). Note also that the spaces $\mathcal{M}_k(\mathbb{R}^n, \mathcal{G}_n)$ are eigenspaces for the Gamma operator Γ_∇, which follows from the polar decomposition (5.2.154):

$$\Gamma_\nabla(M_k(\underline{x})) = -kM_k(\underline{x}) \qquad (\forall M_k(\underline{x}) \in \mathcal{M}_k(\mathbb{R}^n, \mathcal{G}_n)) . \tag{5.2.157}$$

As a result, one also has that

$$Sc(M_k(\underline{x})) = -\left(k + \frac{n-1}{2}\right) M_k(\underline{x}) \qquad (\forall M_k(\underline{x}) \in \mathcal{M}_k(\mathbb{R}^n, \mathcal{G}_n)) . \tag{5.2.158}$$

In the next section, we will also need the following:

Lemma 5.2.68. *If $f(r) = f(||\underline{x}||)$ denotes a scalar radial function on $\mathbb{R}^n$, one has for all $M_k(\underline{x}) \in \mathcal{M}_k(\mathbb{R}^n, \mathcal{G}_n)$ that*

$$Sc[M_k(\underline{x})f(r)] = -\left(k + \frac{n-1}{2}\right) M_k(\underline{x})f(r) . \tag{5.2.159}$$

Proof. This follows from the fact that

$$\begin{aligned} Sc[M_k(\underline{x})f(r)] &= \left(\Gamma_\nabla - \frac{n-1}{2}\right)[M_k(\underline{x})f(r)] \\ &= [\Gamma_\nabla M_k(\underline{x})]f(r) - \frac{n-1}{2}M_k(\underline{x})f(r) \\ &= -\left(k + \frac{n-1}{2}\right) M_k(\underline{x})f(r) . \end{aligned} \tag{5.2.160}$$

We hereby used the fact that $\Gamma_\nabla[f(r)] = 0$, which follows from $dH_{ij}^{\underline{x}}(r) = 0$ for all $i < j$. □

5.2.3.3 *The Clifford Fourier transform*

Let us recall the following definition, for which we refer to [185]:

Definition 5.2.69.
The Clifford Fourier transform (CFT) of a function $f(\underline{x}) : \mathbb{R}^n \to \mathcal{G}_n$, with $n = 2, 3 \pmod 4$, is the function

$$\begin{aligned} &\mathcal{F}\{f\} : \mathbb{R}^n \to \mathcal{G}_n \\ &\qquad \boldsymbol{\omega} \mapsto \mathcal{F}\{f\}(\boldsymbol{\omega}) := \frac{1}{(2\pi)^{\frac{n}{2}}}\int_{\mathbb{R}^n} f(\underline{x})e^{-i_n\boldsymbol{\omega}\cdot\underline{x}}d^n\underline{x} , \end{aligned} \tag{5.2.161}$$

where $\underline{x}, \boldsymbol{\omega} \in \mathbb{R}^n$ and $d^n\underline{x} = (dx_1\mathbf{e_1} \wedge dx_2\mathbf{e_2} \wedge \cdots \wedge dx_n\mathbf{e_n})i_n^{-1}$. Also note that in the formula above we have added the factor $(2\pi)^{-\frac{n}{2}}$, which will simplify our expression for the Gaussian eigenfunction (see below).

For a complete list of the properties[19] of this integral transform, which acts on $\mathcal{G}_n$-valued functions, we refer to Table 5.5 (see [185]).

Table 5.5 Properties of the Clifford Fourier transform (CFT) of Definition 5.2.69 with $n = 2, 3 \pmod 4$. Multivector functions f, g, f_1, f_2 all belong to $L^2(\mathbb{R}^n, \mathcal{G}_n)$, the constants are $\alpha, \beta \in \mathcal{G}_n$, $0 \neq a \in \mathbb{R}$, $\boldsymbol{a}, \boldsymbol{\omega}_0 \in \mathbb{R}^n$ and $m \in \mathbb{N}$. Source: [121, Tab. 1].

Property	Multiv. Funct.	CFT
Left lin.	$\alpha f(\boldsymbol{x})+\beta\, g(\boldsymbol{x})$	$\alpha\mathcal{F}\{f\}(\boldsymbol{\omega})+\,\beta\mathcal{F}\{g\}(\boldsymbol{\omega})$
$\boldsymbol{x}$-Shift	$f(\boldsymbol{x}-\boldsymbol{a})$	$\mathcal{F}\{f\}(\boldsymbol{\omega})\, e^{-i_n\boldsymbol{\omega}\cdot\boldsymbol{a}}$
$\boldsymbol{\omega}$-Shift	$f(\boldsymbol{x})\, e^{i_n\boldsymbol{\omega}_0\cdot\boldsymbol{x}}$	$\mathcal{F}\{f\}(\boldsymbol{\omega}-\boldsymbol{\omega}_0)$
Scaling	$f(a\boldsymbol{x})$	$\frac{1}{\lvert a\rvert^n}\mathcal{F}\{f\}(\frac{\boldsymbol{\omega}}{a})$
Vec. diff.	$(\boldsymbol{a}\cdot\nabla)^m f(\boldsymbol{x})$	$(\boldsymbol{a}\cdot\boldsymbol{\omega})^m\mathcal{F}\{f\}(\boldsymbol{\omega})\; i_n^m$
	$(\boldsymbol{a}\cdot\boldsymbol{x})^m\; f(\boldsymbol{x})$	$(\boldsymbol{a}\cdot\nabla_{\boldsymbol{\omega}})^m\;\mathcal{F}\{f\}(\boldsymbol{\omega})\; i_n^m$
Powers of $\boldsymbol{x}$	$\boldsymbol{x}^m f(\boldsymbol{x})$	$\nabla_{\boldsymbol{\omega}}^m\;\mathcal{F}\{f\}(\boldsymbol{\omega})\; i_n^m$
Vec. deriv.	$\nabla^m f(\boldsymbol{x})$	$\boldsymbol{\omega}^m\;\mathcal{F}\{f\}(\boldsymbol{\omega})\; i_n^m$
Plancherel	$\int_{\mathbb{R}^n} f_1(\boldsymbol{x})\widetilde{f_2(\boldsymbol{x})}\, d^n\boldsymbol{x}$	$\frac{1}{(2\pi)^n}\int_{\mathbb{R}^n}\mathcal{F}\{f_1\}(\boldsymbol{\omega})\widetilde{\mathcal{F}\{f_2\}(\boldsymbol{\omega})}\, d^n\boldsymbol{\omega}$
sc. Parseval	$\int_{\mathbb{R}^n}\lVert f(\boldsymbol{x})\rVert^2\, d^n\boldsymbol{x}$	$\frac{1}{(2\pi)^n}\int_{\mathbb{R}^n}\lVert\mathcal{F}\{f\}(\boldsymbol{\omega})\rVert^2\, d^n\boldsymbol{\omega}$

However, in the present approach, the following properties will play a crucial role.

Proposition 5.2.70.
For functions $f(\underline{x}) : \mathbb{R}^n \to \mathcal{G}_n$, the CFT satisfies the following:

(i) powers of $\underline{x}$ from the left:

$$\mathcal{F}\{\underline{x}^m f\}(\boldsymbol{\omega}) = \nabla_{\boldsymbol{\omega}}^m \mathcal{F}\{f\}(\boldsymbol{\omega}) i_n^m\,, \quad m \in \mathbb{N}\,. \tag{5.2.162}$$

(ii) powers of $\boldsymbol{a}\cdot\underline{x}$:

$$\mathcal{F}\{(\boldsymbol{a}\cdot\underline{x})^m f\}(\boldsymbol{\omega}) = (\boldsymbol{a}\cdot\nabla_{\boldsymbol{\omega}})^m \mathcal{F}\{f\}(\boldsymbol{\omega}) i_n^m\,, \quad m \in \mathbb{N}\,. \tag{5.2.163}$$

(iii) vector derivatives from the left:

$$\mathcal{F}\{\nabla_{\underline{x}}^m f\}(\boldsymbol{\omega}) = \boldsymbol{\omega}^m \mathcal{F}\{f\}(\boldsymbol{\omega}) i_n^m\,, \quad m \in \mathbb{N}\,. \tag{5.2.164}$$

(iv) directional derivatives:

$$\mathcal{F}\{(\boldsymbol{a}\cdot\nabla_{\underline{x}})^m f\}(\boldsymbol{\omega}) = (\boldsymbol{a}\cdot\boldsymbol{\omega})^m \mathcal{F}\{f\}(\boldsymbol{\omega}) i_n^m\,, \quad m \in \mathbb{N}\,. \tag{5.2.165}$$

[19]Note that Theorem 4.33 in [185] also states the CFTs of the Clifford convolution. In the case of $n = 3 \pmod 4$ it is simply the product of the CFTs of the two factor functions. In the case of $n = 2 \pmod 4$ the second factor function is split into even and odd grades, and the sign of the frequency argument of the CFT of the first factor function is reversed when multiplied with the CFT of the odd grade part of the second factor function.

In order to obtain an operator exponential expression for the CFT from above, see Definition 5.2.69, we need to construct a family of eigenfunctions which then moreover serves as a basis for the function space $\mathcal{S}(\mathbb{R}^n, \mathcal{G}_n)$ of rapidly decreasing test functions taking values in $\mathcal{G}_n$. In other words the operator exponential expression acts on functions in the very same was as the Clifford-Fourier transform does. For this we use an expansion into eigenfunctions which form an orthonormal basis of $\mathcal{S}(\mathbb{R}^n, \mathcal{G}_n)$. To do so, we need a series of results:

Proposition 5.2.71.
The Gaussian function $G(\underline{x}) := \exp(-\frac{1}{2}\underline{x}^2) = \exp(-\frac{1}{2}||\underline{x}||^2)$ *on* $\mathbb{R}^n$ *defines an eigenfunction for the CFT:*

$$\mathcal{F}\{G(\underline{x})\}(\boldsymbol{\omega}) := \frac{1}{(2\pi)^{\frac{n}{2}}} \int_{\mathbb{R}^n} \exp(-\frac{1}{2}\underline{x}^2) e^{-i_n \boldsymbol{\omega}\cdot\underline{x}} d^n\underline{x} = G(\boldsymbol{\omega}). \tag{5.2.166}$$

Proof. This follows from a similar property in [185], taking the rescaling factor $(2\pi)^{-\frac{n}{2}}$ into account. □

Next, we prove that the Gaussian $G(\underline{x})$ may be multiplied with arbitrary monogenic polynomials: this will still yield an eigenfunction for $\mathcal{F}$. In order to prove this, we recall formula (5.2.152):

$$M_k(\underline{x}) \in \mathcal{P}_k(\mathbb{R}^n, \mathcal{G}_n) \cap \ker \nabla \implies M_k(\underline{x}) = \sum_A \boldsymbol{e}_A M_k^{(A)}(\underline{x}), \tag{5.2.167}$$

with each $M_k^{(A)}(\underline{x}) \in \mathcal{H}_k(\mathbb{R}^n, \mathbb{R})$ a real-valued harmonic polynomial on $\mathbb{R}^n$. This allows us to focus our attention on *harmonic* polynomials. As is well known, the vector space $\mathcal{H}_k(\mathbb{R}^n, \mathbb{C})$ of k-homogeneous harmonic polynomials in n real variables defines an irreducible module for the special orthogonal group $\mathrm{SO}(n)$ or its Lie algebra $\mathfrak{so}(n)$. This algebra is spanned by the $\binom{n}{2}$ angular momentum operators

$$dH_{ij}^{x} := x_i(\boldsymbol{e_j} \cdot \nabla) - x_j(\boldsymbol{e_i} \cdot \nabla) = x_i\partial_{x_j} - x_j\partial_{x_i} \quad (1 \le i < j \le n), \tag{5.2.168}$$

see Remark 5.2.66. It then follows from general Lie theoretical considerations that the vector space $\mathcal{H}_k(\mathbb{R}^n, \mathbb{C})$ is generated by the repeated action of the negative root vectors in $\mathfrak{so}(m)$ acting on a unique highest weight vector. For the representation space $\mathcal{H}_k(\mathbb{R}^n, \mathbb{C})$, this highest weight vector is given by $h_k(\underline{x}) := (x_1 - ix_2)^k$, see e.g. [82, 141]. Without going into too much detail, it suffices to understand that this implies that arbitrary elements in $\mathcal{H}_k(\mathbb{R}^n, \mathbb{C})$ can always be written as

$$H_k(\underline{x}) = \mathcal{L}(dH_{ij}^{x})h_k(\underline{x}), \tag{5.2.169}$$

where $\mathcal{L}(dH_{ij}^{x})$ denotes some linear combination[20] of products of the angular momentum operators dH_{ij}^{x}. Note that the presence of the complex number field in the argument above has no influence on the fact that we are working with functions taking values in the *real* algebra $\mathcal{G}_n$ in this paper: it suffices to focus on the real (or pure imaginary) part afterward.

[20] In a sense, each combination $\mathcal{L}(dH_{ij}^{x})$ can be seen as some sort of non-commutative polynomial (the factors dH_{ij}^{x} do not commute).

Theorem 5.2.72.
Given an arbitrary element $M_k(\underline{x}) \in \mathcal{M}_k(\mathbb{R}^n, \mathcal{G}_n)$, one has:

$$\mathcal{F}\{M_k(\underline{x})G(\underline{x})\}(\boldsymbol{\omega}) = M_k(\boldsymbol{\omega})G(\boldsymbol{\omega})(-i_n)^k. \tag{5.2.170}$$

Proof. In view of the decomposition (5.2.167), we have that

$$\mathcal{F}\{M_k(\underline{x})G(\underline{x})\}(\boldsymbol{\omega}) = \sum_A e_A \mathcal{F}\{M_k^{(A)}(\underline{x})G(\underline{x})\}(\boldsymbol{\omega}). \tag{5.2.171}$$

If we then write each scalar component $M_k^{(A)}(\underline{x})$ as a linear combination of the form

$$M_k^{(A)}(\underline{x}) = \mathcal{L}^{(A)}(dH_{ij}^{x})h_k(\underline{x}), \tag{5.2.172}$$

we are clearly left with the analysis of terms of the following type (with $1 \leq i < j \leq n$ arbitrary):

$$\mathcal{F}\{\mathcal{L}^{(A)}(dH_{ij}^{x})(x_1 - ix_2)^k G(\underline{x})\}(\boldsymbol{\omega}). \tag{5.2.173}$$

Without loosing generality, we can focus our attention on a single operator dH_{ij}^{x}, since any $\mathcal{L}^{(A)}(dH_{ij}^{x})$ can always be written as a sum of products of these operators. Invoking properties from Proposition 5.2.70, it is clear that

$$\begin{aligned}
\mathcal{F}\{dH_{ij}^{x}h_k G\}(\boldsymbol{\omega}) &= \mathcal{F}\{(x_i(\boldsymbol{e_j} \cdot \nabla_x) - x_j(\boldsymbol{e_i} \cdot \nabla_{\underline{x}}))h_k G\}(\boldsymbol{\omega}) \\
&= (\omega_j(\boldsymbol{e_i} \cdot \nabla_{\boldsymbol{\omega}}) - \omega_i(\boldsymbol{e_j} \cdot \nabla_{\boldsymbol{\omega}}))\mathcal{F}\{h_k G\}(\boldsymbol{\omega})i_n^2 \\
&= dH_{ij}^{\omega}\mathcal{F}\{h_k G\}(\boldsymbol{\omega}),
\end{aligned} \tag{5.2.174}$$

where dH_{ij}^{ω} denotes the angular momentum operator in the variable $\boldsymbol{\omega} \in \mathbb{R}^n$. Next, invoking the same proposition, we also have that

$$\begin{aligned}
\mathcal{F}\{h_k G\}(\boldsymbol{\omega}) &= \mathcal{F}\{(x_1 - ix_2)^k G\}(\boldsymbol{\omega}) \\
&= ((\boldsymbol{e_1} \cdot \nabla_{\boldsymbol{\omega}}) - i(\boldsymbol{e_2} \cdot \nabla_{\boldsymbol{\omega}}))^k \mathcal{F}\{G\}(\boldsymbol{\omega})i_n^k \\
&= ((\boldsymbol{e_1} \cdot \nabla_{\boldsymbol{\omega}}) - i(\boldsymbol{e_2} \cdot \nabla_{\boldsymbol{\omega}}))^k G(\boldsymbol{\omega})i_n^k.
\end{aligned} \tag{5.2.175}$$

It then suffices to note that

$$\begin{aligned}
&((\boldsymbol{e_1} \cdot \nabla_{\boldsymbol{\omega}}) - i(\boldsymbol{e_2} \cdot \nabla_{\boldsymbol{\omega}}))^k \exp(-\frac{1}{2}||\boldsymbol{\omega}||^2) \\
&= (-\omega_1 + i\omega_2)^k \exp(-\frac{1}{2}||\boldsymbol{\omega}||^2)
\end{aligned} \tag{5.2.176}$$

to arrive at $\mathcal{F}\{h_k G\}(\boldsymbol{\omega}) = (-1)^k h_k(\boldsymbol{\omega})G(\boldsymbol{\omega})i_n^k$. Together with what was found above, this then proves the theorem. □

Remark 5.2.73. *There exist alternative ways to prove the fact that harmonic polynomials times a Gaussian kernel define eigenfunctions for the Fourier transform, see e.g. the seminal work [310] by Stein and Weiss, but to our best knowledge the proof above has not appeared in the literature yet.*

In order to arrive at a basis of eigenfunctions for the space $\mathcal{S}(\mathbb{R}^n, \mathcal{G}_n)$, we need more eigenfunctions for the CFT. For that purpose, we introduce the following definition (recall that we have defined d_k in the previous section).

Definition 5.2.74.
For a given monogenic polynomial $M_k^{(b)}(\underline{x}) \in \mathcal{M}_k(\mathbb{R}^n, \mathcal{G}_n)$, where the index $b \in \{1, 2, \ldots, d_k\}$ is used to label a basis for the vector space of monogenics of degree k, we define the Clifford-Hermite eigenfunctions as

$$\varphi_{a,b;k}(\underline{x}) := (\nabla - \underline{x})^a M_k^{(b)}(\underline{x}) G(\underline{x}). \tag{5.2.177}$$

Hereby, the index $a \geq 0$ denotes an arbitrary non-negative integer.

Theorem 5.2.75.
For all indice $(a, b; k) \in \mathbb{N} \times \{1, \ldots, d_k\} \times \mathbb{N}$, one has that

$$\mathcal{F}\{\varphi_{a,b;k}\}(\boldsymbol{\omega}) = \varphi_{a,b;k}(\boldsymbol{\omega})(-i_n)^{a+k} . \tag{5.2.178}$$

Proof. This can again be proved using Proposition 5.2.70. Indeed, we clearly have that

$$\begin{aligned} \mathcal{F}\{\varphi_{a,b;k}\}(\boldsymbol{\omega}) &= \mathcal{F}\{(\nabla_{\underline{x}} - \underline{x})^a M_k^{(b)}(\underline{x}) G(\underline{x})\}(\boldsymbol{\omega}) \\ &= (\boldsymbol{\omega} - \nabla_{\boldsymbol{\omega}})^a \mathcal{F}\{M_k^{(b)}(\underline{x}) G(\underline{x})\}(\boldsymbol{\omega}) i_n^a \\ &= (\nabla_{\boldsymbol{\omega}} - \boldsymbol{\omega})^a M_k^{(b)}(\boldsymbol{\omega}) G(\boldsymbol{\omega})(-i_n)^k (-i_n)^a \\ &= \varphi_{a,b;k}(\boldsymbol{\omega})(-i_n)^{a+k}, \end{aligned} \tag{5.2.179}$$

where we have made use of Theorem 5.2.72. □

Before we come to an exponential operator form for the CFT, we prove a few additional results:

Lemma 5.2.76.
For all indices $(a, b; k) \in \mathbb{N} \times \{1, \ldots, d_k\} \times \mathbb{N}$, one has:

$$\begin{aligned} (\nabla + \underline{x})\varphi_{2a,b;k}(\underline{x}) &= -4a\varphi_{2a-1,b;k}(\underline{x}) \\ (\nabla + \underline{x})\varphi_{2a+1,b;k}(\underline{x}) &= -2(n + 2k + 2a)\varphi_{2a,b;k}(\underline{x}). \end{aligned} \tag{5.2.180}$$

Proof. This lemma can easily be proved by induction on the parameter $a \in \mathbb{N}$, hereby taking into account that

$$[\nabla + \underline{x}, \nabla - \underline{x}] = 2[\underline{x}, \nabla] = 4Sc - 2, \tag{5.2.181}$$

with $Sc \in \mathfrak{osp}(1, 2)$ the Scasimir operator defined in (5.2.155). Indeed, for $a = 0$ we immediately get that

$$\begin{aligned} (\nabla + \underline{x})\varphi_{0,b;k} &= \dot{\nabla} G(\dot{\underline{x}}) M_k^{(b)}(\underline{x}) + \underline{x}\varphi_{0,b;k} \\ &= -\underline{x} G(\underline{x}) M_k^{(b)}(\underline{x}) + \underline{x}\varphi_{0,b;k} = 0. \end{aligned} \tag{5.2.182}$$

For $a = 1$, we get that

$$\begin{aligned}(\nabla + \underline{x})\varphi_{1,b;k} &= \big((\nabla - \underline{x})(\nabla + \underline{x}) + 2(2Sc - 1)\big)\varphi_{0,b;k} \\ &= -2(n + 2k)\varphi_{0,b;k}\ . \end{aligned} \tag{5.2.183}$$

Here we have used the fact (5.2.158) that monogenic homogeneous polynomials are eigenfunctions for the Scasimir operator, together with lemma 5.2.68. Let us then for example consider a general odd index $2a + 1$ (the case of an even index $2a$ is completely similar):

$$\begin{aligned}(\nabla + \underline{x})\varphi_{2a+1,b;k} &= \big((\nabla - \underline{x})(\nabla + \underline{x}) + 2(2Sc - 1)\big)\varphi_{2a,b;k} \\ &= -4a(\nabla - \underline{x})\varphi_{2a-1,b;k} + 2(\nabla - \underline{x})^{2a}(2Sc - 1)\varphi_{0,b;k} \\ &= -2(2a + n + 2k)\varphi_{2a,b;k}. \end{aligned} \tag{5.2.184}$$

Here, we have used the induction hypothesis and the fact that the operator Sc commutes with an even power $(\nabla - \underline{x})$, as it anti-commutes with each individual factor. □

This lemma will now be used to construct another operator, for which our Clifford-Hermite functions from Definition 5.2.74 are again eigenfunctions.

Theorem 5.2.77.
For all indices $(a, b; k) \in \mathbb{N} \times \{1, \ldots, d_k\} \times \mathbb{N}$, *one has:*

$$(\nabla^2 - \underline{x}^2)\varphi_{a,b;k}(\underline{x}) = -(n + 2a + 2k)\varphi_{a,b;k}(\underline{x}). \tag{5.2.185}$$

Proof. First of all, we note that

$$(\nabla + \underline{x})(\nabla - \underline{x}) = \nabla^2 - \underline{x}^2 + [\underline{x}, \nabla], \tag{5.2.186}$$

from which we note that the operator appearing in the theorem can also be written as

$$\nabla^2 - \underline{x}^2 = (\nabla + \underline{x})(\nabla - \underline{x}) - (2Sc - 1). \tag{5.2.187}$$

Using lemma 5.2.76 and properties of the Scasimir operator, we get:

$$\begin{aligned}(\nabla^2 - \underline{x}^2)\varphi_{2a,b;k} &= \big((\nabla + \underline{x})(\nabla - \underline{x}) - (2Sc - 1)\big)\varphi_{2a,b;k} \\ &= (\nabla + \underline{x})\varphi_{2a+1,b;k} - (\nabla - \underline{x})^{2a}(2Sc - 1)\varphi_{0,b;k} \\ &= -2(n + 2k + 2a)\varphi_{2a,b;k} + (n + 2k)\varphi_{2a,b;k} \\ &= -(n + 2k + 4a)\varphi_{2a,b;k} \end{aligned} \tag{5.2.188}$$

for the case of an even index $2a$, and

$$\begin{aligned}(\nabla^2 - \underline{x}^2)\varphi_{2a+1,b;k} &= \big((\nabla + \underline{x})(\nabla - \underline{x}) - (2Sc - 1)\big)\varphi_{2a+1,b;k} \\ &= (\nabla + \underline{x})\varphi_{2a+2,b;k} + (\nabla - \underline{x})^{2a+1}(2Sc + 1)\varphi_{0,b;k} \\ &= -2(2a + 2)\varphi_{2a+1,b;k} - (n + 2k - 2)\varphi_{2a+1,b;k} \\ &= -(n + 2k + 4a + 2)\varphi_{2a+1,b;k} \end{aligned} \tag{5.2.189}$$

for odd indices $2a + 1$. Together, this proves the theorem. □

In order to compare the eigenvalues of the Clifford-Hermite functions as eigenfunctions for the CFT and the operator from the theorem above (which is nothing but the Hamiltonian of the harmonic oscillator), we mention the following remarkable property:

$$\begin{aligned}\varphi_{a,b;k}(\dot{\underline{x}})e^{-\frac{\pi}{4}(\underline{x}^2-\dot{\nabla}^2-n)i_n} &= \sum_{j=0}^{\infty}\frac{1}{j!}(\underline{x}^2-\nabla^2-n)^j\varphi_{a,b;k}(\underline{x})\left(-\frac{\pi}{4}i_n\right)^j \\ &= \varphi_{a,b;k}(\underline{x})\sum_{j=0}^{\infty}\frac{1}{j!}\left(-(a+k)\frac{\pi}{2}i_n\right)^j \\ &= \varphi_{a,b;k}(\underline{x})(-i_n)^{a+k} . \end{aligned} \tag{5.2.190}$$

In the first line above, we have used the Hestenes' overdot notation, to stress the fact that the operator acts on $\underline{x}$ from the right. This is due to the fact that the pseudoscalar i_n does *not* necessarily commute with the Clifford-Hermite eigenfunction.

It thus suffices to prove that these eigenfunctions provide a basis for the function space $\mathcal{S}(\mathbb{R}^n, \mathcal{G}_n)$[21] in order to arrive at the main result of our paper, which is the exponential operator form for the CFT.

Proposition 5.2.78.
The space $\mathcal{S}(\mathbb{R}^n, \mathcal{G}_n)$ is spanned by the countable basis

$$\mathcal{B} := \{\varphi_{a,b;k}(\underline{x}) : (a, b; k) \in \mathbb{N} \times \{1, \ldots, d_k\} \times \mathbb{N}\} . \tag{5.2.191}$$

Proof. In order to prove this, it suffices to show that we can express arbitrary elements of $\mathcal{P}(\mathbb{R}^n, \mathcal{G}_n) \otimes G(\underline{x})$ as a linear combination of the Clifford-Hermite eigenfunctions. To do so, we can use the fact that we know the structure of the space of $\mathcal{G}_n$-valued polynomials in terms of the Fischer decomposition, see (5.2.153). It thus suffices to note that

$$(\nabla - \underline{x})^a M_k^{(b)}(\underline{x})G(\underline{x}) = (-2)^a \underline{x}^a M_k^{(b)}(\underline{x})G(\underline{x}) + \text{L.O.T.} , \tag{5.2.192}$$

where L.O.T. = lower order terms, i.e. it refers to lower powers in $\underline{x}$ times the product of $M_k^{(b)}(\underline{x})$ and the Gaussian function. For $a \in \{0, 1\}$, this is trivial, as we for example have that

$$\begin{aligned}(\nabla - \underline{x})M_k^{(b)}(\underline{x})G(\underline{x}) &= -\underline{x}M_k^{(b)}(\underline{x})G(\underline{x}) + \dot{\nabla}G(\dot{\underline{x}})M_k^{(b)}(\underline{x}) \\ &= -2\underline{x}M_k^{(b)}(\underline{x})G(\underline{x}) , \end{aligned} \tag{5.2.193}$$

and the rest follows from an easy induction argument. In case of an odd index $2a+1$ for example, the induction hypothesis gives:

$$(\nabla - \underline{x})^{2a+1} M_k^{(b)}(\underline{x})G(\underline{x}) = (\nabla - \underline{x})(4^a \underline{x}^{2a} M_k^{(b)}(\underline{x})G(\underline{x}) + \text{L.O.T.}) , \tag{5.2.194}$$

[21] More precisely: for a dense subspace thereof.

where the first power in $\underline{x}$ appearing in L.O.T. is equal to $(2a-2)$. This is due to the fact that the operator $(\nabla - \underline{x})$ can only raise by one, either from the multiplication by $\underline{x}$ or the action of ∇ on $G(\underline{x})$, or lower by one, which comes from the action of ∇ on a power in $\underline{x}$. Using the relation

$$\nabla_{\underline{x}}\underline{x} = \underline{x}\nabla_{\underline{x}} - (2Sc - 1), \tag{5.2.195}$$

it is then easily seen that we indeed arrive at the constant $(-2)^{2a+1}$ for the leading term in $\underline{x}$. As all the leading terms are different, it follows that any function of the form $\underline{x}^a M_k^{(b)}(\underline{x})G(\underline{x})$ can indeed be expressed as a unique linear combination of the Clifford-Hermite eigenfunctions. □

Bringing everything together, we have obtained from Theorem 5.2.75, equation (5.2.190), and Proposition 5.2.78, the following final result:

Theorem 5.2.79. *CFT as exponential operator*
The Clifford Fourier transform $\mathcal{F}\{f\} : \mathbb{R}^n \to \mathcal{G}_n$, as an operator on the function space $\mathcal{S}(\mathbb{R}^n, \mathcal{G}_n)$, can be defined by means of

$$\mathcal{F}\{f\}(\boldsymbol{\omega}) = f(\dot{\boldsymbol{\omega}})e^{-\frac{\pi}{4}(\boldsymbol{\omega}^2 - \dot{\nabla}_{\boldsymbol{\omega}}^2 - n)i_n}. \tag{5.2.196}$$

In other words, the CFT can be written as an exponential operator acting from the right, involving the Hamiltonian operator $\boldsymbol{\omega}^2 - \dot{\nabla}_{\boldsymbol{\omega}}^2$, for the harmonic oscillator, and a phase factor $e^{\frac{n\pi}{4}i_n}$ which reduces to $(i_3 - 1)/\sqrt{2}$ for $n = 3$ and to i_2 for $n = 2$.

5.3 TWO-SIDED CLIFFORD FOURIER TRANSFORMS

We now generalize to two-sided CFTs for multivector valued functions in $Cl(p, q)$, including multiple kernel factors.

5.3.1 Two-sided Clifford Fourier transform with two square roots of -1 in $Cl(p, q)$

5.3.1.1 Background

In this section, we generalize quaternion and Clifford Fourier transforms to general *two-sided* Clifford Fourier transforms (CFT), and study their properties (from linearity to convolution). Two general *multivector square roots* $\in Cl(p, q)$ *of* -1 are used to split multivector signals, and to construct the left and right CFT kernel factors. This section is based on [209].

We therefore tap into the new results on square roots of -1 in Clifford algebras (see Section 2.7) and fully general construct CFTs, with two general square roots of -1 in $Cl(p, q)$. Our new CFTs form therefore a more general class of CFTs, subsuming and generalizing previous results. A further benefit is, that these new CFTs become *fully steerable* within the continuous Clifford algebra submanifolds of square roots of -1. We thus obtain a comprehensive *new mathematical framework* for the investigation and application of Clifford Fourier transforms together with *new properties* (full steerability). Regarding the question of the *most suitable* CFT for a certain application, we are only just beginning to leave the terra cognita of familiar transforms to map out the vast array of possible CFTs in $Cl(p, q)$.

This treatment is organized as follows. In Section 5.3.1.2, we show how the $\pm$ split (of Section 4.1.1) or orthogonal 2D planes split (of Section 4.1.2) of quaternions can be generalized to *split multivector signal functions* with respect to a general pair of square roots of -1 in Clifford algebra. Next, in Section 5.3.1.3, we define the central notion of *general two-sided Clifford Fourier transforms* with respect to any two square roots of -1 in Clifford algebra. Finally, in Section 5.3.1.4 we investigate the *properties* of these new CFTs: linearity, shift, modulation, dilation, moments, inversion, derivatives, Plancherel and Parseval formulas, as well as a convolution theorem.

For a systematic introduction to Clifford's geometric algebras, we refer to Chapter 2. For vector differential calculus, we refer to Chapter 3. In this section, we will use the *principal reverse* $\widetilde{N}$ of Section 2.1.1.6.

For $M, N \in Cl(p,q)$ we get $M * \widetilde{N} = \sum_A M_A N_A$. Two multivectors $M, N \in Cl(p,q)$ are *orthogonal* if and only if $M * \widetilde{N} = 0$. The modulus $|M|$ of a multivector $M \in Cl(p,q)$ is defined as

$$|M|^2 = M * \widetilde{M} = \sum_A M_A^2. \tag{5.3.1}$$

5.3.1.2 The $\pm$ split with respect to two square roots of -1

Here we generalize the quaternionic split of Sections 4.1.1 and 4.1.2. We note that with respect to any square root $i \in Cl(p,q)$ of -1, $i^2 = -1$, every multivector $A \in Cl(p,q)$ can be split into *commuting* A_{+i} and *anticommuting* A_{-i} parts, see Lemma 2.7.5 and [203].

For $f, g \in Cl(p,q)$ an arbitrary pair of square roots of -1, $f^2 = g^2 = -1$, the map $f(\,)g$ is an involution, because $f^2 x g^2 = (-1)^2 x = x, \forall x \in Cl(p,q)$. In Section 4.1.1, a split of quaternions by means of the pure unit quaternion basis elements $\boldsymbol{i}, \boldsymbol{j} \in \mathbb{H}$ was introduced, and generalized to a general pair of pure unit quaternions in Section 4.1.2. We now *generalize the split* to $Cl(p,q)$.

Definition 5.3.1 ($\pm$ split with respect to two square roots of -1). *Let $f, g \in Cl(p,q)$ be an arbitrary pair of square roots of -1, $f^2 = g^2 = -1$, including the cases $f = \pm g$. The general $\pm$ split is then defined with respect to the two square roots f, g of -1 as*

$$x_\pm = \frac{1}{2}(x \pm f x g), \qquad \forall x \in Cl(p,q). \tag{5.3.2}$$

Note that the split of Lemma 2.7.5 is a special case of Definition 5.3.1 with $g = -f$.

We observe from (5.3.2), that $fxg = x_+ - x_-$, i.e. under the map $f(\,)g$ the x_+ part is invariant, but the x_- part changes sign

$$f x_\pm g = \frac{1}{2}(fxg \pm f^2 x g^2) = \frac{1}{2}(fxg \pm x) = \pm\frac{1}{2}(x \pm fxg) = \pm x_\pm. \tag{5.3.3}$$

The two parts $x_\pm$ can be represented with Lemma 2.7.5 as linear combinations of x_{+f} and x_{-f}, or of x_{+g} and x_{-g}

$$\begin{aligned} x_\pm &= \frac{1}{2}(x_{+f} + x_{-f} \pm f(x_{+f} + x_{-f})g) = x_{+f}\,\frac{1 \pm fg}{2} + x_{-f}\,\frac{1 \mp fg}{2} \\ &= \frac{1}{2}(x_{+g} + x_{-g} \pm f(x_{+g} + x_{-g})g) = \frac{1 \pm fg}{2}\,x_{+g} + \frac{1 \mp fg}{2}\,x_{-g}. \end{aligned} \tag{5.3.4}$$

For $Cl(p,q) \cong \mathcal{M}(2d,\mathbb{C})$ or $\mathcal{M}(d,\mathbb{H})$ or $\mathcal{M}(d,\mathbb{H}^2)$, or for both f,g being blades in $Cl(p,q) \cong \mathcal{M}(2d,\mathbb{R})$ or $\mathcal{M}(2d,\mathbb{R}^2)$, we have $\widetilde{f} = -f$, $\widetilde{g} = -g$. We therefore obtain under this assumption the following lemma.

Lemma 5.3.2 (Orthogonality of two $\pm$ split parts). *Given any two multivectors $x, y \in Cl(p,q)$ and applying the $\pm$ split (5.3.2) with respect to two square roots of -1, assuming that $\widetilde{f} = -f$ and $\widetilde{g} = -g$, we get zero for the scalar part of the mixed products*

$$Sc(x_+\widetilde{y_-}) = 0, \qquad Sc(x_-\widetilde{y_+}) = 0. \tag{5.3.5}$$

We prove the first identity, the second follows from $Sc(x) = Sc(\widetilde{x})$.

$$\begin{aligned} Sc(x_+\widetilde{y_-}) &= \frac{1}{4}Sc((x + fxg)(\widetilde{y} - g\widetilde{y}f)) = \frac{1}{4}Sc(x\widetilde{y} - fxgg\widetilde{y}f + fxg\widetilde{y} - xg\widetilde{y}f) \\ &= \frac{1}{4}Sc(x\widetilde{y} - x\widetilde{y} + xg\widetilde{y}f - xg\widetilde{y}f) = 0, \end{aligned} \tag{5.3.6}$$

where the symmetry $Sc(xy) = Sc(yx)$ was used for the third equality.

We will now establish the *general identity*

$$e^{\alpha f} x_\pm e^{\beta g} = x_\pm e^{(\beta \mp \alpha) g} = e^{(\alpha \mp \beta) f} x_\pm. \tag{5.3.7}$$

First, we prove

$$\begin{aligned} e^{\alpha f} x_+ &= (\cos\alpha + f\sin\alpha)\frac{1}{2}(x + fxg) = \frac{1}{2}(x + fxg)\cos\alpha + \frac{1}{2}f(x + fxg)\sin\alpha \\ &= \frac{1}{2}(x + fxg)\cos\alpha + \frac{1}{2}(fxg(-g) - xg)\sin\alpha \\ &= \frac{x + fxg}{2}\cos\alpha + \frac{fxg + x}{2}(-g)\sin\alpha = \frac{x + fxg}{2}(\cos\alpha - g\sin\alpha) \\ &= x_+ e^{-\alpha g}. \end{aligned} \tag{5.3.8}$$

Similarly we can prove that $e^{\alpha f} x_- = x_- e^{+\alpha g}$ by replacing $g \to -g$ ($\Rightarrow x_+ \to x_-$) in (5.3.8).

5.3.1.3 General two-sided Clifford Fourier transforms

The *general two-sided Clifford Fourier transform* (CFT), to be introduced now, can both be understood as a generalization of known one-sided CFTs [185], or of the two-sided quaternion Fourier transformation (QFT) [184,194] to a general Clifford algebra setting. Most previously known CFTs use in their kernels specific square roots of -1, like bivectors, pseudoscalars, unit pure quaternions or sets of coorthogonal blades (commuting or anticommuting blades) [61]. We will *remove all these restrictions* on the square roots of -1 used in a CFT. Note that if the left or right phase angle is identical to zero, we get one-sided right or left sided CFTs, respectively.

Definition 5.3.3 (CFT with respect to two square roots of -1). *Let $f, g \in Cl(p,q)$, $f^2 = g^2 = -1$, be any two square roots of -1. The general two-sided Clifford Fourier transform*[22] *(CFT) of $h \in L^1(\mathbb{R}^{p,q}, Cl(p,q))$, with respect to f, g is*

$$\mathcal{F}^{f,g}\{h\}(\boldsymbol{\omega}) = \int_{\mathbb{R}^{p,q}} e^{-fu(\boldsymbol{x},\boldsymbol{\omega})} h(\boldsymbol{x})\, e^{-gv(\boldsymbol{x},\boldsymbol{\omega})} d^n\boldsymbol{x}, \tag{5.3.9}$$

where $d^n\boldsymbol{x} = dx_1 \dots dx_n$, $\boldsymbol{x}, \boldsymbol{\omega} \in \mathbb{R}^{p,q}$, and $u, v : \mathbb{R}^{p,q} \times \mathbb{R}^{p,q} \to \mathbb{R}$.

Since square roots of -1 in $Cl(p,q)$ populate *continuous submanifolds* in $Cl(p,q)$, the CFT of Definition 5.3.3 is generically *steerable* within these manifolds. In Definition 5.3.3, the two square roots $f, g \in Cl(p,q)$ of -1 may be from the same (or different) conjugacy class and component, respectively. A further generalization to a two-sided CFT on signal functions $h \in L^1(\mathbb{R}^{p,q}; Cl_{p',q'})$ is given later in the context of studying new concepts of convolution, see Definition 5.4.13.

Linearity of the CFT integral (5.3.9) allows us to use the $\pm$ split $h = h_- + h_+$ of Definition 5.3.1 to obtain

$$\mathcal{F}^{f,g}\{h\}(\boldsymbol{\omega}) = \mathcal{F}^{f,g}\{h_-\}(\boldsymbol{\omega}) + \mathcal{F}^{f,g}\{h_+\}(\boldsymbol{\omega}) = \mathcal{F}^{f,g}_-\{h\}(\boldsymbol{\omega}) + \mathcal{F}^{f,g}_+\{h\}(\boldsymbol{\omega}), \tag{5.3.10}$$

since by their construction the operators of the Clifford Fourier transformation $\mathcal{F}^{f,g}$, and of the $\pm$ split with respect to f, g commute. From (5.3.7) follows

Theorem 5.3.4 (CFT of $h_\pm$). *The CFT of the $\pm$ split parts $h_\pm$, with respect to two square roots $f, g \in Cl(p,q)$ of -1 , of a Clifford module function $h \in L^1(\mathbb{R}^{p,q}; Cl(p,q))$ have the quasi-complex forms*

$$\begin{aligned}\mathcal{F}^{f,g}_\pm\{h\} = \mathcal{F}^{f,g}\{h_\pm\} &= \int_{\mathbb{R}^{p,q}} h_\pm\, e^{-g(v(\boldsymbol{x},\boldsymbol{\omega}) \mp u(\boldsymbol{x},\boldsymbol{\omega}))} d^n x \\ &= \int_{\mathbb{R}^{p,q}} e^{-f(u(\boldsymbol{x},\boldsymbol{\omega}) \mp v(\boldsymbol{x},\boldsymbol{\omega}))} h_\pm\, d^n x. \end{aligned} \tag{5.3.11}$$

Remark 5.3.5. *The quasi-complex forms in Theorem 5.3.4 allow to establish discretized and fast versions of the general two-sided CFT of Definition 5.3.3 as sums of complex discretized and fast Fourier transformations (FFT), respectively.*

5.3.1.4 Properties of the general two-sided CFT

We now study important properties of the general two-sided CFT of Definition 5.3.3.

---

[22]For the Fourier transform on L^1 and on L^2, see Footnote 2 on page 144.

5.3.1.4.1 Linearity, shift, modulation, dilation and powers of f, g Regarding *left and right linearity* of the general two-sided CFT of Definition 5.3.3 we can establish with the help of Lemma 2.7.5 that for $h_1, h_2 \in L^1(\mathbb{R}^{p,q}; Cl(p,q))$, and constants $\alpha, \beta \in Cl(p,q)$

$$\begin{aligned}\mathcal{F}^{f,g}\{\alpha h_1 + \beta h_2\}(\boldsymbol{\omega}) &= \alpha_{+f}\mathcal{F}^{f,g}\{h_1\}(\boldsymbol{\omega}) + \alpha_{-f}\mathcal{F}^{-f,g}\{h_1\}(\boldsymbol{\omega}) \\ &\quad + \beta_{+f}\mathcal{F}^{f,g}\{h_2\}(\boldsymbol{\omega}) + \beta_{-f}\mathcal{F}^{-f,g}\{h_2\}(\boldsymbol{\omega}), \qquad (5.3.12)\\ \mathcal{F}^{f,g}\{h_1\alpha + h_2\beta\}(\boldsymbol{\omega}) &= \mathcal{F}^{f,g}\{h_1\}(\boldsymbol{\omega})\alpha_{+g} + \mathcal{F}^{f,-g}\{h_1\}(\boldsymbol{\omega})\alpha_{-g} \\ &\quad + \mathcal{F}^{f,g}\{h_2\}(\boldsymbol{\omega})\beta_{+g} + \mathcal{F}^{f,-g}\{h_2\}(\boldsymbol{\omega})\beta_{-g}. \qquad (5.3.13)\end{aligned}$$

Regarding the *$\boldsymbol{x}$-shifted* function $h_0(\boldsymbol{x}) = h(\boldsymbol{x} - \boldsymbol{x}_0)$ we obtain with constant $\boldsymbol{x}_0 \in \mathbb{R}^{p,q}$, assuming linearity of $u(\boldsymbol{x},\boldsymbol{\omega}), v(\boldsymbol{x},\boldsymbol{\omega})$ in their vector space argument $\boldsymbol{x}$,

$$\mathcal{F}^{f,g}\{h_0\}(\boldsymbol{\omega}) = e^{-fu(\boldsymbol{x}_0,\boldsymbol{\omega})}\mathcal{F}^{f,g}\{h\}(\boldsymbol{\omega})\, e^{-gv(\boldsymbol{x}_0,\boldsymbol{\omega})}. \qquad (5.3.14)$$

For the purpose of *modulation* we make the special assumption, that the functions $u(\boldsymbol{x},\boldsymbol{\omega}), v(\boldsymbol{x},\boldsymbol{\omega})$ are both linear in their frequency argument $\boldsymbol{\omega}$. Then we obtain for $h_m(\boldsymbol{x}) = e^{-fu(\boldsymbol{x},\boldsymbol{\omega}_0)}h(x)\, e^{-gv(\boldsymbol{x},\boldsymbol{\omega}_0)}$, and constant $\boldsymbol{\omega}_0 \in \mathbb{R}^{p,q}$ the modulation formula

$$\mathcal{F}^{f,g}\{h_m\}(\boldsymbol{\omega}) = \mathcal{F}^{f,g}\{h\}(\boldsymbol{\omega} + \boldsymbol{\omega}_0). \qquad (5.3.15)$$

Regarding *dilations*, we make the special assumption, that for constants $a_1, \ldots, a_n \in \mathbb{R} \setminus \{0\}$, and $\boldsymbol{x}' = \sum_{k=1}^n a_k x_k \boldsymbol{e}_k$, we have $u(\boldsymbol{x}',\boldsymbol{\omega}) = u(\boldsymbol{x},\boldsymbol{\omega}')$, and $v(\boldsymbol{x}',\boldsymbol{\omega}) = v(\boldsymbol{x},\boldsymbol{\omega}')$, with $\boldsymbol{\omega}' = \sum_{k=1}^n a_k \omega_k \boldsymbol{e}_k$. We then obtain for $h_d(\boldsymbol{x}) = h(\boldsymbol{x}')$ that

$$\mathcal{F}^{f,g}\{h_d\}(\boldsymbol{\omega}) = \frac{1}{|a_1 \ldots a_n|}\mathcal{F}^{f,g}\{h\}(\boldsymbol{\omega}_d), \qquad \boldsymbol{\omega}_d = \sum_{k=1}^n \frac{1}{a_k}\omega_k \boldsymbol{e}_k. \qquad (5.3.16)$$

For $a_1 = \ldots = a_n = a \in \mathbb{R} \setminus \{0\}$ this simplifies under the same special assumption to

$$\mathcal{F}^{f,g}\{h_d\}(\boldsymbol{\omega}) = \frac{1}{|a|^n}\mathcal{F}^{f,g}\{h\}(\frac{1}{a}\boldsymbol{\omega}). \qquad (5.3.17)$$

Note, that the above assumption would, e.g., be fulfilled for $u(\boldsymbol{x},\boldsymbol{\omega}) = v(\boldsymbol{x},\boldsymbol{\omega}) = \boldsymbol{x} * \widetilde{\boldsymbol{\omega}}$.

For f, g *power factors* in $h_{p,q}(\boldsymbol{x}) = f^p h(\boldsymbol{x}) g^q$, $p, q \in \mathbb{Z}$, we obtain

$$\mathcal{F}^{f,g}\{h_{p,q}\}(\boldsymbol{\omega}) = f^p \mathcal{F}^{f,g}\{h\}(\boldsymbol{\omega}) g^q. \qquad (5.3.18)$$

5.3.1.4.2 CFT inversion, moments, derivatives, Plancherel, Parseval For establishing an inversion formula, moment and derivative properties, Plancherel and Parseval identities, certain *assumptions* about the phase functions $u(\boldsymbol{x},\boldsymbol{\omega})$, $v(\boldsymbol{x},\boldsymbol{\omega})$ need to be made. One possibility is, e.g. to arbitrarily partition the scalar product $\boldsymbol{x} * \widetilde{\boldsymbol{\omega}} = \sum_{l=1}^n x_l \omega_l = u(\boldsymbol{x},\boldsymbol{\omega}) + v(\boldsymbol{x},\boldsymbol{\omega})$, with

$$u(\boldsymbol{x},\boldsymbol{\omega}) = \sum_{l=1}^k x_l \omega_l, \qquad v(\boldsymbol{x},\boldsymbol{\omega}) = \sum_{l=k+1}^n x_l \omega_l, \qquad (5.3.19)$$

for any arbitrary but fixed $1 \leq k \leq n$. We could also include any subset $A_u \subseteq \{1, \ldots, n\}$ of coordinates in $u(\boldsymbol{x}, \boldsymbol{\omega})$ and the complementary set $A_v = \{1, \ldots, n\} \setminus A_u$ of coordinates in $v(\boldsymbol{x}, \boldsymbol{\omega})$, etc. (5.3.19) will be assumed for the current subsection.

We then get the following *inversion* formula

$$h(\boldsymbol{x}) = \mathcal{F}^{f,g}_{-1}\{\mathcal{F}^{f,g}\{h\}\}(\boldsymbol{x}) = \int_{\mathbb{R}^{p,q}} e^{fu(\boldsymbol{x},\boldsymbol{\omega})}\mathcal{F}^{f,g}\{h\}(\boldsymbol{\omega})\, e^{gv(\boldsymbol{x},\boldsymbol{\omega})} d^n\boldsymbol{\omega}, \tag{5.3.20}$$

where $d^n\boldsymbol{\omega} = d\omega_1 \ldots d\omega_n$, $\boldsymbol{x}, \boldsymbol{\omega} \in \mathbb{R}^{p,q}$. For the existence of (5.3.20) we need $\mathcal{F}^{f,g}\{h\} \in L^1(\mathbb{R}^{p,q}; Cl(p,q))$.

Additionally, we get the transformation law for *partial derivatives* $h'_l(\boldsymbol{x}) = \partial_{x_l} h(\boldsymbol{x})$, $1 \leq l \leq n$, for h piecewise smooth and integrable, and $h, h'_l \in L^1(\mathbb{R}^{p,q}; Cl(p,q))$ as

$$\mathcal{F}^{f,g}\{h'_l\}(\boldsymbol{\omega}) = \begin{cases} f\omega_l\, \mathcal{F}^{f,g}\{h\}(\boldsymbol{\omega}), & \text{for } l \leq k \\ \mathcal{F}^{f,g}\{h\}(\boldsymbol{\omega})\, g\,\omega_l, & \text{for } l > k \end{cases}. \tag{5.3.21}$$

For the transformation of the *spatial moments* with $h_l(\boldsymbol{x}) = x_l h(\boldsymbol{x})$, $1 \leq l \leq n$, $h, h_l \in L^1(\mathbb{R}^{p,q}; Cl(p,q))$, we obtain

$$\mathcal{F}^{f,g}\{h_l\}(\boldsymbol{\omega}) = \begin{cases} f\, \partial_{\omega_l} \mathcal{F}^{f,g}\{h\}(\boldsymbol{\omega}), & \text{for } l \leq k \\ \partial_{\omega_l} \mathcal{F}^{f,g}\{h\}(\boldsymbol{\omega})\, g, & \text{for } l > k \end{cases}. \tag{5.3.22}$$

Moreover, for the functions $h_1, h_2, h \in L^2(\mathbb{R}^{p,q}; Cl(p,q))$ we obtain the *Plancherel* identity

$$\langle h_1, h_2 \rangle = \frac{1}{(2\pi)^n} \langle \mathcal{F}^{f,g}\{h_1\}, \mathcal{F}^{f,g}\{h_2\} \rangle, \tag{5.3.23}$$

as well as the *Parseval* identity

$$\|h\| = \frac{1}{(2\pi)^{n/2}} \left\| \mathcal{F}^{f,g}\{h\} \right\|. \tag{5.3.24}$$

5.3.1.4.3 Convolution We define the *convolution* of two multivector signals $a, b \in L^1(R^{p,q}; Cl(p,q))$ as

$$(a \star b)(\boldsymbol{x}) = \int_{\mathbb{R}^{p,q}} a(\boldsymbol{y}) b(\boldsymbol{x} - \boldsymbol{y}) d^n\boldsymbol{y}. \tag{5.3.25}$$

For establishing the general two-sided CFT of the convolution, we need the identity

$$e^{\alpha f} e^{\beta g} = e^{\beta g} e^{\alpha f} + [f, g] \sin(\alpha) \sin(\beta), \qquad [f, g] = fg - gf. \tag{5.3.26}$$

We further define the following two mixed exponential-sine transforms

$$\mathcal{F}^{f,\pm s}\{h\}(\boldsymbol{\omega}) = \int_{\mathbb{R}^{p,q}} e^{-fu(\boldsymbol{x},\boldsymbol{\omega})} h(\boldsymbol{x}) (\pm 1) \sin(-v(\boldsymbol{x}, \boldsymbol{\omega})) d^n\boldsymbol{x}, \tag{5.3.27}$$

$$\mathcal{F}^{\pm s,g}\{h\}(\boldsymbol{\omega}) = \int_{\mathbb{R}^{p,q}} (\pm 1) \sin(-u(\boldsymbol{x}, \boldsymbol{\omega})) h(\boldsymbol{x}) e^{-gv(\boldsymbol{x},\boldsymbol{\omega})} d^n\boldsymbol{x}. \tag{5.3.28}$$

We assume that the functions u, v are both linear with respect to their first argument. The *general two-sided CFT of the convolution* (5.3.25) can then be expressed as

$$
\begin{aligned}
&\mathcal{F}^{f,g}\{a \star b\}(\boldsymbol{\omega}) = \\
&\quad \mathcal{F}^{f,g}\{a_{+f}\}(\boldsymbol{\omega})\mathcal{F}^{f,g}\{b_{+g}\}(\boldsymbol{\omega}) + \mathcal{F}^{f,-g}\{a_{+f}\}(\boldsymbol{\omega})\mathcal{F}^{f,g}\{b_{-g}\}(\boldsymbol{\omega}) \\
&\quad +\mathcal{F}^{f,g}\{a_{-f}\}(\boldsymbol{\omega})\mathcal{F}^{-f,g}\{b_{+g}\}(\boldsymbol{\omega}) + \mathcal{F}^{f,-g}\{a_{-f}\}(\boldsymbol{\omega})\mathcal{F}^{-f,g}\{b_{-g}\}(\boldsymbol{\omega}) \\
&\quad +\mathcal{F}^{f,s}\{a_{+f}\}(\boldsymbol{\omega})[f,g]\mathcal{F}^{s,g}\{b_{+g}\}(\boldsymbol{\omega}) + \mathcal{F}^{f,-s}\{a_{+f}\}(\boldsymbol{\omega})[f,g]\mathcal{F}^{s,g}\{b_{-g}\}(\boldsymbol{\omega}) \\
&\quad +\mathcal{F}^{f,s}\{a_{-f}\}(\boldsymbol{\omega})[f,g]\mathcal{F}^{-s,g}\{b_{+g}\}(\boldsymbol{\omega}) + \mathcal{F}^{f,-s}\{a_{-f}\}(\boldsymbol{\omega})[f,g]\mathcal{F}^{-s,g}\{b_{-g}\}(\boldsymbol{\omega}).
\end{aligned}
\tag{5.3.29}
$$

5.3.1.5 Reflection on two-sided Clifford Fourier transform with two square roots of -1 in $Cl(p,q)$

We have by now established a comprehensive *new mathematical framework* for the investigation and application of Clifford Fourier transforms (CFTs) together with *new properties*. Our new CFTs form a more general class of CFTs, subsuming and generalizing previous results. We have applied new results on square roots of -1 in Clifford algebras to fully general construct CFTs, with two general square roots of -1 in Clifford algebras $Cl(p,q)$, see Section 2.7. The new CFTs are *fully steerable* within the continuous Clifford algebra submanifolds of square roots of -1. We have thus left the terra cognita of familiar transforms to outline the vast array of possible two-sided CFTs in $Cl(p,q)$, even more general then the one-sided CFTs of Section 5.2.2. We showed how the $\pm$ split or orthogonal 2D planes split of quaternions can be generalized to *split multivector signal functions* with respect to a general pair of square roots of -1 in Clifford algebra. Next, we defined the central notion of *general two-sided Clifford Fourier transforms* with respect to any two square roots of -1 in Clifford algebra. Finally, we investigated important *properties* of these new CFTs: linearity, shift, modulation, dilation, moments, inversion, derivatives, Plancherel and Parseval formulas, as well as a convolution theorem.

Regarding numerical implementations, Theorem 5.3.4 shows that 2^n complex Fourier transformations (FTs) are sufficient. In some cases this can be reduced to $2^{(n-1)}$ complex FTs, e.g., when one of the two square roots of -1 is a pseudoscalar. Further algebraic studies may widen the class of CFTs, where $2^{(n-1)}$ complex FTs are sufficient. Numerical implementation is then possible with 2^n (or $2^{(n-1)}$) discrete complex FTs, which can also be fast Fourier transforms (FFTs), leading to fast CFT implementations.

A well-known example of a CFT with two square roots of -1 are the quaternion FTs (QFTs) [56, 58, 124, 127, 184, 194, 291], compare Chapter 4, which are particularly used in applications to partial differential systems, color image processing, filtering, disparity estimation (two images differ by local translations) and texture segmentation. Another example is the spacetime FT, which leads to a multivector wave packet analysis of spacetime signals (e.g. electro-magnetic signals), applicable even to relativistic signals [184, 187], see Section 4.4.

Depending on the choice of the phase functions $u(\boldsymbol{x}, \boldsymbol{\omega})$ and $v(\boldsymbol{x}, \boldsymbol{\omega})$, the multivector basis coefficient functions of the CFT result carry information on the symmetry of the signal, similar to the special case of the QFT [56].

The convolution theorem allows to design and apply multivector valued filters to multivector valued signals.

5.3.2 Two-sided (Clifford) geometric Fourier transform with multiple square roots of -1 in $Cl(p,q)$

5.3.2.1 *Background*

Here we introduce one single straightforward definition of a general geometric Fourier transform covering many versions in the literature. We show which constraints are additionally necessary to obtain certain features such as linearity or a shift theorem. The section is based on [61].

The kernel of the Fourier transform consists of the complex exponential function. With the square root of minus one, the imaginary unit i, as part of the argument it is periodic and therefore suitable for the analysis of oscillating systems.

Geometric algebras usually contain continuous submanifolds of geometric square roots of -1 [193, 203], see Section 2.7. Each multivector has a natural geometric interpretation so the generalization of the Fourier transform to multivector valued functions in the geometric algebras is very reasonable. It helps to interpret the transform, apply it in a target-oriented way to the specific underlying problem and it allows a new point of view in fields like fluid mechanics.

For a historical overview of the development of a wide variety of Clifford Fourier transforms, we refer to Chapter 1. The analysis of their similarities reveals a lot about their qualities, too. We concentrate on this matter and try to summarize most of them in one general definition.

Here we focus on continuous geometric Fourier transforms over flat spaces $\mathbb{R}^{p,q}$ in their integral representation. That way their finite, regular discrete equivalents as used in computational signal and image processing can be intuitively constructed and direct applicability to the existing practical issues and easy numerical manageability are ensured.

5.3.2.2 *Definition of the (Clifford) geometric Fourier transform*

For background on Clifford geometric algebra, we refer to Chapter 2.

For the sake of brevity we now want to refer to arbitrary multivectors

$$\boldsymbol{A} = \sum_{k=0}^{n} \sum_{1 \le j_1 < \ldots < j_k \le n} a_{j_1 \ldots j_k} \boldsymbol{e}_{j_1} \ldots \boldsymbol{e}_{j_k} \in \mathcal{G}^{p,q} = Cl(p,q), \tag{5.3.30}$$

where $a_{j_1 \ldots j_k} \in \mathbb{R}$, as

$$\boldsymbol{A} = \sum_{\boldsymbol{j}} a_{\boldsymbol{j}} \boldsymbol{e}_{\boldsymbol{j}}, \tag{5.3.31}$$

where each of the 2^n multi-indices $\boldsymbol{j} \subseteq \{1, \ldots, n\}$ indicates a basis vector of $Cl(p,q)$ by $\boldsymbol{e}_{\boldsymbol{j}} = \boldsymbol{e}_{j_1} \ldots \boldsymbol{e}_{j_k}$, $1 \le j_1 < \ldots < j_k \le n$, $\boldsymbol{e}_{\emptyset} = \boldsymbol{e}_0 = 1$ and its associated coefficient $a_{\boldsymbol{j}} = a_{j_1 \ldots j_k} \in \mathbb{R}$.

Definition 5.3.6. *The* ***exponential function*** *of a multivector $\boldsymbol{A} \in Cl(p,q)$ is defined by the power series*

$$e^{\boldsymbol{A}} : = \sum_{j=0}^{\infty} \frac{\boldsymbol{A}^j}{j!}. \tag{5.3.32}$$

Lemma 5.3.7. *For two multivectors $\boldsymbol{AB} = \boldsymbol{BA}$ that commute we have*

$$e^{\boldsymbol{A}+\boldsymbol{B}} = e^{\boldsymbol{A}} e^{\boldsymbol{B}}. \tag{5.3.33}$$

Proof. Analogous to the exponent rule of real matrices. □

Notation 5.3.8. *For each geometric algebra $Cl(p,q)$ we will write $\mathscr{I}^{p,q} = \{i \in Cl(p,q), i^2 \in \mathbb{R}^-\}$ to denote the real multiples of all geometric square roots of -1 , compare [193] and [203]. Here we choose the symbol $\mathscr{I}$ to be reminiscent of the imaginary numbers.*

Definition 5.3.9. *Let $Cl(p,q)$ be a geometric algebra, $\boldsymbol{A} : \mathbb{R}^m \to Cl(p,q)$ be a multivector field[23] and $\boldsymbol{x}, \boldsymbol{u} \in \mathbb{R}^m$ vectors. A* ***Geometric Fourier Transform*** *(GFT) $\mathcal{F}_{F_1,F_2}(\boldsymbol{A})$ is defined by two ordered finite sets $F_1 = \{f_1(\boldsymbol{x},\boldsymbol{u}), \dots, f_{\mu}(\boldsymbol{x},\boldsymbol{u})\}$, $F_2 = \{f_{\mu+1}(\boldsymbol{x},\boldsymbol{u}), \dots, f_{\nu}(\boldsymbol{x},\boldsymbol{u})\}$ of mappings $f_k\ (\boldsymbol{x},\boldsymbol{u}) : \mathbb{R}^m \times \mathbb{R}^m \to \mathscr{I}^{p,q}, \forall k = 1, \dots, \boldsymbol{\nu}$ and the calculation rule*

$$\mathcal{F}_{F_1,F_2}(\boldsymbol{A})(\boldsymbol{u}) := \int_{\mathbb{R}^m} \prod_{f \in F_1} e^{-f(\boldsymbol{x},\boldsymbol{u})} \boldsymbol{A}(\boldsymbol{x}) \prod_{f \in F_2} e^{-f(\boldsymbol{x},\boldsymbol{u})} \, d^m \boldsymbol{x}. \tag{5.3.34}$$

This definition combines many Fourier transforms into a single general one. It enables us to prove the well-known theorems which depend only on the properties of the chosen mappings.

Exercise 5.3.10. *Depending on the choice of F_1 and F_2 we obtain previously published transforms.*

1. *In the case of $\boldsymbol{A} : \mathbb{R}^n \to Cl(n,0), n = 2 \pmod 4$ or $n = 3 \pmod 4$, we can reproduce the Clifford Fourier transform introduced by Jancewicz [222] for $n = 3$ and expanded by Ebling and Scheuermann [119] for $n = 2$ and Hitzer and Mawardi [185] for $n = 2 \pmod 4$ or $n = 3 \pmod 4$ using the configuration*

$$F_1 = \emptyset, \quad F_2 = \{f_1\}, \quad f_1(\boldsymbol{x},\boldsymbol{u}) = 2\pi i_n \boldsymbol{x} \cdot \boldsymbol{u}, \tag{5.3.35}$$

   with i_n being the pseudoscalar of $Cl(n,0)$.

2. *Choosing multivector fields $\mathbb{R}^n \to Cl(0,n)$,*

$$\begin{aligned} F_1 &= \emptyset, \quad F_2 = \{f_1, \dots, f_n\}, \\ f_k(\boldsymbol{x},\boldsymbol{u}) &= 2\pi \boldsymbol{e}_k x_k u_k, \quad \forall k = 1, \dots, n \end{aligned} \tag{5.3.36}$$

   we have the Sommen Bülow Clifford Fourier transform from [56, 306].

[23]For the Fourier transform with $\boldsymbol{A} \in L^1(\mathbb{R}^m; Cl(p,q))$ and $\boldsymbol{A} \in L^2(\mathbb{R}^m; Cl(p,q))$, see Footnote 2 on page 144 regarding the roles of L^1 and L^2.

3. *For* $\boldsymbol{A} : \mathbb{R}^2 \to Cl(0,2) \approx \mathbb{H}$ *the quaternion Fourier transform [56, 124, 184] is generated by*

$$\begin{aligned} F_1 = \{f_1\}, \quad F_2 = \{f_2\}, \\ f_1(\boldsymbol{x},\boldsymbol{u}) = 2\pi i x_1 u_1, \quad f_2(\boldsymbol{x},\boldsymbol{u}) = 2\pi j x_2 u_2. \end{aligned} \tag{5.3.37}$$

4. *Using* $Cl(3,1)$ *we can build the spacetime, respectively the volume-time, Fourier transform from [184][24] with the* $\mathcal{G}^{3,1}$*-pseudoscalar* i_4 *as follows*

$$\begin{aligned} F_1 = \{f_1\}, \quad F_2 = \{f_2\}, \\ f_1(\boldsymbol{x},\boldsymbol{u}) = \boldsymbol{e}_4 x_4 u_4, \quad f_2(\boldsymbol{x},\boldsymbol{u}) = \epsilon_4 \boldsymbol{e}_4 i_4 (x_1 u_1 + x_2 u_2 + x_3 u_3). \end{aligned} \tag{5.3.38}$$

5. *The Clifford Fourier transform for colour images by Batard, Berthier and Saint-Jean [18] for* $m = 2, n = 4$, $\boldsymbol{A} : \mathbb{R}^2 \to Cl(4,0)$, *a fixed bivector* $\boldsymbol{B}$, *and the pseudoscalar* i *can intuitively be written as*

$$\begin{aligned} F_1 &= \{f_1\}, \quad F_2 = \{f_2\}, \\ f_1(\boldsymbol{x},\boldsymbol{u}) &= \frac{1}{2}(x_1 u_1 + x_2 u_2)(\boldsymbol{B} + i\boldsymbol{B}), \\ f_2(\boldsymbol{x},\boldsymbol{u}) &= -\frac{1}{2}(x_1 u_1 + x_2 u_2)(\boldsymbol{B} + i\boldsymbol{B}), \end{aligned} \tag{5.3.39}$$

but $(\boldsymbol{B} + i\boldsymbol{B})$ *does not square to a negative real number, see [193]. The special property that* $\boldsymbol{B}$ *and* $i\boldsymbol{B}$ *commute allows us to express the formula using*

$$\begin{aligned} F_1 &= \{f_1, f_2\}, \quad F_2 = \{f_3, f_4\}, \\ f_1(\boldsymbol{x},\boldsymbol{u}) &= \frac{1}{2}(x_1 u_1 + x_2 u_2)\boldsymbol{B}, \quad f_2(\boldsymbol{x},\boldsymbol{u}) = \frac{1}{2}(x_1 u_1 + x_2 u_2) i\boldsymbol{B}, \\ f_3(\boldsymbol{x},\boldsymbol{u}) &= -\frac{1}{2}(x_1 u_1 + x_2 u_2)\boldsymbol{B}, \quad f_4(\boldsymbol{x},\boldsymbol{u}) = -\frac{1}{2}(x_1 u_1 + x_2 u_2) i\boldsymbol{B}, \end{aligned} \tag{5.3.40}$$

which fulfills the conditions of Definition 5.3.9.

6. *Using* $Cl(0,n)$ *and*

$$F_1 = \{f_1\}, \quad F_2 = \emptyset, \quad f_1(\boldsymbol{x},\boldsymbol{u}) = -\boldsymbol{x} \wedge \boldsymbol{u}, \tag{5.3.41}$$

produces the cylindrical Fourier transform as introduced by Brackx, De Schepper and Sommen in [49].

5.3.2.3 General GFT properties

First we prove general properties valid for arbitrary sets F_1, F_2.

Theorem 5.3.11 (Existence). *The geometric Fourier transform exists for all integrable multivector fields* $\boldsymbol{A} \in L_1(\mathbb{R}^n)$.

[24]Please note that Hitzer uses a different notation in [184]. His $\boldsymbol{x} = t\boldsymbol{e}_0 + x_1\boldsymbol{e}_1 + x_2\boldsymbol{e}_2 + x_3\boldsymbol{e}_3$ corresponds to our $\boldsymbol{x} = x_1\boldsymbol{e}_1 + x_2\boldsymbol{e}_2 + x_3\boldsymbol{e}_3 + x_4\boldsymbol{e}_4$, with $\boldsymbol{e}_0\boldsymbol{e}_0 = \epsilon_0 = -1$ being equivalent to our $\boldsymbol{e}_4\boldsymbol{e}_4 = \epsilon_4 = -1$.

Proof. The property

$$f_k^2(\boldsymbol{x},\boldsymbol{u}) \in \mathbb{R}^- \tag{5.3.42}$$

of the mappings f_k for $k = 1, \ldots, \boldsymbol{\nu}$ leads to

$$\frac{f_k^2(\boldsymbol{x},\boldsymbol{u})}{|f_k^2(\boldsymbol{x},\boldsymbol{u})|} = -1 \tag{5.3.43}$$

for all $f_k(\boldsymbol{x},\boldsymbol{u}) \neq 0$. So using the decomposition

$$f_k(\boldsymbol{x},\boldsymbol{u}) = \frac{f_k(\boldsymbol{x},\boldsymbol{u})}{|f_k(\boldsymbol{x},\boldsymbol{u})|}|f_k(\boldsymbol{x},\boldsymbol{u})|, \tag{5.3.44}$$

we can write $\forall j \in \mathbb{N}$

$$f_k^j(\boldsymbol{x},\boldsymbol{u}) = \begin{cases} (-1)^l |f_k(\boldsymbol{x},\boldsymbol{u})|^j & \text{for } j = 2l, l \in \mathbb{N}_0 \\ (-1)^l \dfrac{f_k(\boldsymbol{x},\boldsymbol{u})}{|f_k(\boldsymbol{x},\boldsymbol{u})|}|f_k(\boldsymbol{x},\boldsymbol{u})|^j & \text{for } j = 2l+1, l \in \mathbb{N}_0 \end{cases}, \tag{5.3.45}$$

which results in

$$\begin{aligned} e^{-f_k(\boldsymbol{x},\boldsymbol{u})} &= \sum_{j=0}^{\infty} \frac{(-f_k(\boldsymbol{x},\boldsymbol{u}))^j}{j!} = \sum_{j=0}^{\infty} \frac{(-1)^j |f_k(\boldsymbol{x},\boldsymbol{u})|^{2j}}{(2j)!} \\ &\quad - \frac{f_k(\boldsymbol{x},\boldsymbol{u})}{|f_k(\boldsymbol{x},\boldsymbol{u})|} \sum_{j=0}^{\infty} \frac{(-1)^j |f_k(\boldsymbol{x},\boldsymbol{u})|^{2j+1}}{(2j+1)!} \\ &= \cos(|f_k(\boldsymbol{x},\boldsymbol{u})|) - \frac{f_k(\boldsymbol{x},\boldsymbol{u})}{|f_k(\boldsymbol{x},\boldsymbol{u})|} \sin(|f_k(\boldsymbol{x},\boldsymbol{u})|). \end{aligned} \tag{5.3.46}$$

Because of

$$\begin{aligned} |e^{-f_k(\boldsymbol{x},\boldsymbol{u})}| &= \left| \cos(|f_k(\boldsymbol{x},\boldsymbol{u})|) - \frac{f_k(\boldsymbol{x},\boldsymbol{u})}{|f_k(\boldsymbol{x},\boldsymbol{u})|} \sin(|f_k(\boldsymbol{x},\boldsymbol{u})|) \right| \\ &\leq |\cos(|f_k(\boldsymbol{x},\boldsymbol{u})|)| + \left| \frac{f_k(\boldsymbol{x},\boldsymbol{u})}{|f_k(\boldsymbol{x},\boldsymbol{u})|} \right| |\sin(|f_k(\boldsymbol{x},\boldsymbol{u})|)| \\ &\leq 2, \end{aligned} \tag{5.3.47}$$

the magnitude of the improper integral

$$\begin{aligned} |\mathcal{F}_{F_1,F_2}(\boldsymbol{A})(\boldsymbol{u})| &= \left| \int_{\mathbb{R}^m} \prod_{f \in F_1} e^{-f(\boldsymbol{x},\boldsymbol{u})} \boldsymbol{A}(\boldsymbol{x}) \prod_{f \in F_2} e^{-f(\boldsymbol{x},\boldsymbol{u})} \, \mathrm{d}^m \boldsymbol{x} \right| \\ &\leq \int_{\mathbb{R}^m} \prod_{f \in F_1} \left| e^{-f(\boldsymbol{x},\boldsymbol{u})} \right| |\boldsymbol{A}(\boldsymbol{x})| \prod_{f \in F_2} \left| e^{-f(\boldsymbol{x},\boldsymbol{u})} \right| \, \mathrm{d}^m \boldsymbol{x} \\ &\leq \int_{\mathbb{R}^m} \prod_{f \in F_1} 2 \, |\boldsymbol{A}(\boldsymbol{x})| \prod_{f \in F_2} 2 \, \mathrm{d}^m \boldsymbol{x} = 2^{\boldsymbol{\nu}} \int_{\mathbb{R}^m} |\boldsymbol{A}(\boldsymbol{x})| \, \mathrm{d}^m \boldsymbol{x} \end{aligned} \tag{5.3.48}$$

is finite and therefore the geometric Fourier transform exists. □

Theorem 5.3.12 (Scalar linearity). *The geometric Fourier transform is linear with respect to scalar factors. Let $b, c \in \mathbb{R}$ and $\boldsymbol{A}, \boldsymbol{B}, \boldsymbol{C} : \mathbb{R}^m \to Cl(p, q)$ be three multivector fields that satisfy $\boldsymbol{A}(\boldsymbol{x}) = b\boldsymbol{B}(\boldsymbol{x}) + c\boldsymbol{C}(\boldsymbol{x})$, then*

$$\mathcal{F}_{F_1,F_2}(\boldsymbol{A})(\boldsymbol{u}) = b\mathcal{F}_{F_1,F_2}(\boldsymbol{B})(\boldsymbol{u}) + c\mathcal{F}_{F_1,F_2}(\boldsymbol{C})(\boldsymbol{u}). \tag{5.3.49}$$

Proof. The assertion is an easy consequence of the distributivity of the geometric product over addition, the commutativity of scalars and the linearity of the integral. □

5.3.2.4 Bilinearity

All geometric Fourier transforms from the introductory example can also be expressed in terms of a stronger claim. The mappings $f_1, \ldots, f_{\boldsymbol{\nu}}$, with the first $\boldsymbol{\mu}$ terms to the left of the argument function and the $\boldsymbol{\nu} - \boldsymbol{\mu}$ others on the right of it, are all bilinear and therefore take the form

$$f_k(\boldsymbol{x}, \boldsymbol{u}) = f_k\left(\sum_{j=1}^{m} x_j \boldsymbol{e}_j, \sum_{l=1}^{m} u_l \boldsymbol{e}_l\right) = \sum_{j,l=1}^{m} x_j f_k(\boldsymbol{e}_j, \boldsymbol{e}_l) u_l = \boldsymbol{x}^T M_k \boldsymbol{u}, \tag{5.3.50}$$

$\forall k = 1, \ldots, \boldsymbol{\nu}$, where $M_k \in (\mathscr{I}^{p,q})^{m\times m}, (M_k)_{jl} = f_k(\boldsymbol{e}_j, \boldsymbol{e}_l)$ according to Notation 5.3.8.

Exercise 5.3.13. *Ordered in the same way as in the previous example, the geometric Fourier transforms expressed in the way of (5.3.50) take the following shapes:*

1. *In the Clifford Fourier transform f_1 can be written with*
$$M_1 = 2\pi i_n \,\mathrm{Id}\,. \tag{5.3.51}$$

2. *The $\boldsymbol{\nu} = m = n$ mappings $f_k, k = 1, \ldots, n$ of the Bülow Clifford Fourier transform can be expressed using*
$$(M_k)_{lj} = \begin{cases} 2\pi \boldsymbol{e}_k & \textit{for } k = l = j, \\ 0 & \textit{otherwise.} \end{cases} \tag{5.3.52}$$

3. *Similarly the quaternionic Fourier transform is generated using*
$$(M_1)_{l\iota} = \begin{cases} 2\pi i & \textit{for } l = \iota = 1, \\ 0 & \textit{otherwise,} \end{cases}$$
$$(M_2)_{l\iota} = \begin{cases} 2\pi j & \textit{for } l = \iota = 2, \\ 0 & \textit{otherwise.} \end{cases} \tag{5.3.53}$$

4. *We can build the spacetime Fourier transform with*
$$(M_1)_{lj} = \begin{cases} \boldsymbol{e}_4 & \textit{for } l = j = 1, \\ 0 & \textit{otherwise,} \end{cases}$$
$$(M_2)_{lj} = \begin{cases} \epsilon_4 \boldsymbol{e}_4 i_4 & \textit{for } l = j \in \{2, 3, 4\}, \\ 0 & \textit{otherwise.} \end{cases} \tag{5.3.54}$$

5. *The Clifford Fourier transform for colour images can be described by*

$$\begin{aligned} M_1 &= \frac{1}{2}\boldsymbol{B}\,\mathrm{Id}, \quad M_2 = \frac{1}{2}i\boldsymbol{B}\,\mathrm{Id}, \\ M_3 &= -\frac{1}{2}\boldsymbol{B}\,\mathrm{Id}, \quad M_4 = -\frac{1}{2}i\boldsymbol{B}\,\mathrm{Id}. \end{aligned} \tag{5.3.55}$$

6. *The cylindrical Fourier transform can also be reproduced with mappings satisfying* (5.3.50) *because we can write*

$$\begin{aligned} \boldsymbol{x} \wedge \boldsymbol{u} = \boldsymbol{e}_1\boldsymbol{e}_2 x_1 u_2 - \boldsymbol{e}_1\boldsymbol{e}_2 x_2 u_1 + \ldots + \\ + \boldsymbol{e}_{m-1}\boldsymbol{e}_m x_{m-1} u_m - \boldsymbol{e}_{m-1}\boldsymbol{e}_m x_m u_{m-1} \end{aligned} \tag{5.3.56}$$

and set

$$(M_1)_{lj} = \begin{cases} 0 & \text{for } l = j, \\ \boldsymbol{e}_l\boldsymbol{e}_j & \text{otherwise.} \end{cases} \tag{5.3.57}$$

Theorem 5.3.14 (Scaling). *Let* $0 \neq a \in \mathbb{R}$ *be a real number,* $\boldsymbol{A}(\boldsymbol{x}) = \boldsymbol{B}(a\boldsymbol{x})$ *two multivector fields and all* F_1, F_2 *be bilinear mappings then the geometric Fourier transform satisfies*

$$\mathcal{F}_{F_1,F_2}(\boldsymbol{A})(\boldsymbol{u}) = |a|^{-m}\mathcal{F}_{F_1,F_2}(\boldsymbol{B})\left(\frac{\boldsymbol{u}}{a}\right). \tag{5.3.58}$$

Proof. A change of coordinates together with the bilinearity proves the assertion by

$$\begin{aligned} \mathcal{F}_{F_1,F_2}(\boldsymbol{A})(\boldsymbol{u}) &= \int_{\mathbb{R}^m} \prod_{f\in F} e^{-f(\boldsymbol{x},\boldsymbol{u})}\boldsymbol{B}(a\boldsymbol{x}) \prod_{f\in B} e^{-f(\boldsymbol{x},\boldsymbol{u})}\,\mathrm{d}^m\boldsymbol{x} \\ &\overset{a\boldsymbol{x}=\boldsymbol{y}}{=} \int_{\mathbb{R}^m} \prod_{f\in F} e^{-f(\frac{\boldsymbol{y}}{a},\boldsymbol{u})}\boldsymbol{B}(\boldsymbol{y}) \prod_{f\in B} e^{-f(\frac{\boldsymbol{y}}{a},\boldsymbol{u})}|a|^{-m}\,\mathrm{d}^m\boldsymbol{y} \\ &\overset{f \text{ bilin.}}{=} |a|^{-m}\int_{\mathbb{R}^m} \prod_{f\in F} e^{-f(\boldsymbol{y},\frac{\boldsymbol{u}}{a})}\boldsymbol{B}(\boldsymbol{y}) \prod_{f\in B} e^{-f(\boldsymbol{y},\frac{\boldsymbol{u}}{a})}\,\mathrm{d}^m\boldsymbol{y} \\ &= |a|^{-m}\mathcal{F}_{F_1,F_2}(\boldsymbol{B})\left(\frac{\boldsymbol{u}}{a}\right). \end{aligned} \tag{5.3.59}$$

□

5.3.2.5 *Products with invertible factors*

To obtain properties of the GFT like linearity with respect to arbitrary multivectors or a shift theorem we will have to change the order of multivectors and products of exponentials. Since the geometric product usually is neither commutative nor anticommutative this is not trivial. In this section, we provide useful Lemmata that allow a swap if at least one of the factors is invertible. For more information see [161] and [203].

Remark 5.3.15. *Every multiple of a square root of* -1, $i \in \mathcal{G}^{p,q}$ *is invertible, since from* $i^2 = -r, r \in \mathbb{R}\setminus\{0\}$ *follows* $i^{-1} = -i/r$. *Because of that, for all* $\boldsymbol{u}, \boldsymbol{x} \in \mathbb{R}^m$ *a function* $f_k(\boldsymbol{x},\boldsymbol{u}) : \mathbb{R}^m \times \mathbb{R}^m \to \mathcal{G}^{p,q}$ *is pointwise invertible.*

Definition 5.3.16. *For an invertible multivector $\boldsymbol{B} \in Cl(p,q)$ and an arbitrary multivector $\boldsymbol{A} \in Cl(p,q)$ we define*

$$\boldsymbol{A}_{\boldsymbol{c}^0(\boldsymbol{B})} = \frac{1}{2}(\boldsymbol{A} + \boldsymbol{B}^{-1}\boldsymbol{A}\boldsymbol{B}), \quad \boldsymbol{A}_{\boldsymbol{c}^1(\boldsymbol{B})} = \frac{1}{2}(\boldsymbol{A} - \boldsymbol{B}^{-1}\boldsymbol{A}\boldsymbol{B}). \tag{5.3.60}$$

Lemma 5.3.17. *Let $\boldsymbol{B} \in Cl(p,q)$ be invertible with the unique inverse $\boldsymbol{B}^{-1} = \bar{\boldsymbol{B}}/\boldsymbol{B}^2$, $\boldsymbol{B}^2 \in \mathbb{R} \setminus \{0\}$. Every multivector $\boldsymbol{A} \in Cl(p,q)$ can be expressed unambiguously by the sum of $\boldsymbol{A}_{\boldsymbol{c}^0(\boldsymbol{B})} \in Cl(p,q)$ that commutes and $\boldsymbol{A}_{\boldsymbol{c}^1(\boldsymbol{B})} \in Cl(p,q)$ that anticommutes with respect to $\boldsymbol{B}$. That means*

$$\begin{aligned} \boldsymbol{A} &= \boldsymbol{A}_{\boldsymbol{c}^0(\boldsymbol{B})} + \boldsymbol{A}_{\boldsymbol{c}^1(\boldsymbol{B})}, \\ \boldsymbol{A}_{\boldsymbol{c}^0(\boldsymbol{B})}\boldsymbol{B} &= \boldsymbol{B}\boldsymbol{A}_{\boldsymbol{c}^0(\boldsymbol{B})}, \quad \boldsymbol{A}_{\boldsymbol{c}^1(\boldsymbol{B})}\boldsymbol{B} = -\boldsymbol{B}\boldsymbol{A}_{\boldsymbol{c}^1(\boldsymbol{B})}. \end{aligned} \tag{5.3.61}$$

Proof. We will only prove the assertion for $\boldsymbol{A}_{\boldsymbol{c}^0(\boldsymbol{B})}$.
Existence: With Definition 5.3.16 we get

$$\boldsymbol{A}_{\boldsymbol{c}^0(\boldsymbol{B})} + \boldsymbol{A}_{\boldsymbol{c}^1(\boldsymbol{B})} = \frac{1}{2}(\boldsymbol{A} + \boldsymbol{B}^{-1}\boldsymbol{A}\boldsymbol{B} + \boldsymbol{A} - \boldsymbol{B}^{-1}\boldsymbol{A}\boldsymbol{B}) = \boldsymbol{A}, \tag{5.3.62}$$

and considering

$$\boldsymbol{B}^{-1}\boldsymbol{A}\boldsymbol{B} = \frac{\bar{\boldsymbol{B}}\boldsymbol{A}\boldsymbol{B}}{\boldsymbol{B}^2} = \boldsymbol{B}\boldsymbol{A}\boldsymbol{B}^{-1}, \tag{5.3.63}$$

we also get

$$\begin{aligned} \boldsymbol{A}_{\boldsymbol{c}^0(\boldsymbol{B})}\boldsymbol{B} &= \frac{1}{2}(\boldsymbol{A} + \boldsymbol{B}^{-1}\boldsymbol{A}\boldsymbol{B})\boldsymbol{B} = \frac{1}{2}(\boldsymbol{A} + \boldsymbol{B}\boldsymbol{A}\boldsymbol{B}^{-1})\boldsymbol{B} \\ &= \frac{1}{2}(\boldsymbol{A}\boldsymbol{B} + \boldsymbol{B}\boldsymbol{A}) = \boldsymbol{B}\frac{1}{2}(\boldsymbol{B}^{-1}\boldsymbol{A}\boldsymbol{B} + \boldsymbol{A}) = \boldsymbol{B}\boldsymbol{A}_{\boldsymbol{c}^0(\boldsymbol{B})}. \end{aligned} \tag{5.3.64}$$

Uniqueness: From the first claim in (5.3.61) we get

$$\boldsymbol{A}_{\boldsymbol{c}^1(\boldsymbol{B})} = \boldsymbol{A} - \boldsymbol{A}_{\boldsymbol{c}^0(\boldsymbol{B})}, \tag{5.3.65}$$

together with the third one this leads to

$$\begin{aligned} (\boldsymbol{A} - \boldsymbol{A}_{\boldsymbol{c}^0(\boldsymbol{B})})\boldsymbol{B} &= -\boldsymbol{B}(\boldsymbol{A} - \boldsymbol{A}_{\boldsymbol{c}^0(\boldsymbol{B})}), \\ \boldsymbol{A}\boldsymbol{B} - \boldsymbol{A}_{\boldsymbol{c}^0(\boldsymbol{B})}\boldsymbol{B} &= -\boldsymbol{B}\boldsymbol{A} + \boldsymbol{B}\boldsymbol{A}_{\boldsymbol{c}^0(\boldsymbol{B})}, \\ \boldsymbol{A}\boldsymbol{B} + \boldsymbol{B}\boldsymbol{A} &= \boldsymbol{A}_{\boldsymbol{c}^0(\boldsymbol{B})}\boldsymbol{B} + \boldsymbol{B}\boldsymbol{A}_{\boldsymbol{c}^0(\boldsymbol{B})}. \end{aligned} \tag{5.3.66}$$

and from the second claim finally follows

$$\boldsymbol{A}\boldsymbol{B} + \boldsymbol{B}\boldsymbol{A} = 2\boldsymbol{B}\boldsymbol{A}_{\boldsymbol{c}^0(\boldsymbol{B})}\frac{1}{2}(\boldsymbol{B}^{-1}\boldsymbol{A}\boldsymbol{B} + \boldsymbol{A}) = \boldsymbol{A}_{\boldsymbol{c}^0(\boldsymbol{B})}. \tag{5.3.67}$$

The derivation of the expression for $\boldsymbol{A}_{\boldsymbol{c}^1(\boldsymbol{B})}$ works analogously. □

Corollary 5.3.18 (Decomposition w.r.t. commutativity)**.** *Let $\boldsymbol{B} \in Cl(p,q)$ be invertible, then $\forall \boldsymbol{A} \in Cl(p,q)$*

$$\boldsymbol{B}\boldsymbol{A} = (\boldsymbol{A}_{\boldsymbol{c}^0(\boldsymbol{B})} - \boldsymbol{A}_{\boldsymbol{c}^1(\boldsymbol{B})})\boldsymbol{B}. \tag{5.3.68}$$

Definition 5.3.19. *For $d \in \mathbb{N}, \boldsymbol{A} \in Cl(p,q)$, the ordered set $B = \{\boldsymbol{B}_1, \ldots, \boldsymbol{B}_d\}$ of invertible multivectors and any multi-index $\boldsymbol{j} \in \{0,1\}^d$ we define*

$$\begin{aligned}
\boldsymbol{A}_{\mathbf{c}^{\boldsymbol{j}}(\overrightarrow{B})} &:= ((\boldsymbol{A}_{\mathbf{c}^{j_1}(\boldsymbol{B}_1)})_{\mathbf{c}^{j_2}(\boldsymbol{B}_2)} \cdots)_{\mathbf{c}^{j_d}(\boldsymbol{B}_d)}, \\
\boldsymbol{A}_{\mathbf{c}^{\boldsymbol{j}}(\overleftarrow{B})} &:= ((\boldsymbol{A}_{\mathbf{c}^{j_d}(\boldsymbol{B}_d)})_{\mathbf{c}^{j_{d-1}}(\boldsymbol{B}_{d-1})} \cdots)_{\mathbf{c}^{j_1}(\boldsymbol{B}_1)},
\end{aligned} \tag{5.3.69}$$

recursively with $\mathbf{c}^0, \mathbf{c}^1$ as in Definition 5.3.16.

Exercise 5.3.20. *Let $\boldsymbol{A} = a_0 + a_1\boldsymbol{e}_1 + a_2\boldsymbol{e}_2 + a_{12}\boldsymbol{e}_{12} \in \mathcal{G}^{2,0}$ then, for example*

$$\begin{aligned}
\boldsymbol{A}_{\mathbf{c}^0(\boldsymbol{e}_1)} &= \frac{1}{2}(\boldsymbol{A} + \boldsymbol{e}_1^{-1}\boldsymbol{A}\boldsymbol{e}_1) \\
&= \frac{1}{2}(\boldsymbol{A} + a_0 + a_1\boldsymbol{e}_1 - a_2\boldsymbol{e}_2 - a_{12}\boldsymbol{e}_{12}) = a_0 + a_1\boldsymbol{e}_1,
\end{aligned} \tag{5.3.70}$$

and further

$$\boldsymbol{A}_{\mathbf{c}^{0,0}(\overrightarrow{\boldsymbol{e}_1,\boldsymbol{e}_2})} = (\boldsymbol{A}_{\mathbf{c}^0(\boldsymbol{e}_1)})_{\mathbf{c}^0(\boldsymbol{e}_2)} = (a_0 + a_1\boldsymbol{e}_1)_{\mathbf{c}^0(\boldsymbol{e}_2)} = a_0. \tag{5.3.71}$$

The computation of the other multi-indices with $d = 2$ works analogously and therefore

$$\begin{aligned}
\boldsymbol{A} &= \sum_{\boldsymbol{j} \in \{0,1\}^d} \boldsymbol{A}_{\mathbf{c}^{\boldsymbol{j}}(\boldsymbol{e}_1,\boldsymbol{e}_2)} \\
&= \boldsymbol{A}_{\mathbf{c}^{00}(\overrightarrow{\boldsymbol{e}_1,\boldsymbol{e}_2})} + \boldsymbol{A}_{\mathbf{c}^{01}(\overrightarrow{\boldsymbol{e}_1,\boldsymbol{e}_2})} + \boldsymbol{A}_{\mathbf{c}^{10}(\overrightarrow{\boldsymbol{e}_1,\boldsymbol{e}_2})} + \boldsymbol{A}_{\mathbf{c}^{11}(\overrightarrow{\boldsymbol{e}_1,\boldsymbol{e}_2})} \\
&= a_0 + a_1\boldsymbol{e}_1 + a_2\boldsymbol{e}_2 + a_{12}\boldsymbol{e}_{12}.
\end{aligned} \tag{5.3.72}$$

Lemma 5.3.21. *Let $d \in \mathbb{N}, B = \{\boldsymbol{B}_1, \ldots, \boldsymbol{B}_d\}$ be invertible multivectors and for $\boldsymbol{j} \in \{0,1\}^d$ let $|\boldsymbol{j}| := \sum_{k=1}^d j_k$, then $\forall \boldsymbol{A} \in Cl(p,q)$*

$$\begin{aligned}
\boldsymbol{A} &= \sum_{\boldsymbol{j} \in \{0,1\}^d} \boldsymbol{A}_{\mathbf{c}^{\boldsymbol{j}}(\overrightarrow{B})}, \\
\boldsymbol{A}\boldsymbol{B}_1 \ldots \boldsymbol{B}_d &= \boldsymbol{B}_1 \ldots \boldsymbol{B}_d \sum_{\boldsymbol{j} \in \{0,1\}^d} (-1)^{|\boldsymbol{j}|} \boldsymbol{A}_{\mathbf{c}^{\boldsymbol{j}}(\overrightarrow{B})}, \\
\boldsymbol{B}_1 \ldots \boldsymbol{B}_d\boldsymbol{A} &= \sum_{\boldsymbol{j} \in \{0,1\}^d} (-1)^{|\boldsymbol{j}|} \boldsymbol{A}_{\mathbf{c}^{\boldsymbol{j}}(\overleftarrow{B})} \boldsymbol{B}_1 \ldots \boldsymbol{B}_d.
\end{aligned} \tag{5.3.73}$$

Proof. Apply Lemma 5.3.17 repeatedly. □

Remark 5.3.22. *The distinction of the two directions can be omitted using the equality*

$$\boldsymbol{A}_{\mathbf{c}^{\boldsymbol{j}}(\overrightarrow{\boldsymbol{B}_1,\ldots,\boldsymbol{B}_d})} = \boldsymbol{A}_{\mathbf{c}^{\boldsymbol{j}}(\overleftarrow{\boldsymbol{B}_d,\ldots,\boldsymbol{B}_1})}. \tag{5.3.74}$$

We established it for the sake of notational brevity and will not formulate nor prove every assertion for both directions.

Lemma 5.3.23. *Let* $F = \{f_1(\boldsymbol{x},\boldsymbol{u}),\ldots,f_d(\boldsymbol{x},\boldsymbol{u})\}$ *be a set of pointwise invertible functions then the ordered product of their exponentials and an arbitrary multivector* $\boldsymbol{A} \in Cl(p,q)$ *satisfies*

$$\prod_{k=1}^{d} e^{-f_k(\boldsymbol{x},\boldsymbol{u})}\boldsymbol{A} = \sum_{\boldsymbol{j}\in\{0,1\}^d} \boldsymbol{A}_{\boldsymbol{c}^{\boldsymbol{j}}(\overleftarrow{F})}(\boldsymbol{x},\boldsymbol{u}) \prod_{k=1}^{d} e^{-(-1)^{j_k} f_k(\boldsymbol{x},\boldsymbol{u})}, \tag{5.3.75}$$

where $A_{\boldsymbol{c}^{\boldsymbol{j}}(\overleftarrow{F})}(\boldsymbol{x},\boldsymbol{u}) := A_{\boldsymbol{c}^{\boldsymbol{j}}(\overleftarrow{F(\boldsymbol{x},\boldsymbol{u})})}$ *is a multivector valued function* $\mathbb{R}^m \times \mathbb{R}^m \to Cl(p,q)$.

Proof. For all $\boldsymbol{x},\boldsymbol{u} \in \mathbb{R}^m$ the commutation properties of $f_k(\boldsymbol{x},\boldsymbol{u})$ dictate the properties of $e^{-f_k(\boldsymbol{x},\boldsymbol{u})}$ by

$$\begin{aligned} e^{-f_k(\boldsymbol{x},\boldsymbol{u})}\boldsymbol{A} &\overset{\text{Def. 5.3.6}}{=} \sum_{l=0}^{\infty} \frac{(-f_k(\boldsymbol{x},\boldsymbol{u}))^l}{l!}\boldsymbol{A} \\ &\overset{\text{Lem. 5.3.17}}{=} \sum_{l=0}^{\infty} \frac{(-f_k(\boldsymbol{x},\boldsymbol{u}))^l}{l!}\left(\boldsymbol{A}_{\boldsymbol{c}^0(f_k(\boldsymbol{x},\boldsymbol{u}))} + \boldsymbol{A}_{\boldsymbol{c}^1(f_k(\boldsymbol{x},\boldsymbol{u}))}\right) \end{aligned} \tag{5.3.76}$$

The shape of this decomposition of $\boldsymbol{A}$ may depend on $\boldsymbol{x}$ and $\boldsymbol{u}$. To stress this fact we will interpret $\boldsymbol{A}_{\boldsymbol{c}^0(f_k(\boldsymbol{x},\boldsymbol{u}))}$ as a multivector function and write $\boldsymbol{A}_{\boldsymbol{c}^0(f_k)}(\boldsymbol{x},\boldsymbol{u})$. According to Lemma 5.3.17 we can move $\boldsymbol{A}_{\boldsymbol{c}^0(f_k)}(\boldsymbol{x},\boldsymbol{u})$ through all factors, because it commutes. Analogously swapping $\boldsymbol{A}_{\boldsymbol{c}^1(f_k)}(\boldsymbol{x},\boldsymbol{u})$ will change the sign of each factor because it anticommutes. Hence we get

$$\begin{aligned} &= \boldsymbol{A}_{\boldsymbol{c}^0(f_k)}(\boldsymbol{x},\boldsymbol{u}) \sum_{l=0}^{\infty} \frac{(-f_k(\boldsymbol{x},\boldsymbol{u}))^l}{l!} + \boldsymbol{A}_{\boldsymbol{c}^1(f_k)}(\boldsymbol{x},\boldsymbol{u}) \sum_{l=0}^{\infty} \frac{(f_k(\boldsymbol{x},\boldsymbol{u}))^l}{l!} \\ &= \boldsymbol{A}_{\boldsymbol{c}^0(f_k)}(\boldsymbol{x},\boldsymbol{u}) e^{-f_k(\boldsymbol{x},\boldsymbol{u})} + \boldsymbol{A}_{\boldsymbol{c}^1(f_k)}(\boldsymbol{x},\boldsymbol{u}) e^{f_k(\boldsymbol{x},\boldsymbol{u})}. \end{aligned} \tag{5.3.77}$$

Applying this repeatedly to the product we can deduce

$$\begin{aligned} \prod_{k=1}^{d} e^{-f_k(\boldsymbol{x},\boldsymbol{u})}\boldsymbol{A} &= \prod_{k=1}^{d-1} e^{-f_k(\boldsymbol{x},\boldsymbol{u})} \begin{pmatrix} \boldsymbol{A}_{\boldsymbol{c}^0(f_d)}(\boldsymbol{x},\boldsymbol{u}) e^{-f_d(\boldsymbol{x},\boldsymbol{u})} \\ +\boldsymbol{A}_{\boldsymbol{c}^1(f_d)}(\boldsymbol{x},\boldsymbol{u}) e^{f_d(\boldsymbol{x},\boldsymbol{u})} \end{pmatrix} \\ &= \prod_{k=1}^{d-2} e^{-f_k(\boldsymbol{x},\boldsymbol{u})} \begin{pmatrix} \boldsymbol{A}_{\boldsymbol{c}^{0,0}(\overleftarrow{f_{d-1},f_d})}(\boldsymbol{x},\boldsymbol{u}) e^{-f_{d-1}(\boldsymbol{x},\boldsymbol{u})} e^{-f_d(\boldsymbol{x},\boldsymbol{u})} \\ +\boldsymbol{A}_{\boldsymbol{c}^{1,0}(\overleftarrow{f_{d-1},f_d})}(\boldsymbol{x},\boldsymbol{u}) e^{f_{d-1}(\boldsymbol{x},\boldsymbol{u})} e^{-f_d(\boldsymbol{x},\boldsymbol{u})} \\ +\boldsymbol{A}_{\boldsymbol{c}^{0,1}(\overleftarrow{f_{d-1},f_d})}(\boldsymbol{x},\boldsymbol{u}) e^{-f_{d-1}(\boldsymbol{x},\boldsymbol{u})} e^{f_d(\boldsymbol{x},\boldsymbol{u})} \\ +\boldsymbol{A}_{\boldsymbol{c}^{1,1}(\overleftarrow{f_{d-1},f_d})}(\boldsymbol{x},\boldsymbol{u}) e^{f_{d-1}(\boldsymbol{x},\boldsymbol{u})} e^{f_d(\boldsymbol{x},\boldsymbol{u})} \end{pmatrix} \\ &\quad\vdots \qquad\qquad \vdots \qquad\qquad \vdots \\ &= \sum_{\boldsymbol{j}\in\{0,1\}^d} \boldsymbol{A}_{\boldsymbol{c}^{\boldsymbol{j}}(\overleftarrow{F})}(\boldsymbol{x},\boldsymbol{u}) \prod_{k=1}^{d} e^{-(-1)^{j_k} f_k(\boldsymbol{x},\boldsymbol{u})}. \end{aligned} \tag{5.3.78}$$

□

5.3.2.6 *Separable geometric Fourier transform*

From now on we want to restrict ourselves to an important group of geometric Fourier transforms whose square roots of -1 are independent from the first argument.

Definition 5.3.24. *We call a GFT* ***left (right) separable****, if*

$$f_l = |f_l\,(\boldsymbol{x}, \boldsymbol{u})|\, i_l(\boldsymbol{u}), \tag{5.3.79}$$

$\forall l = 1, \ldots, \boldsymbol{\mu}$, $(l = \boldsymbol{\mu} + 1, \ldots, \boldsymbol{\nu})$, where $|f_l(\boldsymbol{x}, \boldsymbol{u})| : \mathbb{R}^m \times \mathbb{R}^m \to \mathbb{R}$ is a real function and $i_l : \mathbb{R}^m \to \mathscr{I}^{p,q}$ a function that does not depend on $\boldsymbol{x}$.

Exercise 5.3.25. *The first five transforms from the introductory example are separable, while the cylindrical transform (vi) can not be expressed as in* (5.3.79) *except for the two dimensional case.*

We have seen in the proof of Lemma 5.3.23 that the decomposition of a constant multivector $\boldsymbol{A}$ with respect to a product of exponentials generally results in multivector valued functions $\boldsymbol{A}_{\boldsymbol{c}^{\boldsymbol{j}}(F)}(\boldsymbol{x}, \boldsymbol{u})$ of $\boldsymbol{x}$ and $\boldsymbol{u}$. Separability guarantees independence from $\boldsymbol{x}$ and therefore allows separation from the integral.

Corollary 5.3.26 (Decomposition independent from $\boldsymbol{x}$)**.** *Consider a set of functions $F = \{f_1(\boldsymbol{x}, \boldsymbol{u}), \ldots, f_d(\boldsymbol{x}, \boldsymbol{u})\}$ satisfying condition* (5.3.79) *then the ordered product of their exponentials and an arbitrary multivector $\boldsymbol{A} \in Cl(p, q)$ satisfies*

$$\prod_{k=1}^{d} e^{-f_k(\boldsymbol{x},\boldsymbol{u})} \boldsymbol{A} = \sum_{\boldsymbol{j}\in\{0,1\}^d} \boldsymbol{A}_{\boldsymbol{c}^{\boldsymbol{j}}(\overleftarrow{F})}(\boldsymbol{u}) \prod_{k=1}^{d} e^{-(-1)^{j_k} f_k(\boldsymbol{x},\boldsymbol{u})}. \tag{5.3.80}$$

Remark 5.3.27. *If a GFT can be expressed as in 5.3.79 but with multiples of square roots of -1, $i_k \in \mathscr{I}^{p,q}$, which are independent from $\boldsymbol{x}$ and $\boldsymbol{u}$, the parts $\boldsymbol{A}_{\boldsymbol{c}^{\boldsymbol{j}}(\overleftarrow{F})}$ of $\boldsymbol{A}$ will be constants. Note that the first five GFTs from the reference example satisfy this stronger condition, too.*

Definition 5.3.28. *For a set of functions $F = \{f_1(\boldsymbol{x}, \boldsymbol{u}), \ldots, f_d(\boldsymbol{x}, \boldsymbol{u})\}$ and a multi-index $\boldsymbol{j} \in \{0, 1\}^d$, we define the set of functions $F(\boldsymbol{j})$ by*

$$F(\boldsymbol{j}) := \left\{(-1)^{j_1} f_1(\boldsymbol{x}, \boldsymbol{u}), \ldots, (-1)^{j_d} f_d(\boldsymbol{x}, \boldsymbol{u})\right\}. \tag{5.3.81}$$

Theorem 5.3.29 (Left and right products)**.** *Let $\boldsymbol{C} \in Cl(p, q)$ and $\boldsymbol{A}, \boldsymbol{B} : \mathbb{R}^m \to Cl(p, q)$ be two multivector fields with $\boldsymbol{A}(\boldsymbol{x}) = \boldsymbol{C}\boldsymbol{B}(\boldsymbol{x})$ then a left separable geometric Fourier transform obeys*

$$\mathcal{F}_{F_1,F_2}(\boldsymbol{A})(\boldsymbol{u}) = \sum_{\boldsymbol{j}\in\{0,1\}^{\mu}} \boldsymbol{C}_{\boldsymbol{c}^{\boldsymbol{j}}(\overleftarrow{F_1})}(\boldsymbol{u}) \mathcal{F}_{F_1(\boldsymbol{j}),F_2}(\boldsymbol{B})(\boldsymbol{u}). \tag{5.3.82}$$

If $\boldsymbol{A}(\boldsymbol{x}) = \boldsymbol{B}(\boldsymbol{x})\boldsymbol{C}$ we analogously get

$$\mathcal{F}_{F_1,F_2}(\boldsymbol{A})(\boldsymbol{u}) = \sum_{\boldsymbol{k}\in\{0,1\}^{(\nu-\mu)}} \mathcal{F}_{F_1,F_2(\boldsymbol{k})}(\boldsymbol{B})(\boldsymbol{u}) \boldsymbol{C}_{\boldsymbol{c}^{\boldsymbol{k}}(\overrightarrow{F_2})}(\boldsymbol{u}) \tag{5.3.83}$$

for a right separable GFT.

Proof. We restrict ourselves to the proof of the first assertion.

$$\begin{aligned}
\mathcal{F}_{F_1,F_2}(\boldsymbol{A})(\boldsymbol{u}) &= \int_{\mathbb{R}^m} \prod_{f\in F_1} e^{-f(\boldsymbol{x},\boldsymbol{u})}\boldsymbol{C}\boldsymbol{B}(\boldsymbol{x}) \prod_{f\in F_2} e^{-f(\boldsymbol{x},\boldsymbol{u})}\,\mathrm{d}^m\boldsymbol{x} \\
&\overset{\text{Lem. 5.3.23}}{=} \int_{\mathbb{R}^m} \left(\sum_{\boldsymbol{j}\in\{0,1\}^{\mu}} \boldsymbol{C}_{\boldsymbol{c}^{\boldsymbol{j}}(\overleftarrow{F_1})}(\boldsymbol{u}) \prod_{l=1}^{\mu} e^{-(-1)^{j_l} f_l(\boldsymbol{x},\boldsymbol{u})} \right) \\
&\qquad\qquad \boldsymbol{B}(\boldsymbol{x}) \prod_{f\in F_2} e^{-f(\boldsymbol{x},\boldsymbol{u})}\,\mathrm{d}^m\boldsymbol{x} \\
&= \sum_{\boldsymbol{j}\in\{0,1\}^{\mu}} \boldsymbol{C}_{\boldsymbol{c}^{\boldsymbol{j}}(\overleftarrow{F_1})}(\boldsymbol{u}) \int_{\mathbb{R}^m} \prod_{l=1}^{\mu} e^{-(-1)^{j_l} f_l(\boldsymbol{x},\boldsymbol{u})} \\
&\qquad\qquad \boldsymbol{B}(\boldsymbol{x}) \prod_{f\in F_2} e^{-f(\boldsymbol{x},\boldsymbol{u})}\,\mathrm{d}^m\boldsymbol{x} \\
&= \sum_{\boldsymbol{j}\in\{0,1\}^{\mu}} \boldsymbol{C}_{\boldsymbol{c}^{\boldsymbol{j}}(\overleftarrow{F_1})}(\boldsymbol{u}) \mathcal{F}_{F_1(\boldsymbol{j}),F_2}(\boldsymbol{B})(\boldsymbol{u}).
\end{aligned} \tag{5.3.84}$$

The second one follows in the same way. □

Corollary 5.3.30 (Uniform constants). *Let the claims from Theorem 5.3.29 hold. If the constant* $\boldsymbol{C}$ *satisfies* $\boldsymbol{C} = \boldsymbol{C}_{\boldsymbol{c}^{\boldsymbol{j}}(\overleftarrow{F_1})}(\boldsymbol{u})$ *for a multi-index* $\boldsymbol{j} \in \{0,1\}^{\mu}$ *then the theorem simplifies to*

$$\mathcal{F}_{F_1,F_2}(\boldsymbol{A})(\boldsymbol{u}) = \boldsymbol{C}\mathcal{F}_{F_1(\boldsymbol{j}),F_2}(\boldsymbol{B})(\boldsymbol{u}) \tag{5.3.85}$$

for $\boldsymbol{A}(\boldsymbol{x}) = \boldsymbol{C}\boldsymbol{B}(\boldsymbol{x})$, *respectively,*

$$\mathcal{F}_{F_1,F_2}(\boldsymbol{A})(\boldsymbol{u}) = \mathcal{F}_{F_1,F_2(\boldsymbol{k})}(\boldsymbol{B})(\boldsymbol{u})\boldsymbol{C} \tag{5.3.86}$$

for $\boldsymbol{A}(\boldsymbol{x}) = \boldsymbol{B}(\boldsymbol{x})\boldsymbol{C}$ *and* $\boldsymbol{C} = \boldsymbol{C}_{\boldsymbol{c}^{\boldsymbol{k}}(\overrightarrow{F_2})}(\boldsymbol{u})$ *for a multi-index* $\boldsymbol{k} \in \{0,1\}^{(\boldsymbol{\nu}-\boldsymbol{\mu})}$.[25]

Corollary 5.3.31 (Left and right linearity). *The geometric Fourier transform is left (respectively right) linear if* F_1 *(respectively* F_2*) consists only of functions* f_k *with values in the center of* $Cl(p,q)$*, that means* $\forall \boldsymbol{x}, \boldsymbol{u} \in \mathbb{R}^m, \forall \boldsymbol{A} \in Cl(p,q) : \boldsymbol{A}f_k(\boldsymbol{x},\boldsymbol{u}) = f_k(\boldsymbol{x},\boldsymbol{u})\boldsymbol{A}$.

Remark 5.3.32. *Note that for empty sets* F_1 *(or* F_2*) necessarily all elements satisfy commutativity and therefore the condition in Corollary 5.3.31.*

The different appearances of Theorem 5.3.29 are summarized in Table 5.6 and Table 5.7.

We have seen how to change the order of a multivector and a product of exponentials in the previous section. To get a shift theorem we will have to separate sums appearing in the exponent and sort the resulting exponentials with respect to the summands. Note that Corollary 5.3.26 can be applied in two ways here, because exponentials appear on both sides.

[25] Corollary 5.3.30 follows directly from $(\boldsymbol{C}_{\boldsymbol{c}^{\boldsymbol{j}}(\overleftarrow{F_1})})_{\boldsymbol{c}^{\boldsymbol{k}}(\overleftarrow{F_1})} = 0$ for all $\boldsymbol{k} \neq \boldsymbol{j}$ because no non-zero component of $\boldsymbol{C}$ can commute and anticommute with respect to a function in F_1.

Table 5.6 Theorem 5.3.29 (Left products) applied to the GFTs of the first example enumerated in the same order.
Notation: on the LHS $\mathcal{F}_{F_1,F_2} = \mathcal{F}_{F_1,F_2}(\boldsymbol{A})(\boldsymbol{u})$, on the RHS $\mathcal{F}_{F_1',F_2'} = \mathcal{F}_{F_1',F_2'}(\boldsymbol{B})(\boldsymbol{u})$.
Source: [61, Tab. 1].

	GFT	$\boldsymbol{A}(\boldsymbol{x}) = \boldsymbol{C}\boldsymbol{B}(\boldsymbol{x})$
1.	Clifford	$\mathcal{F}_{f_1} = \boldsymbol{C}\mathcal{F}_{f_1}$
2.	Bülow	$\mathcal{F}_{f_1,\dots,f_n} = \boldsymbol{C}\mathcal{F}_{f_1,\dots,f_n}$
3.	Quaternionic	$\mathcal{F}_{f_1,f_2} = \boldsymbol{C}_{\boldsymbol{c}^0(i)}\mathcal{F}_{f_1,f_2} + \boldsymbol{C}_{\boldsymbol{c}^1(i)}\mathcal{F}_{-f_1,f_2}$
4.	Spacetime	$\mathcal{F}_{f_1,f_2} = \boldsymbol{C}_{\boldsymbol{c}^0(\boldsymbol{e}_4)}\mathcal{F}_{f_1,f_2} + \boldsymbol{C}_{\boldsymbol{c}^1(\boldsymbol{e}_4)}\mathcal{F}_{-f_1,f_2}$
5.	Colour Image	$\mathcal{F}_{f_1,f_2,f_3,f_4} = \boldsymbol{C}_{\boldsymbol{c}^{00}(\overleftarrow{\boldsymbol{B},i\boldsymbol{B}})}\mathcal{F}_{f_1,f_2,f_3,f_4}$
		$+\boldsymbol{C}_{\boldsymbol{c}^{10}(\overleftarrow{\boldsymbol{B},i\boldsymbol{B}})}\mathcal{F}_{-f_1,f_2,f_3,f_4}$
		$+\boldsymbol{C}_{\boldsymbol{c}^{01}(\overleftarrow{\boldsymbol{B},i\boldsymbol{B}})}\mathcal{F}_{f_1,-f_2,f_3,f_4}$
		$+\boldsymbol{C}_{\boldsymbol{c}^{11}(\overleftarrow{\boldsymbol{B},i\boldsymbol{B}})}\mathcal{F}_{-f_1,-f_2,f_3,f_4}$
6.	Cylindrical $n = 2$	$\mathcal{F}_{f_1} = \boldsymbol{C}_{\boldsymbol{c}^0(\boldsymbol{e}_{12})}\mathcal{F}_{f_1} + \boldsymbol{C}_{\boldsymbol{c}^1(\boldsymbol{e}_{12})}\mathcal{F}_{-f_1}$
	Cylindrical $n \neq 2$	-

Table 5.7 Theorem 5.3.29 (Right products) applied to the GFTs of the first example, enumerated in the same order.
Notation: on the LHS $\mathcal{F}_{F_1,F_2} = \mathcal{F}_{F_1,F_2}(\boldsymbol{A})(\boldsymbol{u})$, on the RHS $\mathcal{F}_{F_1',F_2'} = \mathcal{F}_{F_1',F_2'}(\boldsymbol{B})(\boldsymbol{u})$
Source: [61, Tab. 2].

	GFT	$\boldsymbol{A}(\boldsymbol{x}) = \boldsymbol{B}(\boldsymbol{x})\boldsymbol{C}$
1.	Clif. $n = 2 \pmod 4$	$\mathcal{F}_{f_1} = \mathcal{F}_{f_1}\boldsymbol{C}_{\boldsymbol{c}^0(i)} + \mathcal{F}_{-f_1}\boldsymbol{C}_{\boldsymbol{c}^1(i)}$
	Clif. $n = 3 \pmod 4$	$\mathcal{F}_{f_1} = \mathcal{F}_{f_1}\boldsymbol{C}$
2.	Bülow	$\mathcal{F}_{f_1,\dots,f_n}$
		$= \sum_{\boldsymbol{k}\in\{0,1\}^n} \mathcal{F}_{(-1)^{k_1}f_1,\dots,(-1)^{k_n}f_n}\boldsymbol{C}_{\boldsymbol{c}^{\boldsymbol{k}}(\overrightarrow{f_1,\dots,f_n})}$
3.	Quaternionic	$\mathcal{F}_{f_1,f_2} = \mathcal{F}_{f_1,f_2}\boldsymbol{C}_{\boldsymbol{c}^0(j)} + \mathcal{F}_{f_1,-f_2}\boldsymbol{C}_{\boldsymbol{c}^1(j)}$
4.	Spacetime	$\mathcal{F}_{f_1,f_2} = \mathcal{F}_{f_1,f_2}\boldsymbol{C}_{\boldsymbol{c}^0(\boldsymbol{e}_4 i_4)} + \mathcal{F}_{f_1,-f_2}\boldsymbol{C}_{\boldsymbol{c}^1(\boldsymbol{e}_4 i_4)}$
5.	Colour Image	$\mathcal{F}_{f_1,f_2,f_3,f_4} = \mathcal{F}_{f_1,f_2,f_3,f_4}\boldsymbol{C}_{\boldsymbol{c}^{00}(\overrightarrow{\boldsymbol{B},i\boldsymbol{B}})}$
		$+\mathcal{F}_{f_1,f_2,-f_3,f_4}\boldsymbol{C}_{\boldsymbol{c}^{10}(\overrightarrow{\boldsymbol{B},i\boldsymbol{B}})}$
		$+\mathcal{F}_{f_1,f_2,f_3,-f_4}\boldsymbol{C}_{\boldsymbol{c}^{01}(\overrightarrow{\boldsymbol{B},i\boldsymbol{B}})}$
		$+\mathcal{F}_{f_1,f_2,-f_3,-f_4}\boldsymbol{C}_{\boldsymbol{c}^{11}(\overrightarrow{\boldsymbol{B},i\boldsymbol{B}})}$
6.	Cylindrical	$\mathcal{F}_{f_1} = \mathcal{F}_{f_1}\boldsymbol{C}$

Not every factor will need to be swapped with every other. So, to keep things short, we will make use of the notation $\boldsymbol{c}^{(J)_l}(f_1,\dots,f_l,0,\dots,0)$ for $l \in \{1,\dots,d\}$ instead of distinguishing between differently sized multi-indices for every l that appears. The zeros at the end substitutionally indicate real numbers. They commute with every multivector. That implies, that for the last $d-l$ factors no swap and therefore no separation needs to be made. It would also be possible to use the notation $\boldsymbol{c}^{(J)_l}(f_1,\ldots,f_{l-1},0,\ldots,0)$ for $l\in\{1,\ldots,d\}$, because every function commutes with itself. The choice we have made means that no exceptional treatment of f_1 is necessary. But please note that the multivectors $(J)_l$ indicating the commutative and anticommutative parts will all have zeros from l to d and therefore form a strictly triangular matrix.

Lemma 5.3.33. *Let a set of functions $F=\{f_1(\boldsymbol{x},\boldsymbol{u}),\ldots,f_d(\boldsymbol{x},\boldsymbol{u})\}$ fulfill (5.3.79) and be linear with respect to $\boldsymbol{x}$. Further let $J\in\{0,1\}^{d\times d}$ be a strictly lower triangular matrix, that is associated column by column with a multi-index $\boldsymbol{j}\in\{0,1\}^d$ by $\forall k=1,\ldots,d:(\sum_{l=1}^d J_{l,k}) \bmod 2=j_k$, with $(J)_l$ being its l-th row, then*

$$\prod_{l=1}^{d} e^{-f_l(\boldsymbol{x}+\boldsymbol{y},\boldsymbol{u})}$$
$$=\sum_{\boldsymbol{j}\in\{0,1\}^d}\ \sum_{\substack{J\in\{0,1\}^{d\times d},\\ \sum_{l=1}^d (J)_l \bmod 2=\boldsymbol{j}}}\ \prod_{l=1}^{d} e^{-f_l(\boldsymbol{x},\boldsymbol{u})}_{\boldsymbol{c}^{(J)_l}(\overleftarrow{f_1,\ldots,f_l,0,\ldots,0})}\prod_{l=1}^{d} e^{-(-1)^{j_l}f_l(\boldsymbol{y},\boldsymbol{u})}, \tag{5.3.87}$$

or alternatively with strictly upper triangular matrices J:

$$\prod_{l=1}^{d} e^{-f_l(\boldsymbol{x}+\boldsymbol{y},\boldsymbol{u})}$$
$$=\sum_{\boldsymbol{j}\in\{0,1\}^d}\ \sum_{\substack{J\in\{0,1\}^{d\times d},\\ \sum_{l=1}^d (J)_l \bmod 2=\boldsymbol{j}}}\ \prod_{l=1}^{d} e^{-(-1)^{j_l}f_l(\boldsymbol{x},\boldsymbol{u})}\prod_{l=1}^{d} e^{-f_l(\boldsymbol{y},\boldsymbol{u})}_{\boldsymbol{c}^{(J)_l}(\overrightarrow{0,\ldots,0,f_l,\ldots,f_d})}. \tag{5.3.88}$$

We do not explicitly indicate the dependence of the partition on $\boldsymbol{u}$ as in Corollary 5.3.26, because the functions in the exponents already contain this dependence. Please note that the decomposition is pointwise.

Proof. We will only prove the first assertion. The second one follows analogously by applying Corollary 5.3.26 the other way around.

$$\prod_{l=1}^{d} e^{-f_l(\boldsymbol{x}+\boldsymbol{y},\boldsymbol{u})} \overset{F\text{ lin.}}{=} \prod_{l=1}^{d} e^{-f_l(\boldsymbol{x},\boldsymbol{u})-f_l(\boldsymbol{y},\boldsymbol{u})} \tag{5.3.89}$$
$$\overset{\text{Lem. 5.3.7}}{=} \prod_{l=1}^{d} e^{-f_l(\boldsymbol{x},\boldsymbol{u})}e^{-f_l(\boldsymbol{y},\boldsymbol{u})} \tag{5.3.90}$$
$$= e^{-f_1(\boldsymbol{x},\boldsymbol{u})}e^{-f_1(\boldsymbol{y},\boldsymbol{u})}\prod_{l=2}^{d} e^{-f_l(\boldsymbol{x},\boldsymbol{u})}e^{-f_l(\boldsymbol{y},\boldsymbol{u})} \tag{5.3.91}$$
$$\overset{\text{cor. 5.3.26}}{=} e^{-f_1(\boldsymbol{x},\boldsymbol{u})}\Big(e^{-f_2(\boldsymbol{x},\boldsymbol{u})}_{\boldsymbol{c}^0(f_1)}e^{-f_1(\boldsymbol{y},\boldsymbol{u})}e^{-f_2(\boldsymbol{y},\boldsymbol{u})}$$
$$+\,e^{-f_2(\boldsymbol{x},\boldsymbol{u})}_{\boldsymbol{c}^1(f_1)}e^{f_1(\boldsymbol{y},\boldsymbol{u})}e^{-f_2(\boldsymbol{y},\boldsymbol{u})}\Big)\prod_{l=3}^{d} e^{-f_l(\boldsymbol{x},\boldsymbol{u})}e^{-f_l(\boldsymbol{y},\boldsymbol{u})} \tag{5.3.92}$$

Now we use Corollary 5.3.26 to step by step rearrange the order of the product.

$$
\begin{aligned}
&\overset{\text{cor. } 5.3.26}{=} e^{-f_1(\boldsymbol{x},\boldsymbol{u})}\Big(e_{\mathbf{c}^0(f_1)}^{-f_2(\boldsymbol{x},\boldsymbol{u})} e_{\mathbf{c}^{00}(\overleftarrow{f_1,f_2})}^{-f_3(\boldsymbol{x},\boldsymbol{u})} e^{-f_1(\boldsymbol{y},\boldsymbol{u})} e^{-f_2(\boldsymbol{y},\boldsymbol{u})} e^{-f_3(\boldsymbol{y},\boldsymbol{u})} \\
&\quad + e_{\mathbf{c}^0(f_1)}^{-f_2(\boldsymbol{x},\boldsymbol{u})} e_{\mathbf{c}^{01}(\overleftarrow{f_1,f_2})}^{-f_3(\boldsymbol{x},\boldsymbol{u})} e^{-f_1(\boldsymbol{y},\boldsymbol{u})} e^{f_2(\boldsymbol{y},\boldsymbol{u})} e^{-f_3(\boldsymbol{y},\boldsymbol{u})} \\
&\quad + e_{\mathbf{c}^0(f_1)}^{-f_2(\boldsymbol{x},\boldsymbol{u})} e_{\mathbf{c}^{10}(\overleftarrow{f_1,f_2})}^{-f_3(\boldsymbol{x},\boldsymbol{u})} e^{f_1(\boldsymbol{y},\boldsymbol{u})} e^{-f_2(\boldsymbol{y},\boldsymbol{u})} e^{-f_3(\boldsymbol{y},\boldsymbol{u})} \\
&\quad + e_{\mathbf{c}^0(f_1)}^{-f_2(\boldsymbol{x},\boldsymbol{u})} e_{\mathbf{c}^{11}(\overleftarrow{f_1,f_2})}^{-f_3(\boldsymbol{x},\boldsymbol{u})} e^{f_1(\boldsymbol{y},\boldsymbol{u})} e^{f_2(\boldsymbol{y},\boldsymbol{u})} e^{-f_3(\boldsymbol{y},\boldsymbol{u})} \\
&\quad + e_{\mathbf{c}^1(f_1)}^{-f_2(\boldsymbol{x},\boldsymbol{u})} e_{\mathbf{c}^{00}(\overleftarrow{f_1,f_2})}^{-f_3(\boldsymbol{x},\boldsymbol{u})} e^{f_1(\boldsymbol{y},\boldsymbol{u})} e^{-f_2(\boldsymbol{y},\boldsymbol{u})} e^{-f_3(\boldsymbol{y},\boldsymbol{u})} \\
&\quad + e_{\mathbf{c}^1(f_1)}^{-f_2(\boldsymbol{x},\boldsymbol{u})} e_{\mathbf{c}^{01}(\overleftarrow{f_1,f_2})}^{-f_3(\boldsymbol{x},\boldsymbol{u})} e^{f_1(\boldsymbol{y},\boldsymbol{u})} e^{f_2(\boldsymbol{y},\boldsymbol{u})} e^{-f_3(\boldsymbol{y},\boldsymbol{u})} \\
&\quad + e_{\mathbf{c}^1(f_1)}^{-f_2(\boldsymbol{x},\boldsymbol{u})} e_{\mathbf{c}^{10}(\overleftarrow{f_1,f_2})}^{-f_3(\boldsymbol{x},\boldsymbol{u})} e^{-f_1(\boldsymbol{y},\boldsymbol{u})} e^{-f_2(\boldsymbol{y},\boldsymbol{u})} e^{-f_3(\boldsymbol{y},\boldsymbol{u})} \\
&\quad + e_{\mathbf{c}^1(f_1)}^{-f_2(\boldsymbol{x},\boldsymbol{u})} e_{\mathbf{c}^{11}(\overleftarrow{f_1,f_2})}^{-f_3(\boldsymbol{x},\boldsymbol{u})} e^{-f_1(\boldsymbol{y},\boldsymbol{u})} e^{f_2(\boldsymbol{y},\boldsymbol{u})} e^{-f_3(\boldsymbol{y},\boldsymbol{u})}\Big) \\
&\quad \prod_{l=4}^{d} e^{-f_l(\boldsymbol{x},\boldsymbol{u})} e^{-f_l(\boldsymbol{y},\boldsymbol{u})}
\end{aligned}
\tag{5.3.93}
$$

There are only 2^δ ways of distributing the signs of δ exponents, so some of the summands can be combined.

$$
\begin{aligned}
&= e^{-f_1(\boldsymbol{x},\boldsymbol{u})}\Bigg(\Big(e_{\mathbf{c}^0(f_1)}^{-f_2(\boldsymbol{x},\boldsymbol{u})} e_{\mathbf{c}^{00}(\overleftarrow{f_1,f_2})}^{-f_3(\boldsymbol{x},\boldsymbol{u})} + e_{\mathbf{c}^1(f_1)}^{-f_2(\boldsymbol{x},\boldsymbol{u})} e_{\mathbf{c}^{10}(\overleftarrow{f_1,f_2})}^{-f_3(\boldsymbol{x},\boldsymbol{u})}\Big) \\
&\quad e^{-f_1(\boldsymbol{y},\boldsymbol{u})} e^{-f_2(\boldsymbol{y},\boldsymbol{u})} e^{-f_3(\boldsymbol{y},\boldsymbol{u})} \\
&+ \Big(e_{\mathbf{c}^0(f_1)}^{-f_2(\boldsymbol{x},\boldsymbol{u})} e_{\mathbf{c}^{01}(\overleftarrow{f_1,f_2})}^{-f_3(\boldsymbol{x},\boldsymbol{u})} + e_{\mathbf{c}^1(f_1)}^{-f_2(\boldsymbol{x},\boldsymbol{u})} e_{\mathbf{c}^{11}(\overleftarrow{f_1,f_2})}^{-f_3(\boldsymbol{x},\boldsymbol{u})}\Big) \\
&\quad e^{-f_1(\boldsymbol{y},\boldsymbol{u})} e^{f_2(\boldsymbol{y},\boldsymbol{u})} e^{-f_3(\boldsymbol{y},\boldsymbol{u})} \\
&+ \Big(e_{\mathbf{c}^0(f_1)}^{-f_2(\boldsymbol{x},\boldsymbol{u})} e_{\mathbf{c}^{10}(\overleftarrow{f_1,f_2})}^{-f_3(\boldsymbol{x},\boldsymbol{u})} + e_{\mathbf{c}^1(f_1)}^{-f_2(\boldsymbol{x},\boldsymbol{u})} e_{\mathbf{c}^{00}(\overleftarrow{f_1,f_2})}^{-f_3(\boldsymbol{x},\boldsymbol{u})}\Big) \\
&\quad e^{f_1(\boldsymbol{y},\boldsymbol{u})} e^{-f_2(\boldsymbol{y},\boldsymbol{u})} e^{-f_3(\boldsymbol{y},\boldsymbol{u})} \\
&+ \Big(e_{\mathbf{c}^0(f_1)}^{-f_2(\boldsymbol{x},\boldsymbol{u})} e_{\mathbf{c}^{11}(\overleftarrow{f_1,f_2})}^{-f_3(\boldsymbol{x},\boldsymbol{u})} + e_{\mathbf{c}^1(f_1)}^{-f_2(\boldsymbol{x},\boldsymbol{u})} e_{\mathbf{c}^{01}(\overleftarrow{f_1,f_2})}^{-f_3(\boldsymbol{x},\boldsymbol{u})}\Big) e^{f_1(\boldsymbol{y},\boldsymbol{u})} \\
&\quad e^{f_2(\boldsymbol{y},\boldsymbol{u})} e^{-f_3(\boldsymbol{y},\boldsymbol{u})}\Bigg) \prod_{l=4}^{d} e^{-f_l(\boldsymbol{x},\boldsymbol{u})} e^{-f_l(\boldsymbol{y},\boldsymbol{u})}
\end{aligned}
\tag{5.3.94}
$$

To get a compact notation we expand all multi-indices by adding zeros until they have the same length. Note that the last non-zero argument in terms like $\mathbf{c}^{000}(\overleftarrow{f_1, 0, 0})$ always coincides with the exponent of the corresponding factor. Because of that it will always commute and could also be replaced by a zero.

$$
\begin{aligned}
&= e^{-f_1(\boldsymbol{x},\boldsymbol{u})}_{\boldsymbol{c}^{000}(\overleftarrow{f_1,0,0})} \\
&\left(\left(e^{-f_2(\boldsymbol{x},\boldsymbol{u})}_{\boldsymbol{c}^{000}(\overleftarrow{f_1,f_2,0})} e^{-f_3(\boldsymbol{x},\boldsymbol{u})}_{\boldsymbol{c}^{000}(\overleftarrow{f_1,f_2,f_3})} + e^{-f_2(\boldsymbol{x},\boldsymbol{u})}_{\boldsymbol{c}^{100}(\overleftarrow{f_1,f_2,0})} e^{-f_3(\boldsymbol{x},\boldsymbol{u})}_{\boldsymbol{c}^{100}(\overleftarrow{f_1,f_2,f_3})}\right)\right. \\
&e^{-f_1(\boldsymbol{y},\boldsymbol{u})} e^{-f_2(\boldsymbol{y},\boldsymbol{u})} e^{-f_3(\boldsymbol{y},\boldsymbol{u})} \\
&+ \left(e^{-f_2(\boldsymbol{x},\boldsymbol{u})}_{\boldsymbol{c}^{000}(\overleftarrow{f_1,f_2,0})} e^{-f_3(\boldsymbol{x},\boldsymbol{u})}_{\boldsymbol{c}^{010}(\overleftarrow{f_1,f_2,f_3})} + e^{-f_2(\boldsymbol{x},\boldsymbol{u})}_{\boldsymbol{c}^{100}(\overleftarrow{f_1,f_2,0})} e^{-f_3(\boldsymbol{x},\boldsymbol{u})}_{\boldsymbol{c}^{110}(\overleftarrow{f_1,f_2,f_3})}\right) \\
&e^{-f_1(\boldsymbol{y},\boldsymbol{u})} e^{f_2(\boldsymbol{y},\boldsymbol{u})} e^{-f_3(\boldsymbol{y},\boldsymbol{u})} \\
&+ \left(e^{-f_2(\boldsymbol{x},\boldsymbol{u})}_{\boldsymbol{c}^{000}(\overleftarrow{f_1,f_2,0})} e^{-f_3(\boldsymbol{x},\boldsymbol{u})}_{\boldsymbol{c}^{100}(\overleftarrow{f_1,f_2,f_3})} + e^{-f_2(\boldsymbol{x},\boldsymbol{u})}_{\boldsymbol{c}^{100}(\overleftarrow{f_1,f_2,0})} e^{-f_3(\boldsymbol{x},\boldsymbol{u})}_{\boldsymbol{c}^{000}(\overleftarrow{f_1,f_2,f_3})}\right) \\
&e^{f_1(\boldsymbol{y},\boldsymbol{u})} e^{-f_2(\boldsymbol{y},\boldsymbol{u})} e^{-f_3(\boldsymbol{y},\boldsymbol{u})} \\
&+ \left(e^{-f_2(\boldsymbol{x},\boldsymbol{u})}_{\boldsymbol{c}^{000}(\overleftarrow{f_1,f_2,0})} e^{-f_3(\boldsymbol{x},\boldsymbol{u})}_{\boldsymbol{c}^{110}(\overleftarrow{f_1,f_2,f_3})} + e^{-f_2(\boldsymbol{x},\boldsymbol{u})}_{\boldsymbol{c}^{100}(\overleftarrow{f_1,f_2,0})} e^{f_3(\boldsymbol{x},\boldsymbol{u})}_{\boldsymbol{c}^{010}(\overleftarrow{f_1,f_2,f_3})}\right) \\
&\left. e^{f_1(\boldsymbol{y},\boldsymbol{u})} e^{f_2(\boldsymbol{y},\boldsymbol{u})} e^{-f_3(\boldsymbol{y},\boldsymbol{u})}\right) \prod_{l=4}^{d} e^{-f_l(\boldsymbol{x},\boldsymbol{u})} e^{-f_l(\boldsymbol{y},\boldsymbol{u})}
\end{aligned}
\tag{5.3.95}
$$

For $\delta = 3$ we look at all strictly lower triangular matrices $J \in \{0,1\}^{\delta\times\delta}$ with the property

$$
\forall k = 1,\ldots,\delta : \left(\sum_{l=1}^{\delta} (J)_{l,k}\right) \bmod 2 = j_k. \tag{5.3.96}
$$

That means the l-th row $(J)_l$ of J contains a multi-index $(J)_l \in \{0,1\}^\delta$, with the last $\delta - l - 1$ entries being zero and the k-th column sum being even when $j_k = 0$ and odd when $j_k = 1$. For example, the first multi-index is $\boldsymbol{j} = (0,0,0)$. There are only two different strictly lower triangular matrices that have columns summing up to even numbers:

$$
J = \begin{pmatrix} 0 & 0 & 0 \\ 0 & 0 & 0 \\ 0 & 0 & 0 \end{pmatrix} \text{ and } J = \begin{pmatrix} 0 & 0 & 0 \\ 1 & 0 & 0 \\ 1 & 0 & 0 \end{pmatrix}. \tag{5.3.97}
$$

The first row of each contains the multi-index that belongs to $e^{-f_1(\boldsymbol{x},\boldsymbol{u})}$, the second one belongs to $e^{-f_2(\boldsymbol{x},\boldsymbol{u})}$ and so on. So the summands with exactly these multi-indices are the ones assigned to the product of exponentials whose signs are invariant during the reordering. With this notation and all $J \in \{0,1\}^{3\times 3}$ that satisfy the property (5.3.96) we can write

$$
\begin{aligned}
\prod_{l=1}^{d} e^{-f_l(\boldsymbol{x}+\boldsymbol{y},\boldsymbol{u})} = \sum_{\boldsymbol{j}\in\{0,1\}^3} \sum_{J} \prod_{l=1}^{3} e^{-f_l(\boldsymbol{x},\boldsymbol{u})}_{\boldsymbol{c}^{(J)_l}(\overleftarrow{f_1,\ldots,f_l,0,\ldots,0})} \prod_{l=1}^{3} e^{-(-1)^{j_l} f_l(\boldsymbol{y},\boldsymbol{u})} \\
\prod_{l=4}^{d} e^{-f_l(\boldsymbol{x},\boldsymbol{u})} e^{-f_l(\boldsymbol{y},\boldsymbol{u})}.
\end{aligned}
\tag{5.3.98}
$$

Using mathematical induction with matrices $J \in \{0,1\}^{\delta\times\delta}$ as introduced above for growing δ and Corollary 5.3.26 repeatedly until we reach $\delta = d$ we get

$$= \sum_{j\in\{0,1\}^d} \sum_J \prod_{l=1}^{d} e^{-f_l(\boldsymbol{x},\boldsymbol{u})}_{\boldsymbol{c}^{(J)_l}(\overleftarrow{f_1,\ldots,f_l,0,\ldots,0})} \prod_{l=1}^{d} e^{-(-1)^{j_l} f_l(\boldsymbol{y},\boldsymbol{u})}. \tag{5.3.99}$$

□

Remark 5.3.34. *The number of summands actually appearing is usually much smaller than in Theorem 5.3.35. It is determined by the number of distinct strictly lower (upper) triangular matrices J with entries being either zero or one, namely:*

$$2^{\frac{d(d-1)}{2}}. \tag{5.3.100}$$

Theorem 5.3.35 (Shift). *Let $\boldsymbol{A}(\boldsymbol{x}) = \boldsymbol{B}(\boldsymbol{x}-\boldsymbol{x}_0)$ be multivector fields, F_1, F_2, linear with respect to $\boldsymbol{x}$, and let $\boldsymbol{j} \in \{0,1\}^{\mu}, \boldsymbol{k} \in \{0,1\}^{(\nu-\mu)}$ be multi-indices, and $F_1(\boldsymbol{j}), F_2(\boldsymbol{k})$ be as introduced in Definition 5.3.28, then a separable GFT suffices*

$$\begin{aligned}\mathcal{F}_{F_1,F_2}(\boldsymbol{A})(\boldsymbol{u}) = &\sum_{\boldsymbol{j},\boldsymbol{k}} \sum_{J,K} \prod_{l=1}^{\mu} e^{-f_l(\boldsymbol{x}_0,\boldsymbol{u})}_{\boldsymbol{c}^{(J)_l}(\overleftarrow{f_1,\ldots,f_l,0,\ldots,0})} \mathcal{F}_{F_1(\boldsymbol{j}),F_2(\boldsymbol{k})}(\boldsymbol{B})(\boldsymbol{u}) \\ &\prod_{l=\mu+1}^{\nu} e^{-f_l(\boldsymbol{x}_0,\boldsymbol{u})}_{\boldsymbol{c}^{(K)_{l-\mu}}(\overrightarrow{0,\ldots,0,f_l,\ldots,f_\nu})},\end{aligned} \tag{5.3.101}$$

where $J \in \{0,1\}^{\mu\times\mu}$ and $K \in \{0,1\}^{(\nu-\mu)\times(\nu-\mu)}$ are the strictly lower, respectively upper, triangular matrices with rows $(J)_l, (K)_{l-\mu}$ summing up to $(\sum_{l=1}^{\mu}(J)_l) \bmod 2 = \boldsymbol{j}$ respectively $\left(\sum_{l=\mu+1}^{\nu}(K)_{l-\mu}\right) \bmod 2 = \boldsymbol{k}$ as in Lemma 5.3.33.

Proof. First we rewrite the transformed function in terms of $\boldsymbol{B}(\boldsymbol{y})$ using a change of coordinates.

$$\begin{aligned}\mathcal{F}_{F_1,F_2}(\boldsymbol{A})(\boldsymbol{u}) &= \int_{\mathbb{R}^m} \prod_{l=1}^{\mu} e^{-f_l(\boldsymbol{x},\boldsymbol{u})} \boldsymbol{A}(\boldsymbol{x}) \prod_{l=\mu+1}^{\nu} e^{-f_l(\boldsymbol{x},\boldsymbol{u})}\, \mathrm{d}^m\boldsymbol{x} \\ &= \int_{\mathbb{R}^m} \prod_{l=1}^{\mu} e^{-f_l(\boldsymbol{x},\boldsymbol{u})} \boldsymbol{B}(\boldsymbol{x}-\boldsymbol{x}_0) \prod_{l=\mu+1}^{\nu} e^{-f_l(\boldsymbol{x},\boldsymbol{u})}\, \mathrm{d}^m\boldsymbol{x} \\ &\overset{\boldsymbol{y}=\boldsymbol{x}-\boldsymbol{x}_0}{=} \int_{\mathbb{R}^m} \prod_{l=1}^{\mu} e^{-f_l(\boldsymbol{y}+\boldsymbol{x}_0,\boldsymbol{u})} \boldsymbol{B}(\boldsymbol{y}) \prod_{l=\mu+1}^{\nu} e^{-f_l(\boldsymbol{y}+\boldsymbol{x}_0,\boldsymbol{u})}\, \mathrm{d}^m\boldsymbol{y}\end{aligned} \tag{5.3.102}$$

Now we separate and sort the factors using Lemma 5.3.33.

$$
\begin{aligned}
&\overset{\text{Lem. 5.3.33}}{=} \int_{\mathbb{R}^m} \sum_{\boldsymbol{j}\in\{0,1\}^{\mu}} \sum_{\substack{J\in\{0,1\}^{\mu\times\mu}\\ \sum(J)_l \bmod 2=\boldsymbol{j}}} \\
&\prod_{l=1}^{\mu} e^{-f_l(\boldsymbol{x}_0,\boldsymbol{u})}_{\boldsymbol{c}^{(J)_l}(\overleftarrow{f_1,\ldots,f_l,0,\ldots,0})} \prod_{l=1}^{\mu} e^{-(-1)^{j_l} f_l(\boldsymbol{y},\boldsymbol{u})} \boldsymbol{B}(\boldsymbol{y}) \\
&\sum_{\boldsymbol{k}\in\{0,1\}^{(\nu-\mu)}} \sum_{\substack{K\in\{0,1\}^{(\nu-\mu)\times(\nu-\mu)}\\ \sum(K)_l \bmod 2=\boldsymbol{k}}} \\
&\prod_{l=\mu+1}^{\nu} e^{-(-1)^{k_{l-\mu}} f_l(\boldsymbol{y},\boldsymbol{u})} \prod_{l=\mu+1}^{\nu} e^{-f_l(\boldsymbol{x}_0,\boldsymbol{u})}_{\boldsymbol{c}^{(K)_{l-\mu}}(\overrightarrow{0,\ldots,0,f_l,\ldots,f_\nu})} \, \mathrm{d}^m\boldsymbol{y} \\
&= \sum_{\boldsymbol{j},\boldsymbol{k}} \sum_{J,K} \prod_{l=1}^{\mu} e^{-f_l(\boldsymbol{x}_0,\boldsymbol{u})}_{\boldsymbol{c}^{(J)_l}(\overleftarrow{f_1,\ldots,f_l,0,\ldots,0})} \\
&\mathcal{F}_{F_1(\boldsymbol{j}),F_2(\boldsymbol{k})}(\boldsymbol{B})(\boldsymbol{u}) \prod_{l=\mu+1}^{\nu} e^{-f_l(\boldsymbol{x}_0,\boldsymbol{u})}_{\boldsymbol{c}^{(K)_{l-\mu}}(\overrightarrow{0,\ldots,0,f_l,\ldots,f_\nu})}.
\end{aligned}
\tag{5.3.103}
$$

□

Corollary 5.3.36 (Shift). *Let $\boldsymbol{A}(\boldsymbol{x}) = \boldsymbol{B}(\boldsymbol{x}-\boldsymbol{x}_0)$ be multivector fields, F_1 and F_2 each consisting of mutually commutative functions*[26] *being linear with respect to $\boldsymbol{x}$, then the GFT obeys*

$$
\mathcal{F}_{F_1,F_2}(\boldsymbol{A})(\boldsymbol{u}) = \prod_{l=1}^{\mu} e^{-f_l(\boldsymbol{x}_0,\boldsymbol{u})} \mathcal{F}_{F_1,F_2}(\boldsymbol{B})(\boldsymbol{u}) \prod_{l=\mu+1}^{\nu} e^{-f_l(\boldsymbol{x}_0,\boldsymbol{u})}. \tag{5.3.104}
$$

Remark 5.3.37. *For sets F_1, F_2 that each consist of less than two functions the condition of Corollary 5.3.36 is necessarily satisfied, compare e.g. the Clifford-Fourier transform, the quaternionic transform or the spacetime Fourier transform listed in the preceding examples.*

The specific forms taken by our standard examples are summarized in Table 5.8. As expected they are often shorter than what could be expected from Remark 5.3.34.

5.3.2.7 Summary on two-sided (Clifford) geometric Fourier transform with multiple square roots of -1 in $Cl(p,q)$

For multivector fields over $\mathbb{R}^{p,q}$ with values in any geometric algebra $Cl(p,q)$ we have successfully defined a general geometric Fourier transform in this section. It covers many popular Fourier transforms from current literature in the introductory example. Its existence, independent of the specific choice of functions F_1, F_2, can be proven for all integrable multivector fields, see Theorem 5.3.11. Theorem 5.3.12 shows that our geometric Fourier transform is generally linear over the field of real numbers. All

[26]Cross commutativity between F_1 and F_2 is not necessary.

Table 5.8 Theorem 5.3.35 (Shift) applied to the GFTs of the first example, enumerated in the same order.
Notations: on the LHS $\mathcal{F}_{F_1,F_2} = \mathcal{F}_{F_1,F_2}(\boldsymbol{A})(\boldsymbol{u})$, on the RHS $\mathcal{F}_{F_1',F_2'} = \mathcal{F}_{F_1',F_2'}(\boldsymbol{B})(\boldsymbol{u})$. In the second row K represents all strictly upper triangular matrices $\in \{0,1\}^{n\times n}$ with rows $(K)_{l-\boldsymbol{\mu}}$ summing up to $\left(\sum_{l=\boldsymbol{\mu}+1}^{\boldsymbol{\nu}}(K)_{l-\boldsymbol{\mu}}\right) \bmod 2 = \boldsymbol{k}$. The simplified shape of the colour image FT results from the commutativity of $\boldsymbol{B}$ and $i\boldsymbol{B}$ and application of Lemma 5.3.7. Source: [61, Tab. 3].

	GFT	$\boldsymbol{A}(\boldsymbol{x}) = \boldsymbol{B}(\boldsymbol{x} - \boldsymbol{x}_0)$
1.	Clifford	$\mathcal{F}_{f_1} = \mathcal{F}_{f_1} e^{-2\pi i \boldsymbol{x}_0 \cdot \boldsymbol{u}}$
2.	Bülow	$\mathcal{F}_{f_1,\ldots,f_n} = \sum_{\boldsymbol{k}\in\{0,1\}^n} \sum_K \mathcal{F}_{(-1)^{k_1} f_1,\ldots,(-1)^{k_n} f_n}$ $\prod_{l=1}^{n} e^{-2\pi x_{0k} u_k}_{\boldsymbol{c}^{(K)_l}(\overrightarrow{0,\ldots,0,f_l,\ldots,f_n})}$
3.	Quaternionic	$\mathcal{F}_{f_1,f_2} = e^{-2\pi i x_{01} u_1} \mathcal{F}_{f_1,f_2} e^{-2\pi j x_{02} u_2}$
4.	Spacetime	$\mathcal{F}_{f_1,f_2} = e^{-\boldsymbol{e}_4 x_{04} u_4} \mathcal{F}_{f_1,f_2} e^{-\epsilon_4 \boldsymbol{e}_4 i_4 (x_1 u_1 + x_2 u_2 + x_3 u_3)}$
5.	Colour Image	$\mathcal{F}_{f_1,f_2,f_3,f_4} = e^{-\frac{1}{2}(x_{01}u_1 + x_{02}u_2)(\boldsymbol{B}+i\boldsymbol{B})} \mathcal{F}_{f_1,f_2,f_3,f_4}$ $e^{\frac{1}{2}(x_{01}u_1 + x_{02}u_2)(\boldsymbol{B}+i\boldsymbol{B})}$
6.	Cyl. $n = 2$	$\mathcal{F}_{f_1} = e^{\boldsymbol{x}_0 \wedge \boldsymbol{u}} \mathcal{F}_{f_1}$
	Cyl. $n \neq 2$	-

transforms from the reference example consist of bilinear F_1 and F_2. We proved that this property is sufficient to ensure the scaling property of Theorem 5.3.14.

If a general geometric Fourier transform is separable as introduced in Definition 5.3.24, then Theorem 5.3.29 (Left and right products) guarantees that constant factors can be separated from the vector field to be transformed. As a consequence general linearity is achieved by choosing F_1, F_2 with values in the center of the geometric algebra $Cl(p,q)$, compare Corollary 5.3.31. All examples except for the cylindrical Fourier transform [49] satisfy this claim.

Under the condition of linearity with respect to the first argument of the functions of the sets F_1 and F_2 additionally to the separability property just mentioned, we have also proven a shift property (Theorem 5.3.35).

5.3.3 Geometric Fourier transform and trigonometric transform with multiple square roots of -1 in $Cl(p,q)$

5.3.3.1 *Background*

As it will turn out in this section, many Clifford Fourier transforms that are relevant for applications are separable and these transforms can be decomposed into a sum of real valued transforms with constant multivector factors. This fact greatly helps their interpretation, their analysis, and their implementation. This section is based on [66], see also [60].

Basically, three different approaches to hypercomplex Fourier transforms can been distinguished. As in [62], we can identify them as follows:

- A: Eigenfunction approach
- B: Generalized roots of -1 approach
- C: Characters of spin group approach

Approach A is studied in many papers like [40,50,91], approach B in Section 5.3.2, see also [61]. The third approach is followed in [18,19]. In this section, we analyze the generalized square roots of -1 approach, because even though the concept of approach C differs very much from approach B, the resulting transforms can be expressed as special cases of approach B. We refer to Section 5.3.2 for a deeper understanding. We will only work with transforms as defined in Section 5.3.2, Definition 5.3.9, throughout this section. That covers all known Fourier transforms over Clifford algebras that belong to approaches B and C.

Remark 5.3.38. *Depending on the choice of F_1 and F_2 in the definition of the GFT, we get already existing transforms. Six examples have already been chosen as standard examples in Example 5.3.10. We will use them again to visualize how the general propositions reduce to special applications.*

In the following section, we will introduce a powerful tool for the analysis of the geometric Fourier transforms: the trigonometric transform. Then, we will first use it to reveal the true nature of the subclasses of the separable and the explicitly invertible GFTs, which cover almost every applied Clifford Fourier transform. In the next two sections, we will enjoy the advantages of this insight and show how it leads to a convolution theorem that is superior to the one in [59] because it requires less restrictions.

5.3.3.2 The trigonometric transform (TT)

For the GFT convolution theorem in [59], one makes use of geometric trigonometric transforms. These are generalized GFTs that may also contain the cosine of the functions F_1, F_2 or their sine paired with the square root of minus one instead of only their exponentials. In contrast to the hypercomplex definition in [59], we want to define real valued general trigonometric transforms, that only consist of scalar appearances of sines and cosines. Therefore we use the following notation.

Notation 5.3.39. *A multivector valued function $f(\boldsymbol{x},\boldsymbol{u}) : \mathbb{R}^2 \to \mathscr{I}_{p,q} = \{B \in Cl(p,q), B^2 \in \mathbb{R}^-\} \subset Cl(p,q)$ that squares to a negative real number, can be separated into a unit part and a real part. This separation is not unique. There are two functions $\pm i(\boldsymbol{x},\boldsymbol{u}) : \mathbb{R}^2 \to \{B \in Cl(p,q), B^2 = -1\} \subset I_{p,q} \subset Cl(p,q)$ that square to minus one*

$$\pm i(\boldsymbol{x},\boldsymbol{u}) = \pm \frac{f(\boldsymbol{x},\boldsymbol{u})}{\|f(\boldsymbol{x},\boldsymbol{u})\|}, \tag{5.3.105}$$

and leave the positive real valued function $\|f(\boldsymbol{x},\boldsymbol{u})\| = \sqrt{-f(\boldsymbol{x},\boldsymbol{u})^2} : \mathbb{R}^2 \to \mathbb{R}^+$. We can choose one of them and define the function $|f(\boldsymbol{x},\boldsymbol{u})| : \mathbb{R}^2 \to \mathbb{R}$, which can also take negative values, by

$$|f(\boldsymbol{x},\boldsymbol{u})| = \frac{f(\boldsymbol{x},\boldsymbol{u})}{i(\boldsymbol{x},\boldsymbol{u})}. \tag{5.3.106}$$

Now, with $j \in \{0,1\}$, we define the decomposition

$$e_j^{f(\boldsymbol{x},\boldsymbol{u})} := \begin{cases} \cos(|f(\boldsymbol{x},\boldsymbol{u})|), & \text{if } j=0, \\ \sin(|f(\boldsymbol{x},\boldsymbol{u})|), & \text{if } j=1. \end{cases} \tag{5.3.107}$$

Lemma 5.3.40. *The exponential of a multivector valued function $f(\boldsymbol{x},\boldsymbol{u}) : \mathbb{R}^2 \to \mathscr{I}_{p,q} = \{B \in Cl(p,q), B^2 \in \mathbb{R}^-\} \subset Cl(p,q)$ that squares to a negative real number satisfies*

$$e^{f(\boldsymbol{x},\boldsymbol{u})} = \sum_{j\in\{0,1\}} (\frac{f(\boldsymbol{x},\boldsymbol{u})}{|f(\boldsymbol{x},\boldsymbol{u})|})^j e_j^{f(\boldsymbol{x},\boldsymbol{u})}. \tag{5.3.108}$$

Proof. Because of $i(\boldsymbol{x},\boldsymbol{u}) = \frac{f(\boldsymbol{x},\boldsymbol{u})}{|f(\boldsymbol{x},\boldsymbol{u})|}$ squares to -1 and $|f(\boldsymbol{x},\boldsymbol{u}) \in \mathbb{R}$, the hypercomplex equivalent to the Euler equation as in [162] leads to

$$\begin{aligned} e^{f(\boldsymbol{x},\boldsymbol{u})} &= e^{i(\boldsymbol{x},\boldsymbol{u})|f(\boldsymbol{x},\boldsymbol{u})|} = \cos(|f(\boldsymbol{x},\boldsymbol{u})|) + i(\boldsymbol{x},\boldsymbol{u})\sin(|f(\boldsymbol{x},\boldsymbol{u})|) \\ &\overset{\text{Not. 5.3.39}}{=} i(\boldsymbol{x},\boldsymbol{u})^0 e_0^{f(\boldsymbol{x},\boldsymbol{u})} + i(\boldsymbol{x},\boldsymbol{u})^1 e_1^{f(\boldsymbol{x},\boldsymbol{u})} \\ &= \sum_{j\in\{0,1\}} i(\boldsymbol{x},\boldsymbol{u})^j e_j^{|f(\boldsymbol{x},\boldsymbol{u})|} = \sum_{j\in\{0,1\}} (\frac{f(\boldsymbol{x},\boldsymbol{u})}{|f(\boldsymbol{x},\boldsymbol{u})|})^j e_j^{f(\boldsymbol{x},\boldsymbol{u})}, \end{aligned} \tag{5.3.109}$$

which proves the assertion □

Remark 5.3.41. *The minus sign that appears in the Fourier transform can be left with the real valued function*

$$\begin{aligned} e^{-f(\boldsymbol{x},\boldsymbol{u})} &= \sum_{j\in\{0,1\}} (\frac{-f(\boldsymbol{x},\boldsymbol{u})}{|-f(\boldsymbol{x},\boldsymbol{u})|})^j e_j^{-f(\boldsymbol{x},\boldsymbol{u})} \\ &= \sum_{j\in\{0,1\}} i(\boldsymbol{x},\boldsymbol{u})^j e_j^{-f(\boldsymbol{x},\boldsymbol{u})}. \end{aligned} \tag{5.3.110}$$

Definition 5.3.42 (Trigonometric transform)**.** *Let $\boldsymbol{A} : \mathbb{R}^m \to Cl(p,q)$ be a multivector field*[27]*, $\boldsymbol{x},\boldsymbol{u} \in \mathbb{R}^m$ vectors, F a finite set of $\boldsymbol{\nu}$, mappings $\mathbb{R}^m \times \mathbb{R}^m \to Cl(p,q)$, and $\boldsymbol{j} \in \{0,1\}^{\boldsymbol{\nu}}$ a multi-index. The* ***Trigonometric transform*** *(TT) $\mathscr{F}_F$ is defined by*

$$\mathscr{F}_F(\boldsymbol{A})(\boldsymbol{u}) := \int_{\mathbb{R}^m} \boldsymbol{A}(\boldsymbol{x}) \prod_{l=1}^{\boldsymbol{\nu}} e_{j_l}^{-f_l(\boldsymbol{x},\boldsymbol{u})} \,\mathrm{d}^m\boldsymbol{x}, \tag{5.3.111}$$

with $e_{j_l}^{-f_l(\boldsymbol{x},\boldsymbol{u})} \in \mathbb{R}$ from Notation 5.3.39.

Remark 5.3.43. *The $e_{j_l}^{-f(\boldsymbol{x},\boldsymbol{u})} \in \mathbb{R}$ are in the center of the geometric algebra. Therefore, there is no need to distinguish the order of their appearances. But it may be*

---

[27]For the Fourier type transform with $\boldsymbol{A} \in L^1(\mathbb{R}^m; Cl(p,q))$ and $\boldsymbol{A} \in L^2(\mathbb{R}^m; Cl(p,q))$, see Footnote 2 on page 144 regarding the roles of L^1 and L^2.

helpful though in order to stress their relation to the GFT. Let $\mathscr{F}_{F_1^{\boldsymbol{j}},F_2^{\boldsymbol{k}}}$ be a geometric Fourier transform and $\boldsymbol{j} \in \{0,1\}^{\mu}, \boldsymbol{k} \in \{0,1\}^{(\nu-\mu)}$ multi-indices. Then,

$$\mathscr{F}_{F_1^{\boldsymbol{j}},F_2^{\boldsymbol{k}}}(\boldsymbol{A})(\boldsymbol{u}) := \int_{\mathbb{R}^m} \prod_{l=1}^{\mu} e_{j_l}^{-f_l(\boldsymbol{x},\boldsymbol{u})} \boldsymbol{A}(\boldsymbol{x}) \prod_{l=\mu+1}^{\nu} e_{k_l}^{-f_l(\boldsymbol{x},\boldsymbol{u})} \,\mathrm{d}^m\boldsymbol{x}, \tag{5.3.112}$$

is a trigonometric transform.

Notation 5.3.44. *The standard sine transform*

$$\mathscr{F}_s(\boldsymbol{A})(\boldsymbol{u}) = \int_{\mathbb{R}^m} \sin(-2\pi\boldsymbol{x}\cdot\boldsymbol{u})\boldsymbol{A}(\boldsymbol{x}) \,\mathrm{d}^m\boldsymbol{x} = \mathscr{F}_{(2\pi i\boldsymbol{x}\cdot\boldsymbol{u})^1}(\boldsymbol{A})(\boldsymbol{u}), \tag{5.3.113}$$

the standard cosine transform

$$\mathscr{F}_c(\boldsymbol{A})(\boldsymbol{u}) = \int_{\mathbb{R}} \cos(-2\pi\boldsymbol{x}\cdot\boldsymbol{u})\boldsymbol{A}(\boldsymbol{x}) \,\mathrm{d}^m\boldsymbol{x} = \mathscr{F}_{(2\pi i\boldsymbol{x}\cdot\boldsymbol{u})^0}(\boldsymbol{A})(\boldsymbol{u}), \tag{5.3.114}$$

and their compositions in higher dimensions are special cases of the trigonometric transform. For them, we will sometimes write the letters s and c as lower index of the transform for the sake of brevity and to stress their special shape. As an example, we would write $\mathscr{F}_{sc}$ instead of $\mathscr{F}_{(i_1x_1u_1)^1,(i_2x_2u_2)^0}$, which means

$$\mathscr{F}_{sc}(\boldsymbol{A})(\boldsymbol{u}) = \int_{\mathbb{R}^2} \sin(-x_1u_1)\cos(-x_2u_2)\boldsymbol{A}(\boldsymbol{x}) \,\mathrm{d}\boldsymbol{x}. \tag{5.3.115}$$

5.3.3.3 The true nature of separable geometric Fourier transforms

The definition of separability has already been introduced in [61].

Definition 5.3.45. *We call a mapping $f : \mathbb{R}^m \times \mathbb{R}^m \to Cl(p,q)$ **x-separable** or separable with respect to its first argument, if it suffices*

$$f = |f(\boldsymbol{x},\boldsymbol{u})|i(\boldsymbol{u}), \tag{5.3.116}$$

*where $i : \mathbb{R}^m \to Cl(p,q)$ is a function that does not depend on $\boldsymbol{x}$ and $|f(\boldsymbol{x},\boldsymbol{u})| : \mathbb{R}^2 \to \mathbb{R}$ a real valued function as in Notation 5.3.39. Analogously we call it **separable** or separable with respect to both arguments, if it suffices*

$$f = |f(\boldsymbol{x},\boldsymbol{u})|i, \tag{5.3.117}$$

with constant $i \in Cl(p,q), i^2 = -1$.

Analogously, a geometric Fourier transform that consists of separable mappings F_1, F_2 is called separable. Separability is a central quality for multiplication, shift and convolution properties of GFTs. Almost every transform from approach B and C is separable.

Exercise 5.3.46. *Every transform of our standard examples from Example 5.3.10, except for the cylindrical Fourier transform for dimensions higher than two, is separable.*

In this section, we want to take a closer look at this vast class of GFTs. By expressing them by means of the trigonometric transforms, we will be able to reveal their true nature: they are combinations of simple real-valued transforms.

Theorem 5.3.47 (GFT decomposition into TT). *Any geometric Fourier transform $\mathscr{F}_{F_1,F_2}$ with x-separable mappings $\forall l = 1, ..., \boldsymbol{\nu} : f_l(\boldsymbol{x}, \boldsymbol{u}) = |f_l(\boldsymbol{x}, \boldsymbol{u})|\, i_l(\boldsymbol{u})$ of a multivector field $\boldsymbol{A}(\boldsymbol{x}) = \sum_{\boldsymbol{r}} a_{\boldsymbol{r}}(\boldsymbol{x})\boldsymbol{e}_{\boldsymbol{r}}$ is the sum of real valued trigonometric transforms $\mathscr{F}_{F_1^{\boldsymbol{j}},F_2^{\boldsymbol{k}}}(a_{\boldsymbol{r}})(\boldsymbol{u}) \in \mathbb{R}$ with multivector factors*

$$\begin{aligned}&\mathscr{F}_{F_1,F_2}(\boldsymbol{A})(\boldsymbol{u})\\&= \sum_{\boldsymbol{r}} \sum_{\substack{\boldsymbol{j}\in\{0,1\}^{\boldsymbol{\mu}},\\ \boldsymbol{k}\in\{0,1\}^{(\boldsymbol{\nu}-\boldsymbol{\mu})}}} \mathscr{F}_{F_1^{\boldsymbol{j}},F_2^{\boldsymbol{k}}}(a_{\boldsymbol{r}})(\boldsymbol{u}) \prod_{l=1}^{\boldsymbol{\mu}} i_l(\boldsymbol{u})^{j_l} \boldsymbol{e}_{\boldsymbol{r}} \prod_{l=\boldsymbol{\mu}+1}^{\boldsymbol{\nu}} i_l(\boldsymbol{u})^{k_l}.\end{aligned} \tag{5.3.118}$$

Proof. Using Notation 5.3.39, we can write any GFT as

$$\begin{aligned}\mathscr{F}_{F_1,F_2}(\boldsymbol{A})(\boldsymbol{u}) &= \int_{\mathbb{R}^m} \prod_{l=1}^{\boldsymbol{\mu}} e^{-f_l(\boldsymbol{x},\boldsymbol{u})} \boldsymbol{A}(\boldsymbol{x}) \prod_{l=\boldsymbol{\mu}+1}^{\boldsymbol{\nu}} e^{-f_l(\boldsymbol{x},\boldsymbol{u})}\, \mathrm{d}^m\boldsymbol{x}\\ &\overset{\text{Lem. 5.3.40}}{=} \int_{\mathbb{R}^m} \prod_{l=1}^{\boldsymbol{\mu}} \sum_{j_l\in\{0,1\}} \left(\frac{f_l(\boldsymbol{x},\boldsymbol{u})}{|f_l(\boldsymbol{x},\boldsymbol{u})|}\right)^{j_l} e_{j_l}^{-f_l(\boldsymbol{x},\boldsymbol{u})} \boldsymbol{A}(\boldsymbol{x})\\ &\qquad \prod_{l=\boldsymbol{\mu}+1}^{\boldsymbol{\nu}} \sum_{k_l\in\{0,1\}} \left(\frac{f_l(\boldsymbol{x},\boldsymbol{u})}{|f_l(\boldsymbol{x},\boldsymbol{u})|}\right)^{k_l} e_{k_l}^{-f_l(\boldsymbol{x},\boldsymbol{u})}\, \mathrm{d}^m\boldsymbol{x}.\end{aligned} \tag{5.3.119}$$

Since the GFT has x-separable mappings, we can replace $\frac{f_l(\boldsymbol{x},\boldsymbol{u})}{|f_l(\boldsymbol{x},\boldsymbol{u})|}$ by $i_l(\boldsymbol{u})$ with $i_l(\boldsymbol{u})^2 = -1$ and write

$$\begin{aligned}\mathscr{F}_{F_1,F_2}(\boldsymbol{A})(\boldsymbol{u}) = \int_{\mathbb{R}^m} &\prod_{l=1}^{\boldsymbol{\mu}} \sum_{j_l\in\{0,1\}} i_l(\boldsymbol{u})^{j_l} e_{j_l}^{-f(\boldsymbol{x},\boldsymbol{u})} \boldsymbol{A}(\boldsymbol{x})\\ &\prod_{l=\boldsymbol{\mu}+1}^{\boldsymbol{\nu}} \sum_{k_l\in\{0,1\}} i_l(\boldsymbol{u})^{k_l} e_{k_l}^{-f(\boldsymbol{x},\boldsymbol{u})}\, \mathrm{d}^m\boldsymbol{x}.\end{aligned} \tag{5.3.120}$$

We collect the indices $j_l, l = 1, ..., \boldsymbol{\mu}$ into the multi-index $\boldsymbol{j} \in \{0,1\}^{\boldsymbol{\mu}}$ and $k_l, l = \boldsymbol{\mu}+1, ..., \boldsymbol{\nu}$ into the multi-index $\boldsymbol{k} \in \{0,1\}^{\boldsymbol{\nu}-\boldsymbol{\mu}}$ and get

$$\begin{aligned}\mathscr{F}_{F_1,F_2}(\boldsymbol{A})(\boldsymbol{u}) = \sum_{\substack{\boldsymbol{j}\in\{0,1\}^{\boldsymbol{\mu}},\\ \boldsymbol{k}\in\{0,1\}^{(\boldsymbol{\nu}-\boldsymbol{\mu})}}} &\int_{\mathbb{R}^m} \prod_{l=1}^{\boldsymbol{\mu}} i_l(\boldsymbol{u})^{j_l} e_{j_l}^{-f_l(\boldsymbol{x},\boldsymbol{u})} \boldsymbol{A}(\boldsymbol{x})\\ &\prod_{l=\boldsymbol{\mu}+1}^{\boldsymbol{\nu}} e_{j_l}^{-f_l(\boldsymbol{x},\boldsymbol{u})} i_l(\boldsymbol{u})^{k_l}\, \mathrm{d}^m\boldsymbol{x}.\end{aligned} \tag{5.3.121}$$

Since the $e_{j_l}^{f_l(\boldsymbol{x},\boldsymbol{u})} \in \mathbb{R}$, this leads to

$$\mathscr{F}_{F_1,F_2}(\boldsymbol{A})(\boldsymbol{u}) = \sum_{\substack{\boldsymbol{j}\in\{0,1\}^{\boldsymbol{\mu}},\\ \boldsymbol{k}\in\{0,1\}^{(\boldsymbol{\nu}-\boldsymbol{\mu})}}} \prod_{l=1}^{\boldsymbol{\mu}} i_l(\boldsymbol{u})^{j_l} \int_{\mathbb{R}^m} \prod_{l=1}^{\boldsymbol{\mu}} e_{j_l}^{-f_l(\boldsymbol{x},\boldsymbol{u})} \boldsymbol{A}(\boldsymbol{x}) \prod_{l=\boldsymbol{\mu}+1}^{\boldsymbol{\nu}} e_{j_l}^{-f_l(\boldsymbol{x},\boldsymbol{u})} \,\mathrm{d}^m\boldsymbol{x} \prod_{l=\boldsymbol{\mu}+1}^{\boldsymbol{\nu}} i_l(\boldsymbol{u})^{k_l}, \tag{5.3.122}$$

and together with Definition 5.3.42 to

$$\mathscr{F}_{F_1,F_2}(\boldsymbol{A})(\boldsymbol{u}) = \sum_{\substack{\boldsymbol{j}\in\{0,1\}^{\boldsymbol{\mu}},\\ \boldsymbol{k}\in\{0,1\}^{(\boldsymbol{\nu}-\boldsymbol{\mu})}}} \prod_{l=1}^{\boldsymbol{\mu}} i_l(\boldsymbol{u})^{j_l} \mathscr{F}_{F_1^{\boldsymbol{j}},F_2^{\boldsymbol{k}}}(\boldsymbol{A})(\boldsymbol{u}) \prod_{l=\boldsymbol{\mu}+1}^{\boldsymbol{\nu}} i_l(\boldsymbol{u})^{k_l}. \tag{5.3.123}$$

The trigonometric transform itself does not contain any multivector and therefore preserves the basis blades of the multivector field $\boldsymbol{A} = \sum_{\boldsymbol{r}} a_{\boldsymbol{r}} \boldsymbol{e}_{\boldsymbol{r}}$, i.e.

$$\mathscr{F}_{F_1^{\boldsymbol{j}},F_2^{\boldsymbol{k}}}(\boldsymbol{A})(\boldsymbol{u}) = \mathscr{F}_{F_1^{\boldsymbol{j}},F_2^{\boldsymbol{k}}} \sum_{\boldsymbol{r}} a_{\boldsymbol{r}} \boldsymbol{e}_{\boldsymbol{r}}(\boldsymbol{u}) = \sum_{\boldsymbol{r}} \mathscr{F}_{F_1^{\boldsymbol{j}},F_2^{\boldsymbol{k}}}(a_{\boldsymbol{r}})(\boldsymbol{u}) \boldsymbol{e}_{\boldsymbol{r}}, \tag{5.3.124}$$

with $\mathscr{F}_{F_1^{\boldsymbol{j}},F_2^{\boldsymbol{k}}}(a_{\boldsymbol{r}})(\boldsymbol{u}) \in \mathbb{R}$. As a result the x-separable GFT can be written as

$$\begin{aligned}
\mathscr{F}_{F_1,F_2}(\boldsymbol{A})(\boldsymbol{u}) &= \sum_{\substack{\boldsymbol{j}\in\{0,1\}^{\boldsymbol{\mu}},\\ \boldsymbol{k}\in\{0,1\}^{(\boldsymbol{\nu}-\boldsymbol{\mu})}}} \prod_{l=1}^{\boldsymbol{\mu}} i_l(\boldsymbol{u})^{j_l} \sum_{\boldsymbol{r}} \mathscr{F}_{F_1^{\boldsymbol{j}},F_2^{\boldsymbol{k}}}(a_{\boldsymbol{r}})(\boldsymbol{u}) \boldsymbol{e}_{\boldsymbol{r}} \prod_{l=\boldsymbol{\mu}+1}^{\boldsymbol{\nu}} i_l(\boldsymbol{u})^{k_l} \\
&= \sum_{\substack{\boldsymbol{j}\in\{0,1\}^{\boldsymbol{\mu}},\\ \boldsymbol{k}\in\{0,1\}^{(\boldsymbol{\nu}-\boldsymbol{\mu})}}} \sum_{\boldsymbol{r}} \mathscr{F}_{F_1^{\boldsymbol{j}},F_2^{\boldsymbol{k}}}(a_{\boldsymbol{r}})(\boldsymbol{u}) \prod_{l=1}^{\boldsymbol{\mu}} i_l(\boldsymbol{u})^{j_l} \boldsymbol{e}_{\boldsymbol{r}} \prod_{l=\boldsymbol{\mu}+1}^{\boldsymbol{\nu}} i_l(\boldsymbol{u})^{k_l},
\end{aligned} \tag{5.3.125}$$

which leads to the assertion. □

Remark 5.3.48. *The formula (5.3.118) can be written more beautifully as*

$$\begin{aligned}
&\mathscr{F}_{F_1,F_2}(\boldsymbol{A})(\boldsymbol{u}) \\
&= \sum_{\substack{\boldsymbol{j}\in\{0,1\}^{\boldsymbol{\mu}},\\ \boldsymbol{k}\in\{0,1\}^{(\boldsymbol{\nu}-\boldsymbol{\mu})}}} \prod_{l=1}^{\boldsymbol{\mu}} i_l(\boldsymbol{u})^{j_l} \mathscr{F}_{F_1^{\boldsymbol{j}},F_2^{\boldsymbol{k}}}(\boldsymbol{A})(\boldsymbol{u}) \prod_{l=\boldsymbol{\mu}+1}^{\boldsymbol{\nu}} i_l(\boldsymbol{u})^{k_l},
\end{aligned} \tag{5.3.126}$$

but we prefer the first version, because it stresses how the transform is done completely in the real numbers.

Corollary 5.3.49. *A geometric Fourier transform $\mathscr{F}_{F_1,F_2}$ with separable mappings $\forall l = 1,...,\boldsymbol{\nu} : f_l(\boldsymbol{x},\boldsymbol{u}) = i_l|f_l(\boldsymbol{x},\boldsymbol{u})|, i_l^2 \in \mathbb{R}^-$ of a multivector field $\boldsymbol{A}(\boldsymbol{x}) = \sum_{\boldsymbol{r}} a_{\boldsymbol{r}}(\boldsymbol{x})\boldsymbol{e}_{\boldsymbol{r}}$ is the sum of real valued trigonometric transforms $\mathscr{F}_{F_1^{\boldsymbol{j}},F_2^{\boldsymbol{k}}}(a_{\boldsymbol{r}})(\boldsymbol{u}) \in \mathbb{R}$ with constant multivector factors*

$$\mathscr{F}_{F_1,F_2}(\boldsymbol{A})(\boldsymbol{u}) = \sum_{\boldsymbol{r}} \sum_{\substack{\boldsymbol{j}\in\{0,1\}^{\boldsymbol{\mu}},\\ \boldsymbol{k}\in\{0,1\}^{(\boldsymbol{\nu}-\boldsymbol{\mu})}}} \mathscr{F}_{F_1^{\boldsymbol{j}},F_2^{\boldsymbol{k}}}(a_{\boldsymbol{r}})(\boldsymbol{u}) \prod_{l=1}^{\boldsymbol{\nu}} i_l^{j_l} \boldsymbol{e}_{\boldsymbol{r}} \prod_{l=\boldsymbol{\mu}+1}^{\boldsymbol{\nu}} i_l^{k_l}. \tag{5.3.127}$$

That means, if we interpret a multivector valued signal as many signals, saved in 2^n channels, the separable geometric Fourier transforms can be interpreted as real valued transforms, that work one after another on each of the channels, get added in a certain way and written into certain channels depending on the multivector factor. As a result they can be interpreted, analyzed and implemented with the same tools as the classical real-valued transforms.

This corollary shows that the separable geometric Fourier transforms may look difficult, but really are structures hardly more complicated than the classical Fourier transform. They are linear combinations of the real valued trigonometric transforms in a Clifford algebra.

Exercise 5.3.50. *The restrictions of Corollary 5.3.49 are fulfilled by all transforms from Example 5.3.10, except for the cylindrical transform for dimensions higher that two. For these examples it takes the following shapes.*

1. *For the Clifford Fourier transform from [119, 185, 222] for multivector fields $A : \mathbb{R}^n \to Cl(n,0)$, $n = 2 \pmod 4$ or $n = 3 \pmod 4$ the Corollary takes the form*

$$\begin{aligned} \mathscr{F}_{f_1}(\boldsymbol{A}) &= \sum_{\boldsymbol{r}} \sum_{k \in \{0,1\}} \mathscr{F}_{f_1^k}(a_{\boldsymbol{r}}) \boldsymbol{e}_{\boldsymbol{r}} i_n^k \\ &= \sum_{\boldsymbol{r}} \mathscr{F}_c(a_{\boldsymbol{r}}) \boldsymbol{e}_{\boldsymbol{r}} + \mathscr{F}_s(a_{\boldsymbol{r}}) \boldsymbol{e}_{\boldsymbol{r}} i_n, \end{aligned} \tag{5.3.128}$$

with n-dimensional sine and cosine transforms.

2. *The transform by Sommen [56, 306] satisfies*

$$\mathscr{F}_{f_1,\dots,f_n}(\boldsymbol{A}) = \sum_{\boldsymbol{r}} \sum_{\boldsymbol{k} \in \{0,1\}^n} \mathscr{F}_{f_1^{k_1},\dots,f_n^{k_n}}(a_{\boldsymbol{r}}) \boldsymbol{e}_{\boldsymbol{r}} \prod_{l=1}^{n} \boldsymbol{e}_l^{k_l}. \tag{5.3.129}$$

3. *The quaternionic Fourier transform [124] takes the shape*

$$\begin{aligned} \mathscr{F}_{f_1,f_2}(\boldsymbol{A}) &= \sum_{\boldsymbol{r}} \sum_{j,k \in \{0,1\}} i^j \mathscr{F}_{f_1^j,f_2^k}(a_{\boldsymbol{r}}) \boldsymbol{e}_{\boldsymbol{r}} j^k \\ &= \sum_{\boldsymbol{r}} \mathscr{F}_{cc}(a_{\boldsymbol{r}}) \boldsymbol{e}_{\boldsymbol{r}} + \mathscr{F}_{cs}(a_{\boldsymbol{r}}) \boldsymbol{e}_{\boldsymbol{r}} j + i \mathscr{F}_{sc}(a_{\boldsymbol{r}}) \boldsymbol{e}_{\boldsymbol{r}} + i \mathscr{F}_{ss}(a_{\boldsymbol{r}}) \boldsymbol{e}_{\boldsymbol{r}} j. \end{aligned} \tag{5.3.130}$$

4. *Analogously, the spacetime Fourier transform [184] takes the shape*

$$\begin{aligned} \mathscr{F}_{f_1,f_2}(\boldsymbol{A}) &= \sum_{\boldsymbol{r}} \sum_{j,k \in \{0,1\}} \boldsymbol{e}_4^j \mathscr{F}_{f_1^j,f_2^k}(a_{\boldsymbol{r}}) \boldsymbol{e}_{\boldsymbol{r}} (\epsilon_4 \boldsymbol{e}_4 i_4)^k \\ &= \sum_{\boldsymbol{r}} \mathscr{F}_{cc}(a_{\boldsymbol{r}}) \boldsymbol{e}_{\boldsymbol{r}} + \mathscr{F}_{cs}(a_{\boldsymbol{r}}) \boldsymbol{e}_{\boldsymbol{r}} \epsilon_4 \boldsymbol{e}_4 i_4 + \boldsymbol{e}_4 \mathscr{F}_{sc}(a_{\boldsymbol{r}}) \boldsymbol{e}_{\boldsymbol{r}} \\ &\quad + \boldsymbol{e}_4 \mathscr{F}_{ss}(a_{\boldsymbol{r}}) \boldsymbol{e}_{\boldsymbol{r}} \epsilon_4 \boldsymbol{e}_4 i_4. \end{aligned} \tag{5.3.131}$$

5. *The Clifford Fourier transform for color images [18] with bivector B takes the form*

$$\mathscr{F}_{f_1,f_2,f_3,f_4}(\boldsymbol{A}) = \sum_{\boldsymbol{r}} \sum_{\boldsymbol{j},\boldsymbol{k}\in\{0,1\}^2} B^{j_1}(iB)^{j_2} \mathscr{F}_{f_1^{j_1},f_2^{j_2},f_3^{k_1},f_4^{j_3}}(\boldsymbol{a_r})\boldsymbol{e_r} B^{k_1}(iB)^{k_2}.$$

$$\begin{aligned}
=\sum_{\boldsymbol{r}} &\mathscr{F}_{cccc}(\boldsymbol{a_r})\boldsymbol{e_r} + \mathscr{F}_{cccs}(\boldsymbol{a_r})\boldsymbol{e_r} iB + \mathscr{F}_{ccsc}(\boldsymbol{a_r})\boldsymbol{e_r} B \\
&+ \mathscr{F}_{ccss}(\boldsymbol{a_r})\boldsymbol{e_r} BiB + iB\mathscr{F}_{cscc}(\boldsymbol{a_r})\boldsymbol{e_r} + iB\mathscr{F}_{cscs}(\boldsymbol{a_r})\boldsymbol{e_r} iB \\
&+ iB\mathscr{F}_{cssc}(\boldsymbol{a_r})\boldsymbol{e_r} B + iB\mathscr{F}_{csss}(\boldsymbol{a_r})\boldsymbol{e_r} BiB + B\mathscr{F}_{sccc}(\boldsymbol{a_r})\boldsymbol{e_r} \\
&+ B\mathscr{F}_{sccs}(\boldsymbol{a_r})\boldsymbol{e_r} iB + B\mathscr{F}_{scsc}(\boldsymbol{a_r})\boldsymbol{e_r} B + B\mathscr{F}_{scss}(\boldsymbol{a_r})\boldsymbol{e_r} BiB \\
&+ BiB\mathscr{F}_{sscc}(\boldsymbol{a_r})\boldsymbol{e_r} + BiB\mathscr{F}_{sscs}(\boldsymbol{a_r})\boldsymbol{e_r} iB + BiB\mathscr{F}_{sssc}(\boldsymbol{a_r})\boldsymbol{e_r} B \\
&+ BiB\mathscr{F}_{ssss}(\boldsymbol{a_r})\boldsymbol{e_r} BiB.
\end{aligned} \tag{5.3.132}$$

Please note that here expressions like $\mathscr{F}_{cccs}, \ldots$ refer to two-dimensional sine and cosine transforms with respect to the variables $x_1u_1 + x_2u_2$ each.

6. *The cylindrical Fourier transform [49] is not separable except for the case $n = 2$. Here, it satisfies*

$$\begin{aligned}
\mathscr{F}_{f_1}(\boldsymbol{A}) &= \sum_{\boldsymbol{r}} \sum_{j\in\{0,1\}} (\boldsymbol{e}_{12})^{j_l} \mathscr{F}_{f_1^j}(\boldsymbol{a_r})\boldsymbol{e_r} \\
&= \sum_{\boldsymbol{r}} \mathscr{F}_c(\boldsymbol{a_r})\boldsymbol{e_r} + \boldsymbol{e}_{12}\mathscr{F}_s(\boldsymbol{a_r})\boldsymbol{e_r},
\end{aligned} \tag{5.3.133}$$

with sine and cosine transforms of the variables $|\boldsymbol{x} \wedge \boldsymbol{u}|$. But for all other dimensions, no closed formula can be constructed in a similar way.

5.3.3.4 The explicitly invertible geometric Fourier transform

There is another important class of GFTs, which we want to take a closer look at.

Definition 5.3.51. *Let $F_1 := \{i_1, \ldots, i_\mu\}, F_2 := \{i_{\mu+1}, \ldots, i_m\}$ be two ordered finite sets of square roots of minus one, $i_k \in Cl(p,q), \forall k = 1, \ldots, m : i_k^2 = -1$, then we denote the transform*

$$\begin{aligned}
&\mathscr{F}_{F_1,F_2}(\boldsymbol{A})(\boldsymbol{u}) \\
&:= (2\pi)^{-\frac{m}{2}} \int_{\mathbb{R}^m} \left(\prod_{k=1}^{\mu} e^{-i_k x_k u_k}\right) \boldsymbol{A}(\boldsymbol{x}) \left(\prod_{k=\mu+1}^{m} e^{-i_k x_k u_k}\right) \mathrm{d}^m \boldsymbol{x}
\end{aligned} \tag{5.3.134}$$

of a function[28] $\boldsymbol{A} : \mathbb{R}^m \to Cl(p,q)$ by ***explicitly invertible geometric Fourier transform*** *(EIGFT).*

This definition is a special case of the general geometric Fourier transform from Definition 5.3.9 and of the separable GFT from the previous section. Please note that

[28]For Fourier transforms with $\boldsymbol{A} \in L^1(\mathbb{R}^m; Cl(p,q))$ and $\boldsymbol{A} \in L^2(\mathbb{R}^m; Cl(p,q))$, see Footnote 2 on page 144 regarding the roles of L^1 and L^2.

it also covers many transforms that can be brought into this form by renaming the components of $\boldsymbol{x}$ and $\boldsymbol{u}$ or by substitution of the integration variables. For example the factor (2π) can be removed from the front into the exponentials by substitution. We chose to use this shape because it has played a special role in [196] and [62], where this shape was used. This restricted version of Definition 5.3.9 is so popular because the additional claims guarantee that the inverse transform of any EIGFT is an EIGFT itself, namely

$$\mathscr{F}_{F_1,F_2}^{-1} = \mathscr{F}_{\{-i_\mu,\dots,-i_1\},\{-i_m,\dots,-i_{\mu+1}\}}. \tag{5.3.135}$$

Even though we do no know yet if the restrictions in Definition 5.3.51 are necessary to guarantee the existence of a GFT inverse, so far it is the only restrictions that we have. Therefore, this class of GFTs is very important because the applications of a transform that puts a function into a space from which it may never return are rather sparse. We were able to construct a bijective transform, that that does not satisfy the Definition 5.3.51. But if there exists a bijective GFT whose inverse is a GFT itself that does not satisfy Definition 5.3.51 is a matter of current research.

Exercise 5.3.52. *Our standard examples of geometric Fourier transforms form Example 5.3.10 satisfy the restricted definition of a EIGFT, except for the Clifford Fourier transform for color images [18] and the cylindrical Fourier transform [49].*

It is easy to show that the color image transform is not bijective if we look at real valued functions $a(\boldsymbol{x}) : \mathbb{R}^m \to \mathbb{R}$. Since they are in the center of the Clifford algebra, the exponentials commute with each other and

$$\begin{aligned}
\mathscr{F}_{f_1,f_2,f_3,f_4}(A)(\boldsymbol{u}) &= \int_{\mathbb{R}^2} e^{\frac{1}{2}(x_1u_1+x_2u_2)B} e^{\frac{1}{2}(x_1u_1+x_2u_2)iB} a(\boldsymbol{x}) e^{-\frac{1}{2}(x_1u_1+x_2u_2)iB} \\
&\quad e^{-\frac{1}{2}(x_1u_1+x_2u_2)B} \,\mathrm{d}^2\boldsymbol{x} \\
&= \int_{\mathbb{R}^2} a(\boldsymbol{x}) e^{\frac{1}{2}(x_1u_1+x_2u_2)B} e^{\frac{1}{2}(x_1u_1+x_2u_2)iB} e^{-\frac{1}{2}(x_1u_1+x_2u_2)iB} \\
&\quad e^{-\frac{1}{2}(x_1u_1+x_2u_2)B} \,\mathrm{d}^2\boldsymbol{x} \\
&= \int_{\mathbb{R}^2} a(\boldsymbol{x}) \,\mathrm{d}^2\boldsymbol{x}.
\end{aligned} \tag{5.3.136}$$

The functions are transformed to one constant real number, their integral. So, this GFT is clearly not bijective. Batard et al. [18] use the transform for vector valued functions only, for which it is invertible.

Whether or not the cylindrical Fourier transform for dimensions higher than two can be inverted is not that easy to decide.

Theorem 5.3.53 (EIGFT decomposition into classic TT)**.** *An explicitly invertible geometric Fourier transform of a multivector field is the sum of the composition of standard 1D sine and cosine transforms with constant multivector factors*

$$\begin{aligned}
\mathscr{F}_{F_1,F_2}(\boldsymbol{A})(\boldsymbol{u}) = &\sum_{\boldsymbol{r}} \sum_{\boldsymbol{j}\in\{0,1\}^m} \mathscr{F}_{(i_1x_1u_1)^{j_1}}\Big(\dots\mathscr{F}_{(i_mx_mu_m)^{j_m}}(a_{\boldsymbol{r}}(\boldsymbol{x}))(\boldsymbol{u})\Big)\Big) \\
&\prod_{l=1}^{\mu} i_l^{j_l} \boldsymbol{e_r} \prod_{l=\mu+1}^{m} i_l^{j_l}.
\end{aligned} \tag{5.3.137}$$

Proof. We start out with Theorem 5.3.47 .

$$\begin{aligned}\mathscr{F}_{F_1,F_2}(\boldsymbol{A})(\boldsymbol{u}) =&\sum_{\boldsymbol{r}} \sum_{\substack{\boldsymbol{j}\in\{0,1\}^{\mu},\\ \boldsymbol{k}\in\{0,1\}^{(\nu-\mu)}}} \mathscr{F}_{F_1^{\boldsymbol{j}},F_2^{\boldsymbol{k}}}(a_{\boldsymbol{r}})(\boldsymbol{u}) \prod_{l=1}^{\mu} i_l(\boldsymbol{u})^{j_l} \boldsymbol{e}_{\boldsymbol{r}} \prod_{l=\mu+1}^{\nu} i_l(\boldsymbol{u})^{k_l}\\ =&\sum_{\boldsymbol{r}} \sum_{\substack{\boldsymbol{j}\in\{0,1\}^{\mu},\\ \boldsymbol{k}\in\{0,1\}^{(\nu-\mu)}}} \int_{\mathbb{R}^m} \prod_{l=1}^{\mu} e_{j_l}^{-f_l(\boldsymbol{x},\boldsymbol{u})} a_{\boldsymbol{r}}(\boldsymbol{x}) \prod_{l=\mu+1}^{\nu} e_{k_l}^{-f_l(\boldsymbol{x},\boldsymbol{u})} \,\mathrm{d}^m\boldsymbol{x}\\ &\prod_{l=1}^{\mu} i_l(\boldsymbol{u})^{j_l} \boldsymbol{e}_{\boldsymbol{r}} \prod_{l=\mu+1}^{\nu} i_l(\boldsymbol{u})^{k_l}\end{aligned} \tag{5.3.138}$$

and insert the special shape of the EIGFT. Here $\boldsymbol{\nu} = m$ and the factor $(2\pi)^{-\frac{m}{2}}$ appears in front of the integral and not in the exponentials

$$\begin{aligned}\mathscr{F}_{F_1,F_2}(\boldsymbol{A})(\boldsymbol{u}) =&(2\pi)^{-\frac{m}{2}} \sum_{\boldsymbol{r}} \sum_{\substack{\boldsymbol{j}\in\{0,1\}^{\mu},\\ \boldsymbol{k}\in\{0,1\}^{(m-\mu)}}} \int_{\mathbb{R}^m} \prod_{l=1}^{\mu} e_{j_l}^{-f_l(\boldsymbol{x},\boldsymbol{u})} a_{\boldsymbol{r}}(\boldsymbol{x})\\ &\prod_{l=\mu+1}^{m} e_{k_l}^{-f_l(\boldsymbol{x},\boldsymbol{u})} \,\mathrm{d}^m\boldsymbol{x} \prod_{l=1}^{\mu} i_l^{j_l} \boldsymbol{e}_{\boldsymbol{r}} \prod_{l=\mu+1}^{m} i_l^{k_l}.\end{aligned} \tag{5.3.139}$$

Comprehension of the multi-indices $\boldsymbol{j}$ and $\boldsymbol{k}$ into just one multi-index $\boldsymbol{j}$ and integration over each of the coordinates of $\boldsymbol{x}$ separately reveals a composition of one-dimensional transforms

$$\begin{aligned}\mathscr{F}_{F_1,F_2}(\boldsymbol{A})(\boldsymbol{u}) =&(2\pi)^{-\frac{m}{2}} \sum_{\boldsymbol{r}} \sum_{\boldsymbol{j}\in\{0,1\}^m} \int_{\mathbb{R}^m} \prod_{l=1}^{m} e_{j_l}^{-f_l(\boldsymbol{x},\boldsymbol{u})} a_{\boldsymbol{r}}(\boldsymbol{x}) \,\mathrm{d}^m\boldsymbol{x} \prod_{l=1}^{\mu} i_l^{j_l} \boldsymbol{e}_{\boldsymbol{r}}\\ &\prod_{l=\mu+1}^{m} i_l^{j_l}\\ =&(2\pi)^{-\frac{m}{2}} \sum_{\boldsymbol{r}} \sum_{\boldsymbol{j}\in\{0,1\}^m} \int_{\mathbb{R}} e_{j_1}^{-i_1 x_1 u_1} \dots \int_{\mathbb{R}} e_{j_m}^{-i_m x_m u_m} a_{\boldsymbol{r}}(\boldsymbol{x})\\ &\mathrm{d}x_m \dots \mathrm{d}x_1 \prod_{l=1}^{\mu} i_l^{j_l} \boldsymbol{e}_{\boldsymbol{r}} \prod_{l=\mu+1}^{m} i_l^{j_l}\\ =&\sum_{\boldsymbol{r}} \sum_{\boldsymbol{j}\in\{0,1\}^m} \mathscr{F}_{(i_1 x_1 u_1)^{j_1}} \Big(\dots \mathscr{F}_{(i_m x_m u_m)^{j_m}} (a_{\boldsymbol{r}}(\boldsymbol{x}))(\boldsymbol{u}) \Big)\Big)\\ &\prod_{l=1}^{\mu} i_l^{j_l} \boldsymbol{e}_{\boldsymbol{r}} \prod_{l=\mu+1}^{m} i_l^{j_l}.\end{aligned} \tag{5.3.140}$$

Depending on whether the multi-indices take the value one or zero, each appearing transform is now a classical one-dimensional sine transform $\mathscr{F}_s$ or cosine transform $\mathscr{F}_c$ of a real valued function. $\square$

This decomposition directly shows how a classical FFT algorithm can be applied to efficiently calculate each of the EIGFTs.

Exercise 5.3.54. *Theorem 5.3.53 can be applied to most of the standard examples from Example 5.3.10.*

1. *The decomposition into the 1D sine and cosine transforms for the Clifford Fourier transform from [119, 185, 222] can be written as*

$$\begin{aligned}\mathscr{F}_{f_1}(\boldsymbol{A}) &= \int_{\mathbb{R}^n} \boldsymbol{A}(\boldsymbol{x}) e^{2\pi i_n \boldsymbol{x}\cdot\boldsymbol{u}}\, \mathrm{d}^n\boldsymbol{x} \\ &= \int_{\mathbb{R}^n} \boldsymbol{A}(\boldsymbol{x}) \prod_{j=1}^{n} e^{2\pi i_n x_j u_j}\, \mathrm{d}^n\boldsymbol{x} \\ &= \sum_{\boldsymbol{r}} \sum_{j\in\{0,1\}^n} \mathscr{F}_{(i_n x_1 u_1)^{j_1}} \ldots \mathscr{F}_{(i_n x_n u_n)^{j_n}} (a_{\boldsymbol{r}}) \boldsymbol{e}_{\boldsymbol{r}} i_n^{|j|}.\end{aligned} \tag{5.3.141}$$

But obviously, the equivalent decomposition into higher dimensional sine and cosine transforms like in Example 5.3.50 is much more beautiful.

2. *The formula for the the Sommen Fourier transform was already shown in Example 5.3.50.*

3. *The formula for the quaternionic Fourier transform was already shown in Example 5.3.50.*

4. *The formula for the spacetime Fourier transform was already shown in Example 5.3.50.*

5. *The color image transform does not suffice the constraints of Theorem 5.3.53. It can only be decomposed as shown in Example 5.3.50.*

6. *The shape for the cylindrical Fourier transform with dimension $n = 2$ can be given using a little trick.*

$$\begin{aligned}\mathscr{F}_{f_1}(\boldsymbol{A})(\boldsymbol{u}) &= \int_{\mathbb{R}^2} A(\boldsymbol{x}) e^{\boldsymbol{x}\wedge\boldsymbol{u}}\, \mathrm{d}^2\boldsymbol{x} \\ &= \int_{\mathbb{R}}\int_{\mathbb{R}} A(x_1, x_2) e^{\boldsymbol{e}_{12}(x_1 u_2 - x_2 u_1)}\, \mathrm{d}x_1\, \mathrm{d}x_2.\end{aligned} \tag{5.3.142}$$

We define the vector $\boldsymbol{u}^* = \boldsymbol{u}\boldsymbol{e}_{12}^{-1} = \boldsymbol{u}(-\boldsymbol{e}_{12}) = \boldsymbol{e}_{12}\boldsymbol{u} = u_2\boldsymbol{e}_1 - u_1\boldsymbol{e}_2$, which satisfies $x \cdot \boldsymbol{u}^* = x_1 u_2 - x_2 u_1$ and get

$$\begin{aligned}\mathscr{F}_{f_1}(\boldsymbol{A})(\boldsymbol{u}) &= \int_{\mathbb{R}}\int_{\mathbb{R}} A(x_1, x_2) e^{\boldsymbol{e}_{12} x\cdot\boldsymbol{u}^*}\, \mathrm{d}x_1\, \mathrm{d}x_2 \\ &= \sum_{\boldsymbol{r}} \mathscr{F}_c(a_{\boldsymbol{r}})(\boldsymbol{u}^*)\boldsymbol{e}_{\boldsymbol{r}} + \mathscr{F}_s(a_{\boldsymbol{r}})(\boldsymbol{u}^*)\boldsymbol{e}_{\boldsymbol{r}}\boldsymbol{e}_{12} \\ &= \sum_{\boldsymbol{r}} \mathscr{F}_c(a_{\boldsymbol{r}})(\boldsymbol{e}_{12}\boldsymbol{u})\boldsymbol{e}_{\boldsymbol{r}} + \mathscr{F}_s(a_{\boldsymbol{r}})(\boldsymbol{e}_{12}\boldsymbol{u})\boldsymbol{e}_{\boldsymbol{r}}\boldsymbol{e}_{12}.\end{aligned} \tag{5.3.143}$$

5.3.3.5 Summary on GFT and TT with multiple square roots of -1 in $Cl(p, q)$

In this section, we have studied separable geometric Fourier transforms. This has two big consequences.

The decomposition of separable geometric Fourier transforms (GFTs) into real valued trigonometric transforms in Theorem 5.3.47 deepens their understanding. Even though the definition of the GFT covers more than the separable transforms, these are the most popular ones. When it comes to real applications, even the stronger restriction to EIGFTs is essential. For them, the decomposition into classical one-dimensional, real-valued sine transforms and cosine transforms with constant multi-vector factors in Theorem 5.3.53 displays their ease of use.

This insight also makes geometric Fourier transforms easier to understand, analyze and implement. One advantage is that the decomposition into one-dimensional sine and cosine transforms directly shows how a classical FFT algorithm can be applied to efficiently calculate the transforms.

5.4 CLIFFORD FOURIER TRANSFORM AND CONVOLUTION

In this section, we continue to explore CFTs and convolution theory.

5.4.1 One-sided CFT and convolution

5.4.1.1 Background and motivation

In this section, we use the general steerable one-sided Clifford Fourier transform (CFT), and relate the classical convolution of Clifford algebra-valued signals over $\mathbb{R}^{p,q}$ with the (equally steerable) Mustard convolution. A Mustard convolution can be expressed in the spectral domain as the point wise product of the CFTs of the factor functions. In full generality do we express the classical convolution of Clifford algebra signals in terms of a linear combination of Mustard convolutions, and vice versa the Mustard convolution of Clifford algebra signals in terms of a linear combination of classical convolutions. This section is based on [217].

The steerable one-sided Clifford Fourier transformation (CFT) was introduced in [201], compare Section 5.2. It generalizes related transforms, like the classical complex Fourier transform, the one-sided single kernel quaternion Fourier transform [129], and the Clifford Fourier transforms with pseudoscalar kernels [117, 185] to higher dimensions. These CFTs essentially replace the imaginary unit $i \in \mathbb{C}$ by a general multivector square root of -1, which usually populate continuous Clifford algebra submanifolds [193,203]. The classical complex Fourier transform needs only one fully commuting kernel factor, due to the commutativity of complex numbers. To have a non-commutative kernel factor under the transform integral on one side of the signal function is meaningful due to the inherent non-commutativity in Clifford algebras. An extensive discussion of the historical development and the application relevance of the CFTs can be found in [51] and [207], see also Chapter 1.

A key strength of the classical complex Fourier transform is its easy and fast application to filtering problems. The convolution of a signal with its filter function becomes in the spectral domain a point wise product of the respective Fourier transformations. This is generally not the case for the convolution of two Clifford algebra-valued signals (Clifford signals) over $\mathbb{R}^{p,q}$, due to Clifford algebra non-commutativity. Yet it is possible to define from the point wise product of the CFTs of two Clifford signals a new type of steerable convolution, called Mustard convolution [62, 270]. This Mustard convolution can be expressed in terms of sums of classical convolutions and vice versa. For the left-sided QFT this has been carried out in [97], for the two-sided QFT in [213], for the space-time Fourier transform in [218] and for the two-sided CFT in [215]. Here we extend this approach in full generality to the steerable one-sided CFT for signal functions which map non-degenerate quadratic form vector spaces to Clifford algebras in all dimensions.

This treatment is organized as follows. We first briefly explain notions of Clifford algebra valued multivector signal functions. Then, we give some background on the steerable one-sided CFT and recalls some of its properties. Section 5.4.1.2 defines the classical convolution of two Clifford signal functions, as well as the steerable Mustard convolution. The rest of the section is devoted to representing the classical convolution in terms of a sum of Mustard convolutions (Theorem 5.4.7) and dually to expressing the Mustard convolution in terms of a sum of classical convolutions (Theorem 5.4.8). Furthermore direct single convolution product identities between classical and Mustard convolutions are established (Theorem 5.4.10), together with the theoretical equivalence (for general Clifford signal convolution product factor functions) of expressing the classical convolution in terms of the Mustard convolution and the reverse (Equation (5.4.27)).

In this treatment, we use the *principal reverse* of Section 2.1.1.6.

With respect to any square root $i \in Cl(p,q)$ of -1, $i^2 = -1$, every multivector $A \in Cl(p,q)$ can be split into *commuting* A_{+i} and *anticommuting* A_{-i} parts, see Lemma 2.7.5 and [203].

A multivector valued function $h : \mathbb{R}^{p,q} \to C\ell(p',q')$, has $2^{n'}$ blade components, $n' = p' + q'$ $(h_A : \mathbb{R}^{p,q} \to \mathbb{R})$

$$h(\boldsymbol{x}) = \sum_A h_A(\boldsymbol{x})\boldsymbol{e}_A. \tag{5.4.1}$$

We define the *inner product* of two functions $h, m : \mathbb{R}^{p,q} \to C\ell(p',q')$ by

$$(h,m) = \int_{\mathbb{R}^{p,q}} h(\boldsymbol{x})\widetilde{m(\boldsymbol{x})}\, d^n\boldsymbol{x} = \sum_{A,B} \boldsymbol{e}_A\widetilde{\boldsymbol{e}_B} \int_{\mathbb{R}^{p,q}} h_A(\boldsymbol{x})m_B(\boldsymbol{x})\, d^n\boldsymbol{x}, \tag{5.4.2}$$

with the *symmetric scalar part*

$$\langle h,m\rangle = \int_{\mathbb{R}^{p,q}} h(\boldsymbol{x}) * \widetilde{m(\boldsymbol{x})}\, d^n\boldsymbol{x} = \sum_A \int_{\mathbb{R}^{p,q}} h_A(\boldsymbol{x})m_A(\boldsymbol{x})\, d^n\boldsymbol{x}, \tag{5.4.3}$$

and the $L^2(\mathbb{R}^{p,q}; C\ell(p',q'))$*-norm*

$$\|h\|^2 = \langle(h,h)\rangle = \int_{\mathbb{R}^{p,q}} |h(\boldsymbol{x})|^2 d^n\boldsymbol{x} = \sum_A \int_{\mathbb{R}^{p,q}} h_A^2(\boldsymbol{x})\, d^n\boldsymbol{x}, \tag{5.4.4}$$

$$L^2(\mathbb{R}^{p,q}; C\ell(p',q')) = \{h : \mathbb{R}^{p,q} \to C\ell(p',q') \mid \|h\| < \infty\}. \tag{5.4.5}$$

Notation 5.4.1 (Argument reflection). *For a function* $h : \mathbb{R}^{p,q} \to Cl(p', q')$ *we set*[29]

$$h^1(\boldsymbol{x}) := h(-\boldsymbol{x}). \tag{5.4.6}$$

Note that we obviously have

$$(h^1)^1(\boldsymbol{x}) = h^1(-\boldsymbol{x}) = h(\boldsymbol{x}). \tag{5.4.7}$$

The definition we will use is *more general* than Definition 3.1 given in [201], because we generalize to multivector signal functions in $L^1(\mathbb{R}^{p,q}; C\ell(p', q'))$ and not only in $L^1(\mathbb{R}^{p,q}; C\ell(p, q))$.

Definition 5.4.2 (Steerable CFT with respect to one square root of -1). *Let* $i \in C\ell(p', q')$, $i^2 = -1$, *be any square root of* -1. *The general Clifford Fourier transform*[30] *(CFT) of* $f \in L^1(\mathbb{R}^{p,q}; C\ell(p', q'))$, *with respect to* i *is*

$$\mathcal{F}^i\{f\}(\boldsymbol{\omega}) = \int_{\mathbb{R}^{p,q}} f(\boldsymbol{x})\, e^{-iu(\boldsymbol{x},\boldsymbol{\omega})} d^n\boldsymbol{x}, \tag{5.4.8}$$

where $d^n\boldsymbol{x} = dx_1 \dots dx_n$, $\boldsymbol{x}, \boldsymbol{\omega} \in \mathbb{R}^{p,q}$, *and* $u : \mathbb{R}^{p,q} \times \mathbb{R}^{p,q} \to \mathbb{R}$.

Since square roots of -1 in $C\ell(p', q')$ populate *continuous submanifolds* in $C\ell(p', q')$, the CFT of Definition 5.4.2 is generically *steerable* within these manifolds, see (5.4.10). In Definition 5.4.2, the square roots $i \in C\ell(p', q')$ of -1 may be from any component of any conjugacy class. The choice of the Clifford's geometric product between multivector signal function f and the multivector kernel $e^{-iu(\boldsymbol{x},\boldsymbol{\omega})}$, in the integrand of (5.4.8) is very important. Because only this choice allowed, e.g. in [117], to define and apply a holistic vector field convolution, without loss of information.

Note that two-sided CFTs can be decomposed to pairs of one-sided CFTs [209].

Remark 5.4.3. *In order to avoid clutter we often drop the upper index* i *as in* $\mathcal{F}\{h\} = \mathcal{F}^i\{h\}$, *but in principle the one-sided CFT always depends on the particular choice* i *of the multivector square root of* -1. *Since square roots of* -1 *in* $C\ell(p', q')$ *populate continuous submanifolds in* $C\ell(p', q')$, *the CFT of Definition 5.4.2 is generically steerable within these submanifolds. In Definition 5.4.2, the square root* $i \in C\ell(p', q')$ *of* -1, *may be from any conjugacy class and component, respectively.*

Within the same conjugacy class of square roots of -1 the CFTs of Definition 5.4.2 are related by the following equation, and therefore steerable. Let $i, i' \in Cl(p', q')$ be any two square roots of -1 in the same conjugacy class, i.e. $i' = a^{-1}ia$, $a \in Cl(p', q')$, a being invertible. As a consequence of this relationship we also have

$$e^{-i'u} = a^{-1}e^{-iu}a, \quad \forall u \in \mathbb{R}. \tag{5.4.9}$$

[29] We are aware that this notation could be confused with an ordinary taking to the power of 1, but as will be seen in the current context no danger of confusion is likely to arise.

[30] For Fourier transforms with $f \in L^1(\mathbb{R}^{p,q}; C\ell(p', q'))$ and $f \in L^2(\mathbb{R}^{p,q}; C\ell(p', q'))$, see Footnote 2 on page 144 regarding the roles of L^1 and L^2.

This in turn leads to the following *steerability relationship* of all CFTs with square roots of -1 from the same conjugacy class:

$$\mathcal{F}^{i'}\{h\}(\boldsymbol{\omega}) = \mathcal{F}^{i}\{ha^{-1}\}(\boldsymbol{\omega})a, \tag{5.4.10}$$

where ha^{-1} means to multiply the signal function h by the constant multivector $a^{-1} \in Cl(p', q')$.

For establishing an *inversion* formula and other properties of the CFT in Definition 5.4.2, certain *assumptions* about the phase function $u(\boldsymbol{x}, \boldsymbol{\omega})$ need to be made. In principle these assumptions could be made based on the desired properties of the resulting CFT. One possibility is, e.g., to assume

$$u(\boldsymbol{x}, \boldsymbol{\omega}) = \boldsymbol{x} * \widetilde{\boldsymbol{\omega}} = \sum_{l=1}^{n} x^l \omega^l = \sum_{l=1}^{n} x_l \omega_l, \tag{5.4.11}$$

which will be assumed in the rest of this paper.

We then get the following *inversion* theorem.[31]

Theorem 5.4.4 (Inversion of one-sided CFT). *For $\mathcal{F}^i\{h\} \in L^1(\mathbb{R}^{p,q}; C\ell(p', q'))$ we have*

$$h(\boldsymbol{x}) = \mathcal{F}^{i}_{-1}\{\mathcal{F}^{i}\{h\}\}(\boldsymbol{x}) = \frac{1}{(2\pi)^n} \int_{\mathbb{R}^n} \mathcal{F}^{i}\{h\}(\boldsymbol{\omega})\, e^{iu(\boldsymbol{x},\boldsymbol{\omega})} d^n\boldsymbol{\omega}, \tag{5.4.12}$$

where $d^n\boldsymbol{\omega} = d\omega_1 \dots d\omega_n$, $\boldsymbol{x}, \boldsymbol{\omega} \in \mathbb{R}^{p,q}$.

The proof of Theorem 5.4.4 is strictly analogous to the proof of equation (4.8) on page 231 of [201], and therefore left as an exercise to the reader.

We further note the following useful relationship using the argument reflection of Notation 5.4.1

$$\mathcal{F}^{-i}\{h\} = \mathcal{F}^{i}\{h^1\} = \mathcal{F}\{h^1\}. \tag{5.4.13}$$

The main properties of the CFT of Definition 5.4.2 have been studied for the special case of multivector signal functions $f \in L^1(\mathbb{R}^{p,q}; C\ell(p, q))$ in detail in [201], and can easily be generalized to the more general case of $f \in L^1(\mathbb{R}^{p,q}; C\ell(p', q'))$.

5.4.1.2 *Convolution and steerable Mustard convolution*

We define the *convolution* of two Clifford (algebra) signals $a, b \in L^1(\mathbb{R}^{p,q};\ Cl(p', q'))$ as

$$(a \star b)(\boldsymbol{x}) = \int_{\mathbb{R}^n} a(\boldsymbol{y}) b(\boldsymbol{x} - \boldsymbol{y}) d^2\boldsymbol{y}, \tag{5.4.14}$$

provided that the integral exists.

---

[31] Note, that we show the inversion symbol -1 as lower index in $\mathcal{F}^{i}_{-1}$, in order to avoid a possible confusion by using two upper indice. The inversion could also be written with the help of the CFT itself as $\mathcal{F}^{i}_{-1} = \frac{1}{(2\pi)^n}\mathcal{F}^{-i}$.

Note that the real continuous Clifford geometric algebra wavelet transform can be written as a convolution of the multivector signal function with the daughter wavelet (a rotated, dilated and translated mother wavelet), essentially evaluated at the center of the daughter wavelet, see [188].

The *Mustard* convolution [62,270] of two Clifford signals $a, b \in L^1(\mathbb{R}^{p,q}; Cl(p', q'))$ is defined as

$$(a \star_M b)(\boldsymbol{x}) = (\mathcal{F}^i)^{-1}(\mathcal{F}^i\{a\}\mathcal{F}^i\{b\})(\boldsymbol{x}), \tag{5.4.15}$$

provided that the integral exists.

Remark 5.4.5. *The Mustard convolution has the conceptual and computational advantage to simply yield, independent of the particular Clifford algebra $Cl(p', q')$ involved and of the particular multivector square root of -1 in the CFT kernel, as spectrum in the CFT Fourier domain the point wise product of the CFTs of the two signals, just as for the classical complex Fourier transform. On the other hand, by its very definition, the Mustard convolution itself depends on the choice of i, i.e. of the multivector square root of -1, used in the Definition 5.4.2 of the CFT. The Mustard convolution* (5.4.15) *is therefore a steerable operator, dependent on the choice of i.*

In the following two subsections, we will express the convolution (5.4.14) in terms of the Mustard convolution (5.4.15), and vice versa, and study the mutual relations of these expressions.

5.4.1.2.1 Expressing the convolution in terms of the Mustard convolution In this section we assume the use of the one-sided CFT with a general multivector square roots of -1, $i \in Cl(p', q')$. The definition of the classical convolution (5.4.14) is independent of the application of a CFT. The Mustard convolution of (5.4.15) depends on the definition of the CFT and in particular on the choice of the multivector square root i of -1.

In our approach we generalize equation (4.17) on page 233 of [201], which expresses the convolution of two Clifford signal functions in the Clifford Fourier domain with the help of the CFT of Definition 5.4.2. We generalize this equation to the case of multivector signal functions $a, b \in L^1(\mathbb{R}^{p,q}; Cl(p', q'))$, and to the CFT of Definition 5.4.2. Nevertheless the proof works perfectly analogous to the one given in [201], we therefore leave this as an exercise to the reader.

Theorem 5.4.6 (CFT of convolution). *We assume that the function u is linear with respect to its first argument. The CFT of the convolution* (5.4.14) *of two multivector signals $a, b \in L^1(\mathbb{R}^{p,q}; Cl(p', q')$ can then be expressed as*

$$\begin{aligned}\mathcal{F}^i\{a \star b\} &= \mathcal{F}^{-i}\{a\}\mathcal{F}^i\{b_{-i}\} + \mathcal{F}^i\{a\}\mathcal{F}^i\{b_{+i}\} \\ &= \mathcal{F}^i\{a^1\}\mathcal{F}^i\{b_{-i}\} + \mathcal{F}^i\{a\}\mathcal{F}^i\{b_{+i}\}.\end{aligned} \tag{5.4.16}$$

We can now easily express the convolution of two multivector signals $\mathcal{F}^i\{a \star b\}(\boldsymbol{\omega})$ in terms of only two Mustard convolutions (5.4.15), by applying the inverse CFT.

Theorem 5.4.7 (Convolution in terms of Mustard convolution). *Assuming a general multivector square root i of -1, the convolution (5.4.14) of two Clifford functions $a, b \in L^1(\mathbb{R}^{p,q}; Cl(p', q'))$ can be expressed in terms of two Mustard convolutions (5.4.15) as*

$$a \star b = a^1 \star_M b_{-i} + a \star_M b_{+i}. \tag{5.4.17}$$

5.4.1.2.2 Expressing the Mustard convolution in terms of the convolution Now we will first simply write out the Mustard convolution (5.4.15) and simplify it until only standard convolutions (5.4.14) remain.

We begin by writing the Mustard convolution (5.4.15) of two multivector functions $a, b \in L^2(\mathbb{R}^{p,q}; Cl(p', q'))$

$$\begin{aligned}
&a \star_M b(\boldsymbol{x}) \\
&= \frac{1}{(2\pi)^n} \int_{\mathbb{R}^{p,q}} \mathcal{F}\{a\}(\boldsymbol{\omega}) \mathcal{F}\{b\}(\boldsymbol{\omega}) e^{iu(\boldsymbol{x},\boldsymbol{\omega})} d^n\boldsymbol{\omega} \\
&= \frac{1}{(2\pi)^n} \int_{\mathbb{R}^{p,q}} \int_{\mathbb{R}^{p,q}} a(\boldsymbol{y}) e^{-iu(\boldsymbol{y},\boldsymbol{\omega})} d^n\boldsymbol{y} \int_{\mathbb{R}^{p,q}} b(\boldsymbol{z}) e^{-iu(\boldsymbol{z},\boldsymbol{\omega})} d^n\boldsymbol{z} e^{iu(\boldsymbol{x},\boldsymbol{\omega})} d^n\boldsymbol{\omega} \\
&= \frac{1}{(2\pi)^n} \iiint a(\boldsymbol{y}) e^{-iu(\boldsymbol{y},\boldsymbol{\omega})} [b_{+i}(\boldsymbol{z}) + b_{-i}(\boldsymbol{z})] e^{iu(\boldsymbol{x}-\boldsymbol{z},\boldsymbol{\omega})} d^n\boldsymbol{y} d^n\boldsymbol{z} d^n\boldsymbol{\omega} \\
&= \frac{1}{(2\pi)^n} \iiint a(\boldsymbol{y}) b_{+i}(\boldsymbol{z}) e^{-iu(\boldsymbol{y},\boldsymbol{\omega})} e^{iu(\boldsymbol{x}-\boldsymbol{z},\boldsymbol{\omega})} d^n\boldsymbol{y} d^n\boldsymbol{z} d^n\boldsymbol{\omega} \\
&\quad + \frac{1}{(2\pi)^n} \iiint a(\boldsymbol{y}) b_{-i}(\boldsymbol{z}) e^{iu(\boldsymbol{y},\boldsymbol{\omega})} e^{iu(\boldsymbol{x}-\boldsymbol{z},\boldsymbol{\omega})} d^n\boldsymbol{y} d^n\boldsymbol{z} d^n\boldsymbol{\omega} \\
&= \frac{1}{(2\pi)^n} \iiint a(\boldsymbol{y}) b_{+i}(\boldsymbol{z}) e^{iu(\boldsymbol{x}-\boldsymbol{y}-\boldsymbol{z},\boldsymbol{\omega})} d^n\boldsymbol{y} d^n\boldsymbol{z} d^n\boldsymbol{\omega} \\
&\quad + \frac{1}{(2\pi)^n} \iiint a(\boldsymbol{y}) b_{-i}(\boldsymbol{z}) e^{iu(\boldsymbol{x}+\boldsymbol{y}-\boldsymbol{z},\boldsymbol{\omega})} d^n\boldsymbol{y} d^n\boldsymbol{z} d^n\boldsymbol{\omega} \\
&= \iint a(\boldsymbol{y}) b_{+i}(\boldsymbol{z}) \delta(\boldsymbol{x} - \boldsymbol{y} - \boldsymbol{z}) d^n\boldsymbol{y} d^n\boldsymbol{z} \\
&\quad + \iint a(\boldsymbol{y}) b_{-i}(\boldsymbol{z}) \delta(\boldsymbol{x} + \boldsymbol{y} - \boldsymbol{z}) d^n\boldsymbol{y} d^n\boldsymbol{z} \\
&= \int_{\mathbb{R}^{p,q}} a(\boldsymbol{y}) b_{+i}(\boldsymbol{x} - \boldsymbol{y}) d^n\boldsymbol{y} + \int_{\mathbb{R}^{p,q}} a(\boldsymbol{y}) b_{-i}(\boldsymbol{x} + \boldsymbol{y}) d^n\boldsymbol{y} \\
&= \int_{\mathbb{R}^{p,q}} a(\boldsymbol{y}) b_{+i}(\boldsymbol{x} - \boldsymbol{y}) d^n\boldsymbol{y} + \int_{\mathbb{R}^{p,q}} a(\boldsymbol{y}) b_{-i}(-(-\boldsymbol{x} - \boldsymbol{y})) d^n\boldsymbol{y} \\
&= a \star b_{+i}(\boldsymbol{x}) + a \star b^1_{-i}(-\boldsymbol{x}) \\
&= a \star b_{+i}(\boldsymbol{x}) + a^1 \star b_{-i}(\boldsymbol{x}).
\end{aligned} \tag{5.4.18}$$

We have abbreviated $\int_{\mathbb{R}^{p,q}} \int_{\mathbb{R}^{p,q}}$ to $\iint$ and $\int_{\mathbb{R}^{p,q}} \int_{\mathbb{R}^{p,q}} \int_{\mathbb{R}^{p,q}}$ to $\iiint$. For the third equality we applied the split of Lemma 2.7.5 to $b(\boldsymbol{x})$ and used the linearity of u with respect to its first argument. For the fourth equality we used the linearity of Clifford's geometric product, the linearity of the triple integral, and we used the commutation and anti-commutation properties of $b_{\pm i}(\boldsymbol{x})$ with the multivector square root $i \in C\ell(p', q')$, which produces the sign change $e^{-iu(\boldsymbol{y},\boldsymbol{\omega})} \to e^{+iu(\boldsymbol{y},\boldsymbol{\omega})}$ in the case of anti-commutation.

For the fifth equality we again applied the linearity of u with respect to its first argument. The integrations $\frac{1}{(2\pi)^n}\int e^{iu(\boldsymbol{x}\pm\boldsymbol{y}-\boldsymbol{z},\boldsymbol{\omega})}d^n\boldsymbol{\omega}$ produce the n-dimensional Dirac delta functions $\delta(\boldsymbol{x}\pm\boldsymbol{y}-\boldsymbol{z})$, giving the sixth equality.
We illustrate the last identity of (5.4.18), $a\star b^1_{-i}(-\boldsymbol{x}) = a^1\star b_{-i}(\boldsymbol{x})$, in the one-dimensional case $\mathbb{R}^{p,q}=\mathbb{R}$, the generalization to $\mathbb{R}^{p,q}$ is then straightforward

$$\begin{aligned} a\star b^1(-x) &= \int_{\mathbb{R}} a(y)b(-(-x-y))dy = \int_{-\infty}^{+\infty} a(y)b(x+y)dy \\ &= \int_{+\infty}^{-\infty} a(-g)b(x-g)(-1)dg = \int_{-\infty}^{+\infty} a(-g)b(x-g)dg \\ &= \int_{\mathbb{R}} a^1(g)b(x-g)dg = a^1\star b(x). \end{aligned} \tag{5.4.19}$$

where we have substituted $g=-y$, $dg=-dy$, including substitution of the integration boundaries for the third equality. The interchange of the integration boundaries eliminates the overall minus sign in the fourth equality of (5.4.19).

Note that in (5.4.18), $a\star b^1_{-i}(-\boldsymbol{x})$, means to first apply the convolution to the pair of functions a and b^1_{-i}, and only then to evaluate the result of the convolution integral with the argument $(-\boldsymbol{x})$. So in general $a\star b^1_{-i}(-\boldsymbol{x}) \neq a\star b_{-i}(+\boldsymbol{x})$.

We finally obtain the desired decomposition of the Mustard convolution (5.4.15) in terms of the classical convolution.

Theorem 5.4.8 (Mustard convolution in terms of standard convolution). *The Mustard convolution* (5.4.15) *of two multivector signal functions* $a,b\in L^1(\mathbb{R}^{p,q};$ $Cl(p',q'))$ *can be expressed in terms of two standard convolutions* (5.4.14) *as*

$$a\star_M b(\boldsymbol{x}) = a^1\star b_{-i}(\boldsymbol{x}) + a\star b_{+i}(\boldsymbol{x}). \tag{5.4.20}$$

Remark 5.4.9 (Theorem duality). *Comparing Theorems 5.4.7 and 5.4.8 we notice an interesting duality: interchanging convolution and Mustard convolution in either theorem yields the other, independent over which vector space* $\mathbb{R}^{p,q}$ *the multivector signals are defined, independent from the signal value Clifford algebra* $C\ell(p',q')$ *and independent from the particular choice of multivector square root of* -1, $i\in C\ell(p',q')$. *The last form of independence also means, that the observed duality is stable with respect to steering the CFT and the Mustard convolution by changing* $i\in C\ell(p',q')$. *Note further, that a corresponding duality will be valid for the left-sided version of the CFT in Definition 5.4.2, by placing the kernel factor on the left side and going analogously through all arguments up to Theorem 5.4.8.*

Yet, it is an interesting non-trivial question, whether a similar duality may hold for other forms of the CFT, e.g. with more than one kernel factor, see e.g. [209, 215].

5.4.1.2.3 Single convolution product identities for classical and Mustard convolutions

Let us now apply Theorem 5.4.7 to the three functions $a, b_{\pm i}$, observing that

$$(b_{+i})_{-i} = (b_{-i})_{+i} = 0, \quad (b_{+i})_{+i} = b_{+i}, \quad (b_{-i})_{-i} = b_{-i}. \tag{5.4.21}$$

Then we obtain

$$a \star b_{+i} = a^1 \star_M (b_{+i})_{-i} + a \star_M (b_{+i})_{+i} = 0 + a \star_M b_{+i} = a \star_M b_{+i}, \tag{5.4.22}$$

and similarly,

$$a \star b_{-i} = a^1 \star_M b_{-i} \quad \Longleftrightarrow \quad a^1 \star b_{-i} = a \star_M b_{-i}, \tag{5.4.23}$$

since double reflection of the argument returns the function itself (5.4.7). Note, that the very same identities are easily obtained by analogously applying Theorem 5.4.8 to $a, b_{\pm i}$. We therefore summarize them in the following theorem.

Theorem 5.4.10 (Partial identities between convolutions and Mustard convolutions). *For pairs of functions* (a, b_{-i}) *and* (a, b_{+i}) *with* $a, b_{\pm i} \in L^1(\mathbb{R}^{p,q}; C\ell(p', q'))$, *where the second factor either commutes or anti-commutes with the multivector square root of* -1, $i \in C\ell(p', q')$ *of the Definition 5.4.2, the following convolution product identities between convolution* (5.4.14) *and Mustard convolution* (5.4.15) *hold*

$$\begin{aligned} a^1 \star b_{-i} = a \star_M b_{-i} \quad &\Longleftrightarrow \quad a \star b_{-i} = a^1 \star_M b_{-i}, \\ a \star b_{+i} &= a \star_M b_{+i}. \end{aligned} \tag{5.4.24}$$

Theorem 5.4.10 can therefore either be derived from Theorem 5.4.7 or from Theorem 5.4.8. Moreover, Theorem 5.4.10 can also be established independently by *direct* computation. Then adding two convolution terms would give the Mustard convolution

$$a^1 \star b_{-i} + a \star b_{+i} \overset{\text{Th. 5.4.10}}{=} a \star_M b_{-i} + a \star_M b_{+i} = a \star_M b. \tag{5.4.25}$$

And conversely adding two Mustard convolution terms would give the convolution

$$a^1 \star_M b_{-i} + a \star_M b_{+i} \overset{\text{Th. 5.4.10}}{=} a \star b_{-i} + a \star b_{+i} = a \star b. \tag{5.4.26}$$

This establishes the following important *threefold theorem equivalence*

$$\text{Theorem 5.4.7} \quad \Longleftrightarrow \quad \text{Theorem 5.4.10} \quad \Longleftrightarrow \quad \text{Theorem 5.4.8}. \tag{5.4.27}$$

Remark 5.4.11. *Note that the need to always decompose the right convolution product factor function* $b = b_{-i} + b_{+i}$ *is manifestly due to the kernel in Definition 5.4.2 being placed on the right side. Using a corresponding left side kernel CFT, would lead to analogous results with decomposing the left convolution product factor* $a = a_{-i} + a_{+i}$.

Furthermore, we can ask under what conditions we get a *full direct single convolution product identity* of the two convolution products $a \star b = a \star_M b$? This identity holds under any of the following conditions:

1. For all functions $a, b \in L^1(\mathbb{R}^{p,q}; C\ell(p', q'))$, with $b_{-i} \equiv 0$. This condition depends on the choice of i.

2. For central multivector square roots $i \in C\ell(p', q')$ of -1 and *all* functions $a, b \in L^1(\mathbb{R}^{p,q}; C\ell(p', q'))$. An important practical example is $i = e_1 e_2 e_3 \in C\ell 3, 0$ [117].

3. For all functions $a, b \in L^1(\mathbb{R}^{p,q}; C\ell(p', q'))$, with reflection symmetry $a^1 = a$. This condition does not depend on the choice of i, and poses no restriction on b.

5.4.1.3 Summary on one-sided CFT and convolution

In this section, we used the Clifford algebra decomposition with respect to a pair of square roots of -1, and the general steerable one-sided Clifford Fourier transform. We defined the notions of (classical non-steerable) convolution of two Clifford algebra valued functions over $\mathbb{R}^{p,q}$, and the steerable Mustard convolution (with its CFT as the point wise product of the CFTs of the factor functions).

The main results are: A decomposition of the classical convolution of Clifford algebra signals in terms of two Mustard convolutions. Next, we showed how in a dual way to fully generally express the Mustard convolution of two Clifford algebra signals in terms of two classical convolutions. Finally, we studied direct single convolution product identities between classical and Mustard convolutions, and showed how even for general Clifford signal factor functions the dual convolution product decompositions are theoretically fully equivalent, including equivalence to pairs of single convolution product identities.

In view of the many potential applications of the CFT [51], including already its lower-dimensional realizations as QFT [213, introduction], and space-time FT [218, introduction], the results may be of great interest in physics, pure and applied mathematics, and engineering, c.g., for filter design and feature extraction in multi-dimensional signal and (color) image processing. Finally, the CFT and all convolutions described above can also be implemented for simulations and real data applications in the Clifford Multivector Toolbox (for MATLAB) [295, 296].

5.4.2 Two-sided CFT with two square roots of -1 in $Cl(p, q)$ and convolution

5.4.2.1 Background

In this section, we use the general steerable two-sided Clifford Fourier transform (CFT), see Section 5.3.1 and [209], and we relate the classical convolution of Clifford algebra-valued signals over $\mathbb{R}^{p,q}$ with the (equally steerable) Mustard convolution. A Mustard convolution can be expressed in the spectral domain as the point wise product of the CFTs of the factor functions. In full generality we express the classical convolution of Clifford algebra signals in terms of finite linear combinations of Mustard convolutions, and vice versa the Mustard convolution of Clifford algebra signals in terms of finite linear combinations of classical convolutions. This Section is based on [215].

A key strength of the classical complex Fourier transform is its easy and fast application to filtering problems. The convolution of a signal with its filter function becomes in the spectral domain a point wise product of the respective Fourier transformations. This is not the case for the convolution of two Clifford algebra-valued signals (Clifford signals) over $\mathbb{R}^{p,q}$, due to Clifford algebra non-commutativity. Yet it is possible to define from the point wise product of the CFTs of two Clifford signals a new type of steerable convolution, called Mustard convolution [270]. This Mustard convolution can be expressed in terms of sums of classical convolutions and vice versa. For the left-sided QFT this has recently been carried out in [97], see also Section 5.4.1, for two-sided QFT in [213] and for the space-time Fourier transform in [218]. Here we extend this approach in full generality to the steerable two-sided CFT for signal functions which map non-degenerate quadratic form vector spaces to Clifford algebras in all dimensions.

This treatment is organized as follows. We first introduce Clifford algebra, multivector signal functions, and briefly reviews an important decomposition (split) of multivectors with respect to a pair of multivector square roots of -1. Then, we give some background on the steerable two-sided CFT and newly define two related (steerable) exponential-sine Clifford Fourier transforms. Finally, Section 5.4.2.2 defines the classical convolution of two Clifford signal functions, as well as two types of steerable Mustard convolutions. The rest of the section is devoted to representing the classical convolution in terms of finite sums of Mustard convolutions. First is the general case in terms of the two types of Mustard convolutions in Theorem 5.4.22. Second, Corollary 5.4.24 expresses for a commuting pair of square roots of -1 in the CFT, the convolution in terms of the standard Mustard convolution. Third, Theorem 5.4.25 generally expresses the classical convolution in terms of the standard Mustard convolution. Fourth, for a pair of anticommuting square roots of -1 in the CFT, Theorem 5.4.26 gives the classical convolution in terms of standard Mustard convolutions. At the end, Theorem 5.4.27 states the Mustard convolution in terms of classical convolutions.

For an introduction to Clifford's geometric algebras, we refer the reader to Chapter 2. We will use the *principal reverse* of $M \in Cl(p,q)$ defined as Section 2.1.1.6.

For $M, N \in Cl(p,q)$ we get $M * \widetilde{N} = \sum_A M_A N_A$. Two multivectors $M, N \in Cl(p,q)$ are *orthogonal* if and only if $M * \widetilde{N} = 0$. The modulus $|M|$ of a multivector $M \in Cl(p,q)$ is defined as

$$|M|^2 = M * \widetilde{M} = \sum_A M_A^2. \tag{5.4.28}$$

Notation 5.4.12 (Argument reflection). *For a function $h : \mathbb{R}^{p,q} \to Cl(p',q')$ and a multi-index $\phi = (\phi_1, \phi_2)$ with $\phi_1, \phi_2 \in \{0,1\}$ we set*

$$h^{\phi} = h^{(\phi_1,\phi_2)}(\boldsymbol{x}) := h((-1)^{\phi_1}\boldsymbol{x}_k, (-1)^{\phi_2}\boldsymbol{x}_{(n-k)}), \tag{5.4.29}$$

where $\boldsymbol{x}_k = x_1 e_1 + \ldots + x_k e_k$, $\boldsymbol{x}_{(n-k)} = \boldsymbol{x} - \boldsymbol{x}_k$, $1 \leq k \leq n$, for arbitrary but fixed k.

Remember that respect to any square root $i \in Cl(p,q)$ of -1, $i^2 = -1$, every multivector $A \in Cl(p,q)$ can be split into *commuting* A_{+i} and *anticommuting* A_{-i} parts, see Lemma 2.7.5 and [203]. We recommend readers to study Section 5.3.1.2 for the role of pairs of square roots of -1, before continuing.

We use the *general steerable two-sided Clifford Fourier transform* (CFT) [209] that can both be understood as a generalization of known one-sided CFTs [185, 201], or of the two-sided quaternion Fourier transformation (QFT) [184, 194], or two-sided space-time Fourier transform [184, 194, 218] to a general Clifford algebra setting. Most known CFTs (prior to [209]) used in their kernels specific square roots of -1, like bivectors, pseudoscalars, unit pure quaternions, or sets of coorthogonal blades (commuting or anticommuting blades) [61]. All those restrictions on the square roots of -1 used in a CFT do not apply in our definition below. Note that if the left or right phase angle is identical to zero, we get one-sided right or left sided CFTs, respectively.

Note that $L^1(\mathbb{R}^{p,q};\, Cl_{p',q'})$ is more general than $L^1(\mathbb{R}^{p,q};\, Cl_{p,q})$ in Section 5.3.1.2, Definition 5.3.3.

Definition 5.4.13 (Steerable CFT with respect to two square roots of -1 [209]). *Let $f, g \in Cl_{p',q'}$, $f^2 = g^2 = -1$, be any two square roots of -1. The general steerable two-sided Clifford Fourier transform*[32] *(CFT) of $h \in L^1(\mathbb{R}^{p,q};\, Cl_{p',q'})$, with respect to f, g is*

$$\mathcal{F}\{h\}(\boldsymbol{\omega}) = \mathcal{F}^{f,g}\{h\}(\boldsymbol{\omega}) = \int_{\mathbb{R}^{p,q}} e^{-fu(\boldsymbol{x},\boldsymbol{\omega})} h(\boldsymbol{x})\, e^{-gv(\boldsymbol{x},\boldsymbol{\omega})} d^n\boldsymbol{x}, \tag{5.4.30}$$

where $d^n\boldsymbol{x} = dx_1 \dots dx_n$, $\boldsymbol{x}, \boldsymbol{\omega} \in \mathbb{R}^{p,q}$, and $u, v : \mathbb{R}^{p,q} \times \mathbb{R}^{p,q} \to \mathbb{R}$.

Remark 5.4.14. *In order to avoid clutter we often drop the upper indexes f, g as in $\mathcal{F}\{h\} = \mathcal{F}^{f,g}\{h\}$, but in principle the two-sided CFT always depends on the particular choice f, g of the two square roots of -1. Since square roots of -1 in $Cl_{p',q'}$ populate continuous submanifolds, the CFT of Definition 5.4.13 is generically steerable within these submanifolds. In Definition 5.4.13, the two square roots $f, g \in Cl_{p',q'}$ of -1, may be from the same (or different) conjugacy class and component, respectively.*

Linearity of the CFT integral (5.4.13) allows us to use the $\pm$ split $h = h_- + h_+$ of Definition 5.3.1 to obtain, see [209] and Section 5.3.1,

$$\begin{aligned}\mathcal{F}^{f,g}\{h\}(\boldsymbol{\omega}) &= \mathcal{F}^{f,g}\{h_-\}(\boldsymbol{\omega}) + \mathcal{F}^{f,g}\{h_+\}(\boldsymbol{\omega}) \\ &= \mathcal{F}_-^{f,g}\{h\}(\boldsymbol{\omega}) + \mathcal{F}_+^{f,g}\{h\}(\boldsymbol{\omega}),\end{aligned} \tag{5.4.31}$$

since by their construction the operators of the Clifford Fourier transformation $\mathcal{F}^{f,g}$, and of the $\pm$ split with respect to f, g commute. From (5.3.7) follows the next theorem.

Theorem 5.4.15 (CFT of $h_\pm$, [209] and Section 5.3.1). *The CFT of the $\pm$ split parts $h_\pm$, with respect to two square roots $f, g \in Cl_{p',q'}$ of -1 , of a Clifford function $h \in L^1(\mathbb{R}^{p,q};\; Cl_{p',q'})$ have the quasi-complex forms*

$$\begin{aligned}\mathcal{F}_\pm^{f,g}\{h\} &= \mathcal{F}^{f,g}\{h_\pm\} \\ &= \int_{\mathbb{R}^{p,q}} h_\pm\, e^{-g(v(\boldsymbol{x},\boldsymbol{\omega})\mp u(\boldsymbol{x},\boldsymbol{\omega}))} d^n\boldsymbol{x} = \int_{\mathbb{R}^{p,q}} e^{-f(u(\boldsymbol{x},\boldsymbol{\omega})\mp v(\boldsymbol{x},\boldsymbol{\omega}))} h_\pm\, d^n\boldsymbol{x}\,.\end{aligned} \tag{5.4.32}$$

Remark 5.4.16. *Theorem 5.4.15 establishes in combination with (5.4.31) a general method for how to compute a two-sided CFT in terms of two one-sided CFTs*[33]*. For special relations of two-sided and one-sided quaternionic Fourier transforms see [184, 187, 194, 205].*

---

[32] The image Clifford algebra $Cl_{p',q'}$ can be identical to $Cl_{p,q}$ over the domain vector space $\mathbb{R}^{p,q}$, but this is not necessary, and completely depends on the application context. For Fourier transforms with $h \in L^1(\mathbb{R}^{p,q}; C\ell(p',q'))$ and $h \in L^2(\mathbb{R}^{p,q}; C\ell(p',q'))$, see Footnote 2 on page 144 regarding the roles of L^1 and L^2.

[33] For a general study of one-sided CFTs see [201] and Section 5.2.

Remark 5.4.17. *The quasi-complex forms in Theorem 5.4.15 allow to establish discretized and fast versions of the general two-sided CFT of Definition 5.4.13 as sums of complex discretized and fast Fourier transformations (FFT), respectively.*

For establishing an inversion formula, and other important CFT properties, certain *assumptions* about the phase functions $u(\boldsymbol{x},\boldsymbol{\omega})$, $v(\boldsymbol{x},\boldsymbol{\omega})$ need to be made. One possibility is, e.g. to arbitrarily partition the scalar product $\boldsymbol{x} * \widetilde{\boldsymbol{\omega}} = \sum_{l=1}^{n} x_l \omega_l = u(\boldsymbol{x},\boldsymbol{\omega}) + v(\boldsymbol{x},\boldsymbol{\omega})$, with

$$u(\boldsymbol{x},\boldsymbol{\omega}) = \sum_{l=1}^{k} x_l \omega_l, \qquad v(\boldsymbol{x},\boldsymbol{\omega}) = \sum_{l=k+1}^{n} x_l \omega_l, \tag{5.4.33}$$

for any arbitrary but fixed $1 \leq k \leq n$. We could also include any subset $B_u \subseteq \{1,\ldots,n\}$ of coordinates in $u(\boldsymbol{x},\boldsymbol{\omega})$ and the complementary set $B_v = \{1,\ldots,n\} \setminus B_u$ of coordinates in $v(\boldsymbol{x},\boldsymbol{\omega})$, etc. (5.4.33) will be assumed whenever the inverse CFT (5.4.34) is applied. We then get the following *inversion* theorem.

Theorem 5.4.18 (CFT inversion [209]). *With the assumption* (5.4.33) *for* u, v, *we get for* $h \in L^1(\mathbb{R}^{p,q}; Cl_{p',q'})$, *and that in any finite interval* h *and the partial coordinate derivatives of* h *are piecewise continuous, and have at most a finite number of extrema and discontinuities, then for* h *being continuous at* $\boldsymbol{x} \in \mathbb{R}^{p,q}$, *we have*

$$\begin{aligned} h(\boldsymbol{x}) &= \mathcal{F}_{-1}\{\mathcal{F}\{h\}\}(\boldsymbol{x}) \\ &= \frac{1}{(2\pi)^n} \int_{\mathbb{R}^{p,q}} e^{fu(\boldsymbol{x},\boldsymbol{\omega})} \mathcal{F}^{f,g}\{h\}(\boldsymbol{\omega})\, e^{gv(\boldsymbol{x},\boldsymbol{\omega})} d^n\boldsymbol{\omega}, \end{aligned} \tag{5.4.34}$$

where $d^n\boldsymbol{\omega} = d\omega_1 \ldots d\omega_n$, $\boldsymbol{x}, \boldsymbol{\omega} \in \mathbb{R}^{p,q}$. *For the existence of* (5.4.34) *we further need* $\mathcal{F}^{f,g}\{h\} \in L^1(\mathbb{R}^{p,q}; Cl_{p',q'})$.

We further define for later use the following two mixed (steerable) *exponential-sine* Fourier transforms

$$\mathcal{F}^{f,\pm s}\{h\}(\boldsymbol{\omega}) = \int_{\mathbb{R}^{p,q}} e^{-fu(\boldsymbol{x},\boldsymbol{\omega})} h(\boldsymbol{x})(\pm 1) \sin(-v(\boldsymbol{x},\boldsymbol{\omega})\,) d^n\boldsymbol{x}, \tag{5.4.35}$$

$$\mathcal{F}^{\pm s,g}\{h\}(\boldsymbol{\omega}) = \int_{\mathbb{R}^{p,q}} (\pm 1) \sin(-u(\boldsymbol{x},\boldsymbol{\omega})) h(\boldsymbol{x}) e^{-gv(\boldsymbol{x},\boldsymbol{\omega})}\, d^n\boldsymbol{x}. \tag{5.4.36}$$

With the help of

$$\begin{aligned} \sin(-u(\boldsymbol{x},\boldsymbol{\omega})) &= \tfrac{f}{2}(e^{-fu(\boldsymbol{x},\boldsymbol{\omega})} - e^{ftu(\boldsymbol{x},\boldsymbol{\omega})}), \\ \sin(-v(\boldsymbol{x},\boldsymbol{\omega})) &= \tfrac{g}{2}(e^{-gv(\boldsymbol{x},\boldsymbol{\omega})} - e^{gv(\boldsymbol{x},\boldsymbol{\omega})}), \end{aligned} \tag{5.4.37}$$

we can rewrite the above mixed exponential-sine Fourier transforms in terms of the CFT of Definition 5.4.13 as

$$\mathcal{F}^{f,\pm s}\{h\} = \pm\tfrac{1}{2}(\mathcal{F}^{f,g}\{hg\} - \mathcal{F}^{f,-g}\{hg\}), \tag{5.4.38}$$

$$\mathcal{F}^{\pm s,g}\{h\} = \pm\tfrac{1}{2}(\mathcal{F}^{f,g}\{fh\} - \mathcal{F}^{-f,g}\{fh\}). \tag{5.4.39}$$

We further note the following useful relationships using the argument reflection of Notation 5.4.12

$$\mathcal{F}^{-f,g}\{h\} = \mathcal{F}^{f,g}\{h^{(1,0)}\} = \mathcal{F}\{h^{(1,0)}\}, \quad \mathcal{F}^{f,-g}\{h\} = \mathcal{F}\{h^{(0,1)}\}, \tag{5.4.40}$$

and similarly

$$\mathcal{F}^{f,-s}\{h\} = \mathcal{F}^{f,s}\{h^{(0,1)}\}, \quad \mathcal{F}^{-s,g}\{h\} = \mathcal{F}^{s,g}\{h^{(1,0)}\}. \tag{5.4.41}$$

The main properties of the CFT of Definition 5.4.13 have been studied in detail in [209], see also Section 5.3.1.

5.4.2.2 *Convolution and steerable Mustard convolution*

We define the *convolution* of two Clifford (algebra) signals $a, b \in L^1(\mathbb{R}^{p,q}; Cl(p',q'))$ as

$$(a \star b)(\boldsymbol{x}) = \int_{\mathbb{R}^{p,q}} a(\boldsymbol{y})b(\boldsymbol{x}-\boldsymbol{y})d^2\boldsymbol{y}, \tag{5.4.42}$$

provided that the integral exists.

The *Mustard* convolution [270] of two Clifford signals $a, b \in L^1(\mathbb{R}^{p,q}; Cl(p',q'))$ is defined as

$$(a \star_M b)(\boldsymbol{x}) = (\mathcal{F}^{f,g})^{-1}(\mathcal{F}^{f,g}\{a\}\mathcal{F}^{f,g}\{b\}), \tag{5.4.43}$$

provided that the integral exists.

Remark 5.4.19. *The Mustard convolution has the conceptual and computational advantage to simply yield as spectrum in the CFT Fourier domain the point wise product of the CFTs of the two signals, just as for the classical complex Fourier transform. On the other hand, by its very definition, the Mustard convolution depends on the choice of the pair f, g, of multivector square roots of -1, used in the Definition 5.4.13 of the CFT. The Mustard convolution* (5.4.43) *is therefore a steerable operator, depending on the choice*[34] *of the pair f, g.*

We additionally define a further type of (steerable) *exponential-sine* Mustard convolution as

$$(a \star_{Ms} b)(\boldsymbol{x}) = (\mathcal{F}^{f,g})^{-1}(\mathcal{F}^{f,s}\{a\}\mathcal{F}^{s,g}\{b\}), \tag{5.4.44}$$

provided that the integral exists.

In the following two subsections we will express the convolution (5.4.42) in terms of the Mustard convolution (5.4.43) and vice versa.

5.4.2.2.1 Expressing the convolution in terms of the Mustard convolution

In this subsection, we assume the use of the two-sided CFT with two general multivector square roots of -1, $f, g \in Cl(p',q')$. The definition of the classical convolution (5.4.42) is independent of the application of a CFT. The Mustard convolutions of (5.4.43) and (5.4.44) depend on the definition of the CFT and in particular on the choice of the two multivector square roots of -1, f, g. Therefore it is meaningful in the following to distinguish[35] the expression of the classical convolution in terms of Mustard

[34]For an example particularly relevant to relativistic physics see [218].

[35]This distinction is a direct consequence of the *steerability* of the Mustard convolutions (5.4.43) and (5.4.44) inherited from the two-sided CFT of Definition 5.4.13.

convolutions for the three cases of general pairs f, g, of commuting f, g (i.e. $[f, g] = 0$), and of anticommuting f, g (i.e. $fg = -gf$), which we consequently state in different theorems and corollaries.

In [97] Theorem 4.1 on page 584 expresses the classical convolution of two quaternion functions with the help of the general left-sided QFT as a sum of 40 Mustard convolutions. Corresponding expressions have been established for the general two-sided QFT in [209], and the space-time Fourier transform in [218]. In our approach we use Theorem 5.12 on page 327 of [209], which expresses the convolution of two Clifford signal functions in the Clifford Fourier domain with the help of the CFT of Definition 5.4.13. Because of its importance for our current study, we recall this theorem here again.

Theorem 5.4.20 (CFT of convolution [209]). *Assuming a general pair of square roots of* -1*,* f, g*, the general two-sided CFT of the convolution* (5.4.42) *of two functions* $a, b \in L^1(\mathbb{R}^{p,q}; Cl(p', q')$ *can then be expressed as*

$$\begin{aligned}\mathcal{F}^{f,g}\{a \star b\} &= \mathcal{F}^{f,g}\{a_{+f}\}\mathcal{F}^{f,g}\{b_{+g}\} + \mathcal{F}^{f,-g}\{a_{+f}\}\mathcal{F}^{f,g}\{b_{-g}\} \\ &\quad + \mathcal{F}^{f,g}\{a_{-f}\}\mathcal{F}^{-f,g}\{b_{+g}\} + \mathcal{F}^{f,-g}\{a_{-f}\}\mathcal{F}^{-f,g}\{b_{-g}\} \\ &\quad + \mathcal{F}^{f,s}\{a_{+f}\}[f,g]\mathcal{F}^{s,g}\{b_{+g}\} + \mathcal{F}^{f,-s}\{a_{+f}\}[f,g]\mathcal{F}^{s,g}\{b_{-g}\} \\ &\quad + \mathcal{F}^{f,s}\{a_{-f}\}[f,g]\mathcal{F}^{-s,g}\{b_{+g}\} + \mathcal{F}^{f,-s}\{a_{-f}\}[f,g]\mathcal{F}^{-s,g}\{b_{-g}\}.\end{aligned} \tag{5.4.45}$$

Note that due to the commutation properties of (5.4.35) and (5.4.36) we can place the commutator $[f, g]$ also inside the exponential-sine transform terms as e.g. in

$$\begin{aligned}\mathcal{F}^{f,s}\{a_{+f}\}[f,g]\mathcal{F}^{s,g}\{b_{+g}\} &= \mathcal{F}^{f,s}\{a_{+f}[f,g]\}\mathcal{F}^{s,g}\{b_{+g}\} \\ &= \mathcal{F}^{f,s}\{a_{+f}\}\mathcal{F}^{s,g}\{[f,g]b_{+g}\}.\end{aligned} \tag{5.4.46}$$

For the special case of a commuting pair of square roots of -1, $[f, g] = 0$, we obtain a much simpler equation.

Corollary 5.4.21 (CFT of convolution with commuting f, g: $fg = gf$). *Assuming a commuting pair of square roots of* -1*,* $[f, g] = 0$*, the general two-sided CFT of the convolution* (5.4.42) *of two functions* $a, b \in L^1(\mathbb{R}^{p,q}; Cl(p', q'))$ *can be expressed as*

$$\begin{aligned}\mathcal{F}^{f,g}\{a \star b\} &= \mathcal{F}^{f,g}\{a_{+f}\}\mathcal{F}^{f,g}\{b_{+g}\} + \mathcal{F}^{f,-g}\{a_{+f}\}\mathcal{F}^{f,g}\{b_{-g}\} \\ &\quad + \mathcal{F}^{f,g}\{a_{-f}\}\mathcal{F}^{-f,g}\{b_{+g}\} + \mathcal{F}^{f,-g}\{a_{-f}\}\mathcal{F}^{-f,g}\{b_{-g}\}.\end{aligned} \tag{5.4.47}$$

We can now easily express the convolution of two Clifford signals $\mathcal{F}^{f,g}\{a \star b\}(\boldsymbol{\omega})$ in terms of only eight Mustard convolutions (5.4.43) and (5.4.44).

Theorem 5.4.22 (Convolution in terms of two types of Mustard convolution). *Assuming a general pair of square roots of* -1*,* f, g*, the convolution* (5.4.42) *of two Clifford functions* $a, b \in L^1(\mathbb{R}^{p,q}; Cl(p', q'))$ *can be expressed in terms of four Mustard convolutions* (5.4.43) *and four exponential-sine Mustard convolutions* (5.4.44) *as*

$$\begin{aligned}a \star b &= a_{+f} \star_M b_{+g} + a_{+f}^{(0,1)} \star_M b_{-g} + a_{-f} \star_M b_{+g}^{(1,0)} + a_{-f}^{(0,1)} \star_M b_{-g}^{(1,0)} \\ &\quad + a_{+f} \star_{Ms} [f,g]b_{+g} + a_{+f}^{(0,1)} \star_{Ms} [f,g]b_{-g} \\ &\quad + a_{-f} \star_{Ms} [f,g]b_{+g}^{(1,0)} + a_{-f}^{(0,1)} \star_{Ms} [f,g]b_{-g}^{(1,0)}.\end{aligned} \tag{5.4.48}$$

Remark 5.4.23. *We use the convention, that terms such as* $a_{+f}\star_{Ms}[f,g]\, b_{+g}$, *should be understood with brackets* $a_{+f} \star_{Ms} ([f,g]b_{+g})$, *which are omitted to avoid clutter.*

Assuming commutation, $[f,g]=0$, the standard Mustard convolution is sufficient to express the classical convolution.

Corollary 5.4.24 (Convolution in terms of Mustard convolution with commuting f,g). *Assuming a commuting pair of square roots of* -1, $[f,g]=0$, *the convolution* (5.4.42) *of two Clifford functions* $a,b \in L^1(\mathbb{R}^{p,q}; Cl(p',q'))$ *can be expressed in terms of four Mustard convolutions* (5.4.43) *as*

$$a \star b = a_{+f} \star_M b_{+f} + a_{+f}^{(0,1)} \star_M b_{-f} + a_{-f} \star_M b_{+f}^{(1,0)} + a_{-f}^{(0,1)} \star_M b_{-f}^{(1,0)} \qquad (5.4.49)$$

Furthermore, applying (5.4.38) and (5.4.39), we can expand the terms in (5.4.45) with exponential-sine transforms into sums of products of CFTs. For example, the first term gives

$$\begin{aligned}
&\mathcal{F}^{f,s}\{a_{+f}\}[f,g]\mathcal{F}^{s,g}\{b_{+g}\} \\
&= \frac{1}{4}\left(\mathcal{F}^{f,g}\{a_{+f}g\} - \mathcal{F}^{f,-g}\{a_{+f}g\}\right)\left(\mathcal{F}^{f,g}\{f[f,g]b_{+g}\} - \mathcal{F}^{-f,g}\{f[f,g]b_{+g}\}\right) \\
&= \frac{1}{4}\Big(\mathcal{F}\{a_{+f}g\}\mathcal{F}\{f[f,g]b_{+g}\} - \mathcal{F}\{a_{+f}g\}\mathcal{F}\{f[f,g]b_{+g}^{(1,0)}\} \\
&\qquad -\mathcal{F}\{a_{+f}^{(0,1)}g\}\mathcal{F}\{f[f,g]b_{+g}\} + \mathcal{F}\{a_{+f}^{(0,1)}g\}\mathcal{F}\{f[f,g]b_{+g}^{(1,0)}\}\Big)\,.
\end{aligned} \qquad (5.4.50)$$

For the important special case of anticommuting square roots of -1, $fg=-gf$, e.g. in quaternion algebra $f=\boldsymbol{i}$, $g=\boldsymbol{k}$ [124, 184, 187] or in the generalization to space-time $f=e_4=e_t$, $g=e_1e_2e_3=i_3$ in $Cl(3,1)$ [184, 187], equation (5.4.50) can be further simplified. Assuming $fg=-gf$, we have

$$[f,g]=2fg, \qquad f[f,g]=2ffg=-2g, \qquad (5.4.51)$$

and we can simplify (5.4.50) to

$$\begin{aligned}
&\mathcal{F}^{f,s}\{a_{+f}\}[f,g]\mathcal{F}^{s,g}\{b_{+g}\} = 2\mathcal{F}^{f,s}\{a_{+f}\}fg\mathcal{F}^{s,g}\{b_{+g}\} \\
&= \frac{2}{4}\Big(\mathcal{F}\{a_{+f}g\}\mathcal{F}\{(-g)b_{+g}\} - \mathcal{F}\{a_{+f}g\}\mathcal{F}\{(-g)b_{+g}^{(1,0)}\} \\
&\qquad -\mathcal{F}\{a_{+f}^{(0,1)}g\}\mathcal{F}\{(-g)b_{+g}\} + \mathcal{F}\{a_{+f}^{(0,1)}g\}\mathcal{F}\{(-g)b_{+g}^{(1,0)}\}\Big) \\
&= \frac{1}{2}\Big(\mathcal{F}\{a_{+f}\}\mathcal{F}\{b_{+g}^{(1,0)}\} - \mathcal{F}\{a_{+f}\}\mathcal{F}\{b_{+g}\} \\
&\qquad -\mathcal{F}\{a_{+f}^{(0,1)}\}\mathcal{F}\{b_{+g}^{(1,0)}\} + \mathcal{F}\{a_{+f}^{(0,1)}\}\mathcal{F}\{b_{+g}\}\Big)\,,
\end{aligned} \qquad (5.4.52)$$

because

$$\begin{aligned}
\mathcal{F}\{a_{+f}g\}\mathcal{F}\{(-g)b_{+g}\} &= \mathcal{F}\{a_{+f}\}g(-g)\mathcal{F}\{b_{+g}^{(1,0)}\} \\
&= \mathcal{F}\{a_{+f}\}\mathcal{F}\{b_{+g}^{(1,0)}\}, \quad \text{etc.}
\end{aligned} \qquad (5.4.53)$$

where we applied for the first equality that for $fg = -gf$,

$$e^{\alpha f} g = g e^{-\alpha f}. \tag{5.4.54}$$

In general equation (5.4.50) allows us in turn to express the Clifford signal convolution purely in terms of standard Mustard convolutions.

Theorem 5.4.25 (Convolution in terms of Mustard convolution). *Assuming a general pair of multivector square roots of* -1, f, g, *the convolution* (5.4.42) *of two Clifford functions* $a, b \in L^1(\mathbb{R}^{p,q}; Cl(p', q'))$ *can be expressed in terms of twenty standard Mustard convolutions* (5.4.43) *as*

$$\begin{aligned} a \star b = {} & a_{+f} \star_M b_{+g} + a_{+f}^{(0,1)} \star_M b_{-g} + a_{-f} \star_M b_{+g}^{(1,0)} + a_{-f}^{(0,1)} \star_M b_{-g}^{(1,0)} \\ & + \tfrac{1}{4}\big(a_{+f} g \star_M f c b_{+g} - a_{+f} g \star_M f c b_{+g}^{(1,0)} - a_{+f}^{(0,1)} g \star_M f c b_{+g} + a_{+f}^{(0,1)} g \star_M f c b_{+g}^{(1,0)} \\ & + a_{+f}^{(0,1)} g \star_M f c b_{-g} - a_{+f}^{(0,1)} g \star_M f c b_{-g}^{(1,0)} - a_{+f} g \star_M f c b_{-g} + a_{+f} g \star_M f c b_{-g}^{(1,0)} \\ & + a_{-f} g \star_M f c b_{+g}^{(1,0)} - a_{-f} g \star_M f c b_{+g} - a_{-f}^{(0,1)} g \star_M f c b_{+g}^{(1,0)} + a_{-f}^{(0,1)} g \star_M f c b_{+g} \\ & + a_{-f}^{(0,1)} g \star_M f c b_{-g}^{(1,0)} - a_{-f}^{(0,1)} g \star_M f c b_{-g} - a_{-f} g \star_M f c b_{-g}^{(1,0)} + a_{-f} g \star_M f c b_{-g} \big), \end{aligned} \tag{5.4.55}$$

with the abbreviation $c = [f, g]$.

Assuming anticommutation, $fg = -gf$, we can eliminate in Theorem 5.4.25 the commutators $c = [f, g]$ with the help of (5.4.52), which after cancellations leaves only sixteen terms.

$$\begin{aligned} & a \star b = \\ & \tfrac{1}{2}\big(a_{+f} \star_M b_{+g}^{(1,0)} + a_{+f} \star_M b_{+g} - a_{+f}^{(0,1)} \star_M b_{+g}^{(1,0)} + a_{+f}^{(0,1)} \star_M b_{+g} \\ & + a_{+f}^{(0,1)} \star_M b_{-g}^{(1,0)} + a_{+f}^{(0,1)} \star_M b_{-g} - a_{+f} \star_M b_{-g}^{(1,0)} + a_{+f} \star_M b_{-g} \\ & + a_{-f} \star_M b_{+g} + a_{-f} \star_M b_{+g}^{(1,0)} - a_{-f}^{(0,1)} \star_M b_{+g} + a_{-f}^{(0,1)} \star_M b_{+g}^{(1,0)} \\ & + a_{-f}^{(0,1)} \star_M b_{-g} + a_{-f}^{(0,1)} \star_M b_{-g}^{(1,0)} - a_{-f} \star_M b_{-g} + a_{-f} \star_M b_{-g}^{(1,0)} \big). \end{aligned} \tag{5.4.56}$$

Furthermore, we can combine by Definition 5.3.1 four pairs of $\pm$ split terms, e.g.,

$$a_{+f} \star_M b_{+g} + a_{+f} \star_M b_{-g} = a_{+f} \star_M b, \text{ etc.} \tag{5.4.57}$$

Assuming $fg = -gf$, this leaves only twelve terms for expressing a classical convolution in terms of a Mustard convolution,

$$\begin{aligned} a \star b = {} & \tfrac{1}{2}\big(a_{+f} \star_M b_{+g}^{(1,0)} + a_{+f} \star_M b - a_{+f}^{(0,1)} \star_M b_{+g}^{(1,0)} + a_{+f}^{(0,1)} \star_M b \\ & + a_{+f}^{(0,1)} \star_M b_{-g}^{(1,0)} - a_{+f} \star_M b_{-g}^{(1,0)} \\ & + a_{-f} \star_M b_{+g} + a_{-f} \star_M b^{(1,0)} - a_{-f}^{(0,1)} \star_M b_{+g} + a_{-f}^{(0,1)} \star_M b^{(1,0)} \\ & + a_{-f}^{(0,1)} \star_M b_{-g} - a_{-f} \star_M b_{-g} \big). \end{aligned} \tag{5.4.58}$$

Moreover, we can combine with the help of the involution $f()g$ of (5.3.3) four pairs of terms like

$$a_{+f} \star_M b^{(1,0)}_{+g} - a_{+f} \star_M b^{(1,0)}_{-g} = a_{+f} \star_M [b^{(1,0)}_{+g} - b^{(1,0)}_{-g}]$$
$$= a_{+f} \star_M (f[b^{(1,0)}_{+g} + b^{(1,0)}_{-g}]g) = a_{+f} \star_M fb^{(1,0)}g, \tag{5.4.59}$$

where in the final result we omit the round brackets, i.e. we understand $a_{+f} \star_M fb^{(1,0)}g = a_{+f} \star_M (fb^{(1,0)}g)$. This in turn leaves only eight terms for expressing a classical convolution in terms of a Mustard convolution, assuming $fg = -gf$.

$$a \star b = \tfrac{1}{2}\big(a_{+f} \star_M fb^{(1,0)}g + a_{+f} \star_M b - a^{(0,1)}_{+f} \star_M fb^{(1,0)}g + a^{(0,1)}_{+f} \star_M b$$
$$+ a_{-f} \star_M fbg + a_{-f} \star_M b^{(1,0)}$$
$$- a^{(0,1)}_{-f} \star_M fbg + a^{(0,1)}_{-f} \star_M b^{(1,0)} \big). \tag{5.4.60}$$

Finally, we note, that (5.4.60) contains pairs of functions $a_{\pm f}$ with unreflected and reflected second argument. Adding these pairs leads to even $\oplus$ or odd $\ominus$ symmetry in the second argument. That is we combine

$$a^{\oplus}_{+f} = \tfrac{1}{2}(a_{+f} + a^{(0,1)}_{+f}), \qquad a^{\ominus}_{+f} = \tfrac{1}{2}(a_{+f} - a^{(0,1)}_{+f}), \qquad \text{etc.} \tag{5.4.61}$$

This allows us for $fg = -gf$, to write the classical convolution in terms of just four Mustard convolutions.

Theorem 5.4.26 (Convolution in terms of Mustard convolution with anticommuting f, g). *Assuming an anticommuting pair f, g, of multivector square roots of -1, with $fg = -gf$, the convolution* (5.4.42) *of two Clifford functions $a, b \in L^1(\mathbb{R}^{p,q}; Cl(p', q'))$ can be expressed in terms of four standard Mustard convolutions* (5.4.43) *as*

$$a \star b = \tfrac{1}{2}\big(a^{\ominus}_{+f} \star_M fb^{(1,0)}g + a^{\oplus}_{+f} \star_M b + a^{\ominus}_{-f} \star_M fbg + a^{\oplus}_{-f} \star_M b^{(1,0)} \big). \tag{5.4.62}$$

5.4.2.2.2 Expressing the Mustard convolution in terms of the convolution Now we will simply write out the Mustard convolution (5.4.43) and simplify it until only standard convolutions (5.4.42) remain. In this subsection, we will use the general $\pm$ split of Definition 5.3.1. Our result should be compared, e.g, in the special case of the left-sided QFT with the Theorem 2.5 on page 584 of [97] with 32 classical convolutions for expressing the Mustard convolution of quaternion functions. Similar results to ours can be found in [209] for the two-sided QFT, and in [218] for the space-time Fourier transform.

We begin by writing the Mustard convolution (5.4.43) of two Clifford functions $a, b \in L^2(\mathbb{R}^{p,q}; Cl(p', q'))$

$$a \star_M b(\boldsymbol{x}) = \frac{1}{(2\pi)^n} \int_{\mathbb{R}^{p,q}} e^{fu(\boldsymbol{x},\boldsymbol{\omega})} \mathcal{F}\{a\}(\boldsymbol{\omega}) \mathcal{F}\{b\}(\boldsymbol{\omega}) e^{gv(\boldsymbol{x},\boldsymbol{\omega})} d^n\boldsymbol{\omega}$$
$$= \frac{1}{(2\pi)^n} \int_{\mathbb{R}^{p,q}} e^{-fu(\boldsymbol{x},\boldsymbol{\omega})} \int_{\mathbb{R}^2} e^{-fu(\boldsymbol{y},\boldsymbol{\omega})} a(\boldsymbol{y}) e^{-gv(\boldsymbol{y},\boldsymbol{\omega})} d^n\boldsymbol{y}$$
$$\int_{\mathbb{R}^{p,q}} e^{-fu(\boldsymbol{z},\boldsymbol{\omega})} b(\boldsymbol{z}) e^{-gv(\boldsymbol{z},\boldsymbol{\omega})} d^2\boldsymbol{z} e^{gv(\boldsymbol{x},\boldsymbol{\omega})} d^n\boldsymbol{\omega}$$
$$= \frac{1}{(2\pi)^n} \int_{\mathbb{R}^{p,q}} \int_{\mathbb{R}^{p,q}} \int_{\mathbb{R}^{p,q}} e^{fu(\boldsymbol{x}-\boldsymbol{y},\boldsymbol{\omega})} (a_+(\boldsymbol{y}) + a_-(\boldsymbol{y})) e^{-gv(\boldsymbol{y},\boldsymbol{\omega})}$$
$$e^{-fu(\boldsymbol{z},\boldsymbol{\omega})} (b_+(\boldsymbol{z}) + b_-(\boldsymbol{z})) e^{gv(\boldsymbol{x}-\boldsymbol{z},\boldsymbol{\omega})} d^n\boldsymbol{y} d^n\boldsymbol{z} d^n\boldsymbol{\omega}. \tag{5.4.63}$$

Next, we use the identities (5.3.7) in order to shift the inner factor $e^{-gv(\boldsymbol{y},\boldsymbol{\omega})}$ to the left and $e^{-fu(\boldsymbol{z},\boldsymbol{\omega})}$ to the right, respectively. We abbreviate $\int_{\mathbb{R}^{p,q}}\int_{\mathbb{R}^{p,q}}\int_{\mathbb{R}^{p,q}}$ to $\iiint$.

$$\begin{aligned}
&a \star_M b(\boldsymbol{x}) \qquad (5.4.64)\\
&= \frac{1}{(2\pi)^n}\iiint e^{fu(\boldsymbol{x}-\boldsymbol{y},\boldsymbol{\omega})}e^{fv(\boldsymbol{y},\boldsymbol{\omega})}a_+(\boldsymbol{y})b_+(\boldsymbol{z})e^{gu(\boldsymbol{z},\boldsymbol{\omega})}e^{gv(\boldsymbol{x}-\boldsymbol{z},\boldsymbol{\omega})}d^n\boldsymbol{y}d^n\boldsymbol{z}d^n\boldsymbol{\omega}\\
&+ \frac{1}{(2\pi)^n}\iiint e^{fu(\boldsymbol{x}-\boldsymbol{y},\boldsymbol{\omega})}e^{fv(\boldsymbol{y},\boldsymbol{\omega})}a_+(\boldsymbol{y})b_-(\boldsymbol{z})e^{-gu(\boldsymbol{z},\boldsymbol{\omega})}e^{gv(\boldsymbol{x}-\boldsymbol{z},\boldsymbol{\omega})}d^n\boldsymbol{y}d^n\boldsymbol{z}d^n\boldsymbol{\omega}\\
&+ \frac{1}{(2\pi)^n}\iiint e^{fu(\boldsymbol{x}-\boldsymbol{y},\boldsymbol{\omega})}e^{-fv(\boldsymbol{y},\boldsymbol{\omega})}a_-(\boldsymbol{y})b_+(\boldsymbol{z})e^{gu(\boldsymbol{z},\boldsymbol{\omega})}e^{gv(\boldsymbol{x}-\boldsymbol{z},\boldsymbol{\omega})}d^n\boldsymbol{y}d^n\boldsymbol{z}d^n\boldsymbol{\omega}\\
&+ \frac{1}{(2\pi)^n}\iiint e^{fu(\boldsymbol{x}-\boldsymbol{y},\boldsymbol{\omega})}e^{-fv(\boldsymbol{y},\boldsymbol{\omega})}a_-(\boldsymbol{y})b_-(\boldsymbol{z})e^{-gu(\boldsymbol{z},\boldsymbol{\omega})}e^{gv(\boldsymbol{x}-\boldsymbol{z},\boldsymbol{\omega})}d^n\boldsymbol{y}d^n\boldsymbol{z}d^n\boldsymbol{\omega}.
\end{aligned}$$

Furthermore, we abbreviate the inner function products as $ab_{\pm\pm}(\boldsymbol{y},\boldsymbol{z}) := a_\pm(\boldsymbol{y})b_\pm(\boldsymbol{z})$, and apply the general $\pm$ split of Definition 5.3.1 once again to obtain $ab_{\pm\pm}(\boldsymbol{y},\boldsymbol{z}) = [ab_{\pm\pm}(\boldsymbol{y},\boldsymbol{z})]_+ + [ab_{\pm\pm}(\boldsymbol{y},\boldsymbol{z})]_- = ab_{\pm\pm}(\boldsymbol{y},\boldsymbol{z})_+ + ab_{\pm\pm}(\boldsymbol{y},\boldsymbol{z})_-$. We omit the square brackets and use the convention that the final $\pm$ split indicated by the final $\pm$ index should be performed last. This allows to apply (5.3.7) again in order to shift the factors $e^{\pm gu(\boldsymbol{z},\boldsymbol{\omega})}e^{gv(\boldsymbol{x}-\boldsymbol{z},\boldsymbol{\omega})}$ to the left. We end up with the following eight terms

$$\begin{aligned}
&a \star_M b(\boldsymbol{x}) \qquad (5.4.65)\\
&= \frac{1}{(2\pi)^n}\iiint e^{fu(\boldsymbol{x}-\boldsymbol{y}-\boldsymbol{z},\boldsymbol{\omega})}e^{fv(\boldsymbol{y}-(\boldsymbol{x}-\boldsymbol{z}),\boldsymbol{\omega})}ab_{++}(\boldsymbol{y},\boldsymbol{z})_+d^n\boldsymbol{y}d^n\boldsymbol{z}d^n\boldsymbol{\omega}\\
&+ \frac{1}{(2\pi)^n}\iiint e^{fu(\boldsymbol{x}-\boldsymbol{y}+\boldsymbol{z},\boldsymbol{\omega})}e^{fv(\boldsymbol{y}+(\boldsymbol{x}-\boldsymbol{z}),\boldsymbol{\omega})}ab_{++}(\boldsymbol{y},\boldsymbol{z})_-d^n\boldsymbol{y}d^n\boldsymbol{z}d^n\boldsymbol{\omega}\\
&+ \frac{1}{(2\pi)^n}\iiint e^{fu(\boldsymbol{x}-\boldsymbol{y}+\boldsymbol{z},\boldsymbol{\omega})}e^{fv(\boldsymbol{y}-(\boldsymbol{x}-\boldsymbol{z}),\boldsymbol{\omega})}ab_{+-}(\boldsymbol{y},\boldsymbol{z})_+d^n\boldsymbol{y}d^n\boldsymbol{z}d^n\boldsymbol{\omega}\\
&+ \frac{1}{(2\pi)^n}\iiint e^{fu(\boldsymbol{x}-\boldsymbol{y}-\boldsymbol{z},\boldsymbol{\omega})}e^{fv(\boldsymbol{y}+(\boldsymbol{x}-\boldsymbol{z}),\boldsymbol{\omega})}ab_{+-}(\boldsymbol{y},\boldsymbol{z})_-d^n\boldsymbol{y}d^n\boldsymbol{z}d^n\boldsymbol{\omega}\\
&+ \frac{1}{(2\pi)^n}\iiint e^{fu(\boldsymbol{x}-\boldsymbol{y}-\boldsymbol{z},\boldsymbol{\omega})}e^{fv(-\boldsymbol{y}-(\boldsymbol{x}-\boldsymbol{z}),\boldsymbol{\omega})}ab_{-+}(\boldsymbol{y},\boldsymbol{z})_+d^n\boldsymbol{y}d^n\boldsymbol{z}d^n\boldsymbol{\omega}\\
&+ \frac{1}{(2\pi)^n}\iiint e^{fu(\boldsymbol{x}-\boldsymbol{y}+\boldsymbol{z},\boldsymbol{\omega})}e^{fv(-\boldsymbol{y}+(\boldsymbol{x}-\boldsymbol{z}),\boldsymbol{\omega})}ab_{-+}(\boldsymbol{y},\boldsymbol{z})_-d^n\boldsymbol{y}d^n\boldsymbol{z}d^n\boldsymbol{\omega}\\
&+ \frac{1}{(2\pi)^n}\iiint e^{fu(\boldsymbol{x}-\boldsymbol{y}+\boldsymbol{z},\boldsymbol{\omega})}e^{fv(-\boldsymbol{y}-(\boldsymbol{x}-\boldsymbol{z}),\boldsymbol{\omega})}ab_{--}(\boldsymbol{y},\boldsymbol{z})_+d^n\boldsymbol{y}d^n\boldsymbol{z}d^n\boldsymbol{\omega}\\
&+ \frac{1}{(2\pi)^n}\iiint e^{fu(\boldsymbol{x}-\boldsymbol{y}-\boldsymbol{z},\boldsymbol{\omega})}e^{fv(-\boldsymbol{y}+(\boldsymbol{x}-\boldsymbol{z}),\boldsymbol{\omega})}ab_{--}(\boldsymbol{y},\boldsymbol{z})_-d^n\boldsymbol{y}d^n\boldsymbol{z}d^n\boldsymbol{\omega}.
\end{aligned}$$

We now only show explicitly how to simplify the second triple integral, the others follow the same pattern.

$$\begin{aligned}
&\frac{1}{(2\pi)^n}\iiint e^{fu(\boldsymbol{x}-\boldsymbol{y}+\boldsymbol{z},\boldsymbol{\omega})}e^{fv(\boldsymbol{y}+(\boldsymbol{x}-\boldsymbol{z}),\boldsymbol{\omega})}[a_+(\boldsymbol{y})b_+(\boldsymbol{z})]_-d^n\boldsymbol{y}d^n\boldsymbol{z}d^n\boldsymbol{\omega}\\
&= \iint \frac{1}{(2\pi)^n}\int_{\mathbb{R}^{p,q}} e^{\sum_{l=1}^k (x_l-y_l+z_l)\omega_l}e^{\sum_{m=k+1}^n (y_m+(x_m-z_m))\omega_m}[a_+(\boldsymbol{y})b_+(\boldsymbol{z})]_-\\
&\quad d^n\boldsymbol{\omega}d^n\boldsymbol{y}d^n\boldsymbol{z}\\
&= \iint \frac{1}{(2\pi)^n}\int_{\mathbb{R}^{p,q}} \Pi_{l=1}^k e^{f(x_l-y_l+z_l)\omega_l}\Pi_{m=k+1}^n e^{f(y_m+(x_m-z_m))\omega_m}[a_+(\boldsymbol{y})b_+(\boldsymbol{z})]_-\\
&\quad d^n\boldsymbol{\omega}d^n\boldsymbol{y}d^n\boldsymbol{z}
\end{aligned}$$

$$= \iint \Pi_{l=1}^{k}\delta(x_l - y_l + z_l)\,\Pi_{m=k+1}^{n}\delta(y_m + (x_m - z_m))[a_+(\boldsymbol{y})b_+(\boldsymbol{z}_k, \boldsymbol{z}_{(n-k)})]_- \, d^n\boldsymbol{y}d^n\boldsymbol{z}$$
$$= \int_{\mathbb{R}^{p,q}} [a_+(\boldsymbol{y})b_+(-(\boldsymbol{x}_k - \boldsymbol{y}_k), \boldsymbol{x}_{(n-k)} + \boldsymbol{y}_{(n-k)})]_- d^n\boldsymbol{y}$$
$$= \int_{\mathbb{R}^{p,q}} [a_+(\boldsymbol{y})b_+(-(\boldsymbol{x}_k - \boldsymbol{y}_k), -(-\boldsymbol{x}_{(n-k)} - \boldsymbol{y}_{(n-k)}))]_- d^n\boldsymbol{y}$$
$$= \int_{\mathbb{R}^{p,q}} [a_+(\boldsymbol{y})b_+^{(1,1)}(\boldsymbol{x}_k - \boldsymbol{y}_k, -\boldsymbol{x}_{(n-k)} - \boldsymbol{y}_{(n-k)})]_- d^n\boldsymbol{y}$$
$$= [a_+ \star b_+^{(1,1)}(\boldsymbol{x}_k, -\boldsymbol{x}_{(n-k)})]_-. \tag{5.4.66}$$

Note that $a_+ \star b_+^{(1,1)}(\boldsymbol{x}_k, -\boldsymbol{x}_{(n-k)})$ means to first apply the convolution to the pair of functions a_+ and $b_+^{(1,1)}$, and only then to evaluate them with the argument $(-\boldsymbol{x}_k, \boldsymbol{x}_{(n-k)})$. So in general $a_+ \star b_+^{(1,1)}(\boldsymbol{x}_k, -\boldsymbol{x}_{(n-k)}) \neq a_+ \star b_+(-\boldsymbol{x}_k, \boldsymbol{x}_{(n-k)})$. Simplifying the other seven triple integrals similarly we finally obtain the desired decomposition of the Mustard convolution (5.4.43) in terms of the classical convolution.

Theorem 5.4.27 (Mustard convolution in terms of standard convolution). *The Mustard convolution* (5.4.43) *of two Clifford functions* $a, b \in L^1(\mathbb{R}^{p,q}; Cl(p', q'))$ *can be expressed in terms of eight standard convolutions* (5.4.42) *as*

$$\begin{aligned} a \star_M b(\boldsymbol{x}) = {} & [a_+ \star b_+(\boldsymbol{x})]_+ + [a_+ \star b_+^{(1,1)}(x_1, -x_2)]_- \\ & + [a_+ \star b_-^{(1,0)}(\boldsymbol{x})]_+ + [a_+ \star b_-^{(0,1)}(x_1, -x_2)]_- \\ & + [a_- \star b_+^{(0,1)}(x_1, -x_2)]_+ + [a_- \star b_+^{(1,0)}(\boldsymbol{x})]_- \\ & + [a_- \star b_-^{(1,1)}(x_1, -x_2)]_+ + [a_- \star b_-(\boldsymbol{x})]_-. \end{aligned} \tag{5.4.67}$$

This Mustard convolution in terms of conventional convolutions can be numerically applied to color edge detection in color images, as discussed in detail in Section VII of [127], by applying the QFT (identical to the CFT for signals and filters $\in L^1(\mathbb{R}^2; Cl(0,2))$ there to the split parts (there named simplex and perplex parts) of the color image data and color edge filters, and summing over the resulting eight terms. Alternatively the same result could be obtained by directly applying the CFT to the (unsplit) image data and filter function, followed by applying the inverse CFT to the point wise product of the CFTs of the image data and the filter function. Since this color edge filter approach has been similarly implemented in other Clifford algebras, e.g. in $Cl(3,0)$ [263], analogous computations could be performed there as well. A detailed real data example unfortunately goes beyond the scope of our current treatment.

Remark 5.4.28. *If we would explicitly insert according Definition 5.3.1* $a_\pm = \frac{1}{2}(a \pm fag)$ *and* $b_\pm = \frac{1}{2}(b \pm fbg)$, *and similarly explicitly insert the second level* $\pm$ *split* $[\ldots]_\pm$, *we would obtain up to a maximum of 64 terms. It is therefore obvious how significant and efficient the use of the general* $\pm$ *split is in this context.*

5.4.2.3 *Summary on two-sided CFT with two square roots of -1 in $Cl(p,q)$ and convolution*

In this treatment we use the general steerable two-sided Clifford Fourier transform, and introduced a pair of related steerable mixed exponential-sine Clifford Fourier transforms. We defined the notions of (classical non-steerable) convolution of two Clifford algebra valued functions over $\mathbb{R}^{p,q}$, the steerable Mustard convolution (with its CFT as the point wise product of the CFTs of the factor functions), and a special steerable Mustard convolution involving the point wise products of mixed exponential-sine CFTs.

The main results are: an efficient decomposition of the classical convolution of Clifford algebra signals in terms of eight Mustard type convolutions. For the special cases of two commuting (or anticommuting) multivector square roots of -1 axis in the CFT, only four terms of the standard Mustard convolution prove to be sufficient. Even in the case of two general multivector square roots of -1 axis in the CFT, the classical convolution of two Clifford algebra signals can always be fully expanded in terms of standard Mustard convolutions. Finally we showed how to fully generally expand the Mustard convolution of two Clifford algebra signals in terms of eight classical convolutions.

Outlook: In view of the many potential applications of the CFT [51], including already its lower-dimensional realizations as QFT [213, introduction] and Chapter 4, and space-time FT [218, introduction] and Section 4.4, we expect our results to be of great interest in physics, pure and applied mathematics, and engineering, e.g., for filter design and feature extraction in multi-dimensional signal and (color) image processing. Furthermore, because higher dimensional Clifford algebras like $Cl(3,3)$ for projective geometry [245], conformal geometric algebra $Cl(4,1)$ for modeling Euclidean 3D geometry [244], $Cl(6,3)$ for quadric surfaces [334], double conformal geometric algebra $Cl(8,2)$ for Darboux cyclides (including tori, quartic surfaces, Dupin cyclides and quadrics) [114, 116], and double conformal space time algebra $Cl(8,4)$ [115] become increasingly popular in higher order surface modeling, lighting, computing applications, etc., we expect a growing need for efficient Fourier analytic tools for functions taking values in these Clifford algebras. Finally, the CFT and all convolutions described above can again be implemented for simulations and real data applications in the Clifford Multivector Toolbox (for MATLAB) [295].

5.4.3 Two-sided CFT with multiple square roots of -1 in $Cl(p,q)$ and convolution

5.4.3.1 *Background and notation*

The large variety of Fourier transforms in geometric algebras inspired the straight forward definition of *A General Geometric Fourier Transform* in Section 5.3.2 based on [61], covering many versions in the literature. We saw which constraints are additionally necessary to obtain certain features like linearity, a scaling or a shift theorem. In this section, we extend the former results by a convolution theorem. This treatment is based on [59]. Because the techniques of proof employed in Section 5.3.2 are very similar to the techniques we need in the present section, we omit all proofs, but encourage the interested reader to either try proofs by himself, or to simply look them up in [59].

We included many Clifford algebra Fourier transforms in one general definition in Section 5.3.2, based on [61]. There we analyzed the separation of constant factors from the transform, the scaling theorem and shift properties. Now we want to derive a convolution theorem. Proofs of some of the lemmata that we need here can be found in Section 5.3.2, based on [61].

Regarding notation, we examine geometric algebras $\mathcal{G}^{p,q} = Cl(p,q), p+q = n \in \mathbb{N}$ over $\mathbb{R}^{p,q}$ [161], see also Chapter 2, generated by the associative, bilinear geometric product with neutral element 1 satisfying

$$\boldsymbol{e}_j\boldsymbol{e}_k + \boldsymbol{e}_k\boldsymbol{e}_j = \epsilon_j\delta_{jk}, \tag{5.4.68}$$

for all $j,k \in \{1,...,n\}$ with the Kronecker symbol δ and

$$\epsilon_j = \begin{cases} 1 & \forall j = 1,...,p, \\ -1 & \forall j = p+1,...,n. \end{cases} \tag{5.4.69}$$

For the sake of brevity we want to refer to arbitrary multivectors

$$\boldsymbol{A} = \sum_{k=0}^{n} \sum_{1\le j_1<...<j_k\le n} a_{j_1...j_k}\boldsymbol{e}_{j_1}...\boldsymbol{e}_{j_k} \in \mathcal{G}^{p,q}, \tag{5.4.70}$$

$a_{j_1...j_k} \in \mathbb{R}$, as

$$\boldsymbol{A} = \sum_{\boldsymbol{j}} a_{\boldsymbol{j}}\boldsymbol{e}_{\boldsymbol{j}}. \tag{5.4.71}$$

where each of the 2^n multi-indices $\vec{j} \subseteq \{1,...,n\}$ indicates a basis multivector of dimension k of $\mathcal{G}^{p,q}$ by $\boldsymbol{e}_{\boldsymbol{j}} = \boldsymbol{e}_{j_1}...\boldsymbol{e}_{j_k}$, $1 \le j_1 < ... < j_k \le n$, $\boldsymbol{e}_\emptyset = \boldsymbol{e}_0 = 1$ and its associated coefficient $a_{\boldsymbol{j}} = a_{j_1...j_k} \in \mathbb{R}$. For each geometric algebra $\mathcal{G}^{p,q}$ we will write $\mathscr{I}^{p,q} = \{i \in \mathcal{G}^{p,q}, i^2 \in \mathbb{R}^-\}$ to denote the real multiples of all geometric square roots of minus one, compare [193] and [203]. We chose the symbol $\mathscr{I}$ to be reminiscent of the imaginary numbers.

Throughout this section, we analyze multivector fields $\boldsymbol{A}: \mathbb{R}^{p',q'} \to \mathcal{G}^{p,q}$, $p'+q' = m \in \mathbb{N}, p+q = n \in \mathbb{N}$. To keep notations short we will often denote the argument vector space by just $\mathbb{R}^m$, so please keep in mind, that it has a signature p', q', too.

We defined in Definition 5.3.9 [61], the general **geometric Fourier transform** (GFT) $\mathscr{F}_{F_1,F_2}(\boldsymbol{A})$ of a multivector field[36] $\boldsymbol{A}: \mathbb{R}^{p',q'} \to \mathcal{G}^{p,q}, p'+q' = m \in \mathbb{N}, p+q = n \in \mathbb{N}$ by the calculation rule

$$\mathscr{F}_{F_1,F_2}(\boldsymbol{A})(\boldsymbol{u}) := \int_{\mathbb{R}^m} \prod_{f\in F_1} e^{-f(\boldsymbol{x},\boldsymbol{u})}\boldsymbol{A}(\boldsymbol{x}) \prod_{f\in F_2} e^{-f(\boldsymbol{x},\boldsymbol{u})}\,\mathrm{d}^m\boldsymbol{x}, \tag{5.4.72}$$

with $\boldsymbol{x},\boldsymbol{u} \in \mathbb{R}^m$ and two ordered finite sets $F_1 = \{f_1(\boldsymbol{x},\boldsymbol{u}),...,f_{\boldsymbol{\mu}}(\boldsymbol{x},\boldsymbol{u})\}$, $F_2 = \{f_{\boldsymbol{\mu}+1}(\boldsymbol{x},\boldsymbol{u}),...,f_{\boldsymbol{\nu}}(\boldsymbol{x},\boldsymbol{u})\}$ of mappings $f_l(\boldsymbol{x},\boldsymbol{u}): \mathbb{R}^m \times \mathbb{R}^m \to \mathscr{I}^{p,q}, \forall l = 1,...,\boldsymbol{\nu}$. We proved some fundamental theorems in dependence on properties of the functions f_l, like existence, linearity, shift and scaling.

[36]For Fourier transforms with $\boldsymbol{A} \in L^1(\mathbb{R}^{p',q'};\mathcal{G}^{p,q})$ and $\boldsymbol{A} \in L^2(\mathbb{R}^{p',q'};\mathcal{G}^{p,q})$, see Footnote 2 on page 144 regarding the roles of L^1 and L^2.

5.4.3.2 Convolution theorem for not coorthogonal exponents

In Section 5.3.3, the description of the GFTs by means of the trigonometric transform (TT) showed, that many GFTs do not differ very much from real valued trigonometric transforms. Now we exploit this simplicity to derive new properties, that could not easily be found otherwise.

In [59], there is a convolution theorem for general geometric Fourier transforms with F_1, F_2 being coorthogonal, separable and linear with respect to the first argument. Coorthogonality can be interpreted as mutual commutation or anticommutation among the functions.

Exercise 5.4.29. *Every GFT in Example 5.3.10 is coorthogonal.*

Although coorthogonality is fulfilled by almost every popular geometric Fourier transform, we want to deduce a convolution theorem that holds for functions that are separable and linear with respect to the first argument but have arbitrary commutation properties. This formulation of the theorem is especially useful for the treatment of steerable Fourier transforms over the manifolds of square roots of minus one, as in [196, 203] and Section 2.7, which are generally not coorthogonal.

Definition 5.4.30 (convolution). *Let $\boldsymbol{A}(\boldsymbol{x}), \boldsymbol{B}(\boldsymbol{x}) : \mathbb{R}^m \to Cl(p,q)$ be two multivector fields. Their* ***convolution*** *$(\boldsymbol{A} * \boldsymbol{B})(\boldsymbol{x})$ is defined as*

$$(\boldsymbol{A} * \boldsymbol{B})(\boldsymbol{x}) = \int_{\mathbb{R}^m} \boldsymbol{A}(\boldsymbol{y})\boldsymbol{B}(\boldsymbol{x}-\boldsymbol{y})\, d^m \boldsymbol{y}. \tag{5.4.73}$$

Theorem 5.4.31 (convolution). *Let $\boldsymbol{A}, \boldsymbol{B}, \boldsymbol{C} : \mathbb{R}^m \to Cl(p,q)$ be multivector fields with $\boldsymbol{A}(\boldsymbol{x}) = (\boldsymbol{C} * \boldsymbol{B})(\boldsymbol{x})$ and F_1, F_2 be separable and linear with respect to the first argument, then*

$$\mathscr{F}_{F_1,F_2}(\boldsymbol{A})(\boldsymbol{u}) = \sum_{\substack{\boldsymbol{j},\boldsymbol{j}'\in\{0,1\}^{\mu}\\ \boldsymbol{k},\boldsymbol{k}'\in\{0,1\}^{(\nu-\mu)}}} \prod_{l=1}^{\mu} i_l(\boldsymbol{u})^{j_l+j_l'} \mathscr{F}_{F_1^{\boldsymbol{j}},F_2^{\boldsymbol{k}}}(\boldsymbol{C})(\boldsymbol{u}) \mathscr{F}_{F_1^{\boldsymbol{j}'},F_2^{\boldsymbol{k}'}}(\boldsymbol{B})(\boldsymbol{u}) \prod_{l=\mu+1}^{\nu} i_l(\boldsymbol{u})^{k_l+k_l'}. \tag{5.4.74}$$

Corollary 5.4.32 (convolution). *Let $\boldsymbol{A}, \boldsymbol{B}, \boldsymbol{C} : \mathbb{R}^m \to Cl(p,q)$ be multivector fields with $\boldsymbol{A}(\boldsymbol{x}) = (\boldsymbol{C} * \boldsymbol{B})(\boldsymbol{x})$ and F_1, F_2 each consist of functions in the center of $Cl(p,q)$, then the the GFT satisfy the simple product formula*

$$\mathscr{F}_{F_1,F_2}(\boldsymbol{A})(\boldsymbol{u}) = \mathscr{F}_{F_1,F_2}(\boldsymbol{C})(\boldsymbol{u})\mathscr{F}_{F_1,F_2}(\boldsymbol{B})(\boldsymbol{u}). \tag{5.4.75}$$

Remark 5.4.33. *For all $Cl(p,q), p+q=n$ even, the center is trivial $\{1\}$. For n odd, the center has two elements $\{1, i_n\}$, with i_n the pseudoscalar. In general, the scalar part of every square root of -1 is zero. So, square roots exist in the center of $Cl(p,q)$ only for $p+q=n$ odd and they will be proportional to i_n .*

The convolution theorem Theorem 5.4.31 is not simply a generalization of the convolution theorem from [59]. That means in the case of coorthogonal functions F_1, F_2, it will not reduce to the other theorem. Here, the exponentials are decomposed, while the other theorem decomposes the multivector functions.

Exercise 5.4.34. *Theorem 5.4.31 can take various shapes, depending on different GFTs. We shorten the formulae and stress the simplicity using Notation 5.3.44 for the standard sine and cosine transforms.*

1. *The Clifford Fourier transform from [119, 185, 222] takes the form*

$$\begin{aligned}\mathscr{F}_{f_1}(\boldsymbol{A}) &= \sum_{j,j'\in\{0,1\}} i^{j+j'} \mathscr{F}_{f_1^j}(\boldsymbol{C}) \mathscr{F}_{f_1^{j'}}(\boldsymbol{B}) \\ &= \mathscr{F}_c(\boldsymbol{C})\mathscr{F}_c(\boldsymbol{B}) - \mathscr{F}_s(\boldsymbol{C})\mathscr{F}_s(\boldsymbol{B}) \\ &\quad + i(\mathscr{F}_c(\boldsymbol{C})\mathscr{F}_s(\boldsymbol{B}) + \mathscr{F}_s(\boldsymbol{C})\mathscr{F}_c(\boldsymbol{B}))\end{aligned} \tag{5.4.76}$$

for $n = 2 \pmod 4$. For $n = 3 \pmod 4$ we can apply Corollary 5.4.32

$$\mathscr{F}_{f_1}(\boldsymbol{A}) = \mathscr{F}_{f_1}(\boldsymbol{C}) \mathscr{F}_{f_1}(\boldsymbol{B}). \tag{5.4.77}$$

Because in this case, the pseudoscalar is in the center of $C\ell_{n,0}(n, 0)$.

2. *The transform by Sommen [56, 306] satisfies*

$$\begin{aligned}\mathscr{F}_{f_1,\dots,f_n}(\boldsymbol{A}) = \sum_{\boldsymbol{k},\boldsymbol{k}'\in\{0,1\}^n} &\mathscr{F}_{f_1^{k_1},\dots,f_n^{k_n}}(\boldsymbol{C}) \mathscr{F}_{f_1^{k'_1},\dots,f_n^{k'_n}}(\boldsymbol{B}) \\ &\prod_{l=1}^n \boldsymbol{e}_l^{k_l+k'_l}.\end{aligned} \tag{5.4.78}$$

3. *The quaternion Fourier transform [56, 124, 184] has the shape*

$$\mathscr{F}_{f_1,f_2}(\boldsymbol{A}) = \sum_{j,j',k,k'\in\{0,1\}} i^{j+j'} \mathscr{F}_{f_1^j,f_2^k}(\boldsymbol{C}) \mathscr{F}_{f_1^{j'},f_2^{k'}}(\boldsymbol{B}) j^{k+k'}, \tag{5.4.79}$$

which can be explicitly written as

$$\begin{aligned}\mathscr{F}_{f_1,f_2}(\boldsymbol{A}) &= \mathscr{F}_{cc}(\boldsymbol{C})\mathscr{F}_{cc}(\boldsymbol{B}) - \mathscr{F}_{sc}(\boldsymbol{C})\mathscr{F}_{sc}(\boldsymbol{B}) - \mathscr{F}_{cs}(\boldsymbol{C})\mathscr{F}_{cs}(\boldsymbol{B}) \\ &\quad + \mathscr{F}_{ss}(\boldsymbol{C})\mathscr{F}_{ss}(\boldsymbol{B}) + i(\mathscr{F}_{sc}(\boldsymbol{C})\mathscr{F}_{cc}(\boldsymbol{B}) + \mathscr{F}_{cc}(\boldsymbol{C})\mathscr{F}_{sc}(\boldsymbol{B}) \\ &\quad - \mathscr{F}_{cs}(\boldsymbol{C})\mathscr{F}_{ss}(\boldsymbol{B}) - \mathscr{F}_{ss}(\boldsymbol{C})\mathscr{F}_{cs}(\boldsymbol{B})) + (\mathscr{F}_{cc}(\boldsymbol{C})\mathscr{F}_{cs}(\boldsymbol{B}) \\ &\quad + \mathscr{F}_{cs}(\boldsymbol{C})\mathscr{F}_{cc}(\boldsymbol{B}) - \mathscr{F}_{sc}(\boldsymbol{C})\mathscr{F}_{ss}(\boldsymbol{B}) - \mathscr{F}_{ss}(\boldsymbol{C})\mathscr{F}_{sc}(\boldsymbol{B}))j \\ &\quad + i(\mathscr{F}_{cc}(\boldsymbol{C})\mathscr{F}_{ss}(\boldsymbol{B}) + \mathscr{F}_{cs}(\boldsymbol{C})\mathscr{F}_{sc}(\boldsymbol{B}) + \mathscr{F}_{sc}(\boldsymbol{C})\mathscr{F}_{cs}(\boldsymbol{B}) \\ &\quad + \mathscr{F}_{ss}(\boldsymbol{C})\mathscr{F}_{cc}(\boldsymbol{B}))j.\end{aligned} \tag{5.4.80}$$

4. *The convolution theorem for the spacetime Fourier transform [184] takes the same shape as for the quaternionic transform*

$$\begin{aligned}\mathscr{F}_{f_1,f_2}(\boldsymbol{A}) = \sum_{j,j',\boldsymbol{k},k'\in\{0,1\}} &\boldsymbol{e}_4^{j+j'} \mathscr{F}_{f_1^j,f_2^k}(\boldsymbol{C}) \\ &\mathscr{F}_{f_1^{j'},f_2^{k'}}(\boldsymbol{B})(\epsilon_4 \boldsymbol{e}_4 i_4)^{k+k'}.\end{aligned} \tag{5.4.81}$$

5. *The Clifford Fourier transform for color images [18] with bivector B takes the form*

$$\mathscr{F}_{f_1,f_2,f_3,f_4}(\boldsymbol{A}) = \sum_{\boldsymbol{j},\boldsymbol{k}\in\{0,1\}^2} (B)^{j_1+j_1'}(iB)^{j_2+j_2'}\mathscr{F}_{f_1^{j_1},f_2^{j_2},f_3^{k_1},f_4^{k_2}}(\boldsymbol{C})$$
$$\mathscr{F}_{f_1^{j_1'},f_2^{j_2'},f_3^{k_1'},f_4^{k_2'}}(\boldsymbol{B})(-B)^{k_1+k_1'}(-iB)^{k_2+k_2'}, \tag{5.4.82}$$

which we will not write down summand by summand, because it comprises 256 terms.

6. *The cylindrical Fourier transform [49] is not separable except for the case $n = 2$. Here it satisfies*

$$\begin{aligned}\mathscr{F}_{f_1}(\boldsymbol{A}) &= \sum_{j,j'\in\{0,1\}} \boldsymbol{e}_{12}^{j+j'}\mathscr{F}_{f_1^j}(\boldsymbol{C})\mathscr{F}_{f_1^{j'}}(\boldsymbol{B})\\ &= \mathscr{F}_{f_1^0}(\boldsymbol{C})\mathscr{F}_{f_1^0}(\boldsymbol{B}) - \mathscr{F}_{f_1^1}(\boldsymbol{C})\mathscr{F}_{f_1^1}(\boldsymbol{B}) + \boldsymbol{e}_{12}(\mathscr{F}_{f_1^0}(\boldsymbol{C})\mathscr{F}_{f_1^1}(\boldsymbol{B})\\ &\quad + \mathscr{F}_{f_1^1}(\boldsymbol{C})\mathscr{F}_{f_1^0}(\boldsymbol{B})),\end{aligned} \tag{5.4.83}$$

but for all other no closed formula can be constructed in a similar way.

5.4.3.3 Short convolution theorem for not coorthogonal exponents

The number of summands in Theorem 5.4.31 is 4^ν. It is possible to formulate versions of the theorem that consist of only 2^ν summands. This formula still holds for GFT with mappings of arbitrary commutation properties. The main difference to the previous formula is that the summands are not real valued. The elementary transform terms are half GFTs, half TTs and multivector valued.

Theorem 5.4.35 (convolution, short). *Let $\boldsymbol{A}, \boldsymbol{B}, \boldsymbol{C} : \mathbb{R}^m \to Cl(p,q)$ be multivector fields with $\boldsymbol{A}(\boldsymbol{x}) = (\boldsymbol{C} * \boldsymbol{B})(\boldsymbol{x})$ and F_1, F_2 be separable and linear with respect to the first argument, then the geometric Fourier transform of $\boldsymbol{A}$ satisfies the convolution property*

$$\mathscr{F}_{F_1,F_2}(\boldsymbol{A})(\boldsymbol{u}) = \sum_{\substack{\boldsymbol{j}\in\{0,1\}^{\mu}\\ \boldsymbol{k}\in\{0,1\}^{(\nu-\mu)}}} \mathscr{F}_{-i_l(\boldsymbol{u})\frac{\pi}{2}j_l+f_l(\boldsymbol{y},\boldsymbol{u}),F_2^{\boldsymbol{k}}}(\boldsymbol{C})(\boldsymbol{u})$$
$$\mathscr{F}_{F_1^{\boldsymbol{j}},-i_l(\boldsymbol{u})\frac{\pi}{2}k_l+f_l(\boldsymbol{y},\boldsymbol{u})}(\boldsymbol{B})(\boldsymbol{u}). \tag{5.4.84}$$

with the transforms

$$\begin{aligned}&\mathscr{F}_{F_1^{\boldsymbol{j}},-i_l(\boldsymbol{u})\frac{\pi}{2}k_l+f_l(\boldsymbol{y},\boldsymbol{u})}(\boldsymbol{B})(\boldsymbol{u})\\ &= \int_{\mathbb{R}^m} \prod_{l=1}^{\mu} e_{j_l}^{-f_l(\boldsymbol{x},\boldsymbol{u})}\boldsymbol{B}(\boldsymbol{x}) \prod_{l=\mu+1}^{\nu} e^{i_l(\boldsymbol{u})\frac{\pi}{2}k_l-f_l(\boldsymbol{y},\boldsymbol{u})}\,\mathrm{d}^m\boldsymbol{x}\\ &= \int_{\mathbb{R}^m} \prod_{l=1}^{\mu} e_{j_l}^{-f_l(\boldsymbol{x},\boldsymbol{u})}\boldsymbol{B}(\boldsymbol{x}) \prod_{l=\mu+1}^{\nu} (i_l(\boldsymbol{u})^{k_l}e^{-f_l(\boldsymbol{y},\boldsymbol{u})})\,\mathrm{d}^m\boldsymbol{x},\end{aligned} \tag{5.4.85}$$

$$
\begin{aligned}
&\mathscr{F}_{-i_l(\boldsymbol{u})\frac{\pi}{2}j_l+f_l(\boldsymbol{y},\boldsymbol{u}),F_2^{\boldsymbol{k}}}(\boldsymbol{C})(\boldsymbol{u})\\
&= \int_{\mathbb{R}^m} \prod_{l=1}^{\mu} e^{i_l(\boldsymbol{u})\frac{\pi}{2}j_l-f_l(\boldsymbol{y},\boldsymbol{u})}\boldsymbol{C}(\boldsymbol{x}) \prod_{l=\mu+1}^{\nu} e_{k_l}^{-f_l(\boldsymbol{x},\boldsymbol{u})}\,\mathrm{d}^m\boldsymbol{x}\\
&= \int_{\mathbb{R}^m} \prod_{l=1}^{\mu} (i_l(\boldsymbol{u})^{j_l}e^{-f_l(\boldsymbol{y},\boldsymbol{u})})\boldsymbol{C}(\boldsymbol{x}) \prod_{l=\mu+1}^{\nu} e_{k_l}^{-f_l(\boldsymbol{x},\boldsymbol{u})}\,\mathrm{d}^m\boldsymbol{x}
\end{aligned} \tag{5.4.86}
$$

for multi-indices $\boldsymbol{j} \in \{0,1\}^{\boldsymbol{\mu}}$ *or* $\boldsymbol{k} \in \{0,1\}^{(\boldsymbol{\nu}-\boldsymbol{\mu})}$ *and* $e_j^{-f(\boldsymbol{x},\boldsymbol{u})}$ *from Notation 5.3.39.*

Corollary 5.4.36. *Let* $\boldsymbol{A}, \boldsymbol{B}, \boldsymbol{C} : \mathbb{R}^m \to Cl(p,q)$ *be multivector fields with* $\boldsymbol{A}(\boldsymbol{x}) = (\boldsymbol{C}*\boldsymbol{B})(\boldsymbol{x})$ *and* F_1, F_2 *be separable, linear with respect to the first argument and mutually commutative, then the geometric Fourier transform of* $\boldsymbol{A}$ *satisfies the convolution property*

$$
\mathscr{F}_{F_1,F_2}(\boldsymbol{A})(\boldsymbol{u}) = \sum_{\substack{\boldsymbol{j}\in\{0,1\}^{\boldsymbol{\mu}}\\ \boldsymbol{k}\in\{0,1\}^{(\boldsymbol{\nu}-\boldsymbol{\mu})}}} \prod_{l=1}^{\mu} i_l(\boldsymbol{u})^{j_l}\mathscr{F}_{F_1,F_2^{\boldsymbol{k}}}(\boldsymbol{C})(\boldsymbol{u})\mathscr{F}_{F_1^{\boldsymbol{j}},F_2}(\boldsymbol{B})(\boldsymbol{u}) \prod_{l=\mu+1}^{\nu} i_l(\boldsymbol{u})^{k_l}. \tag{5.4.87}
$$

with $e_j^{-f(\boldsymbol{x},\boldsymbol{u})}$ *from Notation 5.3.39.*

Exercise 5.4.37. *Theorem 5.4.35 can take various shapes, depending on the different GFTs. We show this for the same examples as before.*

1. *The Clifford Fourier transform from [119, 185, 222] takes the form*

$$
\mathscr{F}_{f_1}(\boldsymbol{A}) = \mathscr{F}_c(\boldsymbol{C})\mathscr{F}_{f_1}(\boldsymbol{B}) + \mathscr{F}_s(\boldsymbol{C})\mathscr{F}_{f_1}(\boldsymbol{B})i \tag{5.4.88}
$$

   for $n = 2 \pmod 4$. *For* $n = 3 \pmod 4$ *we can apply Corollary 5.4.32*

$$
\mathscr{F}_{f_1}(\boldsymbol{A}) = \mathscr{F}_{f_1}(\boldsymbol{C})\mathscr{F}_{f_1}(\boldsymbol{B}). \tag{5.4.89}
$$

   Because in this case, the pseudoscalar is in the center of $C\ell_{n,0}(n,0)$.

2. *The transform by Sommen [56, 306] is the only one of our examples that fulfills all constraints for Theorem 5.4.35 but not for Corollary 5.4.36. It satisfies*

$$
\begin{aligned}
&\mathscr{F}_{f_1,\ldots,f_n}(\boldsymbol{A})\\
&\quad= \sum_{\boldsymbol{k}\in\{0,1\}^n} \mathscr{F}_{F_2^{\boldsymbol{k}}}(\boldsymbol{C})\mathscr{F}_{-i_1\frac{\pi}{2}k_1+f_1,\ldots,-i_n\frac{\pi}{2}k_n+f_n}(\boldsymbol{B}).
\end{aligned} \tag{5.4.90}
$$

3. *The quaternion Fourier transform [56, 124] has the shape*

$$
\begin{aligned}
\mathscr{F}_{f_1,f_2}(\boldsymbol{A}) &= \sum_{j,k\in\{0,1\}} i_1^j\mathscr{F}_{f_1,f_2^k}(\boldsymbol{C})\mathscr{F}_{f_1^j,f_2}(\boldsymbol{B})i_2^k\\
&= \mathscr{F}_{f_1,c}(\boldsymbol{C})\mathscr{F}_{c,f_2}(\boldsymbol{B}) + \mathscr{F}_{f_1,s}(\boldsymbol{C})\mathscr{F}_{c,f_2}(\boldsymbol{B})j\\
&\quad + i\mathscr{F}_{f_1,c}(\boldsymbol{C})\mathscr{F}_{s,f_2}(\boldsymbol{B}) + i\mathscr{F}_{f_1,s}(\boldsymbol{C})\mathscr{F}_{s,f_2}(\boldsymbol{B})j.
\end{aligned} \tag{5.4.91}
$$

4. *And the spacetime Fourier transform [184] has principally the same shape*

$$\begin{aligned}\mathscr{F}_{f_1,f_2}(\boldsymbol{A}) &= \sum_{j,k\in\{0,1\}} \boldsymbol{e}_4^j \mathscr{F}_{f_1,f_2^k}(\boldsymbol{C})\mathscr{F}_{f_1^j,f_2}(\boldsymbol{B})(\epsilon_4\boldsymbol{e}_4 i_4)^k \\ &= \mathscr{F}_{f_1,c}(\boldsymbol{C})\mathscr{F}_{c,f_2}(\boldsymbol{B}) + \mathscr{F}_{f_1,s}(\boldsymbol{C})\mathscr{F}_{c,f_2}(\boldsymbol{B})\epsilon_4\boldsymbol{e}_4 i_4 \\ &\quad + \boldsymbol{e}_4\mathscr{F}_{f_1,c}(\boldsymbol{C})\mathscr{F}_{s,f_2}(\boldsymbol{B}) + \boldsymbol{e}_4\mathscr{F}_{f_1,s}(\boldsymbol{C})\mathscr{F}_{s,f_2}(\boldsymbol{B})\epsilon_4\boldsymbol{e}_4 i_4. \end{aligned} \tag{5.4.92}$$

5. *The Clifford Fourier transform for color images [18] with bivector B takes the form*

$$\begin{aligned}\mathscr{F}_{f_1,f_2,f_3,f_4}(\boldsymbol{A}) = \sum_{\boldsymbol{j},\boldsymbol{k}\in\{0,1\}^2} &(B)^{j_1}(iB)^{j_2}\mathscr{F}_{f_1,f_2,f_3^{k_1},f_4^{k_2}}(\boldsymbol{C}) \\ &\mathscr{F}_{f_1^{j_1},f_2^{j_2},f_3,f_4}(\boldsymbol{B})(-B)^{k_1}(-iB)^{k_2},\end{aligned} \tag{5.4.93}$$

or explicitly

$$\begin{aligned}&\mathscr{F}_{f_1,f_2,f_3,f_4}(\boldsymbol{A}) = \\ &\quad \mathscr{F}_{f_1,f_2,f_3^0,f_4^0}(\boldsymbol{C})\mathscr{F}_{f_1^0,f_2^0,f_3,f_4}(\boldsymbol{B}) \\ &\quad - \mathscr{F}_{f_1,f_2,f_3^0,f_4^1}(\boldsymbol{C})\mathscr{F}_{f_1^0,f_2^0,f_3,f_4}(\boldsymbol{B})iB \\ &\quad - \mathscr{F}_{f_1,f_2,f_3^1 f_4^0}(\boldsymbol{C})\mathscr{F}_{f_1^0,f_2^0,f_3,f_4}(\boldsymbol{B})B \\ &\quad + \mathscr{F}_{f_1,f_2,f_3^1,f_4^1}(\boldsymbol{C})\mathscr{F}_{f_1^0,f_2^0,f_3,f_4}(\boldsymbol{B})BiB \\ &\quad + iB\mathscr{F}_{f_1,f_2,f_3^0,f_4^0}(\boldsymbol{C})\mathscr{F}_{f_1^0,f_2^1,f_3,f_4}(\boldsymbol{B}) \\ &\quad - iB\mathscr{F}_{f_1,f_2,f_3^0,f_4^1}(\boldsymbol{C})\mathscr{F}_{f_1^0,f_2^1,f_3,f_4}(\boldsymbol{B})iB \\ &\quad - iB\mathscr{F}_{f_1,f_2,f_3^1,f_4^0}(\boldsymbol{C})\mathscr{F}_{f_1^0,f_2^1,f_3,f_4}(\boldsymbol{B})B \\ &\quad + iB\mathscr{F}_{f_1,f_2,f_3^1,f_4^1}(\boldsymbol{C})\mathscr{F}_{f_1^0,f_2^1,f_3,f_4}(\boldsymbol{B})BiB \\ &\quad + B\mathscr{F}_{f_1,f_2,f_3^0,f_4^0}(\boldsymbol{C})\mathscr{F}_{f_1^1,f_2^0,f_3,f_4}(\boldsymbol{B}) \\ &\quad - B\mathscr{F}_{f_1,f_2,f_3^0,f_4^1}(\boldsymbol{C})\mathscr{F}_{f_1^1,f_2^0,f_3,f_4}(\boldsymbol{B})iB \\ &\quad - B\mathscr{F}_{f_1,f_2,f_3^1,f_4^0}(\boldsymbol{C})\mathscr{F}_{f_1^1,f_2^0,f_3,f_4}(\boldsymbol{B})B \\ &\quad + B\mathscr{F}_{f_1,f_2,f_3^1,f_4^1}(\boldsymbol{C})\mathscr{F}_{f_1^1,f_2^0,f_3,f_4}(\boldsymbol{B})BiB \\ &\quad + BiB\mathscr{F}_{f_1,f_2,f_3^0,f_4^0}(\boldsymbol{C})\mathscr{F}_{f_1^1,f_2^1,f_3,f_4}(\boldsymbol{B}) \\ &\quad - BiB\mathscr{F}_{f_1,f_2,f_3^0,f_4^1}(\boldsymbol{C})\mathscr{F}_{f_1^1,f_2^1,f_3,f_4}(\boldsymbol{B})iB \\ &\quad - BiB\mathscr{F}_{f_1,f_2,f_3^1,f_4^0}(\boldsymbol{C})\mathscr{F}_{f_1^1,f_2^1,f_3,f_4}(\boldsymbol{B})B \\ &\quad + BiB\mathscr{F}_{f_1,f_2,f_3^1,f_4^1}(\boldsymbol{C})\mathscr{F}_{f_1^1,f_2^1,f_3,f_4}(\boldsymbol{B})BiB.\end{aligned} \tag{5.4.94}$$

6. *The cylindrical Fourier transform [49] is not separable except for the case $n = 2$. Here it satisfies*

$$\mathscr{F}_{f_1}(\boldsymbol{A}) = \mathscr{F}_{f_1^0}(\boldsymbol{C})\mathscr{F}_{f_1}(\boldsymbol{B}) + \boldsymbol{e}_{12}\mathscr{F}_{f_1^1}(\boldsymbol{C})\mathscr{F}_{f_1}(\boldsymbol{B}), \tag{5.4.95}$$

but for all other no closed formula can be constructed in a similar way.

5.4.3.4 Coorthogonality and bases

Definition 5.4.38. *We call two vectors $\boldsymbol{v}, \boldsymbol{w}$* ***orthogonal*** *($\boldsymbol{v} \perp \boldsymbol{w}$) if $\boldsymbol{v} \cdot \boldsymbol{w} = 0$ and* ***colinear*** *($\boldsymbol{v} \parallel \boldsymbol{w}$) if $\boldsymbol{v} \wedge \boldsymbol{w} = 0$.*

Definition 5.4.39. *We call two blades $\boldsymbol{A}, \boldsymbol{B}$* ***orthogonal*** *($\boldsymbol{A} \perp \boldsymbol{B}$) if all of their generating vectors are mutually orthogonal and* ***colinear*** *($\boldsymbol{A} \parallel \boldsymbol{B}$) if all vectors from one blade are colinear to all vectors in the other one.*

For a vector $\boldsymbol{v}$ and a blade $\boldsymbol{B} = \boldsymbol{b}_1 \wedge ... \wedge \boldsymbol{b}_d$ the following equalities hold

$$\begin{aligned} \boldsymbol{v} \perp \boldsymbol{B} &\Leftrightarrow \boldsymbol{v}\boldsymbol{B} = \boldsymbol{v} \wedge \boldsymbol{B} \Leftrightarrow \boldsymbol{v} \cdot \boldsymbol{B} = 0 \Leftrightarrow \boldsymbol{v}\boldsymbol{B} = (-1)^d \boldsymbol{B}\boldsymbol{v}, \\ \boldsymbol{v} \parallel \boldsymbol{B} &\Leftrightarrow \boldsymbol{v}\boldsymbol{B} = \boldsymbol{v} \cdot \boldsymbol{B} \Leftrightarrow \boldsymbol{v} \wedge \boldsymbol{B} = 0 \Leftrightarrow \boldsymbol{v}\boldsymbol{B} = (-1)^{d-1} \boldsymbol{B}\boldsymbol{v}, \end{aligned} \tag{5.4.96}$$

compare [161, 162]. That inspires the next definition.

Definition 5.4.40. *We call two blades $\boldsymbol{A}$ and $\boldsymbol{B}$* ***coorthogonal*** *if $\boldsymbol{A}\boldsymbol{B} = \pm \boldsymbol{B}\boldsymbol{A}$.*

Notation 5.4.41. *A blade can alternatively be written as an outer product of vectors or as a geometric product of orthogonal vectors. For blades $\boldsymbol{A} = \boldsymbol{a}_1 \wedge ... \wedge \boldsymbol{a}_{\boldsymbol{\mu}}$ and $\boldsymbol{B} = \boldsymbol{b}_1 \wedge ... \wedge \boldsymbol{b}_{\boldsymbol{\nu}}$ we will use the notations $\operatorname{span}(\boldsymbol{B}) := \operatorname{span}(\boldsymbol{b}_1, ..., \boldsymbol{b}_{\boldsymbol{\nu}})$, $\boldsymbol{A} \oplus \boldsymbol{B} := \operatorname{span}(\boldsymbol{A}) \oplus \operatorname{span}(\boldsymbol{B}) \subseteq \mathbb{R}^{p,q}$, $\boldsymbol{A} \cap \boldsymbol{B} := \operatorname{span}(\boldsymbol{A}) \cap \operatorname{span}(\boldsymbol{B}) \subseteq \mathbb{R}^{p,q}$, $\beta(\boldsymbol{A}, \boldsymbol{B}) := \dim(\boldsymbol{A} \cap \boldsymbol{B})$ and $\alpha(\boldsymbol{A}, \boldsymbol{B}) := \dim(\boldsymbol{A} \oplus \boldsymbol{B}) = \boldsymbol{\mu} + \boldsymbol{\nu} - \beta(\boldsymbol{A}, \boldsymbol{B})$. For a set of blades $B = \{\boldsymbol{B}_1, ..., \boldsymbol{B}_d\}, d \in \mathbb{N}$ we use the notation $\operatorname{span}(B) = \bigoplus_{k=1}^{d} \operatorname{span}(\boldsymbol{B}_k)$ and $\alpha(B) = \dim(\operatorname{span}(B))$.*

Lemma 5.4.42. *The basis blades $\boldsymbol{e}_{\boldsymbol{k}}$ of $\mathcal{G}^{p,q}$ that are generated from an orthogonal basis of $\mathbb{R}^{p,q}$ are mutually coorthogonal.*

Lemma 5.4.43. *For two coorthogonal blades $\boldsymbol{A}$ and $\boldsymbol{B}$ there is an orthonormal basis $V = \{\boldsymbol{v}_1, ..., \boldsymbol{v}_{\alpha(\boldsymbol{A},\boldsymbol{B})}\}$ of $\boldsymbol{A} \oplus \boldsymbol{B} \subseteq \mathbb{R}^{p,q}$ such that both can be expressed as real multiples of basis blades, that means $\boldsymbol{A} = \operatorname{sgn}(\boldsymbol{A})|\boldsymbol{A}|\boldsymbol{v}_{j_1}...\boldsymbol{v}_{j_{\boldsymbol{\mu}}}$ and $\boldsymbol{B} = \operatorname{sgn}(\boldsymbol{B})|\boldsymbol{B}|\boldsymbol{v}_{k_1}...\boldsymbol{v}_{k_{\boldsymbol{\nu}}}, a, b \in \mathbb{R}$ with the signum function being 1 or -1.*

Lemma 5.4.44. *Let $B = \{\boldsymbol{B}_1, ..., \boldsymbol{B}_d\}, d \in \mathbb{N}$ be non-zero mutually coorthogonal blades. Then there is an orthonormal basis $\boldsymbol{v}_1, ..., \boldsymbol{v}_{\alpha(B)}$ of $\operatorname{span}(B)$ such that every $\boldsymbol{B}_k, k = 1, ..., d$ can be written as a real multiple of a basis blade, that means $\boldsymbol{B}_k = \operatorname{sgn}(\boldsymbol{B}_k)|\boldsymbol{B}_k|\boldsymbol{v}_{\boldsymbol{j}(k)}$ with $\boldsymbol{v}_{\boldsymbol{j}(k)} = \boldsymbol{v}_{j_1(k),...,j_{\boldsymbol{\mu}}(k)}, \boldsymbol{\mu} = \boldsymbol{\mu}(k) = \dim(\boldsymbol{B}_k), |\boldsymbol{B}_k| \in \mathbb{R}$.*

Trivially spoken, coorthogonality of blades can as well be interpreted as coorthogonality of all their generating vectors, that means all their generating vectors are either orthogonal or colinear.

Theorem 5.4.45. *A finite number of blades are coorthogonal if and only if they are real multiples of basis blades of an orthonormal basis of $\mathbb{R}^{p,q}$.*

Notation 5.4.46. *Throughout this paper we will only deal with geometric Fourier transforms whose defining functions $f_1, ..., f_{\boldsymbol{\nu}}$, compare (5.4.72), are mutually coorthogonal blades, that means they satisfy the property $\forall l, k = 1, ..., \boldsymbol{\nu}, \forall \boldsymbol{x}, \boldsymbol{u} \in \mathbb{R}^m$:*

$$f_l(\boldsymbol{x}, \boldsymbol{u}) f_k(\boldsymbol{x}, \boldsymbol{u}) = \pm f_k(\boldsymbol{x}, \boldsymbol{u}) f_l(\boldsymbol{x}, \boldsymbol{u}). \tag{5.4.97}$$

Theorem 5.4.45 allows us to write

$$f_l(\boldsymbol{x}, \boldsymbol{u}) = \text{sgn}(f_l(\boldsymbol{x}, \boldsymbol{u}))|f_l(\boldsymbol{x}, \boldsymbol{u})|\boldsymbol{e}_{\boldsymbol{j}_l(\boldsymbol{x},\boldsymbol{u})}. \tag{5.4.98}$$

for all $l = 1, ..., \boldsymbol{\nu}$ *with a real valued function* $|f_l(\boldsymbol{x}, \boldsymbol{u})| : \mathbb{R}^m \times \mathbb{R}^m \to \mathbb{R}$ *and a function* $\boldsymbol{j}_l(\boldsymbol{x}, \boldsymbol{u}) : \mathbb{R}^m \times \mathbb{R}^m \to \mathcal{P}(\{1, ..., n\})$ *that maps to a multi-index indicating a basis multivector of a certain basis. We will refer to a set of functions with this property simply as a set of basis blade functions.*

Exercise 5.4.47. *This constraint seems strong but all standard examples of geometric Fourier transforms from [61] fulfill it.*

1. *For* $\boldsymbol{A} : \mathbb{R}^n \to Cl(n, 0), n = 2 \pmod 4$ *or* $n = 3 \pmod 4$*, the Clifford Fourier transform introduced by Jancewicz [222] for* $n = 3$ *and expanded by Ebling and Scheuermann [119] for* $n = 2$ *and Hitzer and Mawardi [185] for* $n = 2 \pmod 4$ *or* $n = 3 \pmod 4$ *with*

$$F_1 = \emptyset, \quad F_2 = \{f_1\}, \quad f_1(\boldsymbol{x}, \boldsymbol{u}) = 2\pi i_n \boldsymbol{x} \cdot \boldsymbol{u}, \tag{5.4.99}$$

   clearly fulfills the restriction, since it has only one defining function and i_n *is a basis blade.*

2. *The Sommen Bülow Clifford Fourier transform from [56, 306], defined by*

$$\begin{aligned} F_1 &= \emptyset, \quad F_2 = \{f_1, ..., f_n\}, \\ f_l(\boldsymbol{x}, \boldsymbol{u}) &= 2\pi \boldsymbol{e}_l x_l u_l, \quad \forall l = 1, ..., n, \end{aligned} \tag{5.4.100}$$

   for multivector fields $\mathbb{R}^n \to \mathcal{G}^{0,n}$ *fulfills it, because all basis vectors* $\boldsymbol{e}_k$ *are of course basis blades.*

3. *For* $\boldsymbol{A} : \mathbb{R}^2 \to Cl(0, 2) \approx \mathbb{H}$ *the quaternion Fourier transform [56, 124, 184] is generated by*

$$\begin{aligned} F_1 &= \{f_1\}, \qquad F_2 = \{f_2\}, \\ f_1(\boldsymbol{x}, \boldsymbol{u}) &= 2\pi i x_1 u_1, \qquad f_2(\boldsymbol{x}, \boldsymbol{u}) = 2\pi j x_2 u_2, \end{aligned} \tag{5.4.101}$$

   and satisfies the condition because i *and* j *are basis blades, too.*

4. *The defining functions of the spacetime Fourier transform of [184][37] with the* $Cl(3, 1)$*-pseudoscalar* i_4 *and*

$$\begin{aligned} F_1 &= \{f_1\}, \qquad F_2 = \{f_2\}, \qquad f_1(\boldsymbol{x}, \boldsymbol{u}) = \boldsymbol{e}_4 x_4 u_4, \\ f_2(\boldsymbol{x}, \boldsymbol{u}) &= \epsilon_4 \boldsymbol{e}_4 i_4 (x_1 u_1 + x_2 u_2 + x_3 u_3), \end{aligned} \tag{5.4.102}$$

   fulfill coorthogonality of blades, because of $\boldsymbol{e}_4 \parallel i_4 \Rightarrow \boldsymbol{e}_4 \perp \boldsymbol{e}_4 i_4$.

[37] Please note that [184] uses a different notation. His $\boldsymbol{x} = t\boldsymbol{e}_0 + x_1\boldsymbol{e}_1 + x_2\boldsymbol{e}_2 + x_3\boldsymbol{e}_3$ corresponds to our $\boldsymbol{x} = x_1\boldsymbol{e}_1 + x_2\boldsymbol{e}_2 + x_3\boldsymbol{e}_3 + x_4\boldsymbol{e}_4$, with $\boldsymbol{e}_0\boldsymbol{e}_0 = \epsilon_0 = -1$ being equivalent to our $\boldsymbol{e}_4\boldsymbol{e}_4 = \epsilon_4 = -1$.

5. *The Clifford Fourier transform for color images by Batard, Berthier and Saint-Jean [18] for* $m = 2, n = 4, \boldsymbol{A} : \mathbb{R}^2 \to Cl(4,0)$*, a fixed bivector* $\boldsymbol{B}$*, and the pseudoscalar* i *can be written as*

$$\begin{aligned} F_1 &= \{f_1, f_2\}, \qquad F_2 = \{f_3, f_4\}, \\ f_1(\boldsymbol{x},\boldsymbol{u}) &= \frac{1}{2}(x_1u_1 + x_2u_2)\boldsymbol{B}, \qquad f_2(\boldsymbol{x},\boldsymbol{u}) = \frac{1}{2}(x_1u_1 + x_2u_2)i\boldsymbol{B}, \\ f_3(\boldsymbol{x},\boldsymbol{u}) &= -f_1(\boldsymbol{x},\boldsymbol{u}), \qquad f_4(\boldsymbol{x},\boldsymbol{u}) = -f_2(\boldsymbol{x},\boldsymbol{u}). \end{aligned} \tag{5.4.103}$$

There are bivectors in $\mathcal{G}^{4,0}$ *that are not blades. But since Batard et al. start from* $\mathcal{G}^{3,0}$ *we may assume* $\boldsymbol{B}$ *to be a blade. So the transform fulfills condition (5.4.98), because* $\boldsymbol{B}$ *and* $i\boldsymbol{B}$ *commute. Let* $\boldsymbol{B}$ *consist of the two orthogonal vectors* $\boldsymbol{v}_1\boldsymbol{v}_2 = \boldsymbol{B}$*, then a basis as in Theorem 5.4.45 could be constructed by orthogonal basis completion of* $\boldsymbol{v}_1\boldsymbol{v}_2$ *to a basis* $B = \{\boldsymbol{v}_1, \boldsymbol{v}_2, \boldsymbol{v}_3, \boldsymbol{v}_4\}$ *and normalization. Because from* $\boldsymbol{v}_1\boldsymbol{v}_2\boldsymbol{v}_3\boldsymbol{v}_4 = ci, c \in \mathbb{R}$ *follows that* $i\boldsymbol{B} = -c^{-1}\boldsymbol{v}_3\boldsymbol{v}_4$ *is a basis blade, too.*

6. *The cylindrical Fourier transform as introduced by Brackx, De Schepper and Sommen in [49] with*

$$F_1 = \{f_1\}, \qquad F_2 = \emptyset, \qquad f_1(\boldsymbol{x},\boldsymbol{u}) = -\boldsymbol{x} \wedge \boldsymbol{u}, \tag{5.4.104}$$

satisfies the restriction because it has only one defining function, too. We will see that in contrast to the other transforms the basis guaranteed by Theorem 5.4.45 depends locally on $\boldsymbol{x}$ *and* $\boldsymbol{u}$ *here.*

Remark 5.4.48. *Theorem 5.4.45 guarantees, that there is an orthonormal basis of* $\mathbb{R}^{p,q}$ *such that* $\forall l = 1, ..., \boldsymbol{\nu}, \forall \boldsymbol{x}, \boldsymbol{u} \in \mathbb{R}^m$ *: the values of the functions* $f_l(\boldsymbol{x},\boldsymbol{u}) = \operatorname{sgn}(f_l(\boldsymbol{x},\boldsymbol{u}))|f_l(\boldsymbol{x},\boldsymbol{u})|\boldsymbol{e}_{\boldsymbol{k}(l)}$ *are real multiples of basis blades of* $\mathcal{G}^{p,q}$*. We assume that this basis is the one we use and we call the basis vectors simply* $\boldsymbol{e}_1, ..., \boldsymbol{e}_n$*. Therefore we can use the terms coorthogonal blades and basis blades as synonyms up to a real multiple, especially in terms of commutativity properties they can be used equivalently.*

5.4.3.5 Products with basis blades

From [61] we already know the following facts about products with invertible multivectors. Please note that every square root of minus one $i \in \mathcal{I}^{p,q}$ is invertible and that therefore the functions $f_l : \mathbb{R}^m \times \mathbb{R}^m \to \mathcal{I}^{p,q}$ from (5.4.72) are pointwise invertible, too.

Definition 5.4.49. *For an invertible multivector* $\boldsymbol{B} \in \mathcal{G}^{p,q} = Cl(p,q)$ *and an arbitrary multivector* $\boldsymbol{A} \in Cl(p,q)$ *we define*

$$\boldsymbol{A}_{c^0(\boldsymbol{B})} = \frac{1}{2}(\boldsymbol{A} + \boldsymbol{B}^{-1}\boldsymbol{A}\boldsymbol{B}), \quad \boldsymbol{A}_{c^1(\boldsymbol{B})} = \frac{1}{2}(\boldsymbol{A} - \boldsymbol{B}^{-1}\boldsymbol{A}\boldsymbol{B}). \tag{5.4.105}$$

Definition 5.4.50. *For $d \in \mathbb{N}$, $\boldsymbol{A} \in Cl(p,q)$, the ordered set $B = \{\boldsymbol{B}_1, ..., \boldsymbol{B}_d\}$ of invertible multivectors of $Cl(p,q)$ and any multi-index $\boldsymbol{j} \in \{0,1\}^d$ we define*

$$\begin{aligned} \boldsymbol{A}_{\boldsymbol{c}^{\boldsymbol{j}}(\overrightarrow{B})} &:= ((\boldsymbol{A}_{\boldsymbol{c}^{j_1}(\boldsymbol{B}_1)})_{\boldsymbol{c}^{j_2}(\boldsymbol{B}_2)} \cdots)_{\boldsymbol{c}^{j_d}(\boldsymbol{B}_d)}, \\ \boldsymbol{A}_{\boldsymbol{c}^{\boldsymbol{j}}(\overleftarrow{B})} &:= ((\boldsymbol{A}_{\boldsymbol{c}^{j_d}(\boldsymbol{B}_d)})_{\boldsymbol{c}^{j_{d-1}}(\boldsymbol{B}_{d-1})} \cdots)_{\boldsymbol{c}^{j_1}(\boldsymbol{B}_1)}, \end{aligned} \tag{5.4.106}$$

recursively with $\mathbf{c}^0, \mathbf{c}^1$ of Definition 5.4.49.

Lemma 5.4.51. *Let $d \in \mathbb{N}$, $B = \{\boldsymbol{B}_1, ..., \boldsymbol{B}_d\}$ be invertible multivectors and for $\boldsymbol{j} \in \{0,1\}^d$ let $|\boldsymbol{j}| := \sum_{k=1}^{d} j_k$, then $\forall \boldsymbol{A} \in Cl(p,q)$*

$$\begin{aligned} \boldsymbol{A} &= \sum_{\boldsymbol{j} \in \{0,1\}^d} \boldsymbol{A}_{\boldsymbol{c}^{\boldsymbol{j}}(\overrightarrow{B})}, \\ \boldsymbol{A}\boldsymbol{B}_1 ... \boldsymbol{B}_d &= \boldsymbol{B}_1 ... \boldsymbol{B}_d \sum_{\boldsymbol{j} \in \{0,1\}^d} (-1)^{|\boldsymbol{j}|} \boldsymbol{A}_{\boldsymbol{c}^{\boldsymbol{j}}(\overrightarrow{B})}, \\ \boldsymbol{B}_1 ... \boldsymbol{B}_d \boldsymbol{A} &= \sum_{\boldsymbol{j} \in \{0,1\}^d} (-1)^{|\boldsymbol{j}|} \boldsymbol{A}_{\boldsymbol{c}^{\boldsymbol{j}}(\overleftarrow{B})} \boldsymbol{B}_1 ... \boldsymbol{B}_d. \end{aligned} \tag{5.4.107}$$

Now we use the concept of coorthogonality to simplify and enhance the preliminary findings. For $d \in \mathbb{N}$ we take a closer look at sets of coorthogonal blades $B = \{\boldsymbol{B}_1, ..., \boldsymbol{B}_d\}$.

Lemma 5.4.52. *Let $B = \{\boldsymbol{B}_1, ..., \boldsymbol{B}_d\}$, $d \in \mathbb{N}$, be a set of mutually coorthogonal blades with the unique inverse $\boldsymbol{B}_k^{-1} = \boldsymbol{B}_k \boldsymbol{B}_k^{-2}, \boldsymbol{B}_k^2 \in \mathbb{R} \setminus \{0\}$. Further let $\boldsymbol{A} \in Cl(p,q)$ and $\boldsymbol{j} \in \{0,1\}^d$ be arbitrary, then $\boldsymbol{A}_{\boldsymbol{c}^{\boldsymbol{j}}(\overrightarrow{B})}$ and $\boldsymbol{A}_{\boldsymbol{c}^{\boldsymbol{j}}(\overleftarrow{B})}$ are independent from the order of B.*

Corollary 5.4.53. *For $d \in \mathbb{N}$, $\boldsymbol{A} \in Cl(p,q)$, the ordered set $B = \{\boldsymbol{B}_1, ..., \boldsymbol{B}_d\}$ of mutually coorthogonal blades and any multi-index $\boldsymbol{j} \in \{0,1\}^d$ we have*

$$\boldsymbol{A}_{\boldsymbol{c}^{\boldsymbol{j}}(\overrightarrow{B})} = \boldsymbol{A}_{\boldsymbol{c}^{\boldsymbol{j}}(\overleftarrow{B})}. \tag{5.4.108}$$

Notation 5.4.54. *Because of Corollary 5.4.53 we will not distinguish between $\boldsymbol{A}_{\boldsymbol{c}^{\boldsymbol{j}}(\overrightarrow{B})}$ and $\boldsymbol{A}_{\boldsymbol{c}^{\boldsymbol{j}}(\overleftarrow{B})}$ but just refer to the expression as $\boldsymbol{A}_{\boldsymbol{c}^{\boldsymbol{j}}(B)}$.*

Exercise 5.4.55. *There are simple partitions of a multivector into commutative and anticommutative parts, like for example for $\boldsymbol{A} = a_0\boldsymbol{e}_0 + a_1\boldsymbol{e}_1 + a_2\boldsymbol{e}_2 + a_{12}\boldsymbol{e}_{12} \in Cl(2,0)$ we get*

$$\begin{aligned} \boldsymbol{A}_{\boldsymbol{c}^0(\boldsymbol{e}_1)} &= \frac{1}{2}(\boldsymbol{A} + \boldsymbol{e}_1^{-1}\boldsymbol{A}\boldsymbol{e}_1) \\ &= \frac{1}{2}(\boldsymbol{A} + a_0 + a_1\boldsymbol{e}_1 - a_2\boldsymbol{e}_2 - a_{12}\boldsymbol{e}_{12}) = a_0 + a_1\boldsymbol{e}_1 \end{aligned} \tag{5.4.109}$$

and therefore $\boldsymbol{A} = \boldsymbol{A}_{\boldsymbol{c}^0(\boldsymbol{e}_1)} + \boldsymbol{A}_{\boldsymbol{c}^1(\boldsymbol{e}_1)} = a_0\boldsymbol{e}_0 + a_1\boldsymbol{e}_1 + a_2\boldsymbol{e}_2 + a_{12}\boldsymbol{e}_{12}$. But a decompositions can not always be achieved by just splitting up the multivector into its blades *with respect to a given basis. Sometimes the expressions of these parts are even longer than the multivector itself, for example* $\boldsymbol{A} = \boldsymbol{e}_1$ *satisfies*

$$
\begin{aligned}
(\boldsymbol{e}_1)_{\boldsymbol{c}^0(\boldsymbol{e}_1+\boldsymbol{e}_2)} &= \frac{1}{2}(\boldsymbol{e}_1 + (\boldsymbol{e}_1 + \boldsymbol{e}_2)^{-1}\boldsymbol{e}_1(\boldsymbol{e}_1 + \boldsymbol{e}_2)) \\
&= \frac{1}{2}(\boldsymbol{e}_1 + \frac{1}{2}(\boldsymbol{e}_1 + \boldsymbol{e}_2)\boldsymbol{e}_1(\boldsymbol{e}_1 + \boldsymbol{e}_2)) \\
&= \frac{1}{2}(\boldsymbol{e}_1 + \frac{1}{2}(\boldsymbol{e}_1 + \boldsymbol{e}_2 - \boldsymbol{e}_1\boldsymbol{e}_2\boldsymbol{e}_1 - \boldsymbol{e}_1\boldsymbol{e}_2\boldsymbol{e}_2)) \\
&= \frac{1}{2}(\boldsymbol{e}_1 + \boldsymbol{e}_2)
\end{aligned}
\tag{5.4.110}
$$

and gets decomposed into $\boldsymbol{e}_1 = (\boldsymbol{e}_1)_{\boldsymbol{c}^0(\boldsymbol{e}_1+\boldsymbol{e}_2)} + (\boldsymbol{e}_1)_{\boldsymbol{c}^1(\boldsymbol{e}_1+\boldsymbol{e}_2)} = \frac{1}{2}(\boldsymbol{e}_1 + \boldsymbol{e}_2) + \frac{1}{2}(\boldsymbol{e}_1 - \boldsymbol{e}_2).$

We will show that the decomposition of a multivector into commutative and anticommutative parts with respect to basis blades always is a decomposition into its blades along this basis. First consider one basis blade $\boldsymbol{e}_k$, here $\boldsymbol{c}^0(\boldsymbol{e}_k)$ can be interpreted as a mapping $\boldsymbol{c}^0 : \mathcal{G}^{p,q} \to \mathcal{P}(\{\boldsymbol{j} \subset \{1, ..., n\}, 1 \leq j_1, < ..., < j_\iota \leq n\})$ of the multivector argument into the power set of all multi-indices $\boldsymbol{j}$ as in (5.4.71) which indicate the basis blades of $\mathcal{G}^{p,q}$. The mapping $\boldsymbol{c}^0$ returns the blades of any multivector that commute with its argument $\boldsymbol{e}_k$ and its counterpart $\boldsymbol{c}^1 : \mathcal{G}^{p,q} \to \mathcal{P}(\{\boldsymbol{j} \subset \{1, ..., n\}, 1 \leq j_1, < ..., < j_\iota \leq n\})$ returns the blades that anticommute. The next Lemma will justify this interpretation, but for better understanding we start with a motivational example.

Exercise 5.4.56. *In the previous example the value of* $\boldsymbol{c}^0(\boldsymbol{e}_1)$ *would be* $\{\{0\}, \{1\}\}$ *and* $\boldsymbol{c}^1(\boldsymbol{e}_1) = \{\{2\}, \{12\}\}$*, so we could write*

$$
\begin{aligned}
\boldsymbol{A}_{\boldsymbol{c}^0(\boldsymbol{e}_1)} &= \sum_{\boldsymbol{j} \in \boldsymbol{c}^0(\boldsymbol{e}_1)} a_{\boldsymbol{j}}\boldsymbol{e}_{\boldsymbol{j}} - \sum_{\boldsymbol{j} \in \{\{0\},\{1\}\}} a_{\boldsymbol{j}}\boldsymbol{e}_{\boldsymbol{j}} = a_0\boldsymbol{e}_0 + a_1\boldsymbol{e}_1, \\
\boldsymbol{A}_{\boldsymbol{c}^1(\boldsymbol{e}_1)} &= \sum_{\boldsymbol{j} \in \boldsymbol{c}^1(\boldsymbol{e}_1)} a_{\boldsymbol{j}}\boldsymbol{e}_{\boldsymbol{j}} = \sum_{\boldsymbol{j} \in \{\{2\},\{12\}\}} a_{\boldsymbol{j}}\boldsymbol{e}_{\boldsymbol{j}} = a_2\boldsymbol{e}_2 + a_{12}\boldsymbol{e}_{12}.
\end{aligned}
\tag{5.4.111}
$$

Lemma 5.4.57. *We denote the length of the multi-indices* $\boldsymbol{j}, \boldsymbol{k}$ *by* ι *and* κ*. For a basis blade* $\boldsymbol{e}_k$ *and an arbitrary element* $\boldsymbol{A} = \sum_{\boldsymbol{j}} a_{\boldsymbol{j}}\boldsymbol{e}_{\boldsymbol{j}}$ *of* $\mathcal{G}^{p,q}$ *the multivectors* $\boldsymbol{A}_{\boldsymbol{c}^0(\boldsymbol{e}_k)}$ *and* $\boldsymbol{A}_{\boldsymbol{c}^1(\boldsymbol{e}_k)}$ *are a decomposition of* $\boldsymbol{A}$ *along the basis blades, that means*

$$
\boldsymbol{A}_{\boldsymbol{c}^0(\boldsymbol{e}_k)} = \sum_{\boldsymbol{j} \in \boldsymbol{c}^0(\boldsymbol{e}_k)} a_{\boldsymbol{j}}\boldsymbol{e}_{\boldsymbol{j}}, \qquad \boldsymbol{A}_{\boldsymbol{c}^1(\boldsymbol{e}_k)} = \sum_{\boldsymbol{j} \in \boldsymbol{c}^1(\boldsymbol{e}_k)} a_{\boldsymbol{j}}\boldsymbol{e}_{\boldsymbol{j}},
\tag{5.4.112}
$$

with $\boldsymbol{c}^0(\boldsymbol{e}_k) \cup \boldsymbol{c}^1(\boldsymbol{e}_k) = \{\boldsymbol{j} \subset \{1, ..., n\}, 1 \leq j_1, < ..., < j_\iota \leq n\}$ *as in (5.4.71) and* $\boldsymbol{c}^0(\boldsymbol{e}_k) \cap \boldsymbol{c}^1(\boldsymbol{e}_k) = \emptyset$ *and the index sets* $\boldsymbol{c}^0(\boldsymbol{e}_k), \boldsymbol{c}^1(\boldsymbol{e}_k)$ *take the forms*

$$
\begin{aligned}
\boldsymbol{c}^0(\boldsymbol{e}_k) &= \{\boldsymbol{j} \subset \{1, ..., n\}, \ 1 \leq j_1, < ..., < j_\iota \leq n, \ \iota\kappa - \beta(\boldsymbol{e}_{\boldsymbol{j}}, \boldsymbol{e}_{\boldsymbol{k}}) \text{ even } \}, \\
\boldsymbol{c}^1(\boldsymbol{e}_k) &= \{\boldsymbol{j} \subset \{1, ..., n\}, \ 1 \leq j_1, < ..., < j_\iota \leq n, \\
&\qquad \iota\kappa - \beta(\boldsymbol{e}_{\boldsymbol{j}}, \boldsymbol{e}_{\boldsymbol{k}}) \text{ odd } \}.
\end{aligned}
\tag{5.4.113}
$$

Remark 5.4.58. *An alternative way of describing the decomposition would be* $\boldsymbol{A}_{\boldsymbol{c}^0(\boldsymbol{e_k})} = \sum_{\boldsymbol{j}} a_{0\boldsymbol{j}}\boldsymbol{e_j}$ *and* $\boldsymbol{A}_{\boldsymbol{c}^1(\boldsymbol{e_k})} = \sum_{\boldsymbol{j}} a_{1\boldsymbol{j}}\boldsymbol{e_j}$ *with*

$$a_{0\boldsymbol{j}} = \begin{cases} a_{\boldsymbol{j}} & \text{for } \iota\kappa - \beta(\boldsymbol{j},\boldsymbol{k}) \text{ even,} \\ 0 & \text{else.} \end{cases} \tag{5.4.114}$$

The proof works analogously for $\boldsymbol{A}_{\boldsymbol{c}^1}(\boldsymbol{e_k}) = \sum_{\boldsymbol{j}} a_{1\boldsymbol{j}}\boldsymbol{e_j}$ *with*

$$a_{1\boldsymbol{j}} = \begin{cases} a_{\boldsymbol{j}} & \text{for } \iota\kappa - \beta(\boldsymbol{j},\boldsymbol{k}) \text{ odd,} \\ 0 & \text{else.} \end{cases} \tag{5.4.115}$$

and $\boldsymbol{A} = \sum_{\boldsymbol{j}}(a_{0\boldsymbol{j}} + a_{1\boldsymbol{j}})\boldsymbol{e_j}$ *with* $\forall \boldsymbol{j} : (a_{0\boldsymbol{j}} = a_{\boldsymbol{j}}, a_{1\boldsymbol{j}} = 0)$ *or* $(a_{0\boldsymbol{j}} = 0, a_{1\boldsymbol{j}} = a_{\boldsymbol{j}})$.

Lemma 5.4.59. *For basis blades* $B = \{\boldsymbol{e}_{\boldsymbol{k}(1)}, ..., \boldsymbol{e}_{\boldsymbol{k}(d)}\}$ *and multi-indices* $\vec{l} \in \{0,1\}^d$ *the* $\boldsymbol{A}_{\boldsymbol{c}^{\vec{l}}(B)}$ *form a decomposition of the multivector* $\boldsymbol{A}$ *along the basis blades, that means*

$$\boldsymbol{A}_{\boldsymbol{c}^{\vec{l}}(B)} = \sum_{\boldsymbol{j}\in \boldsymbol{c}^{\vec{l}}(B)} a_{\boldsymbol{j}}\boldsymbol{e_j}, \tag{5.4.116}$$

with $\bigcup_{l\in\{0,1\}^d} \boldsymbol{c}^{\vec{l}}(B) = \{\boldsymbol{j} \subset \{1,...,n\}, 1 \leq j_1, < ..., < j_\iota \leq n\}$ *as in (5.4.71) and* $\forall \vec{l} \neq \vec{l}' \in \{0,1\}^d : \boldsymbol{c}^{\vec{l}}(B) \cap \boldsymbol{c}^{\vec{l}'}(B) = \emptyset$ *and the index set* $\boldsymbol{c}^{\vec{l}}(B)$ *takes the form*

$$\boldsymbol{c}^{\vec{l}}(\boldsymbol{e}_{\boldsymbol{k}(1)}, ..., \boldsymbol{e}_{\boldsymbol{k}(d)}) = \bigcap_{\nu=1}^{d} \boldsymbol{c}^{l_\nu}(\boldsymbol{e}_{\boldsymbol{k}(\nu)})\,. \tag{5.4.117}$$

Remark 5.4.60. *For* $d \in \mathbb{N}$ *basis blades we can use the mapping* $\boldsymbol{c}^{\vec{l}} : (\mathcal{G}^{p,q})^d \to \mathcal{P}(\{\boldsymbol{j} \subset \{1,...,n\}, 1 \leq j_1, < ..., < j_\iota \leq n\})$ *to express the decomposition of a multivector. Compared to Definition 5.4.50 it can be computed much faster using the formula (5.4.117), which by the way again shows very clearly that the partition does not depend on the order of the blades.*

Exercise 5.4.61. *Like in the two preceding examples we look at* $\boldsymbol{A} \in \mathcal{G}^{2,0}$ *but this time we use (5.4.117). From*

$$\boldsymbol{c}^{0,0}(\boldsymbol{e}_1,\boldsymbol{e}_2) = \boldsymbol{c}^0(\boldsymbol{e}_1) \cap \boldsymbol{c}^0(\boldsymbol{e}_2) = \{\{0\},\{1\}\} \cap \{\{0\},\{2\}\} = \{0\}, \tag{5.4.118}$$

follows

$$\boldsymbol{A}_{\boldsymbol{c}^{0,0}(\boldsymbol{e}_1,\boldsymbol{e}_2)} = \sum_{\boldsymbol{j}\in \boldsymbol{c}^{0,0}(\boldsymbol{e}_1,\boldsymbol{e}_2)} a_{\boldsymbol{j}}\boldsymbol{e_j} = a_0\,, \tag{5.4.119}$$

and computing the other parts analogously we get

$$\begin{aligned} \boldsymbol{A} &= \boldsymbol{A}_{\boldsymbol{c}^{0,0}(\boldsymbol{e}_1,\boldsymbol{e}_2)} + \boldsymbol{A}_{\boldsymbol{c}^{1,0}(\boldsymbol{e}_1,\boldsymbol{e}_2)} + \boldsymbol{A}_{\boldsymbol{c}^{0,1}(\boldsymbol{e}_1,\boldsymbol{e}_2)} + \boldsymbol{A}_{\boldsymbol{c}^{1,1}(\boldsymbol{e}_1,\boldsymbol{e}_2)} \\ &= a_0 + a_1\boldsymbol{e}_1 + a_2\boldsymbol{e}_2 + a_{12}\boldsymbol{e}_{12} = \sum_{\boldsymbol{j}} a_{\boldsymbol{j}}\boldsymbol{e_j}\,. \end{aligned} \tag{5.4.120}$$

Remark 5.4.62. *The decomposition with respect to basis blades is independent from the multivector* $\boldsymbol{A}$ *and the total amount of parts that occur is limited by* $\min\{2^d, 2^n\}$, *where* d *is the number of blades in* B *and* $n = p+q$ *the dimension of the underlying vector space. In the case of the previous example this means that a higher* d *would not result in a finer segmentation of* $\boldsymbol{A}$. *Certain combinations of commutation properties will just remain empty, for instance*

$$\mathbf{c}^{1,1,1}(\mathbf{e}_1, \mathbf{e}_2, \mathbf{e}_{12}) = \emptyset. \tag{5.4.121}$$

Now we take a look at the decomposition of exponentials of functions as they appear in (5.4.72) that satisfy condition (5.4.98) and show that they take a very simple form.

Lemma 5.4.63. *Let the value of* $f(\boldsymbol{x},\boldsymbol{u}) : \mathbb{R}^m \times \mathbb{R}^m \to \mathscr{I}^{p,q}$ *be a real multiple of a basis blade* $\forall \boldsymbol{x},\boldsymbol{u} \in \mathbb{R}^m$, $f(\boldsymbol{x},\boldsymbol{u}) = \operatorname{sgn}(f(\boldsymbol{x},\boldsymbol{u}))|f(\boldsymbol{x},\boldsymbol{u})|\mathbf{e}_{\boldsymbol{k}(\boldsymbol{x},\boldsymbol{u})}$ *like in (5.4.98). The decompositions* $e^{-f(\boldsymbol{x},\boldsymbol{u})}_{\mathbf{c}^0(\mathbf{e}_{\vec{l}})}, e^{-f(\boldsymbol{x},\boldsymbol{u})}_{\mathbf{c}^1(\mathbf{e}_{\vec{l}})}$ *of the exponential with respect to any basis blade* $\mathbf{e}_{\vec{l}} \in \mathcal{G}^{p,q}$ *can only take two different shapes:*

$$\begin{aligned}
e^{-f(\boldsymbol{x},\boldsymbol{u})}_{\mathbf{c}^0(\mathbf{e}_{\vec{l}})} &= \begin{cases} e^{-f(\boldsymbol{x},\boldsymbol{u})} & \text{if } \boldsymbol{k}(\boldsymbol{x},\boldsymbol{u}) \in \mathbf{c}^0(\mathbf{e}_{\vec{l}}), \\ \cos(|f(\boldsymbol{x},\boldsymbol{u})|) & \text{if } \boldsymbol{k}(\boldsymbol{x},\boldsymbol{u}) \notin \mathbf{c}^0(\mathbf{e}_{\vec{l}}), \end{cases} \\
e^{-f(\boldsymbol{x},\boldsymbol{u})}_{\mathbf{c}^1(\mathbf{e}_{\vec{l}})} &= \begin{cases} -\frac{f(\boldsymbol{x},\boldsymbol{u})}{|f(\boldsymbol{x},\boldsymbol{u})|}\sin(|f(\boldsymbol{x},\boldsymbol{u})|) & \text{if } \boldsymbol{k}(\boldsymbol{x},\boldsymbol{u}) \in \mathbf{c}^1(\mathbf{e}_{\vec{l}}), \\ 0 & \text{if } \boldsymbol{k}(\boldsymbol{x},\boldsymbol{u}) \notin \mathbf{c}^1(\mathbf{e}_{\vec{l}}). \end{cases}
\end{aligned} \tag{5.4.122}$$

Lemma 5.4.64. *Let* $f(\boldsymbol{x},\boldsymbol{u}) : \mathbb{R}^m \times \mathbb{R}^m \to \mathscr{I}^{p,q}$ *satisfy property (5.4.98)* $\forall \boldsymbol{x},\boldsymbol{u} \in \mathbb{R}^m$, $\vec{l} \in \{0,1\}^d$ *a multi-index and* $B = \{\mathbf{e}_{\boldsymbol{k}(1)}, ..., \mathbf{e}_{\boldsymbol{k}(d)}\}$ *be a set of basis blades. The decompositions of the exponential with respect to* B *can only take four different shapes:*

$$e^{-f(\boldsymbol{x},\boldsymbol{u})}_{\mathbf{c}^{\vec{l}}(B)} = \begin{cases} e^{-f(\boldsymbol{x},\boldsymbol{u})} & \text{if } \vec{l} = 0 \text{ and } \boldsymbol{k}(\boldsymbol{x},\boldsymbol{u}) \in \mathbf{c}^{\vec{l}}(B), \\ \cos(|f(\boldsymbol{x},\boldsymbol{u})|) & \text{if } \vec{l} = 0 \text{ and } \boldsymbol{k}(\boldsymbol{x},\boldsymbol{u}) \notin \mathbf{c}^{\vec{l}}(B), \\ -\frac{f(\boldsymbol{x},\boldsymbol{u})}{|f(\boldsymbol{x},\boldsymbol{u})|}\sin(|f_l(\boldsymbol{x},\boldsymbol{u})|) & \text{if } \vec{l} \neq 0 \text{ and } \boldsymbol{k}(\boldsymbol{x},\boldsymbol{u}) \in \mathbf{c}^{\vec{l}}(B), \\ 0 & \text{if } \vec{l} \neq 0 \text{ and } \boldsymbol{k}(\boldsymbol{x},\boldsymbol{u}) \notin \mathbf{c}^{\vec{l}}(B). \end{cases} \tag{5.4.123}$$

Remark 5.4.65. *The shift theorem introduced in [61] takes a simpler form for GFTs satisfying property (5.4.98), that means for all transforms in the first example. The simplification results from the predictable shape of the decomposition of exponentials with respect to basis blades from Lemma 5.4.64.*

5.4.3.6 Geometric convolution theorem

We have seen that coorthogonal blade functions can be expressed as real multiples of basis blades $f(\boldsymbol{x},\boldsymbol{u}) = \operatorname{sgn}(f(\boldsymbol{x},\boldsymbol{u}))|f(\boldsymbol{x},\boldsymbol{u})|\mathbf{e}_{\boldsymbol{j}(\boldsymbol{x},\boldsymbol{u})} : \mathbb{R}^m \times \mathbb{R}^m \to \mathscr{I}^{p,q}$ in Notation 5.4.46. Lemma 5.4.59 guarantees for basis blade functions, that during the decomposition into commutative and anticommutative parts of a multivector no additional terms appear in the sum over the basis blades (5.4.71). Each part is a real fragment of the multivector along the basis blades of the orthogonal basis from Theorem 5.4.45. Because of that an exponential can only become decomposed into four different shapes: itself, a cosine, a basis blade multiplied with a sine or zero, compare Lemma 5.4.64. This motivates the generalization of geometric Fourier transforms (5.4.72) to trigonometric transforms. We will just use it as an auxiliary construction here and analyze its properties and applications in a future paper.

Definition 5.4.66 (Geometric Trigonometric Transform). *Let $\boldsymbol{A} : \mathbb{R}^m \to \mathcal{G}^{p,q} = Cl(p,q)$ be a multivector field and $\boldsymbol{x}, \boldsymbol{u} \in \mathbb{R}^m$ vectors, F_1, F_2 two ordered finite sets of $\boldsymbol{\mu}$, respectively $\boldsymbol{\nu}-\boldsymbol{\mu}$, mappings $\mathbb{R}^m \times \mathbb{R}^m \to \mathscr{I}^{p,q}$, G_1, G_2 two ordered finite sets of $\boldsymbol{\mu}$, respectively $\boldsymbol{\nu}-\boldsymbol{\mu}$, mappings $(\mathbb{R}^m \times \mathbb{R}^m \to \mathscr{I}^{p,q}) \to \mathcal{G}^{p,q}$ with each $g_l(-f_l(\boldsymbol{x},\boldsymbol{u}))\forall l = 1, ..., \boldsymbol{\nu}$ having one of the shapes from (5.4.123):*

$$g_l(-f_l(\boldsymbol{x},\boldsymbol{u})) = \begin{cases} e^{-f_l(\boldsymbol{x},\boldsymbol{u})}, \\ \cos(|f_l(\boldsymbol{x},\boldsymbol{u})|), \\ -\frac{f_l(\boldsymbol{x},\boldsymbol{u})}{|f_l(\boldsymbol{x},\boldsymbol{u})|}\sin(|f_l(\boldsymbol{x},\boldsymbol{u})|), \\ 0. \end{cases} \tag{5.4.124}$$

The ***Geometric Trigonometric Transform*** *(GTT) $\mathscr{F}_{G_1(F_1),G_2(F_2)}(\boldsymbol{A})$ is defined by*

$$\mathscr{F}_{G_1(F_1),G_2(F_2)}(\boldsymbol{A})(\boldsymbol{u}) := \int_{\mathbb{R}^m} \prod_{l=1}^{\boldsymbol{\mu}} g_l(-f_l(\boldsymbol{x},\boldsymbol{u}))\boldsymbol{A}(\boldsymbol{x}) \prod_{l=\boldsymbol{\mu}+1}^{\boldsymbol{\nu}} g_l(-f_l(\boldsymbol{x},\boldsymbol{u})). \tag{5.4.125}$$

Notation 5.4.67. *We have seen in Lemma 5.4.64 that the decomposition of an exponential with respect to basis blades takes the same shape like the functions G_1, G_2 of a GTT (5.4.124). Therefore for a geometric Fourier transform with basis blade functions F_1, F_2, two sets of basis blades $B_1 = \{\boldsymbol{e}_{\boldsymbol{k}(1)}, ..., \boldsymbol{e}_{\boldsymbol{k}(\eta)}\}$, $B_2 = \{\boldsymbol{e}_{\boldsymbol{k}(\eta+1)}, ..., \boldsymbol{e}_{\boldsymbol{k}(\theta-\eta)}\}$ and strictly lower and upper triangular matrices[38] $J \in \{0,1\}^{\boldsymbol{\mu}\times\eta}$, $K \in \{0,1\}^{(\boldsymbol{\nu}-\boldsymbol{\mu})\times\theta}$ whose rows are $\boldsymbol{\mu}$ and $\boldsymbol{\nu}-\boldsymbol{\mu}$ multi-indices $(J)_l \in \{0,1\}^\eta$ respectively $(K)_l \in \{0,1\}^\theta$, we can construct a geometric trigonometric transform $\mathscr{F}_{G_1(F_1),G_2(F_2)}(\boldsymbol{A})$ by setting $g_l(-f_l(\boldsymbol{x},\boldsymbol{u})) = e_{\boldsymbol{c}^{(J)_l}}^{-f_l(\boldsymbol{x},\boldsymbol{u})}$ for $l = 1, ..., \boldsymbol{\mu}$ and $g_l(-f_l(\boldsymbol{x},\boldsymbol{u})) = e_{\boldsymbol{c}^{(K)_l}}^{-f_l(\boldsymbol{x},\boldsymbol{u})}$ for $l = \boldsymbol{\mu}+1, ..., \boldsymbol{\nu}$. We refer to it shortly as*

$$\mathscr{F}_{(F_1)_{\boldsymbol{c}^J(B_1)},(F_2)_{\boldsymbol{c}^K(B_2)}}(\boldsymbol{A})(\boldsymbol{u}) := \int_{\mathbb{R}^m} \prod_{l=1}^{\boldsymbol{\mu}} e_{\boldsymbol{c}^{(J)_l}(B_1)}^{-f_l(\boldsymbol{x},\boldsymbol{u})} \boldsymbol{A}(\boldsymbol{x}) \prod_{l=\boldsymbol{\mu}+1}^{\boldsymbol{\nu}} e_{\boldsymbol{c}^{(K)_{l-\boldsymbol{\mu}}}(B_2)}^{-f_l(\boldsymbol{x},\boldsymbol{u})} \, \mathrm{d}^m\boldsymbol{x}. \tag{5.4.126}$$

In the case of $\mathscr{F}_{(F_1)_{\boldsymbol{c}^J(F_1)},(F_2)_{\boldsymbol{c}^K(F_2)}}$ we will only write $\mathscr{F}_{(F_1)_{\boldsymbol{c}^J},(F_2)_{\boldsymbol{c}^K}}$.

[38] These matrices were introduced originally in Lemma 6.8 in [61]. It is repeated in this work as Lemma 5.4.72. The proof can be found in [61].

The geometric trigonometric transform is a generalization of the geometric Fourier transform from (5.4.72). It can be used to prove the convolution theorem of the GFT. To accomplish this we additionally need the following facts shown in [61]. Please note that for the proofs of all Lemmata from [61] the claim for the set of functions to be basis blades functions is not necessary.

Definition 5.4.68. *We call a GFT* ***left (right) separable****, if*

$$f_l = |f_l(\boldsymbol{x}, \boldsymbol{u})| i_l(\boldsymbol{u}), \tag{5.4.127}$$

$\forall l = 1, ..., \boldsymbol{\mu}$, $(l = \boldsymbol{\mu} + 1, ..., \boldsymbol{\nu})$*, where* $|f_l(\boldsymbol{x}, \boldsymbol{u})| : \mathbb{R}^m \times \mathbb{R}^m \to \mathbb{R}$ *is a real function and* $i_l : \mathbb{R}^m \to \mathscr{I}^{p,q}$ *a function that does not depend on* $\boldsymbol{x}$*.*

Lemma 5.4.69. *Let* $F = \{f_1(\boldsymbol{x}, \boldsymbol{u}), ..., f_d(\boldsymbol{x}, \boldsymbol{u})\}$ *be a set of pointwise invertible functions then the ordered product of their exponentials and an arbitrary multivector* $\boldsymbol{A} \in \mathcal{G}^{p,q}$ *satisfies*

$$\prod_{l=1}^{d} e^{-f_l(\boldsymbol{x},\boldsymbol{u})} \boldsymbol{A} = \sum_{\boldsymbol{j} \in \{0,1\}^d} \boldsymbol{A}_{\boldsymbol{c}^{\boldsymbol{j}}(\overleftarrow{F})}(\boldsymbol{x}, \boldsymbol{u}) \prod_{l=1}^{d} e^{-(-1)^{j_l} f_l(\boldsymbol{x},\boldsymbol{u})}, \tag{5.4.128}$$

where $\boldsymbol{A}_{\boldsymbol{c}^{\boldsymbol{j}}(\overleftarrow{F})}(\boldsymbol{x}, \boldsymbol{u}) := \boldsymbol{A}_{\boldsymbol{c}^{\boldsymbol{j}}(\overleftarrow{F(\boldsymbol{x},\boldsymbol{u})})}$ *is a multivector valued function* $\mathbb{R}^m \times \mathbb{R}^m \to \mathcal{G}^{p,q}$*.*

Lemma 5.4.70. *Let* $F = \{f_1(\boldsymbol{x}, \boldsymbol{u}), ..., f_d(\boldsymbol{x}, \boldsymbol{u})\}$ *be a set of separable functions that are linear with respect to* $\boldsymbol{x}$*. Further let* $J \in \{0,1\}^{d \times d}$ *be a strictly lower triangular matrix, that is associated column by column with a multi-index* $\boldsymbol{j} \in \{0,1\}^d$ *by* $\forall k = 1, ..., d : (\sum_{l=1}^{d} J_{l,k}) \bmod 2 = j_k$*, with* $(J)_l$ *being its* l*-th row, then*

$$\begin{aligned}
&\prod_{l=1}^{d} e^{-f_l(\boldsymbol{x}+\boldsymbol{y},\boldsymbol{u})} \\
&= \sum_{\boldsymbol{j} \in \{0,1\}^d} \sum_{\substack{J \subset \{0,1\}^{d \times d}, \\ \sum_{l=1}^{d} (J)_l \bmod 2 = \boldsymbol{j}}} \prod_{l=1}^{d} e^{-f_l(\boldsymbol{x},\boldsymbol{u})}_{\boldsymbol{c}^{(J)_l}(\overleftarrow{f_1,...,f_l,0,...,0})} \prod_{l=1}^{d} e^{-(-1)^{j_l} f_l(\boldsymbol{y},\boldsymbol{u})},
\end{aligned} \tag{5.4.129}$$

or alternatively with strictly upper triangular matrices J

$$\begin{aligned}
&\prod_{l=1}^{d} e^{-f_l(\boldsymbol{x}+\boldsymbol{y},\boldsymbol{u})} \\
&= \sum_{\boldsymbol{j} \in \{0,1\}^d} \sum_{\substack{J \in \{0,1\}^{d \times d}, \\ \sum_{l=1}^{d} (J)_l \bmod 2 = \boldsymbol{j}}} \prod_{l=1}^{d} e^{-(-1)^{j_l} f_l(\boldsymbol{x},\boldsymbol{u})} \prod_{l=1}^{d} e^{-f_l(\boldsymbol{y},\boldsymbol{u})}_{\boldsymbol{c}^{(J)_l}(\overrightarrow{0,...,0,f_l,...,f_d})}.
\end{aligned} \tag{5.4.130}$$

Definition 5.4.71. *For a set of functions* $F = \{f_1(\boldsymbol{x}, \boldsymbol{u}), ..., f_d(\boldsymbol{x}, \boldsymbol{u})\}$ *and a multi-index* $\boldsymbol{j} \in \{0,1\}^d$*, we define the set of functions* $F(\boldsymbol{j})$ *by*

$$F(\boldsymbol{j}) := \{(-1)^{j_1} f_1(\boldsymbol{x}, \boldsymbol{u}), ..., (-1)^{j_d} f_d(\boldsymbol{x}, \boldsymbol{u})\}. \tag{5.4.131}$$

We also need a generalization of Lemma 5.4.69 that allows us to swap the order of partial exponentials and multivectors.

Lemma 5.4.72. *For sets of functions $F = \{f_1(\boldsymbol{x}, \boldsymbol{u}), ..., f_d(\boldsymbol{x}, \boldsymbol{u})\}, G = \{g_1, ..., g_d\}$ like in (5.4.124) we get analogously to Lemma 5.4.69*

$$\prod_{l=1}^{d} g_l(-f_l(\boldsymbol{x}, \boldsymbol{u}))\boldsymbol{A} = \sum_{\boldsymbol{j}\in\{0,1\}^d} \boldsymbol{A}_{\boldsymbol{c}^{\boldsymbol{j}}(F)} \prod_{l=1}^{d} g_l(-(-1)^{j_l} f_l(\boldsymbol{x}, \boldsymbol{u})). \tag{5.4.132}$$

Definition 5.4.73. *Let $\boldsymbol{A}(\boldsymbol{x}), \boldsymbol{B}(\boldsymbol{x}) : \mathbb{R}^m \to \mathcal{G}^{p,q}$ be two multivector fields. Their* ***convolution*** *$(\boldsymbol{A} * \boldsymbol{B})(\boldsymbol{x})$ is defined as*

$$(\boldsymbol{A} * \boldsymbol{B})(\boldsymbol{x}) := \int_{\mathbb{R}^m} \boldsymbol{A}(\boldsymbol{y})\boldsymbol{B}(\boldsymbol{x} - \boldsymbol{y}) \, \mathrm{d}^m\boldsymbol{y}. \tag{5.4.133}$$

Theorem 5.4.74 (Convolution)**.** *Let $\boldsymbol{A}, \boldsymbol{B}, \boldsymbol{C} : \mathbb{R}^m \to \mathcal{G}^{p,q}$ be multivector fields with $\boldsymbol{A}(\boldsymbol{x}) = (\boldsymbol{C} * \boldsymbol{B})(\boldsymbol{x})$ and F_1, F_2 be coorthogonal, separable and linear with respect to the first argument, $\boldsymbol{j}, \boldsymbol{j}' \in \{0,1\}^{\mu}, \boldsymbol{k}, \boldsymbol{k}' \in \{0,1\}^{(\nu-\mu)}$ and $J \in \{0,1\}^{\mu\times\mu}$ and $K \in \{0,1\}^{(\nu-\mu)\times(\nu-\mu)}$ are the strictly lower, respectively upper, triangular matrices with rows $(J)_l, (K)_{l-\mu}$ summing up to $(\sum_{l=1}^{\mu}(J)_l) \bmod 2 = \boldsymbol{j}$ respectively $(\sum_{l=\mu+1}^{\nu}(K)_{l-\mu}) \bmod 2 = \boldsymbol{k}$ as in Lemma 5.4.70, then the geometric Fourier transform of $\boldsymbol{A}$ satisfies the convolution property*

$$\begin{aligned}\mathscr{F}_{F_1,F_2}(\boldsymbol{A})(\boldsymbol{u}) = \sum_{\boldsymbol{j},\boldsymbol{j}',\boldsymbol{k},\boldsymbol{k}'} \sum_{J,K} &\left(\mathscr{F}_{F_1(\boldsymbol{j}),F_2(\boldsymbol{k}+\boldsymbol{k}')}(\boldsymbol{C})(\boldsymbol{u})\right)_{\boldsymbol{c}^{\boldsymbol{j}'}(F_1)} \\ &\mathscr{F}_{(F_1(\boldsymbol{j}'))_{\boldsymbol{c}^J},(F_2)_{\boldsymbol{c}^K}}(\boldsymbol{B}_{\boldsymbol{c}^{\boldsymbol{k}'}(F_2)})(\boldsymbol{u}).\end{aligned} \tag{5.4.134}$$

Remark 5.4.75. *The formula in the convolution theorem can take various other shapes depending on the way Lemma 5.4.70 is applied. In Theorem 5.4.74 we used Lemma 5.4.70 in its first version (5.4.129) on F_1 and in its second version (5.4.130) on F_2. This has the advantage that the GTT is needed only on one side. We get the same effect with its second version (5.4.130) on F_1 and its first version (5.4.130) on F_2 by*

$$\begin{aligned}\mathscr{F}_{F_1,F_2}(\boldsymbol{A})(\boldsymbol{u}) = \sum_{\boldsymbol{j},\boldsymbol{j}',\boldsymbol{k},\boldsymbol{k}'} \sum_{J,K} &\left(\mathscr{F}_{(F_1)_{\boldsymbol{c}^J},(F_2(\boldsymbol{k}'))_{\boldsymbol{c}^K}}(\boldsymbol{C})(\boldsymbol{u})\right)_{\boldsymbol{c}^{\boldsymbol{j}'}(F_1)} \\ &\mathscr{F}_{F_1(\boldsymbol{j}+\boldsymbol{j}'),F_2(\boldsymbol{k})}(\boldsymbol{B}_{\boldsymbol{c}^{\boldsymbol{k}'}(F_2)})(\boldsymbol{u}).\end{aligned} \tag{5.4.135}$$

Using the first version twice leads to

$$\begin{aligned}\mathscr{F}_{F_1,F_2}(\boldsymbol{A})(\boldsymbol{u}) = \sum_{\boldsymbol{j},\boldsymbol{j}',\boldsymbol{k},\boldsymbol{k}'} \sum_{J,K} &\left(\mathscr{F}_{F_1(\boldsymbol{j}),(F_2(\boldsymbol{k}'))_{\boldsymbol{c}^K}}(\boldsymbol{C})(\boldsymbol{u})\right)_{\boldsymbol{c}^{\boldsymbol{j}'}(F_1)} \\ &\mathscr{F}_{(F_1(\boldsymbol{j}'))_{\boldsymbol{c}^J},F_2(\boldsymbol{k})}(\boldsymbol{B}_{\boldsymbol{c}^{\boldsymbol{k}'}(F_2)})(\boldsymbol{u}),\end{aligned} \tag{5.4.136}$$

and using the second twice to

$$\begin{aligned}\mathscr{F}_{F_1,F_2}(\boldsymbol{A})(\boldsymbol{u}) = \sum_{\boldsymbol{j},\boldsymbol{j}',\boldsymbol{k},\boldsymbol{k}'} \sum_{J,K} &\left(\mathscr{F}_{(F_1)_{\boldsymbol{c}^J},F_2(\boldsymbol{k}+\boldsymbol{k}')}(\boldsymbol{C})(\boldsymbol{u})\right)_{\boldsymbol{c}^{\boldsymbol{j}'}(F_1)} \\ &\mathscr{F}_{F_1(\boldsymbol{j}+\boldsymbol{j}'),(F_2)_{\boldsymbol{c}^K}}(\boldsymbol{B}_{\boldsymbol{c}^{\boldsymbol{k}'}(F_2)})(\boldsymbol{u}).\end{aligned} \tag{5.4.137}$$

These versions have the advantage of being a bit more symmetric. During the proof of Theorem 5.4.74 one can start by recomposing the transform around $\boldsymbol{C}$. *Each of the four formulae obtained has a counterpart that is constructed by restructuring around* $\boldsymbol{B}$ *first. Listed in the analog order they take the shapes*

$$
\begin{aligned}
&\mathscr{F}_{F_1,F_2}(\boldsymbol{A})(\boldsymbol{u})\\
&= \sum_{\boldsymbol{j},\boldsymbol{j}',\boldsymbol{k},\boldsymbol{k}'} \sum_{J,K} \mathscr{F}_{F_1(\boldsymbol{j}),F_2(\boldsymbol{k}+\boldsymbol{k}')}(\boldsymbol{C}_{\boldsymbol{c}^{\boldsymbol{j}'}(F_1)})(\boldsymbol{u}) \big(\mathscr{F}_{(F_1(\boldsymbol{j}'))_{\boldsymbol{c}^J},(F_2)_{\boldsymbol{c}^K}}(\boldsymbol{B})(\boldsymbol{u})\big)_{\boldsymbol{c}^{\boldsymbol{k}'}(F_2)},\\
&\mathscr{F}_{F_1,F_2}(\boldsymbol{A})(\boldsymbol{u})\\
&= \sum_{\boldsymbol{j},\boldsymbol{j}',\boldsymbol{k},\boldsymbol{k}'} \sum_{J,K} \mathscr{F}_{(F_1)_{\boldsymbol{c}^J},(F_2(\boldsymbol{k}'))_{\boldsymbol{c}^K}}(\boldsymbol{C}_{\boldsymbol{c}^{\boldsymbol{j}'}(F_1)})(\boldsymbol{u}) \big(\mathscr{F}_{F_1(\boldsymbol{j}+\boldsymbol{j}'),F_2(\boldsymbol{k})}(\boldsymbol{B})(\boldsymbol{u})\big)_{\boldsymbol{c}^{\boldsymbol{k}'}(F_2)},\\
&\mathscr{F}_{F_1,F_2}(\boldsymbol{A})(\boldsymbol{u})\\
&= \sum_{\boldsymbol{j},\boldsymbol{j}',\boldsymbol{k},\boldsymbol{k}'} \sum_{J,K} \mathscr{F}_{F_1(\boldsymbol{j}),(F_2(\boldsymbol{k}'))_{\boldsymbol{c}^K}}(\boldsymbol{C}_{\boldsymbol{c}^{\boldsymbol{j}'}(F_1)})(\boldsymbol{u}) \big(\mathscr{F}_{(F_1(\boldsymbol{j}'))_{\boldsymbol{c}^J},F_2(\boldsymbol{k})}(\boldsymbol{B})(\boldsymbol{u})\big)_{\boldsymbol{c}^{\boldsymbol{k}'}(F_2)},\\
&\mathscr{F}_{F_1,F_2}(\boldsymbol{A})(\boldsymbol{u})\\
&= \sum_{\boldsymbol{j},\boldsymbol{j}',\boldsymbol{k},\boldsymbol{k}'} \sum_{J,K} \mathscr{F}_{(F_1)_{\boldsymbol{c}^J},F_2(\boldsymbol{k}+\boldsymbol{k}')}(\boldsymbol{C}_{\boldsymbol{c}^{\boldsymbol{j}'}(F_1)})(\boldsymbol{u}) \big(\mathscr{F}_{F_1(\boldsymbol{j}+\boldsymbol{j}'),(F_2)_{\boldsymbol{c}^K}}(\boldsymbol{B})(\boldsymbol{u})\big)_{\boldsymbol{c}^{\boldsymbol{k}'}(F_2)}.
\end{aligned}
\tag{5.4.138}
$$

Depending on the application one or some of these might be preferred compared to the others, because of savings in memory or runtime.

Corollary 5.4.76 (Convolution). *Let* $\boldsymbol{A}, \boldsymbol{B}, \boldsymbol{C} : \mathbb{R}^m \to \mathcal{G}^{p,q}$ *be multivector fields with* $\boldsymbol{A}(\boldsymbol{x}) = (\boldsymbol{C} * \boldsymbol{B})(\boldsymbol{x})$ *and* F_1, F_2 *each consist of mutually commutative functions*[39], *being separable and linear with respect to the first argument and* $\boldsymbol{j}' \in \{0,1\}^{\mu}$, $\boldsymbol{k}' \in \{0,1\}^{(\nu-\mu)}$ *multi-indices, then the geometric Fourier transforms satisfy the convolution property*

$$
\begin{aligned}
\mathscr{F}_{F_1,F_2}(\boldsymbol{A})(\boldsymbol{u}) = \sum_{\boldsymbol{j}',\boldsymbol{k}'} \big(\mathscr{F}_{F_1,F_2(\boldsymbol{k}')}(\boldsymbol{C})(\boldsymbol{u})\big)_{\boldsymbol{c}^{\boldsymbol{j}'}(F_1)}\\
\mathscr{F}_{F_1(\boldsymbol{j}'),F_2}(\boldsymbol{B}_{\boldsymbol{c}^{\boldsymbol{k}'}(F_2)})(\boldsymbol{u}),
\end{aligned}
\tag{5.4.139}
$$

or

$$
\begin{aligned}
\mathscr{F}_{F_1,F_2}(\boldsymbol{A})(\boldsymbol{u}) = \sum_{\boldsymbol{j}',\boldsymbol{k}'} \mathscr{F}_{F_1,F_2(\boldsymbol{k}')}(\boldsymbol{C}_{\boldsymbol{c}^{\boldsymbol{j}'}(F_1)})(\boldsymbol{u})\\
\big(\mathscr{F}_{F_1(\boldsymbol{j}'),F_2}(\boldsymbol{B})(\boldsymbol{u})\big)_{\boldsymbol{c}^{\boldsymbol{k}'}(F_2)}.
\end{aligned}
\tag{5.4.140}
$$

If the values of the functions in F_1 *and* F_2 *are in the center of* $\mathcal{G}^{p,q}$ *it even satisfies the simple product formula*

$$
\mathscr{F}_{F_1,F_2}(\boldsymbol{A})(\boldsymbol{u}) = \mathscr{F}_{F_1,F_2}(\boldsymbol{C})(\boldsymbol{u})\mathscr{F}_{F_1,F_2}(\boldsymbol{B})(\boldsymbol{u}). \tag{5.4.141}
$$

Exercise 5.4.77. *We summarize the exact shape of the convolution of multivector fields under the transforms from the first Example using the same order.*

[39]Cross commutativity is not necessary.

1. *The Clifford Fourier transform from [119, 185, 222] takes the form*

$$\mathscr{F}_{f_1}(\boldsymbol{A}) = \mathscr{F}_{f_1}(\boldsymbol{C}_{\boldsymbol{c}^0(i)})\mathscr{F}_{f_1}(\boldsymbol{B}) + \mathscr{F}_{-f_1}(\boldsymbol{C}_{\boldsymbol{c}^1(i)})\mathscr{F}_{f_1}(\boldsymbol{B}) \tag{5.4.142}$$

   for $n = 2 \pmod 4$ *and for* $n = 3 \pmod 4$ *the even simpler one*

$$\mathscr{F}_{f_1}(\boldsymbol{A}) = \mathscr{F}_{f_1}(\boldsymbol{C})\mathscr{F}_{f_1}(\boldsymbol{B}) \tag{5.4.143}$$

   because in this case the pseudoscalar is in the center of $\mathcal{G}^{n,0}$.

2. *The Sommen Bülow Clifford Fourier transform [56, 306] is the only one of our examples that does not fulfill the constraints of Corollary 5.4.76 but the ones of Theorem 5.4.74*

$$\begin{aligned}\mathscr{F}_{f_1,\dots,f_n}(\boldsymbol{A}) = \sum_{\boldsymbol{k},\boldsymbol{k}'\in\{0,1\}^n} \sum_{K} &\mathscr{F}_{(-1)^{k_1+k'_1}f_1,\dots,(-1)^{k_n+k'_n}f_n}(\boldsymbol{C}) \\ &\mathscr{F}_{(f_1,\dots,f_n)_{\boldsymbol{c}K}}(\boldsymbol{B}_{\boldsymbol{c}^{\boldsymbol{k}'}(f_1,\dots,f_n)})\end{aligned} \tag{5.4.144}$$

   with strictly upper triangular matrices in $\{0,1\}^{n\times n}$ *with rows* $(K)_{l-\boldsymbol{\mu}}$ *summing up to* $(\sum_{l=\boldsymbol{\mu}+1}^{\boldsymbol{\nu}}(K)_{l-\boldsymbol{\mu}}) \bmod 2 = \boldsymbol{k}$.

3. *The quaternion Fourier transform [56, 124, 184] has the shape*

$$\begin{aligned}\mathscr{F}_{f_1,f_2}(\boldsymbol{A}) =&\Big(\mathscr{F}_{f_1,f_2}(\boldsymbol{C})\Big)_{\boldsymbol{c}^0(f_1)}\mathscr{F}_{f_1,f_2}(\boldsymbol{B}_{\boldsymbol{c}^0(f_2)}) \\ &+\Big(\mathscr{F}_{f_1,f_2}(\boldsymbol{C})\Big)_{\boldsymbol{c}^1(f_1)}\mathscr{F}_{-f_1,f_2}(\boldsymbol{B}_{\boldsymbol{c}^0(f_2)})\end{aligned} \tag{5.4.145}$$

4. *And the spacetime Fourier transform [184] has exactly the same shape*

$$\begin{aligned}\mathscr{F}_{f_1,f_2}(\boldsymbol{A}) =&\Big(\mathscr{F}_{f_1,f_2}(\boldsymbol{C})\Big)_{\boldsymbol{c}^0(f_1)}\mathscr{F}_{f_1,f_2}(\boldsymbol{B}_{\boldsymbol{c}^0(f_2)}) \\ &+\Big(\mathscr{F}_{f_1,f_2}(\boldsymbol{C})\Big)_{\boldsymbol{c}^1(f_1)}\mathscr{F}_{-f_1,f_2}(\boldsymbol{B}_{\boldsymbol{c}^0(f_2)}).\end{aligned} \tag{5.4.146}$$

5. *The Clifford Fourier transform for color images [18] takes the rather long form*

$$\begin{aligned}\mathscr{F}_{f_1,f_2,f_3,f_4}(\boldsymbol{A}) = &\Big(\mathscr{F}_{f_1,f_2,f_3,f_4}(\boldsymbol{C})\Big)_{\boldsymbol{c}^{00}(f_1,f_2)}\mathscr{F}_{f_1,f_2,f_3,f_4}(\boldsymbol{B}_{\boldsymbol{c}^{00}(f_3,f_4)}) \\ &+\Big(\mathscr{F}_{f_1,f_2,f_3,f_4}(\boldsymbol{C})\Big)_{\boldsymbol{c}^{01}(f_1,f_2)}\mathscr{F}_{f_1,-f_2,f_3,f_4}(\boldsymbol{B}_{\boldsymbol{c}^{00}(f_3,f_4)}) \\ &+\Big(\mathscr{F}_{f_1,f_2,f_3,f_4}(\boldsymbol{C})\Big)_{\boldsymbol{c}^{10}(f_1,f_2)}\mathscr{F}_{-f_1,f_2,f_3,f_4}(\boldsymbol{B}_{\boldsymbol{c}^{00}(f_3,f_4)}) \\ &+\Big(\mathscr{F}_{f_1,f_2,f_3,f_4}(\boldsymbol{C})\Big)_{\boldsymbol{c}^{11}(f_1,f_2)}\mathscr{F}_{-f_1,-f_2,f_3,f_4}(\boldsymbol{B}_{\boldsymbol{c}^{00}(f_3,f_4)}) \\ &+\Big(\mathscr{F}_{f_1,f_2,f_3,-f_4}(\boldsymbol{C})\Big)_{\boldsymbol{c}^{00}(f_1,f_2)}\mathscr{F}_{f_1,f_2,f_3,f_4}(\boldsymbol{B}_{\boldsymbol{c}^{01}(f_3,f_4)}) \\ &+\Big(\mathscr{F}_{f_1,f_2,f_3,-f_4}(\boldsymbol{C})\Big)_{\boldsymbol{c}^{01}(f_1,f_2)}\mathscr{F}_{f_1,-f_2,f_3,f_4}(\boldsymbol{B}_{\boldsymbol{c}^{01}(f_3,f_4)}) \\ &+\Big(\mathscr{F}_{f_1,f_2,f_3,-f_4}(\boldsymbol{C})\Big)_{\boldsymbol{c}^{10}(f_1,f_2)}\mathscr{F}_{-f_1,f_2,f_3,f_4}(\boldsymbol{B}_{\boldsymbol{c}^{01}(f_3,f_4)})\end{aligned}$$

$$
\begin{aligned}
&+ \left(\mathscr{F}_{f_1,f_2,f_3,-f_4}(\boldsymbol{C})\right)_{\mathbf{c}^{11}(f_1,f_2)} \mathscr{F}_{-f_1,-f_2,f_3,f_4}(\boldsymbol{B}_{\mathbf{c}^{01}(f_3,f_4)}) \\
&+ \left(\mathscr{F}_{f_1,f_2,-f_3,f_4}(\boldsymbol{C})\right)_{\mathbf{c}^{00}(f_1,f_2)} \mathscr{F}_{f_1,f_2,f_3,f_4}(\boldsymbol{B}_{\mathbf{c}^{10}(f_3,f_4)}) \\
&+ \left(\mathscr{F}_{f_1,f_2,-f_3,f_4}(\boldsymbol{C})\right)_{\mathbf{c}^{01}(f_1,f_2)} \mathscr{F}_{f_1,-f_2,f_3,f_4}(\boldsymbol{B}_{\mathbf{c}^{10}(f_3,f_4)}) \\
&+ \left(\mathscr{F}_{f_1,f_2,-f_3,f_4}(\boldsymbol{C})\right)_{\mathbf{c}^{10}(f_1,f_2)} \mathscr{F}_{-f_1,f_2,f_3,f_4}(\boldsymbol{B}_{\mathbf{c}^{10}(f_3,f_4)}) \\
&+ \left(\mathscr{F}_{f_1,f_2,-f_3,f_4}(\boldsymbol{C})\right)_{\mathbf{c}^{11}(f_1,f_2)} \mathscr{F}_{-f_1,-f_2,f_3,f_4}(\boldsymbol{B}_{\mathbf{c}^{10}(f_3,f_4)}) \\
&+ \left(\mathscr{F}_{f_1,f_2,-f_3,-f_4}(\boldsymbol{C})\right)_{\mathbf{c}^{00}(f_1,f_2)} \mathscr{F}_{f_1,f_2,f_3,f_4}(\boldsymbol{B}_{\mathbf{c}^{11}(f_3,f_4)}) \\
&+ \left(\mathscr{F}_{f_1,f_2,-f_3,-f_4}(\boldsymbol{C})\right)_{\mathbf{c}^{01}(f_1,f_2)} \mathscr{F}_{f_1,-f_2,f_3,f_4}(\boldsymbol{B}_{\mathbf{c}^{1}(f_3,f_4)}) \\
&+ \left(\mathscr{F}_{f_1,f_2,-f_3,-f_4}(\boldsymbol{C})\right)_{\mathbf{c}^{10}(f_1,f_2)} \mathscr{F}_{-f_1,f_2,f_3,f_4}(\boldsymbol{B}_{\mathbf{c}^{11}(f_3,f_4)}) \\
&+ \left(\mathscr{F}_{f_1,f_2,-f_3,-f_4}(\boldsymbol{C})\right)_{\mathbf{c}^{11}(f_1,f_2)} \mathscr{F}_{-f_1,-f_2,f_3,f_4}(\boldsymbol{B}_{\mathbf{c}^{11}(f_3,f_4)}).
\end{aligned}
\tag{5.4.147}
$$

6. *The cylindrical Fourier transform [49] is not separable except for the case* $n = 2$. *Here the convolution corollary holds*

$$\mathscr{F}_{f_1}(\boldsymbol{C}_{\mathbf{c}^0(f_1)})\mathscr{F}_{f_1}(\boldsymbol{B}) + \mathscr{F}_{-f_1}(\boldsymbol{C}_{\mathbf{c}^1(f_1)})\mathscr{F}_{f_1}(\boldsymbol{B}), \tag{5.4.148}$$

but for all other no closed formula can be constructed in a similar way.

5.4.3.7 Summary of two-sided CFT with multiple square roots of -1 and convolution

In this section, we introduced the concept of coorthogonality as the property of commutation or anticommutation of blades. We showed that it is equivalent to the claim for blades to be real multiples of basis blades for an orthonormal basis and presented an algorithm to compute this basis. The Lemmata 5.4.51, 5.4.69 and 5.4.70 about multiplication with invertible factors, that were primarily stated and proved in Section 5.3.2 (following [61]), become simplified for coorthogonal blades. We saw that in this case the partition of the multivector $\boldsymbol{A}$ takes place along the basis blades and that it is independent from the relative order of the factors it is exchanged with. A consequence of this is, that every exponential can only have four simple predictable shapes after decomposition: itself, a sine, a cosine or zero. That fact inspired the definition of the geometric trigonometric transform, whose properties should be studied further.

As examples of the advantages of the point of view on the GFTs related to TTs (Section 5.3.3), we presented two convolution theorems. Until this equivalence to the trigonometric transforms was shown, the convolution theorems of the GFT were only valid for the subclass of the coorthogonal GFT. Now Theorems 5.4.31 and 5.4.35 give closed expressions for the convolution of any GFT with mappings that are separable and linear in the first argument, but have arbitrary commutation properties.

By means of the GTT we were able to show a convolution theorem (Theorem 5.4.74) for the general geometric Fourier transform introduced in Section 5.3.2 [61].

It highlights the rich consequences of the geometric structure created by utilizing general geometric square roots of minus one in sets F_1, F_2. The information contained in the multivector fields, appears now finely segmented and related term by term. The choice of F_1, F_2 determines this segmentation. Because convolution appears in wavelet theory and is closely related to correlation, the GTF convolution theorem may have interesting consequences for multidimensional geometric pattern matching and neural network type learning algorithms as well as for geometric algebra wavelet theory.

5.5 SPECIAL CLIFFORD FOURIER TRANSFORMS

5.5.1 Windowed Clifford Fourier transform (CWFT)

5.5.1.1 Background

We have already discussed in previous sections several generalizations to higher dimension of the classical Fourier transform (FT) using Clifford geometric algebra, including the two-dimensional (2D) Clifford Fourier transform (CFT), $n = 2$, also a special case of Section 5.2.1. Based on the 2D CFT, we now establish the two-dimensional Clifford windowed Fourier transform (CWFT). Using the spectral representation of the CFT, we derive several important properties such as shift, modulation, a reproducing kernel, isometry and an orthogonality relation. Finally, we discuss examples of the CWFT and compare the CFT and the CWFT. This section is based on [254].

One of the basic problems encountered in signal representations using the conventional Fourier transform (FT) is the ineffectiveness of the Fourier kernel to represent and compute location information. One method to overcome such a problem is the windowed Fourier transform (WFT). Some authors [148, 318] have extensively studied the WFT and its properties from a mathematical point of view. In [234, 337] they applied the WFT as a tool of spatial-frequency analysis which is able to characterize the local frequency at any location in a fringe pattern.

On the other hand, we know that Clifford geometric algebra leads to the consequent generalization of real and harmonic analysis to higher dimensions. Clifford algebra accurately treats geometric entities depending on their dimension as scalars, vectors, bivectors (oriented plane area elements) and tri-vectors (oriented volume elements), etc. Motivated by the above facts, we generalize the WFT in the framework of Clifford geometric algebra.

In this section, we study the two-dimensional (2D) Clifford windowed Fourier transform (CWFT). A complementary motivation for studying this topic comes from the understanding that the 2D CWFT is in fact intimately related with Clifford Gabor filters [42] and quaternionic Gabor filters [56, 58]. This generalization also enables us to establish the two-dimensional Clifford Gabor filters.

Let us consider an orthonormal vector basis $\{\boldsymbol{e}_1, \boldsymbol{e}_2\}$ of the real two-dimensional (2D) Euclidean vector space $\mathbb{R}^2 = \mathbb{R}^{2,0}$. The geometric algebra over $\mathbb{R}^2$ denoted by $\mathcal{G}_2 = Cl(2,0)$ then has the graded 4-dimensional basis

$$\{1, \boldsymbol{e}_1, \boldsymbol{e}_2, \boldsymbol{e}_{12}\}, \tag{5.5.1}$$

where 1 is the real scalar identity element (grade 0), $\boldsymbol{e}_1, \boldsymbol{e}_2 \in \mathbb{R}^2$ are vectors (grade 1), and $\boldsymbol{e}_{12} = \boldsymbol{e}_1\boldsymbol{e}_2 = i_2$ defines the unit oriented pseudoscalar[40] (grade 2), i.e. the highest grade blade element in $\mathcal{G}_2$.

The general elements of a geometric algebra are called multivectors. Every multivector $f \in \mathcal{G}_2$ can be expressed as

$$f = \underbrace{\alpha_0}_{\text{scalar part}} + \underbrace{\alpha_1\boldsymbol{e}_1 + \alpha_2\boldsymbol{e}_2}_{\text{vector part}} + \underbrace{\alpha_{12}\boldsymbol{e}_{12}}_{\text{bivector part}}, \forall\alpha_0, \alpha_1, \alpha_2, \alpha_{12} \in \mathbb{R}. \tag{5.5.2}$$

The grade selector is defined as $\langle f \rangle_k$ for the k-vector part of f. We often write $\langle\ldots\rangle = \langle\ldots\rangle_0$. Then equation (5.5.2) can be expressed as[41]

$$f = \langle f \rangle + \langle f \rangle_1 + \langle f \rangle_2. \tag{5.5.3}$$

The multivector f is called a parabivector if the vector part of (5.5.3) is zero, i.e.

$$f = \alpha_0 + \alpha_{12}\boldsymbol{e}_{12}. \tag{5.5.4}$$

The reverse $\tilde{f}$ of a multivector $f \in \mathcal{G}_2$ is an anti-automorphism given by

$$\tilde{f} = \langle f \rangle + \langle f \rangle_1 - \langle f \rangle_2, \tag{5.5.5}$$

which fulfills $\widetilde{fg} = \tilde{g}\tilde{f}$ for every $f, g \in \mathcal{G}_2$. In particular $\tilde{i}_2 = -i_2$.

The scalar product of two multivectors $f, \tilde{g}$ is defined as the scalar part of the geometric product $f\tilde{g}$

$$f * \tilde{g} = \langle f\tilde{g} \rangle = \alpha_0\beta_0 + \alpha_1\beta_1 + \alpha_2\beta_2 + \alpha_{12}\beta_{12}, \tag{5.5.6}$$

which leads to a cyclic product symmetry

$$\langle pqr \rangle = \langle qrp \rangle, \quad \forall p, q, r \in \mathcal{G}_2. \tag{5.5.7}$$

For $f = g$ in (5.5.6) we obtain the modulus (or magnitude) $|f|$ of a multivector $f \in \mathcal{G}_2$ defined as

$$|f|^2 = f * \tilde{f} = \alpha_0^2 + \alpha_1^2 + \alpha_2^2 + \alpha_{12}^2. \tag{5.5.8}$$

It is convenient to introduce an inner product for two multivector valued functions $f, g : \mathbb{R}^2 \to \mathcal{G}_2$ as follows:

$$(f, g)_{L^2(\mathbb{R}^2;\mathcal{G}_2)} = \int_{\mathbb{R}^2} f(\boldsymbol{x})\widetilde{g(\boldsymbol{x})}\, d^2\boldsymbol{x}. \tag{5.5.9}$$

---

[40]Other names in use are *bivector* or *oriented area element*.

[41]Note that (5.5.3) and (5.5.5) show grade selection and not component selection.

One can check that this inner product satisfies the following rules:

$$\begin{aligned}
(f, g+h)_{L^2(\mathbb{R}^2;\mathcal{G}_2)} &= (f,g)_{L^2(\mathbb{R}^2;\mathcal{G}_2)} + (f,h)_{L^2(\mathbb{R}^2;\mathcal{G}_2)}, \\
(f, \lambda g)_{L^2(\mathbb{R}^2;\mathcal{G}_2)} &= (f,g)_{L^2(\mathbb{R}^2;\mathcal{G}_2)}\tilde{\lambda}, \\
(f\lambda, g)_{L^2(\mathbb{R}^2;\mathcal{G}_2)} &= (f, g\tilde{\lambda})_{L^2(\mathbb{R}^2;\mathcal{G}_2)}, \\
(f, g)_{L^2(\mathbb{R}^2;\mathcal{G}_2)} &= \widetilde{(g, f)}_{L^2(\mathbb{R}^2;\mathcal{G}_2)},
\end{aligned} \tag{5.5.10}$$

where $f, g \in L^2(\mathbb{R}^2; \mathcal{G}_2)$, and $\lambda \in \mathcal{G}_2$ is a multivector constant. The scalar part of the inner product gives the L^2-norm

$$\|f\|^2_{L^2(\mathbb{R}^2;\mathcal{G}_2)} = \left\langle (f,f)_{L^2(\mathbb{R}^2;\mathcal{G}_2)} \right\rangle. \tag{5.5.11}$$

Definition 5.5.1 (Clifford module). *Let $\mathcal{G}_2$ be the real Clifford algebra of 2D Euclidean space $\mathbb{R}^2$. A Clifford algebra module $L^2(\mathbb{R}^2; \mathcal{G}_2)$ is defined by*

$$L^2(\mathbb{R}^2; \mathcal{G}_2) = \{ f : \mathbb{R}^2 \longrightarrow \mathcal{G}_2 \mid \|f\|_{L^2(\mathbb{R}^2;\mathcal{G}_2)} < \infty \}. \tag{5.5.12}$$

We will use the following definition of 2D CFT.

Definition 5.5.2. *The CFT[42] of $f \in L^2(\mathbb{R}^2; \mathcal{G}_2) \bigcap L^1(\mathbb{R}^2; \mathcal{G}_2)$ is the function $\mathcal{F}\{f\}$: $\mathbb{R}^2 \to \mathcal{G}_2$ given by*

$$\mathcal{F}\{f\}(\boldsymbol{\omega}) = \int_{\mathbb{R}^2} f(\boldsymbol{x}) e^{-i_2 \boldsymbol{\omega} \cdot \boldsymbol{x}} d^2\boldsymbol{x}, \tag{5.5.13}$$

where we can write $\boldsymbol{\omega} = \omega_1 \boldsymbol{e}_1 + \omega_2 \boldsymbol{e}_2$ and $\boldsymbol{x} = x_1 \boldsymbol{e}_1 + x_2 \boldsymbol{e}_2$. Note that

$$d^2\boldsymbol{x} = \frac{d\boldsymbol{x}_1 \wedge d\boldsymbol{x}_2}{i_2} \tag{5.5.14}$$

is scalar valued ($d\boldsymbol{x}_k = dx_k \boldsymbol{e}_k$, $k = 1, 2$, no summation). Notice that the Clifford Fourier kernel $e^{-i_2 \boldsymbol{\omega} \cdot \boldsymbol{x}}$ does not commute with every element of the Clifford algebra $\mathcal{G}_2$. Furthermore, the product has to be performed in a fixed order.

Theorem 5.5.3. *Suppose that $f \in L^2(\mathbb{R}^2; \mathcal{G}_2)$ and $\mathcal{F}\{f\} \in L^1(\mathbb{R}^2; \mathcal{G}_2)$. Then the CFT is an invertible transform and its inverse is calculated by*

$$\mathcal{F}^{-1}[\mathcal{F}\{f\}(\boldsymbol{\omega})](\boldsymbol{x}) = f(\boldsymbol{x}) = \frac{1}{(2\pi)^2} \int_{\mathbb{R}^2} \mathcal{F}\{f\}(\boldsymbol{\omega})\, e^{i_2 \boldsymbol{\omega} \cdot \boldsymbol{x}} d^2\boldsymbol{\omega}. \tag{5.5.15}$$

5.5.1.2 2D Clifford windowed Fourier transform

In [42, 185] the 2D CFT has been introduced, compare Section 5.2.1 and the above Definition 5.5.2. This now enables us to establish the 2D CWFT. We will see that several properties of the WFT can be established in the new construction with some modifications. We begin with the definition of the 2D CWFT.

[42]For Fourier transforms with $f \in L^1(\mathbb{R}^2; \mathcal{G}_2)$ and $f \in L^2(\mathbb{R}^2; \mathcal{G}_2)$, see Footnote 2 on page 144 regarding the roles of L^1 and L^2.

5.5.1.2.1 Definition of the CWFT

Definition 5.5.4. *A Clifford window function is a function* $\phi \in L^2(\mathbb{R}^2; \mathcal{G}_2) \setminus \{0\}$ *so that* $|\boldsymbol{x}|^{1/2}\phi(\boldsymbol{x}) \in L^2(\mathbb{R}^2; \mathcal{G}_2)$.

$$\phi_{\boldsymbol{\omega},\boldsymbol{b}}(\boldsymbol{x}) = \frac{e^{i_2\boldsymbol{\omega}\cdot\boldsymbol{x}}\phi(\boldsymbol{x}-\boldsymbol{b})}{(2\pi)^2}, \tag{5.5.16}$$

denote the so-called Clifford window daughter functions.

Definition 5.5.5 (Clifford windowed Fourier transform). *The Clifford windowed Fourier transform (CWFT)* $G_\phi f$ *of* $f \in L^2(\mathbb{R}^2; \mathcal{G}_2)$ *is defined by*

$$\begin{aligned} f(\boldsymbol{x}) \quad \longrightarrow \quad & G_\phi f(\boldsymbol{\omega}, \boldsymbol{b}) = (f, \phi_{\boldsymbol{\omega},\boldsymbol{b}})_{L^2(\mathbb{R}^2;\mathcal{G}_2)} \\ = \quad & \frac{1}{(2\pi)^2}\int_{\mathbb{R}^2} f(\boldsymbol{x})\ \{e^{i_2\boldsymbol{\omega}\cdot\boldsymbol{x}}\phi(\boldsymbol{x}-\boldsymbol{b})\}^{\sim}\, d^2\boldsymbol{x} \\ = \quad & \frac{1}{(2\pi)^2}\int_{\mathbb{R}^2} f(\boldsymbol{x})\ \widetilde{\phi(\boldsymbol{x}-\boldsymbol{b})}\, e^{-i_2\boldsymbol{\omega}\cdot\boldsymbol{x}} d^2\boldsymbol{x}. \end{aligned} \tag{5.5.17}$$

This shows that the CWFT can be regarded as the CFT of the product of a Clifford-valued function f and a shifted and reversed Clifford window function ϕ, or as an inner product (5.5.9) of a Clifford-valued function f and the Clifford window daughter functions $\phi_{\boldsymbol{\omega},\boldsymbol{b}}$.

Taking the Gaussian function as the window function of (5.5.16), with $\boldsymbol{\omega} = \boldsymbol{\omega}_0 = \omega_{0,1}\boldsymbol{e}_1 + \omega_{0,2}\boldsymbol{e}_2$ fixed we obtain Clifford Gabor filters, i.e.

$$g_c(\mathbf{x}, \sigma_1, \sigma_2) = \frac{1}{(2\pi)^2} e^{i_2\boldsymbol{\omega}_0\cdot\boldsymbol{x}} e^{-\left[(x_1/\sigma_1)^2+(x_2/\sigma_2)^2\right]/2}, \tag{5.5.18}$$

where σ_1 and σ_2 are standard deviations of the Gaussian functions and the translation parameters are $b_1 = b_2 = 0$.

In terms of the $\mathcal{G}_2$ Clifford Fourier transform equation (5.5.18) can be expressed as

$$\mathcal{F}\{g_c\}(\boldsymbol{\omega}) = \frac{1}{\pi\sigma_1\sigma_2} e^{-\frac{1}{2}\left[(\sigma_1^2(\omega_1-\omega_{0,1})^2+\sigma_2^2(\omega_2-\omega_{0,2})^2\right]}. \tag{5.5.19}$$

From equations (5.5.18) and (5.5.19) we see that Clifford Gabor filters are well localized in the spatial and Clifford Fourier domains.

The energy density is defined as the square modulus of the CWFT (5.5.17) given by

$$|G_\phi f(\boldsymbol{\omega}, \boldsymbol{b})|^2 = \frac{1}{(2\pi)^4}\left|\int_{\mathbb{R}^2} f(\boldsymbol{x})\ \widetilde{\phi(\boldsymbol{x}-\boldsymbol{b})}e^{-i_2\boldsymbol{\omega}\cdot\boldsymbol{x}}\, d^2\boldsymbol{x}\right|^2. \tag{5.5.20}$$

Equation (5.5.20) is often called a spectrogram which measures the energy of a Clifford-valued function f in the position-frequency neighborhood of $(\boldsymbol{b}, \boldsymbol{\omega})$.

In particular, when the Gaussian function (5.5.18) is chosen as the Clifford window function, the CWFT (5.5.17) is called the Clifford Gabor transform.

5.5.1.2.2 Properties of the CWFT We will discuss the properties of the CWFT. We find that many of the properties of the WFT are still valid for the CWFT, however with certain modifications.

Theorem 5.5.6 (Left linearity). *Let $\phi \in L^2(\mathbb{R}^2;\mathcal{G}_2)$ be a Clifford window function. The CWFT of $f, g \in L^2(\mathbb{R}^2;\mathcal{G}_2)$ is a left linear operator*[43]*, which means*

$$[G_\phi(\lambda f + \boldsymbol{\mu} g)](\boldsymbol{\omega}, \boldsymbol{b}) = \lambda G_\phi f(\boldsymbol{\omega}, \boldsymbol{b}) + \boldsymbol{\mu} G_\phi g(\boldsymbol{\omega}, \boldsymbol{b}), \tag{5.5.21}$$

with Clifford constants $\lambda, \boldsymbol{\mu} \in \mathcal{G}_2$.

Proof. Using definition of the CWFT, the proof is obvious. □

Remark 5.5.7. *Since the geometric multiplication is non-commutative, the right linearity property of the CWFT does not hold in general.*

Theorem 5.5.8 (Reversion). *Let $f \in L^2(\mathbb{R}^2;\mathcal{G}_2^+)$ be a parabivector-valued function. For a parabivector-valued window function ϕ we have*

$$G_{\tilde{\phi}}\tilde{f}(\boldsymbol{\omega}, \boldsymbol{b}) = \{G_\phi f(-\boldsymbol{\omega}, \boldsymbol{b})\}^\sim. \tag{5.5.22}$$

Proof. Application of Definition 5.5.5 to the left-hand side of (5.5.22) gives

$$\begin{aligned} G_{\tilde{\phi}}\tilde{f}(\boldsymbol{\omega}, \boldsymbol{b}) &= \frac{1}{(2\pi)^2}\int_{\mathbb{R}^2} \widetilde{f(\boldsymbol{x})}\phi(\boldsymbol{x}-\boldsymbol{b})\, e^{-i_2\boldsymbol{\omega}\cdot\boldsymbol{x}}\, d^2\boldsymbol{x} \\ &= \frac{1}{(2\pi)^2}\{\int_{\mathbb{R}^2} e^{i_2\boldsymbol{\omega}\cdot\boldsymbol{x}}\, \widetilde{\phi(\boldsymbol{x}-\boldsymbol{b})} f(\boldsymbol{x})\, d^2\boldsymbol{x}\}^\sim \\ &= \frac{1}{(2\pi)^2}\{\int_{\mathbb{R}^2} f(\boldsymbol{x})\, \widetilde{\phi(\boldsymbol{x}-\boldsymbol{b})}\, e^{i_2\boldsymbol{\omega}\cdot\boldsymbol{x}}\, d^2\boldsymbol{x}\}^\sim. \end{aligned} \tag{5.5.23}$$

This finishes the proof of the theorem. □

Theorem 5.5.9 (Switching). *If $|\boldsymbol{x}|^{1/2} f(\boldsymbol{x}) \in L^2(\mathbb{R}^2;\mathcal{G}_2)$ and $|\boldsymbol{x}|^{1/2}\phi(\boldsymbol{x}) \in L^2(\mathbb{R}^2;\mathcal{G}_2)$ are parabivector-valued functions, then we obtain*

$$G_\phi f(\boldsymbol{\omega}, \boldsymbol{b}) = e^{-i_2\boldsymbol{\omega}\cdot\boldsymbol{b}}\, \{G_f\phi(-\boldsymbol{\omega}, -\boldsymbol{b})\}^\sim. \tag{5.5.24}$$

Proof. We have, by the CWFT definition,

$$\begin{aligned} G_\phi f(\boldsymbol{\omega}, \boldsymbol{b}) &= \frac{1}{(2\pi)^2}\int_{\mathbb{R}^2} f(\boldsymbol{x})\widetilde{\phi(\boldsymbol{x}-\boldsymbol{b})}\, e^{-i_2\boldsymbol{\omega}\cdot\boldsymbol{x}}\, d^2\boldsymbol{x} \\ &= \frac{1}{(2\pi)^2}\{\int_{\mathbb{R}^2} \phi(\boldsymbol{x}-\boldsymbol{b})\widetilde{f(\boldsymbol{x})}\, e^{i_2\boldsymbol{\omega}\cdot\boldsymbol{x}}\, d^2\boldsymbol{x}\}^\sim. \end{aligned} \tag{5.5.25}$$

The substitution $\boldsymbol{y} = \boldsymbol{x} - \boldsymbol{b}$ in the above expression gives

$$\begin{aligned} G_\phi f(\boldsymbol{\omega}, \boldsymbol{b}) &= \frac{1}{(2\pi)^2}\{\int_{\mathbb{R}^2} \phi(\boldsymbol{y})\widetilde{f(\boldsymbol{y}+\boldsymbol{b})}\, e^{i_2\boldsymbol{\omega}\cdot(\boldsymbol{y}+\boldsymbol{b})}\, d^2\boldsymbol{y}\}^\sim \\ &= \frac{1}{(2\pi)^2} e^{-i_2\boldsymbol{\omega}\cdot\boldsymbol{b}}\{\int_{\mathbb{R}^2} \phi(\boldsymbol{y})\widetilde{f(\boldsymbol{y}+\boldsymbol{b})}\, e^{i_2\boldsymbol{\omega}\cdot\boldsymbol{y}}\, d^2\boldsymbol{y}\}^\sim \\ &= \frac{1}{(2\pi)^2} e^{-i_2\boldsymbol{\omega}\cdot\boldsymbol{b}}\{\int_{\mathbb{R}^2} \phi(\boldsymbol{y})\widetilde{f(\boldsymbol{y}-(-\boldsymbol{b}))}\, e^{-i_2(-\boldsymbol{\omega})\cdot\boldsymbol{y}}\, d^2\boldsymbol{y}\}^\sim, \end{aligned} \tag{5.5.26}$$

which proves the theorem. □

[43]The CWFT of f is a *linear* operator for real constants $\boldsymbol{\mu}, \lambda \in \mathbb{R}$.

Theorem 5.5.10 (Parity). *Let $\phi \in L^2(\mathbb{R}^2; \mathcal{G}_2)$ be a Clifford window function. If P is the parity operator defined as $P\phi(\boldsymbol{x}) = \phi(-\boldsymbol{x})$, then we have*

$$G_{P\phi}\{Pf\}(\boldsymbol{\omega}, \boldsymbol{b}) = G_\phi f(-\boldsymbol{\omega}, -\boldsymbol{b}). \tag{5.5.27}$$

Proof. Direct calculations give for every $f \in L^2(\mathbb{R}^2; \mathcal{G}_2)$

$$\begin{aligned} G_{P\phi}\{Pf\}(\boldsymbol{\omega}, \boldsymbol{b}) &= \frac{1}{(2\pi)^2} \int_{\mathbb{R}^2} f(-\boldsymbol{x})\{\phi(-\boldsymbol{x} + \boldsymbol{b})\}^\sim e^{-i_2(-\boldsymbol{\omega})\cdot(-\boldsymbol{x})}\, d^2\boldsymbol{x} \\ &= \frac{1}{(2\pi)^2} \int_{\mathbb{R}^2} f(-\boldsymbol{x})\{\phi(-\boldsymbol{x} - (-\boldsymbol{b}))\}^\sim e^{-i_2(-\boldsymbol{\omega})\cdot(-\boldsymbol{x})}\, d^2\boldsymbol{x} \\ &= \frac{1}{(2\pi)^2} \int_{\mathbb{R}^2} f(\boldsymbol{x})\{\phi(\boldsymbol{x} - (-\boldsymbol{b}))\}^\sim e^{-i_2(-\boldsymbol{\omega})\cdot\boldsymbol{x}}\, d^2\boldsymbol{x}, \end{aligned} \tag{5.5.28}$$

which completes the proof. □

Theorem 5.5.11 (Shift in space domain, delay). *Let ϕ be a Clifford window function. Introducing the translation operator $T_{\boldsymbol{x}_0} f(\boldsymbol{x}) = f(\boldsymbol{x} - \boldsymbol{x}_0)$, we obtain*

$$G_\phi\{T_{\boldsymbol{x}_0} f\}(\boldsymbol{\omega}, \boldsymbol{b}) = (G_\phi f(\boldsymbol{\omega}, \boldsymbol{b} - \boldsymbol{x}_0))\, e^{-i_2\boldsymbol{\omega}\cdot\boldsymbol{x}_0}. \tag{5.5.29}$$

Proof. We have by using (5.5.17)

$$G_\phi\{T_{\boldsymbol{x}_0} f\}(\boldsymbol{\omega}, \boldsymbol{b}) = \frac{1}{(2\pi)^2} \int_{\mathbb{R}^2} f(\boldsymbol{x} - \boldsymbol{x}_0)\widetilde{\phi(\boldsymbol{x} - \boldsymbol{b})}\, e^{-i_2\boldsymbol{\omega}\cdot\boldsymbol{x}}\, d^2\boldsymbol{x}. \tag{5.5.30}$$

We substitute $\boldsymbol{t} = \boldsymbol{x} - \boldsymbol{x}_0$ in the above expression and get, with $d^2\boldsymbol{x} = d^2\boldsymbol{t}$,

$$\begin{aligned} G_\phi\{T_{\boldsymbol{x}_0} f\}(\boldsymbol{\omega}, \boldsymbol{b}) &= \frac{1}{(2\pi)^2} \int_{\mathbb{R}^2} f(\boldsymbol{t})\{\phi(\boldsymbol{t} - (\boldsymbol{b} - \boldsymbol{x}_0))\}^\sim e^{-i_2\boldsymbol{\omega}\cdot(\boldsymbol{t}+\boldsymbol{x}_0)}\, d^2\boldsymbol{t} \qquad (5.5.31) \\ &= \frac{1}{(2\pi)^2} \int_{\mathbb{R}^2} \left[f(\boldsymbol{t})\{\phi(\boldsymbol{t} - (\boldsymbol{b} - \boldsymbol{x}_0))\}^\sim e^{-i_2\boldsymbol{\omega}\cdot\boldsymbol{t}}\right] d^2\boldsymbol{t}\, e^{-i_2\boldsymbol{\omega}\cdot\boldsymbol{x}_0}. \end{aligned}$$

This ends the proof of (5.5.29). □

Theorem 5.5.12 (Shift in frequency domain, modulation). *Let ϕ be a parabivector valued Clifford window function. If $\boldsymbol{\omega}_0 \in \mathbb{R}^2$ and $f_0(\boldsymbol{x}) = f(\boldsymbol{x})e^{i_2\boldsymbol{\omega}_0\cdot\boldsymbol{x}}$, then*

$$G_\phi f_0(\boldsymbol{\omega}, \boldsymbol{b}) = G_\phi f(\boldsymbol{\omega} - \boldsymbol{\omega}_0, \boldsymbol{b}). \tag{5.5.32}$$

Proof. Using Definition 5.5.5 and simplifying it we get

$$\begin{aligned} G_\phi f_0(\boldsymbol{\omega}, \boldsymbol{b}) &= \frac{1}{(2\pi)^2} \int_{\mathbb{R}^2} f(\boldsymbol{x})e^{i_2\boldsymbol{\omega}_0\cdot\boldsymbol{x}}\, \widetilde{\phi(\boldsymbol{x} - \boldsymbol{b})}\, e^{-i_2\boldsymbol{\omega}\cdot\boldsymbol{x}}\, d^2\boldsymbol{x} \\ &= \frac{1}{(2\pi)^2} \int_{\mathbb{R}^2} f(\boldsymbol{x})\widetilde{\phi(\boldsymbol{x} - \boldsymbol{b})}\, e^{-i_2(\boldsymbol{\omega} - \boldsymbol{\omega}_0)\cdot\boldsymbol{x}}\, d^2\boldsymbol{x}, \end{aligned} \tag{5.5.33}$$

which proves the theorem. □

Theorem 5.5.13 (Reconstruction formula). *Let ϕ be a Clifford window function. Then every 2D Clifford signal $f \in L^2(\mathbb{R}^2; \mathcal{G}_2)$ can be fully reconstructed by*

$$f(\boldsymbol{x}) = (2\pi)^2 \int_{\mathbb{R}^2} \int_{\mathbb{R}^2} G_\phi f(\boldsymbol{\omega}, \boldsymbol{b}) \phi_{\boldsymbol{\omega},\boldsymbol{b}}(\boldsymbol{x}) \, (\tilde{\phi}, \tilde{\phi})^{-1}_{L^2(\mathbb{R}^2;\mathcal{G}_2)} d^2\boldsymbol{b} \, d^2\boldsymbol{\omega}. \tag{5.5.34}$$

Proof. It follows from the CWFT defined by (5.5.17) that

$$G_\phi f(\boldsymbol{\omega}, \boldsymbol{b}) = \frac{1}{(2\pi)^2} \mathcal{F}\{f(\boldsymbol{x}) \widetilde{\phi(\boldsymbol{x} - \boldsymbol{b})}\}(\boldsymbol{\omega}). \tag{5.5.35}$$

Taking the inverse CFT of both sides of (5.5.35) we obtain

$$\begin{aligned} f(\boldsymbol{x}) \widetilde{\phi(\boldsymbol{x} - \boldsymbol{b})} &= (2\pi)^2 \mathcal{F}^{-1}\{G_\phi f(\boldsymbol{\omega}, \boldsymbol{b})\}(\boldsymbol{x}) \\ &= \frac{(2\pi)^2}{(2\pi)^2} \int_{\mathbb{R}^2} G_\phi f(\boldsymbol{\omega}, \boldsymbol{b}) \, e^{i_2 \boldsymbol{\omega} \cdot \boldsymbol{x}} \, d^2\boldsymbol{\omega}. \end{aligned} \tag{5.5.36}$$

Multiplying both sides of (5.5.36) by $\phi(\boldsymbol{x} - \boldsymbol{b})$ and then integrating with respect to $d^2\boldsymbol{b}$ we get

$$\begin{aligned} f(\boldsymbol{x}) &\int_{\mathbb{R}^2} \widetilde{\phi(\boldsymbol{x} - \boldsymbol{b})} \phi(\boldsymbol{x} - \boldsymbol{b}) d^2\boldsymbol{b} \\ &= \int_{\mathbb{R}^2} \int_{\mathbb{R}^2} G_\phi f(\boldsymbol{\omega}, \boldsymbol{b}) \, e^{i_2 \boldsymbol{\omega} \cdot \boldsymbol{x}} \phi(\boldsymbol{x} - \boldsymbol{b}) \, d^2\boldsymbol{\omega} \, d^2\boldsymbol{b}. \end{aligned} \tag{5.5.37}$$

Or, equivalently,

$$f(\boldsymbol{x})(\tilde{\phi}, \tilde{\phi})_{L^2(\mathbb{R}^2;\mathcal{G}_2)} = (2\pi)^2 \int_{\mathbb{R}^2} \int_{\mathbb{R}^2} G_\phi f(\boldsymbol{\omega}, \boldsymbol{b}) \, \phi_{\boldsymbol{\omega},\boldsymbol{b}}(\boldsymbol{x}) \, d^2\boldsymbol{\omega} \, d^2\boldsymbol{b}, \tag{5.5.38}$$

which gives (5.5.34). □

It is worth noting here that if the Clifford window function is a parabivector-valued function, then the reconstruction formula (5.5.34) can be written in the following form

$$f(\boldsymbol{x}) = \frac{(2\pi)^2}{\|\phi\|^2_{L^2(\mathbb{R}^2;\mathcal{G}_2)}} \int_{\mathbb{R}^2} \int_{\mathbb{R}^2} G_\phi f(\boldsymbol{\omega}, \boldsymbol{b}) \phi_{\boldsymbol{\omega},\boldsymbol{b}}(\boldsymbol{x}) \, d^2\boldsymbol{b} \, d^2\boldsymbol{\omega}. \tag{5.5.39}$$

Theorem 5.5.14 (Orthogonality relation). *Assume that the Clifford window function ϕ is a parabivector-valued function. If two Clifford functions $f, g \in L^2(\mathbb{R}^2; \mathcal{G}_2)$, then we have*

$$\begin{aligned} \int_{\mathbb{R}^2} \int_{\mathbb{R}^2} &(f, \phi_{\boldsymbol{\omega},\boldsymbol{b}})_{L^2(\mathbb{R}^2;\mathcal{G}_2)} \widetilde{(g, \phi_{\boldsymbol{\omega},\boldsymbol{b}})}_{L^2(\mathbb{R}^2;\mathcal{G}_2)} d^2\boldsymbol{\omega} \, d^2\boldsymbol{b} \\ &= \frac{\|\phi\|^2_{L^2(\mathbb{R}^2;\mathcal{G}_2)}}{(2\pi)^2} \, (f, g)_{L^2(\mathbb{R}^2;\mathcal{G}_2)}. \end{aligned} \tag{5.5.40}$$

Proof. By inserting (5.5.17) into the left side of (5.5.40), we obtain

$$\begin{aligned}
&\int_{\mathbb{R}^2}\int_{\mathbb{R}^2}(f,\phi_{\boldsymbol{\omega},\boldsymbol{b}})_{L^2(\mathbb{R}^2;\mathcal{G}_2)}\widetilde{(g,\phi_{\boldsymbol{\omega},\boldsymbol{b}})}_{L^2(\mathbb{R}^2;\mathcal{G}_2)}d^2\boldsymbol{\omega}\,d^2\boldsymbol{b}\\
&=\int_{\mathbb{R}^2}\int_{\mathbb{R}^2}(f,\phi_{\boldsymbol{\omega},\boldsymbol{b}})_{L^2(\mathbb{R}^2;\mathcal{G}_2)}\left(\int_{\mathbb{R}^2}\frac{1}{(2\pi)^2}e^{i_2\boldsymbol{\omega}\cdot\boldsymbol{x}}\phi(\boldsymbol{x}-\boldsymbol{b})\widetilde{g(\boldsymbol{x})}d^2\boldsymbol{x}\right)d^2\boldsymbol{\omega}\,d^2\boldsymbol{b}\\
&=\int_{\mathbb{R}^2}\int_{\mathbb{R}^2}\left(\int_{\mathbb{R}^2}\int_{\mathbb{R}^2}\frac{1}{(2\pi)^4}f(\boldsymbol{x}')\,\widetilde{\phi(\boldsymbol{x}'-\boldsymbol{b})}\,e^{i_2\boldsymbol{\omega}\cdot(\boldsymbol{x}-\boldsymbol{x}')}d^2\boldsymbol{\omega}\,d^2\boldsymbol{x}'\right)\\
&\qquad\phi(\boldsymbol{x}-\boldsymbol{b})\widetilde{g(\boldsymbol{x})}d^2\boldsymbol{x}d^2\boldsymbol{b}\\
&=\frac{1}{(2\pi)^2}\int_{\mathbb{R}^2}\int_{\mathbb{R}^2}\left(\int_{\mathbb{R}^2}f(\boldsymbol{x}')\widetilde{\phi(\boldsymbol{x}'-\boldsymbol{b})}\delta(\boldsymbol{x}-\boldsymbol{x}')\phi(\boldsymbol{x}-\boldsymbol{b})d^2\boldsymbol{x}'\right)\widetilde{g(\boldsymbol{x})}\,d^2\boldsymbol{b}d^2\boldsymbol{x}\\
&=\frac{1}{(2\pi)^2}\int_{\mathbb{R}^2}f(\boldsymbol{x})\int_{\mathbb{R}^2}\underbrace{\widetilde{\phi(\boldsymbol{x}-\boldsymbol{b})}\phi(\boldsymbol{x}-\boldsymbol{b})}_{\phi\text{ parabiv. funct.}}\,d^2\boldsymbol{b}\,\widetilde{g(\boldsymbol{x})}\,d^2\boldsymbol{x}\\
&=\frac{1}{(2\pi)^2}\|\phi\|^2_{L^2(\mathbb{R}^2;\mathcal{G}_2)}\int_{\mathbb{R}^2}f(\boldsymbol{x})\widetilde{g(\boldsymbol{x})}\,d^2\boldsymbol{x}, \qquad (5.5.41)
\end{aligned}$$

which completes the proof of (5.5.40). □

Theorem 5.5.15 (Reproducing kernel). *For a parabivector valued Clifford window function $|\boldsymbol{x}|^{1/2}\phi \in L^2(\mathbb{R}^2;\mathcal{G}_2)$ if*

$$\mathbb{K}_\phi(\boldsymbol{\omega},\boldsymbol{b};\boldsymbol{\omega}',\boldsymbol{b}')=\frac{(2\pi)^2}{\|\phi\|^2_{L^2(\mathbb{R}^2;\mathcal{G}_2)}}(\phi_{\boldsymbol{\omega},\boldsymbol{b}},\phi_{\boldsymbol{\omega}',\boldsymbol{b}'})_{L^2(\mathbb{R}^2;\mathcal{G}_2)}, \qquad (5.5.42)$$

then $\mathbb{K}_\phi(\boldsymbol{\omega},\boldsymbol{b};\boldsymbol{\omega}',\boldsymbol{b}')$ is a reproducing kernel, i.e.

$$G_\phi f(\boldsymbol{\omega}',\boldsymbol{b}')=\int_{\mathbb{R}^2}\int_{\mathbb{R}^2}G_\phi f(\boldsymbol{\omega},\boldsymbol{b})\mathbb{K}_\phi(\boldsymbol{\omega},\boldsymbol{b};\boldsymbol{\omega}',\boldsymbol{b}')\,d^2\boldsymbol{\omega}\,d^2\boldsymbol{b}. \qquad (5.5.43)$$

Proof. By inserting the inverse CWFT (5.5.39) into the definition of the CWFT (5.5.17) we easily obtain

$$\begin{aligned}
G_\phi f(\boldsymbol{\omega}',\boldsymbol{b}')&=\int_{\mathbb{R}^2}f(\boldsymbol{x})\,\widetilde{\phi_{\boldsymbol{\omega}',\boldsymbol{b}'}(\boldsymbol{x})}\,d^2\boldsymbol{x}\\
&=\int_{\mathbb{R}^2}\left(\frac{(2\pi)^2}{\|\phi\|^2_{L^2(\mathbb{R}^2;\mathcal{G}_2)}}\int_{\mathbb{R}^2}\int_{\mathbb{R}^2}G_\phi f(\boldsymbol{\omega},\boldsymbol{b})\,\phi_{\boldsymbol{\omega},\boldsymbol{b}}(\boldsymbol{x})d^2\boldsymbol{b}d^2\boldsymbol{\omega}\right)\widetilde{\phi_{\boldsymbol{\omega}',\boldsymbol{b}'}(\boldsymbol{x})}d^2\boldsymbol{x}\\
&=\int_{\mathbb{R}^2}\int_{\mathbb{R}^2}G_\phi f(\boldsymbol{\omega},\boldsymbol{b})\frac{(2\pi)^2}{\|\phi\|^2_{L^2(\mathbb{R}^2;\mathcal{G}_2)}}\left(\int_{\mathbb{R}^2}\phi_{\boldsymbol{\omega},\boldsymbol{b}}(\boldsymbol{x})\widetilde{\phi_{\boldsymbol{\omega}',\boldsymbol{b}'}(\boldsymbol{x})}\,d^2\boldsymbol{x}\right)d^2\boldsymbol{b}d^2\boldsymbol{\omega}\\
&=\int_{\mathbb{R}^2}\int_{\mathbb{R}^2}G_\phi f(\boldsymbol{\omega},\boldsymbol{b})\mathbb{K}_\phi(\boldsymbol{\omega},\boldsymbol{b};\boldsymbol{\omega}',\boldsymbol{b}')\,d^2\boldsymbol{b}d^2\boldsymbol{\omega}, \qquad (5.5.44)
\end{aligned}$$

which finishes the proof. □

Remark 5.5.16. *Formulas (5.5.39), (5.5.40) and (5.5.42) also hold if the Clifford window function is a vector-valued function, i.e. $\phi(\boldsymbol{x})=\phi_1(\boldsymbol{x})\boldsymbol{e}_1+\phi_2(\boldsymbol{x})\boldsymbol{e}_2$.*

The above properties of the CWFT are summarized in Table 5.9.

Table 5.9 Properties of the CWFT of $f, g \in L^2(\mathbb{R}^2;\mathcal{G}_2)$, $L^2 = L^2(\mathbb{R}^2;\mathcal{G}_2)$, where $\lambda, \boldsymbol{\mu} \in \mathcal{G}_2$ are constants, $\boldsymbol{\omega}_0 = \omega_{0,1}\boldsymbol{e}_1 + \omega_{0,2}\boldsymbol{e}_2 \in \mathbb{R}^2$ and $\boldsymbol{x}_0 = x_0\boldsymbol{e}_1 + y_0\boldsymbol{e}_2 \in \mathbb{R}^2$. Source: [254, Tab. 1].

Property	Clifford Valued Function	2D CWFT
Left linearity	$\lambda f(\mathbf{x}) + \boldsymbol{\mu} g(\mathbf{x})$	$\lambda G_\phi f(\boldsymbol{\omega}, \boldsymbol{b}) + \boldsymbol{\mu} G_\phi g(\boldsymbol{\omega}, \boldsymbol{b})$
Delay	$f(\boldsymbol{x} - \boldsymbol{x}_0)$	$(G_\phi f(\boldsymbol{\omega}, \boldsymbol{b} - \boldsymbol{x}_0))\ e^{-i_2 \boldsymbol{\omega}\cdot\boldsymbol{x}_0}$
Modulation	$f(\boldsymbol{x}) e^{i_2 \boldsymbol{\omega}_0 \cdot \boldsymbol{x}}$	$G_\phi f(\boldsymbol{\omega} - \boldsymbol{\omega}_0, \boldsymbol{b})$, if ϕ parabivector valued
Formulas		
Reversion	$G_{\tilde{\phi}} \tilde{f}(\boldsymbol{\omega}, \boldsymbol{b}) =$	$\{G_\phi f(-\boldsymbol{\omega}, \boldsymbol{b})\}^{\sim}$, if f and ϕ are parabivector-valued functions
Switching	$G_\phi f(\boldsymbol{\omega}, \boldsymbol{b}) =$	$e^{-i_2 \boldsymbol{\omega}\cdot\boldsymbol{b}} \{G_f \phi(-\boldsymbol{\omega}, -\boldsymbol{b})\}^{\sim}$, if f and ϕ are parabivector-valued functions
Parity	$G_{P\phi}\{Pf\}(\boldsymbol{\omega}, \boldsymbol{b}) =$	$G_\phi f(-\boldsymbol{\omega}, -\boldsymbol{b})$
Orthogonality	$\frac{1}{(2\pi)^2}\lVert\phi\rVert^2_{L^2}(f, g)_{L^2} =$	$\int_{\mathbb{R}^2}\int_{\mathbb{R}^2} (f, \phi_{\boldsymbol{\omega},\boldsymbol{b}})_{L^2(\mathbb{R}^2;\mathcal{G}_2)} \widetilde{(g, \phi_{\boldsymbol{\omega},\boldsymbol{b}})}_{L^2(\mathbb{R}^2;\mathcal{G}_2)}$ $d^2\boldsymbol{\omega}\, d^2\boldsymbol{b}$, if ϕ parabivector valued
Reconstruction	$f(\boldsymbol{x}) =$	$(2\pi)^2 \int_{\mathbb{R}^2}\int_{\mathbb{R}^2} G_\phi f(\boldsymbol{\omega}, \boldsymbol{b}) \phi_{\boldsymbol{\omega},\boldsymbol{b}}(\boldsymbol{x}) \times (\tilde{\phi}, \tilde{\phi})^{-1}_{L^2(\mathbb{R}^2;\mathcal{G}_2)} d^2\boldsymbol{b} d^2\boldsymbol{\omega}$, $\frac{(2\pi)^2}{\lVert\phi\rVert^2_{L^2(\mathbb{R}^2;\mathcal{G}_2)}} \int_{\mathbb{R}^2}\int_{\mathbb{R}^2} G_\phi f(\boldsymbol{\omega}, \boldsymbol{b}) \phi_{\boldsymbol{\omega},\boldsymbol{b}}(\boldsymbol{x})$ $d^2\boldsymbol{b} d^2\boldsymbol{\omega}$, if ϕ parabivector valued
Reproducing kernel	$G_\phi f(\boldsymbol{\omega}', \boldsymbol{b}') =$	$\int_{\mathbb{R}^2}\int_{\mathbb{R}^2} G_\phi f(\boldsymbol{\omega}, \boldsymbol{b}) \mathbb{K}_\phi(\boldsymbol{\omega}, \boldsymbol{b}; \boldsymbol{\omega}', \boldsymbol{b}') d^2\boldsymbol{\omega} d^2\boldsymbol{b}$, $\mathbb{K}_\phi(\boldsymbol{\omega}, \boldsymbol{b}; \boldsymbol{\omega}', \boldsymbol{b}') = \frac{(2\pi)^2}{\lVert\phi\rVert^2_{L^2(\mathbb{R}^2;\mathcal{G}_2)}} (\phi_{\boldsymbol{\omega},\boldsymbol{b}}, \phi_{\boldsymbol{\omega}',\boldsymbol{b}'})_{L^2(\mathbb{R}^2;\mathcal{G}_2)}$, if ϕ parabivector valued.

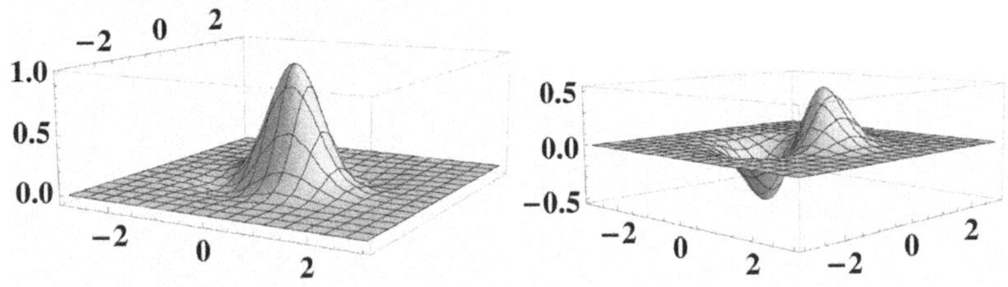

Figure 5.1 The real part (*left*) and bivector part(*right*) of Clifford Gabor filter for the parameters $\omega_{0,1} = \omega_{0,2} = 1, b_1 = b_2 = 0, \sigma_1 = \sigma_2 = 1/\sqrt{2}$ in the spatial domain using Mathematica 6.0. Source: [254, Fig. 1].

5.5.1.2.3 Examples of the CWFT For illustrative purposes, we shall discuss examples of the CWFT. We then compute their energy densities.

Example 5.5.17. *Consider Clifford Gabor filters (see Figure 5.1) defined by* $(\sigma_1 = \sigma_2 = 1/\sqrt{2})$

$$f(\boldsymbol{x}) = \frac{1}{(2\pi)^2} e^{-\boldsymbol{x}^2 + i_2\boldsymbol{\omega}_0\cdot\boldsymbol{x}}, \tag{5.5.45}$$

Obtain the CWFT of f with respect to the Gaussian window function $\phi(\boldsymbol{x}) = e^{-\boldsymbol{x}^2}$.

By definition of the CWFT (5.5.17), we have

$$G_\phi f(\boldsymbol{\omega}, \boldsymbol{b}) = \frac{1}{(2\pi)^4} \int_{\mathbb{R}^2} e^{-\boldsymbol{x}^2 + i_2\boldsymbol{\omega}_0\cdot\boldsymbol{x}} e^{-(\boldsymbol{x}-\boldsymbol{b})^2} e^{-i_2\boldsymbol{\omega}\cdot\boldsymbol{x}} d^2\boldsymbol{x}. \tag{5.5.46}$$

Substituting $\boldsymbol{x} = \boldsymbol{y} + \boldsymbol{b}/2$ we can rewrite (5.5.46) as

$$\begin{aligned} G_\phi f(\boldsymbol{\omega}, \boldsymbol{b}) &= \frac{1}{(2\pi)^4} \int_{\mathbb{R}^2} e^{-(\boldsymbol{y}+\boldsymbol{b}/2)^2 + i_2\boldsymbol{\omega}_0\cdot(\boldsymbol{y}+\boldsymbol{b}/2)} e^{-(\boldsymbol{y}-\boldsymbol{b}/2)^2} e^{-i_2\boldsymbol{\omega}\cdot(\boldsymbol{y}+\boldsymbol{b}/2)}\, d^2\boldsymbol{y} \\ &= \frac{e^{-\boldsymbol{b}^2/2}}{(2\pi)^4} \int_{\mathbb{R}^2} e^{-2\boldsymbol{y}^2} e^{-i_2\boldsymbol{\omega}\cdot\boldsymbol{y}}\, e^{i_2\boldsymbol{\omega}_0\cdot\boldsymbol{y}} d^2\boldsymbol{y}\, e^{-i_2(\boldsymbol{\omega}-\boldsymbol{\omega}_0)\cdot\boldsymbol{b}/2} \\ &= \frac{e^{-\boldsymbol{b}^2/2}}{(2\pi)^4} \int_{\mathbb{R}^2} e^{-2\boldsymbol{y}^2} e^{-i_2(\boldsymbol{\omega}-\boldsymbol{\omega}_0)\cdot\boldsymbol{y}} d^2\boldsymbol{y}\, e^{-i_2(\boldsymbol{\omega}-\boldsymbol{\omega}_0)\cdot\boldsymbol{b}/2} \\ &= \frac{e^{-\boldsymbol{b}^2/2}}{(2\pi)^4} \frac{\pi}{2} e^{-(\boldsymbol{\omega}-\boldsymbol{\omega}_0)^2/8} e^{-i_2(\boldsymbol{\omega}-\boldsymbol{\omega}_0)\cdot\boldsymbol{b}/2} \\ &= \frac{e^{-\boldsymbol{b}^2/2}}{32\pi^3} e^{-(\boldsymbol{\omega}-\boldsymbol{\omega}_0)^2/8} e^{-i_2(\boldsymbol{\omega}-\boldsymbol{\omega}_0)\cdot\boldsymbol{b}/2}. \end{aligned} \tag{5.5.47}$$

The energy density is given by

$$|G_\phi f(\boldsymbol{\omega}, \boldsymbol{b})|^2 = \frac{e^{-\boldsymbol{b}^2}}{(32\pi^3)^2} e^{-(\boldsymbol{\omega}-\boldsymbol{\omega}_0)^2/4}. \tag{5.5.48}$$

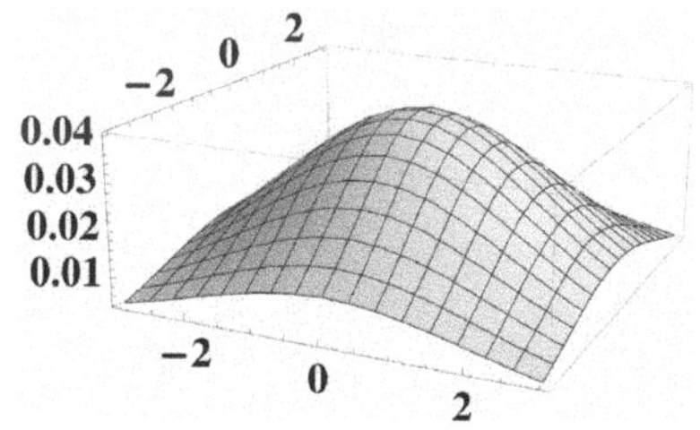

Figure 5.2 Plot of the CWFT of Clifford Gabor filter of Example 5.5.17 using Mathematica 6.0. Note that it is real valued for the parameters $b_1 = b_2 = 0$. Source: [254, Fig. 2].

Example 5.5.18. *Consider the first order two-dimensional B-spline window function defined by*

$$\phi(\boldsymbol{x}) = \begin{cases} 1, & \text{if } 0 \leq x_1 \leq 1 \text{ and } 0 \leq x_2 \leq 1, \\ 0, & \text{otherwise.} \end{cases} \tag{5.5.49}$$

Obtain the CWFT of the function defined as follows:

$$f(\boldsymbol{x}) = \begin{cases} \boldsymbol{x}, & \text{if } 0 \leq x_1 \leq 1 \text{ and } 0 \leq x_2 \leq 1, \\ 0, & \text{otherwise.} \end{cases} \tag{5.5.50}$$

Applying Definition 5.5.5 and simplifying it we obtain

$$\begin{aligned} G_\phi f(\boldsymbol{\omega}, \boldsymbol{b}) &= \frac{1}{(2\pi)^2} \int_{b_1}^{1+b_1} \int_{b_2}^{1+b_2} \boldsymbol{x}\, e^{-i_2 \boldsymbol{\omega}\cdot\boldsymbol{x}} dx_1 dx_2 \\ &= \frac{1}{(2\pi)^2} \int_{b_1}^{1+b_1} \int_{b_2}^{1+b_2} (x_1 \boldsymbol{e}_1 + x_2 \boldsymbol{e}_2)\left(e^{-i_2\omega_1 x_1} e^{-i_2 \omega_2 x_2}\right) dx_1 dx_2 \\ &= \frac{1}{(2\pi)^2} \boldsymbol{e}_1 \int_{b_1}^{1+b_1} x_1 e^{-i_2\omega_1 x_1} dx_1 \int_{b_2}^{1+b_2} e^{-i_2\omega_2 x_2}\, dx_2 \\ &+ \frac{1}{(2\pi)^2} \boldsymbol{e}_2 \int_{b_1}^{1+b_1} e^{-i_2\omega_1 x_1} dx_1 \int_{b_2}^{1+b_2} x_2 e^{-i_2\omega_2 x_2}\, dx_2 \\ &= \Big\{ \boldsymbol{e}_2 \omega_2 [(1 + i_2\omega_1 b_1)(e^{-i_2\omega_1} - 1) + i_2\omega_1 e^{-i_2\omega_1}](e^{-i_2\omega_2} - 1) \\ &- \boldsymbol{e}_1 \omega_1 [(1 + i_2 \omega_2 b_2)(e^{-i_2\omega_2} - 1) \\ &\qquad + i_2\omega_2 e^{-i_2\omega_2}](e^{-i_2\omega_1} - 1)\Big\} \frac{e^{-i_2 \boldsymbol{\omega}\cdot\boldsymbol{b}}}{(2\pi\omega_1\omega_2)^2}, \end{aligned} \tag{5.5.51}$$

with $\boldsymbol{b} = b_1 \boldsymbol{e}_1 + b_2 \boldsymbol{e}_2$.

5.5.1.3 Comparison of CFT and CWFT

Since the Clifford Fourier kernel $e^{-i_2 \boldsymbol{\omega}\cdot\boldsymbol{x}}$ is a global function, the CFT basis has an infinite spatial extension as shown in Figure 5.3. In contrast the CWFT basis $\phi(\boldsymbol{x} - \boldsymbol{b})\, e^{-i_2 \boldsymbol{\omega}\cdot\boldsymbol{x}}$ has a limited spatial extension due to the local Clifford window function $\phi(\boldsymbol{x} - \boldsymbol{b})$ (see Figure 5.4). It means that the CFT analysis can not provide

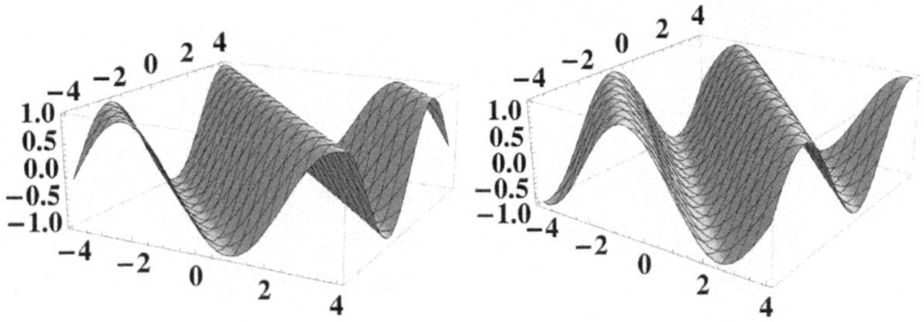

Figure 5.3 Representation of the CFT basis for $\omega_1 = \omega_2 = 1$ with scalar part (*left*) and bivector part (*right*) using Mathematica 6.0. Source: [254, Fig. 3].

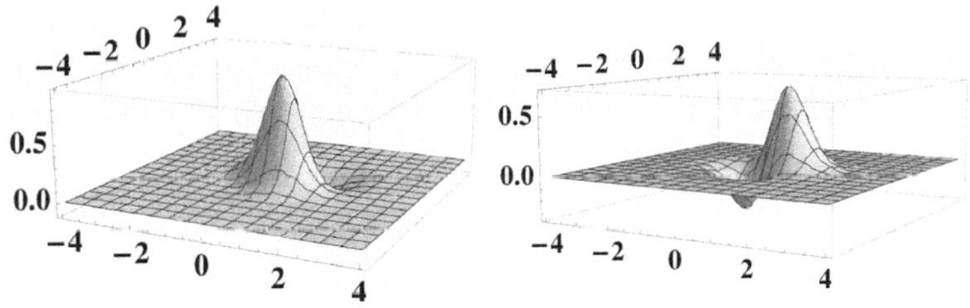

Figure 5.4 Representation of the CWFT basis of a Gaussian window function for the parameters $\omega_{0,1} = \omega_{0,2} = 1, b_1 = b_2 = 0.2$ with scalar part (*left*) and bivector part (*right*) using Mathematica 6.0. Source: [254, Fig. 4].

information about the signal with respect to position and frequency so that we need the CWFT to fully describe the characteristics of the signal simultaneously in both spatial and frequency domains.

5.5.1.4 Summary on windowed CFT

Using the basic concepts of Clifford geometric algebra and the CFT for $n = 2$, we introduced the CWFT. Important properties of the CWFT were demonstrated. This generalization enables us to work with 2D Clifford Gabor filters, which can extend the applications of the 2D complex Gabor filters.

5.5.2 Clifford Fourier Mellin transform

5.5.2.1 Background

In the following we describe a non-commutative generalization of the complex Fourier-Mellin transform to Clifford algebra valued signal functions over the domain $\mathbb{R}^{p,q}$ taking values in $Cl(p, q)$, $p + q = 2$. This Section is based on [199].

The Fourier-Mellin transform is an excellent tool in order to achieve translation, rotation and scale invariant shape recognition. In Section 4.3.2, we already dealt with its extension to the algebra of quaternions $\mathbb{H}$ in the form of the quaternionic

Fourier-Mellin transform [198], which in principle also allows a non-marginal processing of color images. Now we make a first attempt to generalize the Fourier-Mellin transform to Clifford algebras $Cl(p,q)$, $p+q=2$, using a set of two real square roots of -1, $f, g \in Cl(p,q)$, $f^2 = g^2 = -1$. This includes the case of quaternions, because of the isomorphism $\mathbb{H} \equiv Cl(0,2)$, but in addition the algebras $Cl(2,0)$ and $Cl(1,1)$ are also included. First, for the quaternion Fourier transform a *split* of quaternions with respect to any two pure unit quaternions has been developed [184, 194]. It has also been applied to the quaternionic Fourier-Mellin transform, see Section 4.3.2, and it has been generalized to Clifford algebras [196]. We will again recall and apply this split.

For a systematic introduction to Clifford geometric algebras, we refer to Chapter 2. We will employ the *principal reverse* of $M \in Cl(p,q)$ as defined in Section 2.1.1.6. Details of vector differential calculus can be found in Chapter 3. To familiarize oneself with the notation used below, we recommend the reader to refer to Section 5.2.2.1. For some necessary background on the $\pm$ split with respect to two multivector square roots of -1, we refer to Section 5.3.1.2.

A multivector valued function $f : \mathbb{R}^{p,q} \to Cl(p,q)$, has 2^n blade components $(f_A : \mathbb{R}^{p,q} \to \mathbb{R})$

$$f(\boldsymbol{x}) = \sum_A f_A(\boldsymbol{x})\boldsymbol{e}_A, \qquad \boldsymbol{x} = \sum_{l=1}^{n} x_l e^l = \sum_{l=1}^{n} x^l e_l. \tag{5.5.52}$$

We define the *inner product* of two functions $f, g : \mathbb{R}^{p,q} \to Cl(p,q)$ by

$$(f,g) = \int_{\mathbb{R}^{p,q}} f(\boldsymbol{x})\widetilde{g(\boldsymbol{x})}\, d^n\boldsymbol{x} = \sum_{A,B} \boldsymbol{e}_A \widetilde{\boldsymbol{e}_B} \int_{\mathbb{R}^{p,q}} f_A(\boldsymbol{x}) g_B(\boldsymbol{x})\, d^n\boldsymbol{x}, \tag{5.5.53}$$

with the *symmetric scalar part*

$$\langle f, g\rangle = \int_{\mathbb{R}^{p,q}} f(\boldsymbol{x}) * \widetilde{g(\boldsymbol{x})}\, d^n\boldsymbol{x} = \sum_A \int_{\mathbb{R}^{p,q}} f_A(\boldsymbol{x}) g_A(\boldsymbol{x})\, d^n\boldsymbol{x}, \tag{5.5.54}$$

and the $L^2(\mathbb{R}^{p,q}; Cl(p,q))$-*norm*

$$\|f\|^2 = \langle (f,f)\rangle = \int_{\mathbb{R}^{p,q}} |f(\boldsymbol{x})|^2 d^n\boldsymbol{x} = \sum_A \int_{\mathbb{R}^{p,q}} f_A^2(\boldsymbol{x})\, d^n\boldsymbol{x}, \tag{5.5.55}$$

$$L^2(\mathbb{R}^{p,q}; Cl(p,q)) = \{f : \mathbb{R}^{p,q} \to Cl(p,q) \mid \|f\| < \infty\}. \tag{5.5.56}$$

Notation: The *vector derivative* ∇ of a function $f : \mathbb{R}^{p,q} \to Cl(p,q)$ can be expanded in a basis of $\mathbb{R}^{p,q}$ as [304]

$$\nabla = \sum_{l=1}^{n} e^l \partial_l \quad \text{with} \quad \partial_l = \partial_{x_l} = \frac{\partial}{\partial x_l}, \quad 1 \le l \le n. \tag{5.5.57}$$

5.5.2.2 The Clifford Fourier Mellin transformations (CFMT)

We recall the Fourier Mellin transform.

Definition 5.5.19 (Classical Fourier Mellin transform (FMT)).

$$\forall (v,k) \in \mathbb{R} \times \mathbb{Z}, \quad \mathcal{M}\{h\}(v,k) = \frac{1}{2\pi}\int_0^\infty \int_0^{2\pi} h(r,\theta) r^{-iv} e^{-ik\theta} d\theta \frac{dr}{r}, \tag{5.5.58}$$

where $h : \mathbb{R}^2 \to \mathbb{R}$ denotes a function representing, e.g., a gray level image defined over a compact set of $\mathbb{R}^2$.

Well known applications are to shape recognition (independent of rotation and scale), image registration and similarity.

For $Cl(0,2) \cong \mathcal{M}(2d,\mathbb{C})$, or for both f,g being blades in $Cl(2,0)$ or $Cl(1,1)$, we have $\widetilde{f} = -f$, $\widetilde{g} = -g$. We therefore obtain the following Pythagorean modulus identity for the $L^2(\mathbb{R}^2; Cl(p,q))$-norm, $p+q=2$, i.e.,

$$\|h\|^2 = \|h_+\|^2 + \|h_-\|^2 . \tag{5.5.59}$$

We now define the generalization of the FMT to Clifford-valued signals.

Definition 5.5.20 (Clifford FMT (CFMT)). *Let $f,g \in Cl(p,q) : f^2 = g^2 = -1$, $p+q=2$, be any pair of real square roots of -1 in $Cl(p,q)$, $p+q=2$. The Clifford Fourier Mellin transform (CFMT) is given by*

$$\begin{aligned} &\forall (v,k) \in \mathbb{R} \times \mathbb{Z}, \quad \hat{h}(v,k) = \mathcal{M}\{h\}(v,k) = \\ &= \frac{1}{2\pi}\int_0^\infty \int_0^{2\pi} r^{-fv} h(r,\theta) e^{-gk\theta} d\theta \frac{dr}{r}, \end{aligned} \tag{5.5.60}$$

where $h : \mathbb{R}^2 \to Cl(p,q)$ denotes a function from $\mathbb{R}^2$ into the real Clifford $Cl(p,q)$, $p+q=2$, such that $|h|$ is summable over $\mathbb{R}^*_+ \times \mathbb{S}^1$ under the measure $d\theta\frac{dr}{r}$. $\mathbb{R}^*_+$ is the multiplicative group of positive and non-zero real numbers.

For special pure unit quaternion (isomorphic to $Cl(0,2)$) values $f = \boldsymbol{i}$, $g = \boldsymbol{j}$ we have the special case

$$\begin{aligned} &\forall (v,k) \in \mathbb{Z} \times \mathbb{R}, \quad \hat{h}(v,k) = \mathcal{M}\{h\}(v,k) = \\ &= \frac{1}{2\pi}\int_0^\infty \int_0^{2\pi} r^{-\boldsymbol{i}v} h(r,\theta) e^{-\boldsymbol{j}k\theta} d\theta \frac{dr}{r}, \end{aligned} \tag{5.5.61}$$

Note, that the $\pm$ split and the CFMT commute:

$$\mathcal{M}\{h_\pm\} = \mathcal{M}\{h\}_\pm .$$

Theorem 5.5.21 (Inverse CFMT). *The CFMT can be inverted by*

$$h(r,\theta) = \mathcal{M}^{-1}\{h\}(r,\theta) = \frac{1}{2\pi}\int_{-\infty}^\infty \sum_{k\in\mathbb{Z}} r^{fv}\, \hat{h}(v,k)\, e^{gk\theta} dv. \tag{5.5.62}$$

The proof uses

$$\frac{1}{2\pi}\sum_{k\in\mathbb{Z}} e^{gk(\theta-\theta')} = \delta(\theta-\theta'), \qquad r^{fv} = e^{fv\ln r},$$
$$\frac{1}{2\pi}\int_0^{2\pi} e^{fv(\ln(r)-s)}dv = \delta(\ln(r)-s). \tag{5.5.63}$$

We now investigate the basic properties of the CFMT. First, left linearity: For $\alpha,\beta \in \{q \mid q = q_r + q_f f,\ q_r, q_f \in \mathbb{R}\}$,

$$m(r,\theta) = \alpha h_1(r,\theta) + \beta h_2(r,\theta)$$
$$\Longrightarrow\ \hat{m}(v,k) = \alpha\hat{h}_1(v,k) + \beta\hat{h}_2(v,k). \tag{5.5.64}$$

Second, right linearity: For $\alpha',\beta' \in \{q \mid q = q_r + q_g g,\ q_r, q_g \in \mathbb{R}\}$,

$$m(r,\theta) = h_1(r,\theta)\alpha' + h_2(r,\theta)\beta'$$
$$\Longrightarrow\ \hat{m}(v,k) = \hat{h}_1(v,k)\alpha' + \hat{h}_2(v,k)\beta'. \tag{5.5.65}$$

The linearity of the CFMT leads to

$$\begin{aligned}\mathcal{M}\{h\}(v,k) &= \mathcal{M}\{h_- + h_+\}(v,k)\\ &= \mathcal{M}\{h_-\}(v,k) + \mathcal{M}\{h_+\}(v,k),\end{aligned} \tag{5.5.66}$$

which gives rise to the following theorem about *the quasi-complex FMT like forms for CFMT of* $h_\pm$.

Theorem 5.5.22 (Quasi-complex forms for CFMT). *The CFMT of $h_\pm$ parts of $h \in L^2(\mathbb{R}^2,\mathbb{H})$ have simple quasi-complex forms*

$$\begin{aligned}\mathcal{M}\{h_\pm\} &= \frac{1}{2\pi}\int_0^\infty\int_0^{2\pi} h_\pm r^{\pm gv} e^{-gk\theta}\, d\theta\frac{dr}{r}\\ &= \frac{1}{2\pi}\int_0^\infty\int_0^{2\pi} r^{-fv} e^{\pm fk\theta} h_\pm\, d\theta\frac{dr}{r}\ .\end{aligned} \tag{5.5.67}$$

Theorem 5.5.22 allows to use *discrete* and *fast* software to compute the CFMT based on a pair of complex FMT transformations.

For $Cl(0,2) \cong \mathcal{M}(2d,\mathbb{C})$, or for both f,g being blades in $Cl(2,0)$ or $Cl(1,1)$, we have $\widetilde{f} = -f$, $\widetilde{g} = -g$, and under these conditions we have for the two split parts of the CFMT, the following lemma.

Lemma 5.5.23 (Modulus identities). *Due to $|x|^2 = |x_-|^2 + |x_+|^2$, for $Cl(0,2) \cong \mathcal{M}(2d,\mathbb{C})$, or for both f,g being blades in $Cl(2,0)$ or $Cl(1,1)$, we get for $f:\mathbb{R}^2 \to Cl(p,q)$, $p+q=2$, the following identities*

$$|h(r,\theta)|^2 = |h_-(r,\theta)|^2 + |h_+(r,\theta)|^2,$$
$$|\mathcal{M}\{h\}(v,k)|^2 = |\mathcal{M}\{h_-\}(v,k)|^2 + |\mathcal{M}\{h_+\}(v,k)|^2. \tag{5.5.68}$$

Further properties are *scaling* and *rotation*: For $m(r,\theta) = h(ar,\theta+\phi),\ a>0,\ 0 \leq \phi \leq 2\pi$,

$$\widehat{m}(v,k) = a^{fv}\hat{h}(v,k)e^{gk\phi}. \tag{5.5.69}$$

Moreover, we have the following magnitude identity:

$$|\widehat{m}(v,k)| = |\hat{h}(v,k)|, \tag{5.5.70}$$

i.e. the magnitude of the CFMT of a scaled and rotated quaternion signal $m(r,\theta) = h(ar,\theta+\phi)$ is identical to the magnitude of the CFMT of h. Equation (5.5.70) forms the basis for applications to rotation and scale invariant shape recognition and image registration. This may now be extended to color images, since quaternions can encode colors RGB in their $\boldsymbol{i},\boldsymbol{j},\boldsymbol{k}$ components, and to signals with values in $Cl(2,0)$ and $Cl(1,1)$.

The *reflection at the unit circle* $(r \to \frac{1}{r})$ leads to

$$m(r,\theta) = h(\frac{1}{r},\theta) \quad \Longrightarrow \quad \widehat{m}(v,k) = \hat{h}(-v,k). \tag{5.5.71}$$

Reversing the sense of sense of *rotation* $(\theta \to -\theta)$ yields

$$m(r,\theta) = h(r,-\theta) \quad \Longrightarrow \quad \widehat{m}(v,k) = \hat{h}(v,-k). \tag{5.5.72}$$

Regarding radial and rotary modulation we assume

$$m(r,\theta) = r^{fv_0}\, h(r,\theta)\, e^{gk_0\theta}, \qquad v_0 \in \mathbb{R},\ k_0 \in \mathbb{Z}. \tag{5.5.73}$$

Then we get

$$\widehat{m}(v,k) = \hat{h}(v-v_0,k-k_0). \tag{5.5.74}$$

5.5.2.2.1 CFMT derivatives and power scaling We note for the logarithmic derivative that $\frac{d}{d\ln r} = r\frac{d}{dr} = r\partial_r$,

$$\mathcal{M}\{(r\partial_r)^n h\}(v,k) = (fv)^n\hat{h}(v,k), \qquad n \in \mathbb{N}. \tag{5.5.75}$$

Applying the angular derivative with respect to θ we obtain

$$\mathcal{M}\{\partial_\theta^n h\}(v,k) = \hat{h}(v,k)(gk)^n, \qquad n \in \mathbb{N}. \tag{5.5.76}$$

Finally, power scaling with $\ln r$ and θ leads for all $m,n \in \mathbb{N}$, to

$$\mathcal{M}\{(\ln r)^m\theta^n h\}(v,k) = f^m\, \partial_v^m\partial_k^n\hat{h}(v,k)\, g^n. \tag{5.5.77}$$

5.5.2.2.2 CFMT Plancherel and Parseval theorems For the CFMT we have the following two theorems.

Theorem 5.5.24 (CFMT Plancherel theorem). *The scalar part of the inner product of two functions* $h,m : \mathbb{R}^2 \to Cl(p,q),\ p+q=2$, *is*

$$\langle h,m\rangle = \langle \hat{h},\widehat{m}\rangle. \tag{5.5.78}$$

Theorem 5.5.25 (CFMT Parseval theorem). *Let* $h : \mathbb{R}^2 \to Cl(p,q),\ p+q=2$, *and assume* $Cl(0,2) \cong \mathcal{M}(2d,\mathbb{C})$, *or for both* f,g *being blades in* $Cl(2,0)$ *or* $Cl(1,1)$, *then*

$$\|h\| = \|\hat{h}\|, \qquad \|h\|^2 = \|\hat{h}\|^2 = \|\hat{h}_+\|^2 + \|\hat{h}_-\|^2. \tag{5.5.79}$$

5.5.2.3 *Symmetry of the CFMT*

The CFMT of real signals analyzes symmetry. The following notation will be used[44]. The function h_{ee} is *even* with respect to (w.r.t.) $r \to \frac{1}{r} \Longleftrightarrow \ln r \to -\ln r$, i.e. w.r.t. the reflection at the unit circle, and *even* w.r.t. $\theta \to -\theta$, i.e. w.r.t. reversing the sense of rotation (reflection at the $\theta = 0$ line of polar coordinates in the (r, θ)-plane). Similarly we denote by h_{eo} even-odd symmetry, by h_{oe} odd-even symmetry, and by h_{oo} odd-odd symmetry.

Let h be a real valued function $\mathbb{R}^2 \to \mathbb{R}$. The CFMT of h results in

$$\hat{h}(v,k) = \underbrace{\hat{h}_{ee}(v,k)}_{\text{real part}} + \underbrace{\hat{h}_{eo}(v,k)}_{f\text{-part}} + \underbrace{\hat{h}_{oe}(v,k)}_{g\text{-part}} + \underbrace{\hat{h}_{oo}(v,k)}_{fg\text{-part}} \,. \tag{5.5.80}$$

The CFMT of a real signal therefore automatically separates components with different combinations of symmetry w.r.t. reflection at the unit circle and reversal of the sense of rotation. The four components of the CFMT kernel differ by radial and angular phase shifts.

5.5.2.4 *Summary on Clifford Fourier Mellin transform*

We have generalized the Fourier-Mellin transform to a Clifford Fourier Mellin transform (CFMT) acting on $Cl(p,q)$-valued signals, with $p+q=2$, which includes the previously treated case of quaternions [198] of Section 4.3.2, as well as the algebras $Cl(2,0)$ and $Cl(1,1)$. We have derived several properties of this transform: inversion, linearity, quasi-complex (split) forms, and a (split) modulus identity. Beyond this for scaling, rotation, the total modulus remains invariant. This forms the foundation for translation, rotation and scale invariant shape recognition. We have further studied symmetry properties of the CFMT, derivatives and power scaling, Plancherel and Parseval theorems.

5.5.3 The quest for conformal geometric algebra Fourier transformations

5.5.3.1 *Background*

Conformal geometric algebra is preferred in many applications. Clifford Fourier transforms (CFT) allow holistic signal processing of (multi) vector fields, different from marginal (channel wise) processing: Flow fields, color fields, electro-magnetic fields, ... The Clifford algebra sets (manifolds) of $\sqrt{-1}$ lead to continuous manifolds of CFTs, compare Section 2.7. A frequently asked question is: What does a Clifford Fourier transform of conformal geometric algebra (see Section 2.6.5) look like? We try to give a first answer. This Section is based on [197].

Conformal geometric algebra is widely used in applications [190,202]. Reasons are the elegant representation of geometric objects by products of points. The products of these objects form in turn new objects (via intersection, union, projection, ...). Conformal transformations are products of reflections at hyperplanes (versors). This

[44] In this section, we assume $g \neq \pm f$, but a similar study is possible for $g = \pm f$.

even leads to the linearization of translations. Moreover, it is very interesting for the approximation abilities of conformal geometric algebra neural networks.

Clifford's geometric algebra $Cl(2,0)$ is the geometric algebra of the Euclidean plane and unifies 2D vector algebra, complex numbers and spinors in one algebra. Given an orthonormal vector basis $\{e_1, e_2\}$ of $\mathbb{R}^2$, the 4D ($2^2 = 4$) Clifford algebra $Cl(2,0)$ has a basis of 1 scalar, 2 vectors, and 1 bivector $\{1, e_1, e_2, e_{12} = \boldsymbol{i}\}$, where we define $\boldsymbol{i} = e_{12} = e_1 e_2$. The basis bivector $\boldsymbol{i}$ squares to -1. Rotation by a versor (rotor) R is a product of two reflections

$$\boldsymbol{x}' = \widetilde{R}\boldsymbol{x}R, \qquad R = e^{\frac{1}{2}\varphi \boldsymbol{i}}, \qquad \widetilde{R} = e^{-\frac{1}{2}\varphi \boldsymbol{i}}. \tag{5.5.81}$$

5.5.3.2 Prerequisites from conformal geometric algebra of the Euclidean plane

We briefly recall key notions of the conformal geometric algebra of the *Euclidean plane*. The conformal model (see [190] and its references) of the Euclidean plane in $Cl(3,1)$ extends the basis of $\mathbb{R}^2$ by adding a plane $\{e_+, e_-\}$, $e_+^2 = 1$, $e_-^2 = -1$, $e_+ \cdot e_-1 = 0$ and thus generates $Cl(2+1, 0+1) = Cl(3,1)$. We choose a null-basis, assigning origin and infinity vectors (like in projective geometry):

$$\begin{aligned} e_0 &= \frac{1}{2}(e_- - e_+), \qquad e_\infty = e_- + e_+, \\ e_0^2 &= e_\infty^2 = 0, \qquad e_0 \cdot e_\infty = -1. \end{aligned} \tag{5.5.82}$$

The bivector E of the added origin and infinity plane is

$$E = e_+ \wedge e_- = e_\infty \wedge e_0, \qquad E^2 = 1. \tag{5.5.83}$$

There are the multiplication properties

$$e_0 E = -e_0, \qquad E e_0 = e_0, \qquad e_\infty E = e_\infty, \qquad E e_\infty = -e_\infty. \tag{5.5.84}$$

The full 16D basis of $Cl(3,1)$ is

$$\begin{aligned} &\{1, \ e_1, e_2, e_0, e_\infty, \ e_{12} = \boldsymbol{i}, e_1 e_0, e_2 e_0, e_1 e_\infty, e_2 e_\infty, E, \\ &\boldsymbol{i} e_0, \boldsymbol{i} e_\infty, e_1 E, e_2 E, \ \boldsymbol{i} E\}, \end{aligned} \tag{5.5.85}$$

with scalars (1), vectors (4), bivectors (6), trivectors (4), pseudoscalars (1), and $Cl(2,0) \subset Cl(3,1)$. Note that only bivectors and trivectors change sign under reversion (reversing the order of all vector factors). Note further that the algebra $Cl(3,1)$ is the spacetime algebra (STA) of Section 4.4, albeit with a different geometric and physical interpretation.

We have a set of geometric objects $GO \subset Cl(3,1)$ described in the conformal model [190] as conformal points $P = \boldsymbol{p} + \frac{1}{2}\boldsymbol{p}^2 e_\infty + e_0$ (homogeneous, projective), point pairs $P_1 \wedge P_2$, flat point pairs $P \wedge e_\infty$, circles $C = P_1 \wedge P_2 \wedge P_3$ through three points, and lines $P_1 \wedge P_2 \wedge e_\infty$ (hyperplane), offset from the origin. A point P is on one of these objects (with blade Obj): $P \in Obj \Leftrightarrow P \wedge Obj = 0$. The products of these objects with their reverse give [190]

$$\begin{aligned} &P\widetilde{P} = 0, \ \ Pp\widetilde{Pp} = -r^2, \ \ (P \wedge e_\infty)(\widetilde{P \wedge e_\infty}) = -1, \ \ Line\,\widetilde{Line} = -1, \\ &Circle\,\widetilde{Circle} = r^2 D^2 < 0, \end{aligned} \tag{5.5.86}$$

where D is a scalar multiple of e_{12} with negative square. We therefore assume as norm for these objects $\|Obj\| = \sqrt{-Obj\,\widetilde{Obj}}$. Note that for two conformal points P, Q the number $\sqrt{-P\widetilde{Q}} = |\boldsymbol{p} - \boldsymbol{q}|/\sqrt{2}$ is their distance.

Conformal transformations of $\mathbb{R}^2$ become versors (products of vectors) in $Cl(3,1)$: rotation around origin: $R = e^{\frac{1}{2}\varphi \boldsymbol{i}}$ (see (5.5.81)), translation by $\boldsymbol{t} \in \mathbb{R}^2$: $T = e^{\frac{1}{2}\boldsymbol{t}e_\infty}$, transversion by $\boldsymbol{t} \in \mathbb{R}^2$: $Tv = e^{\frac{1}{2}\boldsymbol{t}e_0}$ (a transversion composes inversion at the unit sphere around the origin, translation, and a second inversion at the unit sphere around the origin), and scaling by $e^\gamma, \gamma \in \mathbb{R}$: $S = e^{\frac{1}{2}\gamma E}$. These transformation versors V are applied to conformal objects A as

$$A \to A' = \widetilde{V}AV. \tag{5.5.87}$$

The inner product of $f, g : \mathbb{R}^{p,q} \to GO$, respectively its symmetric scalar part, are

$$\begin{aligned}(f,g) &= -\int_{\mathbb{R}^{p,q}} f(\boldsymbol{x})\widetilde{g(\boldsymbol{x})}\, d^n\boldsymbol{x}, \\ \langle f,g\rangle &= -\int_{\mathbb{R}^{p,q}} f(\boldsymbol{x}) * \widetilde{g(\boldsymbol{x})}\, d^n\boldsymbol{x}.\end{aligned} \tag{5.5.88}$$

The $L^2(\mathbb{R}^{p,q}; GO)$-quasi-norm (indicating distance in the case of conformal points) is

$$\|f\|^2 = \langle (f,f)\rangle, \qquad L^2(\mathbb{R}^{p,q}; GO) = \{f : \mathbb{R}^{p,q} \to GO \mid \|f\| < \infty\}. \tag{5.5.89}$$

Note, that for ensuring finite basis coefficient values of geometric objects in $GO \subset Cl(3,0)$ in the basis (5.5.85) of $Cl(3,1)$ the principal reverse operation of Clifford algebra can be used (reverse combined with changing the sign of every basis vector with negative square).

The Clifford algebra $Cl(3,1)$ is isomorphic to the (square) matrix algebras $\mathcal{M}(4,\mathbb{R})$. $Sc(f) = 0$ for every $f = \sqrt{-1} \in Cl(3,1)$ [193,203]. All $\sqrt{-1} \in Cl(3,1)$ are computable with the Maple package CLIFFORD [5,6,204]. The square roots f of $-\mathbf{1}$ constitute a unique conjugacy class of dimension 8, with as many connected components as the group $\mathcal{G}(\mathcal{M}(4,\mathbb{R}))$ of invertible elements in $\mathcal{M}(4,\mathbb{R})$. For $\mathcal{M}(4,\mathbb{R})$, the centralizer (all elements in $Cl(3,1)$ commuting with f) and the conjugacy class of a square root f of $-\mathbf{1}$ both have $\mathbb{R}$-dimension 8 with two connected components.

5.5.3.3 *Clifford Fourier transformations in* $Cl(3,1)$

We now consider a generalization of quaternion and Clifford Fourier transforms CFTs [56, 58, 61, 124, 127, 184, 185, 187, 194, 198, 199, 205, 209, 291] to conformal geometric algebra $Cl(3,1)$.

Definition 5.5.26 (CFT with respect to two square roots of -1). *Let $f, g \in Cl(3,1)$, $f^2 = g^2 = -1$, be any two square roots of -1. The general Clifford Fourier transform*

(CFT) of[45] $h \in L^1(\mathbb{R}^{p,q}; GO)$, *with respect to* f, g *is*

$$\mathcal{F}^{f,g}\{h\}(\boldsymbol{\omega}) = \int_{\mathbb{R}^{p,q}} e^{-fu(\boldsymbol{x},\boldsymbol{\omega})} h(\boldsymbol{x})\, e^{-gv(\boldsymbol{x},\boldsymbol{\omega})} d^n\boldsymbol{x}, \tag{5.5.90}$$

where $n = p+q$, $d^n\boldsymbol{x} = dx_1 \dots dx_n$, $\boldsymbol{x}, \boldsymbol{\omega} \in \mathbb{R}^{p,q}$, *and* $u, v : \mathbb{R}^{p,q} \times \mathbb{R}^{p,q} \to \mathbb{R}$.

The square roots $f, g \in Cl(3,1)$ of -1 may be from any component of any conjugacy class. The above CFT is steerable in the continuous submanifolds of $\sqrt{-\mathbf{1}}$ in $Cl(3,1)$. We have the following properties of the general two-sided CFT: a Plancherel identity, respectively a Parseval identity, for functions $h_1, h_2, h \in L^2(\mathbb{R}^{p,q}; GO)$

$$\langle h_1, h_2 \rangle = \frac{1}{(2\pi)^n} \langle \mathcal{F}^{f,g}\{h_1\}, \mathcal{F}^{f,g}\{h_2\} \rangle, \quad \|h\| = \frac{1}{(2\pi)^{n/2}} \left\| \mathcal{F}^{f,g}\{h\} \right\|. \tag{5.5.91}$$

For these identities to hold we need for $\sqrt{-1}$ that $\widetilde{f} = -f$, $\widetilde{g} = -g$.

We now seek to find the most suitable $\sqrt{-1} \in Cl(3,1)$. We can write every real $f = \sqrt{-1}$, $f \in Cl(3,1)$ as

$$\begin{aligned} f = {} & \alpha + \boldsymbol{b} + \beta \boldsymbol{i} + (\alpha_\infty + \boldsymbol{b}_\infty + \beta_\infty \boldsymbol{i}) e_\infty + (\alpha_0 + \boldsymbol{b}_0 + \beta_0 \boldsymbol{i}) e_0 \\ & + (\alpha_E + \boldsymbol{b}_E + \beta_E \boldsymbol{i}) E, \end{aligned} \tag{5.5.92}$$

where $\alpha = 0, \alpha_\infty, \alpha_0, \alpha_E, \beta, \beta_\infty, \beta_0, \beta_E \in \mathbb{R}$, $\boldsymbol{b}, \boldsymbol{b}_\infty, \boldsymbol{b}_0, \boldsymbol{b}_E \in \mathbb{R}^2$.

From $f^2 = -1$ we obtain the root equation (main condition for f)

$$f^2 = \boldsymbol{b}^2 - \beta^2 - 2\alpha_0\alpha_\infty + 2\boldsymbol{b}_0 \cdot \boldsymbol{b}_\infty + 2\beta_0\beta_\infty + \alpha_E^2 + \boldsymbol{b}_E^2 - \beta_E^2 = -1, \tag{5.5.93}$$

plus side conditions for zero non-scalar parts of f^2. By imposing that $\widetilde{f} = -f$ (necessary for Plancherel and Parseval identities), we abandon scalar, vector[46] and pseudoscalar parts of f:

$$f = \beta \boldsymbol{i} + \boldsymbol{b}_\infty e_\infty + \beta_\infty \boldsymbol{i} e_\infty + \boldsymbol{b}_0 e_0 + \beta_0 \boldsymbol{i} e_0 + \alpha_E E + \boldsymbol{b}_E E. \tag{5.5.94}$$

and retain only bivector and trivector parts.

Regarding the trivector $\sqrt{-1}$ in $Cl(3,1)$ we can calculate the following. For $f = \boldsymbol{b}_E E$, $\boldsymbol{t}, \boldsymbol{t}' \in \mathbb{R}^2$, $\boldsymbol{t}' = -\boldsymbol{b}_E \boldsymbol{t} \boldsymbol{b}_E$, we obtain

$$\begin{aligned} & e^{-\boldsymbol{b}_E E} e_0 e^{\boldsymbol{b}_E E} = e_0, \\ & e^{-\boldsymbol{b}_E E} \boldsymbol{t} e^{\boldsymbol{b}_E E} = \mathrm{ch}^2(|\boldsymbol{b}_E|)\boldsymbol{t} + \mathrm{sh}^2(|\boldsymbol{b}_E|)\boldsymbol{t}' + 2\mathrm{ch}(|\boldsymbol{b}_E|)\mathrm{sh}(|\boldsymbol{b}_E|)(\boldsymbol{t} \wedge \frac{\boldsymbol{b}_E}{|\boldsymbol{b}_E|})E. \end{aligned} \tag{5.5.95}$$

For $f = \beta_\infty \boldsymbol{i} e_\infty$ we obtain

$$e^{-\beta_\infty \boldsymbol{i} e_\infty} e_0 e^{\beta_\infty \boldsymbol{i} e_\infty} = e_0 - 2\beta_\infty^2 e_\infty - 2\beta_\infty \boldsymbol{i} E. \tag{5.5.96}$$

[45]For Fourier transforms with $h \in L^1(\mathbb{R}^{p,q}; GO)$ and $h \in L^2(\mathbb{R}^{p,q}; GO)$, see Footnote 2 on page 144 regarding the roles of L^1 and L^2.

[46]Note that in the context of the spacetime Fourier transform of Section 4.4, the vector part plays an essential role.

Like for quaternion Fourier transformations this may help to separate signal symmetry components, but for now we set the trivector parts[47] of $\sqrt{-1}$ in $Cl(3,1)$ aside. We keep only the four bivector parts.

If we only keep the bivector parts of $f = \sqrt{-1}$ in $Cl(3,1)$ we have

$$f = \beta \boldsymbol{i} + \boldsymbol{b}_{\infty} e_{\infty} + \boldsymbol{b}_0 e_0 + \alpha_E E. \tag{5.5.97}$$

We recognize that $e^{-\varphi\beta\boldsymbol{i}} h(\boldsymbol{x}) e^{\varphi\beta\boldsymbol{i}}$ means a local rotation by $2\varphi\beta \in \mathbb{R}$ of the signal function $h : \mathbb{R}^{p,q} \to GO$. $e^{-b\boldsymbol{b}_{\infty}e_{\infty}} h(\boldsymbol{x}) e^{b\boldsymbol{b}_{\infty}e_{\infty}}$, $b \in \mathbb{R}$, means a translation by $2b\boldsymbol{b}_{\infty} \in \mathbb{R}^2$ of the signal function $h : \mathbb{R}^{p,q} \to GO$. $e^{-c\boldsymbol{b}_0 e_0} h(\boldsymbol{x}) e^{c\boldsymbol{b}_0 e_0}$, $c \in \mathbb{R}$, means a transversion over $2c\boldsymbol{b}_0 \in \mathbb{R}^2$ of the signal function $h : \mathbb{R}^{p,q} \to GO$. $e^{-a\alpha_E E} h(\boldsymbol{x}) e^{a\alpha_E E}$, $a \in \mathbb{R}$, means a scaling by the factor $e^{2a\alpha_E}$ of the signal function $h : \mathbb{R}^{p,q} \to GO$. Thus every term in a bivector $\sqrt{-1}$ in $Cl(3,1)$ has a clear geometric transformation interpretation. This could be of great advantage for the choice and application of conformal CFTs in image and signal processing. ixCGA Fourier transform!application

Considering the side conditions for the zero non-scalar components in $f^2 = -1$ let the angle $\Theta = \angle(\boldsymbol{b}_0, \boldsymbol{b}_E)$. For $\Theta \neq 0, \pi$ we find that

$$\beta = \frac{1}{\sin\Theta}[\alpha_E \cos\Theta \pm \sqrt{\alpha_E^2 + \sin^2\Theta}]. \tag{5.5.98}$$

Especially for $\Theta = \pi/2, 3\pi/2$ we obtain

$$\beta = \pm\sqrt{1 + \alpha_E^2}. \tag{5.5.99}$$

For the special case $\Theta = 0$, $\alpha_E = 0$, we have

$$\beta = \pm\sqrt{1 + 2|\boldsymbol{b}_0||\boldsymbol{b}_E|}. \tag{5.5.100}$$

For $\Theta = \pi$, $\alpha_E = 0$, we have

$$\beta = \pm\sqrt{1 - 2|\boldsymbol{b}_0||\boldsymbol{b}_E|}, \tag{5.5.101}$$

whereas for $\Theta = \pi$, $\beta = 0$, we have

$$\alpha_E = \pm\sqrt{2|\boldsymbol{b}_0||\boldsymbol{b}_E| - 1}. \tag{5.5.102}$$

5.5.3.4 Summary on quest for conformal geometric algebra Fourier transformations

We have briefly introduced the conformal geometric algebra $Cl(3,1)$ model of the Euclidean plane. Then we defined Clifford Fourier transforms based on real $f = \sqrt{-1}$ in $Cl(3,1)$, and tried to select a Clifford Fourier transform by studying the manifold of $\sqrt{-1}$ in $Cl(3,1)$. A selection with clear geometric interpretation proved to be the bivector parts of $\sqrt{-1}$ in $Cl(3,1)$ for the construction of a CFT in conformal geometric algebra model of Euclidean plane. Then every term in a bivector $\sqrt{-1}$ in

[47] Note that in the context of the spacetime Fourier transform of Section 4.4, the trivector part plays an essential role.

$Cl(3,1)$ has a geometric transformation interpretation. Setting $g = -f$ leads to conformal rotor transformation CFTs. We completed this by a detailed characterization of bivector $\sqrt{-1}$ in $Cl(3,1)$, taking all side conditions from non-scalar parts of ff, to be zero, into account. We expect that this new conformal CFT with real bivector $\sqrt{-1}$ in $Cl(3,1)$ will be potentially be useful in applications. It still shares many properties of conventional Fourier transform (such as linearity, shift, modulation, partial differential, Plancherel, Parseval, convolution, ...).

CHAPTER 6

On the interrelationship of QFTs and CFTs

6.1 BACKGROUND

Real and complex Fourier transforms have been extended to W.R. Hamilton's algebra of quaternions (Chapter 4) and to W.K. Clifford's geometric algebras (Chapter 5). This was initially motivated by applications in nuclear magnetic resonance and electric engineering. Followed by an ever wider range of applications in color image and signal processing. Clifford's geometric algebras are complete algebras, algebraically encoding a vector space and all its subspace elements. Applications include electromagnetism, and the processing of images, color images, vector field and climate data. Further developments of Clifford Fourier transforms include operator exponential representations, and extensions to wider classes of integral transforms, like Clifford algebra versions of linear canonical transforms and wavelets.

This Chapter is mainly based on [207], see also [206]. Several of the transforms we feature in this discussion have been studied in more detail in Chapters 4 and 5. But we briefly characterize them here once more in order to make it easier to grasp the larger picture of hypercomplex Fourier transforms and their relationships.

We begin by introducing Clifford Fourier transforms, including the important class of quaternion Fourier transforms mainly along the lines of [206] and [51], adding further detail, emphasize and new developments.

There is the alternative operator exponential Clifford Fourier transform (CFT) approach, mainly pursued by the Clifford Analysis Group at the University of Ghent (Belgium) [51]. New work in this direction closely related to the roots of -1 approach explained below is [120], see also Section 5.2.3.

A CFT analyzes scalar, vector and multivector signals[1] in terms of sine and cosine waves with multivector coefficients. Basically, the imaginary unit $i \in \mathbb{C}$ in the transformation kernel $e^{i}\phi = \cos\phi + i\sin\phi$ is replaced by a $\sqrt{-1}$ in $Cl(p,q)$. This produces a host of CFTs, a still incomplete brief overview is sketched in Figure 6.1, see also the historical overview in [51]. Additionally, the $\sqrt{-1}$ in $Cl(p,q)$ allow to

[1]For Fourier transforms of multivector signals in L^1 and L^2, see Footnote 2 on page 144 regarding the roles of L^1 and L^2.

DOI: 10.1201/9781003184478-6

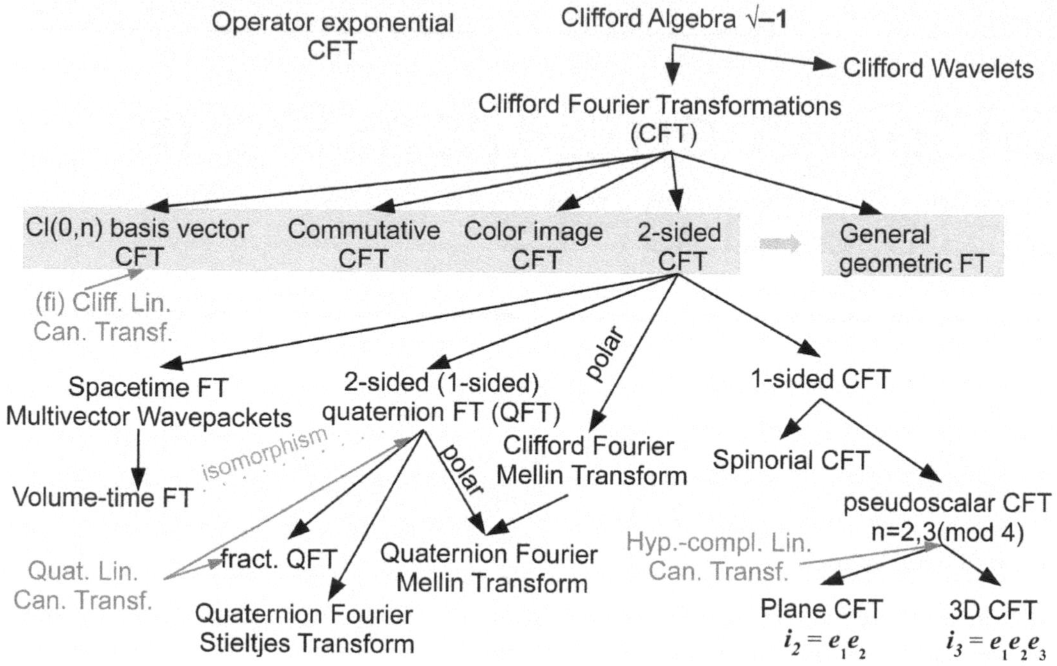

Figure 6.1 Family tree of Clifford Fourier transformations. Source: [207, Fig. 2].

construct further types of integral transformations, notably Clifford wavelets [188, 253].

6.2 GENERAL GEOMETRIC FOURIER TRANSFORM

A rigorous effort was made in [61] to design a *general geometric Fourier transform*, compare also Sections 5.3.2 and 5.3.3, that incorporates most of the previously known CFTs with the help of very general sets of left and right kernel factor products

$$\mathcal{F}_{GFT}\{h\}(\boldsymbol{\omega}) = \int_{\mathbb{R}^{p',q'}} L(\mathbf{x},\omega)h(\mathbf{x})R(\mathbf{x},\omega)d^{n'}\mathbf{x}, \quad L(\mathbf{x},\omega) = \prod_{s\in F_L} e^{-s(\mathbf{x},\omega)}, \tag{6.2.1}$$

with $p'+q' = n'$, $F_L = \{s_1(\mathbf{x},\omega),\dots,s_L(\mathbf{x},\omega)\}$ a set of mappings $\mathbb{R}^{p',q'}\times\mathbb{R}^{p',q'} \to \mathcal{I}^{p,q}$ into the manifold of real multiples of $\sqrt{-1}$ in $Cl(p,q)$. $R(\mathbf{x},\omega)$ is defined similarly, and $h:\mathbb{R}^{p',q'}\to Cl(p,q)$ is the multivector signal function.

6.3 CFT DUE TO SOMMEN AND BÜLOW

Equation (6.2.1) clearly subsumes the *CFT due to Sommen and Bülow* [58]

$$\mathcal{F}_{SB}\{h\}(\boldsymbol{\omega}) = \int_{\mathbb{R}^n} h(\mathbf{x}) \prod_{k=1}^{n} e^{-2\pi x_k\omega_k e_k} d^n\mathbf{x}, \tag{6.3.1}$$

where $\mathbf{x},\omega\in\mathbb{R}^n$ with components x_k,ω_k, and $\{e_1,\dots e_k\}$ is an orthonormal basis of $\mathbb{R}^{0,n}$, $h:\mathbb{R}^n\to Cl(0,n)$.

6.4 COLOR IMAGE CFT

It is further possible [133] to only pick strictly mutually commuting sets of $\sqrt{-1}$ in $Cl(p,q)$, e.g. e_1e_2, $e_3e_4 \in Cl(4,0)$ and construct CFTs with therefore commuting kernel factors in analogy to (6.3.1). Also contained in (6.2.1) is the *color image CFT* of [262]

$$\mathcal{F}_{CI}\{h\}(\boldsymbol{\omega}) = \int_{\mathbb{R}^2} e^{\frac{1}{2}\boldsymbol{\omega}\cdot\mathbf{x}I_4B} e^{\frac{1}{2}\boldsymbol{\omega}\cdot\mathbf{x}B} h(\mathbf{x}) e^{-\frac{1}{2}\boldsymbol{\omega}\cdot\mathbf{x}B} e^{-\frac{1}{2}\boldsymbol{\omega}\cdot\mathbf{x}I_4B} d^2\mathbf{x}, \tag{6.4.1}$$

where $B \in Cl(4,0)$ is a bivector and $I_4B \in Cl(4,0)$ its dual complementary bivector. It is especially useful for the introduction of efficient non-marginal generalized color image Fourier descriptors.

6.5 TWO-SIDED CFT

The main type of CFT, which we will review here is the general *two sided CFT* [209], see Section 5.3.1, with only one kernel factor on each side

$$\mathcal{F}^{f,g}\{h\}(\boldsymbol{\omega}) = \int_{\mathbb{R}^{p',q'}} e^{-fu(\mathbf{x},\omega)} h(\mathbf{x}) e^{-gv(\mathbf{x},\omega)} d^{n'}\mathbf{x}, \tag{6.5.1}$$

with f,g two $\sqrt{-1}$ in $Cl(p,q)$, $u,v : \mathbb{R}^{p',q'} \times \mathbb{R}^{p',q'} \to \mathbb{R}$ and often $\mathbb{R}^{p',q'} = \mathbb{R}^{p,q}$. In the following we will discuss a family of transforms, which belong to this class of CFTs, see the lower half of Figure 6.1.

6.6 QUATERNION FOURIER TRANSFORM (QFT)

One of the nowadays most widely applied CFTs is the *quaternion Fourier transform* (QFT) [184, 205], compare Chapter 4,

$$\mathcal{F}^{f,g}\{h\}(\boldsymbol{\omega}) = \int_{\mathbb{R}^2} e^{-fx_1\omega_1} h(\mathbf{x}) e^{-gx_2\omega_2} d^2\mathbf{x}, \tag{6.6.1}$$

which also has variants were one of the left or right kernel factors is dropped, or both are placed together at the right or left side. It was first described by Ernst, et al, [130, pp. 307–308] (with $f = \boldsymbol{i}, g = \boldsymbol{j}$) for spectral analysis in two-dimensional nuclear magnetic resonance, suggesting to use the QFT as a method to independently adjust phase angles with respect to two frequency variables in two-dimensional spectroscopy. Later Ell [124] independently formulated and explored the QFT for the analysis of linear time-invariant systems of PDEs. The QFT was further applied by Bülow, et al [56] for image, video and texture analysis, by Sangwine et al [51, 291] for color image analysis and analysis of non-stationary improper complex signals, vector image processing and quaternion polar signal representations. It is possible to split every quaternion-valued signal and its QFT into two quasi-complex components [205], which allow the application of complex discretization and fast FT methods. The split can be generalized to the general CFT (6.5.1) [209] in the form

$$x_{\pm} = \frac{1}{2}(x \pm fxg), \quad x \in Cl(p,q). \tag{6.6.2}$$

In the case of quaternions the quaternion coefficient space $\mathbb{R}^4$ is thereby split into two steerable (by the choice of two pure quaternions f, g) orthogonal two-dimensional planes [205]. The geometry of this split appears closely related to the quaternion geometry of rotations [261]. For colors expressed by quaternions, these two planes become chrominance and luminance when $f = g =$ gray line [127].

6.7 QUATERNION FOURIER STIELTJES TRANSFORM

Georgiev and Morais have modified the QFT to a *quaternion Fourier Stieltjes transform* [139].

$$\mathcal{F}_{Stj}(\sigma^1, \sigma^2) = \int_{\mathbb{R}^2} e^{-fx_1\omega_1} d\sigma^1(x_1) d\sigma^2(x_2) e^{-gx_2\omega_2}, \tag{6.7.1}$$

with $f = -\boldsymbol{i}, g = -\boldsymbol{j}$, $\sigma^k : \mathbb{R} \to \mathbb{H}$, $|\sigma^k| \leq \delta_k$ for real numbers $0 < \delta_k < \infty$, $k = 1, 2$.

6.8 QUATERNION FOURIER MELLIN TRANSFORM, CLIFFORD FOURIER MELLIN TRANSFORM

Introducing polar coordinates in $\mathbb{R}^2$ allows to establish a *quaternion Fourier Mellin transform* (QFMT) [198], see Section 4.3.2,

$$\mathcal{F}_{QM}\{h\}(\boldsymbol{\nu}, k) = \frac{1}{2\pi} \int_0^{\infty} \int_0^{2\pi} r^{-f\boldsymbol{\nu}} h(r, \theta) e^{-gk\theta} d\theta dr/r, \quad \forall(\boldsymbol{\nu}, k) \in \mathbb{R} \times \mathbb{Z}, \tag{6.8.1}$$

which can characterize 2D shapes rotation, translation and scale invariant, possibly including color encoded in the quaternion valued signal $h : \mathbb{R}^2 \to \mathbb{H}$ such that $|h|$ is summable over $\mathbb{R}_+^* \times \mathbb{S}^1$ under the measure $d\theta dr/r$, $\mathbb{R}^*$ the multiplicative group of positive non-zero numbers, and $f, g \in \mathbb{H}$ two $\sqrt{-1}$. The QFMT can be generalized straightforward to a *Clifford Fourier Mellin transform* applied to signals $h : \mathbb{R}^2 \to Cl(p, q), p + q = 2$ [199], see Section 5.5.2, with $f, g \in Cl(p, q), p + q = 2$.

6.9 VOLUME-TIME CFT AND SPACETIME CFT

The spacetime algebra $Cl(3, 1)$ of Minkowski space $\mathbb{R}^{3,1}$ with orthonormal vector basis $\{\mathbf{e}_t, \mathbf{e}_1, \mathbf{e}_2, \mathbf{e}_3\}$, $-\mathbf{e}_t^2 = \mathbf{e}_1^2 = \mathbf{e}_2^2 = \mathbf{e}_3^2$, has three blades $\mathbf{e}_t, i_3, i_{st}$ of time vector, unit space volume 3-vector and unit hyperspace volume 4-vector, which are isomorphic to Hamilton's three quaternion units

$$\mathbf{e}_t^2 = -1, \quad i_3 = \mathbf{e}_1\mathbf{e}_2\mathbf{e}_3 = \mathbf{e}_t^* = \mathbf{e}_t i_{st}^{-1}, i_3^2 = -1, \quad i_{st} = \mathbf{e}_t i_3, i_{st}^2 = -1. \tag{6.9.1}$$

The $Cl(3, 1)$ subalgebra with basis $\{1, \mathbf{e}_t, i_3, i_{st}\}$ is therefore isomorphic to quaternions and allows to generalize the two-sided QFT to a *volume-time Fourier transform* (see Section 4.4)

$$\mathcal{F}_{VT}\{h\}(\boldsymbol{\omega}) = \int_{\mathbb{R}^{3,1}} e^{-\mathbf{e}_t\omega_t} h(\mathbf{x}) e^{-\vec{x}\cdot\vec{\omega}} d^4\mathbf{x}, \tag{6.9.2}$$

with $\mathbf{x} = t\mathbf{e}_t + \vec{x} \in \mathbb{R}^{3,1}$, $\vec{x} = x_1\mathbf{e}_1 + x_2\mathbf{e}_2 + x_3\mathbf{e}_3$, $\boldsymbol{\omega} = \omega_t\mathbf{e}_t + \vec{\omega} \in \mathbb{R}^{3,1}$, $\vec{\omega} = \omega_1\mathbf{e}_1 + \omega_2\mathbf{e}_2 + \omega_3\mathbf{e}_3$. The split (6.6.2) with $f = \mathbf{e}_t$, $g = i_3 = \mathbf{e}_t^*$ becomes the spacetime split of special relativity

$$h_{\pm} = \frac{1}{2}(1 \pm \mathbf{e}_t h \mathbf{e}_t^*). \tag{6.9.3}$$

It is most interesting to observe, that the volume-time Fourier transform can indeed be applied to multivector signal functions valued in the whole spacetime algebra $h : \mathbb{R}^{3,1} \to Cl(3,1)$ without changing its form [184, 187], see Section 4.4,

$$\mathcal{F}_{ST}\{h\}(\boldsymbol{\omega}) = \int_{\mathbb{R}^{3,1}} e^{-\mathbf{e}_t \omega_t t} h(\mathbf{x}) e^{-i_3 \vec{x} \cdot \vec{\omega}} d^4\mathbf{x}. \tag{6.9.4}$$

The split (6.9.3) applied to *spacetime Fourier transform* (6.9.4) leads to a relativistic *multivector wavepacket analysis*

$$\mathcal{F}_{ST}\{h\}(\boldsymbol{\omega}) = \int_{\mathbb{R}^{3,1}} h_+(\mathbf{x}) e^{-i_3(\vec{x}\cdot\vec{\omega} - t\omega_t)} d^4\mathbf{x} + \int_{\mathbb{R}^{3,1}} h_-(\mathbf{x}) e^{-i_3(\vec{x}\cdot\vec{\omega} + t\omega_t)} d^4\mathbf{x}, \tag{6.9.5}$$

in terms of right and left propagating spacetime multivector wave packets.

6.10 ONE-SIDED CFTS

Finally, we turn to *one-sided CFTs* [201], see Section 5.2.2, which are obtained by setting the phase function $u = 0$ in (6.5.1). A recent discrete *spinor CFT* used for edge and texture detection is given in [19], where the signal is represented as a spinor and the $\sqrt{-1}$ is a local tangent bivector $B \in Cl(3,0)$ to the image intensity surface ($\mathbf{e}_3$ is the intensity axis).

6.11 PSEUDOSCALAR KERNEL CFTS

The following class of *one-sided CFTs which uses a single pseudoscalar* $\sqrt{-1}$ has been well studied and applied [185], see Section 5.2.1,

$$\mathcal{F}_{PS}\{h\}(\boldsymbol{\omega}) = \int_{\mathbb{R}^n} h(\mathbf{x}) e^{-i_n \mathbf{x}\cdot\boldsymbol{\omega}} d^n\mathbf{x}, \quad i_n = \mathbf{e}_1\mathbf{e}_2 \ldots \mathbf{e}_n, \quad n = 2, 3 (\mathrm{mod}\, 4), \tag{6.11.1}$$

where $h : \mathbb{R}^n \to Cl(n, 0)$, and $\{\mathbf{e}_1, \mathbf{e}_2, \ldots, \mathbf{e}_n\}$ is the orthonormal basis of $\mathbb{R}^n$. Historically the special case of (6.11.1), $n = 3$, was already introduced in 1990 [222] for the processing of electromagnetic fields. This same transform was later applied [132] to two-dimensional images embedded in $Cl(3,0)$ to yield a two-dimensional analytic signal, and in image structure processing. Moreover, the *pseudoscalar CFT* (6.11.1), $n = 3$, was successfully applied to three-dimensional vector field processing in [117, 118] with vector signal convolution based on Clifford's full geometric product of vectors. The theory of the transform has been thoroughly studied in [185].

For embedding one-dimensional signals in $\mathbb{R}^2$, [132] considered in (6.11.1) the special case of $n = 2$, and in [117, 118] this was also applied to the processing of two-dimensional vector fields.

Recent applications of (6.11.1) with $n = 2, 3$, to geographic information systems and climate data can be found in [331–333].

6.12 QUATERNION AND CLIFFORD LINEAR CANONICAL TRANSFORMS

Real and complex linear canonical transforms parametrize a continuum of transforms, which include the Fourier, fractional Fourier, Laplace, fractional Laplace, Gauss-Weierstrass, Bargmann, Fresnel and Lorentz transforms, as well as scaling operations. A Fourier transform transforms multiplication with the space argument x into differentiation with respect to the frequency argument ω. In Schrödinger quantum mechanics this constitutes a rotation in position-momentum phase space. A linear canonical transform transforms the position and momentum operators into linear combinations (with a two-by-two real or complex parameter matrix), preserving the fundamental position-momentum commutator relationship, at the core of the uncertainty principle. The transform operator can be made to act on appropriate spaces of functions, and can be realized in the form of integral transforms, parametrized in terms of the four real (or complex) matrix parameters [329].

KitIan Kou et al [237] introduce the quaternionic linear canonical transform (QLCT). They consider a pair of unit determinant two-by-two matrices

$$A_1 = \begin{pmatrix} a_1 & b_1 \\ c_1 & d_1 \end{pmatrix}, \qquad A_2 = \begin{pmatrix} a_2 & b_2 \\ c_2 & d_2 \end{pmatrix}, \tag{6.12.1}$$

with entries $a_1, a_2, b_1, b_2, c_1, c_2, d_1, d_2 \in \mathbb{R}$, $a_1 d_1 - c_1 b_1 = 1$, $a_2 d_2 - c_2 b_2 = 1$, where they disregard the cases $b_1 = 0$, $b_2 = 0$, for which the LCT is essentially a chirp multiplication.

We now *generalize* the definitions of [237] using the following two kernel functions with two pure unit quaternions $f, g \in \mathbb{H}$, $f^2 = g^2 = -1$, including the cases $f = \pm g$,

$$\begin{aligned} K^f_{A_1}(x_1, \omega_1) &= \frac{1}{\sqrt{f 2\pi b_1}} e^{f(a_1 x_1^2 - 2x_1\omega_1 + d_1\omega_1^2)/(2b_1)}, \\ K^g_{A_2}(x_2, \omega_2) &= \frac{1}{\sqrt{g 2\pi b_2}} e^{g(a_2 x_2^2 - 2x_2\omega_1 + d_2\omega_2^2)/(2b_2)}. \end{aligned} \tag{6.12.2}$$

The *two-sided QLCT* of signals $h \in L^1(\mathbb{R}^2, \mathbb{H})$ can now generally be defined as

$$\mathcal{L}^{f,g}(\boldsymbol{\omega}) = \int_{\mathbb{R}^2} K^f_{A_1}(x_1, \omega_1) h(\mathbf{x}) K^g_{A_2}(x_2, \omega_2) d^2\mathbf{x}. \tag{6.12.3}$$

The *left-sided* and *right-sided QLCTs* can be defined correspondingly by placing the two kernel factors both on the left or on the right[2], respectively. For $a_1 = d_1 = a_2 = d_2 = 0$, $b_1 = b_2 = 1$, the conventional two-sided (left-sided, right-sided) QFT is recovered. We note that it will be of interest to "complexify" the matrices A_1 and A_2,

[2]In [237] the possibility of a more general pair of unit quaternions $f, g \in \mathbb{H}$, $f^2 = g^2 = -1$, is only indicated for the case of the right-sided QLCT, but with the restriction that f, g should be an *orthonormal pair* of pure quaternions, i.e. $Sc(f\overline{g}) = 0$. Otherwise [237] always strictly sets $f = \boldsymbol{i}$ and $g = \boldsymbol{j}$.

by including replacing $a_1 \to a_{1r} + fa_{1f}$, $a_2 \to a_{2r} + ga_{2g}$, etc. In [237] for $f = \boldsymbol{i}$ and $g = \boldsymbol{j}$ the right-sided QLCT and its properties, including an uncertainty principle are studied in some detail.

In [330] a complex Clifford linear canonical transform is defined and studied for signals $f \in L^1(\mathbb{R}^m, C^{m+1})$, where $C^{m+1} = \text{span}\{1, e_1, \ldots, e_m\} \subset Cl(0,m)$ is the subspace of paravectors in $Cl(0,m)$. This includes uncertainty principles. Motivated by Remark 2.2 in [330], we now modify this definition to generalize the one-sided CFT of [201] for real Clifford algebras $Cl(n,0)$ to a *general real Clifford linear canonical transform (CLNT)*. We define the parameter matrix

$$A = \begin{pmatrix} a & b \\ c & d \end{pmatrix}, \quad a, b, c, d \in \mathbb{R}, \quad ad - cb = 1. \tag{6.12.4}$$

We again omit the case $b = 0$ and define the kernel

$$K^f(\mathbf{x}, \boldsymbol{\omega}) = \frac{1}{\sqrt{f(2\pi)^n b}} e^{f(a\mathbf{x}^2 - 2\mathbf{x}\cdot\boldsymbol{\omega} + d\boldsymbol{\omega}^2)/(2b)}, \tag{6.12.5}$$

with the general square root of -1: $f \in Cl(n,0)$, $f^2 = -1$. Then the general real CLNT can be defined for signals $h \in L^1(\mathbb{R}^n; Cl(n,0))$ as

$$\mathcal{L}^f\{h\}(\boldsymbol{\omega}) = \int_{\mathbb{R}^n} h(\mathbf{x}) K^f(\mathbf{x}, \boldsymbol{\omega}) d^n\mathbf{x}. \tag{6.12.6}$$

For $a = d = 0$, $b = 1$, the conventional one-sided CFT of [201] in $Cl(n,0)$ is recovered. It is again of interest to modify the entries of the parameter matrix to $a \to a_0 + fa_f, b \to b_0 + fb_f$, etc.

Similarly a *Clifford version of a linear canonical transform* (CLCT) for signals $h \in L^1(\mathbb{R}^m; \mathbb{R}^{m+1})$ is formulated in [236], using two-by-two parameter matrices $A_1, \ldots, A_m$, which maps $\mathbb{R}^m \to Cl(0,m)$. The Sommen Bülow CFT (6.3.1) is recovered for parameter matrix entries $a_k = d_k = 0$, $b_k = 1$, $1 \leq k \leq m$.

6.13 SUMMARY ON THE INTERRELATIONSHIP OF QFTS AND CFTS

We have briefly reviewed Clifford Fourier transforms which apply the manifolds of $\sqrt{-1} \in Cl(p,q)$ in order to create a rich variety of new Clifford valued Fourier transformations. The history of these transforms spans just over 30 years. Major steps in the development were: $Cl(0,n)$ CFTs, then pseudoscalar CFTs, and quaternion FTs. In the 1990ies especially applications to electromagnetic fields/electronics and in signal/image processing dominated. This was followed by by color image processing and most recently applications in Geographic Information Systems (GIS). This overview could only feature a part of the approaches in CFT research, and only a part of the applications. Omitted were details on operator exponential CFT approach [51], and CFT for conformal geometric algebra. Regarding applications, e.g. CFT Fourier descriptor representations of shape [288] of B. Rosenhahn, et al was omitted. Note that there are further types of Clifford algebra/analysis related integral transforms: Clifford wavelets, Clifford radon transforms, Clifford Hilbert transforms, ... which we did also not discuss.

APPENDIX A

Summary of square roots of -1, Cauchy-Schwarz, uncertainty equality

The appendix contains some tables of multivector square roots of -1 (Appendix A.1), a sample worksheet for their computations can be found in [204], the proof of a multivector Cauchy-Schwarz inequality (Appendix A.2), and of the uncertainty equality for Gaussian multivector functions (Appendix A.3).

A.1 SUMMARY OF ROOTS OF -1 IN $C\ell(p,q) \cong \mathcal{M}(2d,\mathbb{C})$ FOR $d = 1,2,4$

In this appendix we summarize roots of $-\mathbf{1}$ for Clifford algebras $C\ell(p,q) \cong \mathcal{M}(2d,\mathbb{C})$ for $d = 1,2,4$. These roots have been computed with CLIFFORD [5]. Maple [317] worksheets (like the one presented in [204]) written to derive these roots are posted at [204].

k	f_k	$\Delta_k(t)$
1	$\omega = e_{123}$	$(t-i)^2$
0	e_{23}	$(t-i)(t+i)$
-1	$-\omega = -e_{123}$	$(t+i)^2$

Table A.1 Square roots of $-\mathbf{1}$ in $C\ell(3,0) \cong \mathcal{M}(2,\mathbb{C})$, $d = 1$. Source: [203, Tab. 1].

A.2 MULTIVECTOR CAUCHY-SCHWARZ INEQUALITY

We will show that for multivectors $M, N \in Cl(p,q)$, with the principal reverse $\widetilde{N}$ of Section 2.1.1.6, we have

$$|\langle M\widetilde{N}\rangle| \leq \|M\|\,\|N\|. \tag{A1}$$

DOI: 10.1201/9781003184478-A

k	f_k	$\Delta_k(t)$
2	$\omega = e_{12345}$	$(t-i)^4$
1	$\frac{1}{2}(e_{23} + e_{123} - e_{2345} + e_{12345})$	$(t-i)^3(t+i)$
0	e_{123}	$(t-i)^2(t+i)^2$
-1	$\frac{1}{2}(e_{23} + e_{123} + e_{2345} - e_{12345})$	$(t-i)(t+i)^3$
-2	$-\omega = -e_{12345}$	$(t+i)^4$

Table A.2 Square roots of -1 in $C\ell(4,1) \cong \mathcal{M}(4,\mathbb{C})$, $d=2$. Source: [203, Tab. 2].

k	f_k	$\Delta_k(t)$
2	$\omega = e_{12345}$	$(t-i)^4$
1	$\frac{1}{2}(e_3 + e_{12} + e_{45} + e_{12345})$	$(t-i)^3(t+i)$
0	e_{45}	$(t-i)^2(t+i)^2$
-1	$\frac{1}{2}(-e_3 + e_{12} + e_{45} - e_{12345})$	$(t-i)(t+i)^3$
-2	$-\omega = -e_{12345}$	$(t+i)^4$

Table A.3 Square roots of -1 in $C\ell(0,5) \cong \mathcal{M}(4,\mathbb{C})$, $d=2$. Source: [203, Tab. 3].

Proof Note that for any $t \in \mathbb{R}$ holds

$$\begin{aligned} 0 &\leq \|M+tN\|^2 = (M+tN) * (\widetilde{M+tN}) \\ &= M * \widetilde{M} + t(M * \widetilde{N} + N * \widetilde{M}) + t^2 N * \widetilde{N} \\ &= \|M\|^2 + 2t\langle M\widetilde{N}\rangle + t^2\|N\|^2. \end{aligned} \tag{A2}$$

The negative discriminant of this quadratic polynomial implies

$$\langle M\widetilde{N}\rangle^2 - \|M\|^2\|N\|^2 \leq 0. \tag{A3}$$

k	f_k	$\Delta_k(t)$
2	$\omega = e_{12345}$	$(t-i)^4$
1	$\frac{1}{2}(e_3 + e_{134} + e_{235} + \omega)$	$(t-i)^3(t+i)$
0	e_{134}	$(t-i)^2(t+i)^2$
-1	$\frac{1}{2}(-e_3 + e_{134} + e_{235} - \omega)$	$(t-i)(t+i)^3$
-2	$-\omega = -e_{12345}$	$(t+i)^4$

Table A.4 Square roots of -1 in $C\ell(2,3) \cong \mathcal{M}(4,\mathbb{C})$, $d=2$. Source: [203, Tab. 4].

k	f_k	$\Delta_k(t)$
4	$\omega = e_{1234567}$	$(t-i)^8$
3	$\frac{1}{4}(e_{23} - e_{45} + e_{67} - e_{123} + e_{145} - e_{167} + e_{234567} + 3\omega)$	$(t-i)^7(t+i)$
2	$\frac{1}{2}(e_{67} - e_{45} - e_{123} + \omega)$	$(t-i)^6(t+i)^2$
1	$\frac{1}{4}(e_{23} - e_{45} + 3e_{67} - e_{123} + e_{145} + e_{167} - e_{234567} + \omega)$	$(t-i)^5(t+i)^3$
0	$\frac{1}{2}(e_{23} - e_{45} + e_{67} - e_{234567})$	$(t-i)^4(t+i)^4$
−1	$\frac{1}{4}(e_{23} - e_{45} + 3e_{67} + e_{123} - e_{145} - e_{167} - e_{234567} - \omega)$	$(t-i)^3(t+i)^5$
−2	$\frac{1}{2}(e_{67} - e_{45} + e_{123} - \omega)$	$(t-i)^2(t+i)^6$
−3	$\frac{1}{4}(e_{23} - e_{45} + e_{67} + e_{123} - e_{145} + e_{167} + e_{234567} - 3\omega)$	$(t-i)(t+i)^7$
−4	$-\omega = -e_{1234567}$	$(t+i)^8$

Table A.5 Square roots of $-\mathbf{1}$ in $C\ell(7,0) \cong \mathcal{M}(8,\mathbb{C})$, $d = 4$. Source: [203, Tab. 5].

This proves (A1) and (5.2.149):

$$\langle M\widetilde{N}\rangle = M * \widetilde{N} \leq |\langle M\widetilde{N}\rangle| \leq \|M\| \, \|N\|. \tag{A4}$$

Inserting into (5.2.148) and (2.6.35) into the multivector Cauchy-Schwarz inequality (A4) we can express it in a basis (5.5.1) of the geometric algebra as

$$|\sum_A \alpha_A \beta_A| \leq \left(\sum_A \alpha_A^2\right)^{\frac{1}{2}} \left(\sum_B \beta_B^2\right)^{\frac{1}{2}}. \tag{A5}$$

A.3 UNCERTAINTY EQUALITY FOR GAUSSIAN MULTIVECTOR FUNCTIONS

Note that according to line 3 of the proof for Theorem 5.2.55 the uncertainty principle (5.2.103) can be rewritten as

$$(2\pi)^3 \int_{\mathbb{R}^3} (\boldsymbol{a}\cdot\boldsymbol{x})^2 \|f(\boldsymbol{x})\|^2 \, d^3\boldsymbol{x} \int_{\mathbb{R}^3} \|\boldsymbol{a}\cdot\nabla f(\boldsymbol{x})\|^2 d^3\boldsymbol{x} \geq \frac{(2\pi)^3}{4} F^2. \tag{B1}$$

Now we have for Gaussian multivector functions (5.2.104)

$$\boldsymbol{a}\cdot\nabla f = \boldsymbol{a}\cdot\nabla \; C_0 \, e^{-k\boldsymbol{x}^2} = -2k \; \boldsymbol{a}\cdot\boldsymbol{x} \; C_0 \, e^{-k\boldsymbol{x}^2} = -2k \; \boldsymbol{a}\cdot\boldsymbol{x} \; f. \tag{B2}$$

k	f_k	$\Delta_k(t)$
4	$\omega = e_{1234567}$	$(t-i)^8$
3	$\frac{1}{4}(e_4 - e_{23} - e_{56} + e_{1237} + e_{147} + e_{1567} - e_{23456} + 3\omega)$	$(t-i)^7(t+i)$
2	$\frac{1}{2}(-e_{23} - e_{56} + e_{147} + \omega)$	$(t-i)^6(t+i)^2$
1	$\frac{1}{4}(-e_4 - e_{23} - 3e_{56} - e_{1237} + e_{147} + e_{1567} - e_{23456} + \omega)$	$(t-i)^5(t+i)^3$
0	$\frac{1}{2}(e_4 + e_{23} + e_{56} + e_{23456})$	$(t-i)^4(t+i)^4$
-1	$\frac{1}{4}(-e_4 - e_{23} - 3e_{56} + e_{1237} - e_{147} - e_{1567} - e_{23456} - \omega)$	$(t-i)^3(t+i)^5$
-2	$\frac{1}{2}(-e_{23} - e_{56} - e_{147} - \omega)$	$(t-i)^2(t+i)^6$
-3	$\frac{1}{4}(e_4 - e_{23} - e_{56} - e_{1237} - e_{147} - e_{1567} - e_{23456} - 3\omega)$	$(t-i)(t+i)^7$
-4	$-\omega = -e_{1234567}$	$(t+i)^8$

Table A.6 Square roots of $-\mathbf{1}$ in $C\ell(1,6) \simeq \mathcal{M}(8,\mathbb{C})$, $d = 4$. Source: [203, Tab. 6].

so, we get

$$\boldsymbol{a} \cdot \boldsymbol{x}\, f = \frac{-1}{2k}\, \boldsymbol{a} \cdot \nabla f, \tag{B3}$$

and

$$\|\boldsymbol{a} \cdot \nabla f\|^2 = 4k^2 \|(\boldsymbol{a} \cdot \boldsymbol{x}) f\|^2 = 4k^2\, (\boldsymbol{a} \cdot \boldsymbol{x})^2 \|f\|^2 \tag{B4}$$

k	f_k	$\Delta_k(t)$
4	$\omega = e_{1234567}$	$(t-i)^8$
3	$\frac{1}{4}(e_4 + e_{145} + e_{246} + e_{347} - e_{12456} - e_{13457} - e_{23467} + 3\omega)$	$(t-i)^7(t+i)$
2	$\frac{1}{2}(e_{145} - e_{12456} - e_{13457} + \omega)$	$(t-i)^6(t+i)^2$
1	$\frac{1}{4}(-e_4 + e_{145} + e_{246} - e_{347} - 3e_{12456} - e_{13457} - e_{23467} + \omega)$	$(t-i)^5(t+i)^3$
0	$\frac{1}{2}(e_4 + e_{12456} + e_{13457} + e_{23467})$	$(t-i)^4(t+i)^4$
−1	$\frac{1}{4}(-e_4 - e_{145} - e_{246} + e_{347} - 3e_{12456} - e_{13457} - e_{23467} - \omega)$	$(t-i)^3(t+i)^5$
−2	$\frac{1}{2}(-e_{145} - e_{12456} - e_{13457} - \omega)$	$(t-i)^2(t+i)^6$
−3	$\frac{1}{4}(e_4 - e_{145} - e_{246} - e_{347} - e_{12456} - e_{13457} - e_{23467} - 3\omega)$	$(t-i)(t+i)^7$
−4	$-\omega = -e_{1234567}$	$(t+i)^8$

Table A.7 Square roots of $-\mathbf{1}$ in $C\ell(3,4) \cong \mathcal{M}(8,\mathbb{C})$, $d = 4$. Source: [203, Tab. 7].

Substituting (B4) and (B3) in the left side of (B1) we get for $\boldsymbol{a}^2 = 1$

$$
\begin{aligned}
(2\pi)^3 \int_{\mathbb{R}^3} (\boldsymbol{a}\cdot\boldsymbol{x})^2 \|f(\boldsymbol{x})\|^2 \, d^3\boldsymbol{x} \int_{\mathbb{R}^3} \|\boldsymbol{a}\cdot\nabla f(\boldsymbol{x})\|^2 d^3\boldsymbol{x} & \\
\overset{(B4)}{=} \quad & 4k^2(2\pi)^3 \left(\int_{\mathbb{R}^3} \boldsymbol{a}\cdot\boldsymbol{x}\ \boldsymbol{a}\cdot\boldsymbol{x} \|f\|^2 d^3\boldsymbol{x} \right)^2 \\
= \quad & 4k^2(2\pi)^3 \left(\int_{\mathbb{R}^3} \boldsymbol{a}\cdot\boldsymbol{x}\boldsymbol{a}\cdot\boldsymbol{x} \langle f\tilde{f}\rangle d^3\boldsymbol{x} \right)^2 \\
= \quad & 4k^2(2\pi)^3 \left(\int_{\mathbb{R}^3} \boldsymbol{a}\cdot\boldsymbol{x} \langle \boldsymbol{a}\cdot\boldsymbol{x} f\tilde{f}\rangle d^3\boldsymbol{x} \right)^2 \\
\overset{(B3)}{=} \quad & 4k^2(2\pi)^3 \left(\int_{\mathbb{R}^3} \frac{\boldsymbol{a}\cdot\boldsymbol{x}}{-2k} \langle (\boldsymbol{a}\cdot\nabla f)\tilde{f}\rangle d^3\boldsymbol{x} \right)^2 \\
\overset{(5.2.101)}{=} \quad & (2\pi)^3 \left(\int_{\mathbb{R}^3} \boldsymbol{a}\cdot\boldsymbol{x} \frac{1}{2} \boldsymbol{a}\cdot\boldsymbol{\nabla} \|f\|^2 d^3\boldsymbol{x} \right)^2 \\
\overset{\text{P. 5.2.9}}{=} \quad & \frac{(2\pi)^3}{4} \left(\int_{\mathbb{R}^3} \underbrace{(\boldsymbol{a}\cdot\boldsymbol{\nabla}\ \boldsymbol{a}\cdot\boldsymbol{x})}_{=\boldsymbol{a}^2=1} \|f\|^2 d^3\boldsymbol{x} \right)^2 \\
= \quad & \frac{(2\pi)^3}{4} F^2. \qquad \text{(B5)}
\end{aligned}
$$

k	f_k	$\Delta_k(t)$
4	$\omega = e_{1234567}$	$(t-i)^8$
3	$\frac{1}{4}(-e_{23} + e_{123} + e_{2346} + e_{2357} - e_{12346} - e_{12357} + e_{234567} + 3\omega)$	$(t-i)^7(t+i)$
2	$\frac{1}{2}(e_{123} - e_{12346} - e_{12357} + \omega)$	$(t-i)^6(t+i)^2$
1	$\frac{1}{4}(-e_{23} + e_{123} - e_{2346} + e_{2357} - 3e_{12346} - e_{12357} - e_{234567} + \omega)$	$(t-i)^5(t+i)^3$
0	$\frac{1}{2}(e_{23} + e_{12346} + e_{12357} + e_{234567})$	$(t-i)^4(t+i)^4$
−1	$\frac{1}{4}(-e_{23} - e_{123} + e_{2346} - e_{2357} - 3e_{12346} - e_{12357} - e_{234567} - \omega)$	$(t-i)^3(t+i)^5$
−2	$\frac{1}{2}(-e_{123} - e_{12346} - e_{12357} - \omega)$	$(t-i)^2(t+i)^6$
−3	$\frac{1}{4}(-e_{23} - e_{123} - e_{2346} - e_{2357} - e_{12346} - e_{12357} + e_{234567} - 3\omega)$	$(t-i)(t+i)^7$
−4	$-\omega = -e_{1234567}$	$(t+i)^8$

Table A.8 Square roots of $-\mathbf{1}$ in $C\ell(5,2) \cong \mathcal{M}(8,\mathbb{C})$, $d = 4$. Source: [203, Tab. 8].

Bibliography

[1] R. Abłamowicz, Z. Oziewicz and J. Rzewuski, *Clifford algebra approach to twistors*. J. Math. Phys. **23**, pp. 231–242 (1982).

[2] R. Abłamowicz, *Structure of spin groups associated with degenerate Clifford algebras*, J. Math. Phys. **27**(1), pp. 1–6 (1986).

[3] R. Abłamowicz, B. Fauser, K. Podlaski and J. Rembieliński, *Idempotents of Clifford Algebras*. Czechoslovak Journal of Physics, **53**(11), pp. 949–954 (2003).

[4] R. Abłamowicz and G. Sobczyk, Appendix 7.1 of *Lectures on Clifford (Geometric) Algebras and Applications*. Birkhäuser, Boston, 2004.

[5] R. Abłamowicz and B. Fauser, *CLIFFORD – A Maple Package for Clifford Algebra Computations with Bigebra, SchurFkt, GfG - Groebner for Grassmann, Cliplus, Define, GTP, Octonion, SP, SymGroupAlgebra, and code_support.* `http://www.math.tntech.edu/rafal/`, December 2008.

[6] R. Abłamowicz, *Computations with Clifford and Grassmann Algebras*, Adv. Appl. Clifford Algebras **19**, (3–4), pp. 499–545 (2009).

[7] R. Abłamowicz and B. Fauser, *On the transposition anti-involution in real Clifford algebras I: the transposition map*, Linear and Multilinear Algebra, **60**(6), pp. 621–644 (2012), DOI: `https://doi.org/10.1080/03081087.2011.624093`

[8] G. Agarwal, D. Pattanayak and E. Wolf, *Structure of the Electromagnetic Field in a Spatially Dispersive Medium*, Phys. Rev. Lett. **27**, pp. 1022–1025 (1971).

[9] L.V. Ahlfors, *Möbius transformations in* $\mathbb{R}^n$ *expressed through* 2×2 *matrices of Clifford numbers*, Complex Variables, **5**, pp. 215–224 (1986).

[10] N.I. Akhiezer, *Lectures on Integral Transforms*, American Mathematical Society, Rhode Island, 1988.

[11] S.T. Ali, J.P. Antoine and J.P. Gazeau, *Coherent States, Wavelets and Their Generalizations*, Springer, Heidelberg, 2000.

[12] S. Altmann, *Rotations, Quaternions, and Double Groups*, Dover, New York, 1986.

[13] A. Arnaudon, M. Bauer and L. Frappat, *On Casimir's Ghost*, Commun. Math. Phys. **187**, pp. 429–439 (1997).

[14] M. Ashdown, Cambridge GA Package for Maple, http://www.mrao.cam.ac.uk/~maja1/software/GA/, last accessed: 17 Sep. 2020.

[15] Johann Sebastian Bach.

[16] V. Bargmann, *On a Hilbert Space of Analytic Functions and an Associated Integral Transform*, Communications on Pure and Applied Mathematics, **14**, pp. 187–214 (1961).

[17] P. Bas, N. Le Bihan and J.M. Chassery, *Color image watermarking using quaternion Fourier transform.* In Acoustics, Speech, and Signal Processing, 2003. Proceedings (ICASSP'03). 2003 IEEE International Conference on (Vol. 3, pp. III–521).

[18] T. Batard, M. Berthier and C. Saint-Jean. *Clifford Fourier transform for color image processing.* In Bayro-Corrochano and Scheuermann (eds.), Geometric Algebra Computing in Engineering and Computer Science. Springer, London, 2010, pp. 135–162 (2010).

[19] T. Batard and M. Berthier, *Clifford-Fourier Transform and Spinor Representation of Images*, in: E. Hitzer, S.J. Sangwine (eds.), "Quaternion and Clifford Fourier Transforms and Wavelets", TIM **27**, Birkhäuser, Basel, 2013, pp. 177–195.

[20] W. E. Baylis (ed.), *Clifford (Geometric) Algebras with Applications in Physics, Mathematics and Engineering*, Birkhäuser, Basel, 1996.

[21] E. Bayro-Corrochano, *List of Publications*, https://gdl.cinvestav.mx/~edb/publications_crono-1.html, last accessed: 17 Sep. 2020.

[22] E. Bayro-Corrochano and M. Angel de la Torre Gomora, *Image processing using the quaternion wavelet transform*, Proceedings of the Iberoamerican Congress on Pattern Recognition, CIARP'2004, October 2004, Puebla, Mexico, pp. 612–620 (2004).

[23] E. Bayro-Corrochano, *Multi-resolution image analysis using the quaternion wavelet transform*, Numerical algorithms, **39**, pp. 35–55 (2005).

[24] E. Bayro-Corrochano, *The theory and use of the quaternion wavelet transform.* Journal of Mathematical Imaging and Vision, **24**, pp. 19–35 (2006).

[25] E. Bayro-Corrochano, N. Trujillo and M. Naranjo, *Quaternion Fourier descriptors for the preprocessing and recognition of spoken words using images of spatiotemporal representations.* Journal of Mathematical Imaging and Vision, **28**(2), pp. 179–190 (2007).

[26] I. Bell, *Maths for (Games) Programmers, Section 4 - Multivector Methods*, http://www.iancgbell.clara.net/maths/geoalg.htm, last accessed: 17 Sep. 2020.

[27] S. Bernstein, *Cliff. Cont. Wavelet Trans. in $L_{0,2}$ and $L_{0,3}$*, AIP Proceedings of ICNAAM 2008, **1048**, pp. 634–637 (2008).

[28] S. Bernstein, *Spherical Singular Integrals, Monogenic Kernels and Wavelets on the 3D Sphere*, Advances in Applied Clifford Algebras, **19**(2), pp. 173–189 (2009), DOI: 10.1007/s00006-009-0149-4.

[29] S. Bernstein, S. Ebert and R.S. Kraußhar, *Diffusion Wavelets on Conformally Flat Cylinders and Tori* in NUMERICAL ANALYSIS AND APPLIED MATHEMATICS: International Conference on Numerical Analysis and Applied Mathematics 2009: Volume 1 and Volume 2. AIP Conference Proceedings, Volume 1168, pp. 773–776 (2009).

[30] S. Bernstein, J.L. Bouchot, M. Reinhardt, and B. Heise, *Generalized Analytic Signals in Image Processing: Comparison, Theory and Applications*, in: E. Hitzer and S.J. Sangwine (Eds.), Quaternion and Clifford Fourier Transforms and Wavelets, Trends in Mathematics (TIM) **27**, Birkhäuser, 2013, pp. 221–246. DOI: `https://doi.org/10.1007/978-3-0348-0603-9_11`.

[31] A. Beutelspacher, *A Survey of Grassmann's Lineale Ausdehnungslehre*, in G. Schubring (ed.), Hermann Günther Grassmann (1809–1877): Visionary Mathematician, Scientist and Neohumanist Scholar, Kluwer, Dordrecht, 1996.

[32] R. Bracewell, *The Fourier Transform and its Applications*, McGraw-Hill Book Company, New York, 2000.

[33] F. Brackx, R. Delanghe and F. Sommen, *Clifford Analysis*, Vol. **76** of Research Notes in Mathematics, Pitman Advanced Publishing Program, Boston, 1982.

[34] F. Brackx and F. Sommen, *The Continuous Wavelet Transform in Clifford Analysis*, in F. Brackx, J.S.R. Chisholm and V. Souček (eds.), Clifford Analysis and Its Applications, NATO ARW Series, Kluwer Academic Publishers, Dordrecht, 2001, pp. 9–26.

[35] F. Brackx and F. Sommen, *The generalized Clifford–Hermite continuous wavelet transform*, Advances in Applied Clifford Algebras, **11**(S1), pp. 219–231 (2001).

[36] F. Brackx and F. Sommen, *Clifford–Bessel wavelets in Euclidean space*, Mathematical Methods in the Applied Sciences, **25**(16–18), pp. 1479–1491 (2002).

[37] F. Brackx and F. Sommen, *Benchmarking of Three-dimensional Clifford Wavelet Functions*, Complex Variables, **47**(7), pp. 577–588 (2002).

[38] F. Brackx, N. De Schepper and F. Sommen, *The Clifford–Laguerre continuous wavelet transform*, Bulletin of the Belgian Mathematical Society – Simon Stevin, **11**(2), pp. 201–215 (2004).

[39] F. Brackx, N. De Schepper and F. Sommen, *The Clifford–Gegenbauer polynomials and the associated continuous wavelet transform*, Integral Transforms and Special Functions, **15**(5), pp. 387–404 (2004).

[40] F. Brackx, N. De Schepper and F. Sommen, *The Clifford-Fourier Transform*, Journal of Fourier Analysis and Applications, **11**(6), pp. 669–681 (2005).

[41] F. Brackx, H. De Schepper and F. Sommen, *A Hermitian setting for wavelet analysis: the basics*, In Proceedings of the 4th International Conference on Wavelet Analysis and Its Applications, (2005).

[42] F. Brackx, N. De Schepper and F. Sommen, *The Two-Dimensional Clifford Fourier Transform.* Journal of Mathematical Imaging and Vision, **26**(1), pp. 5–18 (2006).

[43] F. Brackx, N. De Schepper and F. Sommen, *Clifford–Jacobi Polynomials and the Associated Continuous Wavelet Transform in Euclidean Space*, T. Qian, V. Mang, and Y. Xu, book*Applied and Numerical Harmonic Analysis*, Birkhäuser Verlag, Basel, pp. 185–198 (2007).

[44] F. Brackx, H. De Schepper and F. Sommen, *A theoretical framework for wavelet analysis in a Hermitean Clifford setting*, Communications on Pure and Applied Analysis, **6**(3), pp. 549–567 (2007).

[45] F. Brackx, H. De Schepper, N. De Schepper and F. Sommen, *The generalized Hermitean Clifford–Hermite continuous wavelet transform*, Numerical Analysis and Applied Mathematics, **936**, pp. 721–725 (2007).

[46] F. Brackx, N. De Schepper and F. Sommen, *The Fourier Transform in Clifford Analysis*, Advances in Imaging and Electron Physics, **156**, pp. 55–302 (2009).

[47] F. Brackx, H. De Schepper, N. De Schepper and F. Sommen, *Generalized Hermitean Clifford–Hermite polynomials and the associated wavelet transform*, Mathematical Methods in the Applied Sciences, **32**(5), pp. 606–630 (2009).

[48] F. Brackx, H. De Schepper, D. Eelbode and V. Souček, *The Howe dual pair in Hermitean Clifford analysis*, Rev. Mat. Iberoam. **26**(2), pp. 449–479 (2010).

[49] F. Brackx, N. De Schepper and F. Sommen. *The Cylindrical Fourier Transform.* In Bayro-Corrochano and Scheuermann (eds.), Geometric Algebra Computing in Engineering and Computer Science. Springer, London, 2010. p. 107–119 (2010).

[50] F. Brackx, N. De Schepper and F. Sommen. *The Clifford-Fourier integral kernel in even dimensional Euclidean space*, Journal of Mathematical Analysis and Applications, **365**(2), pp. 718–728 (2010).

[51] F. Brackx, E. Hitzer and S.J. Sangwine, *History of Quaternion and Clifford-Fourier Transforms*, in: E. Hitzer, S.J. Sangwine (eds.), "Quaternion and Clifford Fourier Transforms and Wavelets", Trends in Mathematics (TIM) Vol.

27, Birkhäuser, Basel, 2013, pp. xi–xxvii. Free online text: http://link.springer.com/content/pdf/bfm\%3A978-3-0348-0603-9\%2F1.pdf .

[52] I.N. Bronstein and K.A. Semendjajew, *Taschenbuch der Mathematik* (German), 22nd ed., Verlag Harri Deutsch, Frankfurt am Main, 1985.

[53] S. Buchholz and G. Sommer, *Introduction to Neural Computation in Clifford Algebra*, in G. Sommer (ed.), Geometric Computing with Clifford Algebras, Springer, Berlin, 2001.

[54] S. Buchholz, K. Tachibana and E. Hitzer, *Optimal Learning Rates for Clifford Neurons*, In: J.M. de Sá, L. A. Alexandre, W. Duch and D. Mandic (eds), Artificial Neural Networks – ICANN 2007, Lecture Notes in Computer Science, Vol. 4668, Springer, Berlin, Heidelberg, pp. 864–873 (2007). DOI: 10.1007/978-3-540-74690-4_88, Preprint: https://www.researchgate.net/publication/221078851_Optimal_Learning_Rates_for_Clifford_Neurons, last accessed: 21 Jan. 2021.

[55] S. Buchholz, E. Hitzer and K. Tachibana, *Coordinate independent update formulas for versor Clifford neurons*, Proceedings of Joint 4th Int. Conf. on Soft Comp. and Intel. Sys., and 9th Int. Symp. on Adv. Intel. Sys., 17–21 Sep. 2008, Japan Society for Fuzzy Theory and intelligent informatics, Nagoya, Japan, pp. 814–819 (2008), DOI: 10.14864/softscis.2008.0.814.0, Open Access: https://www.jstage.jst.go.jp/article/softscis/2008/0/2008_0_814/_article/-char/ja/, Preprint: https://pdfs.semanticscholar.org/10f6/bb1bb2918d830ab5721379910221f93f4668.pdf, accessed 21 Jan. 2021.

[56] T. Bülow, *Hypercomplex Spectral Signal Representations for the Processing and Analysis of Images*, Ph.D. Thesis, Institut für Informatik und Praktische Mathematik, University of Kiel, Germany, Aug. 1999. http://www.uni-kiel.de/journals/servlets/MCRFileNodeServlet/jportal_derivate_00001015/1999_tr03.pdf

[57] T. Bülow and G. Sommer, *Hypercomplex Signals – A Novel Extension of the Analytic Signal to the Multidimensional Case*, (2001), IEEE Transactions on Signal Processing, **49**(11), pp. 2844–2852 (Nov. 2001), DOI: 10.1109/78.960432.

[58] T. Bülow, M. Felsberg and G. Sommer, *Non-commutative Hypercomplex Fourier Transforms of Multidimensional Signals.* In G. Sommer (ed.), "Geometric Computing with Clifford Algebras: Theoretical Foundations and Applications in Computer Vision and Robotics" Springer-Verlag, Berlin, 2001, pp. 187–207.

[59] R. Bujack, G. Scheuermann and E. Hitzer, *A General Geometric Fourier Transform Convolution Theorem*, Adv. Appl. Clifford Algebras **23**(1), pp. 15–38 (2013), Preprint: http://www.informatik.uni-leipzig.de/~bujack/GFTconv.pdf, DOI: 10.1007/s00006-012-0338-4.

[60] R. Bujack, E. Hitzer and G. Scheuermann, *Demystification of the Geometric Fourier Transforms*, In T. Simos, G. Psihoyios and C. Tsitouras (eds.), Numerical Analysis and Applied Mathematics ICNAAM 2013, AIP Conf. Proc. **1558**, pp. 525–528 (2013). DOI: 10.1063/1.4825543, Preprint: http://vixra.org/abs/1310.0255

[61] R. Bujack, G. Scheuermann and E. Hitzer, *A General Geometric Fourier Transform*, in: E. Hitzer, S.J. Sangwine (eds.), "Quaternion and Clifford Fourier Transforms and Wavelets", Trends in Mathematics (TIM) **27**, Birkhäuser, Basel, 2013, pp. 155–176. DOI: 10.1007/978-3-0348-0603-9_8 , Preprint: http://arxiv.org/abs/1306.2184.

[62] R. Bujack, H. De Bie, N. De Schepper and G. Scheuermann, *Convolution products for hypercomplex Fourier transforms*, J. Math. Imaging Vision, **48**, pp. 606–624 (2014), preprint: http://arxiv.org/abs/1303.1752 .

[63] R. Bujack, M. Hlawitschka, G. Scheuermann and E. Hitzer, *Customized TRS Invariants for 2D Vector Fields via Moment Normalization*, Pattern Recognition Letters, **46**, pp. 46–59, online 27 May 2014, DOI: 10.1016/j.patrec.2014.05.005

[64] R. Bujack, I. Holz, G. Scheuermann and E. Hitzer, *Moment Invariants for 2D Flow Fields Using Normalization*, Proceedings of IEEE Pacific Visualization Symposium (PacificVis) 2014, March 4–7, Yokohama, Japan, pp. 41–48, DOI: 10.1109/PacificVis.2014.16.

[65] R. Bujack, I. Hotz, G. Scheuermann and E. Hitzer, *Moment Invariants for 2D Flow Fields via Normalization in Detail*, in IEEE Transactions on Visualization and Computer Graphics, **21**(8), pp. 916–929, 1 Aug. 2015, DOI: 10.1109/TVCG.2014.2369036. Preprint: http://vixra.org/abs/1411.0348.

[66] R. Bujack, E. Hitzer and G. Scheuermann, *Demystification of the Geometric Fourier Transforms and resulting Convolution Theorems*, Mathematical Methods in the Applied Sciences, **39**(7), pp. 1877–1890 (2016) Article first published online: 3 Sep. 2015. DOI: 10.1002/mma.3607.

[67] E.J. Candes, *Ridgelets: Theory and Applications*, PhD Thesis, Stanford University, Stanford, USA, 1998, http://statweb.stanford.edu/~candes/papers/thesis.ps

[68] E. Cartan, *La géométrie des groupes simples*, Ann. Mat. Pura Appl. **4**(4) (1927), pp. 209–256. *Complément au mémoire sur la géométrie des groupes simples*, Ann. Mat. Pura Appl. **5**(4) (1928), pp. 253–260.

[69] F. Catoni, R. Cannata and P. Zampeti, *An Introduction to Commutative Quaternions*, Adv. in App. Cliff. Alg. **16**(1), pp. 1–28 (2006).

[70] C. Chevalley, *The Theory of Lie Groups*, Princeton University Press, Princeton, 1957.

[71] C. Chevalley, *The Algebraic Theory of Spinors and Clifford Algebras*, Springer, Berlin, 1997.

[72] V.M. Chernov, *Discrete orthogonal transforms with data representation in composition algebras*, Proceedings Scandinavian Conference on Image Analysis, Uppsala, Sweden, pp. 357–364, (1995).

[73] J.S.R. Chisholm, *Homepage of John Stephen Roy Chisholm*, `http://www.roychisholm.com/`, last accessed: 23 Feb. 2021.

[74] J.G. Christensen, *Uncertainty Principles*, Master Thesis, University of Copenhagen, 2003.

[75] CitizenGo NGO, *Pétition à l'attention de: The EU and the UN. You Must Recognize the Crime of Genocide Against Christians in Iraq - #CallItGenocide.* `http://www.citizengo.org/en/node/32625`, last accessed 4th Feb. 2016.

[76] C.K. Chui, *An Introduction to Wavelets*, Academic Press, New York, 1992.

[77] W.K. Clifford, *Applications of Grassmann's Extensive Algebra.* American Journal of Mathematics, Pure and Applied **1**(1878), pp. 350–358.

[78] W.K. Clifford, *On the classification of geometric algebras*, 1876, in W. Tucker (ed.), Mathematical Papers by W.K. Clifford, Macmillan, London (1882), pp. 397–401.

[79] S.A. Collins, *Lens–system Diffraction Integral Written in Terms of Matrix Optics*, Journal of the Optical Society of America, **60**, pp. 1168–1177 (1970).

[80] F. Collins, Director of the US National Human Genome Research Institute, in Time Magazine, 5 Nov. 2006.

[81] E.U. Condon, *Immersion of the Fourier Transform in a continuous group of functional transformations*, Proceedings of the National Academy of Sciences USA, **23**, pp. 158–164 (1937).

[82] D. Constales, F. Sommen and P. Van Lancker, *Models for irreducible representations of Spin(m)*, Adv. Appl. Clifford Algebras, **11**(S1), pp. 271–289 (2001).

[83] K. Coulembier and H. De Bie, *Hilbert Space for Quantum Mechanics on Superspace*, Journal of Mathematical Physics, **52**(6), 30 pages (2011), DOI: `https://doi.org/10.1063/1.3592602`.

[84] H.S.M. Coxeter, *Quaternions and Reflections*, The American Mathematical Monthly, **53**(3) (Mar., 1946), pp. 136–146.

[85] H.S.M. Coxeter and W.O.J. Moser, *Generators and Relations for Discrete Groups*, Springer, New York, 4th ed., 1980.

[86] J.G. Daugman, *Complete Discrete 2-D Gabor Transforms by Neural Network for Image Analysis and Compression, IEEE Transaction on Acoustics, Speech, and Signal Processing*, **36**(7), pp. 1169–1179 (1988).

[87] B. Davies *Integral Transforms and Their Applications*, Springer, New York, 2002.

[88] H. De Bie, *Fourier Transform and Related Integral Transforms in Superspace*, Journal of Mathematical Analysis and Applications, **345**, pp. 147–164 (2008).

[89] H. De Bie and F. Sommen, *Vector and bivector Fourier transforms in Clifford analysis*. In K. Gürlebeck and C. Könke (eds.), 18th International Conference on the Application of Computer Science and Mathematics in Architecture and Civil Engineering, p. 11, 2009.

[90] H. De Bie and Y. Xu, *On the Clifford-Fourier transform*, International Mathematics Research Notices **2011**(22), pp. 5123–5163 (2011), DOI: `10.1093/imrn/rnq288`.

[91] H. De Bie, N. De Schepper and F. Sommen, *The Class of Clifford–Fourier Transforms*, Journal of Fourier Analysis and Applications, **17**, pp. 1198–1231 (2011).

[92] H. De Bie and N. De Schepper, *The fractional Clifford-Fourier transform*, Complex Analysis and Operator Theory **6**, pp. 1047–1067 (2012).

[93] H. De Bie, B. Orsted, P. Somberg and V. Souček, *Dunkl Operators and a Family of Realizations of* $\mathfrak{osp}(1|2)$, Transactions of the American Mathematical Society, **364**, pp. 3875–3902 (2012).

[94] H. De Bie, *Clifford Algebras, Fourier Transforms and Quantum Mechanics*, Mathematical Methods in the Applied Sciences, **35**(18), pp. 2198–2228 (2012).

[95] H. De Bie, *The kernel of the radially deformed Fourier transform*, Integral Transforms Spec. Funct. **24**(12), pp. 1000–1008 (2013).

[96] H. De Bie, B. Orsted, P. Somberg and V. Souček, *The Clifford deformation of the Hermite semigroup*, SIGMA 9, 010, 22 pages (2013).

[97] H. De Bie, N. De Schepper, T.A. Ell, K. Rubrecht and S.J. Sangwine, *Connecting spatial and frequency domains for the quaternion Fourier transform*, Applied Mathematics and Computation, **271**, pp. 581–593 (2015).

[98] H. De Bie, *The Kernel of the Radially Deformed Fourier Transform*, INTEGRAL TRANSFORMS AND SPECIAL FUNCTIONS, **24**(12). pp. 1000–1008 (2013).

[99] P.-P. Dechant, *Clifford algebra unveils a surprising geometric significance of quaternionic root systems of Coxeter groups*, Adv. in App. Cliff. Alg., **23**(2), pp. 301–321 (2013), DOI: `https://doi.org/10.1007/s00006-012-0371-3`.

[100] R. Delanghe, F. Sommen and V. Souček, *Clifford analysis and spinor valued functions*, Kluwer Academic Publishers, Dordrecht, 1992.

[101] R. Delanghe, *Clifford analysis: history and perspective*, Comp. Meth. Funct. Theory, **1**, pp. 107–153 (2001).

[102] M.A. Delsuc, *Spectral representation of 2D NMR spectra by hypercomplex numbers*, Journal of magnetic resonance, **77**(1), pp. 119–124, Mar. 1988, (1988).

[103] A. De Martino and K. Diki, *On the quaternionic short-time Fourier and Segal-Bargmann transforms*, Preprint: `https://arxiv.org/pdf/2009.00073.pdf`, accessed 09 Jan. 2021.

[104] P. Denis, P. Carré and C. Fernandez-Maloigne, *Spatial and spectral quaternionic approaches for colour images.* Computer Vision and Image Understanding, **107**, pp. 74–87 (2007).

[105] N. De Schepper, *Multidimensional continuous wavelet transforms and generalized Fourier transforms in Clifford analysis*, Ph.D. Thesis, Ghent University, Belgium (2006).

[106] C. Doran, D. Hestenes, F. Sommen and N. Van Acker, *Lie Groups as Spin Groups*, J. Math. Phys., **34**(8), pp. 3642–2669 (1993).

[107] C. Doran, *Geometric Algebra and its Application to Mathematical Physics*, Ph.D. thesis, University of Cambridge, 181 pages (1994). `http://www.mrao.cam.ac.uk/~clifford/publications/abstracts/chris_thesis.html`

[108] C. Doran and A. Lasenby, *Geometric Algebra for Physicists*, CUP, Cambridge (UK), 2003.

[109] S. Dorrode and F. Ghorbel, *Robust and efficient Fourier-Mellin transform approximations for gray-level image reconstruction and complete invariant description*, Computer Vision and Image Understanding, **83**(1) (2001), pp. 57–78, DOI `10.1006/cviu.2001.0922`.

[110] L. Dorst, *The Inner Products of Geometric Algebra*, In L. Dorst, C. Doran and J. Lasenby (eds.), Applications of Geometric Algebra in Computer Science and Engineering, Birkhäuser, Basel, 2002, pp. 35–46. Preprint: `https://staff.fnwi.uva.nl/l.dorst/clifford/inner.ps`

[111] L. Dorst, D. Fontijne and S. Mann, *Geometric Algebra for Computer Science – An Object-Oriented Approach to Geometry*, Morgan Kaufmann Series in Computer Graphics, Elsevier, San Francisco, 2007.

[112] L. Dorst and J. Lasenby (eds.), *Guide to Geometric Algebra in Practice.* Springer, Berlin, 2011.

[113] C.F. Dunkl, *Differential–difference Operators Associated to Reflection Groups*, Transactions of the American Mathematical Society, **311**, pp. 167–183 (1989)

[114] R.B. Easter, *G8,2 Geometric Algebra, DCGA*, preprint: http://vixra.org/abs/1508.0086, (2015).

[115] R.B. Easter, *Double Conformal Space-Time Algebra*, preprint: http://vixra.org/abs/1602.0114, (2016).

[116] R.B. Easter and E. Hitzer, *Double Conformal Geometric Algebra for Quadrics and Darboux Cyclides*, CGI '16: Proceedings of the 33rd Computer Graphics International, June 2016, pp. 93–96 (2016), DOI: https://doi.org/10.1145/2949035.2949059.

[117] J. Ebling and G. Scheuermann, *Clifford Fourier Transform on Vector Fields.* IEEE Transactions on Visualization and Computer Graphics, **11**(4), pp. 469–479 (July/August 2005). DOI: 10.1109/TVCG.2005.54

[118] J. Ebling and G. Scheuermann, *Clifford convolution and pattern matching on vector fields*, In Proc. IEEE Vis., **3**, IEEE Computer Society, Los Alamitos, CA., 2003. pp. 193–200.

[119] J. Ebling, *Visualization and Analysis of Flow Fields using Clifford Convolution.* Ph.D. thesis, University of Leipzig, Germany, (2006).

[120] D. Eelbode and E. Hitzer, *Operator Exponentials for the Clifford Fourier Transform on Multivector Fields*, Electronic proceedings of ICCA10, Tartu, Estonia, Aug. 2014, 18 pages. Preprint: http://vixra.org/abs/1403.0310 . Online version: http://icca10.ut.ee/SC_FTW_workshops/FTW/Exp_CFT_multivector_ICCA10.pdf

[121] D. Eelbode and E. Hitzer, *Operator Exponentials for the Clifford Fourier Transform on Multivector Fields in Detail*, Adv. Appl. Clifford Algebras, Vol. **26**, pp. 953–968 (2016) Online First: 22 Oct. 2015, DOI: 10.1007/s00006-015-0600-7.

[122] Y. El Haoui, E. Hitzer and S. Fahlaoui, *Heisenberg's and Hardy's uncertainty principles for special relativistic space-time Fourier transformation*, accepted by Adv. Appl. Clifford Algebras, Sep. 2020, 29 pages. DOI: https://doi.org/10.1007/s00006-020-01093-5 .

[123] T.A. Ell, *Hypercomplex Spectral Transformations*, University of Minnesota, Ph.D. Thesis No. AAT 9231031, June 1992.

[124] T.A. Ell, *Quaternion-Fourier Transforms for Analysis of Two-Dimensional Linear Time-Invariant Partial Differential Systems*, in Proc. of the 32nd IEEE Conf. on Decision and Control, December 15–17, San Antonio, Tx, **2** (1993), pp. 1830–1841.

[125] T.A. Ell and S.J. Sangwine, *The Discrete Fourier Transforms of a Colour Image.* In J.M. Blackledge and M.J. Turner (eds.), Image Processing II: Mathematical Methods, Algorithms and Applications, pp. 430–441 (2000).

[126] T.A. Ell and S.J. Sangwine, *Hypercomplex Wiener-Khintchine theorem with application to color image correlation*, Proceedings of International Conference on Image Processing, **2**, pp. 792–795 (2000).

[127] T.A. Ell and S.J. Sangwine, *Hypercomplex Fourier transforms of color images*, IEEE Trans. on Image Processing **16**(1), pp. 22–35 (2007), DOI: 10.1109/TIP.2006.884955.

[128] T.A. Ell, *Quaternion Fourier Transform: Re-tooling Image and Signal Processing Analysis.* In: E. Hitzer and S.J. Sangwine (eds.), Quaternion and Clifford Fourier Transforms and Wavelets, Trends in Mathematics (TIM) **27**, Birkhäuser, Basel, pp. 3–14, (2013). DOI: https://doi.org/10.1007/978-3-0348-0603-9_1

[129] T.A. Ell, N. Le Bihan, S.J. Sangwine, *Quaternion Fourier Transforms for Signal and Image Processing.* Digital Signal and Image Processing, Wiley-ISTE, Hoboken, 2014.

[130] R.R. Ernst, et al, *Princ. of NMR in One and Two Dim.*, Int. Ser. of Monogr. on Chem., Oxford Univ. Press, 1987.

[131] M.I. Falcao and H.R. Malonek, *Generalized Exponentials through Appell sets in* $\mathbb{R}^{n+1}$ *and Bessel functions*, AIP Conference Proceedings, **936**, pp. 738–741 (2007).

[132] M. Felsberg, *Low-Level Image Processing with the Structure Multivector*, Ph.D. Thesis, Institut für Informatik und Praktische Mathematik, University of Kiel, Christian-Albrechts-Universität, Germany, 2002.

[133] M. Felsberg, et al, *Comm. Hypercomplex Fourier Transf. of Multidim. Signals*, in G. Sommer (ed.), "Geom. Comp. with Cliff. Algebras", Springer, Berlin, 2001, pp. 209–229.

[134] O. Forster, *Analysis 1 (Differential- und Integralrechnung einer Veränderlichen)*, Viewig, Wiesbaden & Rowohlt Taschenbuch Verlag, Reinbek bei Hamburg, 1976.

[135] G. Friesecke, *Course Material Fourieranalysis, Lecture 13: The Fourier transform on* L^2, 2013, https://www-m7.ma.tum.de/foswiki/pub/M7/Analysis/Fourier13/lecture13.pdf, accessed 08 Jan. 2021.

[136] Y. Fu, U. Kähler and P. Cerejeiras, *The Balian-Low Theorem for the Windowed Clifford-Fourier Transform.* In: E. Hitzer, S.J. Sangwine (eds) Quaternion and Clifford Fourier Transforms and Wavelets. Trends in Mathematics (TIM) **27**. Birkhäuser, Basel, (2013).

[137] R. Fueter, *Die Funktionentheorie der Differentialgleichungen* $\Delta u = 0$ *und* $\Delta\Delta u = 0$ *mit vier reellen Variablen*, Comment. Math. Helv., **7**(1), pp. 307–330, (1935).

[138] *Genesis chapter 1 verse 1*, in The Holy Bible, English Standard Version. Wheaton (Illinois): Crossway Bibles, Good News Publishers; 2001, `https://www.biblegateway.com/passage/?search=Genesis+1&version=ESV`, last accessed 23 Feb. 2021.

[139] S. Georgiev and J. Morais, *Bochner's Theorems in the Framework of Quaternion Analysis* in: E. Hitzer and S.J. Sangwine (eds.), "Quaternion and Clifford Fourier Transforms and Wavelets", TIM **27**, Birkhäuser, Basel, 2013, pp. 85–104.

[140] P.K. Ghosh and T.V. Sreenivas, *Time-varying filter interpretation of Fourier transform and its variants*, Signal Processing, **11**(86): pp. 3258–3263 (2006), DOI: `https://doi.org/10.1016/j.sigpro.2006.01.005`.

[141] J. Gilbert and M.A.M. Murray, *Clifford algebras and Dirac operators in harmonic analysis*, Cambridge University Press, Cambridge (UK), (1991).

[142] P.R. Girard, *Quaternions, algèbre de Clifford et physique relativiste*, Presses polytechniques, et universitaires romandes, Lausanne, 2004. *Quaternions, Clifford Algebras and Relativistic Physics*, Birkhäuser, Basel, 2007 (English).

[143] P.R. Girard, R. Pujol, P. Clarysse, A. Marion, R. Goutte and P. Delachartre, *Analytic Video* ($2D + t$) *Signals Using Clifford-Fourier Transforms in Multiquaternion Grassmann-Hamilton Clifford Algebras*. In: E. Hitzer, S.J. Sangwine (eds.), Quaternion and Clifford Fourier Transforms and Wavelets. Trends in Mathematics (TIM) **27**. Birkhäuser, Basel (2013).

[144] R.C. Gonzales, R.E. Woods and S.L. Eddins, *Digital Image Processing Using Matlab*, (Pearson Prentice Hall, Upper Saddle River, 2004).

[145] G.H. Granlund and H. Knutsson, *Signal Processing for Computer Vision*, Kluwer, Dordrecht, 1995.

[146] H. Grassmann, (F. Engel, editor) *Die Ausdehnungslehre von 1844 und die Geometrische Analyse, 1*, part 1, Teubner, Leipzig, 1894.

[147] K. Gröchenig, *Foundations of Time-Frequency Analysis*, Birkhäuser, Basel, 2001.

[148] K. Gröchenig and G. Zimmermann, *Hardy's theorem and the short-time Fourier transform of Schwartz functions*, J. London Math. Soc., **2**(63), pp. 205–214 (2001).

[149] K. Gürlebeck and W. Sprößig, *Quaternionic Analysis and Elliptic Boundary Value Problems*, ISNM 89, Birkhäuser, Basel (1990).

[150] K. Gürlebeck and W. Sprößig, *Quaternionic and Clifford Calculus for Physicists and Engineers*, John Wiley and Sons, England, Chichester, 1997.

[151] K. Gürlebeck, K. Habetha and W. Sprößig, *Holomorphic Functions in the Plane and n-dimensional Space*, Birkhäuser, Basel, 2008.

[152] S. Gull, A. Lasenby and C. Doran, *Imaginary Numbers are not Real. - the Geometric Algebra of Spacetime*, Found. Phys. **23**(9), pp. 1175–1201 (1993), DOI: `https://doi.org/10.1007/BF01883676`.

[153] C. Guo and L. Zhang, *A novel multiresolution spatiotemporal saliency detection model and its applications in image and video compression*, IEEE Trans. Image Process, **19**(1), pp. 185–198 (2010), DOI: `doi:10.1109/TIP.2009.2030969`.

[154] S.L. Hahn, *Wigner distributions and ambiguity functions of 2-D quaternionic and monogenic signals*, IEEE Trans. Signal Process., **53**(8), pp. 3111–3128 (2005), DOI: `https://doi.org/10.1109/TSP.2005.851134`

[155] W.R. Hamilton, *On quaternions, or on a new system of imaginaries in algebra*, Philosophical Magazine. **25**(3), pp. 489–495 (1844).

[156] W.R. Hamilton, *Elements of Quaternions*, Longmans Green, London 1866. Chelsea, New York, 1969.

[157] Hebrews, chapter 11, verses 1 to 3, Holy Bible, New International Version. `https://www.biblegateway.com/passage/?search=Hebrews+11\%3A1-3&version=NIV`, last accessed: 12 June 2020.

[158] D. Hestenes, *Space-Time Algebra*. Gordon and Breach, London, 1966.

[159] D. Hestenes, *Space Time Calculus*, `http://geocalc.clas.asu.edu/html/STC.html`, last accessed: 17 Sep. 2020.

[160] D. Hestenes, *Multivector Calculus*, J. Math. Anal. and Appl., **24**(2), pp. 313–325 (1968). `http://geocalc.clas.asu.edu/pdf/MultCalc.pdf`, last accessed: 17 Sep. 2020.

[161] D. Hestenes and G. Sobczyk, *Clifford Algebra to Geometric Calculus: A Unified Language for Mathematics and Physics*, Kluwer, Dordrecht, reprinted with corrections 1992.

[162] D. Hestenes, *New foundations for classical mechanics*, Kluwer, Dordrecht, 1999.

[163] D. Hestenes, *New Foundations for Mathematical Physics*, `http://geocalc.clas.asu.edu/html/NFMP.html`, last accessed: 17 Sep. 2020.

[164] D. Hestenes, H. Li and A. Rockwood, *New Algebraic Tools for Classical Geometry*, in G. Sommer (ed.), Geometric Computing with Clifford Algebras, Springer, Berlin, 2001.

[165] D. Hestenes, *Point Groups and Space Groups in Geometric Algebra*, in L. Dorst et. al. (eds.), Applications of Geometric Algebra in Computer Science and Engineering, Birkhäuser, Basel, 2002.

[166] D. Hildenbrand, J. Pitt and A. Koch, *Gaalop – High Performance Parallel Computing Based on Conformal Geometric Algebra*, In E. Bayro-Corrochano, G. Scheuermann (eds.), Geometric Algebra Computing in Engineering and Computer Science, Springer, Berlin, (2010) pp. 477–494.

[167] D. Hildenbrand, *Foundations of Geometric Algebra Computing*, Springer, Berlin, 2013.

[168] E. Hitzer, *Creative Peace License*, `http://gaupdate.wordpress.com/2011/12/14/the-creative-peace-license-14-dec-2011/`, last accessed: 12 June 2020.

[169] E. Hitzer, *Geometric Calculus International – Software.* `http://sinai.apphy.u-fukui.ac.jp/gcj/gc\_int.html\#software`

[170] E. Hitzer, *Geometric Calculus for Engineers*, Proc. of the Pukyong National University – Fukui University International Symposium 2001 for Promotion of Research Cooperation, Pukyong National University, Busan, Korea, pp. 59–66 (2001). Preprint: `http://vixra.org/abs/1306.0114`

[171] E. Hitzer, *Imaginary eigenvalues and complex eigenvectors explained by real geometry*, in C. Doran, L. Dorst and J. Lasenby (eds.), Applied Geometrical Algebras in Computer Science and Engineering, AGACSE 2001, Birkhäuser, pp. 145–153 (2001). DOI 10.1007/978-1-4612-0089-5_13, Preprint: `http://arxiv.org/abs/1306.0717` .

[172] E. Hitzer, *A real explanation for imaginary eigenvalues and complex eigenvectors*, in T.M. Karade (ed.), Proc. of the Nat. Symp. on Mathematical Sciences, 1-5 March, 2001, Nagpur, India, Einst. Foundation Int. 1, pp. 1–26 (2001). Preprint: `http://vixra.org/abs/1306.0113` .

[173] E. Hitzer, *Antisymmetric Matrices are Real Bivectors*, Mem. Fac. Eng. Fukui Univ. **49**(2), pp. 283–298 (2001). Preprint: `http://vixra.org/abs/1306.0112`, last accessed: 12 June 2020.

[174] E. Hitzer, *Vector Differential Calculus*, Mem. Fac. Eng. Fukui Univ. **50**(1), pp. 109–125 (2002).

[175] E. Hitzer, *Multivector Differential Calculus*, Adv. in Appl. Cliff. Algs., **12**(2), pp. 135–182 (2002). DOI: `https://doi.org/10.1007/BF03161244`, Preprint: `https://arxiv.org/pdf/1306.2278.pdf`, last accessed: 12 June 2020.

[176] E. Hitzer, *Geometric Calculus – Engineering Mathematics for the 21st Century*, Mem. Fac. Eng. Fukui Univ. **50**(1), pp. 127–137 (2002). Preprint: `http://vixra.org/abs/1306.0117` .

[177] E. Hitzer and R. Nagaoka, *Support Website for the Linear Algebra Lectures of the University of the Air Japan 2004–2008.*, `http://erkenntnis.icu.ac.jp/gala2/index.html` . Website opened: 26 July 2016.

[178] E. Hitzer, *What is an imaginary number?*, Support Website for the Linear Algebra Lectures of the University of the Air Japan 2004–2008. Preprint: `http://vixra.org/abs/1306.0173` . Website opened: 26 July 2016.

[179] E. Hitzer, *The geometric product and derived products*, 11 pages, 21 July 2003, preprint: `https://vixra.org/pdf/1306.0175v1.pdf`, last accessed: 17 June 2020.

[180] E. Hitzer, *Axioms of Geometric Algebra*, 16 May 2003, rev., 8 pages, preprint: `https://vixra.org/pdf/1306.0178v1.pdf`, last accessed: 21 Nov. 2020.

[181] E. Hitzer and B. Mawardi, *Uncertainty Principle for the Clifford Geometric Algebra $Cl(3,0)$ based on Clifford Fourier Transform*, in T.E. Simos, G. Sihoyios and C. Tsitouras (eds.), International Conference on Numerical Analysis and Applied Mathematics 2005, Wiley-VCH, Weinheim, 2005, pp. 922–925 (2005).

[182] E. Hitzer and B. Mawardi, Uncertainty Principle for Clifford Geometric Algebras $Cl_{n,0}, n = 3 (\text{mod } 4)$ based on Clifford Fourier Transform, in T. Qian, M.I. Vai and X. Yusheng (eds.), Wavelet Analysis and Applications, Springer (SCI) Book Series Applied and Numerical Harmonic Analysis, Springer, New York, pp. 45–54 (2006).

[183] E. Hitzer, *Tutorial on Fourier Transformations and Wavelet Transformations in Clifford Geometric Algebra*, in K. Tachibana (ed.), Lecture notes of the International Workshop for "Computational Science with Geometric Algebra" (FCSGA2007), Nagoya University, Japan, 14–21 Feb. 2007, pp. 65–87 (2007). Preprint: `http://vixra.org/abs/1306.0133` .

[184] E. Hitzer, *Quaternion Fourier Transform on Quaternion Fields and Generalizations*, Advances in Applied Clifford Algebras, **17**(3), pp. 497–517 (2007). DOI: `10.1007/s00006-007-0037-8`, preprint: `http://arxiv.org/abs/1306.1023` .

[185] E. Hitzer and B. Mawardi, *Clifford Fourier Transform on Multivector Fields and Uncertainty Principles for Dimensions $n = 2(mod\ 4)$ and $n = 3(mod\ 4)$*, in P. Angles (ed.), Adv. App. Cliff. Alg. Vol. **18**, S3,4, pp. 715–736 (2008). DOI: 10.1007/s00006-008-0098-3, Preprint: `http://vixra.org/abs/1306.0127`

[186] E. Hitzer, *Basic Multivector Calculus*, Proc. of 18th Intelligent Systems Symposium (FAN 2008), 23–24 Oct. 2008, Hiroshima, Japan, pp. 185–190 (2008). Preprint: `http://vixra.org/abs/1306.0124` .

[187] E. Hitzer, *Directional Uncertainty Principle for Quaternion Fourier Transforms*, Advances in Applied Clifford Algebras, **20**(2), pp. 271–284 (2010), online since 08 July 2009. DOI: `10.1007/s00006-009-0175-2`, preprint: `http://arxiv.org/abs/1306.1276` .

[188] E. Hitzer, *Clifford (Geometric) Algebra Wavelet Transform*, in V. Skala and D. Hildenbrand (eds.), Proc. of GraVisMa 2009, 02–04 Sep. 2009, Plzen, Czech Republic, pp. 94–101 (2009). Online: `http://gravisma.zcu.cz/GraVisMa-2009/`

Papers_2009/!_2009_GraVisMa_proceedings-FINAL.pdf, Preprint: http://arxiv.org/abs/1306.1620

[189] E. Hitzer, *Real Clifford Algebra* $Cl(n,0), n = 2, 3(mod\,4)$ *Wavelet Transform*, edited by T.E. Simos et al., AIP Proceedings of ICNAAM 2009, No. **1168**, pp. 781–784 (2009).

[190] E. Hitzer, K. Tachibana, S. Buchholz and I. Yu, *Carrier Method for the General Evaluation and Control of Pose, Molecular Conformation, Tracking, and the Like*, Adv. in App. Cliff. Alg., **19**(2), (2009) pp. 339–364. DOI: https://doi.org/10.1007/s00006-009-0160-9. Preprint: https://www.researchgate.net/publication/226288320_Carrier_Method_for_the_General_Evaluation_and_Control_of_Pose_Molecular_Conformation_Tracking_and_the_Like.

[191] E. Hitzer, *Angles Between Subspaces.* In V. Skala (ed.), Workshop Proceedings of Computer Graphics, Computer Vision and Mathematics, Brno University of Technology, Czech Republic, 2010. UNION Agency, Preprint: https://arxiv.org/abs/1306.1629.

[192] E. Hitzer, *Angles between subspaces computed in Clifford algebra.* AIP Conference Proceedings, **1281**(1), pp. 1476–1479 (2010). Preprint: https://arxiv.org/abs/1306.1825.

[193] E. Hitzer, R. Abłamowicz, *Geometric Roots of* -1 *in Clifford Algebras* $Cl(p,q)$ *with* $p + q \leq 4$, Adv. In Appl. Cliff. Algebras, Vol. **21**(1) pp. 121–144, (2011), DOI: 10.1007/s00006-010-0240-x . Preprints: http://arxiv.org/abs/0905.3019 , Tennessee University of Technology, Department of Mathematics, Technical Report 2009-3, http://www.tntech.edu/files/math/reports/TR_2009_3.pdf .

[194] E. Hitzer, *OPS-QFTs: A New Type of Quaternion Fourier Transforms Based on the Orthogonal Planes Split with One or Two General Pure Quaternions.* in Numerical Analysis and Applied Mathematics ICNAAM 2011, AIP Conf. Proc. **1389**, pp. 280–283 (2011); DOI: 10.1063/1.3636721 preprint: http://arxiv.org/abs/1306.1650 .

[195] E. Hitzer, S.J. Sangwine, *The orthogonal planes split of quaternions*, Proceedings of ICCA9, Weimar 2011.

[196] E. Hitzer, *Two-sided Clifford Fourier transform with two square roots of* -1 *in* $Cl(p,q)$, in M. Berthier, L. Fuchs, C. Saint-Jean (eds.), electronic Proceedings of AGACSE 2012, La Rochelle, France, 2–4 July 2012.

[197] E. Hitzer, *The quest for conformal geometric algebra Fourier transformations*, In T. Simos, G. Psihoyios and C. Tsitouras (eds.), Numerical Analysis and Applied Mathematics ICNAAM 2013, AIP Conf. Proc. **1558**, pp. 30–33 (2013). DOI: 10.1063/1.4825413, Preprint: http://vixra.org/abs/1310.0248 .

[198] E. Hitzer, *Quaternionic Fourier-Mellin Transform*, in T. Sugawa (ed.), Proceedings of the The 19th International Conference on Finite or Infinite Dimensional Complex Analysis and Applications (ICFIDCAA 2011), 11–15 December 2011, Hiroshima, Japan, Tohoku Univ. Press, Sendai (2013), ii, pp. 123–131.

[199] E. Hitzer, *Clifford Fourier-Mellin transform with two real square roots of* -1 *in* $Cl(p,q)$, $p+q=2$, 9th ICNPAA 2012, AIP Conf. Proc., **1493**, pp. 480–485 (2012).

[200] E. Hitzer, *Introduction to Clifford's Geometric Algebra*, SICE Journal of Control, Measurement, and System Integration, Vol. **51**, No. 4, pp. 338–350, April 2012, (April 2012). Preprint: `http://arxiv.org/abs/1306.1660`, last accessed: 12 June 2020.

[201] E. Hitzer, *The Clifford Fourier transform in real Clifford algebras*, in E. H., K. Tachibana (eds.), "Session on Geometric Algebra and Applications, IKM 2012", Special Issue of Clifford Analysis, Clifford Algebras and their Applications, Vol. **2**, No. 3, pp. 227–240, (2013). First published in K. Gürlebeck, T. Lahmer and F. Werner (eds.), electronic Proc. of 19th International Conference on the Application of Computer Science and Mathematics in Architecture and Civil Engineering, IKM 2012, Weimar, Germany, 04–06 July 2012. Preprint: `http://vixra.org/abs/1306.0130` .

[202] E. Hitzer, T. Nitta and Y. Kuroe, *Applications of Clifford's Geometric Algebra.* Adv. Appl. Clifford Algebras **23**, pp. 377–404 (2013). `https://doi.org/10.1007/s00006-013-0378-4`. Preprint: `http://arxiv.org/abs/1305.5663` .

[203] E. Hitzer, J. Helmstetter and R. Ablamowitz, *Square roots of* -1 *in real Clifford algebras*, in E. Hitzer, S.J. Sangwine (eds.), "Quaternion and Clifford Fourier Transforms and Wavelets", Trends in Mathematics **27**, Birkhäuser, Basel, 2013, pp. 123–153. DOI: `10.1007/978-3-0348-0603-9_7` , Preprints: `http://arxiv.org/abs/1204.4576` , `http://www.tntech.edu/files/math/reports/TR_2012_3.pdf` . First published in K. Gürlebeck (ed.), Proc. of The 9th Int. Conf. on Clifford Algebras and their Applications, (2011).

[204] E. Hitzer, J. Helmstetter and R. Ablamowicz, Maple worksheets created with `CLIFFORD` for a verification of results presented in this paper, `http://math.tntech.edu/rafal/publications.html` (ľ2012).

[205] E. Hitzer and S.J. Sangwine, *The orthogonal 2D planes split of quaternions and steerable quaternion Fourier transformations*, in: E. Hitzer and S.J. Sangwine (Eds.), Quaternion and Clifford Fourier Transforms and Wavelets, Trends in Mathematics (TIM) **27**, Birkhäuser, 2013, pp. 15–40. DOI: `10.1007/978-3-0348-0603-9_2` , preprint: `http://arxiv.org/abs/1306.2157`.

[206] E. Hitzer, *Extending Fourier transformations to Hamilton's quaternions and Clifford's geometric algebras*, In T. Simos, G. Psihoyios and C. Tsitouras (eds.), Numerical Analysis and Applied Mathematics ICNAAM 2013, AIP

Conf. Proc. **1558**, pp. 529–532 (2013). DOI: 10.1063/1.4825544, Preprint: http://viXra.org/abs/1310.0249.

[207] E. Hitzer, *New Developments in Clifford Fourier Transforms*, in N. E. Mastorakis, P. M. Pardalos, R. P. Agarwal and L. Kocinac (eds.), Advances in Applied and Pure Mathematics, Proceedings of the 2014 International Conference on Pure Mathematics, Applied Mathematics, Computational Methods (PMAMCM 2014), Santorini Island, Greece, July 17–21, 2014, Mathematics and Computers in Science and Engineering Series, **29**, pp. 19–25 (2014). Preprint: http://viXra.org/abs/1407.0169 .

[208] E. Hitzer, *Preprocessing quaternion data in quaternion spaces using the quaternion domain Fourier transform*, Proceedings of SICE Symposium on Systems and Information 2014 (SSI 2014), Session SS13-6, catalog No. SY0009/14/0000-0834Ì2014 SICE, Okayama University, Okayama, Japan, 21-23 Nov. 2014, pp. 834–839.

[209] E. Hitzer, *Two-Sided Clifford Fourier Transform with Two Square Roots of* -1 *in* $Cl(p,q)$ Adv. Appl. Clifford Algebras, Vol. **24** (2014), pp. 313–332, DOI: 10.1007/s00006-014-0441-9 , Preprint: http://arxiv.org/abs/1306.2092.

[210] E. Hitzer, *Quaternion Domain Fourier Transform*, Electronic proceedings of ICCA10, Tartu, Estonia, Aug. 2014. Online version: http://icca10.ut.ee/06_FTW_workshops/FTW/QuaternionDomainFT_rv1.pdf

[211] E. Hitzer, *The Quaternion Domain Fourier Transform and its Properties*, Advances in Applied Clifford Algebras, **26**, pp. 969–984 (2016), Online First: 02 Nov. 2015, DOI: 10.1007/s00006-015-0620-3 . Preprint: http://vixra.org/abs/1511.0302 .

[212] E. Hitzer, *The orthogonal planes split of quaternions and its relation to quaternion geometry of rotations*, in F. Brackx, H. De Schepper and J. Van der Jeugt (eds.), Proceedings of the 30th International Colloquium on Group Theoretical Methods in Physics (group30), 14–18 July 2014, Ghent, Belgium, IOP Jour. of Phys.: Conf. Ser. (JPCS) Vol. **597** (2015) 012042. DOI: 10.1088/1742-6596/597/1/012042. Open Access URL: http://iopscience.iop.org/1742-6596/597/1/012042/pdf/1742-6596_597_1_012042.pdf, Preprint: http://viXra.org/abs/1411.0362.

[213] E. Hitzer *General two-sided quaternion Fourier transform, convolution and Mustard convolution*, Adv. of App. Cliff. Algs. **27**, pp. 381–395 (2017), Online First: 12 May 2016, DOI: 10.1007/s00006-016-0684-8. Preprint: http://vixra.org/abs/1601.0165 .

[214] E. Hitzer, *Quaternionic Wiener-Khinchine theorems and spectral representation of convolution with steerable two-sided quaternion Fourier transform*, Adv.

Appl. Clifford Algebras, **27**, pp. 1313–1328 (2017), DOI: https://doi.org/10.1007/s00006-016-0744-0.

[215] E. Hitzer, *General Steerable Two-Sided Clifford Fourier Transform, Convolution and Mustard Convolution*, Adv. Appl. Clifford Algebras **27**, pp. 2215–2234 (2017), Online First 9 June 2016, DOI: https://doi.org/10.1007/s00006-016-0687-5 Available as preprint: http://vixra.org/abs/1602.0044

[216] E. Hitzer and S.J. Sangwine, *Multivector and multivector matrix inverses in real Clifford algebras*, Appl. Math. and Comp., **311**(C), pp. 375–389 (Oct 2017), DOI: https://doi.org/10.1016/j.amc.2017.05.027. Preprint: Technical Report CES-534, ISSN: 1744-8050, http://repository.essex.ac.uk/17282.

[217] E. Hitzer, *General one-sided Clifford Fourier transform, and convolution products in the spatial and frequency domains*, 16 pages (2016), Preprint: http://vixra.org/pdf/1604.0001v1.pdf

[218] E. Hitzer, *Special relativistic Fourier transformation and convolutions*, Math. Meth. in the Appl. Science, Vol. **42**, Iss. 7, pp. 2244–2255 (2019), DOI: https://doi.org/10.1002/mma.5502, related preprint: http://vixra.org/pdf/1601.0283v3.pdf.

[219] E. Hitzer, W. Benger, M. Niederwieser, R. Baran and F. Steinbacher, *Strip Adjustment of Airborne Laserscanning Data with Conformal Geometric Algebra*, in preparation.

[220] R.A. Horn and C.R. Johnson, Matrix Analysis, Cambridge University Press, Cambridge (UK), 1985.

[221] K. Ito (ed.), *Encyclopedic Dictionary of Mathematics*, 2nd ed., Mathematical Society of Japan, The MIT Press, Cambridge, Massachusetts, 1996.

[222] B. Jancewicz, *Trivector Fourier transformation and electromagnetic field*, Journal of Mathematical Physics, **31**(8), pp. 1847–1852 (1990).

[223] *The Japanese Tea Ceremony*, http://www.holymtn.com/tea/Japanesetea.htm last accessed: 18 Sep. 2020.

[224] Jesus Christ, Gospel according to Matthew, Bible, Today's English Version, https://www.biblegateway.com/passage/?search=Matthew+1&version=GNT, last accessed: 23 Feb. 2021.

[225] L. Jin, H. Liu, X. Xu and E. Song, *Quaternion-based impulse noise removal from color video sequences*. IEEE Transactions on Circuits and Systems for Video Technology, **23**, pp. 741–755 (2013).

[226] John, chapter 14, verse 6, Holy Bible, King James Version. `https://www.biblegateway.com/passage/?search=John+14:6&version=KJV`, last accessed: 12 June 2020.

[227] U. Kähler et al., *Monogenic Wavelets over Unit Ball*, ZAA, **24**(4), pp. 813–824 (2005) .

[228] G. Kaiser, E. Heyman and V. Lomakin, *Physical source realization of complex-source pulsed beams*, The Journal of the Acoustical Society of America, **107**, pp. 1880–1891 (2000), DOI: `https://doi.org/10.1121/1.428469`.

[229] G. Kaiser, *Communications via holomorphic Green functions*, Clifford Analysis and its Applications, Kluwer NATO Science, (2001).

[230] G. Kaiser, *Complex-distance potential theory, wave equations, and physical wavelets*, Mathematical Methods in the Applied Sciences, In F. Sommen and W. Sprößig, **25**, pp. 1577–1588 (2002), Invited paper, Special Issue on Clifford Analysis in Applications.

[231] G. Kaiser, *Physical wavelets and their sources: Real physics in complex space-time*, 56 pages, preprint: `arxiv.org/abs/math-ph/0303027` (2003).

[232] G. Kaiser, *Huygens' principle in classical electrodynamics: a distributional approach*, preprint: `https://arxiv.org/abs/0906.4167`, June 2009, 14 pages (2009).

[233] C. Kalisa and B. Torrésani, *N-Dimensional Affine Weyl-Heisenberg Wavelets*, *Ann. Inst. Henri Poincaré*, pp. 201–236 (1993) .

[234] Q. Kemao, *Two-dimensional windowed Fourier transform for fringe pattern analysis: Principles, applications, and implementations*, Optics and Laser Engineering, **45**, pp. 304–317 (2007).

[235] T. Kobayashi and G. Mano, *Integral Formulas for the Minimal Representation of* $O(p,2)$, Acta Applicandae Mathematicae, **86**, pp. 103–113 (2005).

[236] K. Kou, J. Morais and Y. Zhang, *Generalized prolate spheroidal wave functions for offset linear canonical transform in Clifford Analysis*, Math. Meth. Appl. Sci. **36**, pp. 1028–1041 (2013), DOI: `10.1002/mma.2657`.

[237] K. Kou, J-Y. Ou and J. Morais, *On Uncertainty Principle for Quaternionic Linear Canonical Transform*, Abs. and App. Anal., Vol. **2013**, IC 72592, 14 pp., Open Access, DOI: `https://doi.org/10.1155/2013/725952`.

[238] J. Kuipers, *Quaternions and Rotation Sequences: A Primer With Applications to Orbits, Aerospace, and Virtual Reality*, (reprint edition), Princeton University Press, Princeton (2002).

[239] G. Laville and I.P. Ramadanoff, *Stone-Weierstrass Theorem*, pp. 1–7 (2004,2007), preprint: `https://arxiv.org/pdf/math/0411090.pdf`, accessed 21 Jan. 2021.

[240] N. Le Bihan and S.J. Sangwine, *Quaternionic Spectral Analysis of Non-Stationary Improper Complex Signals*, in: E. Hitzer, S.J. Sangwine (eds.), "Quaternion and Clifford Fourier Transforms and Wavelets", Trends in Mathematics (TIM) Vol. **27**, Birkhäuser, Basel, 2013, pp. 41–56.

[241] T.S. Lee, *Image Representation using 2D Gabor Wavelets*, IEEE Transaction on Pattern Analysis and Machine Intelligence, **18**(10), pp. 1–13 (1996).

[242] J.C. Lennox, *Seven days that divide the world - the beginning according to Genesis and science.* Zondervan: Grand Rapids (Michigan), 2011.

[243] C. Li, A. McIntosh and T. Qian, *Clifford Algebras, Fourier Transform and Singular Convolution Operators On Lipschitz Surfaces.* Revista Matematica Iberoamericana, **10**(3), pp. 665–695 (1994), open access: `http://dmle.icmat.es/pdf/MATEMATICAIBEROAMERICANA_1994_10_03_06.pdf`, accessed 26 Jan. 2021.

[244] H. Li, *Invariant Algebras and Geometric Reasoning*, World Scientific, Singapore, 2008.

[245] H. Li, et al, *Three-Dimensional Projective Geometry with Geometric Algebra*, preprint `http://arxiv.org/abs/1507.06634` (2015).

[246] S. Lie, *On a class of geometric transformations*, PhD thesis, University of Oslo (formerly Christiania), 1871.

[247] P. Lounesto, *Clifford Algebras and Spinors*, London Mathematical Society Lecture Note **239**, Cambridge University Press, Cambridge (UK), 2001.

[248] A. Macdonald, *Linear and Geometric Algebra*, CreateSpace, LaVergne, 2011, latest corrected edition: 2020.

[249] A. Macdonald, *Vector and Geometric Calculus*, CreateSpace Independent Publishing Platform (December 18, 2012).

[250] S. Mallat, *A wavelet tour of signal processing*, Academic Press, Cambridge, Massachusetts, 2001.

[251] B. Mawardi and E. Hitzer, *Clifford Fourier Transformation and Uncertainty Principle for the Clifford Geometric Algebra* $Cl_{3,0}$, Advances in Applied Clifford Algebras, **16**(1), pp. 41–61 (2006). Preprint: `http://vixra.org/abs/1306.0089`, DOI: `https://doi.org/10.1007/s00006-006-0003-x`.

[252] B. Mawardi and E. Hitzer, *Clifford Algebra* $Cl(3,0)$*-valued Wavelets and Uncertainty Inequality for Clifford Gabor Wavelet Transformation*, in Preprints of

Meeting of the Japan Society for Industrial and Applied Mathematics, ISSN: 1345–3378, Tsukuba University, 16–18 Sep. 2006, Tsukuba, Japan, pp. 64–65 (2006).

[253] B. Mawardi and E. Hitzer, *Clifford Algebra $Cl(3,0)$-valued Wavelet Transformation, Clifford Wavelet Uncertainty Inequality and Clifford Gabor Wavelets*, International Journal of Wavelets, Multiresolution and Information Processing, **5**(6), pp. 997–1019 (2007).

[254] B. Mawardi, E. Hitzer and S. Adji, *Two-Dimensional Clifford Windowed Fourier Transform*, in G. Scheuermann, E. Bayro-Corrochano (eds.), Geometric Algebra Computing, Springer, Lectures Notes in Computer Science (LNCS), New York, 2010, pp. 93–106. First published in: Electronic Proc. of AGACSE 3, Leipzig, Germany, 17–19 Aug. 2008 (2008).

[255] B. Mawardi, E. Hitzer, A. Hayashi and R. Ashino, *An Uncertainty Principle for Quaternion Fourier Transform*, Computer & Mathematics with Applications, **56**(9), pp. 2398–2410 (2008).

[256] B. Mawardi, E. Hitzer, R. Ashino and R. Vaillancourt, *Windowed Fourier transform of two-dimensional quaternionic signals*, Appl. Math. and Computation, **216**(8), pp. 2366–2379, 15 June 2010, DOI: `https://doi.org/10.1016/j.amc.2010.03.082`, Preprint: `https://vixra.org/pdf/1306.0096v1.pdf`

[257] B. Mawardi, S. Adji and J. Zhao, *Clifford Algebra-Valued Wavelet Transform on Multivector Fields*, Advances in Applied Clifford Algebras, **21**(1), pp. 13–30 (2011), DOI: `10.1007/s00006-010-0239-3`.

[258] J.C. Maxwell, *A treatise on electricity and magnetism*, Clarendon Press, Oxford 1873.

[259] A. McIntosh, *Clifford Algebras, Fourier Theory, Singular Integrals and Harmonic Functions on Lipschitz Domains.* Chapter 1 of J. Ryan (ed.), Clifford Algebras in Analysis and Related Topics, CRC Press, Boca Raton, 1996.

[260] Robert Hjalmar Mellin (1854–1933). Biography. Homepage of St. Andrews University, Scotland.

[261] L. Meister and H. Schaeben, *A concise quaternion geometry of rotations*, Math. Meth. in the Appl. Sci. **28**(1), pp. 101–126, 2005.

[262] J. Mennesson, et al, *Color Obj. Recogn. Based on a Clifford Fourier Transf.*, in L. Dorst, J. Lasenby, "Guide to Geom. Algebra in Pract.", Springer, New York, pp. 175–191 (2011)

[263] B. Mishra et al, *A Geometric Algebra Co-Processor for Color Edge Detection*, Electronics, **4**, pp. 94–117 (2015), DOI: `10.3390/electronics4010094` .

[264] M. Mitrea, *Clifford wavelets, singular integrals, and Hardy spaces*, Volume **1575** of Lecture notes in mathematics, Springer: Berlin, 1994.

[265] J.P. Morais, S. Georgiev and W. Sprößig, *Real Quaternionic Calculus Handbook*, Birkhäuser, Basel, 2014.

[266] M. Moshinsky and C. Quesne, *Linear Canonical Transformations and their Unitary Representations*, Journal of Mathematical Physics, **12**, pp. 1772–1780 (1971).

[267] C.E. Moxey, T.A. Ell and S.J. Sangwine, *Hypercomplex operators and vector correlation*, EUSIPCO 2002, Eleventh European Signal Processing Conference, 3–6 September 2002, Toulouse, France, III, pp. 247–250 (2002).

[268] C.E. Moxey, S.J. Sangwine and T.A. Ell, *Hypercomplex correlation techniques for vector images*, IEEE Trans. Signal Process, **51**, pp. 1941–1953 (2003).

[269] R. Murenzi, Wavelet transform associated to the n-dimensional Euclidean group with dilation, in Proceedings of *Wavelets: Time-Frequency Methods and Phase Space, Marseille, France, December 14–18, 1987*, eds. J.M. Combes, A. Grossmann and Ph. Tchamitchian, Springer, New York, pp. 239–246 (1989),

[270] D. Mustard, *Fractional convolution*, J. Aust. Math. Soc. Ser. B, **40**, pp. 257–265 (1998), DOI: `10.1017/S0334270000012509` .

[271] M. Nakaoka and A. Hattori, *Introduction to Linear Algebra*, Kinokuniya Shoten, Tokyo 1986 (Japanese).

[272] V. Namias, *The Fractional Order Fourier Transform and its Application to Quantum Mechanics*, IMA Journal of Applied Mathematics, **25**(3), pp. 241–265 (1980), DOI: `10.1093/imamat/25.3.241`.

[273] T. Needham, *Visual Complex Analysis*, Oxford University Press, 2001.

[274] O. Rodrigues, *Des lois géométriques qui régissent des déplacements d'un système solide*, Journal de Mathématiques Pures et Appliquées (Liouville), **5**, p. 380 (1840).

[275] K. Nono, *Hyperholomorphic functions of a quaternion variable*, Bull. of Fukuoka University of Education, **32**, pp. 21–37 (1982).

[276] H. Ozaktas, Z. Zalevsky and M. Kutay, *The Fractional Fourier Transform*, Wiley: Chichester (2001).

[277] A. Papoulis, *The Fourier Integral and Its Applications*, Mc Graw-Hill Book Company, Inc., New York, 1962.

[278] B. Patra, *An Introduction to Integral Transforms*, CRC Press, London, 2018.

[279] Paul, Romans 1:20, The Holy Bible New International Version, IBS, Colorado, 1973. `https://www.biblegateway.com/`, last accessed: 23 Feb. 2021.

[280] S.C. Pei, J.J. Ding and J.H. Chang, *Efficient Implementation of Quat. Fourier Transf., Convolution, and Correlation by 2-D Complex FFT*, IEEE Trans. on Sig. Proc. **49**(11), pp. 2783–2797 (2001).

[281] C. Perwass, *Free software CLUCalc for intuitive 3D visualizations and scientific calculations*. `http://cluviz.de/`, last accessed: 18 Sep. 2020.

[282] C. Perwass, *Geometric Algebra with Applications in Engineering*, Springer, Berlin, 2009.

[283] I. Porteous, *Clifford Algebras and the Classical Groups*, CUP, Cambridge (UK), 1995.

[284] A.P. Prodnikov, Y.A. Brychkov and O.I. Marichev, *Evaluation of integrals and the Mellin transform*, Transl. from Itogi Nauki i Techniki, Seriya Matematicheskii Analiz, **27**, pp. 3–146 (1989).

[285] Psalm 92, verse 5, *New Int. Version of the Bible*. `www.biblegateway.com`

[286] T. Qian, *Paley-Wiener Theorems and Shannon Sampling in the Clifford Analysis Setting*. In R. Ablamowicz (ed.), "Clifford Algebras - Applications to Mathematics, Physics, and Engineering", Birkäuser, Basel, pp. 115–124 (2004).

[287] J. M. Rassias, *On The Heisenberg-Weyl Inequality*, Jour. of Inequalities in Pure and Appl. Math., **6**(1), article 11, pp. 1–18 (2005), `https://www.emis.de/journals/JIPAM/images/169_04_JIPAM/169_04.pdf`, last accessed: 23 Feb. 2021.

[288] B. Rosenhahn and G. Sommer *Pose estimation of free-form objects*, in T. Pajdla, J. Matas (eds.) European Conference on Computer Vision, Prague, 2004, Springer, Berlin, **127**, pp. 414–427 (2004).

[289] T.L. Saaty, *Private communication*, March 2014.

[290] S. Said, N. Le Bihan and S.J. Sangwine, *Fast complexified quaternion Fourier transform*. IEEE Transactions on Signal Processing, **56**(4), pp. 1522–1531 (2008).

[291] S.J. Sangwine, *Fourier transforms of colour images using quaternion, or hypercomplex, numbers*, Electronics Letters, **32**(21) (1996), pp. 1979–1980. DOI: `10.1049/el:19961331`

[292] S.J. Sangwine and T.A. Ell, *The Discrete Fourier Transform of a Colour Image*, In Blackledge, J. M. and Turner, M. J., Image Processing II Mathematical Methods, Algorithms and Applications, Chichester, 2000, Horwood Publishing for Institute of Mathematics and its Applications, pp. 430–441, Proceedings Second IMA Conference on Image Processing, De Montfort University, Leicester, UK, September 1998.

[293] S.J. Sangwine, *Color image edge detector based on quaternion convolution*, Electronics Letters, **34**, pp. 969–971 (1998).

[294] S.J. Sangwine, *Biquaternion (Complexified Quaternion) Roots of* -1, Adv. Appl. Clifford Algebras **16**(1), pp. 63–68 (2006), DOI: `10.1007/s00006-006-0005-8`.

[295] S.J. Sangwine and E. Hitzer, Clifford Multivector Toolbox (for MATLAB) 2015–2016, Software library available at: `http://clifford-multivector-toolbox.sourceforge.net/`, last accessed 18 Sep. 2020.

[296] S.J. Sangwine and E. Hitzer, *Clifford Multivector Toolbox (for MATLAB)*. Adv. Appl. Clifford Algebras **27**, pp. 539–558 (2017). Online First 19 April 2016, DOI: `https://doi.org/10.1007/s00006-016-0666-x`, Preprint: `http://repository.essex.ac.uk/16434/1/author_final.pdf`.

[297] S.J. Sangwine and N. Le Bihan, *Quaternion and octonion toolbox for Matlab*, `http://qtfm.sourceforge.net/`, last accessed 29 Mar. 2016.

[298] B. Schuler, *Zur Theorie der regulären Funktionen einer Quaternionen-Variablen*, Comm. Math. Helv. **10**, pp. 327–342, (1937/1938).

[299] F. Schwabl, *Quantenmechanik*, 2nd ed., Springer, Berlin, 1990.

[300] D.S. Shirokov, *On determinant, other characteristic polynomial coefficients, and inverses in Clifford algebras of arbitrary dimension*, Preprint: `https://arxiv.org/pdf/2005.04015.pdf`, last accessed: 12 June 2020.

[301] P. Singer, *Uncertainty Inequalities for the Continuous Wavelet transform*, IEEE Transaction on Information Theory, **45**, pp. 1039–1042 (1999).

[302] G. Sobczyk, *Simplicial Calculus with Geometric Algebra*, available at: `http://modelingnts.la.asu.edu/html/GeoCalc.html\#Multivector`

[303] G. Sobczyk and O.L. Sanchez, *Fundamental Theorem of Calculus*, Adv. Appl. Cliff. Algs., **21**, pp. 221–231 (2011).

[304] G. Sobczyk, *Conformal Mappings in Geometric Algebra*, Notices of the AMS, **59**(2), pp. 264–273 (2012).

[305] F. Sommen, *A product and an exponential function in hypercomplex function theory*, Applicable Analysis, **12**, pp. 13–26 (1981).

[306] F. Sommen, *Hypercomplex Fourier and Laplace Transforms I*, Illinois Journal of Mathematics, **26**(2), pp. 332–352 (1982).

[307] F. Sommen, *Hypercomplex Fourier and Laplace transforms II*, Complex Variables, **1**(2–3), pp. 209–238 (1983), DOI: `10.1080/17476938308814016`.

[308] R. Soulard and P. Carré, *Quaternionic wavelets for texture classification*, Pattern Recognition Lett., **32**, pp. 1669–1678 (2011).

[309] R. Soulard and P. Carré, *Colour Extension of Monogenic Wavelets with Geometric Algebra: Application to Color Image Denoising*, in E. Hitzer, S.J. Sangwine (eds.), "Quaternion and Clifford Fourier Transforms and Wavelets", Trends in Mathematics **27**, Birkhäuser, Basel, 2013, pp. 247–268. DOI: `https://doi.org/10.1007/978-3-0348-0603-9_12` .

[310] E.M. Stein and G. Weiss, *Introduction to Fourier analysis on Euclidean spaces*, Princeton University Press, Princeton, 1971.

[311] A. Sudbery, *Quaternionic analysis*, Math. Proc. Cambridge Philos. Soc., **85**(2), pp. 199–225, (1979).

[312] *The Center of an Algebra*, The Unapologetic Mathematician, `https://unapologetic.wordpress.com/2010/10/06/the-center-of-an-algebra/`, last accessed: 23 Feb. 2021.

[313] K. Toraichi, M. Kamada, S. Itahashi and R. Mori, *Window functions represented by B-spline functions*, IEEE Transaction on Acoustics, Speech and Signal Processing, **37**(1) pp. 145–147 (1989).

[314] L. Traversoni, *Quaternion Wavelet Problems*, Proceedings of 8th International Symposium on Approximation Theory, Texas A& M University, Jan. 1995. L. Traversoni, *Image Analysis Using Quaternion Wavelets*, in E. Bayro-Corrochano, G. Sobczyk (eds.), Proc. of AGACSE 1999, Birkhäuser, Basel, 2001.

[315] W.K. Tung, *Group Theory in Physics*, World Scientific, Singapore, 1985.

[316] E. Vinberg, *A Course in Algebra*, Graduate Series in Mathematics, **56**, AMS, Providence, Rhode Island, 2003.

[317] Waterloo Maple Incorporated, *Maple, a general purpose computer algebra system.* Waterloo, `http://www.maplesoft.com` (ľ2012).

[318] F. Weisz, *Multiplier theorems for the short-time Fourier transform*, Integral Equation and Operator Theory, **60**(1), pp. 133–149 (2008).

[319] H. Weyl, *The Theory of Groups and Quantum Mechanics*, second ed., Dover, New York, 1950.

[320] Wikipedia, *Dirac equation*, `http://en.wikipedia.org/wiki/Dirac_equation`, accessed 10 June 2014.

[321] Wikipedia article on *Conjugacy class*, `http://en.wikipedia.org/wiki/Conjugacy_class`, opened 19 March 2011.

[322] Wikipedia article on *Euclidean Group*, `https://en.wikipedia.org/wiki/Euclidean_group`, opened 18 July 2020.

[323] Wikipedia article on *Inner automorphism*, `http://en.wikipedia.org/wiki/Inner_automorphism`, opened 19 March 2011.

[324] Wikipedia, *Maxwell Equations*, `http://en.wikipedia.org/wiki/Maxwell_equations`, accessed 10 June 2014.

[325] Wikipedia, *Partial Differential Equations*, `http://en.wikipedia.org/wiki/Partial_differential_equation`, accessed 10 June 2014.

[326] Wikipedia: *Quaternions*, `http://en.wikipedia.org/wiki/Quaternion`, accessed 28 Apr. 2014.

[327] Wikipedia, *Schrödinger equation*, `http://en.wikipedia.org/wiki/Schroedinger_equation`, accessed 10 June 2014.

[328] Wikipedia: *Texture (crystalline)*, `http://en.wikipedia.org/wiki/Texture_(crystalline)`, accessed 15 May 2015.

[329] K.B. Wolf, *Integral Transforms in Science and Engineering*, Chapters 9&10, Plenum Press, New York, 1979. `http://www.fis.unam.mx/~bwolf/integral.html`

[330] Y. Yang and K. Kou, *Uncertainty principles for hypercomplex signals in the linear canoncial transform domains*, Signal Proc., **95**, pp. 67–75 (2014).

[331] Z. Yu, W. Luo, L. Yi, Y. Hu and L. Yuan, *Clifford algebra-based structure filtering analysis for geophysical vector fields*, Nonlin. Processes Geophys., **20**, pp. 563–570, (2013), DOI: `https://doi.org/10.5194/npg-20-563-2013`.

[332] L. Yuan, et al, *Geom. Alg. for Multidim.-Unified Geogr. Inf. System*, Adv. Appl. Cliff. Algs., **23**, pp. 497–518 (2013).

[333] L. Yuan, et al, *Pattern Forced Geophys. Vec. Field Segm. based on Clifford FFT*, Computer & Geoscience, **60**, pp. 63–69 (2013).

[334] J. Zamora-Esquivel, $G_{6,3}$ *Geometric Algebra; Description and Implementation* Advances in Applied Clifford Algebras, **24**(2), pp. 493–514, (2014), DOI: `10.1007/s00006-014-0442-8`.

[335] J.M. Zhao and L.Z. Peng, *Quaternion-valued Admissible Wavelets Associated with the 2-dimensional Euclidean Group with Dilations*, Journal of Natural Geometry, **20**, pp. 21–32 (2001).

[336] J. Zhao, *Clifford Algebra-valued Admissible Wavelets Associated with Admissible Group*, Acta Scientiarium Naturalium Universitatis Pekinensis, **41**(5), pp. 667–670 (2005).

[337] J. Zhong and H. Zeng, *Multiscale windowed Fourier transform for phase extraction of fringe pattern*, Applied Optics, **46**(14), pp. 2670–2675 (2007).

Index

Milton Keynes UK
Ingram Content Group UK Ltd.
UKHW051930141024
449569UK00027B/1424

9 781032 026589